Earth Science

Earth Science

Ninth Edition

Edward J. Tarbuck
Frederick K. Lutgens

Illinois Central College

Illustrated by
Dennis Tasa

Prentice
Hall

PRENTICE HALL
Upper Saddle River, New Jersey 07458

Library of Congress Cataloging-in-Publication Data

Tarbuck, Edward J.
  Earth Science/ Edward J. Tarbuck, Frederick K. Lutgens—9th ed.
    p.   cm.
  Includes index.
  ISBN 0-13-081566-7
  1. Earth sciences.   I. Lutgens, Frederick K.   II. Title
QE26.2.T38 2000   99–32024
550—dc21   CIP

Senior Editor: *Daniel Kaveney*
Editor in Chief: *Paul F. Corey*
Associate Vice President of Production and Manufacturing: *David W. Riccardi*
Executive Managing Editor: *Kathleen Schiaparelli*
Assistant Managing Editor: *Lisa Kinne*
Production Editor: *Edward Thomas*
Marketing Manager: *Leslie Cavaliere*
Creative Director: *Paula Maylahn*
Art Manager: *Gus Vibal*
Art Editor: *Grace Hazeldine*
Art Director: *Joseph Sengotta*
Supplements Production Editors: *Wendy Rivers and Dawn Murrin*
Photo Editors: *Lorinda Morris-Nantz and Melinda Reo*
Photo Researcher: *Toby Zausner*
Copy Editor: *James Tully*
Assistant Editor: *Margaret Ziegler*
Production Assistant: *Nancy Bauer*
Cover Designer: *Steven Gagliostro*
Interior Design: *Judith A Matz-Coniglio*
Manufacturing Manager: *Trudy Pisciotti*
Text Composition: *Molly Pike/Lido Graphics*
Cover Photo: *Double Falls at dawn near Logan Pass, Glacier National Park, Montana. Galen Rowell/ Mountian Light Photography, Inc.*
Title Page Photo: *Sea stacks and surf-rounded stones at Garrapata Beach, California. Carr Clifton Photography.*

©2000, 1997 by Prentice-Hall, Inc.
Upper Saddle River, New Jersey 07458

Earlier editions copyright 1994, 1991 by Macmillan Publishing Company, and copyright 1988, 1985, 1982, 1979, and 1976 by Merrill Publishing Company.

Printed in the United States of America.

10   9   8   7   6   5   4   3

ISBN 0-13-081566-7

Prentice-Hall International (UK) Limited, *London*
Prentice-Hall of Australia Pty. Limited, *Sydney*
Prentice-Hall Canada, Inc., *Toronto*
Prentice-Hall Hispanoamericana, S.A., *Mexico*
Prentice-Hall of India Private Limited, *New Delhi*
Prentice-Hall of Japan, Inc., *Tokyo*
Prentice-Hall, *(Singapore)* Pte Ltd
Editora Prentice-Hall do Brasil, Ltda., *Rio de Janeiro*

Brief Contents

Contents

6 Earthquakes and Earth's Interior   155

7 Plate Tectonics   181

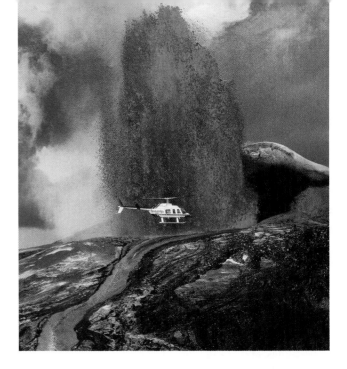

8 Igneous Activity   215

9 Mountain Building   245

Part 3
The Atmosphere   375

14 Composition, Structure, and Temperature   377

15 Moisture, Clouds, and Precipitation   405

Preface

Earth is a very small part of a vast universe, but it is our home. It provides the resources that support our modern society and the ingredients necessary to maintain life. Therefore, a knowledge and understanding of our planet is critical to our social well being and indeed, vital to our survival.

In recent years, media reports have made us increasingly aware of our place in the universe and the forces at work in our physical environment. We are also beginning to learn that human interactions with natural systems can upset delicate balances. News stories inform us of new discoveries in the solar system and beyond. Daily reports remind us of the destruction caused by hurricanes, earthquakes, floods, and landslides. We have been made aware of ozone depletion, potential global warming, and growing environmental concerns about the oceans. To comprehend, prepare for, and solve these and other concerns requires an awareness of how science is done and the scientific principles that influence our planet, its rocks, mountains, atmosphere, and oceans.

Earth Science, ninth edition, like its predecessors, is a college-level text designed for an introductory course that often has the same name as this text. It consists of four units that emphasize broad and up-to-date coverage of basic topics and principles in geology, oceanography, meteorology, and astronomy. The book is intended to be a meaningful, nontechnical survey for undergraduate students with little background in science. Usually these students are taking an Earth science class to meet a portion of their college's or university's general requirements.

In addition to being informative and up-to-date, a major goal of *Earth Science* is to meet the need of beginning students for a readable and user-friendly text, a book that is a highly usable "tool" for learning basic Earth science principles and concepts.

Distinguishing Features

Readability

The language of this book is straightforward and *written to be understood*. Clear, readable discussions with a minimum of technical language are the rule. The frequent headings and subheadings help students follow discussions and identify the important ideas presented in each chapter. In the ninth edition, improved readability was achieved by examining chapter organization and flow, and writing in a more personal style. Large portions of the text were substantially rewritten in an effort to make the material more understandable.

Illustrations and Photographs

The Earth sciences are highly visual. Therefore, photographs and artwork are a very important part of an introductory book. *Earth Science*, ninth edition, contains dozens of new high-quality photographs that were carefully selected to aid understanding, add realism, and heighten the interest of the reader.

The illustrations in each new edition of *Earth Science* keep getting better and better. In the ninth edition more than 100 pieces of line art were redesigned. The new art illustrates ideas and concepts more clearly and realistically than ever before. The art program was carried out by Dennis Tasa, a gifted artist and respected Earth science illustrator.

Focus on Learning

New to the ninth edition: To assist student learning, every chapter now concludes with a *Chapter Summary*. When a chapter has been completed, three useful devices help students review. First, the *Chapter Summary* recaps all of the major points. Next is a checklist of *Key Terms* with page references. Learning the language of Earth science helps students learn the material. This is followed by *Review Questions* that help students examine their knowledge of significant facts and ideas. Each chapter closes with some suggested Web sites for further exploration. Moreover, students are reminded to visit the *all-new* and *greatly improved* Web site for *Earth Science*, ninth edition (http://www.prenhall.com/tarbuck). It contains many excellent opportunities for review and exploration.

Earth as a System

An important occurrence in modern science has been the realization that Earth is a giant multidimensional system. Our planet consists of many separate but interacting parts. A change in any one part can produce changes in any or all of the other parts—often in ways that are neither obvious nor immediately apparent. Although it is not possible to study the entire system at once, it is possible to develop an awareness and appreciation for the concept and for many of the system's important interrelationships. Therefore, beginning with an expanded discussion in the Introduction, the theme of "Earth as a system" keeps recurring through all major units of the book. It is a thread that "weaves" through the chapters and helps tie them together.

Several new and revised special interest boxes relate to *Earth as a system*. To remind the reader of this important theme, the small icon you see at the beginning of this section is used to mark these boxes. Finally, each chapter concludes with a section on *Examining the Earth System*. The questions and problems found here are intended to develop an awareness and appreciation for some of the Earth system's many interrelationships.

People and the Environment

Because knowledge about our planet and how it works is necessary to our survival and well being, the treatment of environmental issues has always been an important part of *Earth Science*. Such discussions serve to illustrate the relevance and application of Earth science knowledge. With each new edition this focus has been given greater emphasis. This is certainly the case with the ninth edition. The text integrates a great deal of information about the relationship between people and the natural environment and explores applications of the Earth sciences to understanding and solving problems that arise from these interactions.

In addition to many basic text discussions, 17 of the text's special interest boxes involve the "People and the Environment" theme and are quickly recognized by the distinctive icon you see at the beginning of this section.

Maintaining a Focus on Basic Principles and Instructor Flexibility

Although many new topical issues are treated in *Earth Science*, ninth edition, it should be emphasized that the main focus of this new edition remains the same as its predecessors—to foster student understanding of basic Earth science principles. Whereas student use of the text is a primary concern, the book's adaptability to the needs and desires of the instructor is equally important. Realizing the broad diversity of Earth science courses in both content and approach, we have continued to use a relatively nonintegrated format to allow maximum flexibility for the instructor. Each of the four major units stands alone; hence, they can be taught in any order. A unit can be omitted entirely without appreciable loss of continuity, and portions of some chapters may be interchanged or excluded at the instructor's discretion.

More About the Ninth Edition

In addition to adding chapter summaries, strengthening the "Earth as a System" and "People and the Environment" themes and making substantial changes to the photography and art programs, much more is new to the ninth edition.

The ninth edition of *Earth Science* represents a thorough revision. *Every* part of the book was examined carefully with the dual goals of keeping topics current and improving the clarity of text discussions. Based on

feedback from reviewers and our students, we believe we have succeeded.

Of special interest is a new section on "Marine Life Zones" in Chapter 12 and a new related special interest box (Box 12.2) on "Hydrothermal Vent Communities". Moreover, Chapter 16 includes a substantial new discussion of "El Niño and Global Weather" and a related new box (Box 16.2), "Tracking El Niño from Space".

Supplements

The authors and publisher have been pleased to work with a number of talented people to produce an excellent supplements package. This package includes the traditional supplements that students and professors have come to expect from authors and publishers, as well as some new kinds of supplements that involve electronic media.

For the Student

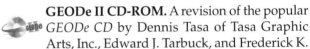**GEODe II CD-ROM.** A revision of the popular *GEODe CD* by Dennis Tasa of Tasa Graphic Arts, Inc., Edward J. Tarbuck, and Frederick K. Lutgens. *GEODe II* is a dynamic program that reinforces key concepts by using animations, tutorials, and interactive exercises. A special *GEODe II* icon appears throughout the book wherever a text discussion has a corresponding *GEODe II* activity. A copy of *GEODe II* has been included with this text. This special offering gives students two valuable products (GEODe II and the textbook) for the price of one.

Internet Support: This site, specific to the text, contains numerous review exercises (from which students get immediate feedback), exercises to expand one's understanding of Earth science, and resources for further exploration. This Web site provides an excellent platform from which to start using the Internet for the study of Earth science. Please visit the site at http://www.pren hall.com/tarbuck.

Study Guide: Written by experienced educator Kenneth Pinzke, the study guide helps students identify the important points from the text, and then provides them with review exercises, study questions, self-check exercises, and vocabulary review.

For the Professor

Transparency Set: More than 150 full-color acetates of illustrations from the text are available free of charge to qualified adopters.

Slides: More than 200 slides of images taken from the text, many of which were taken by the authors, are also available to qualified adopters.

Presentation Manager: This user-friendly navigation software enables professors to custom-build multimedia presentations. *Prentice Hall Presentation Manager 3.0* contains several hundred images from the text, as well as selected animations from GEODe II. The CD-ROM allows professors to organize material in whatever order they choose; preview resources by chapter; search the digital library by keyword; integrate material from their hard drive, a network, or the Internet; or edit lecture notes and annotate images with an overlay tool. This powerful presentation tool is available at no cost to qualified adopters of the text.

The New York Times Themes of the Times—Changing Earth: This unique newspaper-format supplement features recent articles about geology from the pages of the *New York Times*. This supplement, available at no extra charge from your local Prentice Hall representative, encourages students to make connections between the classroom and the world around them.

Instructor's Manual: Written by Kenneth Pinzke of Belleville Area College, the Instructor's Manual is intended as a resource for both new and experienced instructors. It includes a variety of lecture outlines, additional source materials, teaching tips, advice about how to integrate visual supplements (including the Web-based resources), and various other ideas for the classroom.

Test Item File: The Test Item File provides instructors with a wide variety of test questions.

PH Custom Test: Based on the powerful testing technology developed by Engineering Software Associates, Inc. (ESA), *Prentice Hall Custom Test* allows instructors to create and tailor exams to their own needs. With the online testing program, exams can also be administered online and data can then be automatically transferred for evaluation. The comprehensive desk reference guide is included along with online assistance.

For the Laboratory

Applications and Investigations in Earth Science: Written by Ed Tarbuck, Fred Lutgens, and Ken Pinzke, this full-color laboratory manual contains 22 exercises that provide students with hands-on experiences in geology, oceanography, meteorology, astronomy, and Earth science skills.

Acknowledgments

Writing a college textbook requires the talents and cooperation of many individuals. Working with Dennis Tasa, who is responsible for all of the outstanding illustrations, is always special for us. We not only value his outstanding artistic talents and imagination, but his friendship as well. We are also grateful to Professor Ken Pinzke at Belleville Area College. In addition to his many helpful suggestions regarding the manuscript, Ken developed the Web site, prepared the chapter summaries, and the instructor's guide. In addition, he revised the student study guide and the laboratory manual that are available to accompany *Earth Science*. Ken is an important part of our team and a valued friend as well.

Special thanks goes to those colleagues who prepared in-depth reviews. Their critical comments and thoughtful input helped guide our work and clearly strengthened the text. We wish to thank:

Lasty Best; *Palm Beach Community College*
Ted Chamberlain; *Colorado State University*
Dr. Lynn Chyi; *University of Akron*
Chris Hooker; *Waubonsee Community College*
Kris Huysken; *Indiana University NW*
Arthur C. Lee; *Westark Community College*
Amalie Orme; *CSU–Northridge*
Mark Sutherland; *College of Dupage*
Brooke Towery; *Pensacola Junior College*
Matthew Werhner; *Hillsborough Community College*

In addition we would like to acknowledge the aid of our students. Their comments and criticisms sometimes challenge us to search for a deeper understanding and always help us maintain our focus on readability.

We also want to acknowledge the team of professionals at Prentice Hall. Thanks to Editor in Chief Paul Corey. We sincerely appreciate his continuing strong support for excellence and innovation. Thanks also to our editor Dan Kaveney. His strong communication skills and energetic style contributed greatly to the project. The production team, led by Ed Thomas, as always, has done an outstanding job. They are true professionals with whom we are very fortunate to be associated.

Edward J. Tarbuck
Frederick K. Lutgens

Introduction

The spectacular eruption of a volcano, the magnificent scenery of a rocky coast, and the destruction created by a hurricane are all subjects for the Earth scientist. The study of Earth science deals with many fascinating and practical questions about our environment (Figure I.1). What forces produce mountains? Why is our daily weather so variable? Is climate really changing? How old is Earth and how is it related to the other planets in the solar system? What causes ocean tides? What was the Ice Age like? Will there be another? Can a successful well be located at this site?

The subject of this text is **Earth science.** To understand Earth is not an easy task because our planet is not a static and unchanging mass. Rather, it is a dynamic body with many interacting parts and a long and complex history.

Figure I.1 Sunset on Machapuchare Peak in the Himalaya's of Nepal. These mountains are the highest on Earth. Understanding the forces that created this majestic range is just one of the many areas of Earth science study. (Photo by Galen Rowell/Mountain Light Photography, Inc.)

Figure I.2 View of Earth that greeted the *Apollo 8* astronauts as their spacecraft emerged from behind the Moon. (Courtesy of NASA)

A View of Earth

Introduction
A View of Earth

A view such as the one in Figure I.2 provided the *Apollo 8* astronauts as well as the rest of humanity with a unique perspective of our home. Seen from space, Earth is breathtaking in its beauty and startling in its solitude. Such an image reminds us that our home is, after all, a planet—small, self-contained, and in some ways even fragile.

As we look closely at our planet from space, it becomes apparent that Earth is much more than rock and soil. In fact, the most conspicuous features in Figure I.2 are not continents, but swirling clouds suspended above the surface the vast global ocean. These features emphasize the importance of water on our planet.

The closer view of Earth from space shown in Figure I.3 helps us appreciate why the physical environment is traditionally divided into three major spheres: the water portion of our planet, the hydrosphere; Earth's gaseous envelope, the atmosphere; and, of course, the solid Earth. It should be emphasized that our environment is highly integrated and is not dominated by rock, water, or air alone. Rather, it is characterized by continuous interactions as air comes in contact with rock, rock with water, and water with air. Moreover, the biosphere, the totality of life-forms on our planet, extends into each of the three physical realms and is an equally integral part of the planet. Thus, Earth can be thought of as consisting of four major spheres: the hydrosphere, atmosphere, solid Earth, and biosphere.

The interactions among the spheres of Earth's environment are uncountable. Figure I.4 provides us with one easy-to-visualize example. The shoreline is an obvious meeting place for rock, water, and air. In this scene, ocean waves that were created by the drag of air moving across the water are breaking against the rocky shore. The force of the water can be powerful and the erosional work that is accomplished can be great.

Hydrosphere

Earth is sometimes called the *blue* planet. Water more than anything else makes Earth unique. The **hydrosphere** is a dynamic mass of liquid that is continually on the move, from the oceans to the atmosphere, precipitating back to the land, and running back to the ocean again. The global ocean is certainly the most prominent feature of the hydrosphere, blanketing nearly 71 percent of Earth's surface and accounting for about 97 percent of Earth's water. However, the hydrosphere also includes the freshwater found in streams, lakes, and glaciers, as well as that found underground.

Although these latter sources constitute just a tiny fraction of the total, they are much more important than their meager percentage indicates. In addition to providing the freshwater that is so vital to life on the continents, streams, glaciers, and groundwater are responsible for sculpturing and creating many of our planet's varied landforms.

Atmosphere

Earth is surrounded by a life-giving gaseous envelope called the **atmosphere**. This thin blanket of air is an

Figure I.3 Africa and Arabia are prominent in this image of Earth taken from *Apollo 17*. The tan cloud-free zones over the land coincide with the major desert regions where vegetation is relatively sparse. The band of clouds across central Africa is associated with a much wetter climate that in places sustains tropical rain forests. The dark blue of the oceans and the white swirling cloud patterns remind us of the importance of the hydrosphere and the atmosphere. Antarctica, a continent covered by glacial ice, is visible at the South Pole. (Courtesy of NASA/Science Source/Photo Researchers, Inc.)

integral part of the planet. It not only provides the air that we breathe but also acts to protect us from the Sun's intense heat and dangerous radiation. The energy exchanges that continually occur between the atmosphere and the Earth's surface and between the atmosphere and space produce the effects we call weather.

If, like the Moon, Earth had no atmosphere, our planet would not only be lifeless but also many of the processes and interactions that make the surface such a dynamic place could not operate. Without weathering and erosion, the face of our planet might more closely resemble the lunar surface, which has not changed appreciably in nearly 3 billion years.

Solid Earth

Lying beneath the atmosphere and the ocean is the solid Earth. It is divided into three principal units: the dense **core**; the less dense **mantle**; and the **crust**, which is the light and very thin outer skin of Earth (Figure I.5). The crust is not a layer of uniform thickness; rather, it is characterized by many irregularities. It is thinnest beneath the oceans and thickest where continents exist. By examining the most prominent surface features and their global extent, we can obtain clues to the dynamic processes that shape our planet.

The two principal divisions of Earth's surface are the continents and the ocean basins. The most obvious difference between these two diverse provinces is their relative levels. The average elevation of the continents above sea level is about 840 meters (2750 feet), whereas the average depth of the oceans is about 3800 meters (12,500 feet). Thus, the continents stand on the average 4640 meters (about 4.6 kilometers or nearly 3 miles) above the level of the ocean floor.

Biosphere

The **biosphere** includes all life on Earth and consists of the parts of the solid Earth, hydrosphere, and atmosphere in which living organisms can be found. Plants and animals depend on the physical environment for the basics of life. However, it is important to note that organisms do more than just respond to their physical environment. Indeed, through countless interactions, life-forms help maintain and alter their physical environment. Without life, the makeup and nature of solid Earth, hydrosphere, and atmosphere would be very different.

To summarize this brief view of Earth, we have seen that the physical environment consists of the hydrosphere, atmosphere, and solid Earth. These three parts are characterized by countless interactions. In addition, the biosphere extends into each of the spheres of the physical environment, and must also be considered an integral part of our planet.

Figure I.4 The shoreline is one obvious meeting place for the hydrosphere, atmosphere, and solid Earth. In this scene, at Point Lobos, California, ocean waves that were created by the force of moving air break against the rocky shore. The force of the water can be powerful and the erosional work that is accomplished can be great. (Photo by David D. Kean/The Stock Market)

Earth as a System

As we study Earth, it becomes clear that our planet can be viewed as a dynamic body with many separate but interacting parts or *spheres*. The hydrosphere, atmosphere, biosphere, and solid Earth and all of their components can be studied separately. However, the parts are not isolated. Each is related in some way to the others to produce a complex and continuously interacting whole that we call the *Earth system* (Figure I.6).

A **system** is a group of interacting, or interdependent parts that form a complex whole. Most of us hear and use the term frequently. We may service our car's cooling *system*, make use of the city's transportation *system*, and be a participant in the political *system*. A news report might inform us of an approaching weather *system*. Further, we know that Earth is just a small part of a large system known as the *solar system*.

The parts of the Earth system are linked so that a change in one part can produce changes in any or all of the other parts. For example, when a volcano erupts, lava from Earth's interior may flow out at the surface and block a nearby valley. This new obstruction influences the region's drainage system by creating a lake or causing streams to change course. The large quantities of volcanic ash and gases that can be emitted during an eruption might be blown high into the atmosphere and influence the amount of solar energy that can reach Earth's surface (Figure I.7). The result could be a drop in air temperatures over the entire hemisphere.

Where the surface is covered by lava flows or a thick layer of volcanic ash, existing soils are buried. This causes the soil-forming processes to begin anew to transform the new surface material into soil (Figure I.8). The soil that eventually forms will reflect the interactions among many parts of the Earth system—the volcanic

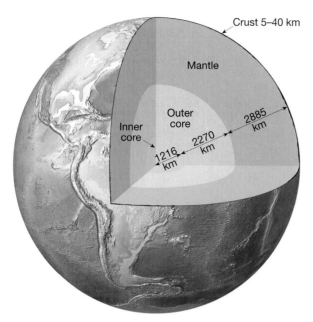

Figure I.5 View of Earth's layered structure. The inner core, outer core, and mantle are drawn to scale but the thickness of the crust is exaggerated by about five times.

parent material, the type and rate of weathering, and the impact of biological activity. Of course, there would also be significant changes in the biosphere. Some organisms and their habitats would be eliminated by the lava and ash, whereas new settings for life, such as the lake, would be created. The potential climate change could also impact sensitive life-forms.

The Earth system is characterized by processes that vary on spatial scales from fractions of millimeters to thousands of kilometers. Time scales for Earth's processes range from milliseconds to billions of years. As we learn about Earth, it becomes increasingly clear that despite significant separations in distance or time, many processes are connected and that a change in one component can influence the entire system.

The Earth system has a nearly endless array of cycles or loops in which matter is recycled over and over again (Figure I.6). One familiar loop is the *hydrologic cycle*. It represents the unending circulation of Earth's water among the hydrosphere, atmosphere, biosphere, and solid Earth. Water enters the atmosphere by evaporation from Earth's surface and by

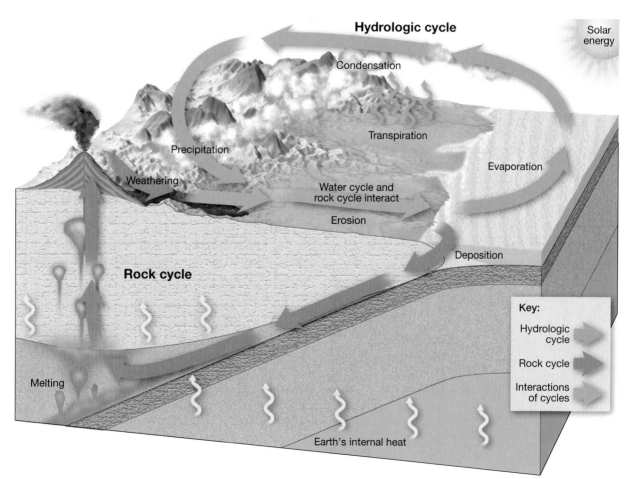

Figure I.6 Each part of the Earth system is related to every other part to produce a complex interacting whole. The Earth system involves many cycles including the hydrologic cycle and the rock cycle. Such cycles are not independent of each other. There are many places where they interface.

Figure I.7 The August 20, 1997, eruption of Soufriere Hills volcano on the Caribbean Island of Montserrat. When large quantities of volcanic dust and gases are blown high into the atmosphere, the amount of solar energy reaching Earth's surface is diminished. (Photo by Kevin West/Liaison Agency, Inc.)

transpiration from plants. Water vapor condenses in the atmosphere to form clouds, which in turn produce precipitation that falls back to Earth' s surface. Some of the rain that falls onto the land sinks in to be taken up by plants or become groundwater and some flows across the surface toward the ocean.

Viewed over long time spans, the rocks of the solid Earth are constantly forming, changing, and reforming (Figure I.6). The loop that involves the processes by which one rock changes to another is called the rock cycle. The cycles of the Earth system, such as the hydrologic and rock cycles, are not independent of one another. To the contrary, there are many places where they interface. An **interface** is a common boundary where different parts of a system come in contact and

interact. For example, in Figure I.6 weathering at the surface gradually disintegrates and decomposes solid rock. Running water may eventually move this material to another place and deposit it. Later, groundwater percolating through the debris may leave behind mineral matter that cements the grains together into solid rock (a rock that is often very different from the rock we started with!). This changing of one rock into another could not have occurred without the movement of water through the hydrologic cycle. There are many places where one cycle or loop in the Earth system interfaces with and is a basic part of another.

The Earth system is powered by energy from two sources. The Sun drives external processes that occur in the atmosphere, hydrosphere, and at Earth's surface. Weather and climate, ocean circulation, and erosional processes are driven by energy from the Sun. Earth's interior is the second source of energy. Heat remaining from when our planet formed and heat that is continuously generated by radioactive decay powers the internal processes that produce volcanoes, earthquakes, and mountains.

Humans are *part of* the Earth system, a system in which the living and nonliving components are entwined and interconnected. Therefore, our actions produce changes in all of the other parts. When we burn gasoline and coal, build breakwaters along the shoreline, dispose of our wastes, and clear the land, we cause other parts of the system to respond, often in unforeseen ways. Throughout this book you will learn about many of Earth's subsystems: the hydrologic system, the tectonic (mountain-building) system, and the climate system, to name a few. Remember that these components *and we humans* are all part of the complex interacting whole we call the Earth system.

The organization of this text involves traditional groupings of chapters that focus on closely related topics. Nevertheless, the theme of *Earth as a system* keeps

Figure I.8 When Mount St. Helens erupted in May 1980, the area shown here was buried by a volcanic mudflow. Now, plants are reestablished and new soil is forming. (Photo by Jack Dykinga)

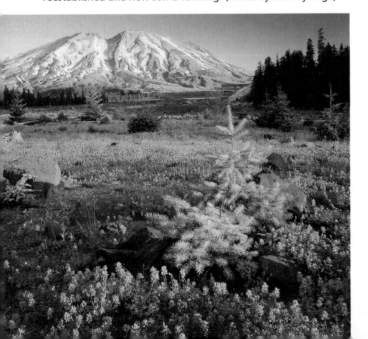

recurring through *all* major units of *Earth Science*. It is a thread that "weaves" through the chapters and helps tie them together. At the end of each chapter there is a section on *Examining the Earth System*. The questions and problems found there are intended to help you develop an awareness and appreciation for some of Earth system's important interrelationships.

People and the Environment

Environment refers to everything that surrounds and influences an organism. Some of these things are biological and social, but others are nonliving. The factors in this latter category are collectively called our *physical environment*. The physical environment encompasses water, air, soil, and rock, as well as conditions such as temperature, humidity, and sunlight. The phenomena and processes studied by the Earth sciences are basic to an understanding of the physical environment. In this sense, most of Earth science can be characterized as environmental science.

However, when the term *environmental* is applied to Earth science today, it is usually reserved for those aspects that focus on the relationships between people and the natural environment.

We humans can dramatically influence natural processes. For example, river flooding is natural, but the magnitude and frequency of flooding can be changed significantly by human activities such as clearing forests, building cities, and constructing dams. Unfortunately, natural systems do not always adjust to artificial changes in ways we can anticipate. Thus, an alteration to the environment that was intended to benefit society may have the opposite effect.

Interactions between people and the natural environment is a second important theme that recurs throughout this text and helps to tie its units together. This book's primary focus is to present basic Earth science principles. But while doing so we will have an opportunity to explore many human-environment issues along the way.

Resources

Resources are an important environmental concern. Resources range from water and soil to metallic and nonmetallic minerals and energy. These materials are the very basis of modern civilization. The mineral and energy resources that are extracted from the crust are the raw materials from which the products used by society are made (Figure I.9).

Few people who live in highly industrialized nations realize the quantity of resources needed to maintain their present standard of living. For example, the annual per capita consumption of metallic and nonmetallic mineral resources for the United States is

Figure I.9 As world population grows, demand for mineral and energy resources climbs. Earth scientists must deal with the search for additional supplies of traditional and alternative resources as well as the environmental impact of their extraction and use. These offshore oil rigs are in California's Santa Barbara channel. (Photo by Glenn Short/Liaison Agency, Inc.)

nearly 10,000 kilograms (11 tons). This is each person's prorated share of the materials required by industry to provide the vast array of products modern society demands. Figures for other highly industrialized countries are comparable.

Resources are commonly divided into two broad categories. Some are classified as **renewable**, which means that they can be replenished over relatively short time spans. Common examples are plants and animals for food, natural fibers for clothing, and forest products for lumber and paper. Energy from flowing water, wind, and the Sun are also considered renewable.

By contrast, many other basic resources are classified as **nonrenewable**. Important metals such as iron, aluminum, and copper fall into this category, as do our most important fuels: oil, natural gas, and coal. Although these and other resources continue to form, the processes that create them are so slow that significant deposits take millions of years to accumulate. In essence, Earth contains fixed quantities of these substances. When the present supplies are mined or

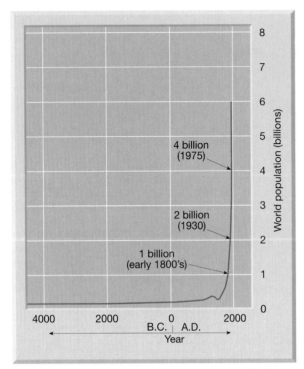

Figure I.10 Growth of world population. It took until 1800 for the number to reach 1 billion. By the year 2010, nearly 7 billion people may inhabit the planet. The demand for basic resources is growing faster than the rate of population increase. (Data from the Population Reference Bureau)

pumped from the ground, there will be no more. Although some nonrenewable resources, such as aluminum, can be used over and over again, others, such as oil, cannot be recycled.

Population

A glance at Figure I.10 shows that the population of our planet is growing rapidly. Although it took until the beginning of the nineteenth century for the number to reach 1 billion, just 130 years were needed for the population to double to 2 billion. Between 1930 and 1975 the figure doubled again to 4 billion, and by about 2010 as many as 7 billion people may inhabit the planet. Clearly, as population grows, the demand for resources expands as well. However, the rate of mineral and energy resource usage has climbed more rapidly than the overall growth of population. This fact results from an increasing standard of living. This is certainly true in the United States, where about 6 percent of the world's population uses approximately 30 percent of the annual production of mineral and energy resources.

How long will the remaining supplies of basic resources last? How long can we sustain the rising standard of living in today's industrialized countries and still provide for the growing needs of developing regions? How much environmental deterioration are we willing to accept in pursuit of basic resources? Can alternatives be found? If we are to cope with an increasing per capita demand and a growing world population, it is important that we have some understanding of our present and potential resources.

Environmental Problems

In addition to the quest for adequate mineral and energy resources, the Earth sciences must also deal with a broad array of other environmental problems. Some are local, some are regional, and still others are global in extent. Developed and developing nations alike face serious difficulties. Urban air pollution, acid rain, ozone depletion, and global warming are just a few concerns that pose significant threats (Figure I.11). Other problems involve the loss of fertile soils to erosion, the disposal of toxic wastes, and the contamination and depletion of water resources. The list continues to grow.

In addition to human-induced and human-accentuated problems, people must also cope with the many

Figure I.11 Air pollution in downtown Los Angeles. Air quality problems affect many cities. Fuel combustion by motor vehicles and power plants provide a high proportion of the pollutants. Meteorological factors determine whether pollutants remain "trapped" in the city or are dispersed. (Photo by Ted Spiegel/Black Star)

Figure I.12 There are many Earth processes and phenomena that are hazardous to people, including volcanoes, earthquakes, landslides, floods, hurricanes, and tornadoes. Heavy rains from Hurricane Mitch in November 1998 created devastating mudflows in Central America. (Photo by Noel Quidu/Liaison Agency, Inc.)

natural hazards posed by the physical environment (Figure I.12). Earthquakes, landslides, floods, and hurricanes are just four of the many risks. Others such as drought, although not as spectacular, are nevertheless equally important environmental concerns. Of course, environmental hazards are simply *natural* processes. They become hazards only when people try to live where these processes occur. In many cases, the threat of natural hazards is aggravated by increases in population as more people crowd into places where an impending danger exists.

It is clear that as world population continues its rapid growth, pressures on the environment will increase as well. Therefore, an understanding of Earth is not only essential for the location and recovery of basic resources, but also for dealing with the human impact on the environment and minimizing the effects of natural hazards. Knowledge about our planet and how it works is necessary to our survival and well-being. Earth is the only suitable habitat we have, and its resources are limited.

The Nature of Scientific Inquiry

As members of a modern society, we are constantly reminded of the benefits derived from science. But what exactly is the nature of scientific inquiry?

All science is based on the assumption that the natural world behaves in a consistent and predictable manner. The overall goal of science is to discover the underlying patterns in the natural world and then to use this knowledge to predict what will or will not happen, given certain facts or circumstances.

The development of new scientific knowledge involves some basic, logical processes that are universally accepted. To determine what is occurring in the natural world, scientists collect scientific facts through observation and measurement (Figure I.13). These data are essential to science and serve as the springboard for the development of scientific theories.

Hypothesis

Once facts have been gathered and principles have been formulated to describe a natural phenomenon, investigators try to explain how or why things happen in the manner observed. They can do this by constructing a preliminary, untested explanation, which we call a scientific **hypothesis**. Often, several different hypotheses are advanced to explain the same facts. Next, scientists think about what will occur or be observed if a hypothesis is correct and devise ways or methods to test the accuracy of predictions drawn from the hypothesis.

If a hypothesis cannot be tested, it is not scientifically useful, no matter how interesting it might seem. Testing usually involves making observations, developing models, and performing experiments. What if test results do not turn out as expected? One possibility is that there were errors in the observations or

Figure I.13 This scientist is surveying a glacier. (Photo by Charles D. Winters/Timeframe Photography, Inc./Photo Researchers, Inc.)

experiments. Of course, another possibility is that the hypothesis is not valid. Before rejecting the hypothesis, the tests might be repeated or new tests might be devised. The more tests the better. The history of science is littered with discarded hypotheses. One of the best known is the idea that Earth was at the center of the universe, a proposal that was supported by the apparent daily motion of the Sun, Moon, and stars around Earth.

Theory

When a hypothesis has survived extensive scrutiny, and when competing hypotheses have been eliminated, a hypothesis may be elevated to the status of a scientific **theory**. In everyday language, we may say "that's only a theory." But a scientific theory is a well-tested and widely accepted view that scientists agree best explains certain observable facts. It is not enough for scientific theories to fit only the data that are already at hand. Theories must also fit additional observations that were not used to formulate them in the first place. Put another way, theories should have predictive power.

Scientific theories, like scientific hypotheses, are accepted only provisionally. It is always possible that a theory which has withstood previous testing may eventually be disproven. As theories survive more testing, they are regarded with higher levels of confidence. Theories that have withstood extensive testing, such as the theory of plate tectonics or the theory of evolution, are held with a very high degree of confidence.

Scientific Methods

The process just described, in which scientists gather facts though observations and formulate scientific hypotheses and theories, is called the *scientific method*. Contrary to popular belief, the scientific method is not a standard recipe that scientists apply in a routine manner to unravel the secrets of our natural world. Rather, it is an endeavor that involves creativity and insight. Rutherford and Ahlgren put it this way: "Inventing hypothesis or theories to imagine how the world works and then figuring out how they can be put to the test of reality is as creative as writing poetry, composing music, or designing skyscrapers."[*]

Modern scientific endeavors frequently incorporate new technologies. With the development of high-speed computers, scientists began creating models that attempted to simulate what was happening in the "real" world. These models have become quite useful when dealing with processes that occur on very long time scales or are physically very large, or take place under extreme conditions. For example, it should be obvious that to design equipment capable of duplicating the conditions found deep within Earth and at the scale required would be virtually impossible. Although not without its own drawbacks, modeling helps narrow the gap between what we can measure and observe and the determination of the processes that produce those observations.

There is not a fixed path that scientists always follow that leads unerringly to scientific knowledge. Nevertheless, many scientific investigations involve the following steps: (1) the collection of scientific facts through observation and measurement (Figure I.14); (2) the development of one or more working hypotheses to explain these facts; (3) development of observations and experiments to test the hypotheses; and (4) the acceptance, modification, or rejection of the hypothesis based on extensive testing. Other scientific discoveries represent purely theoretical ideas, which stand up to extensive examination. Still other scientific advancements have been made when a totally unexpected happening occurred during an experiment. These serendipitous discoveries are more than pure luck; for as Louis Pasteur said, "In the field of observation, chance favors only the prepared mind."

Scientific knowledge is acquired through several avenues, so it might be best to describe the nature of scientific inquiry as the methods of science, rather than the scientific method.

In this book you will discover the results of centuries of scientific work. You will see the end product

---

[*]F. James Rutherford and Andrew Ahlgren, *Science for All Americans* (New York: Oxford University Press, 1990), p. 7.

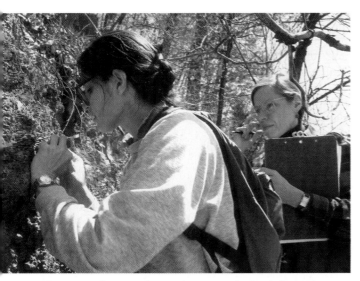

Figure I.14 These Earth scientists are gathering information about an exposure of sedimentary rock. (Photo by Catherine Ursillo/Photo Researchers, Inc..)

of millions of observations, thousands of hypotheses, and hundreds of theories. We have distilled all of this to give you a "briefing" on Earth science.

But realize that our knowledge of Earth is changing daily, as thousands of scientists worldwide make satellite observations, analyze drill cores from the seafloor, measure earthquakes, develop computer models to predict climate, examine the genetic codes of organisms, and discover new facts about our planet's long history. This new knowledge often updates hypotheses and theories. Expect to see many new discoveries and changes in scientific thinking in your lifetime.

The Earth Sciences

Earth science is the name for all the sciences that collectively seek to understand Earth and its neighbors in space. It includes geology, oceanography, meteorology, and astronomy. This text consists of four units, one devoted to each of these traditional Earth science disciplines.

The focus of Part One of this text, *The Solid Earth*, is the science of **geology**, a word that literally means "study of Earth." Geology is traditionally divided into two broad areas—physical and historical.

Physical geology examines the materials composing Earth and seeks to understand the many processes that operate beneath and upon its surface. Earth is a dynamic, ever-changing planet. Internal forces create earthquakes, build mountains, and produce volcanic structures. At the surface, external processes break rock apart and

sculpture a broad array of landforms. The erosional effects of water, wind, and ice result in a great diversity of landscapes. Because rocks and minerals form in response to Earth's internal and external processes, their interpretation is basic to an understanding of our planet.

In contrast to physical geology, the aim of *historical geology* is to understand the origin of Earth and the development of the planet through its 4.5-billion-year history. It strives to establish an orderly chronological arrangement of the multitude of physical and biological changes that have occurred in the geologic past.

Part Two of the text, *The Oceans*, is devoted to **oceanography**. Oceanography is actually not a separate and distinct science. Rather, it involves the application of all sciences in a comprehensive and interrelated study of the oceans in all their aspects and relationships. Oceanography integrates chemistry, physics, geology, and biology. It includes the study of the composition and movements of seawater, as well as coastal processes, seafloor topography, and marine life.

Part Three, *The Atmosphere*, examines the mixture of gases that is held to the planet by gravity and thins rapidly with altitude. Acted on by the combined effects of Earth's motions and energy from the Sun, the formless and invisible atmosphere reacts by producing an infinite variety of weather, which, in turn, creates the basic pattern of global climates. **Meteorology** is the study of the atmosphere and the processes that produce weather and climate. Like oceanography, meteorology involves the application of other sciences in an integrated study of the thin layer of air that surrounds Earth.

Part Four, *Astronomy*, demonstrates that the study of Earth is not confined to investigations of its four interacting "spheres." The Earth sciences also attempt to relate our planet to the larger universe. Because Earth is related to all of the other objects in space, the science of **astronomy**—the study of the universe—is very useful in probing the origins of our own environment. As we are so closely acquainted with the planet on which we live, it is easy to forget that Earth is just a tiny object in a vast universe. Indeed, Earth is subject to the same physical laws that govern the many other objects populating the great expanses of space. Thus, to understand explanations of our planet's origin, it is useful to learn something about the other members of our solar system. Moreover, it is helpful to view the solar system as a part of the great assemblage of stars that comprise our galaxy, which, in turn, is but one of many galaxies (see Box I.1).

Box I.1   Earth as a System: Earth's Place in the Cosmos*

For centuries, those who have gazed at the night sky have wondered about the nature of the universe, Earth's place within it, and whether or not we are "alone." Today, many exciting discoveries in astronomy are beginning to provide some answers concerning the origin of the universe, the formation and evolution of stars, and how Earth and its materials came into existence.

The realization that the universe is immense and orderly began in the early 1900s when Edwin Hubble and other dedicated scientists demonstrated that the Milky Way Galaxy is one of hundreds of billions of galaxies, each of which contains billions of stars. Today we understand that Earth, its materials, and indeed all living things are the consequence of a sequence of events, governed by natural laws, that occurred within the universe during the past 10–15 billion years.

Astronomical evidence supports the theory, called the Big Bang, that the universe began perhaps as long as 20 billion years ago when a dense, hot, supermassive concentration of material

*This box was prepared by Kenneth Pinzke, Belleville Area College.

exploded with cataclysmic force (Figure I.A). Within about one second, the temperature of the expanding universe cooled to approximately 10 billion degrees and fundamental atomic particles called protons and neutrons began to appear. After a few minutes, atoms of the least complex elements—hydrogen and helium—had formed, and the initial conversion of energy to matter in the young universe was completed.

During the first billion years or so, matter (essentially hydrogen and to a lesser extent helium) in the expanding universe became clumpy and fragmented into enormous clouds and groups of clouds that eventually collapsed to become galaxies and clusters of galaxies. Inside these collapsing clouds, smaller concentrations of matter formed into stars. One of the billions of galaxies to form was the one we call the Milky Way.

The Milky Way is a collection of several hundred billion stars, the oldest of which is about 10 billion years. It is one of a cluster of approximately 28 galaxies, called the Local Group, that exist in our region of the universe. Initially, the oldest stars in the Milky Way formed from nearly pure hydrogen. Later, succeeding

generations of younger stars, including our Sun, would have heavier, more complex atoms available for their formation.

After forming, all stars go through a series of changes before finally dying out. During the life of most stars, energy is produced as hydrogen nuclei (protons) fuse with other hydrogen nuclei to form helium. During this process, called *nuclear fusion*, matter is converted to energy. Stars begin to die when their nuclear fuel becomes exhausted. Massive stars often have indescribably explosive deaths. During these violent events, called supernovas, nuclear fusion advances beyond helium and produces heavier, more complex atoms such as oxygen, carbon, and iron. In turn, these products of stellar death may become the materials that make up future generations of stars. It was from the debris scattered during the death of a preexisting star, or stars, that our Sun, as well as the rest of the solar system, formed.

Our star, the Sun, is at the very least a second-generation star. It, along with the planets and other members of the solar system, began forming nearly 5 billion years ago, some 15 billion years after the Big Bang, from a large, interstellar cloud,

Chapter Summary

- Earth's four major spheres are the *hydrosphere, atmosphere, solid Earth,* and *biosphere.* The water of the hydrosphere is continually on the move, from the oceans to the atmosphere, to the land, and back to the ocean again. The atmosphere is the life-giving gaseous envelope that surrounds Earth. Lying beneath the atmosphere and ocean is the solid Earth, which consists of the dense *core,* the less dense *mantle,* and the *crust.* The biosphere, which extends into and influences the other three spheres, includes all life on Earth.

- Although each of Earth's four spheres can be studied separately, they are all related in a complex and continuously interacting whole that we call the *Earth system. Changing one part of the Earth system can produce changes in any or all the other parts.* The Earth system is characterized by processes that vary on spatial scales from fractions of mil-

limeters to thousands of kilometers, and on time scales that range from milliseconds to billions of years. Two of the many cycles that are part of the Earth system are the *hydrologic cycle* (the unending circulation of Earth's water among the hydrosphere, atmosphere, biosphere, and solid Earth) and the *rock cycle* (the cycle that involves the processes by which one rock changes to another). Cycles are not independent of one another. The common boundary where different parts of a system connect and interact is referred to as an *interface.*

- The *two sources of energy that power the Earth system are* (1) *the Sun,* which drives the external processes that occur in the atmosphere, hydrosphere, and at Earth's surface, and (2) *heat from Earth's interior* that powers the internal processes that produce volcanoes, earthquakes, and mountains.

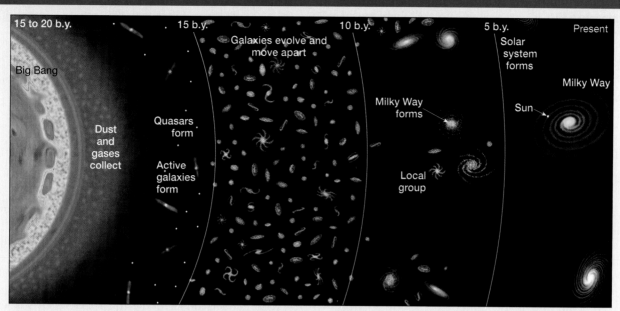

Figure I.A About 20 billion years ago, an incomprehensibly large explosion sent all the matter of the universe flying outward at incredible speeds. After a few billion years, the material cooled and condensed into the first stars and galaxies. Because the universe is expanding, the evolving galaxies continue to move away from one another. About 5 billion years ago, our solar system began forming in one of these galaxies, the Milky Way.

called a *nebula*, which consisted of dust particles and gases enriched in heavy elements from supernova explosions. Gravitational energy caused the nebula to contract and, in so doing, it began to rotate and flatten. Inside, smaller concentrations of matter began condensing to form the planets. At the center of the nebula there was sufficient pressure and heat to initiate hydrogen nuclear fusion, and our Sun was born.

It has been said that all life on Earth is related to the stars. This is true because the very atoms in our bodies, as well as the atoms that make up every other thing on Earth, owe their origin to a supernova event that occurred billions of years ago, trillions of kilometers away.

- Humans are part of the Earth system. Our actions produce changes in all of the other parts of the system.

- *Environment* refers to everything that surrounds and influences an organism. These influences can be biological, social, or physical. When applied to Earth science today, the term *environmental* is usually reserved for those aspects that focus on the relationships between people and the natural environment.

- *Resources* are in important environmental concern. The two broad categories of resources are (1) *renewable*, which means that they can be replenished over relatively short time spans, and (2) *nonrenewable*. As population grows, the demand for resources expands as well.

- Environmental problems can be local, regional, or global. Human-induced problems include urban air pollution, acid rain, ozone depletion, and global warming. Natural hazards include earthquakes, landslides, floods, and hurricanes. As world population grows, pressures on the environment also increase.

- All science is based on the assumption that the natural world behaves in a consistent and predictable manner. The process through which scientists gather facts through observations and formulate scientific *hypotheses* and *theories* is called the *scientific method*. To help determine what is occurring in the natural world, scientists often (1) collect facts, (2) develop a scientific hypothesis, (3) construct experiments to validate the hypothesis, and (4) accept, modify, or reject the hypothesis on the basis of extensive testing. Other discoveries represent purely theoretical ideas that have stood up to extensive examination. Still other scientific advancements have been made when a totally unexpected happening occurred during an experiment.

- *Earth science* is the name for all the sciences that collectively seek to understand Earth and its neighbors in space. It includes *geology, oceanography, meteorology,* and *astronomy*. Geology is traditionally divided into two broad areas—*physical* and *historical*.

Key Terms

astronomy (p. 11)
atmosphere (p. 2)
biosphere (p. 3)
core (p. 3)
crust (p. 3)
geology (p. 11)

hydrosphere (p. 2)
hypothesis (p. 9)
interface (p. 6)
mantle (p. 3)
meteorology (p. 11)
nonrenewable resource (p. 7)

oceanography (p. 11)
renewable resource (p. 7)
system (p. 4)
theory (p. 10)

Review Questions

1. List and briefly define the four "spheres" that constitute our environment.

2. Name the three principal divisions of the solid Earth. Which one is thinnest?

3. The oceans cover nearly _____ percent of Earth's surface and contain about _____ percent of the planet's total water supply.

4. Contrast renewable and nonrenewable resources. Give one or more examples of each.

5. World population may reach nearly ____ billion by the year 2010. How does this compare to the world population near the beginning of the nineteenth century?

6. List at least four phenomena that could be regarded as natural hazards.

7. How is a scientific hypothesis different from a scientific theory?

8. Name the specific Earth science described by each of the following statements.
   (a) The science that deals with the dynamics of the oceans.
   (b) This word literally means "the study of Earth."
   (c) An understanding of the atmosphere is the primary focus of this science.
   (d) This science helps us understand Earth's place in the universe.

Examining the Earth System

1. Examine Figure I.1 closely. Selecting among Earth's four major spheres (atmosphere, solid Earth, hydrosphere, and biosphere) indicate which sphere is represented by each of the following features: (a) mountain, (b) tree, (c) cloud, (d) snow.

2. Referring to Figure I.1, describe two situations in which two or more of the Earth's sphere's (atmosphere, solid Earth, hydrosphere, and biosphere) interface (come in contact) with one another. Indicate which spheres are involved in each case.

Web Resources

For direct links to numerous appropriate Web locations, contact the *Earth Science Ninth Edition* Home Page at:

- **http://www.prenhall.com/tarbuck.**

This site also contains chapter review exercises and relevant topical investigations that utilize Web resources.

Part One

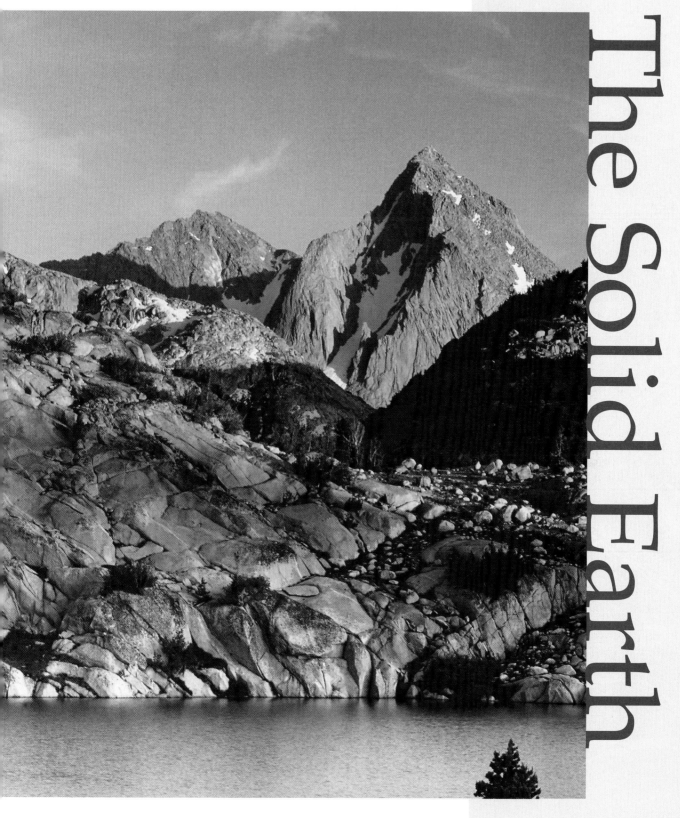

The Solid Earth

Evolution Peaks, Kings Canyon National Park, California.
(Photo by Galen Rowell/Mountain Light Photography, Inc.)

Chapter 1

Minerals: Building Blocks of Rocks

Aerial view of the giant Bingham Canyon copper mine near Salt Lake City, Utah.
Materials that we extract from Earth are the basis of modern civilization. Clearly such
extraction has an impact on our environment. (Photo by Michael Collier)

The study of geology appropriately begins with the study of rocks and minerals, because a knowledge of the materials that make up Earth is basic to our understanding of much of the rest of geology. In this chapter we shall examine atoms and elements, the building blocks of minerals, and then investigate the major mineral groups. Our inquiry shall attempt to answer some important questions: How are minerals identified? How are rocks and minerals similar and how are they different? What are the major rock-forming minerals? What defines a mineral resource?

Earth's crust is the source of a wide variety of minerals, many of which are useful and essential to people. In fact, practically every manufactured product contains materials obtained from minerals. Most people are familiar with the common uses of basic metals, including aluminum in beverage cans, copper in electrical wiring, gold in jewelry, and silicon in computer chips. But fewer are aware that pencil "lead" does not contain lead metal but is really made of the soft black mineral called *graphite*. Baby powder ("talcum powder") is ground-up rock made of the mineral *talc*, with perfume added. Drill bits impregnated with pieces of *diamond* (a mineral) are used by dentists to drill through tooth enamel. The common mineral *quartz* is the main ingredient in ordinary glass and is the source of silicon for computer chips. And on and on. As the material demands of modern society grow, the need to locate additional supplies of useful minerals also grows, and becomes more challenging as the easily mined sources become depleted.

The economic uses of rocks and minerals are important to us. Also important to Earth scientists is that every process they study depends in some way on the properties of these basic Earth materials. Events such as volcanic eruptions, mountain building, weathering and erosion, and even earthquakes involve rocks and minerals (Figure 1.1). Consequently, a basic knowledge of Earth materials is essential to the understanding of all Earth science phenomena.

Figure 1.1 Lone Pine Peak in California's Sierra Nevada. (Photo by Carr Clifton)

Minerals Versus Rocks

Earth Materials
Minerals

Many people think of rocks as nondescript, hard, often dirty objects. Minerals are considered by many to be part of the human diet ("vitamins and minerals"), or possibly rare ores or precious gems. However, these common perceptions are not entirely correct.

Figure 1.2 shows the relation between rocks and minerals. A **rock** can be defined simply as an aggregate of minerals. Minerals are the *building blocks* that make up rocks. In a rock, the minerals occur together as a *mixture,* so each mineral retains its distinctive properties, such as color, as you can see in Figure 1.2.

Although most rocks are composed of more than one mineral, certain minerals are commonly found by themselves in large, relatively pure quantities. In these instances, they are still considered to be rocks. A common example is the mineral *calcite,* which frequently is the dominant constituent in widespread rock layers where it is given the name *limestone.*

Minerals, the building blocks of rocks, have a very precise definition. A **mineral** is a naturally occurring inorganic solid that possesses a definite chemical structure, which gives it a unique set of physical properties.

Thus, for any Earth material to be considered a mineral, it must exhibit the following characteristics:

It must be naturally occurring.

It must be inorganic (was never alive).

It must be a solid.

It must possess a definite chemical structure.

When the term "mineral" is used by geologists, only those substances that fulfill these precise conditions are considered minerals. Consequently, synthetic diamonds, although chemically the same as natural diamonds, are human-made, and therefore are not considered minerals. Further, oil and natural gas, which are not solids, and which once were "living" (they are the remains of ancient organisms), are also not classified as minerals. An exception is material formed by the activity of animals, such as clam shells, which are considered minerals.

Composition and Structure of Minerals

Each of Earth's nearly 4000 minerals is a unique substance that is defined by its chemical composition and internal structure. What elements make up a particular

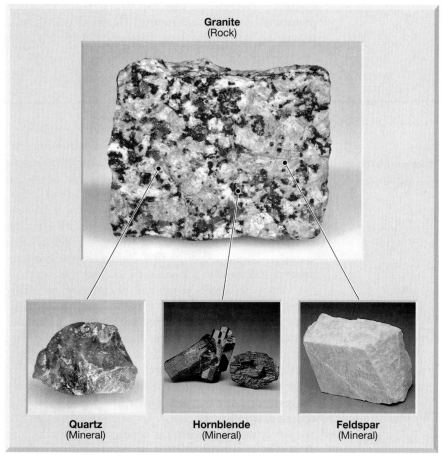

Figure 1.2 Rocks are aggregates of one or more minerals.

mineral? In what pattern are its elements bonded together? And what important properties does this give each mineral?

At this point, let us briefly review the basic building blocks of minerals, the chemical **elements.** At present, 112 elements are known, although 20 of these have been produced only in the laboratory. Some minerals, such as gold and sulfur, are made entirely from one element. But most elements are not stable, and thus most minerals are a combination of two or more elements, joined to form a chemically stable compound.

To understand better how elements combine to form minerals, we must first consider the atom. The **atom** is the smallest particle of matter that has all the characteristics of an element. It is this extremely small particle that does the combining.

How Atoms Are Constructed

Two simplified models of an atom are shown in Figure 1.3 . Each atom has a central region, called the **nucleus,** which contains very dense **protons** (particles with positive electrical charges) and equally dense **neutrons** (particles with neutral electrical charges). Orbiting the

A.

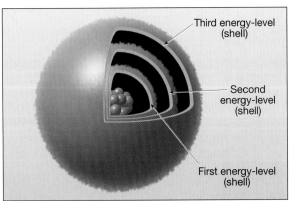

B.

Figure 1.3 Two models of the atom. **A.** A very simplified view of the atom, which consists of a central nucleus, consisting of protons and neutrons, encircled by high-speed electrons. **B.** Another model of the atom showing spherically shaped electron clouds (energy level shells). Note that these models are not drawn to scale. Electrons are minuscule in size compared to protons and neutrons, and the relative space between the nucleus and electron shells is much greater than illustrated.

nucleus are **electrons,** which have negative electrical charges. For convenience, we often diagram atoms to show the electrons orbiting the nucleus, like the orderly orbiting of the planets around the Sun (Figure 1.3A). However, electrons move so rapidly that they create a sphere-shaped negative zone around the nucleus. Hence, a more realistic picture of the positions of electrons can be obtained by envisioning a cloud of negatively charged electrons surrounding the nucleus (Figure 1.3B).

It is also known that individual electrons are located at given distances from the nucleus in regions called **energy levels,** or **shells.** Three such shells are shown in Figure 1.3. As we shall see, an important fact about these shells is that each can have only a specific number of electrons.

The number of protons in an atom's nucleus determines its **atomic number** and name of the element. For example, all atoms with six protons are carbon atoms, all those with eight protons are oxygen atoms, and so forth. "Free" atoms (those not combined with other atoms) have the same number of electrons as protons, so the atomic number also equals the number of electrons surrounding the nucleus. Therefore, carbon has six electrons to match its six protons, and oxygen has eight electrons to match its eight protons. Neutrons have no charge, so the positive charge of the protons is exactly balanced by the negative charge of the electrons. Consequently, uncombined atoms are neutral electrically and have no overall electrical charge. Thus, the *element* is a large collection of electrically neutral atoms, all having the same atomic number.

How Atoms Bond Together

Elements combine with each other to form a wide variety of more complex compounds. A **compound** consists of two or more elements bonded together in definite proportions. When the elements separate, the bonds are broken and the compound no longer exists; only its component elements do. Through experimentation, scientists have learned that the forces bonding atoms together are electrical. Further, it is known that chemical bonding results in a change in the arrangement of electrons in the shells of the bonded atoms.

When an atom combines chemically, it either gains, loses, or shares electrons with another atom. An atom that gains electrons becomes negatively charged because it has more electrons than protons. Atoms that lose electrons become positively charged. Atoms that have an electrical charge because of a gain or loss of electrons are called **ions.** Simply stated, oppositely charged ions attract one another to produce a neutral chemical compound.

An example of chemical bonding involves sodium (Na) atoms and chlorine (Cl) atoms to produce sodium chloride molecules (NaCl). Sodium chloride is common

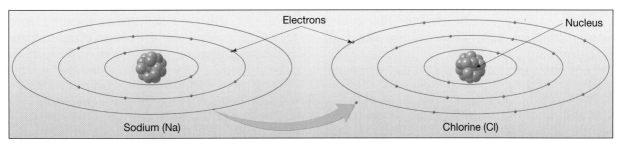

Figure 1.4 Chemical bonding of sodium and chlorine atoms to produce sodium chloride. Through the transfer of one electron in the outer shell of a sodium atom to the outer shell of a chlorine atom, the sodium becomes a positive ion and chlorine a negative ion.

table salt (Figure 1.4). When the sodium atom loses one electron it becomes a positive ion, and when the chlorine atom gains one electron it becomes a negative ion. These opposite charges act as an "electrical glue" that bonds the atoms together.

The properties of a chemical compound are dramatically different from the properties of the elements composing it. For example, chlorine is a green, poisonous gas that is so toxic it was used as a weapon during World War I. Sodium is a soft, silvery metal that reacts vigorously with water and, if held in your hand, could burn it severely. Together, however, these atoms produce the compound sodium chloride (table salt), the familiar clear, crystalline solid that is essential for human life.

This example also illustrates an important difference between a rock and a mineral. When elements join to form a mineral, the constituent elements lose their physical properties. In a *rock,* on the other hand, which is a *mixture* of minerals, each mineral retains its own distinctive physical properties.

Isotopes and Radioactive Decay. Subatomic particles are so incredibly small that a special unit was devised to express their mass. A proton or a neutron has a mass just slightly more than one *atomic mass unit,* whereas an electron is only about one two-thousandth of an atomic mass unit. Thus, although electrons play an active role in chemical reactions, they do not contribute significantly to the mass of an atom.

The **mass number** of an atom is simply the total of its neutrons and protons in the nucleus. Atoms of the same element always have the same number of protons, but commonly have varying numbers of neutrons. This means that an element can have more than one mass number. These variants of the same element are called **isotopes** of that element.

For example, carbon has two well-known isotopes, one having a mass number of 12 (carbon-12), the other a mass number of 14 (carbon-14). All atoms of the same element must have the same number of protons, and carbon always has six (atomic number = 6). Hence, carbon-12 must have six protons *plus six neutrons* to give it a mass number of 12, whereas carbon-14 must have six protons *plus eight neutrons* to give it a mass number of 14.

The *average* atomic mass of any random sample of carbon is much closer to 12 than 14, because carbon-12 is the more common isotope. This average is called *atomic weight.**

In chemical behavior, all isotopes of the same element are nearly identical. To distinguish among them is like trying to differentiate identical twins, with one slightly heavier. Because isotopes of an element react the same chemically, different isotopes can become parts of the same mineral. For example, when calcite is formed from calcium, carbon, and oxygen, some of its carbon atoms are carbon-12 and some are carbon-14.

Although the nuclei of most atoms are stable, many elements do have isotopes in which the nuclei are unstable. "Unstable" means that the isotopes disintegrate through a process called **radioactive decay.** It is so called because a decaying atom actively radiates energy and particles. Radioactive decay occurs when the forces that bind the nucleus are not strong enough to keep it together.

The rate at which the unstable nuclei decay is steady and measurable. This makes such isotopes useful "clocks" for dating the events of Earth history. A discussion of radioactive decay and its applications in dating past geologic events is found in Chapter 10.

Properties of Minerals

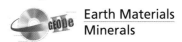

Earth Materials
Minerals

Minerals occur in a fascinating palette of colors, shapes, and lusters. They vary in hardness and may even have a distinctive taste or scent. We will now look at these properties, which are useful to help identify minerals.

Crystal Form

Crystal form is the external expression of a mineral's internal orderly arrangement of atoms. Generally, when a mineral forms without space restrictions, it

---

*The term *weight* is a misnomer that has resulted from long use. The correct term is *atomic mass.*

A.

B.

Figure 1.5 Crystal form is the external expression of a mineral's orderly internal structure. **A.** Pyrite, commonly known as "fool's gold," often forms cubic crystals that may contain parallel lines called striations. (Photo by GeoScience/PH) **B.** Quartz sample that exhibits well-developed hexagonal (six-sided) crystals with pyramidal-shaped ends. (Photo B. by Breck P. Kent)

will develop individual crystals with well-formed crystal faces. Figure 1.5A illustrates this for pyrite, an iron-and-sulfur mineral with cubic crystals. Figure 1.5B shows the distinctive hexagonal crystals of quartz that grow when space and time permit. However, most of the time, crystal growth is severely constrained. It is stunted because of competition for space, resulting in an intergrown mass of small, jammed crystals, none of which exhibits its crystal form. This is what happened to the minerals in the granite in Figure 1.2. Thus, most inorganic solid objects are composed of crystals, but they are not clearly visible to the unaided eye.

Luster

Luster is the appearance or quality of light reflected from the surface of a mineral. Minerals that have the appearance of metals, regardless of color, are said to have a *metallic luster,* like the pyrite crystals in Figure 1.5A. Minerals with a *nonmetallic luster* are described

by various adjectives. These include vitreous (glassy, like the quartz crystals in Figure 1.5B), pearly, silky, resinous, and earthy (dull).

Color

Although **color** is an obvious feature of a mineral, it is often an unreliable diagnostic property. Slight impurities in the common mineral quartz, for example, give it a variety of colors, including pink, purple (amethyst), milky white, and even black.

Streak

Streak is the color of a mineral in its powdered form, which is a much more reliable indication of color. Streak is obtained by rubbing the mineral across a piece of hard, unglazed porcelain, termed a *streak plate.* Whereas the color of a mineral may vary from sample to sample, the streak usually does not, and is therefore

Table 1.1 Mohs scale of mineral hardness.

Relative Scale		Mineral	Hardness of Some Common Objects
Hardest	10	Diamond	
	9	Corundum	
	8	Topaz	
	7	Quartz	
	6	Potassium Feldspar	5.5 Glass, Pocketknife
	5	Apatite	
	4	Fluorite	3.5 Copper Penny
	3	Calcite	2.5 Fingernail
	2	Gypsum	
Softest	1	Talc	

the more consistent property. Streak can also help to distinguish minerals with metallic lusters from those having nonmetallic lusters. Metallic minerals generally have a dense, dark streak, whereas minerals with nonmetallic lusters do not.

Hardness

One of the most useful diagnostic properties is **hardness,** a measure of the resistance of a mineral to abrasion or scratching. This property is determined by rubbing the mineral to be identified against another mineral of known hardness. One will scratch the other (unless they have the same hardness).

Geologists use a standard hardness scale, called the **Mohs scale.** It consists of 10 minerals arranged in order from 10 (hardest) to 1 (softest), as shown in Table 1.1. Any mineral of unknown hardness can be rubbed against these to determine its hardness. In the field, other handy objects work, too. For example, your fingernail has a hardness of 2.5, a copper penny 3, and a piece of glass 5.5. The mineral gypsum, which has a hardness of 2, can be easily scratched by your fingernail. Conversely, the similar-looking mineral calcite, which has a hardness of 3, cannot be scratched by your fingernail. Calcite cannot scratch glass, because its hardness is less than 5.5. Quartz, the hardest of the common minerals at 7, will scratch a glass plate. Diamonds, hardest of all, scratch anything. You can see that by rubbing an unknown mineral against objects of known hardness, you can quickly narrow its hardness range.

Cleavage

In the crystal structure of a mineral, some bonds are weaker than others. These bonds are where a mineral will break when it is stressed. **Cleavage** is the tendency of a mineral to cleave, or break, along planes of

Figure 1.6 Sheet-type cleavage common to the micas. (Photo by Chip Clark)

weak bonding. Not all minerals have definite planes of weak bonding, but those that possess cleavage can be identified by the distinctive smooth surfaces that are produced when the mineral is broken.

The simplest type of cleavage is exhibited by minerals called micas (Figure 1.6). Because the micas have weak bonds in one direction, they cleave to form thin, flat sheets. Some minerals have several cleavage planes that produce smooth surfaces when broken, whereas others exhibit poor cleavage, and still others have no cleavage at all. When minerals break evenly in more than one direction, cleavage is described by the *number of planes* exhibited and the *angles at which they meet* (Figure 1.7).

Do not confuse *cleavage* with *crystal form!* When a mineral exhibits cleavage, it will break into pieces *that have the same geometry as each other.* By contrast, the quartz crystals shown in Figure 1.5B do not have

Figure 1.7 Smooth surfaces produced when a mineral with cleavage is broken. The sample on the left (fluorite) exhibits four planes of cleavage (eight sides), whereas the other two samples exhibit three planes of cleavage (six sides). Also notice that the mineral in the center (halite) has cleavage planes that meet at 90-degree angles, whereas the mineral on the right (calcite) has cleavage planes that meet at 75-degree angles.

← 2 cm →

Figure 1.8 Conchoidal fracture. The smooth curved surfaces result when minerals break in a glasslike manner. (Photo by E. J. Tarbuck)

cleavage. If broken, they fracture into pieces that do not resemble each other or the original crystals.

Fracture

Minerals that do not exhibit cleavage when broken, such as quartz, are said to **fracture**. Even fracturing has variety: Minerals that break into smooth curved surfaces like those seen in broken glass have a *conchoidal fracture* (Figure 1.8). Others break into splinters or fibers, like asbestos, but most minerals fracture irregularly.

Specific Gravity

Specific gravity compares the weight of a mineral to the weight of an equal volume of water. For example, if a cubic centimeter of a mineral weighs three times as much as a cubic centimeter of water, its specific gravity is 3. With a little practice, you can estimate the specific gravity of minerals by hefting them in your hand.

← 5 cm →

Figure 1.9 Galena is lead sulfide. Like other metallic ores, it has a relatively high specific gravity and therefore feels heavy when hefted.

For instance, if a mineral feels as heavy as the common rocks you have handled, its specific gravity will probably be somewhere between 2.5 and 3. Some metallic minerals have a specific gravity noticeably greater than that of common rock-forming minerals. Galena, which is an ore of lead, has a specific gravity of roughly 7.5 (Figure 1.9). The specific gravity of pure 24-karat gold is a very hefty 20 (see Box 1.1).

Other Properties of Minerals

In addition, some minerals can be recognized by other distinctive properties. For example, halite is ordinary salt, so it is quickly identified with your tongue. Thin sheets of mica will bend and elastically snap back. Gold is malleable, which means it can be easily hammered or shaped. Talc and graphite both have distinctive feels. Talc feels soapy. Graphite feels greasy (it is a principal ingredient in dry lubricants).

A few minerals, such as magnetite, have a high iron content and can be picked up with a magnet. Some varieties of magnetite (lodestone) are natural magnets and will pick up small iron-based objects such as pins and paper clips. Some minerals exhibit special optical properties. For example, when a transparent piece of calcite is placed over printed material, the letters appear doubled. This optical property is known as *double refraction.* In addition, the streak of many sulfur-bearing minerals smells like rotten eggs.

One very simple chemical test involves placing a drop of dilute hydrochloric acid from a dropper bottle on a freshly broken mineral surface. Certain minerals, called carbonates, will effervesce (fizz) with hydrochloric acid. This test is useful in identifying the mineral calcite, which is a common carbonate mineral.

In summary, a number of special physical and chemical properties are useful in identifying particular minerals. These include taste, smell, elasticity, malleability, feel, magnetism, double refraction, and chemical reaction to hydrochloric acid. Remember that every one of these properties depends on the composition (elements) of a mineral and its structure (how the atoms are arranged).

Mineral Groups

Earth Materials
Minerals

Nearly 4000 minerals have been named and about 40 to 50 new ones are identified each year. Fortunately, for students who are beginning to study minerals, no more than a few dozen are abundant! Collectively, these few make up most of the rocks of Earth's crust and, as such, are classified as the *rock-forming minerals.* It is also interesting to note that *only eight elements* make up the bulk of these minerals and represent over 98 percent (by weight) of the continental crust (Table 1.2).

Box 1.1 People and the Environment: Asbestos: What Are the Risks?

Once considered safe enough to use in toothpaste, asbestos may have become the most feared contaminant on Earth. Although health concerns arose two decades earlier, the asbestos panic began in 1986 when the Environmental Protection Agency (EPA) instituted the Asbestos Hazard Emergency Response Act. It required inspection of all public and private schools for asbestos. This brought asbestos to public attention and raised parental fears that their children could contract asbestos-related cancers because of high levels of airborne fibers in schools. Since that time, billions of dollars have been spent on asbestos testing and removal.

What Is Asbestos?

Asbestos is not a single material. Rather, asbestos is a general term for a group of silicate minerals that *readily separate into thin, strong fibers* (Figure 1.A). Because these fibers are flexible, heat resistant, and relatively chemically inert, they have many uses. Asbestos has been widely used to strengthen concrete, make fireproof fabrics, and insulate boilers and hot pipes. It is a component in floor tiles and the major ingredient in automobile brake linings. Wall coatings rich in asbestos fibers were used extensively during the U.S. building boom of the 1950s and early 1960s.

Most asbestos comes from three minerals. *Chrysotile* ("white asbestos") is a fibrous form of the mineral serpentine. It is the only type of asbestos mined in North America and once made up 95 percent of world production. *Crocidolite* ("blue asbestos") and *amosite* ("brown asbestos") are currently mined in South Africa and make up about 5 percent of world production.

Exposure and Risk

There is no question that prolonged exposure to air laden with certain types of asbestos dust in an unregulated workplace can be dangerous. When the thin, rodlike fibers are inhaled into the lungs, they are neither easily broken down nor easily expelled, but can remain *for life*. Three lung diseases can result: (1) *asbestosis*, a scarring of tissue, that decreases the lung's ability to absorb oxygen; (2) *mesothelioma*, a rare tumor that develops in the chest or gut; and (3) *lung cancer*.

Evidence that incriminates "blue asbestos" and "brown asbestos" comes from health studies at mines in South Africa and western Australia. Miners and mill workers showed an extremely high incidence of mesothelioma, sometimes after less than a year of exposure.

However, the U.S. Geological Survey concludes that the risks from the most widely used form of asbestos (chrysotile or "white asbestos") are minimal to nonexistent. They cite studies of miners of white asbestos in Canada and in northern Italy, where mortality rates from mesothelioma and lung cancer differ very little from the general public.

Another study was conducted on miner's wives in the area of Thetford Mines, Quebec, once the largest chrysotile mine in the world. For many years there were no dust controls on mining and milling operations, so these women were exposed to extremely high levels of airborne asbestos. Nevertheless, they exhibited *below-normal* levels of the diseases thought to be associated with asbestos exposure.

The various types of asbestos fibers differ in their chemical composition, shape, and durability. The thin, rodlike "blue asbestos" and "brown asbestos" fibers, which can easily penetrate the lining of the lungs, are certainly pathogenic. But chrysotile fibers are curly and can be expelled more easily than can the rodlike fibers. Further, if inhaled, chrysotile fibers break down within a year. This is not the case for the other forms of asbestos, or for fiberglass, which is frequently used as a substitute for asbestos. These differences are thought to explain the fact that the mortality rates for chrysotile workers differ very little from the rates for the general population. Although the EPA has lifted its ban on asbestos, very little of this once exalted mineral is presently used in the United States.

What Risk?

Does asbestos present a risk to the nation's students? Available data indicate that the levels of airborne asbestos in schools are approximately 0.01 of the permissible exposure levels for the U.S. workplace. In addition, the indoor concentrations of the most biologically active fibers are comparable to outdoor

← 2 cm →

Figure 1.A Chrysotile asbestos. This sample is a fibrous form of the mineral serpentine. (Photo by E. J. Tarbuck)

levels. With few exceptions, the type of asbestos found in schools is chrysotile, which in low concentrations has been shown to be relatively harmless.

A comparison of the risk from asbestos exposure in schools to other risks in society is shown in Table 1.A. Playing high school football poses over 100 times the risk as exposure to airborne asbestos. These data clearly demonstrate that the asbestos panic was and is unwarranted. However, it has spawned an entire asbestos-removal industry that now has a life of its own. Misguided public policies, no matter how well intentioned they are, are expensive.

Table 1.A Estimates of Risk from Asbestos Exposure in Schools in Comparison with Other Risks

Cause	Annual Death Rate (per million)
Asbestos exposure in schools	0.005–0.093
Whooping cough vaccination	1–6
Aircraft accidents	6
High school football	10
Drowning (ages 5–14)	27
Motor vehicle accident, Pedestrian (ages 5–14)	32
Home accidents (ages 1–14)	60
Long-term smoking	1200

The two most abundant elements are silicon and oxygen, comprising nearly three-fourths of Earth's continental crust. Silicon and oxygen combine to form the framework of the most common mineral group, the **silicates.** Perhaps the next most common mineral group is the carbonates (carbon plus oxygen plus other elements), of which calcite is the most prominent member. Other common rock-forming minerals include gypsum and halite.

We will first discuss the most common mineral group, the silicates, and then consider other prominent mineral groups.

Rock-Forming Silicates

Each of the silicate minerals contains oxygen and silicon atoms. Except for a few silicate minerals such as quartz, most silicate minerals also contain one or more additional elements in their crystalline structure.

Table 1.2 Relative abundance of the most common elements in Earth's crust.

Element	Approximate Percentage by Weight
Oxygen (O)	46.6
Silicon (Si)	27.7
Aluminum (Al)	8.1
Iron (Fe)	5.0
Calcium (Ca)	3.6
Sodium (Na)	2.8
Potassium (K)	2.6
Magnesium (Mg)	2.1
All others	1.7
Total	100

Source: Data from Brian Mason.

These elements give rise to the great variety of silicate minerals and their varied properties.

All silicates have the same fundamental building block, the **silicon-oxygen tetrahedron.** This structure consists of four oxygen atoms surrounding a much smaller silicon atom as shown in Figure 1.10. Thus, a

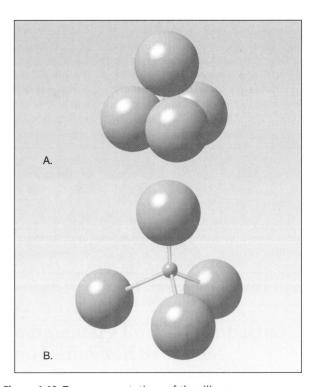

Figure 1.10 Two representations of the silicon-oxygen tetrahedron. **A.** The four large spheres represent oxygen atoms and the blue sphere represents a silicon atom. The spheres are drawn in proportion to the radii of the atoms. **B.** A model of the tetrahedron using rods to depict the bonds that connect the atoms.

typical hand-size silicate mineral specimen contains millions of these silicon-oxygen tetrahedra, joined together in a variety of ways.

In some minerals, the tetrahedra are joined into chains, sheets, or three-dimensional networks by sharing oxygen atoms (Figure 1.11). These larger silicate structures are then connected to one another by other elements. The primary elements that join silicate structures are iron (Fe), magnesium (Mg), potassium (K), sodium (Na), and calcium (Ca).

Major groups of silicate minerals and common examples are given in Figure 1.11. The feldspars are by far the most plentiful group, comprising over 50 percent of Earth's crust. Quartz, the second most abundant mineral in the continental crust, is the only common mineral made completely of silicon and oxygen.

Notice in Figure 1.11 that each mineral *group* has a particular silicate *structure*. A relationship exists between this internal structure of a mineral and the *cleavage* it exhibits. Because the silicon-oxygen bonds are strong, silicate minerals tend to cleave between the silicon-oxygen structures rather than across them. For example, the micas have a sheet structure and thus tend to cleave into flat plates (see muscovite in

Mineral		Idealized Formula	Cleavage	Silicate Structure	
Olivine		$(Mg, Fe)_2SiO_4$	None	Single tetrahedron	
Pyroxene group (Augite)		$(Mg,Fe)SiO_3$	Two planes at right angles	Single chains	
Amphibole group (Hornblende)		$Ca_2(Fe,Mg)_5Si_8O_{22}(OH)_2$	Two planes at 60° and 120°	Double chains	
Micas	Biotite	$K(Mg,Fe)_3AlSi_3O_{10}(OH)_2$	One plane	Sheets	
	Muscovite	$KAl_2(AlSi_3O_{10})(OH)_2$			
Feld-spars	Orthoclase	$KAlSi_3O_8$	Two planes at 90°	Three-dimensional networks	
	Plagioclase	$(Ca,Na)AlSi_3O_8$			
Quartz		SiO_2	None	(Expanded view)	

Figure 1.11 Common silicate minerals. Note that the complexity of the silicate structure increases down the chart.

Figure 1.12 Some common rock-forming minerals. Silicate minerals are common constituents of igneous rocks, while the nonsilicates along with quartz and clay are prominent minerals in many sedimentary rocks.

Figure 1.12). Quartz, which has equally strong silicon-oxygen bonds in all directions, has no cleavage, but fractures instead.

How do silicate minerals form? Most crystallize from molten rock as it cools. This cooling can occur at or near Earth's surface (low temperature and pressure) or at great depths (high temperature and pressure). The *environment* during crystallization and the *chemical composition of the molten rock* mainly determine which minerals are produced. For example, the silicate mineral olivine crystallizes at high temperatures (about 1200°C), whereas quartz crystallizes at much lower temperatures (about 700°C).

In addition, some silicate minerals form at Earth's surface from the weathered products of other silicate minerals. Still other silicate minerals are formed under the extreme pressures associated with mountain building. Each silicate mineral, therefore, has a structure and a chemical composition that *indicate the conditions under which it formed.* Thus, by carefully examining the mineral makeup of rocks, geologists can often determine the circumstances under which the rocks formed.

Important Nonsilicate Minerals

Nonsilicate minerals make up only about one-fourth of the continental crust, so they can be considered scarce compared to the silicates. However, many are important economically. Table 1.3 lists examples of nonsilicate minerals of economic value: oxides, sulfides, sulfates, halides, and "native" elements.

A major rock-forming group worldwide is the *carbonates*, which includes the mineral calcite ($CaCO_3$). This mineral is the major constituent in two well-known rocks: limestone and marble. Limestone is used commercially for road paving, building stone, and as the main ingredient in portland cement. Marble is used decoratively.

Two other nonsilicate minerals frequently found in sedimentary rocks are *halite* and *gypsum*. Both minerals are commonly found in thick layers, which are the last vestiges of ancient seas that long ago evaporated, leaving behind their salts (Figure 1.13). Like limestone, halite and gypsum both are important nonmetallic resources. Halite is the mineral name for common table salt, sodium chloride. Gypsum is the mineral from which plaster and other similar building materials are composed.

In addition, a number of other minerals have economic value (Table 1.3). Ores of metals include hematite (iron), sphalerite (zinc), and galena (lead). Native (free-occurring, not in compounds) elements include gold, silver, copper, and carbon (diamonds and graphite). Other important minerals are fluorite, corundum, and sylvite.

Table 1.3 Common nonsilicate mineral groups.

Mineral Group	Name	Chemical Formula	Economic Use
Oxides	Hematite	Fe_2O_3	Ore of iron, pigment
	Magnetite	Fe_3O_4	Ore of iron
	Corundum	Al_2O_3	Gemstone, abrasive
	Ice	H_2O	Solid form of water
Sulfides	Galena	PbS	Ore of lead
	Sphalerite	ZnS	Ore of zinc
	Pyrite	FeS_2	Sulfuric acid production
	Chalcopyrite	$CuFeS_2$	Ore of copper
	Cinnabar	HgS	Ore of mercury
Sulfates	Gypsum	$CaSO_4 \cdot 2H_2O$	Plaster
	Anhydrite	$CaSO_4$	Plaster
	Barite	$BaSO_4$	Drilling mud
Native elements	Gold	Au	Trade, jewelry
	Copper	Cu	Electrical conductor
	Diamond	C	Gemstone, abrasive
	Sulfur	S	Sulfa drugs, chemicals
	Graphite	C	Pencil lead, dry lubricant
	Silver	Ag	Jewelry, photography
	Platinum	Pt	Catlyst
Halides	Halite	NaCl	Common salt
	Fluorite	CaF_2	Used in steel making
	Sylvite	KCl	Fertilizer
Carbonates	Calcite	$CaCO_3$	Portland cement, lime
	Dolomite	$CaMg(CO_3)_2$	Portland cement, lime

Figure 1.13 Thick beds of halite (salt) at an underground mine near Grand Saline, Texas. (Photo by Tom Bochsler/Pearson Education/PH College)

Mineral Resources

Mineral resources are Earth's storehouse of useful minerals that can be recovered for use. Resources include already identified deposits from which minerals can be extracted profitably, called **reserves,** as well as known deposits that are not yet recoverable under present economic conditions or technology. Deposits inferred to exist but not yet discovered are also considered mineral resources.

The term **ore** denotes useful metallic minerals that can be mined at a profit. In common usage, the term *ore* is also applied to some nonmetallic minerals, such as fluorite and sulfur. However, materials used for such purposes as building stone, road paving, abrasives, ceramics, and fertilizers are not usually called ores; rather, they are classified as *industrial rocks and minerals.*

Recall that more than 98 percent of Earth's crust is composed of only eight elements. Except for oxygen and silicon, all other elements make up a relatively small fraction (about one-fourth) of common crustal rocks (see Table 1.2). Indeed, the natural concentrations of many elements are exceedingly small. A deposit containing only the average crustal percentage of a valuable element like gold is worthless because the cost of extracting it greatly exceeds the value of the material recovered.

Precious stones have been prized by people since antiquity. This tradition survives today. Gems are not just prized keepsakes of the rich but are possessed by people of modest means as well. Misinformation abounds about the nature of gems and the minerals of which they are composed. Part of the misinformation stems from the ancient practice of grouping precious stones by color rather than mineral makeup. For example, the more common red spinels were often passed off to royalty as rubies, which are more valuable gems. Even today, when modern techniques of mineral identification are commonplace, yellow quartz is frequently sold as topaz.

Naming Gemstones

Compounding the confusion is the fact that many gems have names that are different from their mineral names. For example, diamond is composed of the mineral of the same name, whereas sapphire is a form of corundum, an aluminum oxide-rich mineral. Although pure aluminum oxide is colorless, a minute amount of a foreign element can produce a vividly colored gemstone. Hence, depending on the impurity, sapphires of nearly every color are known to exist. Pure aluminum oxide with trace amounts of titanium and iron

Table 1.B Important gemstones.

Gem	Mineral Name	Prized Hues
Precious		
Diamond	Diamond	Colorless, yellows
Emerald	Beryl	Greens
Opal	Opal	Brilliant hues
Ruby	Corundum	Reds
Sapphire	Corundum	Blues
Semiprecious		
Alexandrite	Chrysoberyl	Variable
Amethyst	Quartz	Purples
Cat's-eye	Chrysoberyl	Yellows
Chalcedony	Quartz (agate)	Banded
Citrine	Quartz	Yellows
Garnet	Garnet	Reds, greens
Jade	Jadeite or nephrite	Greens
Moonstone	Feldspar	Transparent blues
Peridot	Olivine	Olive greens
Smoky quartz	Quartz	Browns
Spinel	Spinel	Reds
Topaz	Topaz	Purples, reds
Tourmaline	Tourmaline	Reds, blue-greens
Turquoise	Turquoise	Blues
Zircon	Zircon	Reds

produce the most prized blue sapphires. When the mineral corundum contains sufficient quantities of chromium, it exhibits a brilliant red color, and the gem is called *ruby*. Large, gem-quality rubies are much rarer than diamonds and thus can command a very high price.

To summarize, when a gem-quality sample of corundum exhibits a red hue, it is called *ruby*, but if it exhibits any other color, the gem is called *sapphire*. If the specimen is not suitable as a gem, it simply goes by the mineral name *corundum*. Although common corundum is not a gemstone, it does

---

To be considered of value, an element must be concentrated above the level of its average crustal abundance. For example, copper makes up about 0.0135 percent of the crust. However, for a material to be considered as copper ore, it must contain a concentration that is about 100 times this amount. Aluminum, in contrast, represents 8.13 percent of the crust and must be concentrated to only about four times its average crustal percentage before it can be extracted profitably.

It is important to realize that economic changes may make a deposit profitable to extract, or lose its profitability. If demand for a metal increases and prices rise sufficiently, the status of a previously unprofitable deposit changes, and it becomes an ore.

The status of unprofitable deposits may also change if a technological advance allows the ore to be extracted at a lower cost than before.

Conversely, changing economic factors can turn a once-profitable ore deposit into an unprofitable deposit that can no longer be called an ore. This situation is illustrated by a copper mine at Bingham Canyon, Utah, one of the largest open-pit mines on Earth (Figure 1.14). Mining was halted there in 1985 because outmoded equipment had driven the cost of extracting the copper beyond the current selling price. The owners responded by replacing an antiquated 1000-car railroad with conveyor belts and pipelines for transporting the ore and waste. These devices achieved a cost reduction of nearly 30 percent and returned this mining operation to profitability.

have value as an abrasive material. Whereas two gems, rubies and sapphires, are composed of the mineral corundum, quartz is the parent mineral of more than a dozen gems. Table 1.B lists some well-known gemstones and their mineral names.

What Constitutes a Gemstone?

In essence, certain mineral specimens when cut and polished posses beauty of such quality that they can command a price that makes the process of producing the gem profitable. Gemstones can be divided into two categories: precious and semiprecious. A *precious* gem has beauty, durability, size, and rarity, whereas a *semiprecious* gem generally has only one or two of these qualities. The gems that have traditionally enjoyed the highest esteem are diamonds, rubies, sapphires, emeralds, and some varieties of opal (Table 1.B). All other gemstones are classified as semiprecious. It should be noted, however, that large, high-quality specimens of the so-called semiprecious stones can often command a very high price.

Obviously, beauty is the most important quality that a gem can possess. Today, we prefer translucent stones with evenly tinted colors. The most favored hues appear to be red, blue, green, purple, rose, and yellow.

The most prized stones are pigeon-blood rubies, blue sapphires, grass-green emeralds, and canary-yellow diamonds. Colorless gems are generally less than desirable except in the case of diamonds that display "flashes of color" known as *brilliance*. Note that gemstones in the "rough" are dull and would be passed over by most laypersons as "just another rock." Gemstones must be cut and polished by experienced artisans before their true beauty can be displayed (Figure 1.B).

The durability of a gem depends on its hardness; that is, its resistance to abrasion by objects normally encountered in everyday living. For good durability, gems should be as hard or harder than quartz as defined by the Mohs scale of hardness, which consists of 10 minerals arranged in order from 1 (softest) to 10 (hardest). One notable exception is opal, which is comparatively soft (hardness 5 to 6.5) and brittle. Opal's esteem comes from its "fire," which is a display of a variety of brilliant colors including greens, blues, and reds.

It seems to be human nature to treasure that which is rare. In the case of gemstones, large, high-quality specimens are much rarer than smaller stones. Thus, large rubies, diamonds, and emeralds, which are rare in addition to being beautiful and durable, command the very highest prices.

Figure 1.B Australian sapphires depicting variations in cuts and colors. (Photo by Fred Ward, Black Star)

Figure 1.14 Aerial view of Bingham Canyon copper mine near Salt Lake City, Utah. This huge open-pit mine is about 4 kilometers across and 900 meters deep. Although the amount of copper in the rock is less than 1 percent, the huge volumes of material removed and processed each day (about 200,000 tons) yield significant quantities of metal. (Photo by Michael Collier)

Chapter Summary

- A *mineral* is a naturally occurring inorganic solid possessing a definite chemical structure that gives it a unique set of physical properties. Most *rocks* are aggregates composed of two or more minerals.

- The building blocks of minerals are *elements*. An *atom* is the smallest particle of matter that still retains the characteristics of an element. Each atom has a *nucleus* containing *protons* and *neutrons*. Orbiting the nucleus of an atom are *electrons*. The number of protons in an atom's nucleus determines its *atomic number* and the name of the element. Atoms bond together to form a *compound* by either gaining, losing, or sharing electrons with another atom.

- *Isotopes* are variants of the same element, but with a different *mass number* (the total number of neutrons plus protons found in an atom's nucleus). Some isotopes are unstable and disintegrate naturally through a process called *radioactive decay*.

- The properties of minerals include *crystal form, luster, color, streak, hardness, cleavage, fracture,* and *specific gravity*. In addition, a number of special physical and chemical properties (*taste, smell, elasticity, malleability, feel, magnetism, double refraction,* and *chemical reaction to hydrochloric acid*) are useful in identifying certain minerals. Each mineral has a unique set of properties that can be used for identification.

- The eight most abundant elements found in Earth's continental crust (oxygen, silicon, aluminum, iron, calcium, sodium, potassium, and magnesium) also make up the majority of minerals.

- The most common mineral groups is the *silicates*. All silicate minerals have the *silicon-oxygen tetrahedron* as their fundamental building block. In some silicate minerals the tetrahedra are joined in chains, in others, the tetrahedra are arranged into sheets, or three-dimensional networks. Each silicate mineral has a structure and a chemical composition that indicates the conditions under which it was formed.

- The *nonsilicate* mineral groups include the *oxides* (e.g., magnetite, mined for iron), *sulfides* (e.g., sphalerite, mined for zinc), *sulfates* (e.g., gypsum, used in plaster and frequently found in sedimentary rocks), *native elements* (e.g., graphite, a dry lubricant), *halides* (e.g., halite, common salt and frequently found in sedimentary rocks), and *carbonates* (e.g., calcite, used in portland cement and a major constituent in two well-known rocks: limestone and marble).

- The term *ore* is used to denote useful metallic minerals, like hematite (mined for iron) and galena (mined for lead), that can be mined for a profit, as well as some nonmetallic minerals, such as fluorite and sulfur, that contain useful substances.

Key Terms

atom (p. 20)
atomic number (p. 20)
cleavage (p. 23)
color (p. 22)
compound (p. 20)
crystal form (p. 21)
electron (p. 20)
energy levels (shells) (p. 20)
element (p. 20)
fracture (p. 24)

hardness (p. 23)
ions (p. 20)
isotope (p. 21)
luster (p. 22)
mass number (p.21)
mineral (p. 19)
mineral resource (p. 29)
Mohs hardness scale (p. 23)
neutron (p. 20)
nucleus (p. 20)

ore (p. 29)
proton (p. 20)
radioactive decay (p. 21)
reserve (p. 29)
rock (p. 19)
silicate (p. 26)
silicon-oxygen tetrahedron (p. 26)
specific gravity (p. 24)
streak (p. 22)

Review Questions

1. Define the term *rock*.

2. List the three main particles of an atom and explain how they differ from one another.

3. If the number of electrons in an atom is 35 and its mass number is 80, calculate the following:

   (a) The number of protons.

   (b) The atomic number.

   (c) The number of neutrons.

4. What occurs in an atom to produce an ion?

5. What is an isotope?

6. Although all minerals have an orderly internal arrangement of atoms (crystalline structure), most mineral samples do not visibly demonstrate their crystal form. Why?

7. Why might it be difficult to identify a mineral by its color?

8. If you found a glassy-appearing mineral while rock hunting and had hopes that it was a diamond, what simple test might help you make a determination?

9. Table 1.3 (p. 29) lists a use for corundum as an abrasive. Explain why it makes a good abrasive in terms of Mohs hardness scale.

10. Gold has a specific gravity of almost 20. If a 25-liter pail of water weighs about 25 kilograms, how much would a 25-liter pail of gold weigh?

11. What are the two most common elements in Earth's crust? What is the term used to describe the basic building block of all silicate minerals?

12. What are the two most common silicate minerals? List three nonsilicate minerals that are commonly found in rocks. Contrast a mineral *resource* and a mineral *reserve*. What might cause a mineral deposit that had not been considered an ore to become reclassified as an ore?

Examining the Earth System

1. Perhaps one of the most significant interrelationships between humans and the Earth system involves the extraction, refinement, and distribution of the planet's mineral wealth. To help you understand these associations, begin by thoroughly researching a mineral commodity that is mined in your local region or state. (You might find the information at these United States Geological Survey (USGS) Web sites, **http://minerals.er.usgs.gov/minerals/pubs/state/mo.html** and **http://minerals.er.usgs.gov:80/minerals/ pubs/mcs/**, useful.) What products are made from this mineral? Do you use any of these products? Describe the mining and refining of the mineral and the local impact these processes have on each of Earth's spheres (atmosphere, hydrosphere, solid Earth, and biosphere). Are any of the effects negative? If so, what, if anything, is being done to end or minimize the damage?

2. Referring to the mineral you described above, in your opinion does the environmental impact of extracting this mineral outweigh the benefits derived from the products produced from this mineral?

Web Resources

The following are informative and interesting World Wide Web sites that address topics related to those presented in the chapter:

Minerals on the Internet:

- **http://www.swcp.com/~tasa/minlinks.html**

Mineral Resources Program (United States Geological Survey):

- **http://minerals.er.usgs.gov/**

For direct links to these sites and other appropriate Earth science Web locations, chapter review exercises, and relevant, topical investigations that utilize Web resources, contact the *Earth Science*, 9th Edition Home Page at:

- **http://www.prenhall.com/tarbuck.**

Chapter 2

Rocks: Materials of the Lithosphere

The White Chalk Cliffs near Beachy Head in Sussex, England. The cliffs consist of the
hard parts of tiny (microscopic) marine organisms. Chalk is a sedimentary rock (solid
Earth) formed when organisms (biosphere) extract material dissolved in water
(hydrosphere). (Photo by Adrian Greeman/Aspect Picture Library/The Stock Market)

Why study rocks? You have already learned that rocks and minerals have great economic value. Furthermore, all Earth processes in some way depend on the properties of these basic materials. Events such as volcanic eruptions, mountain building, weathering, erosion, and even earthquakes involve rocks and minerals. Consequently, a basic knowledge of Earth materials is essential in understanding Earth phenomena.

Every rock contains clues about the environments in which it formed (Figure 2.1). For example, some rocks are composed entirely of small shell fragments. This tells Earth scientists that the particles composing the rock originated in a shallow marine environment. Other rocks contain clues that indicate they formed from a volcanic eruption, or deep in Earth during mountain building. Thus, rocks contain a wealth of information about events that have occurred over Earth's long history.

We divide rocks into three groups, based on their mode of origin. The groups are igneous, sedimentary, and metamorphic. Before examining each group, we will view the *rock cycle*, which depicts the interrelationships among these rock groups.

Earth as a System: The Rock Cycle

 Earth Materials
The Rock Cycle

Earth is a system. This means that our planet consists of many interacting parts that form a complex whole. Nowhere is this idea better illustrated than when we examine the rock cycle (Figure 2.2). The **rock cycle** allows us to view many of the interrelationships among different parts of the Earth system. It helps us understand the origin of igneous, sedimentary, and metamorphic rocks and to see that each type is linked to the others by the processes that act upon and within the planet. Learn the rock cycle well; you will be examining its interrelationships in greater detail throughout this chapter and the other chapters in the unit on the solid Earth.

The Basic Cycle

Let us begin at the top of Figure 2.2. **Magma** is molten material that forms inside Earth. Eventually magma cools and solidifies. This process, called **crystallization**, may occur either beneath the surface or, following a

Figure 2.1 Rocks contain information about the processes that produce them. This eroded sandstone monument in Arizona's Vermillion Cliffs Wilderness was once a sand dune. (Photo by Jack W. Dykinga and Associates)

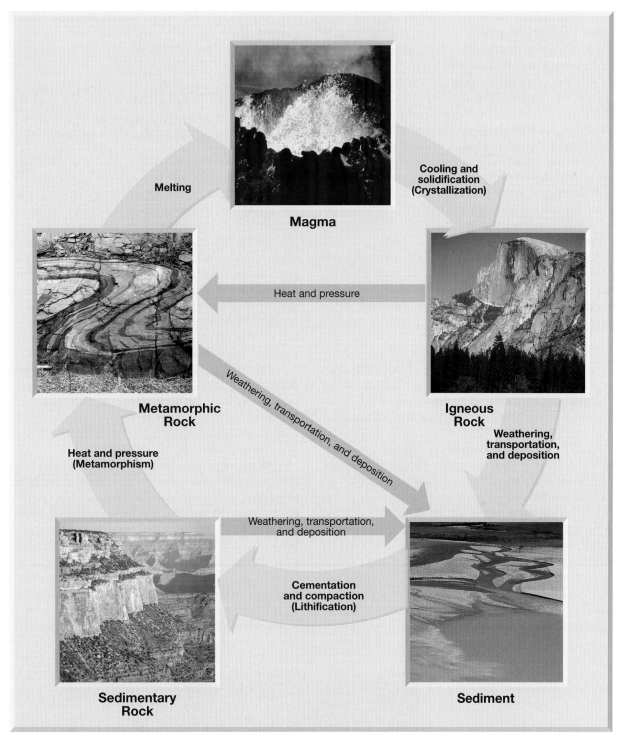

Figure 2.2 The rock cycle is one way of viewing many of the interrelationships among different parts of the Earth system. It shows us that Earth's materials (boxes) are all linked to each other by processes (arrows). Every rock contains clues to the processes that formed it. (Photos by Edward J. Tarbuck)

volcanic eruption, at the surface. In either situation, the resulting rocks are called **igneous rocks**.

If igneous rocks are exposed at the surface, they will undergo **weathering**, in which the day-in and day-out influences of the atmosphere slowly disintegrate and decompose rocks. The materials that result are often moved downslope by gravity before being picked up and transported by any of a number of erosional agents—running water, glaciers, wind, or waves. Eventually these particles and dissolved substances, called **sediment**, are deposited. Although most sediment ultimately comes to rest in the ocean, other sites of deposition include river floodplains, desert basins, swamps, and dunes.

Next the sediments undergo **lithification**, a term meaning "conversion into rock." Sediment is usually lithified into **sedimentary rock** when compacted by the weight of overlying layers or when cemented as percolating groundwater fills the pores with mineral matter.

If the resulting sedimentary rock is buried deep within Earth and involved in the dynamics of mountain building, or intruded by a mass of magma, it will be subjected to great pressures and/or intense heat. The sedimentary rock will react to the changing environment and turn into the third rock type, **metamorphic rock**. When metamorphic rock is subjected to additional pressure changes or to still higher temperatures, it will melt, creating magma, which will eventually crystallize into igneous rock.

Processes driven by heat from Earth's interior are responsible for creating igneous and metamorphic rocks. Weathering and erosion, external processes powered by energy from the Sun, produce the sediment from which sedimentary rocks form.

Alternative Paths

The paths shown in the basic cycle are not the only ones that are possible. To the contrary, other paths are just as likely to be followed as those described in the preceding section. These alternatives are indicated by the blue arrows in Figure 2.2.

Igneous rocks, rather than being exposed to weathering and erosion at Earth's surface, may remain deeply buried. Eventually these masses may be subjected to the strong compressional forces and high temperatures associated with mountain building. When this occurs, they are transformed directly into metamorphic rocks.

Metamorphic and sedimentary rocks, as well as sediment, do not always remain buried. Rather, overlying layers may be stripped away, exposing the once-buried rock. When this happens, the material is attacked by weathering processes and turned into new raw materials for sedimentary rocks.

Although rocks may seem to be unchanging masses, the rock cycle shows that they are not. The changes, however, take time—great amounts of time.

Igneous Rocks: "Formed by Fire"

Earth Materials
Igneous Rocks

In our discussion of the rock cycle, we pointed out that igneous rocks form as magma cools and crystallizes. This molten rock, which originates at depths as great as 200 kilometers within Earth, consists primarily of the elements found in silicate minerals (silicon and oxygen, plus aluminum, iron, calcium, sodium, potassium, magnesium, and others). Magma also contains some gases, particularly water vapor, which are confined within the magma by the surrounding rocks. Because the magma body is less dense than the surrounding rocks, it works its way toward the surface over time spans of thousands to millions of years. On occasion, the magma breaks through the surface, producing a volcanic eruption.

The spectacular explosions that sometimes accompany an eruption are produced by the gases that escape as the confining pressure lessens near the surface. Sometimes blockage of the volcanic vent, coupled with surface-water seepage into the magma chamber, can produce catastrophic explosions.

Along with ejected rock fragments, a volcanic eruption often generates extensive lava flows. **Lava** is similar to magma, except that most of the gaseous component has escaped (Figure 2.3). The rocks that result when lava solidifies are classified as **volcanic** or **extrusive**, because they are *extruded* onto the surface. The magma not able to reach the surface eventually crystallizes at depth. Igneous rocks produced in this manner are termed **plutonic**, or **intrusive**, because they *intrude* existing rocks. These deep rocks would never be observed were it not for erosion stripping away the overlying rock.

Figure 2.3 Fluid basaltic lava moving down the slopes of Hawaii's Kilauea Volcano. (Photo by G. Brad Lewis/Liaison Agency, Inc.)

Magma Crystallizes to Form Igneous Rocks

Magma is basically a very hot, thick fluid, but it also contains solids and gases. The solids are mineral crystals. The liquid portion of a magma body is composed of ions that move about freely. However, as magma cools, the random movements of the ions slow, and the ions begin to arrange themselves into orderly patterns. This process is called *crystallization*. Usually all of the molten material does not solidify at the same time. Rather, as it cools, numerous small crystals develop. In a systematic fashion, ions are added to these centers of crystal growth. When the crystals grow large enough for their edges to meet, their growth ceases for lack of space, and crystallization continues elsewhere. Eventually, all of the liquid is transformed into a solid mass of interlocking crystals.

The rate of cooling strongly influences crystal size. If a magma cools very slowly, relatively few centers of crystal growth develop. Slow cooling also allows ions to migrate over relatively great distances. Consequently, *slow cooling results in the formation of large crystals*. Conversely, if cooling occurs quite rapidly, the ions lose their motion and quickly combine. This results in a large number of tiny crystals that all compete for the available ions. Therefore, the outcome of rapid cooling is the formation of a solid mass of *small intergrown crystals*.

Thus, if a geologist discovers rock containing large, finger-sized crystals, it means the rock cooled very slowly. But if the crystals can be seen only with a microscope, the geologist knows that the magma cooled very quickly.

If the molten material is quenched almost instantly, there is not sufficient time for the ions to arrange themselves into a crystalline network at all. Therefore, solids produced in this manner consist of randomly distributed ions. Such rocks are called *glass* and are quite similar to ordinary manufactured glass. "Instant" quenching occurs during violent volcanic eruptions that produce tiny shards of glass termed "volcanic ash."

Classifying Igneous Rocks

In addition to the rate of cooling, the composition of a magma and the amount of dissolved gases influence crystallization. Because magmas differ in each of these aspects, the physical appearance and mineral composition of igneous rocks vary widely. Nevertheless, it is possible to classify igneous rocks based on their *texture* and *mineral constituents*. We will now look at both features.

Igneous Textures. **Texture** describes the overall appearance of an igneous rock, based on the *size* and *arrangement* of its interlocking crystals. Texture is a very important characteristic, because it reveals a great deal about the environment in which the rock formed. You learned that rapid cooling produces small crystals, whereas very slow cooling produces much larger crystals. As you might expect, the rate of cooling is quite slow in magma chambers lying deep within the crust, whereas a thin layer of lava extruded upon Earth's surface may chill "rock solid" in a matter of hours. Small molten blobs ejected into the air during a violent eruption can solidify almost instantly.

Igneous rocks that form rapidly at the surface or as small masses within the upper crust have a **fine-grained texture**, with the individual crystals too small to be seen with the unaided eye (Figure 2.4A). Common in many fine-grained igneous rocks are voids left by gas bubbles that escape as the magma

A.

B.

Figure 2.4 A. Rhyolite exhibits a fine-grained texture. **B.** Granite is a common igneous rock that has a coarse-grained texture. (Photos by E. J. Tarbuck)

solidifies. These openings are called *vesicles* and are limited to the fast-cooling upper portion of lava flows (Figure 2.5).

When large masses of magma solidify far below the surface, they form igneous rocks that exhibit a **coarse-grained texture**. These coarse-grained rocks have the appearance of a mass of intergrown crystals, which are roughly equal in size and large enough that the individual minerals can be identified with the unaided eye. Granite is a classic example (Figure 2.4B).

A large mass of magma located at depth may require tens of thousands of years to solidify. Because all materials within a magma do not crystallize at the same rate or at the same time during cooling, it is possible for some crystals to become quite large before others even start to form. If magma that already contains some large crystals suddenly erupts at the surface, the remaining molten portion of the lava would cool quickly. The resulting rock, which has large crystals embedded in a matrix of smaller crystals, is said to have a **porphyritic texture** (Figure 2.6).

During some volcanic eruptions, molten rock is ejected into the atmosphere, where it is quenched very quickly. Rapid cooling of this type may generate rock with a **glassy texture**. As was indicated, glass results when the ions do not have sufficient time to unite into an orderly crystalline structure. *Obsidian*, a common type of natural glass, is similar in appearance to a dark chunk of manufactured glass (Figure 2.7A). Another volcanic rock that exhibits a glassy texture is *pumice*. Usually found with obsidian, pumice forms when large amounts of gas escape through lava to generate a gray, frothy mass (Figure 2.7B). In some samples, the vesicles are quite noticeable; in others, the pumice resembles fine shards of intertwined glass. Because of the large volume of air-filled voids, many samples of pumice will float in water.

Mineral Composition. The mineral makeup of an igneous rock depends on the chemical composition of the magma from which it crystallizes. Because a large variety of igneous rocks exists, you might assume an equally large variety of magmas must also exist. However, various eruptive stages of the same volcano often extrude lavas exhibiting somewhat different mineral compositions, particularly if long periods separate the eruptions. Such evidence led geologists to examine the possibility that a single magma might produce rocks of varying mineral content.

N. L. Bowen studied this idea in the first quarter of the twentieth century. Bowen discovered that, as magma cools in the laboratory, certain minerals crystallize first, at very high temperatures (top of Figure 2.8). At successively lower temperatures, other minerals crystallize.

Bowen also demonstrated that if a mineral remains in the molten solution after crystallization, it will react with the remaining liquid to produce the

Figure 2.5 Scoria is a volcanic rock that exhibits vesicles, which form as gas bubbles escape near the top of a lava flow.

next mineral in the sequence shown in Figure 2.8. For this reason, this arrangement of minerals became known as *Bowen's reaction series*.

On the upper left branch of this reaction series, olivine, the first mineral to form, will react with the remaining melt to become pyroxene. This reaction will continue until the last mineral in the series, biotite, is formed. The right branch of the reaction series is a continuum in which the earliest-formed calcium-rich feldspar crystals react with the sodium ions contained in the melt to become progressively more sodium-rich. Ordinarily, these reactions are not complete, so that various amounts of several of these minerals may exist at any given time.

Figure 2.6 Andesite porphyry. Notice the two distinctively different sizes of crystals.(Photo by E. J. Tarbuck)

A.

B.

Figure 2.7 Igneous rocks that exhibit a glassy texture. **A.** Obsidian, a glassy volcanic rock. **B.** Pumice, a glassy rock containing numerous tiny voids.

During the last stage of crystallization, after most of the magma is solidified, the minerals muscovite and potassium feldspar may form. Finally, the remaining molten liquid (if any), which is rich in silica, eventually crystallizes as quartz.

A consequence of this crystallization scheme is that minerals that form at about the same temperature are found together in the same igneous rock. For example, notice that the minerals potassium feldspar, muscovite mica, and quartz are found in the same region of the diagram, and they are the major constituents of the igneous rock granite.

How does Bowen's reaction series account for the great diversity of igneous rocks? During crystallization, a separation of the solid and liquid components of a magma can occur. This happens if the earlier

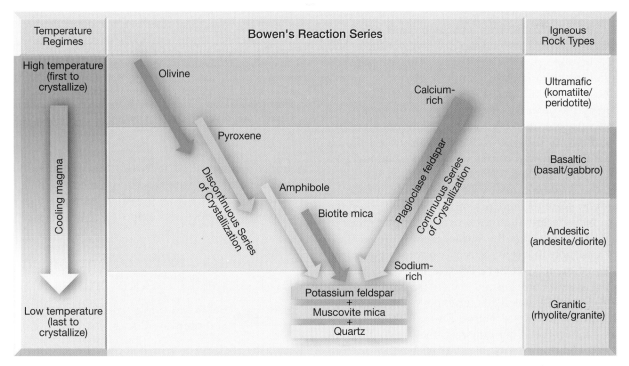

Figure 2.8 Bowen's reaction series shows the sequence in which minerals crystallize from a magma. Compare this figure to the mineral composition of the rock groups in Table 2.1. Note that each rock group consists of minerals that crystallize at the same time.

formed minerals are denser (heavier) than the liquid portion and settle to the bottom, as shown in Figure 2.9A. This *crystal settling* is thought to occur frequently with the dark, dense silicates, such as olivine and pyroxene. When the remaining melt solidifies it will form a rock with a chemical composition much different from the parent magma (Figure 2.9B). Consequently, a wide variety of igneous rocks results.

Naming Igneous Rocks

Igneous rocks are classified by their texture and mineral composition. Various igneous textures result from different cooling histories, while the mineral compositions are a consequence of the chemical makeup of the parent magma and the environment of crystallization. A general classification scheme based on texture and mineral composition is provided in Table 2.1.

The rocks on the right side of Table 2.1 consist of the *first* minerals to crystallize. They contain more iron (Fe) and magnesium (Mg) and less silica (SiO_2) than do later-formed minerals. Their iron content makes them darker in color and slightly more dense than the other rocks. Basalt is a common rock of this composition, so geologists use the term *basaltic* to indicate any igneous rock having this composition. Basalt is the most common extrusive igneous rock. Many volcanic islands, such as the Hawaiian Islands and Iceland, are composed mainly of basalt. Further, the upper layers of the oceanic crust consist primarily of basalt.

The rocks on the left side of Table 2.1 contain the *last* minerals to crystallize and consist mainly of feldspar and quartz. These rocks have a *granitic* composition. *Granite* is perhaps the best-known igneous rock (see Figure 2.4B). This is partly because of its natural beauty, which is enhanced when polished, and partly because of its abundance. Slabs of polished granite are commonly used for tombstones, monuments, and as building stones. Granite is often produced by the processes that generate mountains.

Because granite is a by-product of mountain building and is very resistant to weathering and erosion, it frequently forms the core of eroded mountains: Pikes Peak in the Rockies; Mount Rushmore in the Black Hills; the White Mountains of New Hampshire; Stone Mountain, Georgia; and Yosemite National Park in the Sierra Nevada are all areas where large quantities of granite are exposed at the surface (see Figure 2.10).

Please note that two rocks may have the same mineral constituents, but have *different textures and hence different names*. For example, the coarse-grained intrusive rock granite has a fine-grained volcanic equivalent called *rhyolite* (see Figure 2.4A). Although these rocks contain the same minerals, they have different textures and do not look at all alike. Further, in contrast to granite, rhyolite is rather uncommon.

Note that gradations exist (Table 2.1). For example, an abundant extrusive igneous rock called *andesite* has

A.

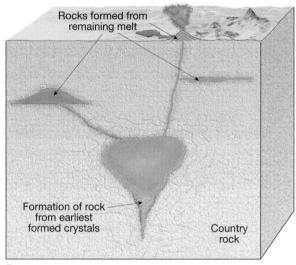

B.

Figure 2.9 :Separation of minerals by crystal settling. **A.** Illustration of how the earliest-formed minerals can be separated from a magma by settling. **B.** The remaining melt could migrate to a number of different locations and, upon further crystallization, generate rocks having compositions much different from that of the parent magma.

a mineral composition between that of granitic and basaltic rocks. Andesite is a fine-grained volcanic rock whose name derives from the Andes Mountains, where numerous volcanoes are composed of this rock type.

Sedimentary Rocks: Compacted and Cemented Sediment

 Earth Materials
Sedimentary Rocks

Recall the rock cycle, which shows the origin of sedimentary rocks. Weathering begins the process. Next, gravity and erosional agents (running water, wind, waves, and glacial ice) remove the products of

Table 2.1 Classification of igneous rocks.

		Granitic (felsic)	Andesitic (intermediate)	Basaltic (mafic)	Ultramafic[*]
Dominant Minerals		Quartz Potassium feldspar	Amphibole Intermediate plagioclase feldspar	Pyroxene Calcium-rich plagioclase feldspar	Olivine Pyroxene
Color		Light-colored Less than 15% dark minerals	Medium-colored 15–40% dark minerals	Dark gray to black More than 40% dark minerals	Dark-green to black Nearly 100% dark minerals
T e x t u r e	**Coarse-grained**	Granite	Diorite	Gabbro	Peridotite
	Fine-grained	Rhyolite	Andesite	Basalt	Komatiite
	Porphyritic	"Porphyry" follows any of the above names whenever there are appreciable phenocrysts			
	Glassy	Obsidian (compact glass) Pumice (frothy glass)			

[*]Ultramafic rocks are dark, dense rocks, composed almost entirely of minerals containing iron and magnesium. Although relatively rare on Earth's surface, these rocks are believed to be major constituents of the upper mantle.

weathering and carry them to a new location where they are deposited. Usually the particles are broken down further during this transport phase. Following deposition, this *sediment* may become lithified, or "turned to rock." Commonly, *compaction* and *cementation* transform the sediment into solid sedimentary rock.

Figure 2.10 El Capitan is a massive granitic structure in Yosemite National Park, California. (Photo by Jeff Gnass/The Stock Market)

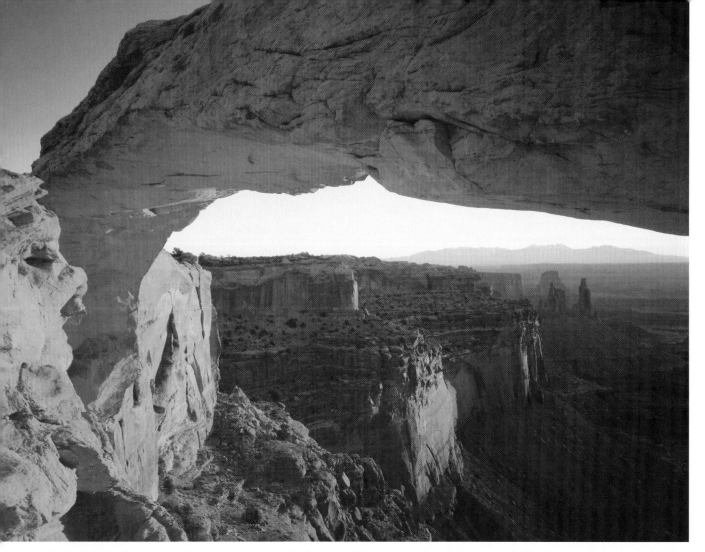

Figure 2.11 Sedimentary rocks exposed in Canyonlands National Park, Utah. About 75 percent of all rock outcrops on the continents are sedimentary rocks. (Photo by Jeff Gnass)

The word *sedimentary* indicates the nature of these rocks, for it is derived from the Latin *sedimentum*, which means "settling," a reference to a solid material settling out of a fluid. Most sediment is deposited in this fashion. Weathered debris is constantly being swept from bedrock and carried away by water, ice, or wind. Eventually the material is deposited in lakes, river valleys, seas, and countless other places. The particles in a desert sand dune, the mud on the floor of a swamp, the gravels in a stream bed, and even household dust are examples of sediment produced by this never-ending process.

The weathering of bedrock and the transport and deposition of the weathering products are continuous. As piles of sediment accumulate, the materials near the bottom are compacted by the weight of the overlying layers. Over long periods, these sediments are cemented together by mineral matter deposited from water in the spaces between particles. This forms solid sedimentary rock.

Geologists estimate that sedimentary rocks account for only about 5 percent (by volume) of Earth's outer 16 kilometers (10 miles). However, the importance of this group of rocks is far greater than

this percentage implies. If you sampled the rocks exposed at Earth's surface, you would find that the great majority are sedimentary (Figure 2.11). Indeed, about 75 percent of all rock outcrops on the continents are sedimentary. Therefore, we can think of sedimentary rocks as comprising a relatively thin and somewhat discontinuous layer in the uppermost portion of the crust. This makes sense because sediment accumulates at the surface.

It is from sedimentary rocks that geologists reconstruct many details of Earth's history. Because sediments are deposited in a variety of different settings at the surface, the rock layers that they eventually form hold many clues to past surface environments. They may also exhibit characteristics that allow geologists to decipher information about the method and distance of sediment transport. Furthermore, it is sedimentary rocks that contains fossils, which are vital evidence in the study of the geologic past.

Finally, many sedimentary rocks are very important economically. Coal, for example, is a sedimentary rock that presently is the source for over half the electric power generated in the United States. Other major energy resources (petroleum and natural gas) occur in pores within sedimentary rocks. Other sedimentary rocks are major sources of iron, aluminum, manganese, fertilizer, and sand and gravel.

Classifying Sedimentary Rocks

Materials accumulating as sediment have two principal sources. First, sediments may originate as solid particles from weathered rocks, such as the igneous rocks earlier described. These particles are called *detritus*, and the sedimentary rocks that they form are called **detrital sedimentary rocks**.

The second major source of sediment is soluble material produced largely by chemical weathering. When these dissolved substances are precipitated back as solids, they are called "chemical sediment," and they form **chemical sedimentary rocks**. We will now look at detrital and chemical sedimentary rocks.

Detrital Sedimentary Rocks. Though a wide variety of minerals and rock fragments may be found in detrital rocks, clay minerals and quartz dominate. Clay minerals are the most abundant product of the chemical weathering of silicate minerals, especially the feldspars. Quartz, in contrast, is abundant because it is extremely durable and very resistant to chemical weathering.

Table 2.2 Particle size classification for detrital rocks.

Size Range (millimeters)	Particle Name	Common Sediment Name	Detrital Rock
>256	Boulder		
64–256	Cobble	Gravel	Conglomerate or breccia
4–64	Pebble		
2–4	Granule		
1/16–2	Sand	Sand	Sandstone
1/256–1/16	Silt	Mud	Shale
<1/256	Clay		

Scale in millimeters
10 20 30 40 50 60

Thus, when igneous rocks such as granite are weathered, individual quartz grains are set free.

Geologists use particle size to distinguish among detrital sedimentary rocks. Table 2.2 presents the four size categories for particles making up detrital rocks. When gravel-sized particles predominate, the rock is called *conglomerate* if the sediment is rounded (Figure 2.12A) and *breccia* if the pieces are angular (Figure 2.12B).

A.

B.

C.

D.

Figure 2.12 Common detrital sedimentary rocks. **A.** Conglomerate (rounded particles). **B.** Breccia (angular particles). **C.** Sandstone **D.** Shale with plant fossil. (Photos by E. J. Tarbuck)

Angular fragments indicate that the particles were not transported very far from their source prior to deposition, and so have not had corners and rough edges abraded. *Sandstone* is the name given rocks when sand-sized grains prevail (Figure 2.12C). *Shale*, the most common sedimentary rock, is made of very fine-grained sediment (Figure 2.12D). *Siltstone*, another rather fine-grained rock, is sometimes difficult to differentiate from rocks such as shale that are composed of even smaller clay-sized sediment.

Particle size is not only a convenient method of dividing detrital rocks; the sizes of the component grains also provide useful information about the environment in which the sediment was deposited. Currents of water or air sort the particles by size. The stronger the current, the larger the particle size carried. Gravels, for example, are moved by swiftly flowing rivers, rockslides, and glaciers. Less energy is required to transport sand; thus, it is common in windblown dunes, river deposits, and beaches. Because silts and clays settle very slowly, accumulations of these materials are generally associated with the quiet waters of a lake, lagoon, swamp, or marine environment.

Although detrital sedimentary rocks are classified by particle size, in certain cases the mineral composition is also part of naming a rock. For example, most sandstones are predominantly quartz-rich, and are often referred to as quartz sandstone. In addition, rocks consisting of detrital sediments are rarely composed of grains of just one size. Consequently, a rock containing quantities of both sand and silt can be correctly classified as sandy siltstone or silty sandstone, depending on which particle size dominates.

Chemical Sedimentary Rocks. In contrast to detrital rocks, which form from the solid products of weathering, chemical sediments are derived from material that is carried in solution to lakes and seas. This material does not remain dissolved in the water indefinitely. When conditions are right, it precipitates to form chemical sediments. This precipitation may occur directly as the result of physical processes, or indirectly through life processes of water-dwelling organisms. Sediment formed in this second way has a *biochemical* origin.

An example of a deposit resulting from physical processes is the salt left behind as a body of salt water evaporates. In contrast, many water-dwelling animals and plants extract dissolved mineral matter to form shells and other hard parts. After the organisms die, their skeletons may accumulate on the floor of a lake or ocean.

Limestone is the most abundant chemical sedimentary rock. It is composed chiefly of the mineral calcite ($CaCO_3$). Ninety percent of limestone is biochemical sediment. The rest precipitates directly from water.

One easily identified biochemical limestone is *coquina*, a coarse rock composed of loosely cemented

Figure 2.13 This rock, called coquina, consists of shell fragments; therefore, it has a biochemical origin. (Photo by E. J. Tarbuck)

shells and shell fragments (Figure 2.13). Another less obvious but familiar example is *chalk*, a soft, porous rock made up almost entirely of the hard parts of microscopic organisms (Figure 2.14).

Inorganic limestones form when chemical changes or high water temperatures increase the concentration of calcium carbonate to the point that it precipitates. *Travertine*, the type of limestone that decorates caverns, is one example. Groundwater is the source of travertine that is deposited in caves. As water drops reach the air in a cavern, some of the carbon dioxide dissolved in the water escapes, causing calcium carbonate to precipitate.

Dissolved silica (SiO_2) precipitates to form varieties of microcrystalline quartz. Minerals composed of microcrystalline quartz include chert (light color), flint (dark), jasper (red), and agate (banded). These chemical sedimentary rocks may have either an inorganic or biochemical origin, but the mode of origin is usually difficult to determine.

Very often, evaporation causes minerals to precipitate from water. Such minerals include halite, the chief component of *rock salt*, and gypsum, the main ingredient of *rock gypsum*. Both materials have significant commercial importance. Halite is familiar to everyone as the common salt used in cooking and seasoning foods. Of course, it has many other uses and has been considered important enough that people have sought, traded, and fought over it for much of human history. Gypsum is the basic ingredient of plaster of Paris. This material is used

Figure 2.14 The White Chalk Cliffs of Dover. (Photo by ZEFA/UK/Chorley & Handford/The Stock Market.)

most extensively in the construction industry for "drywall" and plaster.

In the geologic past, many areas that are now dry land were covered by shallow arms of the sea that had only narrow connections to the open ocean. Under these conditions, water continually moved into the bay to replace water lost by evaporation. Eventually the waters of the bay became saturated and salt deposition began. Today, these arms of the sea are gone, and the remaining deposits are called **evaporites**.

On a smaller scale, evaporite deposits can be seen in such places as Death Valley, California. Here, following rains or periods of snowmelt in the mountains, streams flow from surrounding mountains into an enclosed basin. As the water evaporates, *salt flats* form from dissolved materials left behind as a white crust on the ground (Figure 2.15).

Coal is quite different from other chemical sedimentary rocks. Unlike other rocks in this category, which are calcite- or silica-rich, coal is made mostly of organic matter. Close examination of a piece of coal under a microscope or magnifying glass often reveals plant structures such as leaves, bark, and wood that have been chemically altered but are still identifiable. This supports the conclusion that coal is the end product of the burial of large amounts of plant material over extended periods.

The initial stage in coal formation is the accumulation of large quantities of plant remains. However, special conditions are required for such accumulations, because dead plants normally decompose when exposed to the atmosphere. An ideal environment that allows for the buildup of plant material is a swamp. Because stagnant swamp water is oxygen-deficient, complete decay (oxidation) of the plant material is not possible. At various times during Earth history, such environments have been common. Coal undergoes successive stages of formation. With each successive stage, higher temperatures and pressures drive off impurities and volatiles, as shown in Figure 2.16.

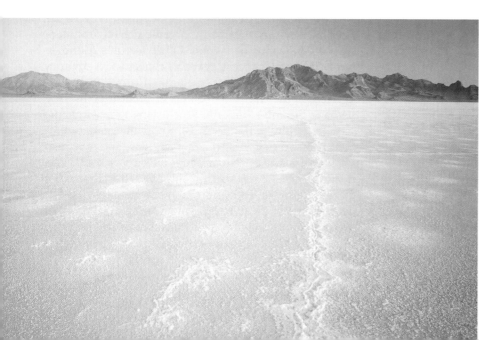

Figure 2.15 Bonneville Salt Flats, Utah. (Photo by Scott T. Smith)

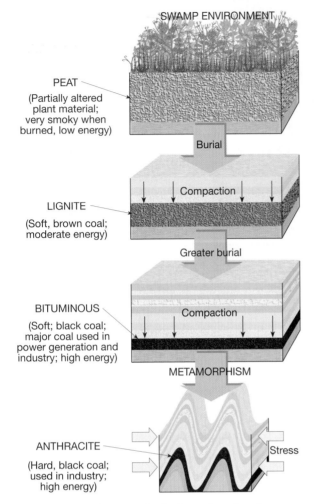

Figure 2.16 Successive stages in the formation of coal.

SWAMP ENVIRONMENT

PEAT
(Partially altered plant material; very smoky when burned, low energy)

Burial

Compaction

LIGNITE
(Soft, brown coal; moderate energy)

Greater burial

BITUMINOUS
(Soft; black coal; major coal used in power generation and industry; high energy)

Compaction

METAMORPHISM

ANTHRACITE
(Hard, black coal; used in industry; high energy)

Stress

grains are pressed closer and closer, pore space is greatly reduced. For example, when clays are buried beneath several thousand meters of material, the volume of the clay might be reduced as much as 40 percent. Compaction is most significant in fine-grained sedimentary rocks such as shale, because sand and other coarse sediments compress little.

Cementation is another important means by which sediments are converted to sedimentary rock. The cementing materials are carried in solution by water percolating through the pore spaces between particles. Through time, the cement precipitates onto the sediment grains, fills the open spaces, and joins the particles. Calcite, silica, and iron oxide are the most common cements. Identification of the cementing material is simple. Calcite cement will effervesce (fizz) with dilute hydrochloric acid. Silica is the hardest cement and thus produces the hardest sedimentary rocks. When a sedimentary rock has an orange or red color, this usually means iron oxide is present.

Features of Sedimentary Rocks

Sedimentary rocks are particularly important evidence of Earth's long history. These rocks form at Earth's surface, and as layer upon layer of sediment accumulates, each records the nature of the environment at the time the sediment was deposited. These layers, called **strata**, or **beds**, are the *single most characteristic feature of sedimentary rocks* (see Figure 2.11).

The thickness of beds range from microscopically thin to dozens of meters thick. Separating the strata are *bedding planes*, flat surfaces along which rocks tend to separate or break. Generally, each bedding plane marks the end of one episode of sedimentation and the beginning of another.

Sedimentary rocks provide geologists with evidence for deciphering past environments. A conglomerate, for example, indicates a high-energy environment, such as a rushing stream, where only the coarse materials can settle out. By contrast, black shale and coal are associated with a low-energy, organic-rich environment such as a swamp or lagoon. Other features found in some sedimentary rocks also give clues to past environments (Figure 2.17).

Fossils, the traces or remains of prehistoric life, are perhaps the most important inclusions found in some sedimentary rock (Figure 2.18). Knowing the nature of the life-forms that existed at a particular time may help to answer many questions about the environment. Was it land or ocean? A lake or swamp? Was the climate hot or cold, rainy or dry? Was the ocean water shallow or deep, turbid or clear? Furthermore, fossils are important time indicators and play a key role in matching up rocks from different places that are the same age. Fossils are important tools used in interpreting the geologic past and will be examined in some detail in Chapter 10.

Lignite and bituminous coals are sedimentary rocks, but anthracite is a metamorphic rock. Anthracite forms when sedimentary layers are subjected to the folding and deformation associated with mountain building.

In summary, we divide sedimentary rocks into two major groups: detrital and chemical. The main criterion for classifying detrital rocks is particle size, whereas chemical rocks are distinguished by their mineral composition. The categories presented here are more rigid than is the actual state of nature. Many detrital sedimentary rocks are a mixture of more than one particle size. Furthermore, many sedimentary rocks classified as chemical also contain at least small quantities of detrital sediment, and practically all detrital rocks are cemented with material that was originally dissolved in water.

Lithification of Sediment

Lithification refers to the processes by which sediments are transformed into solid sedimentary rocks. One of the most common processes is *compaction*. As sediments accumulate through time, the weight of overlying material compresses the deeper sediments. As the

A.

B.

Figure 2.17 A. Ripple marks preserved in sedimentary rocks may indicate a beach or stream channel environment. (Photo by Stephen Trimble) **B.** Mud cracks form when wet mud or clay dries and shrinks, perhaps signifying a tidal flat or desert basin. (Photo by Gary Yeowell/Tony Stone Images)

Metamorphic Rocks: Changed in Form

 Earth Materials
Metamorphic Rocks

Metamorphism literally means to "change form." Thus, metamorphism changes existing rocks in size, shape, texture, and even the minerals they contain. Metamorphic rocks can form from igneous, sedimentary, or even from other metamorphic rocks. The agents of change include heat, pressure, and chemically active fluids.

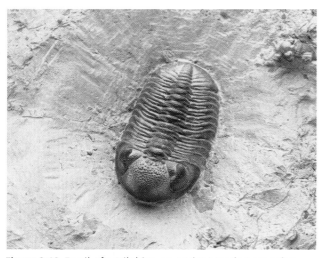

Figure 2.18 Fossil of a trilobite, an ancient marine organism, preserved in shale. Fossils are important tools used to determine past environmental conditions. They are also valuable time indicators.

Metamorphism varies broadly in degree. In some instances, rocks are only slightly changed. For example, under *low-grade metamorphism*, the common sedimentary rock shale becomes the metamorphic rock called *slate*. Although slate is more compact than shale, hand samples of these rocks are sometimes difficult to distinguish. In other cases, the transformation is so complete that the identity of the original rock cannot be determined. In such *high-grade metamorphism*, features like bedding planes, fossils, and vesicles that may have existed in the parent rock are completely obliterated. Further, when rocks at depth are subjected to differential forces, they flow and bend into intricate folds.

In the most extreme metamorphic environments, the temperatures approach those at which rocks melt. However, during metamorphism the deformed material must remain essentially solid, for once complete melting occurs, we enter the realm of igneous activity.

Metamorphism takes place when rock is subjected to conditions unlike those in which it originally formed. As a result, the rock becomes unstable and gradually changes until a state of equilibrium with the new environment is reached. The changes occur at the elevated temperatures and pressures of the region extending from a few kilometers below Earth's surface to the crust–mantle boundary. As the formation of metamorphic rocks is completely hidden from view (which is not the case for many sedimentary and some igneous rocks), metamorphism is one of the most difficult processes for geologists to study.

Metamorphism most often occurs in one of two settings. First, during mountain building, great quantities of rock are subjected to the intense stresses and high temperatures associated with large-scale deformation. The end result may be extensive areas of metamorphic rocks that are said to have undergone **regional metamorphism**. The greatest volume of metamorphic rock is produced in this fashion. Second, when rock is in contact with, or near, a mass of magma, **contact metamorphism** takes place. In this circumstance, the changes are caused primarily by the high temperatures of the molten material, which in effect "bake" the surrounding rock.

Extensive areas of metamorphic rocks are exposed on every continent. Metamorphic rocks are an important component of many mountain belts, where they make up a large portion of a mountain's crystalline core. Even the stable continental interiors, which are generally covered by sedimentary rocks, are underlain by metamorphic basement rocks (Figure 2.19). In all of these settings, the metamorphic rocks are usually highly deformed and intruded by igneous masses. Indeed, significant parts of Earth's continental crust are composed of metamorphic and associated igneous rocks.

Three Metamorphic Agents

The agents of metamorphism include *heat*, *pressure*, and *chemically active fluids*. During metamorphism, rocks are often subjected to all three agents simultaneously. However, the degree of metamorphism and the contribution of each agent vary greatly from one environment to another. In low-grade metamorphism, rocks are subjected to temperatures and pressures only slightly greater than those associated with the lithification of sediments. High-grade metamorphism, in contrast, involves extreme conditions closer to those at which rocks melt.

Moreover, the minerals making up the parent rock determine, to a large extent, the degree to which various metamorphic agents will cause change. For example, when an intruding igneous mass enters a rock unit, hot, ion-rich fluids circulate through the host rock. If the host rock is quartz sandstone, very little alteration may take place. However, if the host rock is limestone, the impact of these fluids can be dramatic and the effects of metamorphism may extend for several kilometers from the igneous mass.

Heat is perhaps the most important agent of metamorphism. It provides the energy to drive chemical reactions that recrystallize minerals. Rocks formed near Earth's surface may be subjected to intense heat when they are intruded by molten material rising from below. Because temperature increases with depth, rocks that originate in a surface environment may also be subjected to extreme temperatures if they are subsequently buried deep within Earth. When buried to a depth of only a few kilometers, certain minerals, such as clay, become unstable and recrystallize into minerals that are stable in this more intense environment. Other minerals, particularly those in crystalline igneous rocks, are stable at relatively high temperatures and pressures and therefore require burial to 20 kilometers or more before metamorphism will occur.

Pressure, like temperature, also increases with depth. Buried rocks are subjected to the force exerted by the load above. This confining pressure is analogous to air pressure, where the force is applied equally in all directions. In addition to the pressure exerted by the load of material above, rocks are also subjected to *differential stresses* during the process of mountain building (Figure 2.20). In this situation the force is unequal in different directions, and the material is squeezed as if it had been placed in a vise. Rock located at great depth is quite warm and behaves plastically during deformation. This accounts for its ability to flow and bend into intricate folds (Figure 2.21).

Figure 2.19 Uplifting and erosion exposed this metamorphic rock in Alaska's Brook Range. (Photo by Jeff Gnass/The Stock Market)

A. Confining pressure

Undeformed strata

Deformed strata

B. Directional stress

Figure 2.20 Pressure (stress) as a metamorphic agent. **A.** In a depositional environment, as confining pressure increases, rocks deform by decreasing in volume. **B.** During mountain building, directional stress shortens and deforms rock strata.

Chemically active fluids also influence the metamorphic process. Most common is the water contained in the pore spaces of virtually every rock. Water that surrounds the crystals acts as a catalyst by aiding the migration of ions. In some instances, the minerals recrystallize into more stable crystal forms. In other cases, ion exchange among minerals results in the formation of completely different minerals.

Figure 2.21 Deformed metamorphic rocks exposed in a road cut in the Eastern Highland of Connecticut. Imagine the tremendous force required to fold rock in this manner. (Photo by Phil Dombrowski)

Metamorphism Changes Texture

The degree of metamorphism is reflected in the rock's texture and mineralogy. When rocks are subjected to low-grade metamorphism, they become more compact and thus more dense. A common example is the metamorphic rock slate, which forms when shale is subjected to temperatures and pressures only slightly greater than those associated with the compaction that lithifies sediment. In this case, differential stress causes the microscopic clay minerals in shale to align into the more compact arrangement found in slate.

Under more extreme conditions, stress causes certain minerals to recrystallize. In general, recrystallization encourages the growth of larger crystals. Consequently, many metamorphic rocks consist of visible crystals, much like coarse-grained igneous rocks. The crystals of some minerals will recrystallize with a preferred orientation, essentially perpendicular to the direction of the compressional force. The resulting mineral alignment usually gives the rock a layered or banded appearance termed **foliated texture** (Figure 2.22). Simply, foliation results whenever the minerals of a rock are brought into parallel alignment.

Not all metamorphic rocks have a foliated texture. Such rocks are said to exhibit a **nonfoliated texture**. Metamorphic rocks composed of only one mineral that forms equidimensional crystals are, as a rule, not visibly foliated. For example, limestone, if pure, is composed of only a single mineral, calcite. When a fine-grained limestone is metamorphosed, the small calcite crystals combine to form larger interlocking crystals. The resulting rock resembles a coarse-grained igneous rock. This nonfoliated metamorphic equivalent of limestone is called *marble*.

In some environments, new materials are actually introduced during the metamorphic process. For example, host rock adjacent to a large intruding magma body may be altered by *hydrothermal* (hot-water) *solutions* released during the latter stages of crystallization. Hydrothermal solutions are rich in ions, including those of valuable metals. Many metallic ore deposits are formed by the precipitation of minerals from hydrothermal solutions.

Classifying Metamorphic Rocks

To review, metamorphic processes cause many changes in existing rocks, including increased density, growth of larger crystals, foliation (reorientation of the mineral grains into a layered or banded appearance), and the transformation of low-temperature minerals into high-temperature minerals. Further, the introduction of ions generates new minerals, some of which are economically important.

Here is a brief look at common rocks produced by metamorphic processes.

Figure 2.22 Under directed pressure, planar minerals, such as the micas, become reoriented or recrystallized so that their surfaces are aligned at right angles to the stress. The resulting planar orientation of mineral grains gives the rock a foliated texture. If the coarse-grained igneous rock (granite) on the left underwent intense metamorphism, it could end up closely resembling the metamorphic rock the right (gneiss). (Photos by E. J. Tarbuck)

Foliated Rocks. Slate is a very fine-grained foliated rock composed of minute mica flakes (Figure 2.23). The most noteworthy characteristic of slate is its excellent rock cleavage, meaning that it splits easily into flat slabs. This property has made slate a most useful rock for roof and

5 cm

Figure 2.23 Slate, a common metamorphic rock produced by the low-grade metamorphism of shale. (Photo by E. J. Tarbuck)

floor tile, chalkboards, and billiard tables. Slate is most often generated by the low-grade metamorphism of shale, although less frequently it forms from the metamorphism of volcanic ash. Slate can be almost any color, depending on its mineral constituents. Black slate contains organic material; red slate gets its color from iron oxide; and green slate is usually composed of chlorite, a micalike mineral.

Schists are strongly foliated rocks, formed by regional metamorphism. They are "platy" and can be readily split into thin flakes or slabs. Like slate, the parent material from which many schists originate is shale, but in the case of schist, the metamorphism is more intense.

The term *schist* describes the *texture* of a rock regardless of composition. For example, schists composed primarily of muscovite and biotite are called *mica schists* (Figure 2.24).

Gneiss (pronounced "nice") is the term applied to banded metamorphic rocks that contain mostly elongated and granular, as opposed to platy, minerals (see Figure 2.22, right). The most common minerals in gneisses are quartz and feldspar, with lesser amounts of muscovite, biotite, and hornblende. Gneisses exhibit

Figure 2.24 Mica schist, a common metamorphic rock composed of shiny mica flakes. (Photo by E. J. Tarbuck)

strong segregation of light and dark silicates, giving them a characteristic banded texture. While in a plastic state, these banded gneisses can be deformed into intricate folds.

Nonfoliated Rocks. *Marble* is a coarse, crystalline rock whose parent rock is limestone (Figure 2.25). Marble is composed of large interlocking calcite crystals, which form from the recrystallization of smaller grains in the parent rock.

Because of its color and relative softness (hardness of only 3 on the Mohs scale) marble is a popular building stone. White marble is particularly prized as a stone from which to carve monuments and statues, such as the famous statue of David by Michelangelo. Often the limestone from which marble forms contains impurities that color the marble. Thus, marble can be pink, gray, green, or even black.

Figure 2.25 Marble, a crystalline rock formed by the metamorphism of limestone. (Photo by E. J. Tarbuck)

Quartzite is a very hard metamorphic rock most often formed from quartz sandstone. Under moderate-to-high-grade metamorphism, the quartz grains in sandstone fuse. Pure quartzite is white, but iron oxide may produce reddish or pinkish stains and dark minerals may impart a gray color.

Resources From Rocks and Minerals

The outer layer of Earth, which we call the crust, is only as thick when compared to the remainder of the Earth as a peach skin is to a peach, yet it is of supreme importance to us. We depend on it for fossil fuels and as a source of such diverse minerals as the talc for baby powder, salt to flavor food, and gold for world trade. In fact, on occasion, the availability or absence of certain Earth materials has altered the course of history. As the material requirements of modern society grow, the need to locate additional supplies of useful minerals also grows, and becomes more challenging as well (see Box 2.1).

Metallic Mineral Resources

Some of the most important accumulations of metals, such as gold, silver, copper, mercury, lead, platinum, and nickel, are produced by igneous and metamorphic processes (Table 2.3). These mineral resources, like

Table 2.3 Ore minerals of important metals.

Mineral	Uses
Aluminum	Bauxite
Chromium	Chromite
Copper	Chalcopyrite
	Bornite
	Chalcocite
Gold	Native gold
Iron	Hematite
	Magnetite
	Limonite
Lead	Galena
Magnesium	Magnesite
	Dolomite
Manganese	Pyrolusite
Mercury	Cinnabar
Molybdenum	Molybdenite
Nickel	Pentlandite
Platinum	Native platinum
Silver	Native silver
	Argentite
Tin	Cassiterite
Titanium	Ilmenite
	Rutile
Tungsten	Wolframite
	Scheelite
Uranium	Uraninite (pitchblende)
Zinc	Sphalerite

Box 2.1   People and the Environment: United States Per Capita Use of Mineral and Energy Resources

Materials that we extract from Earth are the basis of modern civilization. Mineral and energy resources from the crust are the raw materials from which the products used by society are made. Like most people who live in highly industrialized nations, you may not realize the quantity of resources needed to maintain your present standard of living. Figure 2.A shows the approximate annual per capita consumption of several important metallic and nonmetallic mineral resources for the United States. Per capita use of our main energy resources is also shown. This is each person's prorated share of the materials required by industry to provide the vast array of homes, cars, electronics, cosmetics, packaging, and so on that modern society demands. It represents the energy we need to drive cars, fly airplanes, run machinery, and heat buildings. Figures for other highly industrialized countries such as Canada, Australia, and several other nations in Western Europe are comparable.

Every year about 9000 kilograms of stone, sand, and gravel is mined for each person in the United States. Calculate how much stone, sand, and gravel will be needed for an individual during a 75-year lifetime. If 1.5 metric tons (1000 kilograms in a metric ton) occupies 1 cubic meter, how big a hole must be dug to supply an individual with 75 years' worth of stone, sand, and gravel?

Figure 2.A The annual per capita consumption of nonmetallic and metallic mineral resources for the United States is nearly 10,000 kilograms! About 94 percent of the materials used are nonmetallic. The per capita use of oil, coal, and natural gas exceeds 11,000 kilograms.

---

most others, result from processes that concentrate desirable materials to the extent that extraction is economically feasible.

The igneous processes that generate some metal deposits are quite straightforward. For example, as a large magma body cools, the heavy minerals that crystallize early tend to settle to the lower portion of the magma chamber. This type of magmatic segregation is particularly active in large basaltic magmas where chromite (ore of chromium), magnetite, and platinum are occasionally generated. Layers of chromite, interbedded with other heavy minerals, are mined from such deposits in the Bushveld Complex in South Africa, which contains over 70 percent of the world's known reserves of platinum.

Among the best-known and most important ore deposits are those generated from hydrothermal (hot-water) **solutions**. Included in this group are the gold

deposits of the Homestake mine in South Dakota; the lead, zinc, and silver ores near Coeur d'Alene, Idaho; the silver deposits of the Comstock Lode in Nevada; and the copper ores of the Keweenaw Peninsula in Michigan.

The majority of hydrothermal deposits are thought to originate from hot, metal-rich fluids that are associated with cooling magma bodies. During solidification, liquids plus various metallic ions accumulate near the top of the magma chamber. Because these hot fluids are very mobile, they can migrate great distances through the surrounding rock before they are eventually deposited. Some of this fluid moves along fractures or bedding planes, where it cools and precipitates the metallic ions to produce **vein deposits** (Figure 2.26). Many of the most productive deposits of gold, silver, and mercury occur as hydrothermal vein deposits.

Another important type of accumulation generated by hydrothermal activity is called a **disseminated**

Figure 2.26 Light-colored vein deposits emplaced along a series of fractures in dark-colored igneous rock. (Photo by James E. Patterson)

deposit. Rather than being concentrated in narrow veins and dikes, these ores are distributed as minute masses throughout the entire rock mass. Much of the world's copper is extracted from disseminated deposits, including the huge Bingham Canyon copper mine in Utah. Because these accumulations contain only 0.4 to 0.8 percent copper, between 125 and 250 metric tons of ore must be mined for every ton of metal recovered. The environmental impact of these large excavations, including the problems of waste disposal, is significant.

Nonmetallic Mineral Resources

Mineral resources not used as fuels or processed for the metals they contain are referred to as *nonmetallic mineral resources*. These materials are extracted and processed either to make use of the nonmetallic elements they contain or for the physical and chemical properties they possess (Table 2.4). Nonmetallic mineral resources are commonly divided into two broad groups—*building materials* and *industrial minerals*. As some substances have many different uses, they are found in both categories. Limestone, perhaps the most versatile and widely used rock of all, is the best example (Figure 2.27). As a building material, it is used not only as crushed rock and building stone but also in the making of cement. Moreover, as an industrial mineral, it is an ingredient in the manufacture of steel and is used in agriculture to neutralize acidic soils.

Besides aggregate (sand, gravel, and crushed rock) and cut stone, the other important building materials include gypsum for plaster and wallboard, clay for tile and bricks, and cement, which is made from limestone and shale. Cement and aggregate go into the making of concrete, a material that is essential to practically all construction.

Many and various nonmetallic resources are classified as industrial minerals. People often do not realize the importance of industrial minerals because they see only the products that resulted from their use and not the minerals themselves. That is, many nonmetallics are used up in the process of creating other products. Examples include fluorite and limestone, which are part of the steelmaking process; corundum and garnet, which are used as abrasives to make a piece of machinery; and sylvite, which is employed in the production of the fertilizers used to grow a food crop.

Table 2.4 Uses of nonmetallic minerals.

Mineral	Uses
Apatite	Phosphorous fertilizers
Asbestos (chrysotile)	Incombustible fibers
Calcite	Aggregate; steelmaking; soil conditioning; chemicals; cement; building stone
Clay minerals (kaolinite)	Ceramics; china
Corundum	Gemstones; abrasives
Diamond	Gemstones; abrasives
Fluorite	Steelmaking; aluminum refining; glass; chemicals
Garnet	Abrasives; gemstones
Graphite	Pencil lead; lubricant; refractories
Gypsum	Plaster of Paris
Halite	Table salt; chemicals; ice control
Muscovite	Insulator in electrical applications
Quartz	Primary ingredient in glass
Sulfur	Chemicals; fertilizer manufacture
Sylvite	Potassium fertilzers
Talc	Powder used in paints, cosmetics, etc.

Figure 2.27 Limestone quarry, Tivoli, Italy. Limestone has many uses and is both a building material and an industrial mineral. (Photo by Blaine Harrington III/The Stock Market)

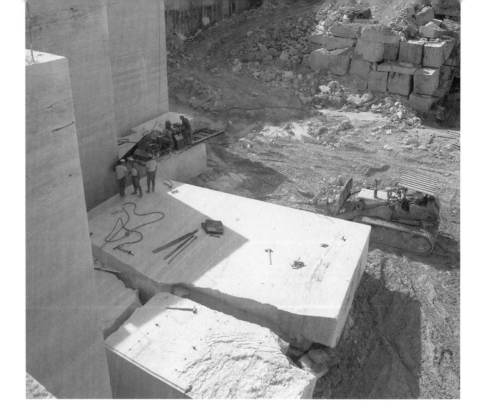

Chapter Summary

- *Igneous rock* forms from *magma* that cools and solidifies in a process called *crystallization*. *Sedimentary rock* forms from the *lithification of sediment*. *Metamorphic rock* forms from rock that has been subjected to great pressure and heat in a process called *metamorphism*.

- The rate of cooling of magma greatly influences the size of mineral crystals in igneous rock. The four basic igneous rock textures are (1) *fine-grained*, (2) *coarse-grained*, (3) *porphyritic*, and (4) *glassy*.

- The mineral makeup of an igneous rock is ultimately determined by the chemical composition of the magma from which it crystallized. N.L. Bowen showed that as magma cools, minerals crystallize in an orderly fashion. *Crystal settling* can change the composition of magma and cause more than one rock type to form from a common parent magma.

- Igneous rocks are classified by their *texture* and *mineral composition*.

- *Detrital sediments* are materials that originate and are transported as solid particles derived from weathering. *Chemical sediments* are soluble materials produced largely by chemical weathering that are precipitated by either inorganic or organic processes. *Detrital sedimentary rocks*, which are classified by particle size, contain a variety of mineral and rock fragments, with clay minerals and quartz the chief constituents. *Chemical sedimentary rocks* often contain the products of biological processes such as shells or mineral crystals that form as water evaporates and minerals precipitate. *Lithification* refers to the processes by which sediments are transformed into solid sedimentary rocks.

- Common detrital sedimentary rocks include *shale* (the most common sedimentary rock), *sandstone*, and *conglomerate*. The most abundant chemical sedimentary rock is *limestone*, composed chiefly of the mineral calcite. *Rock gypsum* and *rock salt* are chemical rocks that form as water evaporates and triggers the deposition of chemical precipates.

- Some of the features of sedimentary rocks that are often used in the interpretation of Earth history and past environments include *strata*, or *beds* (the single most characteristic feature), *bedding planes*, and *fossils*.

- Two types of metamorphism are (1) *regional metamorphism* and (2) *contact metamorphism*. The agents of metamorphism include *heat*, *pressure*, and *chemically active fluids*. Heat is perhaps the most important because it provides the energy to drive the reactions that result in the *recrystallization* of minerals. Metamorphic processes cause many changes in rocks, including *increased density*, growth of *larger mineral crystals*, *reorientation of the mineral grains* into a layered or banded appearance known as *foliation*, and the formation of *new minerals*.

- Some common metamorphic rocks with a *foliated texture* include *slate*, *schist*, and *gneiss*. Metamorphic rocks with a *nonfoliated texture* include *marble* and *quartzite*.

- Some of the most important accumulations of *metallic mineral resources* are produced by igneous and metamorphic processes. *Vein deposits* (deposits in fractures or bedding planes) and *disseminated deposits* (deposits distributed throughout the entire rock mass) are produced from *hydrothermal solutions*—hot, metal-rich fluids associated with cooling magma bodies.

- *Nonmetallic mineral resources* are mined for the nonmetallic elements they contain or for the physical and chemical properties they possess. The two groups of nonmetallic mineral resources are (1) *building materials* (e.g., limestone and gypsum) and (2) *industrial minerals* (e.g., fluorite and corundum).

Key Terms

chemical sedimentary rock (p. 45)
coarse-grained texture (p. 40)
contact metamorphism (p. 50)
crystallization (p. 36)
detrital sedimentary rock (p. 45)
disseminated deposit (p. 54)
evaporite (p. 47)
extrusive (volcanic) (p. 38)
fine-grained texture (p. 39)
foliated texture (p. 51)

glassy texture (p. 40)
hydrothermal solution (p. 54)
igneous rock (p. 37)
intrusive (plutonic) (p. 38)
lava (p. 38)
lithification (p. 38)
magma (p. 36)
metamorphic rock (p. 38)
nonfoliated texture (p. 51)
porphyritic texture (p. 40)

regional metamorphism (p. 50)
rock cycle (p. 36)
sediment (p. 37)
sedimentary rock (p. 38)
strata (beds) (p. 48)
texture (p. 39)
vein deposit (p. 54)
weathering (p. 37)

Review Questions

1. Explain the statement "One rock is the raw material for another" using the rock cycle.

2. If a lava flow at Earth's surface had a basaltic composition, what rock type would the flow likely be (see Table 2.1, p. 43)? What igneous rock would form from the same magma if it did not reach the surface but instead crystallized at great depth?

3. What does a porphyritic texture indicate about the history of an igneous rock?

4. How are granite and rhyolite different? The same? (See Table 2.1, p. 43.)

5. Relate the classification of igneous rocks to Bowen's reaction series.

6. What minerals are most common in detrital sedimentary rocks? Why are these minerals so abundant?

7. What is the primary basis for distinguishing among various detrital sedimentary rocks?

8. Distinguish between the two categories of chemical sedimentary rocks.

9. What are evaporite deposits? Name a rock that is an evaporite.

10. Compaction is an important lithification process with which sediment size?

11. What is probably the single most characteristic feature of sedimentary rocks?

12. What is metamorphism? What are the *agents of change*?

13. Distinguish between regional and contact metamorphism.

14. What feature would easily distinguish schist and gneiss from quartzite and marble?

15. In what ways do metamorphic rocks differ from the igneous and sedimentary rocks from which they formed?

16. List two general types of hydrothermal deposits.

17. Nonmetallic resources are commonly divided into two broad groups. List the two groups and some examples of materials that belong to each.

Examining the Earth System

1. The sedimentary rocks coquina and shale have each formed in response to interactions among two or more of Earth's spheres. List the spheres associated with the formation of each of the rocks and write a short explanation of each of your choices.

2. Of the two main sources of energy that drive the rock cycle, Earth's internal heat and solar energy, which is primarily responsible for each of the three groups of rocks found on and within Earth?

Web Resources

The following are informative and interesting World Wide Web sites that address topics related to those presented in the chapter:

Bob's Rock Shop

• **http://www.rockhounds.com/rockshop/table.html**

Rock Identification and Classification

• **http://www.ei.gov.bc.ca/geosmin/educat/ROCKID/ rockclas.htm**

For direct links to these sites and other appropriate Earth science Web locations, chapter review exercises, and relevant, topical investigations that utilize Web resources, contact the *Earth Science*, 9th Edition Home Page at

• **http://www.prenhall.com/tarbuck.**

Chapter 3

Weathering, Soil, and Mass Wasting

Differential weathering produced the Delicate Arch in Arches National Park, Utah. Weathering plays a critical role in the rock cycle. (Photo by David Muench Photography, Inc.)

Earth's surface is constantly changing. Rock is disintegrated and decomposed, moved to lower elevations by gravity, and carried away by water, wind, or ice. In this manner Earth's physical landscape is sculptured. This chapter focuses on the first two steps of this never-ending process—weathering and mass wasting—probing into how and why rock disintegrates and decomposes and what mechanisms act to move it downslope. Soil, an important product of the weathering process and a vital resource, is also examined.

Earth's External Processes

External Processes
External vs Internal Processes

Weathering, mass wasting, and erosion are called **external processes** because they occur at or near Earth's surface and are powered by energy from the Sun. External processes are a basic part of the rock cycle because they are responsible for transforming solid rock into sediment.

To the casual observer, the face of Earth may appear to be without change, unaffected by time. In fact, less than 200 years ago, most people believed that mountains, lakes, and deserts were permanent features of an Earth that was thought to be no more than a few thousand years old. Today, however, we know that mountains eventually succumb to weathering and erosion, lakes fill with sediment and vegetation or are drained by streams, and deserts come and go as relatively minor climate changes occur.

Earth is a dynamic body. Some parts of Earth's surface are gradually elevated by mountain building and volcanic activity. Meanwhile, opposing processes are continually removing materials from higher elevations and transporting them to lower elevations (Figure 3.1). The latter processes include:

1. **Weathering**—disintegration and decomposition of rock at or near Earth's surface.

2. **Mass wasting**—transfer of rock material downslope under the influence of gravity.

3. **Erosion**—incorporation and transportation of material by a mobile agent, usually water, wind, or ice.

We will first turn our attention to the process of weathering and the products generated by this activity. However, weathering cannot be easily separated

Figure 3.1 Slopes here, as elsewhere, are places where materials are continually moving from higher to lower elevations. Weathering begins the process by attacking the solid rock exposed at the surface. Next, gravity moves the weathered debris downslope. This step, termed mass wasting, may range from a slow and gradual creep to a thundering landslide. Eventually, the material that was once high up the slope reaches the stream at the bottom. The moving water then transports the debris away. (Photo by Carr Clifton)

from the other two processes because, as weathering breaks rocks apart, it facilitates the movement of rock debris by mass wasting and erosion. Conversely, the transport of material by mass wasting and erosion furthers the disintegration and decomposition of rock.

Weathering

All materials are susceptible to weathering. Consider, for example, the synthetic rock we call concrete, which closely resembles the sedimentary rock called conglomerate. A newly poured concrete sidewalk has a smooth, fresh look. However, not many years later, the same sidewalk will appear chipped, cracked, and rough, with pebbles exposed at the surface. If a tree is nearby, its roots may grow under the concrete, heaving and buckling it. The same natural processes that eventually break apart a concrete sidewalk also act to disintegrate natural rocks, regardless of their type or strength.

Why does rock weather? Simply, weathering is the response of Earth materials to a changing environment. For instance, after millions of years of uplift and erosion, the rocks overlying a large body of intrusive igneous rock may be removed. This exposes the rock to a whole new environment at the surface. The mass of crystalline rock, which formed deep below ground where temperatures and pressures are much greater than at the surface, is now subjected to very different and comparatively hostile surface conditions. In response, this rock mass will gradually change until it is once again in equilibrium, or balance, with its new environment. Such transformation of rock is what we call *weathering.*

In the following sections we will discuss two kinds of weathering—mechanical and chemical. Mechanical weathering is the physical breaking up of rocks into smaller pieces. Chemical weathering actually alters a rock's chemistry, changing it into different substances. Although we will consider these two

processes separately, keep in mind that they usually work simultaneously in nature.

Mechanical Weathering

When a rock undergoes **mechanical weathering** it is broken into smaller and smaller pieces, each retaining the characteristics of the original material. The end result is many small pieces from a single large one. Figure 3.2 shows that breaking a rock into smaller pieces increases the surface area available for chemical attack. Hence, by breaking rocks into smaller pieces, mechanical weathering increases the amount of surface area available for chemical weathering.

In nature, three physical processes are especially important in breaking rocks into smaller fragments: frost wedging, expansion resulting from unloading, and biological activity.

Frost Wedging

Alternate freezing and thawing of water is one of the most important processes of mechanical weathering. Water has the unique property of expanding about 9 percent when it freezes. This increase in volume occurs because, as ice forms, the water molecules arrange themselves into a very open crystalline structure. As a result, when water freezes, it expands and exerts a tremendous outward force. This can be verified by completely filling a container with water and freezing it. The formation of ice will rupture the container.

In nature, water works its way into every crack or void in rock and, upon freezing, expands and enlarges the opening. After many freeze-thaw cycles, the rock is broken into pieces. This process is appropriately called **frost wedging** (Figure 3.3). Frost wedging is most pronounced in mountainous regions in the middle latitudes where a daily freeze-thaw cycle often exists. Here, sections of rock are wedged loose and may tumble into large piles called **talus slopes**, which often form at the base of steep rocky cliffs (Figure 3.3).

Figure 3.2 Chemical weathering can occur only to those portions of a rock that are exposed to the elements. Mechanical weathering breaks rock into smaller and smaller pieces, thereby increasing the surface area available for chemical attack.

Figure 3.3 Frost wedging. As water freezes it expands to 109 percent of its original volume, exerting a force great enough to break rock. When frost wedging occurs in a setting such as this, the broken rock fragments fall to the base of the cliff and create a cone-shaped accumulation known as talus. (Photo by Tom Bean)

Unloading

When large masses of igneous rock, particularly those composed of granite, are exposed by erosion, slabs begin to break loose like the layers of an onion. The process is called **sheeting** and probably takes place in response to the great reduction in pressure that occurs when the overlying rock is eroded away. Accompanying this *unloading,* the outer layers expand more than the rock below, and thus separate from the rock body. Continued weathering eventually causes the slabs to separate and spall off, creating **exfoliation domes.** Excellent examples of exfoliation domes include Stone Mountain, Georgia, and Half Dome (Figure 3.4) and Liberty Cap in Yosemite National Park.

Deep underground mining provides us with another example of how rocks behave once the confining pressure is removed. Large rock slabs sometimes explode off the walls of newly cut mine tunnels because of the abruptly reduced pressure. Evidence of this type, plus the fact that fracturing occurs parallel to the floor of a quarry when large blocks of rock are removed, strongly supports the process of unloading as the cause of sheeting.

Although many fractures are created by expansion, others are produced by contraction during the crystallization of magma, and still others by tectonic forces during mountain building. Fractures produced

Figure 3.4 Summit of Half Dome, an exfoliation dome in Yosemite National Park. (Photo by Breck Kent)

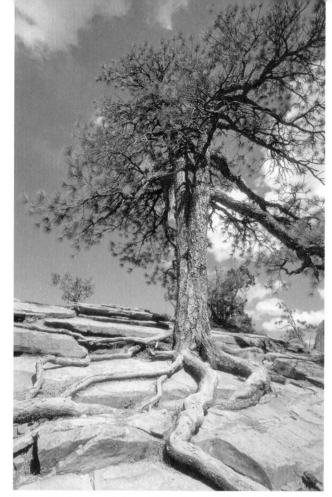

Figure 3.5 Root wedging widens fractures in rock and aids the process of mechanical weathering. (Photo by Tom Bean/DRK Photo)

Chemical Weathering

Chemical weathering involves the complex processes that alter the internal structures of minerals by removing and/or adding elements. During this transformation, the original rock decomposes into substances that are stable in the surface environment. Consequently, the products of chemical weathering will remain essentially unchanged as long as they remain in an environment similar to the one in which they formed.

Water and Carbonic Acid

Water is by far the most important agent of chemical weathering. Although pure water is nonreactive, a small amount of dissolved material is generally all that is needed to activate it. Oxygen dissolved in water will *oxidize* some materials. For example, when an iron nail is found in moist soil, it will have a coating of rust (iron oxide), and if the time of exposure has been long, the nail will be so weak that it can be broken as easily as a toothpick. When rocks containing iron-rich minerals oxidize, a yellow to reddish-brown rust will appear on the surface (Figure 3.6).

Carbon dioxide (CO_2) dissolved in water (H_2O) forms carbonic acid (H_2CO_3), the same weak acid produced when soft drinks are carbonated. Rain dissolves some carbon dioxide as it falls through the atmosphere,

Figure 3.6 Iron reacts with oxygen to form iron oxide as seen on these rusted barrels. (Photo by Stephen J. Krasemann/DRK Photo)

by these activities generally form a definite pattern and are called *joints*. Joints are important rock structures that allow water to penetrate to depth and start the process of weathering long before the rock is exposed at the surface.

Biological Activity

Weathering is also accomplished by the activities of organisms, including plants, burrowing animals, and humans. Plant roots in search of minerals and water grow into fractures, and as the roots grow they wedge the rock apart (Figure 3.5). Burrowing animals further break down the rock by moving fresh material to the surface, where physical and chemical processes can more effectively attack it. Decaying organisms also produce acids, which contribute to chemical weathering. Where rock has been blasted in search of minerals or for road construction, the impact is substantial.

Although usually considered separately from mechanical weathering, the activities of the erosional agents—wind, water, and glaciers—are nonetheless important. For as these mobile agents move rock debris, they relentlessly disintegrate the Earth materials they carry.

and additional amounts released by decaying organic matter are acquired as the water percolates through the soil. Carbonic acid ionizes to form the very reactive hydrogen ion (H^+) and the bicarbonate ion (HCO_3^-).

How Granite Weathers

To illustrate how rock chemically weathers when attacked by carbonic acid, we will consider the weathering of granite, the most abundant continental rock. Recall that granite consists mainly of quartz and potassium feldspar. The weathering of the potassium feldspar component of granite takes place as follows:

$$2KAlSi_3O_8 \ + \ 2(H^+ + HCO_3^-) \ + \ H_2O \ \rightarrow$$
potassium feldspar   carbonic acid   water

$$Al_2Si_2O_5(OH)_4 \ + \ 2KHCO_3 \ + \ 4SiO_2$$
clay mineral   potassium bicarbonite   silica

In this reaction, the hydrogen ions (H^+) attack and replace potassium ions (K^+) in the feldspar structure, thereby disrupting the crystalline network. Once removed, the potassium is available as a nutrient for plants. Or it becomes potassium bicarbonate ($KHCO_3$), a soluble salt that may be incorporated into other minerals or carried to the ocean in dissolved form by streams.

The most abundant products of the chemical breakdown of feldspar are residual clay minerals. Because clay minerals are the end product of weathering, they are very stable under surface conditions. Consequently, clay minerals make up a high percentage of the inorganic material in soils. Moreover, the most abundant sedimentary rock, shale, also contains a high proportion of clay minerals.

In addition to the formation of clay minerals during this reaction, some silica is removed from the feldspar structure and is carried away by groundwater (water beneath Earth's surface). This dissolved silica will eventually precipitate to produce nodules of chert or flint, fill in the pore spaces between sediment grains, or be carried to the ocean, where microscopic animals will remove it to build hard silica shells.

To summarize, the weathering of potassium feldspar generates a residual clay mineral, a soluble salt (potassium bicarbonate), and some silica that enters into solution.

Quartz, the other main component of granite, is *very resistant* to chemical weathering; hence, it remains substantially unaltered when attacked by weakly acidic solutions. As a result, when granite weathers, the feldspar crystals dull and slowly turn to clay, releasing the once-interlocked quartz grains, which still retain their fresh, glassy appearance. Although some quartz remains in the soil, much is transported to the sea or to other sites of deposition, where it

becomes the main constituent of such features as sandy beaches and sand dunes. In time it may become lithified to form the sedimentary rock *sandstone.*

Weathering of Silicate Minerals

Table 3.1 lists the weathered products of some of the most common silicate minerals. Remember that silicate minerals make up most of Earth's crust and that these minerals are composed essentially of only eight elements. When chemically weathered, silicate minerals yield sodium, calcium, potassium, and magnesium ions. These form soluble products that may be removed by groundwater. The element iron combines with oxygen, producing relatively insoluble iron oxides, which give soil a reddish-brown or yellowish color. Under most conditions the three remaining elements, aluminum, silicon, and oxygen, join with water to produce residual clay minerals. However, even the highly insoluble clay minerals are very slowly removed by subsurface water.

Spheroidal Weathering

In addition to altering the internal structure of minerals, chemical weathering causes physical changes as well. For instance, when angular rock masses are attacked by water that enters along joints, the rocks tend to take on a spherical shape. Gradually the corners and edges of the angular blocks become more rounded. The corners are attacked most readily because of their greater surface area, as compared to the edges and faces. This process, called **spheroidal weathering,** gives the weathered rock a more rounded or spherical shape (Figure 3.7).

Sometimes during the formation of spheroidal boulders, successive shells separate from the rock's main body. Eventually the outer shells spall off, allowing the chemical weathering activity to penetrate deeper into the boulder. This spherical scaling results because, as the minerals in the rock weather

Table 3.1 Products of weathering.

Mineral	Residual Products	Material in Solution
Quartz	Quartz grains	Silica
Feldspars	Clay minerals	Silica K^+, Na^+, Ca^{2+}
Amphibole (hornblende)	Clay minerals Limonite Hematite	Silica Ca^{2+}, Mg^{2+}
Olivine	Limonite Hematite	Silica Mg^{2+}

Figure 3.7 Spheroidal weathering is evident in this exposure of granite in California's Joshua Tree National Monument. Because the rocks are attacked more vigorously on the corners and edges, they take on a spherical shape. The lines visible in the rock are called *joints*. Joints are important rock structures that allow water to penetrate and start the weathering process long before the rock is exposed. (Photo by E. J. Tarbuck)

to clay, they increase in size through the addition of water to their structure. This increased bulk exerts an outward force that causes concentric layers of rock to break loose and fall off. Hence, chemical weathering does produce forces great enough to cause mechanical weathering.

This type of spheroidal weathering, in which shells spall off, should not be confused with the phenomenon of sheeting discussed earlier. In sheeting, the fracturing occurs as a result of unloading and the rock layers that separate from the main body are largely unaltered at the time of separation.

Rates of Weathering

Several factors influence the type and rate of rock weathering. We have already seen how mechanical weathering affects the rate of weathering. By breaking rock into smaller pieces, the amount of surface area exposed to chemical weathering is increased. The

presence or absence of joints (cracks) can be significant because they influence the ability of water to penetrate the rock (Figure 3.7). Other important factors include the mineral makeup of rocks and climate.

Mineral Makeup

The variations in weathering rates, attributable to the mineral constituents, can be demonstrated by comparing old headstones made from different rock types. Headstones of granite, which is composed of silicate minerals, are relatively resistant to chemical weathering. We can see this by examining the inscriptions on the headstones shown in Figure 3.8. This is not true of the marble headstone, which shows signs of extensive chemical alteration over a relatively short period. Marble is composed of calcite (calcium carbonate), which readily dissolves even in a weakly acidic solution.

The silicates, the most abundant mineral group, weather in essentially the same sequence as their order of crystallization. By examining Bowen's reaction

Figure 3.8 An examination of headstones reveals the rate of chemical weathering on diverse rock types. The granite headstone (left) was erected three years after the marble headstone (right). The inscription date of 1885 on the marble monument is nearly illegible. (Photos by E. J. Tarbuck)

series (see Figure 2.8, p. 41), we see that olivine crystallizes first and is therefore the least resistant to chemical weathering, whereas quartz, which crystallizes last, is the most resistant.

Climate

Climatic factors, particularly temperature and moisture, are crucial to the rate of rock weathering. These climatic elements largely determine the weathering rate and strongly influence the kind and amount of vegetation present. Regions with lush vegetation generally have a thick mantle of soil rich in decayed organic matter from which chemically active fluids such as carbonic and humic acids are derived.

The optimum environment for chemical weathering is a combination of warm temperatures and abundant moisture. In polar regions chemical weathering is ineffective because frigid temperatures keep the available moisture locked up as ice, whereas in arid regions there is insufficient moisture to foster rapid chemical weathering.

Human activities can influence the composition of the atmosphere, which, in turn, can impact the rate of chemical weathering. Box 3.1 examines one well-known example—acid rain.

Differential Weathering

Many factors determine the type and rate of rock weathering for a given place. There is generally enough variation, even within a relatively small area, for the rocks to exhibit some differential weathering. **Differential weathering** simply relates to the fact that rocks exposed at Earth's surface usually do not weather at the same rate. Because of variations in such factors as mineral makeup, degree of jointing, and exposure to the elements, significant differences occur. Consequently, differential weathering and subsequent erosion create many unusual and often spectacular rock formations and landforms. Included are features such as the natural bridges found in Arches National Park and the sculptured rock pinnacles found in Bryce Canyon National Park (Figure 3.9).

Figure 3.9 Differential weathering is illustrated by these sculpted rock pinnacles in Bryce National Park, Utah. (Photo by Ray Mathis/The Stock Market)

Soil

Soil covers most land surfaces. Along with air and water, it is one of our most indispensable resources (Figure 3.10). Also like air and water, soil is taken for granted by many of us. The following quote helps put this vital layer in perspective.

> Science, in recent years, has focused more and more on the Earth as a planet, one that for all we know is unique—where a thin blanket of air, a thinner film of water, and the thinnest veneer of soil combine to support a web of life of wondrous diversity in continuous change.*

Soil has accurately been called "the bridge between life and the inanimate world." All life—the entire biosphere—owes its existence to a dozen or so elements that must ultimately come from Earth's crust.

---

*Jack Eddy. "A Fragile Seam of Dark Blue Light," in *Proceedings of the Global Change Research Forum*. U.S. Geological Survey Circular 1086, 1993, p. 15.

Box 3.1 Earth as a System: Acid Precipitation: A Human Impact on the Earth

Humans are part of the complex interacting whole we call the Earth system. As such, our actions cause changes to all the other parts of the system. For example, by going about our normal routine, we humans modify the composition of the atmosphere. These atmospheric modifications, in turn, cause unintended and unwanted changes to occur in the hydrosphere, biosphere, and solid Earth. Acid precipitation is one small but significant example.

Decomposed stone monuments and structures are common sights in many cities (Figure 3.A). Although we expect rock to gradually decompose, many of these monuments have succumbed prematurely. An important cause for this accelerated chemical weathering is acid precipitation.

Rain is naturally somewhat acidic. When carbon dioxide from the atmosphere dissolves in water, it becomes weak carbonic acid. However, the term *acid precipitation* refers to precipitation that is much more acidic than natural, unpolluted rain and snow.

As a consequence of burning large quantities of fossil fuels like coal and petroleum products, nearly 40 million tons of sulfur and nitrogen oxides are released into the atmosphere each year in the United States. The major sources of these emissions include power-generating plants, industrial processes, such as ore smelting and petroleum refining, and vehicles of all kinds. Through a series of complex chemical reactions, some of these pollutants are converted into acids that then fall to Earth's surface as rain or snow. Another portion is deposited in dry form and subsequently converted into acid after coming in contact with precipitation, dew, or fog.

Widespread acid rain has been known in northern Europe and eastern North America for some time. Studies

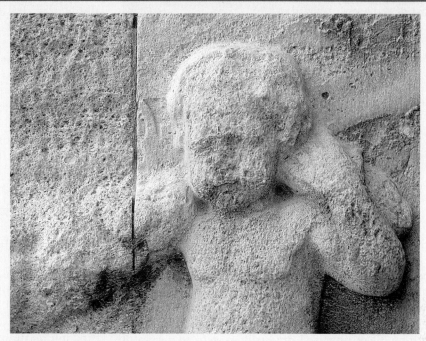

Figure 3.A Acid rain accelerates the chemical weathering of stone monuments and structures, including this building facade in Leipzig, Germany. (Photo by Doug Plummer/Photo Researchers, Inc.)

have also shown that acid rain occurs in many other regions, including western North America, Japan, China, Russia, and South America. In addition to local pollution sources, a portion of the acidity found in the northeastern United States and eastern Canada originates hundreds of kilometers away in industrialized regions to the south and southwest. This situation occurs because many pollutants remain in the atmosphere for periods as long as five days, during which time they may be transported great distances.

The damaging environmental effects of acid rain are thought to be considerable in some areas and imminent in others. The best-known effect is an increased acidity in thousands of lakes in Scandinavia and eastern North America. Accompanying this have been substantial increases in dissolved aluminum that

is leached from the soil by the acidic water and that, in turn, is toxic to fish. Consequently, some lakes are virtually devoid of fish, and others are approaching this condition. Ecosystems are characterized by many interactions at many levels of organization, which means that evaluating the effects of acid precipitation on these complex systems is difficult and expensive, and far from complete.

In addition to the thousands of lakes that can no longer support fish, research indicates that acid precipitation may also reduce agricultural crop yields and impair the productivity of forests. Acid rain not only harms the foliage, but also damages roots and leaches nutrient minerals from the soil. Finally, acid precipitation is known to promote the corrosion of metals and to contribute to the destruction of stone structures.

Once weathering and other processes create soil, plants carry out the intermediary role of assimilating the necessary elements and making them available to animals including humans.

An Interface in the Earth System

When Earth is viewed as a system, soil is referred to as an *interface*—a common boundary where different parts of a system interact. This is an appropriate designation

Figure 3.10 Soil is an essential resource that we often take for granted. Soil is not a living entity, but it contains a great deal of life. Moreover, this complex medium supports nearly all plant life, which, in turn, supports animal life. (Photo by James E. Patterson)

because soil forms where the solid Earth, the atmosphere, the hydrosphere, and the biosphere meet. Soil is a material that develops in response to complex environmental interactions among different parts of the Earth system. Over time, soil gradually evolves to a state of equilibrium or balance with the environment. Soil is dynamic and sensitive to almost every aspect of its surroundings. Thus, when environmental changes occur, such as climate, vegetative cover, and animal (including human) activity, the soil responds. Any such change produces a gradual alteration of soil characteristics until a new balance is reached. Although thinly distributed over the land surface, soil functions as a fundamental interface, providing an excellent example of the integration among many parts of the Earth system.

What Is Soil?

With few exceptions, Earth's land surface is covered by **regolith,** the layer of rock and mineral fragments produced by weathering. Some would call this material soil, but soil is more than an accumulation of weathered debris. **Soil** is a combination of mineral and organic matter, water, and air—that portion of the regolith that supports the growth of plants. Although the proportions of the major components in soil vary, the same four components always are present to some extent (Figure 3.11). About one-half of the total volume of a good-quality surface soil is a mixture of disintegrated and decomposed rock (mineral matter) and **humus,** the decayed remains of animal and plant life (organic matter). The remaining half consists of pore spaces among the solid particles where air and water circulate.

Although the mineral portion of the soil is usually much greater than the organic portion, humus is an essential component. In addition to being an important source of plant nutrients, humus enhances the soil's

ability to retain water. Because plants require air and water to live and grow, the portion of the soil consisting of pore spaces that allow for the circulation of these fluids is as vital as the solid soil constituents.

Soil water is far from "pure" water; instead, it is a complex solution containing many soluble nutrients. Soil water not only provides the necessary moisture for the chemical reactions that sustain life; it also supplies plants with nutrients in a form they can use. The pore spaces not filled with water contain air. This air is the source of necessary oxygen and carbon dioxide for most microorganisms and plants that live in the soil.

Soil Texture and Structure

Most soils are far from uniform, and contain particles of different sizes. **Soil texture** refers to the proportions of different particle sizes. Texture is a very basic soil property because it strongly influences the soil's ability to retain and transmit water and air, both of which are essential to plant growth. Sandy soils may drain too rapidly and dry out quickly. At the opposite extreme, the pore spaces of clay-rich soils may be so small that they inhibit drainage, and long-lasting puddles result. Moreover, when the clay and silt content is very high, plant roots may have difficulty penetrating the soil.

Because soils rarely consist of particles of only one size, *textural categories* have been established, based upon the varying proportions of clay, silt, and sand. The standard system of classes used by the U.S. Department of Agriculture is shown in Figure 3.12. For example, point *A* on this triangular diagram (left center) represents a soil composed of 10 percent silt, 40 percent clay, and 50 percent sand. Such a soil is called

Figure 3.11 Composition (by volume) of a soil in good condition for plant growth. Although the percentages vary, each soil is composed of mineral and organic matter, water, and air.

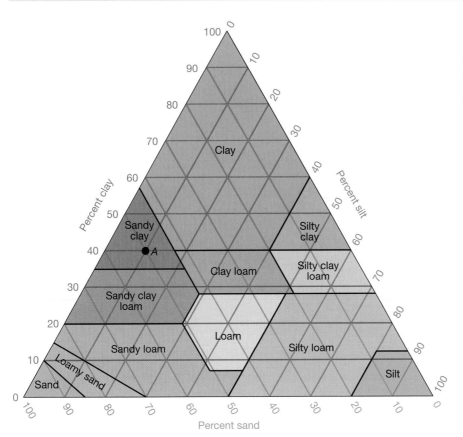

Figure 3.12 The texture of any soil can be represented by a point on this soil texture diagram. Soil texture is one of the most significant factors used to estimate agricultural potential and engineering characteristics. (After U.S. Department of Agriculture)

a *sandy clay.* The soils called *loam,* which occupy the central portion of the diagram, are those in which no single particle size predominates over the other two. Loam soils are best suited to support plant life because they generally have better moisture characteristics and nutrient storage ability than do soils composed predominantly of clay or coarse sand.

Soil particles are seldom completely independent of one another. Rather, they usually form clumps called *peds* that give soils a particular structure. Four basic soil structures are recognized: platy, prismatic, blocky, and spheroidal. Soil structure is important because it influences the ease of a soil's cultivation as well as the susceptibility of a soil to erosion. In addition, soil structure affects the porosity and permeability of soil (that is, the ease with which water can penetrate). This, in turn, influences the movement of nutrients to plant roots. Prismatic and blocky peds usually allow for moderate water infiltration, whereas platy and spheroidal structures are characterized by slower infiltration rates.

Controls of Soil Formation

Soil is the product of the complex interplay of several factors. The most important of these are parent material, time, climate, plants, animals, and slope. Although all of these factors are interdependent, their roles will be examined separately.

Parent Material

The source of the weathered mineral matter from which soils develop is called the **parent material** and is a major factor influencing a newly forming soil. Gradually it undergoes physical and chemical changes as the processes of soil formation progress. Parent material may be the underlying bedrock or it can be a layer of unconsolidated deposits, as in a stream valley. When the parent material is bedrock, the soils are termed *residual soils.* By contrast, those developed on unconsolidated sediment are called *transported soils* (Figure 3.13). Note that transported soils form *in place* on parent materials that have been carried from elsewhere and deposited by gravity, water, wind, or ice.

The nature of the parent material influences soils in two ways. First, the type of parent material affects the rate of weathering, and thus the rate of soil formation. (Consider the weathering rates of granite versus limestone.) Also, because unconsolidated deposits are already partly weathered and provide more surface area for chemical weathering, soil development on such material usually progresses more rapidly. Second, the chemical makeup of the

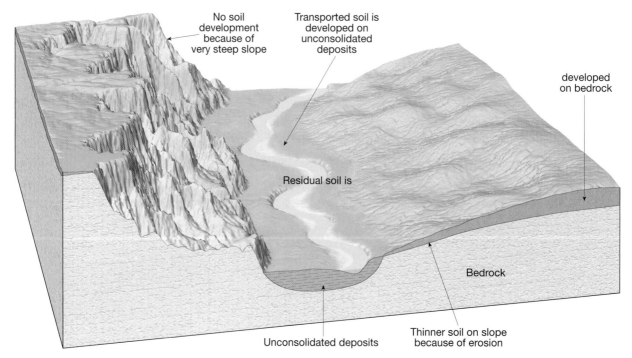

Figure 3.13 The parent material for residual soils is the underlying bedrock, whereas transported soils form on unconsolidated deposits. Also note that as slopes become steeper, soil becomes thinner.

*Labels in figure:*
No soil development because of very steep slope
Transported soil is developed on unconsolidated deposits
developed on bedrock
Residual soil is
Bedrock
Unconsolidated deposits
Thinner soil on slope because of erosion

parent material affects the soil's fertility. This influences the character of the natural vegetation the soil can support.

At one time the parent material was believed to be the primary factor causing differences among soils. Today soil scientists realize that other factors, especially climate, are more important. In fact, it has been found that similar soils are often produced from different parent materials and that dissimilar soils have developed from the same parent material. Such discoveries reinforce the importance of the other soil-forming factors.

Time

Time is an important component of every geological process, and soil formation is no exception. The nature of soil is strongly influenced by the length of time processes have been operating. If weathering has been going on for a comparatively short time, the parent material determines to a large extent the characteristics of the soil. As weathering processes continue, the influence of parent material on soil is overshadowed by the other soil-forming factors, especially climate. The amount of time required for various soils to evolve cannot be specified because the soil-forming processes act at varying rates under different circumstances. However, as a rule the longer a soil has been forming, the thicker it becomes and the less it resembles the parent material.

Climate

Climate is the most influential control of soil formation. Just as temperature and precipitation are the climatic elements that influence people the most, so too are they the elements that exert the strongest impact on soil formation. Variations in temperature and precipitation determine whether chemical or mechanical weathering predominates. They also greatly influence the rate and depth of weathering. For instance, a hot, wet climate may produce a thick layer of chemically weathered soil in the same amount of time that a cold, dry climate produces a thin mantle of mechanically weathered debris. Also, the amount of precipitation influences the degree to which various materials are removed (leached) from the soil, thereby affecting soil fertility. Finally, climatic conditions are important factors controlling the type of plant and animal life present.

Plants and Animals

The biosphere plays a vital role in soil formation. The types and abundance of organisms present have a strong influence on the physical and chemical properties of a soil. In fact, for well-developed soils in many regions, the significance of natural vegetation is frequently implied in the description used by soil scientists. Such phrases as *prairie soil, forest soil,* and *tundra soil* are common.

The chief function of plants and animals is to furnish organic matter to the soil. Certain bog soils are composed almost entirely of organic matter, whereas desert soils may contain only a tiny percentage. Although the quantity of organic matter varies substantially among soils, it is the rare soil that completely lacks it.

The primary source of organic matter is plants, although animals and the uncountable microorganisms also contribute. When organic matter decomposes, important nutrients are supplied to plants, as well as to animals and microorganisms living in the soil. Consequently, soil fertility depends in part on the amount of organic matter present. Furthermore, the decay of plant and animal remains causes the formation of various organic acids. These complex acids hasten the weathering process. Organic matter also has a high water-holding ability and thus aids water retention in a soil.

Microorganisms, including fungi, bacteria, and single-celled protozoa, play an active role in the decay of plant and animal remains. The end product is *humus,* a material that no longer resembles the plants and animals from which it is formed. In addition, certain microorganisms aid soil fertility because they have the ability to *fix* (alter) atmospheric nitrogen gas into soil nitrogen compounds.

Earthworms and other burrowing animals act to mix the mineral and organic portions of a soil. Earthworms, for example, feed on organic matter and thoroughly mix soils in which they live, often moving and enriching many tons per acre each year. Burrows and holes also aid the passage of water and air through the soil.

Slope

The slope of the land can vary greatly over short distances. Such variations, in turn, can lead to widely varied local soil types. Many of the differences exist because slope has a significant impact on the amount of erosion and the water content of soil.

On steep slopes, soils are often poorly developed. In such situations little water can soak in, and as a result, soil moisture may be insufficient for vigorous plant growth. Further, because of accelerated erosion on steep slopes, the soils are thin, or nonexistent (Figure 3.13).

In contrast, poorly drained and waterlogged soils in bottomlands have a much different character. Such soils are usually thick and dark. The dark color results from generous organic matter that accumulates because saturated conditions retard the decay of vegetation. The optimum terrain for soil development is a flat-to-undulating upland surface. Here we find good drainage, minimum erosion, and sufficient infiltration of water into the soil.

Slope orientation, the direction the slope is facing, also is significant. In the mid-latitudes of the Northern Hemisphere, a south-facing slope receives a great deal more sunlight than does a north-facing slope. In fact, a steep north-facing slope may receive no direct sunlight at all. The difference in the amount of solar radiation received causes substantial differences in soil temperature and moisture, which in turn may influence the nature of the vegetation and the character of the soil.

Although we have dealt separately with each of the soil-forming factors, remember that *all work together* to form soil. No single factor is responsible for a soil being as it is. Rather, it is the combined influence of parent material, time, climate, plants, animals, and slope that determines a soil's character.

The Soil Profile

It is important to realize that soil-forming processes *operate from the surface downward.* Thus, variations in composition, texture, structure, and color gradually evolve at varying depths. These vertical differences, which usually become more pronounced as time passes, divide the soil into zones or layers known as **horizons.** If you were to dig a trench in soil, you would see that its walls are layered. Such a vertical section through all of the soil horizons constitutes the **soil profile** (Figure 3.14).

Figure 3.14 A soil profile is a vertical cross-section from the surface down to the parent material. Well-developed soils show distinct layers called horizons. (Photo by E. J. Tarbuck)

Figure 3.15 presents an idealized view of a well-developed soil profile in which five horizons are identified. From the surface downward, they are designated as *O, A, E, B,* and *C,* respectively. These five horizons are common to soils in temperate regions. The characteristics and extent of development of horizons vary in different environments. Thus, different localities exhibit soil profiles that can contrast greatly with one another.

The *O* horizon consists largely of organic material, unlike the layers beneath it that consist mainly of mineral matter. The upper portion of the *O* horizon is primarily plant litter such as loose leaves and other organic debris that are still recognizable. By contrast, the lower portion of the *O* horizon is made up of partly decomposed organic matter (humus) in which plant structures can no longer be identified. In addition to plants, the *O* horizon is teeming with microscopic life including bacteria, fungi, algae, and insects. All of these organisms contribute oxygen, carbon dioxide, and organic acids to the developing soil.

Underlying the organic-rich *O* horizon is the *A* horizon. This zone is largely mineral matter, yet biological activity is high and humus is generally present—up to 30 percent in some instances. Together the *O* and *A* horizons make up what is commonly called *topsoil*. Below the *A* horizon, the *E* horizon is a light-colored layer that contains little organic material. As water percolates downward through this zone, finer particles are carried away. This washing out of the fine soil components is termed **eluviation.** Water percolating downward also dissolves soluble inorganic soil components and carries them to deeper zones. This depletion of soluble materials from the upper soil is termed **leaching.**

Immediately below the *E* horizon is the *B* horizon, or *subsoil*. Much of the material removed from the *E* horizon by eluviation is deposited in the *B* horizon, which is often referred to as the *zone of accumulation*. The accumulation of the fine clay particles enhances water retention in the subsoil. However, in extreme cases, clay accumulation can form a very compact and impermeable layer called *hardpan*. The *O, A, E,* and *B* horizons together constitute the **solum,** or "true soil." It is in the solum that the soil-forming processes are active and that living roots and other plant and animal life are largely confined.

Below the solum and above the unaltered parent material is the *C* horizon, a layer characterized by partially altered parent material. Whereas the *O, A, E,* and

Solum or "true soil"

Topsoil

O horizon
Loose and partly decayed organic matter

A horizon
Mineral matter mixed with some humus

E horizon
Light colored mineral particles. Zone of eluviation and leaching

Subsoil

B horizon
Accumulation of clay transported from above

C horizon
Partially altered parent material

Unweathered parent material

Figure 3.15 Idealized soil profile from a humid climate in the middle latitudes.

B horizons bear little resemblance to the parent material, it is easily identifiable in the *C* horizon. Although this material is undergoing changes that will eventually transform it into soil, it has not yet crossed the threshold that separates regolith from soil.

Soil characteristics and development can vary greatly in different environments. The boundaries between soil horizons can be very distinct, as in Figures 3.14 and 3.15, or the horizons may blend gradually from one to another. A well-developed soil profile indicates that environmental conditions have been relatively stable over an extended time span.

Soil Types

Hundreds of soil types and subtypes exist worldwide, far too many to consider here. Instead, we shall look at just three very generic types—pedalfers, pedocals, and laterites. As you read, notice that the characteristics of each soil primarily reflect the prevailing climatic conditions. A summary of the characteristics is provided in Table 3.2. (For more information on different soil groups and their distribution, see Appendix F.)

In cold or dry climates soils are generally thin and poorly developed. The reasons for this are fairly obvious. Chemical weathering progresses very slowly in such climates, and the scant plant life yields very little organic matter.

Pedalfer

The term **pedalfer** is derived from the Greek **ped***on*, meaning "soil," and the chemical symbols **Al** (aluminum) and **Fe** (Iron). Pedalfers are characterized by an accumulation of iron oxides and aluminum-rich clays in the *B* horizon. In mid-latitude areas where the annual rainfall exceeds 63 centimeters (25 inches), most of the soluble materials, such as calcium carbonate, are leached from the soil and carried away by underground water. The less soluble iron oxides and clays are carried from the *E* horizon and deposited in the *B* horizon, giving it a brown to red-brown color. These soils are best developed under forest vegetation where large quantities of decomposing organic matter provide the acid conditions necessary for leaching. In the United States pedalfers are found east of a line from northwestern Minnesota to south-central Texas.

Pedocal

The term **pedocal** is derived from the Greek **ped***on*, meaning "soil," and the first three letters of **cal***cite* (calcium carbonate). As the name implies, pedocals are characterized by an accumulation of calcium carbonate. This soil type is found in the drier western United States in association with grassland and brush vegetation. Because chemical weathering is less intense in drier areas, pedocals generally contain a smaller percentage of clay materials than pedalfers.

Table 3.2 Summary of soil types.

Climate	Temperate humid (> 63 cm rainfall)	Temperate dry (<63 cm rainfall)	Tropical (heavy rainfall)	Extreme arctic or desert
Vegetation	Forest	Grass and brush	Grass and trees	Almost none, so no humus develops
Typical Area	Eastern U.S.	Western U.S.		
Soil Type	Pedalfer	Pedocal	Laterite	
Topsoil	Sandy, light colored; acid	Commonly enriched in calcite; whitish color	Enriched in iron, (and aluminum); brick red color	No real soil forms because there is no organic material. Chemical weathering is very slow.
Subsoil	Enriched in aluminum, iron and clay; brown color	Enriched in calcite; whitish color	All other elements removed by leaching	
Remarks	Extreme development in conifer forests, because abundant humus makes groundwater very acidic. Produces light gray soil because of removal of iron	*Caliche* is name applied to the accumulation of calcite	Apparently bacteria destroy humus, so no acid is available to remove iron	

(Note: "Zones not developed" appears as a vertical divider between the Tropical column and the adjacent columns.)

In the arid and semiarid western states, a calcite-enriched layer called *caliche* may be present in the soils. In these areas little of the rain that falls penetrates to great depths. Rather, it is held by the soil particles near the surface until it evaporates. As a result, the soluble materials, chiefly calcium carbonate, are removed from the uppermost layer and redeposited below, forming the caliche layer.

Laterite

In hot, wet tropical climates, soils called **laterites** may develop. Chemical weathering is intense under such climate conditions, so these soils are usually deeper than soils developing over a similar period in the midlatitudes. Not only does leaching remove the soluble materials such as calcite, but the great quantities of percolating water also remove much of the silica, with the result that oxides of iron and aluminum become concentrated in the soil. The iron gives the soil a distinctive orange to red color.

Because bacterial activity is great in the tropics, laterites contain practically no humus. This fact, coupled with the highly leached and bricklike nature of these soils, makes laterites poor for growing crops. Their infertility has been demonstrated repeatedly in tropical countries where cultivation has been attempted (see Box 3.2).

Soil Erosion

Soils are just a tiny fraction of all Earth materials, yet they are a vital resource. Because soils are necessary for the growth of rooted plants, they are the very foundation of the human life-support system. Just as human ingenuity can increase the agricultural productivity of soils through fertilization and irrigation, soils can be damaged or destroyed by careless activities. Despite their basic role in providing food, fiber, and other basic materials, soils are among our most abused resources.

Perhaps this neglect and indifference has occurred because a substantial amount of soil seems to remain even where soil erosion is serious. Nevertheless, although the loss of fertile topsoil may not be obvious to the untrained eye, it is a growing problem as human activities expand and disturb more and more of Earth's surface.

How Soil Is Eroded

Soil erosion is a natural process; it is part of the constant recycling of Earth materials that we call the *rock cycle*. Once soil forms, erosional forces, especially water and wind, move soil components from one place to another. Every time it rains, raindrops strike the land with surprising force (Figure 3.16). Each drop acts like a tiny bomb, blasting movable soil particles out of their

Figure 3.16 When it is raining, millions of water drops are falling at velocities approaching 10 meters per second (35 kilometers per hour). When water drops strike an exposed surface, soil particles may splash as high as 1 meter into the air and land more than a meter away from the point of raindrop impact. Soil dislodged by splash erosion is more easily moved by sheet erosion. (Photo courtesy of U.S. Department of Agriculture)

positions in the soil mass. Then, water flowing across the surface carries away the dislodged soil particles. Because the soil is moved by thin sheets of water, this process is termed *sheet erosion*.

After flowing as a thin, unconfined sheet for a relatively short distance, threads of current typically develop and tiny channels called *rills* begin to form. Still deeper cuts in the soil, known as *gullies*, are created as rills enlarge (Figure 3.17). When normal farm cultivation cannot eliminate the channels, we know the rills have grown large enough to be called gullies. Although most dislodged soil particles move only a short distance during each rainfall, substantial quantities eventually leave the fields and make their way downslope to a stream. Once in the stream channel, these soil particles, which can now be called *sediment*, are transported downstream and eventually deposited.

Rates of Erosion

We know that soil erosion is the ultimate fate of practically all soils. In the past, erosion occurred at slower rates than it does today because more of the land surface was covered and protected by trees, shrubs, grasses, and other plants. However, human activities such as farming, logging, and construction, which remove or disrupt the natural vegetation, have greatly accelerated the rate of soil erosion. Without the stabilizing effect of plants, the soil is more easily swept away by the wind or carried downslope by sheet wash.

Box 3.2   People and the Environment: Laterites and the Clearing of the Rain Forest

Laterites are thick red soils that form in the wet tropics and subtropics and are the end product of extreme chemical weathering. Because lush tropical rain forests have lateritic soils, it might seem as though they would have great potential for agriculture. However, this is not the case. In fact, just the opposite is true; laterites are among the poorest soils for farming. Why is this the case?

Because laterites develop under conditions of high temperatures and heavy rainfall, they are severely leached. Leaching leads to low fertility because plant nutrients are removed by the large volume of downward-percolating water. Therefore, even though the vegetation may be dense and luxuriant, the soil itself contains few available nutrients. Most of the nutrients that support the rain forest are locked up in the trees themselves. The forest is maintained by the decay of plant material. That is, as vegetation dies and decomposes, the roots of the rain forest trees quickly absorb the nutrients before they are leached from the soil. The nutrients are continuously recycled as trees die and decompose.

Therefore, when forests are cleared to provide land for farming or to harvest the timber, most of the nutrients are removed as well (Figure 3.B). What remains is a soil that contains little to nourish crop growth.

The clearing of rain forests not only removes the supply of plant nutrients, but also leads to accelerated erosion. When vegetation is present, its roots anchor the soil while its leaves and branches provide a canopy that protects the ground from the full force of the frequent heavy rains. Removal of the vegetation also exposes the ground to strong direct sunlight. When baked by the Sun, laterites can harden to a bricklike consistency and become practically impenetrable to water and crop roots. In only a few years, lateritic soils in a freshly cleared area may no longer be cultivable.

The term *laterite* is derived from the Latin word *latere*, meaning "brick," and was first applied to the use of this material for brick making in India and

Figure 3.B Clearing the Amazon rain forest in Surinam. The thick lateritic soil is highly leached. (Photo by Wesley Bocxe/Photo Researchers)

Cambodia. Laborers simply excavated the soil, shaped it, and allowed it to harden in the Sun. Ancient but still well-preserved structures built of laterite remain standing today in the wet tropics (Figure 3.C). Such structures have withstood centuries of weathering because all of the original soluble materials were already removed from the soil by chemical weathering. Laterites are therefore virtually insoluble and thus very stable.

In summary, we have seen that laterites are highly leached soils that are the products of extreme chemical weathering in the warm, wet tropics. Although they may be associated with lush tropical rain forests, these soils are unproductive when vegetation is removed. Moreover, when cleared of plants, laterites are subject to accelerated erosion and can be baked to bricklike hardness by the Sun.

Figure 3.C This ancient temple at Angor Wat, Cambodia, was built of bricks made of laterite. (Photo by R. Ian Lloyd/The Stock Market)

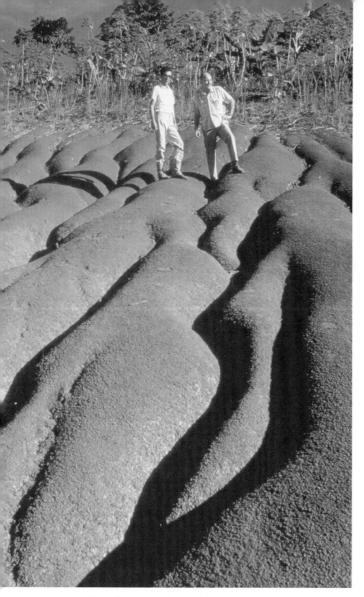

Figure 3.17 Gully erosion in poorly protected soil, southern Colombia (Photo by Carl Purcell/Photo Researchers, Inc.)

Natural rates of soil erosion vary greatly from one place to another and depend on soil characteristics as well as such factors as climate, slope, and type of vegetation. Over a broad area, erosion caused by surface runoff may be estimated by determining the sediment loads of the streams that drain the region. When studies of this kind were made on a global scale they indicated that, prior to the appearance of humans, sediment transport by rivers to the oceans amounted to just over 9 billion metric tons per year (1 metric ton = 1000 kilograms). By contrast, the amount of material currently transported to the sea by rivers is about 24 billion metric tons per year, or more than two and one-half times the earlier rate.

It is more difficult to measure the loss of soil due to wind erosion. However, the removal of soil by wind is generally much less significant than erosion by flowing water except during periods of prolonged drought. When dry conditions prevail, strong winds can remove large quantities of soil from unprotected fields. Such was the case in the 1930s in the portions of the Great Plains that came to be called the Dust Bowl (see Box 5.4, p. 146).

In many regions the rate of soil erosion is significantly greater than the rate of soil formation. This means that a renewable resource has become nonrenewable in these places. At present, it is estimated that topsoil is eroding faster than it forms on more than one-third of the world's croplands. The result is lower productivity, poorer crop quality, reduced agricultural income, and an ominous future.

Sedimentation and Chemical Pollution

Another problem related to excessive soil erosion involves the deposition of sediment. Each year in the United States hundreds of millions of tons of eroded soil are deposited in lakes, reservoirs, and streams. The detrimental impact of this process can be significant. For example, as more and more sediment is deposited in a reservoir, the capacity of the reservoir is reduced, limiting its usefulness for flood control, water supply, and/or hydroelectric power generation. In addition, sedimentation in streams and other waterways can restrict navigation and lead to costly dredging operations.

In some cases soil particles are contaminated with pesticides used in farming. When these chemicals are introduced into a lake or reservoir, the quality of the water supply is threatened and aquatic organisms can be endangered. In addition to pesticides, nutrients found naturally in soils as well as those added by agricultural fertilizers make their way into streams and lakes, where they stimulate the growth of plants. Over a period of time, excessive nutrients accelerate the process by which plant growth leads to the depletion of oxygen and an early death of the lake.

The availability of good soils is critical if the world's rapidly growing population is to be fed. On every continent unnecessary soil loss is occurring because appropriate conservation measures are not being used. Although it is a recognized fact that soil erosion can never be completely eliminated, soil conservation programs can substantially reduce the loss of this basic resource. Windbreaks (rows of trees), and terracing, and plowing along the contours of hills, are some of the effective measures, as are special tillage practices and crop rotation.

Weathering Creates Ore Deposits

Weathering creates many important mineral deposits. It does so by concentrating minor amounts of metals that are scattered through unweathered rock into economically valuable concentrations. Such a transformation is often termed **secondary enrichment** and takes place in one of two ways. In one situation, chemical weathering coupled with downward-percolating water removes undesired materials from decomposing rock, leaving the desired elements enriched in the upper zones of the soil. The second way is basically the reverse of the first. That is, the desirable elements that are found in low concentrations near the surface are removed and carried to lower zones, where they are redeposited and become more concentrated.

Bauxite

The formation of *bauxite,* the principal ore of aluminum, is one important example of an ore created as a result of enrichment by weathering processes (Figure 3.18). Although aluminum is the third most abundant element in Earth's crust, economically valuable concentrations of this important metal are not common, because most aluminum is tied up in silicate minerals, from which it is extremely difficult to extract.

Bauxite forms in rainy tropical climates in association with laterites. (In fact, bauxite is sometimes called aluminum laterite.) When aluminum-rich source rocks are subjected to the intense and prolonged chemical weathering of the tropics, most of the common elements, including calcium, sodium, and silicon, are removed by leaching. Because aluminum is extremely insoluble, it becomes concentrated in the soil (as bauxite, a hydrated aluminum oxide). Thus, the formation of bauxite depends on climatic conditions in which chemical weathering and leaching are pronounced, plus, of course, the presence of aluminum-rich source rock. Important deposits of nickel and cobalt are also found in laterite soils that develop from igneous rocks rich in other silicate minerals.

Figure 3.18 Bauxite is the ore of aluminum and forms as a result of weathering processes under tropical conditions. Its color varies from red or brown to nearly white. (Photo by E. J. Tarbuck)

Other Deposits

Many copper and silver deposits result when weathering processes concentrate metals that are dispersed through a low-grade primary ore. Usually such enrichment occurs in deposits containing pyrite (FeS_2), the most common and widespread sulfide mineral. Pyrite is important because when it chemically weathers, sulfuric acid forms, which enables percolating waters to dissolve the ore metals. Once dissolved, the metals gradually migrate downward through the primary ore body until they are precipitated. Deposition takes place because of changes that occur in the chemistry of the solution when it reaches the groundwater zone (the zone beneath the surface where all pore spaces are filled with water). In this manner, the small percentage of dispersed metal can be removed from a large volume of rock and redeposited as a higher-grade ore in a smaller volume of rock.

Mass Wasting: The Work of Gravity

Earth's surface is never perfectly flat but instead consists of slopes. Some are steep and precipitous; others are moderate or gentle. Some are long and gradual; others are short and abrupt. Some slopes are mantled with soil and covered by vegetation; others consist of barren rock and rubble. Their form and variety are great. Taken together, slopes are the most common elements in our physical landscape. Although most slopes appear to be stable and unchanging, they are not static features because the force of gravity causes material to move downslope. At one extreme, the movement may be gradual and practically imperceptible. At the other extreme, it may consist of a thundering landslide or rockfall.

Rapid forms of mass wasting are significant geologic hazards (Figure 3.19). Each year landslides in the United States are responsible for 25 to 50 deaths and produce more than $1.5 billion in losses. Worldwide, landslides cause an average of 600 deaths annually. Sometimes extreme events occur. Such was the case in November 1998, when mudflows were triggered by heavy rains from Hurricane Mitch. Thousands perished in Nicaragua and Honduras. Thousands more lost their homes. The economies of these two small countries were devastated. (See Figure I.12A, p.9)

Most mass wasting, whether spectacular or subtle, is the result of circumstances that are completely independent of human activities. Very few landslides occur where they cannot be anticipated. In places where mass wasting is a recognized threat, steps can often be taken to control downslope movements or limit the damages that such movements can cause. If the potential for mass wasting goes unrecognized or is ignored, the results can be costly and dangerous. We

Figure 3.19 This home in Pacific Palisades, California, was destroyed by a landslide triggered by the January 1994 Northridge earthquake. In some cases, damages from earthquake-induced mass wasting are greater than damages caused directly by an earthquake's ground vibrations. (Photo by ChromoSohm/Sohm/The Stock Market)

should also note that, although most downslope movements occur whether people are present or not, many occurrences are aggravated or even triggered by human actions.

Mass Wasting and Landform Development

Landslides are spectacular examples of a basic geologic process called *mass wasting*. Mass wasting refers to the downslope movement of rock, regolith, and soil under the direct influence of gravity. It is distinct from the erosional processes that are examined in subsequent chapters because mass wasting does not require a transporting medium.

In the evolution of most landforms, mass wasting is the step that follows weathering. By itself, weathering does not produce significant landforms. Rather, landforms develop as the products of weathering are removed from the places where they originate. Once weathering weakens and breaks rock apart, mass wasting transfers the debris downslope, where a stream, acting as a conveyor belt, usually carries it away (see Figure 3.1). Although there may be many intermediate stops along the way, the sediment is eventually transported to its ultimate destination, the sea.

The combined effects of mass wasting and running water produce stream valleys, which are the most common and conspicuous of Earth's landforms. If streams alone were responsible for creating the valleys in which they flow, the valleys would be very narrow features. However, the fact that most river valleys are much wider than they are deep is a strong indication of the significance of mass wasting processes in supplying material to streams. The walls of a canyon extend far from the river because of the transfer of weathered debris downslope to the river and its tributaries by

mass-wasting processes. In this manner, streams and mass wasting combine to modify and sculpture the surface. Of course, glaciers, groundwater, waves, and wind are also important agents in shaping landforms and developing landscapes.

Controls and Triggers of Mass Wasting

Gravity is the controlling force of mass wasting, but several factors play an important role in overcoming inertia and triggering downslope movements. Among these factors are saturation of material with water, oversteepening of slopes, removal of anchoring vegetation, and ground vibrations from earthquakes.

The Role of Water

When the pores in sediment become filled with water, the cohesion among particles is destroyed, allowing them to slide past one another with relative ease. For example, when sand is slightly moist, it sticks together quite well. However, if enough water is added to fill the openings between the grains, the sand will ooze out in all directions. Thus, saturation reduces the internal resistance of materials, which are then easily set in motion by the force of gravity. When clay is wetted, it becomes very slick—another example of the "lubricating" effect of water. Water also adds considerable weight to a mass of material. The added weight in itself may be enough to cause the material to slide or flow downslope.

Oversteepened Slopes

Oversteepening of slopes is another cause of many mass movements. Unconsolidated, granular sand-sized or coarser particles assume a stable slope called the **angle of repose,** the steepest angle at which material remains stable. Depending on the size and shape of the particles, the angle varies from 25 to 40 degrees. The larger, more angular particles maintain the steepest slopes. If the angle is increased, the rock debris will adjust by moving downslope.

Oversteepening is not just important because it triggers movements of unconsolidated granular materials. Oversteepening also produces unstable slopes and mass movements in cohesive soils, regolith, and bedrock. The response will not be immediate, as with loose, granular material, but sooner or later, one or more mass-wasting processes will eliminate the oversteepening and restore stability to the slope.

There are many situations in nature where this takes place. A stream undercutting a valley wall and waves pounding against the base of a cliff are but two familiar examples. Furthermore, through their activities, people often create oversteepened and unstable slopes that become prime sites for mass wasting.

Vegetation

Plants protect against erosion and contribute to the stability of slopes because their root systems bind soil and regolith together. Where plants are lacking, mass wasting is enhanced, especially if slopes are steep and water is plentiful. When anchoring vegetation is removed by forest fires or by people (for timber, farming, or development), surface materials frequently move downslope.

An unusual but well-known example occurred several decades ago on steep slopes near Menton, France. Farmers replaced olive trees, which have deep roots, with a more profitable, but shallow-rooted crop, carnations. When the less stable slope failed, the landslide took 11 lives.

Earthquakes as Triggers

Conditions favoring mass wasting can exist in an area for a long time without movement occurring. An additional factor is sometimes necessary to trigger the movement. Among the more important and dramatic triggers are earthquakes. An earthquake and its aftershocks can dislodge enormous volumes of rock and unconsolidated material. The mass-wasting events shown in Figures 3.19 and 3.22 were both triggered by earthquakes. In many areas that are jolted by earthquakes, it is not ground vibrations directly but landslides and ground subsidence triggered by the vibrations that cause the greatest damage.

Falls, Slides, and Flows

Geologists include several processes under the name of mass wasting. Four are illustrated in Figure 3.20. Generally, each process is defined by the type of material involved, the kind of motion, and the velocity of the movement.

If soil and regolith dominate, terms such as *debris*, *mud*, or *earth* are used. In contrast, when a mass of bedrock breaks loose and moves downslope, the term *rock* may be part of the description. Generally, the kind of motion is described as either a fall, a slide, or a flow.

When the movement involves the free-fall of detached individual pieces of any size, it is termed a **fall.** Falls are common on slopes that are too steep for loose material to remain on the surface (Figure 3.21). Many falls result when freeze and thaw cycles or the action of plant roots loosen rock to the point that gravity takes over. Rockfall is the primary way in which talus slopes are built and maintained (see Figure 3.3). Sometimes falls may trigger other forms of downslope movement.

Many mass-wasting processes are **slides,** which occur whenever material remains fairly coherent and moves along a well-defined surface that is approximately parallel to the slope. Sometimes the surface is a joint, a fault, or a bedding plane. Sometimes the word *slide* is used as a synonym for the word *landslide*. It should be pointed out that, although many people,

A. Slump

B. Rockside

C. Debris flow

D. Earthflow

Figure 3.20 The four processes illustrated here are all considered to be relatively rapid forms of mass wasting. Because material in slumps **A.** and rockslides **B.** move along well-defined surfaces, they are said to move by sliding. By contrast, when material moves downslope as a viscous fluid, the movement is described as a flow. Debris flow **C.** and earthflow **D.** advance downslope in this manner.

Figure 3.21 Rockfall blocking Montana Highway 2 east of Pipestone Pass, May 1998. (AP/Wide World Photos)

becomes trapped and compressed beneath the falling mass of debris, allowing it to move as a buoyant, flexible sheet across the surface.

Most mass movements, however, do not move with the speed of a rock avalanche. In fact, a great deal of mass wasting is imperceptibly slow. One process that we will examine later, termed *creep*, results in particle movements that are usually measured in millimeters or centimeters per year. Thus, as you can see, rates of movement can be spectacularly sudden or exceptionally gradual. Even the velocity of a single process at a particular site can vary considerably from one time to another.

including geologists, use the term, the word *landslide* has no specific definition in geology. Rather, it should be considered as a popular nontechnical term to describe all relatively rapid forms of mass wasting, including those in which sliding does not occur.

A third type of movement common to mass wasting is termed **flow**. Flow occurs when material moves downslope as a viscous fluid. Most flows are saturated with water and typically move as lobes or tongues.

When mass-wasting events make the news, a large quantity of material has in all likelihood moved rapidly downslope and has had a disastrous effect upon people and property. Indeed, during events called *rock avalanches,* rock and debris can hurtle downslope at speeds exceeding 200 kilometers (125 miles) per hour. Many researchers believe that rock avalanches, such as the one that produced the scene in Figure 3.22, must literally "float on air" as they move downslope. That is, high velocities result when air

Slump

Slump refers to the downward sliding of a mass of rock or unconsolidated material moving as a unit along a *curved* surface (see Figure 3.20A). Usually the slumped material does not travel spectacularly fast nor very far. This is a common form of mass wasting, especially in thick accumulations of cohesive materials such as clay. As the movement occurs, a crescent-shaped scarp (cliff) is created at the head and the block's upper surface is sometimes tilted backward.

Slump commonly occurs because a slope has been oversteepened. The material on the upper portion of a slope is held in place by the material at the bottom of the slope. As this anchoring material at the base is removed, the material above is made unstable and reacts to the pull of gravity. A common example is a valley wall that becomes oversteepened by a meandering river. Another is a coastal area where a cliffed seashore is undercut by wave activity at its base.

Figure 3.22 This 4-kilometer long tongue of rubble was deposited atop Alaska's Sherman Glacier by a rock avalanche. The event was triggered by a tremendous earthquake in March 1964. (Photo by Austin Post, U.S. Geological Survey)

Rock Avalanche Debris

Rockslide

Rockslides frequently occur in high mountain areas such as the Andes, Alps, and Canadian Rockies. They are sudden and rapid movements that happen when detached segments of bedrock break loose and slide downslope (Figure 3.20B). As the moving mass thunders along the surface, it brakes into many smaller pieces. Such events are among the fastest and most destructive mass movements.

Rockslides usually take place where there is an inclined surface of weakness. Such surfaces tend to form where strata are titled or where joints and fractures exist parallel to the slope. When rock in such a setting is undercut at the base of the slope, it loses support and eventually gives way. The Gros Ventre slide described in Box 3.3 is a classic example. Sometimes the rockslide is triggered when rain or melting snow lubricates the underlying surface to the point that friction is no longer sufficient to hold the rock unit in place. As a result, rockslides tend to be more common during the spring, when heavy rains and melting snow are most prevalent.

Earthquakes may trigger rockslides and other mass movements. A tragic example occurred on August 17, 1959, when a severe earthquake west of Yellowstone National Park triggered a massive slide in the canyon of the Madison River in southwestern Montana. In a matter of moments an estimated 27 million cubic meters of rock, soil, and trees slid into the canyon. The debris dammed the river and buried a campground and highway. More than 20 unsuspecting campers perished.

Debris Flow

Debris flow is a relatively rapid type of mass wasting that involves a flow of soil and regolith containing a large amount of water (Figure 3.20C). Debris flows, which are also called *mudflows*, are most characteristic of semiarid mountainous regions and are also common on the slopes of some volcanoes. Because of their fluid properties, debris flows follow canyons and stream channels. In populated areas, debris flows can pose a significant hazard to life and property (Figure 3.23).

Debris Flows in Semiarid Regions

When a cloudburst or rapidly melting mountain snows create a sudden flood in a semiarid region, large quantities of soil and regolith are washed into nearby stream channels because there is usually little vegetation to anchor the surface material. The end product is a flowing tongue of well-mixed mud, soil, rock, and water. Its consistency may range from that of wet concrete to a soupy mixture not much thicker than muddy water. The rate of flow therefore depends not only on the slope but also on the water content. When dense, debris flows are capable of carrying or pushing large boulders, trees, and even houses with relative ease.

Debris flows pose a serious hazard to development in dry mountainous areas such as southern California. The construction of homes on canyon hillsides and the removal of anchoring vegetation by brush fires and other means have increased the frequency of these destructive events (Figure 3.23).

Figure 3.23 On January 25, 1997, a debris flow literally buried this one-story home in Mill Valley, California. Heavy rains from a powerful Pacific storm triggered the event. (Justin Sullivan/AP/Wide World Photos)

Box 3.3 The Gros Ventre Slide: A Classic Case Study

The Gros Ventre River flows west from the northernmost part of the Wind River Range in northwestern Wyoming, through Grand Teton National Park, and eventually empties into the Snake River. On June 23, 1925, a classic rockslide took place in its valley, just east of the small town of Kelly. In the span of just a few minutes a great mass of sandstone, shale, and soil crashed down the south side of the valley, carrying with it a dense pine forest. The volume of debris, estimated at 38 million cubic meters (50 million cubic yards), created a 70-meter-high dam on the Gros Ventre River (Figure 3.D). Because the river was completely blocked, a lake was created. It filled so quickly that a house that had been 18 meters (60 feet) above the river was floated off its foundation 18 hours after the slide. In 1927, the lake overflowed the dam, partially draining the lake and resulting in a devastating flood downstream.

Why did the Gros Ventre rockslide take place? Figure 3.E is a diagrammatic cross-sectional view of the geology of the valley. Notice the following points: (1) the sedimentary strata in this area dip (tilt) 15 to 21 degrees; (2) underlying the bed of sandstone is a relatively thin layer of clay; and (3) at the bottom of the valley the river had cut through much of the sandstone layer. During the spring of 1925, water from heavy rains and melting snow seeped through the sandstone, saturating the clay below. Because much of the sandstone layer had been cut through by the Gros Ventre River, the layer had virtually no support at the bottom of the slope. Eventually the sandstone could no longer hold its position on the

Figure 3.D The scar, 2.4 kilometers (1.5 miles) long and 0.8 kilometers (0.5 miles) wide, left on the side of Sheep Mountain by the Gros Ventre rockslide. (Photo by Stephen Trimble)

wetted clay, and gravity pulled the mass down the side of the valley. The circumstances at this location were such that the event was inevitable.

Figure 3.E Cross-sectional view of the Gros Ventre rockslide. The slide occurred when the tilted and undercut sandstone bed could no longer maintain its position atop the saturated bed of clay. (After W. C. Alden, "Landslide and Flood at Gros Ventre, Wyoming," Transactions (AIME) 76 (1928):348.)

Lahars

Debris flows are also common on the slopes of some volcanoes, in which case they are termed **lahars.** The word originated in Indonesia, a volcanic region that has experienced many of these often destructive events. Lahars result when highly unstable layers of ash and debris become saturated with water and flow down steep volcanic slopes, generally following existing stream channels. Some are initiated when heavy rainfalls erode volcanic deposits. Others are triggered when large volumes of ice and snow are suddenly melted by heat flowing to the surface from within the volcano or by the hot gases and near-molten debris emitted during a violent eruption.

In November 1985, lahars were produced when Nevado del Ruiz, a 5300-meter (17,400-foot) volcano in the Andes Mountains of Colombia erupted. The eruption melted much of the snow and ice that capped the uppermost 600 meters of the peak, producing torrents of hot, thick mud, ash, and debris. The lahars moved outward from the volcano, following the valleys of three rain-swollen rivers that radiate from the peak. The flow that moved down the valley of the Lagunilla River was the most destructive, devastating the town of Armero, 48 kilometers from the mountain. Most of the more than 25,000 deaths caused by the event occurred in this once-thriving agricultural community. Death and property damage also occurred in 13 other villages within the 180-square-kilometer disaster area. Although a great deal of volcanic debris was explosively ejected from Nevado del Ruiz, it was the lahars triggered by this eruption that made this such a devastating natural disaster. In fact, it was the worst volcanic disaster since 28,000 people died following the 1902 eruption of Mount Pelée on the Caribbean island of Martinique. This event is described in Chapter 8.

Earthflow

We have seen that debris flows are frequently confined to channels in semiarid regions. In contrast, **earthflows** most often form on hillsides in humid areas during times of heavy precipitation or snowmelt (see Figure 3.20D). When water saturates the soil and regolith on a hillside, the material may break away, leaving a scar on the slope and forming a tongue- or teardrop-shaped mass that flows downslope (Figure 3.24). The materials most commonly involved are rich in clay and silt and contain only small proportions of sand and coarser particles. Earthflows range in size from bodies a few meters long, a few meters wide, and less than 1 meter deep to masses more than 1 kilometer long, several hundred meters wide, and more than 10 meters deep.

Because earthflows are quite viscous, they generally move at slower rates than the more fluid debris flows. They move slowly and persistently for periods ranging from days to years. Depending on slope steepness and the material's consistency, velocities range from less than 1 millimeter per day up to several meters per day. Movement is typically faster during wet periods. In addition to occurring as isolated hillside phenomena, earthflows commonly take place in association with large slumps. In this situation, they may be seen as tonguelike flows at the base of the slump.

A special type of earthflow, known as *liquefaction*, sometimes occurs in association with earthquakes. Porous, clay- to sand-sized sediments that are saturated with water are most vulnerable. When shaken suddenly, the grains lose cohesion and the ground flows. Liquefaction can cause buildings to sink or tip on

Figure 3.24 This small, tongue-shaped earthflow occurred on a newly formed slope along a recently constructed highway. It formed in clay-rich material following a period of heavy rain. Notice the small slump at the head of the earthflow. (Photo by E. J. Tarbuck)

their sides and underground storage tanks and sewer lines to float upward. To say the least, damage can be substantial. You will learn more about this in Chapter 6.

Slow Movements

Movements such as rockslides, rock avalanches, and lahars are certainly the most spectacular and catastrophic forms of mass wasting. These dangerous events deserve intensive study to enable more effective prediction, timely warnings, and better controls to save lives. However, because of their large size and spectacular nature, they give us a false impression of their importance as a mass-wasting process. Indeed, these sudden movements transport less material than the slow, subtle action of creep. Whereas rapid types of mass wasting are characteristic of mountains and steep hillsides, creep can take place on gentle slopes and is thus much more widespread.

Creep

Creep is a type of mass wasting that involves the gradual downhill movement of soil and regolith. A primary cause is the alternate expansion and contraction of surface material caused by freezing and thawing or wetting and drying. As shown in Figure 3.25, freezing or wetting lifts particles at right angles to the slope, and thawing or drying allows the particles to fall back to a slightly lower level. Each cycle therefore moves the material a short distance downhill. Creep can also be initiated if the ground becomes saturated with water. Following a heavy rain or snowmelt, a waterlogged

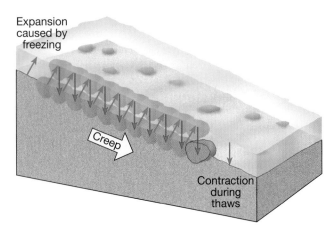

Figure 3.25 The repeated expansion and contraction of the surface material causes a net downslope migration of rock particles—a process called creep.

soil may lose its internal cohesion, allowing gravity to pull the material downslope. Because creep is imperceptibly slow, the process cannot be observed in action. What can be observed, however, are the effects of creep. Creep causes fences and utility poles to tilt and retaining walls to be displaced.

Solifluction

Solifluction is a common form of mass wasting in regions underlain by **permafrost.** Permafrost refers to the permanently frozen ground that occurs in Earth's harsh tundra and ice cap climates. Solifluction is a form of creep in which unconsolidated, water-saturated material gradually moves downslope. It occurs in a zone above the permafrost called the *active layer*, which thaws in summer and refreezes in winter. During summer, water is unable to percolate into the impervious permafrost layer below. As a result, the active layer becomes saturated and slowly flows. The process can occur on slopes as gentle as 2 to 3 degrees. Where there is a well-developed mat of vegetation, a solifluction sheet may move in a series of well-defined lobes or overriding folds (Figure 3.26).

Figure 3.26 Solifluction lobes northeast of Fairbanks, Alaska. (Photo by James E. Patterson)

Chapter Summary

- Earth's external processes include (1) *weathering*—the disintegration and decomposition of rock at or near the surface, (2) *mass wasting*—the transfer of rock material downslope under the influence of gravity, and (3) *erosion*—the incorporation and transportation of material by a mobile agent, usually water, wind, or ice.

- *Mechanical weathering* is the physical breaking up of rock into smaller pieces. *Chemical weathering* alters a rock's chemistry, changing it into different substances. Rocks can be broken into smaller fragments by *frost wedging, unloading,* and *biological activity.* Water is by far the most important agent of chemical weathering. Oxygen in water can *oxidize* some materials, while carbon dioxide (CO_2) dissolved in water forms *carbonic acid.* The chemical weathering of silicate minerals frequently produces (1) soluble products containing sodium, calcium, potassium, and magnesium, (2) insoluble iron oxides, and (3) clay minerals. The type and rate of rock weathering is influenced by a rock's mineral makeup—calcite readily dissolves in mildly acidic solutions and silicate minerals that form first from magma are least resistant to chemical weathering. Climatic factors, particularly temperature and moisture, are crucial to the rate of rock weathering.

- *Soil* is a combination of mineral and organic matter, water, and air—that portion of the *regolith* (the layer of rock and mineral fragments produced by weathering) that supports the growth of plants. *Soil texture* refers to the proportions of different particle sizes (clay, silt, and sand) found in soil. The most important *factors that control soil formation are parent material, time, climate, plants, animals, and slope.*

- Soil-forming processes operate from the surface downward and produce zones or layers in the soil called *horizons.* From the surface downward the horizons are designated as O, A, E, B, and C respectively. Although there are hundreds of soil types and subtypes worldwide, the three very generic types are (1) *pedalfer*—characterized by an accumulation of iron oxides and aluminum-rich clays in the B horizon, (2) *pedocal*—characterized by an accumulation of calcium carbonate, and (3) *laterite*—deep soils that develop in the hot, wet tropics that are poor for growing crops because they are highly leached and bricklike.

- Soil erosion by wind and water is the ultimate fate of practically all soils. *Rates of soil erosion vary* from one place to another and depend on the soil's characteristics as well as such factors as climate, slope, and type of vegetation.

- Weathering creates mineral deposits by concentrating metals into economically valuable deposits. The process, called *secondary enrichment*, is accomplished by either (1) removing undesirable materials and leaving the desired elements enriched in the upper zones of the soil or (2) removing and carrying the desirable elements to lower soil zones where they are redeposited and become more concentrated. *Bauxite*, the principal ore of aluminum, is one important ore created by secondary enrichment.

- In the evolution of most landforms, mass wasting is the step that follows weathering. The combined effect of mass wasting and erosion by running water produce stream valleys. *Gravity is the controlling force of mass wasting.* Other factors that play an important role in overcoming inertia and triggering downslope movements are saturation of the material with water, oversteepening of slopes, removal of anchoring vegetation, and ground vibrations from earthquakes. The various processes included under the name of mass wasting are defined by (1) the type of material involved (debris, mud, earth, or rock), (2) the kind of motion (fall, slide, or flow), and (3) the velocity of the movement (fast, slow). The various kinds of mass wasting include the fast forms called *slump, rockslide, debris flow,* and *earthflow*, as well as the slow movements referred to as *creep* and *solifluction.*

Key Terms

angle of repose (p. 78)
chemical weathering (p. 63)
creep (p. 83)
debris flow (p. 81)
differential weathering (p. 66)
earthflow (p. 83)
eluviation (p. 72)
erosion (p. 60)
exfoliation dome (p. 62)
external process (p. 60)
fall (p. 74)
flow (p. 80)
frost wedging (p. 61)

horizon (p. 71)
humus (p. 68)
lahar (p. 82)
laterite (p. 84)
leaching (p. 72)
mass wasting (p. 60)
mechanical weathering (p. 61)
parent material (p. 69)
pedalfer (p. 73)
pedocal (p. 73)
permafrost (p. 84)
regolith (p. 68)
rockslide (p. 81)

secondary enrichment (p. 76)
sheeting (p. 62)
slide (p. 79)
slump (p. 80)
soil (p. 68)
soil profile (p. 71)
soil texture (p. 68)
solifluction (p. 84)
solum (p. 72)
spheroidal weathering (p. 64)
talus slope (p. 61)
weathering (p. 60)

Review Questions

1. Describe the role of external processes in the rock cycle.

2. If two identical rocks were weathered, one mechanically and the other chemically, how would the products of weathering for the two rocks differ?

3. How does mechanical weathering add to the effectiveness of chemical weathering?

4. Describe the formation of an exfoliation dome. Give an example of such a feature.

5. Granite and basalt are exposed at the surface in a hot, wet region.

(a) Which type of weathering will predominate?

(b) Which of the rocks will weather most rapidly? Why?

6. Heat speeds up a chemical reaction. Why then does chemical weathering proceed slowly in a hot desert?

7. How is carbonic acid (H_2CO_3) formed in nature? What results when this acid reacts with potassium feldspar?

8. List some possible environmental effects of acid precipitation (see Box 3.1).

9. Using the soil texture diagram (Figure 3.12), name the soil that consists of 60 percent sand, 30 percent silt, and 10 percent clay.

10. What factors might cause different soils to develop from the same parent material, or similar soils to form from different parent materials?

11. Which of the controls of soil formation is most important? Explain.

12. How can slope affect the development of soil? What is meant by the term *slope orientation?*

13. List the characteristics associated with each of the horizons in a well-developed soil profile. Which of the horizons constitute the solum? Under what circumstances do soils lack horizons?

14. Distinguish between pedalfers and pedocals.

15. What soil type is associated with tropical rain forests? Since this soil supports the growth of lush natural vegetation, is it also excellent for growing crops? Briefly explain. (See Box 3.2.)

16. Is soil erosion a natural process or primarily the result of inappropriate land use by people? Explain.

17. List three detrimental effects of soil erosion other than the loss of topsoil from croplands.

18. Name the primary ore of aluminum and describe its formation.

19. What role does mass wasting play in sculpting Earth's landscape?

20. What is the controlling force of mass wasting? What other factors are important?

21. Distinguish among fall, slide, and flow.

22. Why can rock avalanches move at such great speeds?

23. What factors led to the massive rockslide at Gros Ventre, Wyoming (See Box 3.3.)?

24. What type of mass wasting event killed thousands of people in the vicinity of the Colombian volcano Nevado del Ruiz in 1985?

25. Compare and contrast debris flow and earthflow.

26. Because creep is an imperceptibly slow process, what evidence might indicate to you that this phenomenon is affecting a slope? Describe the mechanism that creates this slow movement (Figure 3.25).

27. Why is solifluction only a summertime phenomenon?

Examining the Earth System

1. Which of Earth's spheres are associated with the production of carbonic acid?

2. The level of carbon dioxide (CO_2) in the atmosphere has been increasing for more than a century. Should this increase tend to accelerate or slow down the rate of chemical weathering of Earth's surface rocks? Explain how you arrived at your conclusion.

3. In addition to carbon dioxide, list some other chemicals that human activities emit into the atmosphere that aid in the chemical weathering of Earth materials. (You might find the information about acid rain at this United States Environmental Protection Agency (EPA) Web site, **http://www.epa.gov/acidrain/ardhome.html**, both informative and useful.)

4. Discuss the interaction of the atmosphere, solid Earth, biosphere, and hydrosphere in the formation of soil.

Web Resources

The following are informative and interesting World Wide Web sites that address topics related to those presented in the chapter:

Soil Science Society of America:

- **http://www.soils.org/**

WWW Virtual Library: Soils and Substrates:

- **http://www.metla.fi/info/vlib/soils/**

For direct links to these sites and other appropriate Earth science Web locations, chapter review exercises, and relevant, topical investigations that utilize Web resources, contact the *Earth Science*, 9th Edition Home Page at:

- **http://www.prenhall.com/tarbuck.**

Chapter 4

Running Water and Groundwater

There are countless ways in which nature impacts people. This small sinkhole formed suddenly in 1991 when the roof of a cavern collapsed, destroying this home in Frostproof, Florida. (Photo by *St. Petersburg Times*/Liaison Agency, Inc.)

All the rivers run into the sea; yet the sea is not full;
unto the place from whence the rivers come, thither
they return again.

(Ecclesiastes 1:7)

As the perceptive writer of Ecclesiastes indicated, water is continually on the move, from the ocean to the land and back again in an endless cycle. This chapter deals with that part of the hydrologic cycle that returns water to the sea. Some water travels quickly via a rushing stream, and some moves more slowly below the surface. We will examine the factors that influence the distribution and movement of water, as well as look at how water sculptures the landscape. To a great extent, the Grand Canyon, Niagara Falls, Old Faithful, and Mammoth Cave all owe their existence to the action of water on its way to the sea.

Water is everywhere on Earth—in the oceans, glaciers, rivers, lakes, the air, soil, and in living tissue. All of these "reservoirs" constitute Earth's hydrosphere. In all, the water content of the hydrosphere is about 1.36 billion cubic kilometers (326 million cubic miles).

The vast bulk of it, about 97.2 percent, is stored in the global oceans (Figure 4.1). Ice sheets and glaciers account for another 2.15 percent, leaving only 0.65 percent to be divided among lakes, streams, subsurface water, and the atmosphere (Figure 4.1). Although the percentages of Earth's total water found in each of the latter sources is but a small fraction of the total inventory, the absolute quantities are great.

Earth as a System: The Hydrologic Cycle

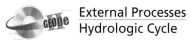

External Processes
Hydrologic Cycle

The water found in each of the reservoirs depicted in Figure 4.1 does not remain in these places indefinitely. Water can readily change from one state of matter (solid, liquid, or gas) to another at the temperatures and pressures occurring at Earth's surface. Therefore, water is constantly moving among the oceans, the atmosphere, the solid Earth, and the biosphere. This unending circulation of Earth's water supply is called the **hydrologic cycle.** The cycle shows us many critical interrelationships among different parts of the Earth system.

The hydrologic cycle is a gigantic worldwide system powered by energy from the Sun in which the atmosphere provides the vital link between the oceans and continents (Figure 4.2). Water evaporates into the atmosphere from the ocean and to a much lesser extent from the continents. Winds transport this moisture-laden air, often great distances, until conditions cause the moisture to condense into clouds, and precipitation to fall. The precipitation that falls into the ocean has completed its cycle and is ready to begin another.

The water that falls on the continents, however, must make its way back to the ocean.

What happens to precipitation once it has fallen on land? A portion of the water soaks into the ground (called **infiltration**), slowly moving downward, then laterally, finally seeping into lakes, streams, or directly into the ocean. When the rate of rainfall exceeds Earth's ability to absorb it, the surplus water flows over the surface into lakes and streams, a process called **runoff**. Much of the water that infiltrates or runs off eventually returns to the atmosphere because of evaporation from the soil, lakes, and streams. Also, some of the water that infiltrates the ground surface is absorbed by plants, which then release it into the atmosphere. This process is called **transpiration**. Each year a field of crops may transpire the equivalent of a water layer 60 centimeters (2 feet) deep over the entire field. The same area of trees may pump twice this amount into the atmosphere.

When precipitation falls in very cold areas—at high elevations or high latitudes—the water may not immediately soak in, run off, or evaporate. Instead, it may become part of a snowfield or a glacier. In this way, glaciers store large quantities of water on land. If present-day glaciers were to melt and release all their water, sea level would rise by several dozen meters. This would submerge many heavily populated coastal areas. As we shall see in Chapter 5, over the past 2 million years, huge ice sheets have formed and melted on several occasions, each time changing the balance of the hydrologic cycle.

Figure 4.2 also shows Earth's overall *water balance,* or the volume of water that passes through each part of the cycle annually. The amount of water vapor in the air at any one time is just a tiny fraction of Earth's total water supply. But the *absolute* quantities that are cycled through

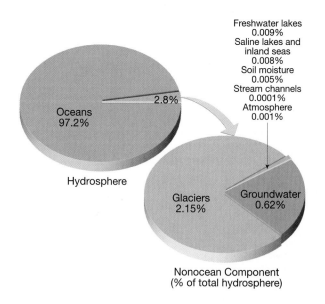

Figure 4.1 Distribution of Earth's water.

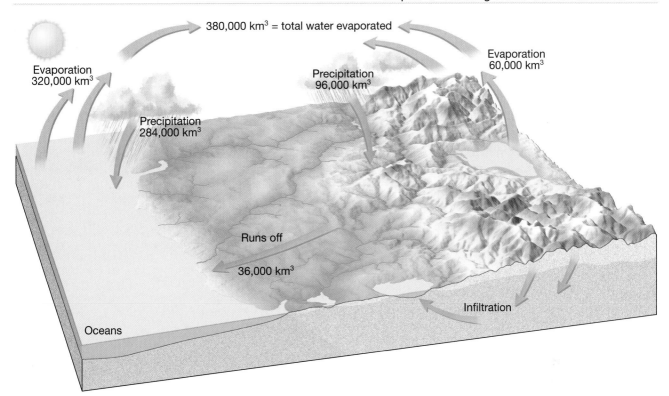

380,000 km³ = total water evaporated

Evaporation
320,000 km³

Evaporation
60,000 km³

Precipitation
96,000 km³

Precipitation
284,000 km³

Runs off

36,000 km³

Infiltration

Oceans

Figure 4.2 Earth's water balance. About 320,000 cubic kilometers of water are evaporated each year from the oceans, while evaporation from the land (including lakes and streams) contributes 60,000 cubic kilometers of water. Of this total of 380,000 cubic kilometers of water, about 284,000 cubic kilometers fall back to the ocean, and the remaining 96,000 cubic kilometers fall on Earth's land surface. Since 60,000 cubic kilometers of water evaporate from the land, 36,000 cubic kilometers of water remain to erode the land during the journey back to the oceans.

the atmosphere over a 1-year period are immense—some 380,000 cubic kilometers—enough to cover Earth's entire surface to a depth of about 1 meter (39 inches). Estimates show that over North America almost six times more water is carried by moving currents of air than is transported by all the continent's rivers.

It is important to know that the hydrologic cycle is *balanced*. Because the total amount of water vapor in the atmosphere remains about the same, the average annual precipitation over Earth must be equal to the quantity of water evaporated. However, for all of the continents taken together, precipitation exceeds evaporation. Conversely, over the oceans, evaporation exceeds precipitation. Because the level of the world ocean is not dropping, the system must be in balance. In Figure 4.2, the 36,000 cubic kilometers of water that annually runs off from the land to the ocean causes enormous erosion. In fact, this immense volume of moving water is the *single most important agent sculpturing Earth's land surface.*

To summarize, the hydrologic cycle is the continuous movement of water from the oceans to the atmosphere, from the atmosphere to the land, and from the land back to the sea. The land-back-to-the-sea step is the primary action that wears down Earth's land surface. In this chapter, we will first observe the work of water running over the surface, including floods, erosion, and the formation of valleys. Then we will look

underground at the slow labors of groundwater as it forms springs and caverns, and provides drinking water on its long migration to the sea.

Running Water

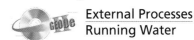

External Processes
Running Water

Running water is of great importance to people. We depend on rivers for energy, transportation, and irrigation. Their fertile floodplains have been favored sites for agriculture and industry since the dawn of civilization. As the dominant agent of erosion, running water has shaped much of our physical environment.

Although we have always depended on running water, its source eluded us for centuries. Not until the 1500s did we realize that streams were supplied by surface runoff and underground water, which ultimately had their sources as rain and snow.

Streamflow

Water makes its way to the sea under the influence of gravity. The time required for the journey depends on the velocity of the stream. Velocity is the distance that water travels in a unit of time. Some sluggish streams flow at less than 1 kilometer per hour, whereas a few rapid ones may exceed 30 kilometers per hour. Velocities are measured at gauging stations (Figure 4.3A). Along straight stretches, the highest velocities are near the center of the channel just below the surface, where friction is lowest (Figure 4.3B). But when a stream curves, its zone of maximum speed shifts toward its outer bank (Figure 4.3C).

The ability of a stream to erode and transport materials depends on its velocity. Even slight variations in velocity can lead to significant changes in how much sediment can be transported by the water. Several factors determine the velocity of a stream, including (1) gradient; (2) shape, size, and roughness of the channel; and (3) discharge.

Gradient. The slope of a stream channel expressed as the vertical drop of a stream over a specified distance is **gradient**. Portions of the lower Mississippi River, for example, have very low gradients of 10 centimeters per kilometer or less. By contrast, some mountain stream channels decrease in elevation at a rate of more than 40 meters per kilometer, or a gradient 400 times steeper than the lower Mississippi (Figure 4.4). Gradient varies not only between streams, but over a stream's length. The steeper the gradient, the more energy available for streamflow. If two streams were identical in every respect except gradient, the stream with the higher gradient would obviously have the greater velocity.

Channel Characteristics. A stream's channel is a conduit that guides the flow of water, but the water encounters friction as it flows. The shape, size, and roughness of the channel affect the amount of friction. Larger channels have more efficient flow because a smaller proportion of water is in contact with the channel. A smooth channel promotes a more uniform flow, whereas an irregular channel filled with boulders creates enough turbulence to slow the stream significantly.

Discharge. The **discharge** of a stream is the volume of water flowing past a certain point in a given unit of time. This is usually measured in cubic meters per second or cubic feet per second. Table 4.1 lists the world's largest rivers in terms of discharge. The largest river in North America, the Mississippi, discharges an average of 17,300 cubic meters (611,000 cubic feet) per second. Although this is a huge quantity of water, it is nevertheless dwarfed by the mighty Amazon in South America, the world's largest river. Draining a rainy region that is nearly three-fourths the size of the conterminous United States, the Amazon discharges 12 times more water than the Mississippi.

The discharges of most rivers are far from constant (see Box 4.1). This is true because of such variables as rainfall and snowmelt. When discharge changes, the factors noted earlier must also change. Indeed, measurements show that when the amount of water in a

A.

B.

C.

Figure 4.3 A. Continuous records of stage and discharge are collected by the U.S. Geological Survey at more than 7000 gauging stations in the United States. Average velocities are determined by using measurements from several spots across the stream. This station is on the Rio Grande south of Taos, New Mexico. (Photo by E.J. Tarbuck) **B.** Along straight stretches, stream velocity is highest at the center of the channel. **C.** When a stream curves, its zone of maximum speed shifts toward the outer bank.

stream increases, the width, depth, and velocity all increase. To handle the additional water, the stream will increase the size of its channel by widening and

Figure 4.4 Rapids can form where a portion of a stream channel has a steep gradient. (Photo by Bruce Gaylord/Visuals Unlimited)

deepening it. The larger the channel, the less friction, and the more swiftly the water will flow.

Changes From Upstream to Downstream

One useful way of studying a stream is to examine its *profile*. A profile is simply a cross-sectional view of a stream from its source area (called the *head* or *headwaters*) to its *mouth*, the point downstream where the river empties into another water body. By examining Figure 4.5, you can see that the most obvious feature of a typical profile is a constantly decreasing gradient from the head to the mouth. Although many local irregularities may exist, the overall profile is a smooth curve.

The profile shows that the gradient decreases downstream. To see how other factors change in a downstream direction, observations and measurements must be made. When data are collected from several gauging stations along a river, they show that discharge increases toward the mouth. This should come as no surprise because, as we move downstream, more and more tributaries contribute water to the main channel (Figure 4.5). Furthermore, in most humid regions, additional water is added from the groundwater supply. Thus, as you move downstream, the stream's width, depth, and velocity change in response to the increased volume of water carried by the stream.

The observed increase in velocity that occurs downstream contradicts our impressions about wild, rushing mountain streams and wide, placid rivers. Although a mountain stream may have the appearance of a raging torrent, its average velocity is often less than that of the river near its mouth. The difference is primarily attributable to the greater efficiency of the larger channel in a downstream direction.

In the headwaters region where the gradient is steep, the water must flow in a relatively small and often boulder-strewn channel. The small channel and rough bed create great friction and inhibit movement

Table 4.1 World's largest rivers ranked by discharge.

| Rank | River | Country | Drainage Area | | Average Discharge | |
			Square kilometers	Square miles	Cubic meters per second	Cubic feet per second
1	Amazon	Brazil	5,778,000	2,231,000	212,400	7,500,000
2	Congo	Zaire	4,014,500	1,550,000	39,650	1,400,000
3	Yangtze	China	1,942,500	750,000	21,800	770,000
4	Brahmaputra	Bangladesh	935,000	361,000	19,800	700,000
5	Ganges	India	1,059,300	409,000	18,700	660,000
6	Yenisei	Russia	2,590,000	1,000,000	17,400	614,000
7	Mississippi	United States	3,222,000	1,244,000	17,300	611,000
8	Orinoco	Venezuela	880,600	340,000	17,000	600,000
9	Lena	Russia	2,424,000	936,000	15,500	547,000
10	Parana	Argentina	2,305,000	890,000	14,900	526,000

Box 4.1 People and the Environment: The Effect of Urbanization on Discharge

When rains occur, stream discharge increases. If the rains are sufficiently heavy, the ability of the channel to contain the discharge is exceeded, and water spills over the banks as a flood. Floods are natural events that should be expected. However, when cities are built, the magnitude and frequency of flooding increases. The top portion of Figure 4.A is a hypothetical hydrograph that shows the time relationship between a rainstorm and the occurrence of flooding. Notice that the water level in the stream does not rise at the onset of precipitation because time is needed for water to move from the place where it fell to the stream. This time difference is called the *lag time*.

When an area changes from being predominantly rural to largely urban, streamflow is affected. The effect of urbanization on streamflow is illustrated by the bottom hydrograph in Figure 4.A. Notice that after urbanization the peak discharge during a flood is greater, and that the lag time between precipitation and flood peak is shorter than before urbanization. The explanation for this effect is relatively simple. The construction of streets, parking lots, and buildings covers over the ground that once soaked up water. Thus, less water infiltrates the ground, and the rate and amount of runoff increase. Further, because much less water soaks into the ground, the low-water (dry-season) flow in urban streams, which is maintained by the seepage of groundwater into the channel, is greatly reduced. As one might expect, the magnitude of these effects is a function of the per-

Figure 4.A When an area changes from rural to urban, the lag time between rainfall and flood peak is shortened. The flood peak is also higher following urbanization. (After L.B. Leopold, U.S. Geological Survey)

centage of land that is covered by impermeable surfaces.

Urbanization is just one example of human interference with streams. There are many other ways that land use inadvertently influences the flow

of streams and the work they carry out. Moreover, there are also many ways by which people intentionally attempt to manipulate and control streams. Some of these are discussed in the section on Floods and Flood Control (p. 103).

by scattering water in all directions, with almost as much backward motion as forward motion. However, downstream, the material on the bed of the stream becomes much smaller, offering less resistance to flow, and the width and depth of the channel increase to accommodate the greater discharge. These factors, especially the wider and deeper channel, permit the water to flow more freely and hence more rapidly.

In summary, you have seen an inverse relationship between gradient and discharge. Where gradient is steep, discharge is small, and where discharge is great, gradient is small. Stated another way, a stream can maintain a higher velocity near its mouth even though it has a lower gradient than upstream because of the greater discharge, larger channel, and smoother bed.

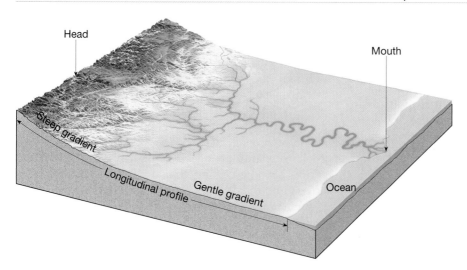

Figure 4.5 A longitudinal profile is a cross-section along the length of a stream. Note the concave-upward curve of the profile, with a steeper gradient upstream and a gentler gradient downstream.

Base Level

External Processes
Running Water

Streams cannot endlessly erode their channels deeper and deeper. There is a lower limit to how deep a stream can erode, and that limit is the stream's base level. **Base level** is the lowest point to which a stream can erode its channel.

Two general types of base level exist. Sea level is considered the *ultimate base level*, because it is the lowest level to which stream erosion could lower the land.

Temporary, or *local*, *base levels* include lakes, resistant layers of rock, and main streams that act as base level for their tributaries. For example, when a stream enters a lake, its velocity quickly approaches zero and its ability to erode decreases. Thus, the lake prevents the stream from eroding below its level at any point upstream from the lake. However, because the outlet of the lake can cut downward and drain the lake, the lake is only a temporary hindrance to the stream's ability to downcut its channel.

In a similar manner, the layer of resistant rock at the lip of the waterfall in Figure 4.6 acts as a temporary

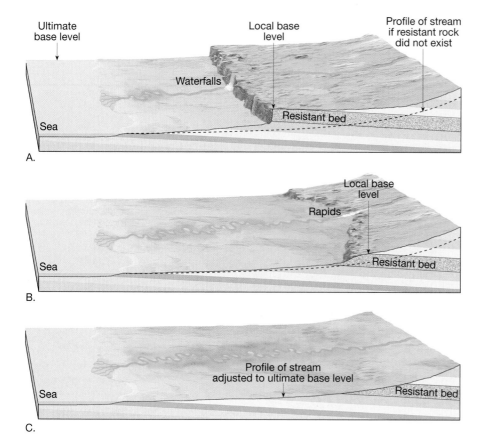

Figure 4.6 A resistant layer of rock can act as a local (temporary) base level. Because the durable layer is eroded more slowly, it limits the amount of downcutting upstream.

base level. Until the ledge of hard rock is eliminated, it will limit the amount of downcutting upstream.

Any change in base level will cause a corresponding readjustment of stream activities. When a dam is built along a stream, the reservoir that forms behind it raises the base level of the stream (Figure 4.7). Upstream from the dam the stream gradient is reduced, lowering its velocity and, hence, its sediment-transporting ability. The stream, now having too little energy to transport all of its load, will deposit sediment. This builds up its channel. This process continues until the stream again has a gradient sufficient to carry its load.

The opposite situation is illustrated in Figure 4.8. A stream in a broad, flat-floored valley that is near base level often develops a course with many bends and turns called *meanders* (Figure 4.8A). What would happen to such a stream if base level dropped or if the land were uplifted? The river, now considerably above base level, would have excess energy and would downcut its channel to establish a balance with its new base level. The result could be *incised meanders*, that is, a winding river in a steep, narrow valley (Figure 4.8B).

Work of Streams

The work of streams includes erosion, transportation, and deposition. These activities go on simultaneously in all stream channels, even though they are presented individually here.

Erosion

Erosion is the removal of rock and soil. Much of the material carried by streams reaches them through underground water, overland flow, and mass wasting. Streams also create some of their load by eroding their own channels. If a channel is composed of bedrock, most of the erosion is accomplished by the abrasive action of water armed with sediment, a process analogous to sandblasting. Pebbles caught in swirling eddies act like cutting tools and bore circular "potholes" into the channel floor. In channels composed of loose material, considerable lifting and removal can be accomplished by the impact of water alone.

Transportation

Streams transport their load of sediment in three ways: (1) in solution (**dissolved load**); (2) in suspension (**suspended load**); and (3) scooting or rolling along the bottom (**bed load**).

Most of the *dissolved load* is brought to the stream by groundwater and to a lesser degree it is acquired by dissolving rock along the stream's course. The quantity of material carried in solution is highly variable and depends on climate and the geologic setting. Usually the dissolved load is expressed as parts of dissolved material per million parts of water (parts per million, or ppm). Although some rivers may have a dissolved load of 1000 ppm or more, the average figure for the world's rivers is estimated at 115 to 120

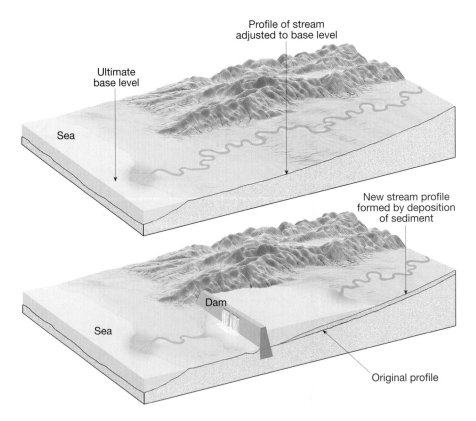

Figure 4.7 When a dam is built and a reservoir forms, the stream's base level is raised. This reduces the stream's velocity and leads to deposition and a reduction of the gradient upstream from the reservoir.

A.   B.

Figure 4.8 A. A river in a broad, flat-floored valley near base level often exhibits a meandering channel. **B.** Incised meanders of the Colorado River in Canyonlands National Park, Utah. Here, as the Colorado Plateau was gradually uplifted, the meandering river adjusts to being higher above base level by downcutting. (Photos by Michael Collier)

ppm. Almost 4 billion metric tons of dissolved mineral matter are supplied to the oceans each year by streams.

Most streams (but not all) carry the largest part of their load in *suspension*. Indeed, the visible cloud of sediment suspended in the water is the most obvious portion of a stream's load. Usually only sand, silt, and clay can be carried this way, but during a flood larger particles are transported as well. Also, during a flood, the total quantity of material carried in suspension increases dramatically, as can be verified by anyone whose home has been a site for the deposition of this material.

A portion of a stream's load of solid material consists of sediment that is too large to be carried in suspension. These coarser particles move along the bottom of the stream and constitute the *bed load*. Unlike the suspended and dissolved loads, which are constantly in motion, the bed load is in motion only intermittently, when the force of the water is sufficient to move the larger particles. In terms of the erosional work accomplished by a downcutting stream, the grinding action of the bed load is of great importance.

Each year, the Mississippi River transports about 750 million tons of material to the Gulf of Mexico. Of this total, it is estimated that approximately 500 million tons are carried in suspension, 200 million tons in solution, and the remaining 50 million tons as bed load; however, such proportions vary widely from stream to stream.

Streams vary in their ability to carry a load. Their ability is determined by two criteria. First, the **competence** of a stream measures the maximum size of particles it is capable of transporting. The stream's velocity determines its competence. If the velocity of a stream doubles, its competence increases four times; if the velocity triples, its competence increases nine times; and so forth. This explains how large boulders

that seem immovable can be transported during a flood, which greatly increases a stream's velocity.

Second, the **capacity** of a stream is the maximum load it can carry. The capacity of a stream is directly related to its discharge. The greater the volume of water flowing in a stream, the greater is its capacity for hauling sediment.

By now it should be clear why the greatest erosion and transportation of sediment occur during floods (Figure 4.9). The increase in discharge results in a greater capacity and the increase in velocity results in greater competence. With rising velocity the water becomes more turbulent, and larger and larger particles are set in motion. In just a few days, or perhaps a few hours, a stream in flood stage can erode and transport more sediment than it does during months of normal flow (see Box 4.2).

Deposition

Whenever a stream slows down, the situation reverses. As its velocity decreases, its competence is reduced and sediment begins to drop out, largest particles first. Each particle size has a *critical settling velocity*. As streamflow drops below the critical settling velocity of a certain particle size, sediment in that category begins to settle out. Thus, stream transport provides a mechanism by which solid particles of various sizes are separated. This process, called **sorting**, explains why particles of similar size are deposited together.

The well-sorted material typically deposited by a stream is called **alluvium**, the general term for any stream-deposited sediment. Many different depositional features are composed of alluvium. Some occur within stream channels, some occur on the valley floor adjacent to the channel, and some exist at the mouth of the stream.

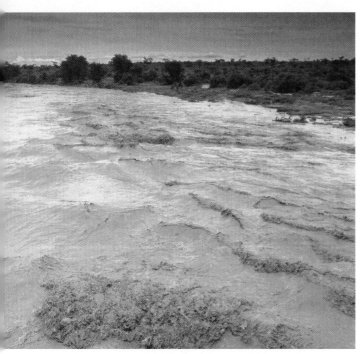

Figure 4.9 The suspended load is clearly visible because it gives this flooding river a "muddy" appearance. During floods both capacity and competency increase. Therefore the greatest erosion and sediment transport occur during these high-water periods. Here we see the sediment-filled floodwaters of Kenya's Mara River. (Photo by Tim Davis/Tony Stone Images)

Deltas. When a stream enters the relatively still waters of an ocean or lake, its velocity drops abruptly, and the resulting deposits form a **delta** (Figure 4.10). As the delta grows outward, the stream's gradient continually lessens. This circumstance eventually causes the channel to become choked with sediment from the slowing water. As a consequence, the river seeks a shorter, higher-gradient route to base level, as illustrated in Figure 4.10B. This illustration shows the main channel dividing into several smaller ones called **distributaries**. Most deltas are characterized by these shifting channels that act in an opposite way to that of tributaries.

Rather than carrying water into the main channel, distributaries carry water away from the main channel. After numerous shifts of the channel, a delta may grow into a rough triangular shape like the Greek letter delta (Δ), for which it is named. Note, however, that many deltas do not exhibit the idealized shape. Differences in the configurations of shorelines and variations in the nature and strength of wave activity result in many shapes.

Many large rivers have deltas extending over thousands of square kilometers. The delta of the Mississippi River is one example. It resulted from the accumulation of huge quantities of sediment derived

Figure 4.10 A. Structure of a simple delta that forms in the relatively quiet waters of a lake. **B.** Growth of a simple delta. As a stream extends its channel, the gradient is reduced. Frequently, during flood stage the river is diverted to a higher-gradient route, forming a new distributary. Old abandoned distributaries are gradually invaded by aquatic vegetation and fill with sediment. (After Ward's Natural Science Establishment, Inc., Rochester, N.Y.)

Box 4.2   People and the Environment: The 1997 Red River Floods

The Red River of the North originates near the point where South Dakota, North Dakota, and Minnesota meet. It flows northward, finally emptying into Manitoba's Lake Winnepeg (Figure 4.B). Along most of its course in the United States, the river serves as the North Dakota–Minnesota boundary.

In April 1997, Grand Forks, North Dakota, and other communities along the Red River experienced disastrous flooding—the worst ever recorded in the region (Figure 4.C). The flood was preceded by an especially snowy winter. As April began, the winter snows were melting and flooding seemed imminent. Then on April 5 and 6 a final winter storm pounded the northern Great Plains. The blizzard rebuilt once-shrinking snowdrifts back to heights of 6 meters (20 feet) in some places. However, the drifts did not last long because rapidly rising temperatures melted the snow in a matter of days, causing rapid runoff into the Red River.

The discharge at Grand Forks jumped from about 70 cubic meters (2500 cubic feet) per second on April 1 to nearly 4200 cubic meters (150,000 cubic feet) per second three weeks later. The floodwaters at Grand Forks crested at 17.7 meters (58.1 feet), almost 8 meters (more than 26 feet) above flood stage! This exceeded the previous record flood stage set in 1897 by 1.2 meters (4 feet). Roughly 4.5 million acres of land were under water, and most of Grand Fork's nearly 50,000 inhabitants had to evacuate. The losses in Grand Forks alone were estimated to exceed $1 billion.

The 1997 Red River flood was truly a record-breaking event—a 500-year flood (that is, the statistical probability of such a flood is once in 500 years). Although such an extraordinary flood is rare, less severe flooding of broad areas along the Red River of the North is common. Why do floods here occur so frequently and over such large areas? Four factors seem to be especially important.*

Figure 4.B The Red River flows north into Manitoba's Lake Winnepeg.

1. The Red River flows northward. When the spring thaw takes place, it proceeds in the same direction the river flows, from south to north. Therefore, runoff from the southern portion of the drainage basin progressively joins with newly created snowmelt from more northerly localities. If the progressive south-to-north thaw is perfectly synchronized, extensive floods occur in the northern portion of the valley.

2. Ice derived from the southern valley progressively meets with freshly broken ice in the central and northern valley. Such ice concentrations can slow or block the water flow, causing river levels to swell.

3. The Red River flows across a broad and extremely flat plain, the former floor of a now-abandoned lake. Therefore, when the river floods onto this plain, the area covered can be dramatic.

4. The river's gradient is very low and decreases downstream. Near Fargo the gradient averages about 8 centimeters per kilometer (5 inches per mile). Farther north, near the Canadian border, the gradient drops to just 2.5 centimeters per kilometer (1.5 inches per mile). During floods in this border region, the river's lack of slope causes it to form what is essentially a massive shallow lake.

*Based on material prepared by Professor Donald P. Schwert, North Dakota State University.

Figure 4.C Floodwaters surround Harwood, North Dakota, during the record-breaking 1997 floods. (Photo by Dan Koeck/Liaison Agency, Inc.)

from the vast region drained by the river and its tributaries. Today, New Orleans rests where there was ocean less than 5000 years ago. Figure 4.11 shows that portion of the Mississippi delta that has been built over the past 5000 to 6000 years. As shown, the delta is actually a series of seven coalescing subdeltas. Each formed when the river left its existing channel in favor of a shorter, more direct path to the Gulf of Mexico. The individual subdeltas interfinger and partially cover one another to produce a very complex structure. The present subdelta, called a *bird-foot* delta because of the configuration of its distributaries, has been built by the Mississippi in the last 500 years.

Natural Levees. Some rivers occupy valleys with broad, flat floors and build **natural levees** that parallel their channels on both banks (Figure 4.12). Natural levees are built by successive floods over many years. When a stream overflows its banks, its velocity immediately diminishes, leaving coarse sediment deposited in strips bordering the channel. As the water spreads out over the valley, a lesser amount of fine sediment is deposited over the valley floor. This uneven distribution of material produces the very gentle slope of the natural levee.

The natural levees of the lower Mississippi rise 6 meters (20 feet) above the valley floor. The area behind the levee is characteristically poorly drained for the obvious reason that water cannot flow up the levee and into the river. Marshes called **backswamp**s result. A tributary stream that cannot enter a river because levees block the way often has to flow parallel to the river until it can breach the levee. Such streams are called **yazoo tributaries** after the Yazoo River, which parallels the Mississippi for over 300 kilometers.

Stream Valleys

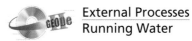

**External Processes
Running Water**

Stream valleys can be divided into two general types. Narrow V-shaped valleys and wide valleys with flat floors exist as the ideal forms, with many gradations between.

Figure 4.11 During the past 5000 to 6000 years, the Mississippi River has built a series of seven coalescing subdeltas. The numbers indicate the order in which the subdeltas were deposited. The present bird-foot delta (number 7) represents the activity of the past 500 years. Without ongoing human efforts, the present course will shift and follow the path of the Atchafalaya River. The inset on left shows the point where the Mississippi may someday break through (arrow) and the shorter path it would take to the Gulf of Mexico. (After C.R. Kolb and J.R. Van Lopik)

Figure 4.12 Natural levees are gently sloping deposits that are created by repeated floods. Because the ground next to the stream channel is higher than the adjacent floodplain, back swamps and yazoo tributaries may develop.

Most stream valleys are much broader at the top than is the width of their channel at the bottom. This would not be the case if the only agent responsible for eroding valleys were the streams flowing through them. The sides of most valleys are shaped by a combination of weathering, overland flow, and mass wasting. In some arid regions, where downcutting is rapid and weathering is slow, and in places where rock is particularly resistant, narrow valleys may not be V-shaped but rather may have nearly vertical walls.

Narrow Valleys

The Yellowstone River provides an excellent example of a narrow valley (Figure 4.13). A narrow V-shaped valley indicates that the primary work of the stream has been downcutting toward base level. The most prominent features of a narrow valley are *rapids* and *waterfalls*. Both occur where the stream profile drops rapidly, a situation usually caused by variations in the erodibility of the bedrock into which a stream channel is cutting. Resistant beds create rapids by acting as a temporary base level upstream while allowing downcutting to continue downstream. Once erosion has eliminated the resistant rock, the stream profile smooths out again. Waterfalls are places where the stream profile makes a vertical drop.

Wide Valleys

Once a stream has cut its channel closer to base level, downward erosion becomes less dominant. At this point more of the stream's energy is directed from side to side. The result is a widening of the valley as the river cuts away first at one bank and then at the other (Figure 4.14).

Floodplains. The side-to-side cutting of a stream eventually produces a flat valley floor, or **floodplain**. It is appropriately named because the river is confined to its channel, except during flood stage, when it overflows its banks and inundates the floodplain.

When a river erodes laterally, creating a floodplain as just described, it is called an *erosional floodplain*. Floodplains can be depositional in nature as well. *Depositional floodplains* are produced by a major fluctuation in conditions, such as a change in base level. The floodplain in California's Yosemite Valley is one such feature, and was produced when a glacier gouged the former stream valley deeper by about 300 meters (1000 feet). After the glacial ice melted, the stream readjusted to its former base level by refilling the valley with alluvium.

Meanders. Streams that flow on floodplains move in sweeping bends called **meanders** (Figure 4.14).

Figure 4.13 V-shaped valley of the Yellowstone River. The rapids and waterfalls indicate that the river is vigorously downcutting. (Photo by Art Wolfe)

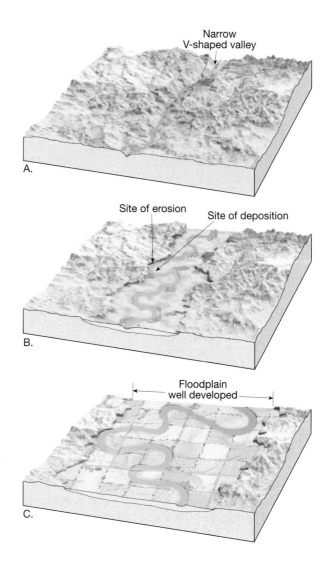

Narrow
V-shaped valley

A.

Site of erosion   Site of deposition

B.

Floodplain
well developed

C.

Figure 4.14 Stream eroding its floodplain.

Deposition
of point bar

Erosion of
cutbank

Figure 4.15 Lateral movement of meanders. By eroding its outer bank and depositing material on the inside of the bend, a stream is able to shift its channel.

Meanders continually change position by eroding sideways and slightly downstream. The sideways movement occurs because the maximum velocity of the stream shifts toward the outside of the bend, causing erosion of the outer bank (Figure 4.15). At the same time, the reduced current at the inside of the meander results in the deposition of coarse sediment, especially sand, called a *point bar*. Thus, by eroding its outer bank and depositing material along its inner bank, a stream moves sideways without changing its channel size.

Because of the slope of the channel, erosion is more effective on the downstream side of a meander. Therefore, in addition to migrating laterally, the bends also gradually migrate down the valley. Sometimes the downstream migration of a meander is slowed when it reaches a more resistant portion of the floodplain. This allows the next meander upstream to overtake it. Gradually the neck of land between the meanders is narrowed. When they get close enough, the river erodes through the narrow neck of land to the next loop (Figure 4.16). The new, shorter channel segment is called a **cutoff** and, because of its shape, the abandoned bend is called an **oxbow lake**.

Floods and Flood Control

When the discharge of a stream becomes so great that it exceeds the capacity of its channel, it overflows its banks as a **flood**. Floods are the most common and most destructive of all geologic hazards. They are, nevertheless, simply part of the *natural* behavior of streams.

Causes of Floods

Rivers flood because of the weather. Rapid melting of snow in the spring, and/or major storms that bring heavy rains over a large region, cause most floods. The extensive 1997 flood along the Red River of the North (described in Box 4.2) is a recent example of an event triggered by rapid snowmelt. Exceptional rains caused the devastating floods in the upper Mississippi River Valley during the summer of 1993 (Figure 4.17).

Human interference with the stream system can worsen or even cause floods. One example is the transformation of a rural landscape to an urban one. Pavement and storm sewers increase runoff and speed water to a nearby river (see Box 4.1). A second example is the failure of a dam. The bursting of a dam in 1889 on the Little Conemaugh River caused the devastating Johnstown, Pennsylvania, flood that took some 3000 lives. A second dam failure occurred there again in 1977 and caused 77 fatalities.

Figure 4.16 Formation of a cutoff and oxbow lake.

Flood Control

Several strategies have been devised to eliminate or lessen the catastrophic effects of floods. Engineering efforts include the construction of artificial levees, the building of flood-control dams, and river channelization.

Artificial Levees. Artificial *levees* are earthen mounds built on the banks of a river to increase the volume of water the channel can hold. These most common of stream-containment structures have been used since ancient times and continue to be used today.

Artificial levees are usually easy to distinguish from natural levees because their slopes are much steeper. When a river is confined by levees during

Figure 4.17 Satellite views of the Missouri River flowing into the Mississippi River. St. Louis is just south of their confluence. The upper image shows the rivers during a drought that occurred in summer 1988. The lower image depicts the peak of the record-breaking 1993 flood. Exceptional rains produced the wettest spring and early summer of the twentieth century in the upper Mississippi River basin. In all, nearly 14 million acres were inundated, displacing at least 50,000 people. (Photos by Space Imaging)

periods of high water, it frequently deposits material in its channel as the discharge diminishes. This is sediment that otherwise would have been dropped on the floodplain. Thus, each time there is a high flow, deposits are left on the river bed and the bottom of the channel is built up. With the buildup of the bed, less water is required to overflow the original levee. As a result, the height of the levee may have to be raised periodically to protect the floodplain. Moreover, many

artificial levees are not built to withstand periods of extreme flooding. For example, levee failures were numerous in the Midwest during the summer of 1993, when the upper Mississippi and many of its tributaries experienced record floods (Figure 4.18).

Flood-Control Dams. *Flood-control dams* are built to store floodwater and then let it out slowly. This lowers the flood crest by spreading it out over a longer time span. Since the 1920s, thousands of dams have been built on nearly every major river in the United States. Many dams have significant nonflood-related functions such as providing water for irrigated agriculture and for hydroelectric power generation. Many reservoirs are also major regional recreational facilities.

Although dams may reduce flooding and provide other benefits, building these structures also has significant costs and consequences. For example, reservoirs created by dams may cover fertile farmland, useful forests, historic sites, and scenic valleys. Of course, dams trap sediment. Therefore, deltas and floodplains downstream erode because they are no longer replenished with silt during floods. Large dams can also cause significant ecological damage to river environments that took thousands of years to establish.

Building a dam is not a permanent solution to flooding. Sedimentation behind a dam means that the volume of its reservoir will gradually diminish, reducing the effectiveness of this flood-control measure.

Channelization. *Channelization* involves altering a stream channel in order to speed the flow of water to prevent it from reaching flood height. This may simply involve clearing a channel of obstructions or dredging a channel to make it wider and deeper.

A more radical alteration involves straightening a channel by creating *artificial cutoffs*. The idea is that by shortening the stream, the gradient and hence the velocity are both increased. By increasing velocity, the larger discharge associated with flooding can be dispersed more rapidly.

Since the early 1930s, the Army Corps of Engineers has created many artificial cutoffs on the Mississippi for the purpose of increasing the efficiency of the channel and reducing the threat of flooding. In all, the river has been shortened more than 240 kilometers (150 miles). The program has been somewhat successful in reducing the height of the river in flood stage. However, because the river's tendency toward meandering still exists, preventing the river from returning to its previous course has been difficult.

A Nonstructural Approach. All of the flood-control measures described so far have involved structural solutions aimed at "controlling" a river. These solutions are expensive and often give people residing on the floodplain a false sense of security.

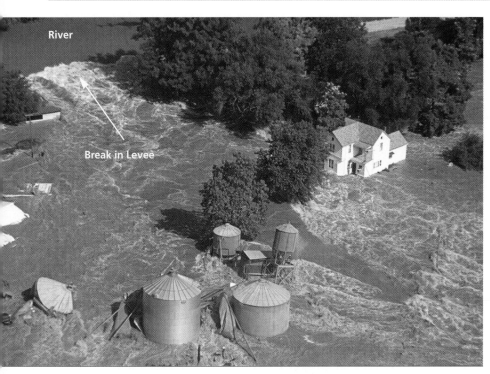

River

Break in Levee

Figure 4.18 Water rushes through a break in an artificial levee in Monroe County, Illinois. During the record-breaking 1993 Midwest floods, many artificial levees could not withstand the force of the floodwaters. Sections of many weakened structures were overtopped or simply collapsed. (Photo by James A. Finley/AP/Wide World Photos)

Today many scientists and engineers advocate a nonstructural approach to flood control. They suggest that an alternative to artificial levees, dams, and channelization is sound floodplain management. By identifying high-risk areas, appropriate zoning regulations can be implemented to minimize development and promote more appropriate land use.

Drainage Basins and Patterns

 External Processes
Running Water

Every stream, no matter how large or small, has a **drainage basin**. This is the land area that contributes water to the stream. The drainage basin of one stream is separated from the drainage basin of another by an imaginary line called a **divide** (Figure 4.19). Divides range in scale from a ridge separating two small gullies on a hillside to a *continental divide*, which splits continents into enormous drainage basins. For example, the continental divide that runs somewhat north—south through the Rocky Mountains separates the drainage that flows west to the Pacific Ocean from that flowing to the Gulf of Mexico. Although divides separate the drainage of two streams, if both streams are tributaries of the same river, they are both a part of the river's drainage system.

Drainage systems are networks of streams that together form distinctive patterns. The nature of a drainage pattern can vary greatly from one type of terrain to another, primarily in response to the kinds of rock on which the streams developed or the structural pattern of faults and folds.

Certainly the most commonly encountered drainage pattern is the **dendritic pattern** (Figure 4.20A). This pattern of irregularly branching tributary streams resembles the branching pattern of a deciduous tree. In fact, the word *dendritic* means "treelike." The dendritic pattern forms where the underlying material is relatively uniform. Because the surface material is essentially uniform in its resistance to erosion, it does not control the pattern of streamflow. Rather, the pattern is determined chiefly by the direction of slope of the land.

When streams diverge from a central area like spokes from the hub of a wheel, the pattern is said to be **radial** (Figure 4.20B). This pattern typically develops on isolated volcanic cones and domal uplifts.

Figure 4.20C illustrates a **rectangular** pattern, in which many right-angle bends can be seen. This pattern develops when the bedrock is crisscrossed by a series of joints and/or faults. Because these structures are eroded more easily than unbroken rock, their geometric pattern guides the directions of valleys.

Figure 4.20D illustrates a **trellis** drainage pattern, a rectangular pattern in which tributary streams are nearly parallel to one another and have the appearance of a garden trellis. This pattern forms in areas underlain by alternating bands of resistant and less-resistant rock.

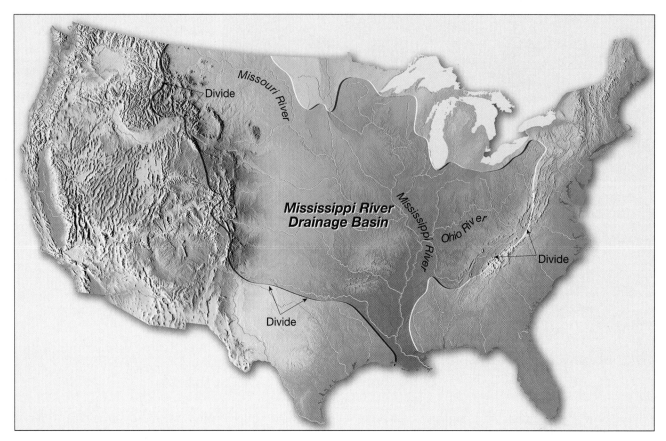

Figure 4.19 A drainage basin is the land area drained by a stream and its tributaries. The drainage basin of the Mississippi River, North America's largest river, covers about 3 million square kilometers. Divides are the boundaries that separate drainage basins from each other. Drainage basins and divides exist for all streams.

Water Beneath the Surface

 External Processes
Groundwater

Groundwater, water that has soaked into the ground, is one of our most important and widely available resources. Yet people's perceptions of groundwater are often unclear and incorrect. The reason is that groundwater is hidden from view except in caves and mines, and the impressions people gain from these subsurface openings are often misleading. Observations on the land surface give an impression that Earth is "solid." This view is not changed very much when we enter a cave and see water flowing in a channel that appears to have been cut into solid rock. Because of such observations many people believe that groundwater occurs only in underground "rivers." But actual rivers underground are extremely rare.

In reality, much of the subsurface environment is not "solid" at all but contains a huge volume of openings that exist as spaces between grains of soil and sediment and occur as narrow joints and fractures in bedrock. It is in these tiny openings that groundwater collects and moves.

The Importance of Groundwater

The importance of groundwater can be demonstrated by comparing its volume with the quantity of water in other parts of the hydrosphere. Of all the world's water, only about six-tenths of 1 percent occurs underground. Nevertheless, the amount of water stored in the rocks and sediments beneath Earth's surface is vast. When the oceans are excluded and only sources of freshwater are considered, the significance of groundwater becomes more apparent. Table 4.2 contains estimates of the distribution of freshwater in the hydrosphere. Clearly, the largest volume occurs as glacial ice. Second in rank is groundwater, with slightly more than 14 percent of the total. However, when ice is excluded and just liquid water is considered, more than 94 percent is groundwater. Without question, *groundwater represents the largest reservoir of freshwater that is readily available to humans.* Its value in terms of economics and human well-being is incalculable.

In many parts of the world, wells and springs provide the water needed not only for great numbers of people, but also for crops, livestock, and industry. In the United States, groundwater is the source of about

A. Dendritic

B. Radial

C. Rectangular

Ridges of
resistant
rock

Valleys cut in
less-resistant
rock

D. Trellis

Figure 4.20 Drainage patterns. **A.** Dendritic. **B.** Radial. **C.** Rectangular. **D.** Trellis.

40 percent of the water used for all purposes except hydroelectric power generation and power plant cooling. Groundwater provides drinking water for more than 50 percent of the population, as well as 40 percent of the water for irrigation and 26 percent of industry's needs. In some areas, however , overuse of this basic resource has caused serious problems, including streamflow depletion, land subsidence, and increased pumping costs. In addition, groundwater

contamination due to human activities is a real and growing threat in many places.

Groundwater's Geological Roles

Geologically, groundwater is important as an erosional agent. The dissolving action of groundwater slowly removes soluble rock, allowing surface depressions known as sinkholes to form as well as creating

Table 4.2 Freshwater of the hydrosphere.

Parts of the Hydrosphere	Volume of Freshwater (km³)	Share of Total Volume of Freshwater (percent)	Rate of Water Exchange
Ice sheets and glaciers	24,000,000	84.945	8000 years
Groundwater	4,000,000	14.158	280 years
Lakes and reservoirs	155,000	0.549	7 years
Soil moisture	83,000	0.294	1 year
Water vapor in the atmosphere	14,000	0.049	9.9 days
River water	1,200	0.004	11.3 days
Total	28,253,200	100.000	

Source: U.S. Geological Survey Water Supply Paper 2220, 1987.

subterranean caverns (Figure 4.21). Groundwater is also an equalizer of streamflow. Much of the water that flows in rivers is not direct runoff from rain and snowmelt. Rather, a large percentage of precipitation soaks in and then moves slowly underground to stream channels. Groundwater is thus a form of storage that sustains streams during periods when rain does not fall. When we observe water flowing in a river during a dry period, it is water from rain that fell at some earlier time and was stored underground.

Distribution and Movement of Groundwater

 External Processes
Groundwater

When rain falls, the water may run off immediately, evaporate, be taken up by plants and transpired back into the air, or soak into the ground. This last path is the primary source of practically all subsurface water. The amount of water that takes each of these paths, however, varies considerably both in time and space. For any location, the steepness of the slope, the nature of the surface material, the intensity of the rainfall, and the type and amount of vegetation are all influential factors. Heavy rains that fall on steep slopes underlain by impervious materials will obviously result in a high percentage of the water running off. Conversely, a gentle, steady rain

that falls on more gradual slopes consisting of materials more easily penetrated by water would result in a much larger percentage of the water soaking into the ground.

Distribution

Some of the water that soaks in does not travel far, because it is held by molecular attraction as a surface film on soil particles. A portion of this moisture evaporates back into the atmosphere. Much of the remainder is used by plants between rains. But water that is not held near the surface penetrates downward until it reaches a zone where all of the open spaces in sediment and rock are completely filled with water. This is called the **zone of saturation**. Water within it is called **groundwater**. The upper limit of this zone is known as the **water table**. The area above the water table where the soil, sediment, and rock are not saturated is called the **zone of aeration** (Figure 4.22). The open spaces here are filled mainly with air.

The water table is rarely level as we might expect a table to be. Instead, its shape is usually a subdued replica of the surface, reaching its highest elevations beneath hills and decreasing in height toward valleys (Figure 4.22). When you see a wetland (swamp), it indicates that the water table is right at the surface. Lakes and streams generally occupy areas where the land surface is below the water table.

Although many factors contribute to the irregular surface of the water table, the most important cause is

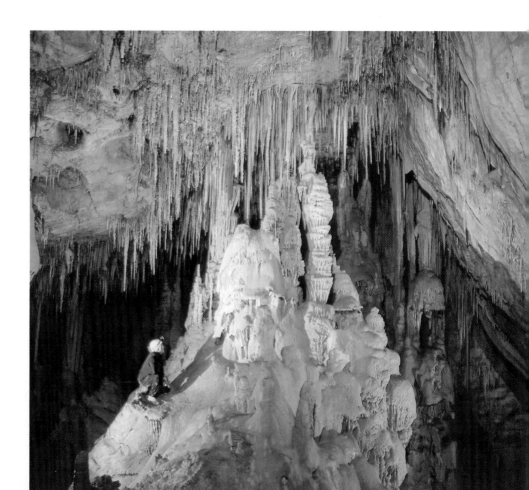

Figure 4.21 The dissolving action of groundwater creates caves. Three Fingers Cave, Lincoln County, New Mexico. (Photo by Harris Photographic/ Tom Stack and Associates)

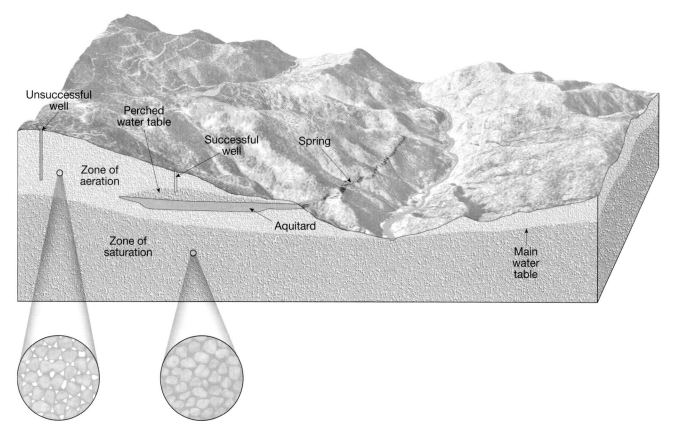

Figure 4.22 This diagram illustrates the relative positions of many features associated with subsurface water.

that groundwater moves very slowly. For this reason, the water tends to "pile up" beneath hills. If rainfall were to cease completely, these water "hills" would slowly subside and gradually approach the level of the valleys. However, new supplies of rainwater are usually added often enough to prevent this. Nevertheless, in times of extended drought, the water table may drop enough to fall below otherwise productive wells.

Movement

Depending on the nature of the subsurface material, the flow of groundwater and the amount of water that can be stored vary greatly. The quantity of groundwater that can be stored depends on the **porosity** of the material; that is, the percentage of the total volume of rock or sediment that consists of pore spaces. Although these openings often consist of spaces between particles of sediment, such features as vesicles (voids left by gases escaping from lava), joints and faults, and cavities formed by the solution of soluble rocks such as limestone are also common.

Rock or sediment may be very porous and still not allow water to move through it. The **permeability** of a material indicates its ability to transmit water through interconnected pore spaces. The smaller the pore spaces, the slower the groundwater moves. If the spaces between particles are too small, water cannot

move at all. Clay exemplifies this circumstance. Although the ability of clay to store water is often high, its pore spaces are so small that water is unable to move through it. Thus, we say that clay is *impermeable*.

Impermeable layers such as clay that hinder or prevent water movement are termed **aquitards**. Larger particles, such as sand or gravel, have larger pore spaces. Therefore, the water can move with relative ease. Such permeable rock strata or sediments that transmit groundwater freely are called **aquifers** ("water carriers"). Aquifers are important because they are the water-bearing layers that well drillers seek. A good aquifer can provide water for numerous wells, homes, cities, and even irrigation and industries.

Springs

 **External Processes
Groundwater**

Springs have aroused the curiosity and wonder of people for thousands of years. The fact that springs were (and to some people still are) rather mysterious phenomena is not difficult to understand, for here is water flowing freely from the ground in all kinds of weather in seemingly inexhaustible supply but with no obvious source. Today we know that the source of springs is water from the zone of saturation and that the ultimate source of this water is precipitation.

Whenever the water table intersects the ground surface, a natural flow of groundwater results, which we call a **spring**. There are many circumstances that create springs. Springs such as the one in Figure 4.23 form when an aquitard blocks the downward movement of groundwater and forces it to move laterally. When the permeable bed (aquifer) outcrops in a valley, a spring or series of springs results.

Another situation that can produce a spring is illustrated in Figure 4.22. Here an aquitard is situated above the main water table. As water percolates downward, a portion accumulates above the aquitard to create a localized zone of saturation and a *perched water table*. Springs, however, are not confined to places where a perched water table creates a flow at the surface. Many geological situations lead to the formation of springs because subsurface conditions vary greatly from place to place.

Hot Springs

By definition, the water in **hot springs** is 6–9°C (10–15°F) warmer than the mean annual air temperature for the localities where they occur. In the United States alone, there are well over 1000 such springs.

Figure 4.23 Spring in Arizona's Marble Canyon. (Photo by Michael Collier)

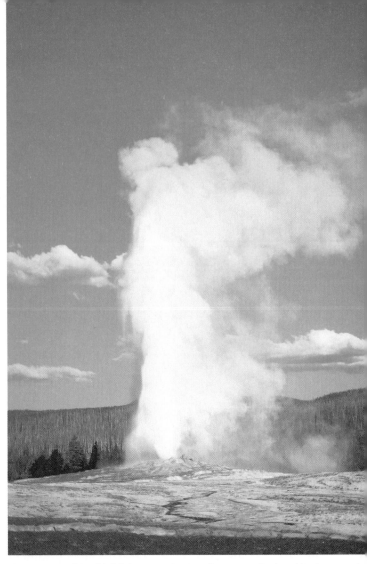

Figure 4.24 Old Faithful, in Wyoming's Yellowstone National Park, one of the world's most famous geysers, emits up to 45,000 liters (almost 12,000 gallons) of hot water and steam about once each hour. (Photo by Kennan Ward/The Stock Market)

Temperatures in deep mines and oil wells usually rise with an increase in depth an average of about 2°C per 100 meters (1°F per 100 feet). Therefore, when groundwater circulates at great depths, it becomes heated, and if it rises to the surface, the water may emerge as a hot spring. The water of some hot springs in the United States, particularly in the East, is heated in this manner. However, the great majority (over 95 percent) of the hot springs (and geysers) in the United States are found in the West. The reason for such a distribution is that the source of heat for most hot springs is cooling igneous rock, and it is in the West that igneous activity has been most recent.

Geysers

Geysers are intermittent hot springs or fountains in which columns of water are ejected with great force at various intervals, often rising 30 to 60 meters (100 to 200 feet; Figure 4.24). After the jet of water ceases, a column of steam rushes out, usually with a thundering roar. Perhaps the most famous geyser in the world is Old Faithful in Yellowstone National Park, which erupts about once each hour. Geysers are also found in other parts of the world, including New Zealand and Iceland. In fact, the Icelandic word *geysa*, to gush, gives us the name *geyser*.

Geysers occur where extensive underground chambers exist within hot igneous rocks. How they operate is shown in Figure 4.25. As relatively cool groundwater enters the chambers, it is heated by the surrounding rock. At the bottom of the chamber, the water is under great pressure because of the weight of the overlying water. This great pressure prevents the water from boiling at the normal surface temperature of 100°C (212°F). For example, at the bottom of a 300-meter (1000-foot) chamber, water must attain a temperature of nearly 230°C (450°F) before it will boil. The heating causes the water to expand, with the result that some is forced out at the surface. This loss of water reduces the pressure on the remaining water in the chamber. The reduced pressure lowers the boiling point and a portion of the water deep within the chamber quickly turns to steam and causes the geyser to erupt. Following the eruption, cool groundwater again seeps into the chamber and the cycle begins anew.

Wells

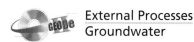

External Processes
Groundwater

The most common method for removing groundwater is the **well**, a hole bored into the zone of saturation (see Figure 4.22). Wells serve as small reservoirs into which groundwater moves and from which it can be pumped to the surface. The use of wells dates back many centuries and continues to be an important method of obtaining water today. By far the single greatest use of this water in the United States is irrigation for agriculture. More than 65 percent of the groundwater used each year is for this purpose. Industrial uses rank a distant second, followed by the amount used in city water systems and rural homes.

The level of the water table may fluctuate considerably during the course of a year, dropping during dry seasons and rising following periods of rain. Therefore, to ensure a continuous supply of water, a well must penetrate far below the water table. Whenever water is withdrawn from a well, the water table around the well is lowered. This effect, termed **drawdown**, decreases with increasing distance from the well. The result is a depression in the water table, roughly conical in shape, known as a **cone of depression** (Figure 4.26). For most small domestic wells, the cone of depression is negligible. However, when wells are used for irrigation or for industrial purposes, the

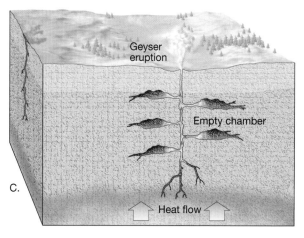

Figure 4.25 Idealized diagrams of a geyser. A geyser can form if the heat is not distributed by convection. **A.** In this figure, the water near the bottom is heated to near its boiling point. The boiling point is higher there than at the surface because the weight of the water above increases the pressure. **B.** The water higher in the geyser system is also heated; therefore, it expands and flows out at the top, reducing the pressure on the water at the bottom. **C.** At the reduced pressure on the bottom, boiling occurs. Some of the bottom water flashes into steam, and the expanding steam causes an eruption.

Figure 4.26 A cone of depression in the water table often forms around a pumping well. If heavy pumping lowers the water table, some wells may be left dry.

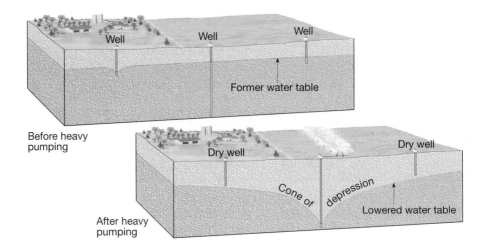

Before heavy pumping

After heavy pumping

withdrawal of water can be great enough to create a very wide and steep cone of depression that may substantially lower the water table in an area and cause nearby shallow wells to become dry. Figure 4.26 illustrates this situation.

Artesian Wells

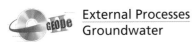

External Processes
Groundwater

In most wells, water does not rise on its own. If water is first encountered at 30 meters of depth, it remains at that level, fluctuating perhaps a meter or two with seasonal wet and dry periods. However, in some wells, water rises, sometimes overflowing at the surface.

The term **artesian** is applied to any situation in which groundwater rises in a well above the level where it was initially encountered. For such a situation to occur, two conditions must exist (Figure 4.27): (1) water must be confined to an aquifer that is inclined so that one end is exposed at the surface, where it can receive water; and (2) aquitards both above and below the aquifer must be present to prevent the water from escaping. When such a layer is tapped, the pressure created by the weight of the water above will force the water to rise. If there were no friction the water in the well would rise to the level of the water at the top of the aquifer. However, friction reduces the height of this pressure surface. The greater the distance from the recharge area (area where water enters the inclined aquifer), the greater the friction and the less the rise of water.

In Figure 4.27, Well 1 is a *nonflowing artesian well*, because at this location the pressure surface is below ground level. When the pressure surface is above the ground and a well is drilled into the aquifer, a *flowing artesian well* is created (Well 2, Figure 4.27).

Artesian systems act as conduits, transmitting water from remote areas of recharge great distances to the points of discharge. In this manner, water that fell in central Wisconsin years ago is now taken from the ground and used by communities many kilometers away in Illinois. In South Dakota, such a system brings water from the Black Hills in the west, eastward across the state.

On a different scale, city water systems may be considered as examples of artificial artesian systems. The water tower, into which water is pumped, may be considered the area of recharge, the pipes the confined aquifer, and the faucets in homes the flowing artesian wells.

Environmental Problems Associated With Groundwater

As with many of our valuable natural resources, groundwater is being exploited at an increasing rate. In some areas, overuse threatens the groundwater supply. In other places, groundwater withdrawal has caused the ground and everything resting upon it to sink. Still other localities are concerned with the possible contamination of their groundwater supply.

Treating Groundwater as a Nonrenewable Resource

For many, groundwater appears to be an endlessly renewable resource, for it is continually replenished by rainfall and melting snow. But in some regions, groundwater has been and continues to be treated as a *nonrenewable* resource. Where this occurs, the amount of water available to recharge the aquifer is significantly less than the amount being withdrawn.

The High Plains provides one example. Here an extensive agricultural economy is largely dependent on irrigation (Figure 4.28). In some parts of the region, where intense irrigation has been practiced for an extended period, depletion of groundwater has been severe. Under these circumstances, it can be said that the groundwater is literally being "mined." Even if

Figure 4.27 Artesian systems occur when an inclined aquifer is surrounded by impermeable beds.

pumping were to cease immediately, it could take hundreds or thousands of years for the groundwater to be fully replenished. Box 4. 3 explores this issue more fully.

Land Subsidence Caused by Groundwater Withdrawal

As you shall see later in this chapter, surface subsidence can result from natural processes related to groundwater. However, the ground may also sink when water is pumped from wells faster than natural recharge processes can replace it. This effect is particularly pronounced in areas underlain by thick layers of loose sediments. As water is withdrawn, the weight of the overburden packs the sediment grains more tightly together and the ground subsides.

Many areas can be used to illustrate land subsidence caused by excessive pumping of groundwater from relatively loose sediment. A classic example in the United States occurred in the San Joaquin Valley of California (Figure 4.29). This important agricultural

region relies heavily on irrigation. Land subsidence due to groundwater withdrawal began in the valley in the mid-1920s and locally exceeded 8 meters (28 feet) by 1970. Then, because of the importation of surface water and a decrease in groundwater pumping, water levels in the aquifer recovered and subsidence ceased. However, during a drought in 1976–1977, heavy groundwater pumping led to renewed subsidence. This time, water levels dropped at a much faster rate than during the previous period because of the reduced storage capacity caused by earlier compaction of material in the aquifer. In all, more than 13,400 square kilometers (5200 square miles) of irrigable land, one-half the entire valley, were affected by subsidence. Damage to structures , including highways, bridges, water lines, and wells, was extensive. Because subsidence changed the gradients of some streams, flooding also became a costly problem. Many other examples of land subsidence due to groundwater pumping occur in the United States and elsewhere in the world.

Figure 4.28 In some agricultural regions water is pumped from the ground faster than it is replenished. In such instances, groundwater is being treated as a nonrenewable resource. (Photo by Steve Welsh/Liaison International)

Figure 4.29 The shaded area on the map shows California's San Joaquin Valley. The marks on the utility pole in the photo indicate the level of the surrounding land in preceding years. Between 1925 and 1975 this part of the San Joaquin Valley subsided almost 9 meters because of the withdrawal of groundwater and the resulting compaction of sediments. (Photo courtesy of U.S. Geological Survey)

Groundwater Contamination

The pollution of groundwater is a serious matter, particularly in areas where aquifers provide a large part of the water supply. One common source of groundwater pollution is sewage. Its sources include an ever-increasing number of septic tanks as well as inadequate or broken sewer systems, and farm wastes.

If sewage water that is contaminated with bacteria enters the groundwater system it may become purified through natural processes. The harmful bacteria can be mechanically filtered by the sediment through which the water percolates, destroyed by chemical oxidation, and/or assimilated by other organisms. For purification to occur, however, the aquifer must be of the correct composition.

For example, extremely permeable aquifers (such as highly fractured crystalline rock, coarse gravel, or cavernous limestone) have such large openings that contaminated groundwater may travel long distances without being cleansed. In this case, the water flows too rapidly and is not in contact with the surrounding material long enough for purification to occur. This is the problem at Well 1 in Figure 4.30A.

Conversely, when the aquifer is composed of sand or permeable sandstone, the water can sometimes be purified after traveling only a few dozen meters through it. The openings between sand grains are large enough to permit water movement, yet the movement of the water is slow enough to allow ample time for its purification (Well 2, Figure 4.30B).

Other sources and types of contamination also threaten groundwater supplies (Figure 4.31). These include widely used substances such as highway salt, fertilizers that are spread across the land surface, and pesticides. In addition, a wide array of chemicals and industrial materials may leak from pipelines, storage tanks, landfills, and holding ponds. Some of these pollutants are classified as *hazardous*, meaning that they are either flammable, corrosive, explosive, or toxic. As rainwater oozes through the refuse, it may dissolve a variety of potential contaminants. If the leached material reaches the water table, it will mix with the groundwater and contaminate the supply.

Because groundwater movement is usually slow, polluted water might go undetected for a long time. In fact, most contamination is discovered only after drinking water has been affected and people become ill. By this time, the volume of polluted water might be very large, and even if the source of contamination is removed immediately, the problem is not solved. Although the sources of groundwater contamination are numerous, the solutions are relatively few.

Once the source of the problem has been identified and eliminated, the most common practice is simply to abandon the water supply and allow the

Box 4.3   People and the Environment: The Ogallala Aquifer— How Long Will the Water Last?

The High Plains extend from the western Dakotas south to Texas. Despite being a land of little rain, this is one of the most important agricultural regions in the United States. The reason is a vast endowment of groundwater that makes irrigation possible throughout most of the region's 450,000 square kilometers (175,000 square miles). The source of most of this water is the Ogallala Formation, the largest aquifer in the United States (Figure 4.D).

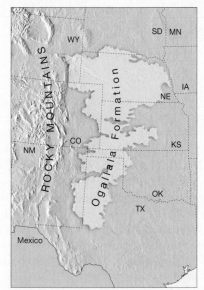

Figure 4.D The Ogallala Formation underlies about 450,000 square kilometers (175,000 square miles) of the High Plains, making it the largest aquifer in the United States.

Geologically, the Ogallala Formation consists of a number of sandy and gravelly rock layers. The sediments were derived from the erosion of the Rocky Mountains and carried eastward by sluggish streams. Erosion has removed much of the formation from eastern Colorado, severing the Ogallala's connection to the Rockies.

The Ogallala Formation averages 60 meters (200 feet) thick and in places it is as much as 180 meters (600 feet) thick. Groundwater in the aquifer originally came downslope from the Rockies as well as from surface precipitation that soaked into the ground over thousands of years. Because of its high porosity and great size, the Ogallala accumulated huge amounts of groundwater—enough freshwater to fill Lake Huron! Today, with the connection between the aquifer and the Rockies severed, all of the Ogallala's recharge must come from the region's meager rainfall.

The Ogallala aquifer was first used for irrigation in the late 1800s, but its use was limited by the capacity of the pumps available at the time. In the 1920s, with the development of large-capacity irrigation pumps, High Plains farmers, particularly in Texas, began tapping the Ogallala for irrigation. Then in the 1950s, improved technology brought large-scale exploitation of the aquifer. Today, nearly 170,000 wells are being used to irrigate more than 65,000 square kilometers (16 million acres) of land.

With the increase in irrigation has come a drastic drop in the Ogallala's water table, especially in the southern High Plains. Declines in the water table of 3 to 15 meters (10 to 50 feet) are common, and in places the water table is now 60 meters (200 feet) below its level prior to irrigation.

Since the 1980s, irrigated acreage has declined in the High Plains. One important reason has been higher energy costs. As water levels have dropped, the costs of pumping groundwater to the surface have increased.

Although the water table decline has slowed in parts of the southern High Plains, substantial pumping continues, often in excess of recharge. The future of irrigated farming in this region is clearly in jeopardy.

> The southern High Plains represent an area of the United States that will return, sooner or later, to dry-land farming. The transition will come sooner and with fewer ecological and economic crises if the agricultural industry is weaned gradually from its dependence on groundwater irrigation. If nothing is done until all the accessible water in the Ogallala aquifer has been removed, the transition will be ecologically dangerous and economically dreadful.[*]

[*]National Research Council. *Solid-Earth Sciences and Society.* Washington, DC: National Academy Press, 1993, p. 148.

pollutants to be flushed away gradually. This is the least costly and easiest solution, but the aquifer must remain unused for many years. To accelerate this process, polluted water is sometimes pumped out and treated. Following removal of the tainted water, the aquifer is allowed to recharge naturally or, in some cases, the treated water or other freshwater is pumped back in. This process is costly, time consuming, and it may be risky because there is no way to be certain that all of the contamination has been removed. Clearly, the most effective solution to groundwater contamination is prevention.

The Geologic Work of Groundwater

Groundwater dissolves rock. This fact is key to understanding how caverns and sinkholes form. Because soluble rocks, especially limestone, underlie millions of square kilometers of Earth's surface, it is here that groundwater carries on its important role as an erosional agent. Limestone is nearly insoluble in pure water, but is quite easily dissolved by water containing small quantities of carbonic acid. Most natural water contains this weak acid because rainwater readily dissolves carbon

Figure 4.30 A. Although the contaminated water has traveled more than 100 meters before reaching Well 1, the water moves too rapidly through the cavernous limestone to be purified. **B.** As the discharge from the septic tank percolates through the permeable sandstone, it is purified in a relatively short distance.

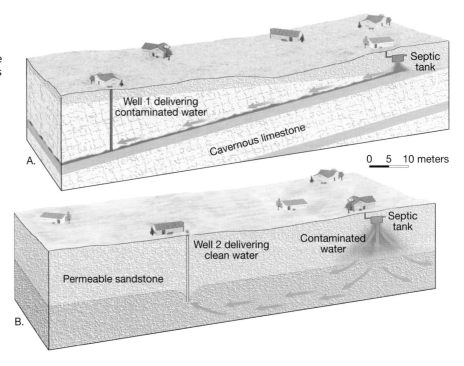

dioxide from the air and from decaying plants. Therefore, when groundwater comes in contact with limestone, the carbonic acid reacts with calcite in the rocks to form calcium bicarbonate, a soluble material that is then carried away in solution.

Caverns

The most spectacular results of groundwater's erosional handiwork are limestone **caverns**. In the United States alone about 17,000 caves have been discovered.

Although most are relatively small, some have spectacular dimensions. Carlsbad Caverns in southeastern New Mexico and Mammoth Cave in Kentucky are famous examples. One chamber in Carlsbad Caverns has an area equivalent to 14 football fields and enough height to accommodate the U.S. Capitol Building. At Mammoth Cave, the total length of interconnected caverns extends for more than 540 kilometers (340 miles).

Most caverns are created at or below the water table in the zone of saturation. Here acidic groundwater follows lines of weakness in the rock, such as joints

A.   **B.**

Figure 4.31 Sometimes agricultural chemicals **A.** and materials leached from landfills **B.** find their way into the groundwater. These are two of the potential sources of groundwater contamination. (Photo A by Roy Morsch/The Stock Market; Photo B by F. Rossotto/The Stock Market)

Of the various dripstone features found in caverns, perhaps the most familiar are **stalactites**. These iciclelike pendants hang from the ceiling of the cavern and form where water seeps through cracks above. When water reaches air in the cave, some of the dissolved carbon dioxide escapes from the drop and calcite begins to precipitate. Deposition occurs as a ring around the edge of the water drop. As drop after drop follows, each leaves an infinitesimal trace of calcite behind, and a hollow limestone tube is created. Water then moves through the tube, remains suspended momentarily at the end, contributes a tiny ring of calcite, and falls to the cavern floor. The stalactite just described is appropriately called a soda straw (Figure 4.33). Often the hollow tube of the *soda straw* becomes plugged or its supply of water increases. In either case, the water is forced to flow and deposit along the outside of the tube. As deposition continues, the stalactite takes on the more common conical shape.

Figure 4.33 "Live" solitary soda straw stalactites. Lehman Caves, Great Basin National Park, Nevada. (Photo by Tom Bean)

Figure 4.32 The dripstone features found in caves are of many types, including stalactites, stalagmites, and columns. Lehman Caves, Great Basin National Park, Nevada. (Photo by Galen Rowell)

and bedding planes. As time passes, the dissolving process slowly creates cavities and gradually enlarges them into caverns. Material that is dissolved by the groundwater is eventually discharged into streams and carried to the ocean.

Certainly the features that arouse the greatest curiosity for most cavern visitors are the stone formations that give some caverns a wonderland appearance (Figure 4.32). These are not erosional features, like the caverns in which they reside, but depositional features. They are created by the seemingly endless dripping of water over great spans of time. The calcite that is left behind produces the limestone we call *travertine*. These cave deposits, however, are also commonly called *dripstone*, an obvious reference to their mode of origin.

Although the formation of caverns takes place in the zone of saturation, the deposition of dripstone is not possible until the caverns are above the water table, in the zone of aeration. This commonly occurs as nearby streams cut their valleys deeper, lowering the water table as the elevation of the rivers drops. As soon as the chamber is filled with air, the conditions are right for the decoration phase of cavern building to begin.

Formations that develop on the floor of a cavern and reach upward toward the ceiling are called **stalagmites**. The water supplying the calcite for stalagmite growth falls from the ceiling and splatters over the surface. As a result, stalagmites do not have a central tube and are usually more massive in appearance and more rounded on their upper ends than stalactites. Given enough time, a downward-growing stalactite and an upward-growing stalagmite may join to form a *column*.

Karst Topography

Many areas of the world have landscapes that, to a large extent, have been shaped by the dissolving power of groundwater. Such areas are said to exhibit **karst topography**, named for the *Krs* region in Slovenia (formerly a part of Yugoslavia) where such topography is strikingly developed. In the United States, karst landscapes occur in many areas that are underlain by limestone, including portions of Kentucky, Tennessee, Alabama, southern Indiana, and central and northern Florida. Generally, arid and semiarid areas do not develop karst topography because there is insufficient groundwater. When solution features exist in such regions, they are likely to be remnants of a time when rainfall was more abundant.

Karst areas typically have irregular terrain punctuated with many depressions called **sinkholes** or, simply, **sinks**. In the limestone areas of Florida, Kentucky, and southern Indiana, there are literally tens of thousands of these depressions varying in depth from just a meter or two to a maximum of more than 50 meters.

Sinkholes commonly form in one of two ways. Some develop gradually over many years without any physical disturbance to the rock. In these situations, the limestone immediately below the soil is dissolved by downward-seeping rainwater that is freshly charged with carbon dioxide. These depressions are usually not deep and are characterized by relatively gentle slopes. By contrast, sinkholes can also form suddenly and without warning when the roof of a cavern collapses under its own weight. Typically, the depressions created in this manner are steep-sided and deep. When they form in populous areas, they may represent a serious geologic hazard (Figure 4.34 and chapter-opening photo).

In addition to a surface pockmarked by sinkholes, karst regions characteristically show a striking lack of surface drainage (streams). Following a rainfall, runoff is quickly funneled below ground through sinks. It then flows through caverns until it finally reaches the water table. Where streams do exist at the surface, their paths are usually short. The names of such streams often give a clue to their fate. In the Mammoth Cave area of Kentucky, for example, there is Sinking Creek, Little Sinking Creek , and Sinking Branch. Some sinkholes become plugged with clay and debris, creating small lakes or ponds.

Figure 4.34 The Winter Park, Florida, sinkhole that grew rapidly for 3 days, swallowing part of a community swimming pool as well as several businesses, houses, and automobiles. (Photo by Leif Skoogfors/Woodfin Camp and Associates)

Chapter Summary

- The *hydrologic cycle* describes the continuous interchange of water among the oceans, atmosphere, and continents. Powered by energy from the Sun, it is a global system in which the atmosphere provides the link between the oceans and continents. The processes involved in the hydrologic cycle include *precipitation, evaporation, infiltration* (the movement of water into rocks or soil through cracks and pore spaces), *runoff* (water that flows over the land, rather than infiltrating into the ground), and *transpiration* (the release of water vapor to the atmosphere by plants). *Running water is the single most important agent sculpturing Earth's land surface.*

- The factors that determine a stream's *velocity* are *gradient* (slope of the stream channel), *shape, size* and *roughness* of the channel, and the stream's *discharge* (amount of water passing a given point per unit of time, frequently measured in cubic feet per second). Most often, the gradient and roughness of a stream decrease downstream, while width, depth, discharge, and velocity increase.

- The two general types of *base level* (the lowest point to which a stream may erode its channel) are (1) *ultimate base level* and (2) *temporary*, or *local base level*. Any change in base level will cause a stream to adjust and establish a new balance. Lowering base level will cause a stream to downcut, whereas raising base level results in deposition of material in the channel.

- The work of a stream includes *erosion* (the incorporation of material), *transportation* (as *dissolved load, suspended load*, and *bed load*), and whenever a stream's velocity decreases, *deposition*.

- Although many gradations exist, the two general types of stream valleys are (1) *narrow V-shaped valleys* and (2) *wide valleys with flat floors*. Because the dominant activity is downcutting toward base level, narrow valleys often contain *waterfalls* and *rapids*. When a stream has cut its channel closer to base level, its energy is directed from side to side, and erosion produces a flat valley floor, or *floodplain*. Streams that flow upon floodplains often move in sweeping bends called *meanders*. Widespread meandering may result in shorter channel segments, called *cutoffs*, and/or abandoned bends, called *oxbow lakes*.

- *Floods* are triggered by heavy rains and/or snowmelt. Sometimes human interference can worsen or even cause floods. Flood-control measures include the building of *artificial levees* and dams, as well as *channelization*, which could involve creating *artificial cutoffs*. Many sci-

entists and engineers advocate a nonstructural approach to flood control that involves more appropriate land use.

- The land area that contributes water to a stream is called a *drainage basin*. Drainage basins are separated by an imaginary line called a *divide*. Common *drainage patterns* produced by streams include (1) *dendritic*, (2) *radial*, (3) *rectangular*, and (4) *trellis*.

- As a resource, *groundwater* represents the largest reservoir of freshwater that is readily available to humans. Geologically, the dissolving action of groundwater produces *caves* and *sinkholes*. Groundwater is also an equalizer of stream flow.

- Groundwater is that water which occupies the pore spaces in sediment and rock in a zone beneath the surface called the *zone of saturation*. The upper limit of this zone is the *water table*. The *zone of aeration* is above the water table where the soil, sediment, and rock are not saturated. The quantity of water that can be stored depends on the *porosity* (the volume of open spaces) of the material. However, the *permeability* (the ability to transmit a fluid through interconnected pore spaces) of a material is the primary factor controlling the movement of groundwater.

- Materials with very small pore spaces (such as clay) hinder or prevent groundwater movement and are called *aquitards*. *Aquifers* consist of materials with larger pore spaces (such as sand) that are permeable and transmit groundwater freely.

- *Springs* occur whenever the water table intersects the land surface and a natural flow of groundwater results. *Wells*, openings drilled into the zone of saturation, withdraw groundwater and create roughly conical depressions in the water table known as *cones of depression*. *Artesian wells* occur when water rises above the level at which it was initially encountered.

- Some of the current environmental problems involving groundwater include (1) *overuse* by intense irrigation, (2) *land subsidence* caused by groundwater withdrawal, and (3) *contamination* by pollutants.

- Most *caverns* form in limestone at or below the water table when acidic groundwater dissolves rock along lines of weakness, such as joints and bedding planes. *Karst topography* exhibits an irregular terrain punctuated with many depressions, called *sinkholes*.

Key Terms

alluvium (p. 98)
aquifer (p. 109)
aquitard (p. 109)
artesian well (p. 112)
backswamp (p. 100)
base level (p. 95)
bed load (p. 95)
capacity (p. 97)
cavern (p. 116)
competence (p. 97)
cone of depression (p. 111)
cutoff (p. 103)
delta (p. 98)
dendritic pattern (p. 105)
discharge (p. 92)
dissolved load (p. 95)
distributary (p. 98)

divide (p. 105)
drainage basin (p. 105)
drawdown (p. 111)
flood (p. 103)
floodplain (p. 101)
geyser (p. 110)
gradient (p. 92)
groundwater (p. 108)
hot spring (p. 110)
hydrologic cycle (p. 90)
infiltration (p. 90)
karst topography (p. 118)
meander (p. 103)
natural levee (p. 100)
oxbow lake (p. 103)
permeability (p. 109)
porosity (p. 109)

radial pattern (p. 105)
rectangular pattern (p. 105)
runoff (p. 90)
sinkhole (sink) (p. 118)
sorting (p. 98)
spring (p. 110)
stalactite (p. 117)
stalagmite (p. 118)
suspended load (p. 95)
transpiration (p. 90)
trellis pattern (p. 105)
water table (p. 108)
well (p. 111)
yazoo tributary (p. 100)
zone of aeration (p. 108)
zone of saturation (p. 108)

Review Questions

1. Once precipitation has fallen on land, what paths are available to it?

2. A stream starts out 2000 meters above sea level and travels 250 kilometers to the ocean. What is its average gradient in meters per kilometer?

3. Suppose that the stream mentioned in Question 2 developed extensive meanders so that its course was lengthened to 500 kilometers. Calculate its new gradient. How does meandering affect gradient?

4. When the discharge of a stream increases, what happens to the stream's velocity?

5. Define base level. Name the main river in your area. For what streams does it act as base level? What is the base level for the Mississippi River? The Missouri River?

6. When an area changes from being predominantly rural to largely urban, how is streamflow affected? (See Box 4.1.)

7. In what three ways does a stream transport its load?

8. If you collect a jar of water from a stream, what part of its load will settle to the bottom of the jar? What portion will remain in the water?

9. Differentiate between competency and capacity.

10. Briefly describe the formation of a natural levee. How is this feature related to backswamps and yazoo tributaries?

11. List and briefly describe three basic flood-control strategies. What are some drawbacks of each?

12. Each of the following statements refers to a particular drainage pattern. Identify the pattern. (a) Streams diverge from a central high area such as a volcano. (b) Streams form a branching, "treelike" pattern. (c) A pat-

tern that develops when bedrock is crisscrossed by joints and faults.

13. What percentage of freshwater is groundwater (see Table 4.2)? If glacial ice is excluded and only liquid freshwater is considered, about what percentage is groundwater?

14. Geologically, groundwater is important as an erosional agent. Name another significant geological role of groundwater.

15. Define groundwater and relate it to the water table.

16. How do porosity and permeability differ?

17. What is the source of heat for most hot springs and geysers? How is this reflected in the distribution of these features?

18. What is meant by the term *artesian*? In order for artesian wells to exist, two conditions must be present. List these conditions.

19. What problem is associated with the pumping of groundwater for irrigation in the southern part of the High Plains? (See Box 4.3.)

20. Briefly explain what happened in the San Joaquin Valley of California as the result of excessive groundwater withdrawal.

21. Which would be most effective in purifying polluted groundwater: an aquifer composed mainly of coarse gravel, or sand, or cavernous limestone?

22. Differentiate between stalactites and stalagmites. How do these features form?

23. Describe two ways in which sinkholes are created.

Examining the Earth System

1. List the process(es) involved in moving water through the hydrologic cycle from the (a) hydrosphere to the atmosphere, (b) atmosphere to the solid Earth, (c) biosphere to the atmosphere, and (d) hydrosphere (land) to the hydrosphere (ocean). Exploring the hydrology and hydrogeology links at this University of Texas Austin site, http://www.geo.utexas.edu/bexton/jge/hydrology.htm, may help you organize your responses.

2. Over the oceans, evaporation exceeds precipitation, yet sea level does not drop. Can you explain why?

3. List at least three specific examples of interactions between humans and the hydrologic cycle (for example, the construction of a dam). Briefly describe the consequence(s) of each of these interactions.

4. Describe the role of the atmosphere, solid Earth, and biosphere in determining the quantity and quality of groundwater available for human consumption.

Web Resources

The following are informative and interesting World Wide Web sites that address topics related to those presented in the chapter:

Water Resources of the United States (United States Geological Survey)

• **http://h2o.usgs.gov/**

National Ground Water Association

• **http://www.h2o-ngwa.org/**

For direct links to these sites and other appropriate Earth science Web locations, chapter review exercises, and relevant, topical investigations that utilize Web resources, contact the Earth Science, 9th Edition Home Page at

• **http://www.prenhall.com/tarbuck.**

Chapter 5
Glaciers, Deserts, and Wind

Desert dunes are closing in on this agricultural land in Egypt. (Photo by Georg Gerster/Photo Researchers, Inc.)

Today, glaciers cover nearly 10 percent of Earth's land surface; however, in the recent geologic past ice sheets were three times more extensive, covering vast areas with ice thousands of meters thick. Many regions still bear the mark of these glaciers. The first part of this chapter examines glaciers and the erosional and depositional features they create. The second part is devoted to dry lands and the geologic work of wind. Because desert and near-desert conditions prevail over an area as large as that affected by the massive glaciers of the Ice Age, the nature of such landscapes is indeed worth investigating.

Glaciers: A Part of Two Basic Cycles

External Processes
Glaciers

Many present-day landscapes were modified by the widespread glaciers of the most recent Ice Age and still strongly reflect the handiwork of ice. The basic character of such diverse places as the Alps, Cape Cod, and Yosemite Valley was fashioned by now-vanished masses of glacial ice. Moreover, Long Island, the Great Lakes, and the fiords of Norway and Alaska all owe their existence to glaciers. Glaciers, of course, are not just a phenomenon of the geologic past. As we shall see, they are still sculpting and depositing debris in many regions today.

Glaciers are a part of two fundamental cycles in the Earth system—the hydrologic cycle and the rock cycle (see Figure I.6, p. 5). Earlier we learned that the water of the hydrosphere is constantly cycled through the atmosphere, biosphere, and solid Earth. Time and time again the same water is evaporated from the oceans into the atmosphere, precipitated upon the land, and carried by rivers and underground flow back to the sea. However, when precipitation falls at high elevations or high latitudes, the water may not immediately make its way toward the sea. Instead, it may become part of a glacier. Although the ice will eventually melt, allowing the water to continue its path to the sea, water can be stored as glacial ice for many tens, hundreds, or even thousands of years.

Figure 5.1 Kennicott Glacier, in Alaska's Wrangell-St. Elias National Park, is an example of a valley glacier. (Photo by Tom Bean)

A **glacier** is a thick ice mass that forms over hundreds or thousands of years. It originates on land from the accumulation, compaction, and recrystallization of snow. A glacier appears to be motionless, but it is not—glaciers move very slowly. Like running water, groundwater, wind, and waves, glaciers are dynamic erosional agents that accumulate, transport, and deposit sediment. As such, glaciers are among the processes that perform a very basic function in the rock cycle. Although glaciers are found in many parts of the world, most are located in remote areas, either near Earth's poles or in high mountains.

Literally thousands of relatively small glaciers exist in lofty mountain areas, where they usually follow valleys originally occupied by streams. Unlike the rivers that previously flowed in these valleys, the glaciers advance slowly, perhaps only a few centimeters each day. Because of their setting, these moving ice masses are termed **valley glaciers** or **alpine glaciers** (Figure 5.1). Each glacier is a stream of ice, bounded by precipitous rock walls, that flows downvalley from a snow accumulation center near its head. Like rivers, valley glaciers can be long or short, wide or narrow, single or with branching tributaries. Generally the widths of alpine glaciers are small compared to their lengths. In length, some extend for just a fraction of a kilometer, whereas others go on for many dozens of kilometers. The west branch of the Hubbard Glacier, for example, runs through 112 kilometers of mountainous terrain in Alaska and the Yukon Territory.

In contrast to valley glaciers, **ice sheets** exist on a *much* larger scale. These enormous masses flow out in all directions from one or more centers and completely obscure all but the highest areas of underlying terrain. Although many ice sheets have existed in the past, just two achieve this status at present (Figure 5.2). In the Northern Hemisphere, Greenland is covered by an imposing ice sheet averaging nearly 1500 meters thick. It occupies 1.7 million square kilometers, or about 80 percent of this large island. In the Southern Hemisphere, the huge Antarctic Ice Sheet attains a maximum thickness of nearly 4300 meters and covers an area of more than 13.9 million square kilometers. Because of the proportions of these huge features, they often are called *continental ice sheets.* Indeed, the combined areas of present-day continental ice sheets represent almost 10 percent of Earth's land area.

In addition to valley glaciers and ice sheets, other types of glaciers are also identified. Covering some uplands and plateaus are masses of glacial ice called **ice caps.** They resemble ice sheets but are much smaller than the continental-scale features. Ice caps occur in many places, including Iceland and several of the large islands in the Arctic Ocean. Another type, known as

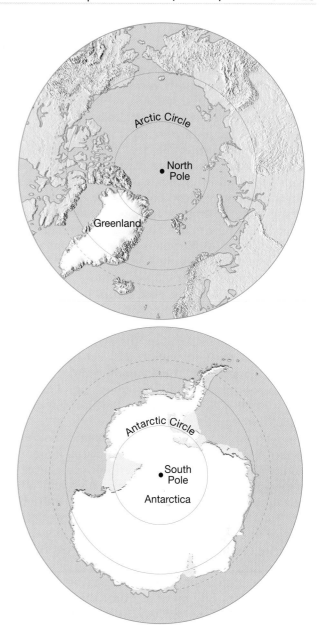

Figure 5.2 The only present-day continental ice sheets are those covering Greenland and Antarctica. Their combined areas represent almost 10 percent of Earth's land area. Greenland's ice sheet occupies 1.7 million square kilometers, or about 80 percent of the island. The area of the Antarctic Ice Sheet is almost 14 million square kilometers.

piedmont glaciers, occupy broad lowlands at the bases of steep mountains and form when one or more valley glaciers emerge from the confining walls of mountain valleys. Here the advancing ice spreads out to form a broad sheet. The size of individual piedmont glaciers varies greatly. Among the largest is the broad Malaspina Glacier along the coast of southern Alaska. It covers more than 5000 square kilometers of the flat coastal plain at the foot of the lofty St. Elias range.

How Glaciers Move

The movement of glacial ice is generally referred to as *flow*. The fact that glacial movement is described in this way seems paradoxical—how can a solid flow? Glacial ice flows in two ways. The first of these, *plastic flow*, involves movement *within* the ice. Ice behaves as a brittle solid until the pressure upon it is equivalent to the weight of about 50 meters (165 feet) of ice. Once that load is surpassed, ice behaves as a plastic material and flow begins. A second and often equally important mechanism of glacial movement consists of the entire ice mass slipping along the ground. The lowest portions of most glaciers are thought to move by this sliding process.

The uppermost 50 meters of a glacier are appropriately referred to as the *zone of fracture*. Because there is not enough overlying ice to cause plastic flow, this upper part of the glacier consists of brittle ice. Consequently, the ice in this zone is carried along piggyback style by the ice below. When the glacier moves over irregular terrain, the zone of fracture is subjected to tension, with cracks called **crevasses** resulting (Figure 5.3). These gaping cracks can make travel across glaciers dangerous and can extend to depths of 50 meters (165 feet). Beyond this depth, plastic flow seals them off.

Unlike streamflow, glacial movement is not obvious (see Box 5.1). How rapidly does glacial ice move? Average rates vary considerably from one glacier to another. Some move so slowly that trees and other vegetation may become well established in the debris that accumulates on the glacier's surface. Others advance up to several meters per day. The movement of some glaciers is characterized by periods of extremely rapid advance followed by periods during which movement is practically nonexistent.

Snow is the raw material from which glacial ice originates. Therefore, glaciers form in areas where more snow falls in winter than can melt during the summer. Glaciers are constantly gaining and losing ice. Snow accumulation and ice formation occur in the **zone of accumulation** (Figure 5.4). Here the addition of snow thickens the glacier and promotes movement. Beyond this area of ice formation is the **zone of wastage**. Here there is a net loss to the glacier as all of the snow from the previous winter melts, as does some of the glacial ice (Figure 5.4).

In addition to melting, glaciers also waste as large pieces of ice break off the front of a glacier in a process called *calving*. Where glaciers reach the sea, calving creates *icebergs* (Figure 5.5). Because icebergs are just slightly less dense than seawater, they float very low in the water, with more than 80 percent of their mass submerged. The margins of the Greenland Ice Sheet produce thousands of icebergs each year. Many drift southward and find their way into the North Atlantic, where they are a hazard to navigation.

Figure 5.3 Crevasses form in the brittle ice of the zone of fracture. They can extend to depths of 50 meters and can obviously make travel across glaciers dangerous. Gasherbrum II expedition, Pakistan. (Photo by Galen Rowell/Mountain Light Photography, Inc.)

Whether the margin of a glacier is advancing, retreating, or remaining stationary depends on the *budget* of the glacier. The glacial budget is the balance or lack of balance between accumulation at the upper end of a glacier and loss at the lower end. If ice accumulation exceeds wastage, the glacial front advances until the two factors balance. At this point, the terminus of the glacier becomes stationary.

If a warming trend increases wastage and/or if a drop in snowfall decreases accumulation, the ice front will retreat. As the terminus of the glacier retreats, the extent of the zone of wastage diminishes. Therefore, in time a new balance will be reached between accumulation and wastage, and the ice front will again become stationary.

Whether the margin of a glacier is advancing, retreating, or stationary, the ice within the glacier

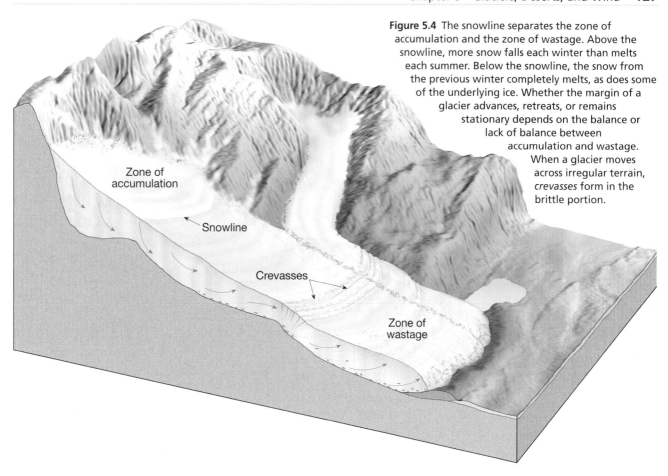

Figure 5.4 The snowline separates the zone of accumulation and the zone of wastage. Above the snowline, more snow falls each winter than melts each summer. Below the snowline, the snow from the previous winter completely melts, as does some of the underlying ice. Whether the margin of a glacier advances, retreats, or remains stationary depends on the balance or lack of balance between accumulation and wastage. When a glacier moves across irregular terrain, *crevasses* form in the brittle portion.

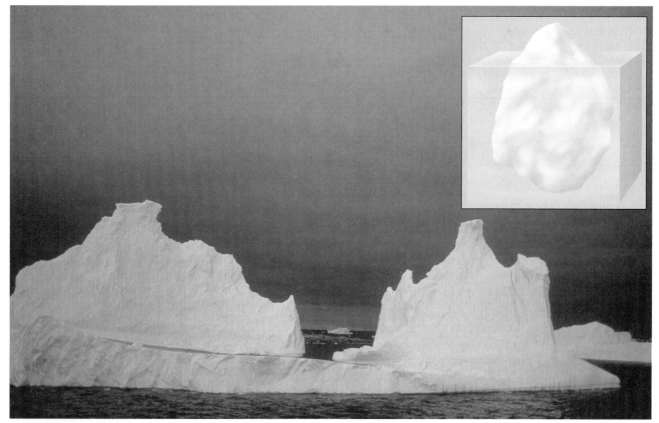

Figure 5.5 Icebergs are created when large pieces break off the front of a glacier after it reaches a water body. These formed in Antarctica's Paradise Bay. (Photo by Galen Rowell/Mountain Light Photography, Inc.)

Box 5.1   Do Glaciers Move? An Application of the Scientific Method

The study of glaciers provides an early application of the scientific method. High in the Alps of Switzerland and France, small glaciers exist in the upper portions of some valleys. In the late eighteenth and early nineteenth centuries, people who farmed and herded animals in these valleys suggested that glaciers in the upper reaches of the valleys had previously been much larger and had occupied downvalley areas. They based their belief on the fact that the valley floors were littered with angular boulders and other rock debris that seemed identical to the materials that they could see in and near the glaciers at the heads of the valleys.

Although the explanation of these observations seemed logical, others did not accept the notion that masses of ice hundreds of meters thick were capable of movement. The disagreement was settled after a simple experiment was designed and carried out to test the hypothesis that glacial ice can move.

Markers were placed in a straight line completely across an alpine glacier. The position of the line was marked on the valley walls so that if the ice moved, the change in position could be detected. After a year or two the results were clear: The markers on the glacier had advanced down the valley, proving that glacial ice indeed moves. In addition, the experiment demonstrated that ice within a glacier does not move

Figure 5.A Ice movement and changes in the terminus at Rhone Glacier, Switzerland. In this classic study of a valley glacier, the movement of stakes clearly showed that ice along the sides of the glacier moves slowest. Also notice that even though the ice front was retreating, the ice within the glacier was advancing.

at a uniform rate, because the markers in the center advanced farther than did those along the margins. Although most glaciers move too slowly for direct visual detection, the experiment succeeded in demonstrating that movement nevertheless occurs. In the years that followed, this experiment was repeated many times with greater accuracy using more modern surveying techniques. Each time, the basic relationships established by earlier attempts were verified.

The experiment illustrated in Figure 5.A was carried out at Switzerland's Rhone Glacier later in the nineteenth century. It not only traced the movement of markers within the ice, but also mapped the position of the glacier's terminus. Notice that even though the ice within the glacier was advancing, the ice front was retreating. As often occurs in science, experiments and observations designed to test one hypothesis yield new information that requires further analysis and explanation.

continues to flow forward. In the case of a receding glacier, the ice still flows forward, but not rapidly enough to offset wastage. This point is illustrated in Figure 5.A in Box 5.1. As the line of stakes within the Rhone Glacier continued to move downvalley, the terminus of the glacier slowly retreated upvalley.

Glacial Erosion

External Processes
Glaciers

Glaciers are capable of great erosion. For anyone who has observed the terminus of an alpine glacier, the evidence of its erosive force is clear. You can witness

firsthand the release of rock fragments of various sizes from the ice as it melts. All signs lead to the conclusion that the ice has scraped, scoured, and torn rock debris from the floor and walls of the valley and carried it downvalley. Indeed, as a transporter of sediment, ice has no equal.

Once rock debris is acquired by the glacier, it cannot settle out as does the load carried by a stream or by the wind. Consequently, glaciers can carry huge blocks that no other erosional agent could possibly budge. Although glaciers are of limited importance as erosional agents today, many landscapes that were modified by the widespread glaciers of the recent Ice Age still reflect to a high degree the work of ice.

How Glaciers Erode

Glaciers erode land primarily in two ways—plucking and abrasion. First, as a glacier flows over a fractured bedrock surface, it loosens and lifts blocks of rock and incorporates them into the ice. This process, known as **plucking,** occurs when meltwater penetrates the cracks and joints along the rock floor of the glacier and refreezes. As the water expands, it exerts tremendous leverage that pries the rock loose. In this manner, sediment of all sizes becomes part of the glacier's load.

The second major erosional process is **abrasion.** As the ice and its load of rock fragments slide over bedrock, they function like sandpaper to smooth and polish the surface below. The pulverized rock produced by the glacial "grist mill" is appropriately called **rock flour.** So much rock flour may be produced that meltwater streams leaving a glacier often have the grayish appearance of skim milk—visible evidence of the grinding power of the ice.

When the ice at the bottom of a glacier contains large fragments, long scratches and grooves called **glacial striations** may be gouged into the bedrock (Figure 5.6). These linear scratches on the bedrock surface provide clues to the direction of glacial movement. By mapping the striations over large areas, glacial flow patterns can often be reconstructed. Conversely, not all abrasive action produces striations. The rock surface over which the glacier moves may also become highly polished by the ice and its load of finer particles.

As is the case with other agents of erosion, the rate of glacial erosion is highly variable. This differential erosion by ice is largely controlled by four factors: (1) rate of glacial movement; (2) thickness of the ice; (3) shape, abundance, and hardness of the rock fragments contained in the ice at the base of the glacier; and (4) the erodibility of the surface beneath the glacier. Variations in any or all of these factors from time to time and/or from place to place mean that the features, effects, and degree of landscape modification in glaciated regions can vary greatly.

Landforms Created by Glacial Erosion

Although the erosional accomplishments of ice sheets can be tremendous, landforms carved by these huge ice masses usually do not inspire the same awe as do the erosional features created by valley glaciers. In regions where the erosional effects of ice sheets are significant, glacially scoured surfaces and subdued terrain are the rule. By contrast, in mountainous areas, erosion by valley glaciers produces many truly spectacular features. Much of the rugged mountain scenery so celebrated for its majestic beauty is the product of erosion by valley glaciers.

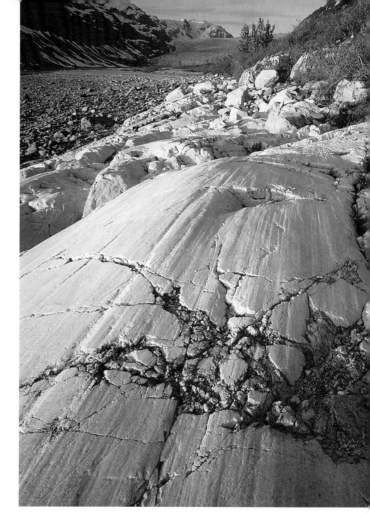

Figure 5.6 Glacial abrasion created the scratches and grooves in this bedrock. Glacier Bay National Park, Alaska. (Photo by Carr Clifton)

Take a moment to study Figure 5.7, which shows a mountain setting before, during, and after glaciation. You will refer to this often in the following discussion.

Prior to glaciation, alpine valleys are characteristically V-shaped because streams are well above base level and are therefore downcutting (Figure 5.7A). However, in mountainous regions that have been glaciated, the valleys are no longer narrow. As a glacier moves down a valley once occupied by a stream, the ice modifies it in three ways: The glacier widens, deepens, and straightens the valley, so that what was once a narrow V-shaped valley is transformed into a U-shaped **glacial trough** (Figures 5.7C and 5.8).

Because the amount of glacial erosion depends in part on the thickness of the ice, the main or trunk glacier cuts its valley deeper than can the smaller tributary glaciers. Thus, after the ice has receded, the valleys of tributary glaciers are left standing above the main trough and are termed **hanging valleys.** Rivers flowing through hanging valleys can produce spectacular waterfalls, such as those in Yosemite National Park, California (Figure 5.9).

At the head of a glacial valley is a characteristic and often imposing feature associated with an alpine

Figure 5.7 Erosional landforms created by alpine glaciers. The unglaciated landscape in part **A** is modified by valley glaciers in part **B**. After the ice recedes, in part **C,** the terrain looks very different than before glaciation.

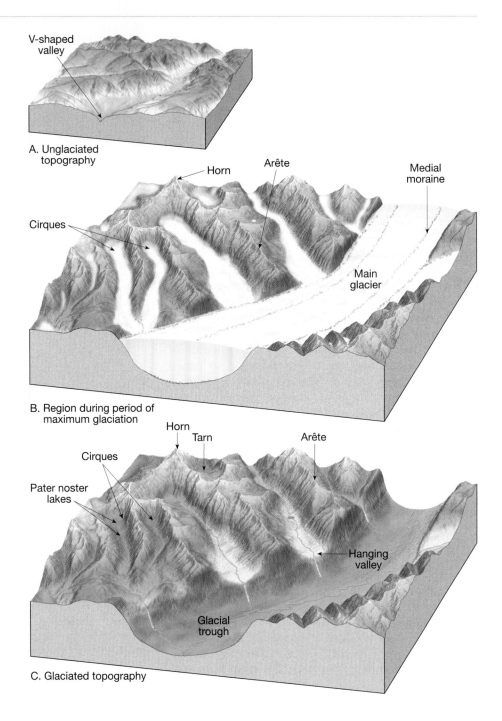

A. Unglaciated topography

B. Region during period of maximum glaciation

C. Glaciated topography

glacier—a **cirque.** As Figure 5.10 illustrates, these hollowed-out, bowl-shaped depressions have precipitous walls on three sides but are open on the downvalley side. The cirque represents the focal point of the glacier's source, that is, the area of snow accumulation and ice formation. Cirques begin as irregularities in the mountainside that are subsequently enlarged by frost wedging and plucking along the sides and bottom of the glacier. The glacier, in turn, acts as a conveyor belt that carries away the debris. After the glacier has melted away, the cirque basin is sometimes occupied by a small lake.

The Alps, Northern Rockies, and many other mountain landscapes carved by valley glaciers reveal more than glacial troughs and cirques. In addition, sinuous, sharp-edged ridges called **arêtes** and sharp, pyramid-like peaks termed **horns** project above the surroundings (Figure 5.7C). Both features can originate from the same basic process—the enlargement of cirques produced by plucking and frost action. Several cirques around a single high mountain create the spires of rock called horns. As the cirques enlarge and converge, an isolated horn is produced. The most famous example is the Matterhorn in the Swiss Alps (Figure 5.11).

Arêtes can form in a similar manner except that the cirques are not clustered around a point but rather exist on opposite sides of a divide. As the cirques

Figure 5.8 Prior to glaciation, a mountain valley is typically narrow and V-shaped. During glaciation, an alpine glacier widens, deepens, and straightens the valley, creating the U-shaped glacial trough seen here. The string of lakes is called pater noster lakes. This valley is in Glacier National Park, Montana. (Photo by John Montagne)

grow, the divide separating them is reduced to a very narrow, knifelike partition. An arête can also be created in another way. When two glaciers occupy parallel valleys, an arête can form when the land separating the moving tongues of ice is progressively narrowed as the glaciers scour and widen their valleys.

Fiords are deep, often spectacular, steep-sided inlets of the sea that exist in many high-latitude areas of the world where mountains are adjacent to the ocean (Figure 5.12). Norway, British Columbia, Greenland, New Zealand, Chile, and Alaska all have coastlines characterized by fiords. They are glacial troughs that became submerged as the ice left the valley and sea level rose following the Ice Age. The depth of some fiords exceeds 1000 meters (3300 feet).

However, the great depths of these flooded troughs are only partly explained by the post-Ice Age rise in sea level. Unlike the situation governing the downward erosional work of rivers, sea level does not act as base level for glaciers. As a consequence, glaciers are capable of eroding their beds far below the surface of the sea. For example, a valley glacier 300 meters (1000 feet) thick can carve its valley floor more than 250 meters (800 feet) below sea level before downward erosion ceases and the ice begins to float.

Glacial Deposits

 External Processes
Glaciers

Glaciers pick up and transport a huge load of debris as they slowly advance across the land. Ultimately these materials are deposited when the ice melts. In regions where glacial sediment is deposited, it can play a truly significant role in forming the physical landscape. For example, in many areas once covered by the ice sheets

of the recent Ice Age, the bedrock is rarely exposed because glacial deposits that are dozens or even hundreds of meters thick completely mantle the terrain. The general effect of these deposits is to reduce the local relief and thus level the topography. Indeed, rural country scenes that are familiar to many of us— rocky pastures in New England, wheat fields in the

Figure 5.9 Bridalveil Falls in Yosemite National Park cascades from a hanging valley into the glacial trough below. (Photo by E. J. Tarbuck)

Figure 5.10 Glacier flowing from a cirque in Glacier Bay National Park, Alaska. These bowl-shaped depressions are found at the heads of glacial valleys and represent a focal point of ice formation. (Photo by Tom Beam)

Dakotas, rolling farmland in the Midwest—result directly from glacial deposition.

Types of Glacial Drift

Long before the theory of an extensive Ice Age was proposed, much of the soil and rock debris covering portions of Europe was recognized as coming from elsewhere. At the time, these foreign materials were believed to have been "drifted" into their present positions by floating ice during an ancient flood. As a consequence, the term *drift* was applied to this sediment. Although rooted in a concept that was not correct, this term was so well established by the time the true glacial origin of the debris became widely recognized that it remained in the glacial vocabulary. Today, **drift** is an all-embracing term for sediments of glacial origin, no matter how, where, or in what form they were deposited.

Glacial drift is divided into two distinct types: (1) materials deposited directly by the glacier, which are known as *till*, and (2) sediments laid down by glacial meltwater, called *stratified drift*. Here is the difference: **till** is deposited as glacial ice melts and drops its load of rock debris. Unlike moving water and wind, ice cannot sort the sediment it carries; therefore, deposits of till are characteristically unsorted mixtures of many particle sizes (Figure 5.13). **Stratified drift** is sorted according to the size and weight of the fragments. As ice is not capable of such sorting activity, these sediments are not deposited directly by the glacier. Rather, they reflect the sorting action of glacial meltwater.

Some deposits of stratified drift are made by streams issuing directly from the glacier. Other stratified deposits involve sediment that was originally laid down as till and later picked up, transported, and redeposited by meltwater beyond the margin of the ice. Accumulations of stratified drift often consist largely of sand and gravel, because the meltwater is not capable of moving larger material and because the finer rock flour remains suspended and is commonly carried far from the glacier. An indication that stratified drift consists primarily of sand and gravel can be seen in many areas where these deposits are actively mined as aggregate for road work and other construction projects.

When boulders are found in the till or lying free on the surface, they are called **glacial erratics** if they are different from the bedrock below. Of course, this means that they must have been derived from a source outside the area where they are found (Figure 5.14). Although

Figure 5.11 Horns are sharp, pyramid-like peaks that are fashioned by alpine glaciers. This example is the famous Matterhorn in the Swiss Alps. (Photo by E. J. Tarbuck)

Figure 5.12 Like other fiords, this one at Tracy Arm, Alaska is a drowned glacial trough. (Photo by Tom Bean)

the locality of origin for most erratics is unknown, the origin of some can be determined. Therefore, by studying glacial erratics as well as the mineral composition of the till, geologists can sometimes trace the path of a lobe of ice. In portions of New England as well as other areas, erratics can be seen dotting pastures and farm fields. In some places, these rocks were cleared from fields and piled to make fences and walls.

Figure 5.14 A large, glacial erratic near Hensler, North Dakota. (Photo by Tom Bean/DRK Photo)

Figure 5.13 Glacial till is an unsorted mixture of many different sediment sizes. A close examination often reveals cobbles that have been scratched as they were dragged along by the glacier. (Photos by E. J. Tarbuck)

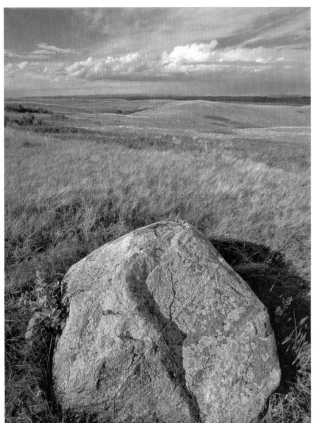

Moraines, Outwash Plains, and Kettles

Perhaps the most widespread features created by glacial deposition are *moraines,* which are simply layers or ridges of till. Several types of moraines are identified; some are common only to mountain valleys, and others are associated with areas affected by either ice sheets or valley glaciers. Lateral and medial moraines fall in the first category, whereas end moraines and ground moraines are in the second.

The sides of a valley glacier accumulate large quantities of debris from the valley walls. When the glacier wastes away, these materials are left as ridges, called **lateral moraines,** along the sides of the valley (Figure 5.15). **Medial moraines** are formed when two valley glaciers coalesce to form a single ice stream. The till that was once carried along the edges of each glacier joins to form a single dark stripe of debris within the newly enlarged glacier. Creation of these dark stripes within the ice stream is one obvious proof that glacial ice moves, because the medial moraine could not form if the ice did not flow down-valley (Figure 5.15).

End moraines, as the name implies, form at the terminus of a glacier. Here, while the ice front is stationary, the glacier continues to carry in and deposit large quantities of rock debris, creating a ridge of till tens to hundreds of meters high. The end moraine marking the farthest advance of the glacier is called the *terminal end moraine.* Those moraines that formed as the ice front periodically became stationary during retreat are termed *recessional end moraines.* As the glacier recedes, a layer of till is laid down, forming a gently undulating surface of **ground moraine.** Ground moraine has a leveling effect, filling in low spots and clogging old stream channels, often leading to a disruption of drainage.

End moraines deposited by the most recent major stage of Ice Age glaciation are prominent features in many parts of the Midwest and Northeast. In Wisconsin, the wooded, hilly terrain of the Kettle Moraine near Milwaukee is a particularly picturesque example. A well-known example in the Northeast is Long Island. This linear strip of glacial sediment that extends northeastward from New York City is part of an end moraine complex that stretches from eastern Pennsylvania to Cape Cod, Massachusetts (Figure 5.16).

At the same time that an end moraine is forming, meltwater emerges from the ice in rapidly moving streams. Often they are choked with suspended material and carry a substantial bed load. As the water leaves the glacier, it rapidly loses velocity and much of its bed load is dropped. In this way a broad, ramplike surface of stratified drift is built adjacent to the downstream edge of most end moraines. When the feature is formed in association with an ice sheet, it is termed an **outwash plain,** and when it is confined to a mountain valley, it is usually referred to as a **valley train.** Figure 5.17 shows an outwash plain and other common depositional features.

Often end moraines, outwash plains, and valley trains are pockmarked with basins or depressions known as **kettles** (Figure 5.17). Kettles form when blocks of stagnant ice become buried in drift and melt,

Figure 5.15 Lateral moraines form from the accumulation of debris along the sides of a valley glacier. Medial moraines form when the lateral moraines of merging valley glaciers join. Medial moraines could not form if the ice did not advance downvalley. Therefore, these dark stripes are proof that glacial ice moves. Kaskawulsh Glacier, Kluane National Park, Yukon, Canada. (Photo by Theo Allofs/Tony Stone Images)

Figure 5.16 End moraines make up substantial parts of Long Island, Cape Cod, Martha's Vineyard, and Nantucket. Although portions are submerged, the Ronkonkoma moraine (a terminal endmoraine) extends through central Long Island, Martha's Vineyard, and Nantucket. It was deposited about 20,000 years ago. The recessional Harbor Hill moraine, which formed about 14,000 years ago, extends along the north shore of Long Island, through southern Rhode Island and Cape Cod.

leaving pits in the glacial sediment. Most kettles do not exceed 2 kilometers in diameter, and the typical depth of most kettles is less than 10 meters (33 feet). Water often fills the depression and forms a pond or lake. One well-known example is Walden Pond near Concord, Massachusetts. It is here that Henry David Thoreau lived alone for two years in the 1840s and about which he wrote *Walden*, his classic of American literature.

Drumlins, Eskers, and Kames

Drumlins are streamlined asymmetrical hills composed of till (Figure 5.17). They range in height from 15 to 60 meters (50 to 200 feet) and average 0.4 to 0.8 kilometer (0.25–0.50 mile) in length. The steep side of the hill faces the direction from which the ice advanced, whereas the gentler slope points in the direction the ice

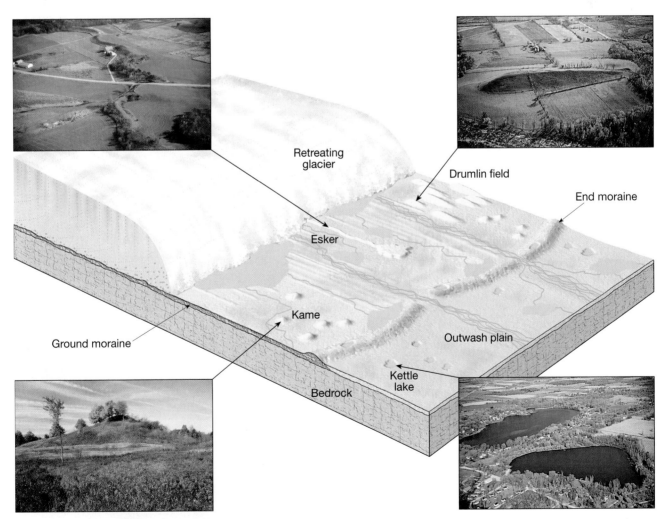

Figure 5.17 This hypothetical area illustrates many common depositional landforms. (Drumlin photo courtesy by Ward's Natural Science Establishment; Kame, Esker, and Kettle photos by Richard P. Jacobs/JLM Visuals)

moved. Drumlins are not found singly, but rather occur in clusters, called *drumlin fields.* One such cluster, east of Rochester, New York, is estimated to contain about 10,000 drumlins. Their streamlined shape indicates that they were molded in the zone of flow within an active glacier. It is thought that drumlins originate when glaciers advance over previously deposited drift and reshape the material.

In some areas that were once occupied by glaciers, sinuous ridges composed largely of sand and gravel might be found. These ridges, called **eskers,** are deposits made by streams flowing in tunnels beneath the ice, near the terminus of a glacier (Figure 5.17). They may be several meters high and extend for many kilometers. In some areas they are mined for sand and gravel, and for this reason, eskers are disappearing in some localities.

Kames are steep-sided hills that, like eskers, are composed of sand and gravel (Figure 5.17). Kames originate when glacial meltwater washes sediment into openings and depressions in the stagnant wasting terminus of a glacier. When the ice eventually melts away, the stratified drift is left behind as mounds or hills.

Glaciers of the Ice Age

At various points in the preceding pages, we mentioned the Ice Age, a time when ice sheets and alpine glaciers were far more extensive than they are today.

There was a time when the most popular explanation for what we now know to be glacial deposits was that the material had been drifted in by means of icebergs or perhaps simply swept across the landscape by a catastrophic flood. However, during the nineteenth century, field investigations by many scientists provided convincing evidence that an extensive Ice Age was responsible for these deposits, and for many other features.

By the beginning of the twentieth century, geologists had largely determined the extent of Ice Age glaciation. Further, they discovered that many glaciated regions had not one layer of drift, but several. Close examination of these older deposits showed well-developed zones of chemical weathering and soil formation as well as the remains of plants that require warm temperatures. The evidence was clear; there had not been just one glacial advance but many, each separated by extended periods when climates were as warm or warmer than at present. The Ice Age had not simply been a time when the ice advanced over the land, lingered for a while, and then receded. Rather, the period was a very complex event characterized by a number of advances and withdrawals of glacial ice.

The glacial record on land is punctuated by many erosional gaps. This makes it difficult to reconstruct clearly the episodes of the Ice Age. But sediment on the ocean floor provides an uninterrupted record of

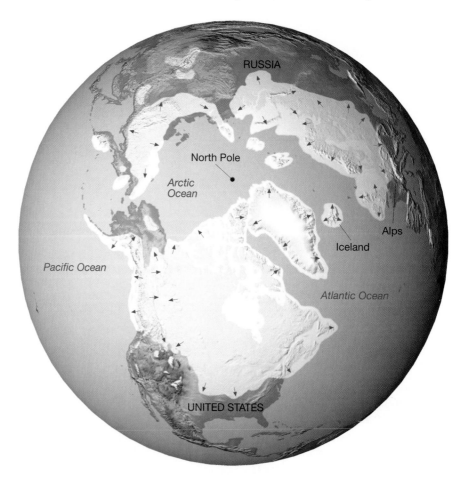

Figure 5.18 Maximum extent of ice sheets in the Northern Hemisphere during the Ice Age.

climate cycles for this period. Studies of cores drilled from these seafloor sediments show that glacial/interglacial cycles have occurred about every 100,000 years. About 20 such cycles of cooling and warming were identified for the span we call the Ice Age.

During the glacial age, ice left its imprint on almost 30 percent of Earth's land area, including about 10 million square kilometers of North America, 5 million square kilometers of Europe, and 4 million square kilometers of Siberia (Figure 5.18). The amount of glacial ice in the Northern Hemisphere was roughly twice that of the Southern Hemisphere. The primary reason is that the Southern Hemisphere has little land in the middle latitudes and therefore the southern polar ice could not spread far beyond the margins of Antarctica. By contrast, North America and Eurasia provided great expanses of land for the spread of ice sheets.

Today we know that the Ice Age began between 2 million and 3 million years ago. This means that most of the major glacial episodes occurred during a division of the geologic time scale called the **Pleistocene epoch.** Although the Pleistocene is commonly used as a synonym for the Ice Age, this epoch does not encompass it all. The Antarctic Ice Sheet, for example, formed at least 14 million years ago, and, in fact, might be much older.

Some Indirect Effects of Ice Age Glaciers

In addition to the massive erosional and depositional work carried on by Pleistocene glaciers, the ice sheets had other, sometimes profound, effects on the landscape. For example, as the ice advanced and retreated, animals and plants were forced to migrate. This led to stresses that some organisms could not tolerate. Furthermore, many present-day stream courses bear little resemblance to their preglacial routes. The Missouri River once flowed northward toward Hudson Bay in Canada. The Mississippi River followed a path through central Illinois, and the head of the Ohio River reached only as far as Indiana. Some rivers that today carry only a trickle of water but occupy broad channels are a testament to the fact that they once carried torrents of glacial meltwater.

In areas that were centers of ice accumulation, such as Scandinavia and northern Canada, the land has been slowly rising for the past several thousand years. The land had downwarped under the tremendous weight of 3-kilometer-thick masses of ice. Following the removal of this immense load, the crust has been adjusting by gradually rebounding upward ever since.

A far-reaching effect of the Ice Age was the worldwide change in sea level that accompanied each advance and retreat of the ice sheets (see Box 5.2). The

Box 5.2 What If the Ice Melted?

How much water is stored as glacial ice? Estimates by the U.S. Geological Survey indicate that only slightly more than 2 percent of the world's water is accounted for by glaciers. But this small figure may be misleading when the actual amounts of water are considered. The total volume of all valley glaciers is about 210,000 cubic kilometers, comparable to the combined volume of the world's largest saline and freshwater lakes. Furthermore, 80 percent of the world's ice and nearly two-thirds of Earth's freshwater are represented by Antarctica's ice sheet, which covers an area almost one and one-half times that of the United States. If this ice melted, sea level would rise an estimated 60 to 70 meters, and the ocean would inundate many densely populated coastal areas (Figure 5.B).

The hydrologic importance of the continent and its ice can be illustrated in another way. If Antarctica's ice sheet were melted at a uniform rate, it could feed (1) the Mississippi River for more than 50,000 years, (2) all the

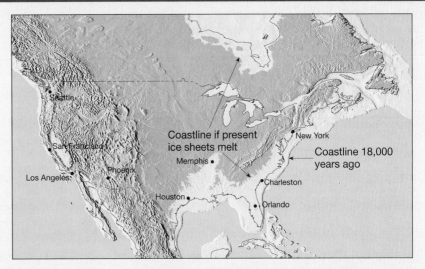

Figure 5.B This map of a portion of North America shows the present-day coastline compared to the coastline that existed during the last ice-age maximum (18,000 years ago) and the coastline that would exist if present ice sheets in Greenland and Antarctica melted. (After R. H. Dott, Jr. and R. L. Battan, *Evolution of the Earth*, New York: McGraw-Hill, 1971.)

rivers in the United States for about 17,000 years, (3) The Amazon River for approximately 5000 years, or (4) all the rivers of the world for about 750 years. As the foregoing discussion illustrates, the quantity of ice on Earth today is truly immense. However, present glaciers occupy only about one-third the area they did in the very recent geologic past.

snow that nourishes glaciers ultimately comes from moisture evaporated from the oceans. Therefore, when the ice sheets increased in size, sea level fell and the shoreline shifted seaward (see Box 5.2, Figure 5.B). Estimates suggest that sea level was as much as 100 meters (330 feet) lower than today. Consequently, the Atlantic Coast of the United States was located more than 100 kilometers (60 miles) to the east of New York City. Moreover, France and Britain were joined where the English Channel is today. Alaska and Siberia were connected across the Bering Strait, and Southeast Asia was tied by dry land to the islands of Indonesia.

The formation and growth of ice sheets was an obvious response to significant changes in climate. But the existence of the glaciers themselves triggered climatic changes in the regions beyond their margins. In arid and semiarid areas on all continents, temperatures were lowered, which meant evaporation rates were also lowered. At the same time, precipitation was moderate. This cooler, wetter climate resulted in the formation of many lakes called **pluvial lakes** (from the Latin term *pluvia* meaning "rain"). In North America, pluvial lakes were concentrated in the vast Basin and Range region of Nevada and Utah (Figure 5.19). Although most are now gone, there are a few remnants, the largest being Utah's Great Salt Lake.

Causes of Glaciation

A great deal is known about glaciers and glaciation. Much has been learned about glacier formation and movement, the extent of glaciers past and present, and the features created by glaciers, both erosional and depositional. However, a widely accepted theory for the causes of glacial ages has not yet been established.

Although widespread glaciation has been rare in Earth's history, the Pleistocene Ice Age is not the only glacial period for which a record exists. Two Precambrian glacial episodes have been identified in the geologic record, the first approximately 2 billion years ago and the second about 600 million years ago. Further, a well-documented record of an earlier glacial age is found in late Paleozoic rocks that are about 250 million years old and which exist on several landmasses.

Any theory that attempts to explain the causes of glacial ages must successfully answer two basic questions. (1) *What causes the onset of glacial conditions?* For continental ice sheets to have formed, average temperatures must have been somewhat lower than at present and perhaps substantially lower than throughout much of geologic time. Thus, a successful theory would have to account for the cooling that finally leads to glacial conditions. (2) *What caused the alternation of glacial and interglacial stages that have been documented for the Pleistocene epoch?* The first question deals with long-term trends in temperature on a scale

Figure 5.19 Pluvial lakes of the Western United States. (After R. F. Flint, *Glacial and Quaternary Geology*, New York: John Wiley & Sons.)

of millions of years, but this second question relates to much shorter-term changes.

Although the literature of science contains a vast array of hypotheses relating to the possible causes of glacial periods, we will discuss only a few major ideas to summarize current thought.

Plate Tectonics

Probably the most attractive proposal for explaining the fact that extensive glaciations have occurred only a few times in the geologic past comes from the theory of plate tectonics.* Not only does this theory provide geologists with explanations about many previously misunderstood processes and features, but it also provides a possible explanation for some hitherto unexplainable climatic changes, including the onset of glacial conditions. Because glaciers can form only on land, we know that landmasses must exist somewhere in the higher latitudes before an ice age can commence. Many scientists suggest that ice ages have

---

*A complete discussion of plate tectonics is presented in Chapter 7.

occurred only when Earth's shifting crustal plates have carried the continents from tropical latitudes to more poleward positions.

Glacial features in present-day Africa, Australia, South America, and India indicate that these regions, which are now tropical or subtropical, experienced an Ice Age near the end of the Paleozoic era, about 250 million years ago. However, there is no evidence that ice sheets existed during this same period in what are today the higher latitudes of North America and Eurasia. For many years this puzzled scientists. Was the climate in these relatively tropical latitudes once like it is today in Greenland and Antarctica? Why did glaciers not form in North America and Eurasia? Until the plate tectonics theory was formulated, there had been no reasonable explanation.

Today, scientists realize that the areas containing these ancient glacial features were joined together as a single supercontinent (Pangaea) located at latitudes far to the south of their present positions. Later, this landmass broke apart, and its pieces, each moving on a different plate, migrated toward their present locations (Figure 5.20). It is now understood that during the geologic past, plate movements accounted for many dramatic climatic changes as landmasses shifted in relation to one another and moved to different latitudinal positions. Changes in oceanic circulation also must have occurred, altering the transport of heat and moisture and consequently the climate as well. Because the rate of plate movement is very slow—a few centimeters per year—appreciable changes in the positions of the continents occur only over great spans of geologic time. Thus, climatic changes brought about by shifting plates are extremely gradual and happen on a scale of millions of years.

Variations in Earth's Orbit

Because climatic changes brought about by moving plates are extremely gradual, the plate tectonics theory cannot be used to explain the alternation between glacial and interglacial climates that occurred during the Pleistocene epoch. Therefore, we must look to some other triggering mechanism that may cause climatic change on a scale of thousands rather than millions of years. Today many scientists strongly suspect that the climatic oscillations that characterized the Pleistocene may be linked to variations in Earth's orbit. This hypothesis was first developed and strongly advocated by the Yugoslavian scientist Milutin Milankovitch and is based on the premise that variations in incoming solar radiation are a principal factor in controlling Earth's climate.

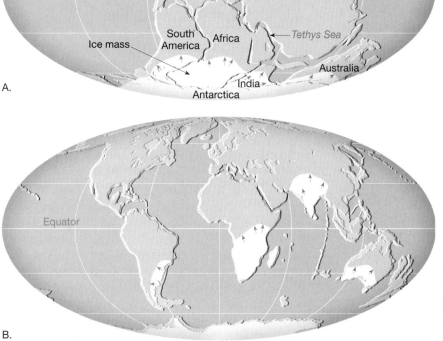

Figure 5.20 A. The supercontinent Pangaea showing the area covered by glacial ice 300 million years ago. **B.** The continents as they are today. The white areas indicate where evidence of the old ice sheets exists.

Milankovitch formulated a comprehensive mathematical model based on the following elements (Figure 5.21):

1. Variations in the shape (*eccentricity*) of Earth's orbit about the Sun;
2. Changes in *obliquity*; that is, changes in the angle that the axis makes with the plane of Earth's orbit; and
3. The wobbling of Earth's axis, called *precession*.

Using these factors, Milankovitch calculated variations in the receipt of solar energy and the corresponding surface temperature of Earth back into time in an attempt to correlate these changes with the climatic fluctuations of the Pleistocene. In explaining climatic changes that result from these three variables, note that they cause little or no variation in the total solar energy reaching the ground. Instead, their impact is felt because they change the degree of contrast between the seasons. Somewhat milder winters in the middle to high latitudes means greater snowfall totals, whereas cooler summers would bring a reduction in snowmelt.

Among the studies that have added credibility to this astronomical hypothesis is one in which deep-sea sediments containing certain climatically sensitive microorganisms were analyzed to establish a chronology of temperature changes going back nearly one-half million years.* This time scale of climatic change was then compared to astronomical calculations of eccentricity, obliquity, and precession to determine if a correlation did indeed exist. Although the study was very involved and mathematically complex, the conclusions were straightforward. The authors found that major variations in climate over the past several hundred thousand years were closely associated with changes in the geometry of Earth's orbit; that is, cycles of climatic change were shown to correspond closely with the periods of obliquity, precession, and orbital eccentricity. More specifically, the authors stated: "It is concluded that changes in the earth's orbital geometry are the fundamental cause of the succession of Quaternary ice ages."†

Let us briefly summarize the ideas that were just described. The theory of plate tectonics provides an explanation for the widely spaced and nonperiodic onset of glacial conditions at various times in the geologic past, whereas the astronomical model proposed by Milankovitch and supported by the work of J.D. Hays and his colleagues furnishes an explanation for the alternating glacial and interglacial episodes of the Pleistocene.

A.

B.

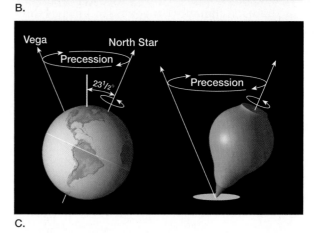

C.

Figure 5.21 Orbital variations. **A.** The shape of Earth's orbit changes during a cycle that spans about 100,000 years. It gradually changes from nearly circular to one that is more elliptical and then back again. This diagram greatly exaggerates the amount of change. **B.** Today the axis of rotation is tilted about 23.5° to the plane of Earth's orbit. During a cycle of 41,000 years, this angle varies from 21.5° to 24.5°. **C.** Precession. Earth's axis wobbles like that of a spinning top. Consequently, the axis points to different spots in the sky during a cycle of about 26,000 years.

---

*J. D. Hays, John Imbrie, and N.J. Shackelton, "Variations in the Earth's Orbit: Pacemaker of the Ice Ages," *Science* 194 (1976): 1121–32.

†J. D. Hays et al., p. 1131. The term *Quaternary* refers to the period on the geologic time scale that encompasses the last 1.6 million years.

In conclusion, we emphasize that the ideas just discussed do not represent the only possible explanations for glacial ages. Although interesting and attractive, these proposals are certainly not without critics; nor are they the only possibilities currently under study. Other factors may be, and probably are, involved.

Deserts

External Processes
Deserts

Climate has a strong influence on the nature and intensity of Earth's external processes. This was clearly demonstrated in the preceding section on glaciers and the Ice Age. Another excellent example of the strong link between climate and geology is seen when we examine the development of arid landscapes.

The dry regions of the world encompass about 42 million square kilometers, nearly 30 percent of Earth's land surface. No other climatic group covers so large a land area.[††] The word *desert* literally means "deserted" or "unoccupied." For many dry regions this is a very appropriate description. Yet where water is available in deserts, many people thrive. Nevertheless, the world's dry regions are among the least familiar land areas on Earth outside of the polar realm (see Box 5.3).

Desert landscapes frequently appear stark. Their profiles are not softened by a carpet of soil and abundant plant life. Instead, barren rocky outcrops with steep, angular slopes are common. At some places the rocks are tinted orange and red. At others they are gray and brown and streaked with black. For many visitors desert scenery exhibits a striking

beauty; to others, the terrain seems bleak. No matter which feeling is elicited, it is clear that deserts are very different from the more humid places where most people live.

As you shall see, arid regions are not dominated by a single geologic process. Rather, the effects of tectonic forces, running water, and wind are all apparent. Because these processes combine in different ways from place to place, the appearance of desert landscapes varies a great deal as well (Figure 5.22).

Geologic Processes in Arid Climates

External Processes
Deserts

The angular rock exposures, the sheer canyon walls, and the pebble- or sand-covered surface of the desert contrast sharply with the rounded hills and curving slopes of more humid places. To a visitor from a humid region, a desert landscape may seem to have been shaped by forces different from those operating in wetter areas. However, although the contrasts might be striking, they do not reflect different processes. They merely disclose the differing effects of the same processes that operate under contrasting climatic conditions.

[††]An examination of dry climates is found in Chapter 18.

Figure 5.22 A scene in north-central New Mexico near Abiquiu. The appearance of desert landscapes varies a great deal from place to place. (Photo by Linda Waidhofer/Liaison Agency, Inc.)

Box 5.3 Common Misconceptions About Deserts

Deserts are hot, lifeless, sand-covered landscapes shaped largely by the force of wind. The preceding statement summarizes the image of arid regions that many people hold, especially those living in more humid places. Is it an accurate view? The answer is no. Although there are clearly elements of reality in such an impression, it is a generalization that contains a number of misconceptions (Figure 5.C).

One common fallacy about deserts is that they are lifeless or almost lifeless. Although reduced in amount and different in character, plant and animal life are usually present. Desert plants all have one characteristic in common: They have developed adaptations that make them highly tolerant of drought. Many have waxy leaves, stems, or branches or a thickened cuticle (outermost protective layer) to reduce water loss. Others have very small leaves or no leaves at all. Also, the roots of some species often extend to great depths to tap the moisture found there, whereas others produce a shallow but widespread root system that enables them to absorb great amounts of moisture quickly from the infrequent desert downpours. Thus, although widely dispersed and providing little ground cover, plants of many kinds flourish in the desert.

A second widely held belief about the world's dry lands is that they are always hot. This fact seems to be reinforced by commonly quoted temperature statistics. The highest accepted temperature record for the United States as well as the entire Western Hemisphere is 57°C (134°F). This long-standing record was set at Death Valley, California, on July 10, 1913. The nearly 59°C (137°F) reading in Azizia, Libya, in North Africa's Sahara Desert on

September 13, 1922, is the world record. Despite these remarkably high figures, cold temperatures are also experienced in desert regions. For example, the average daily minimum in January at Phoenix, Arizona, is 17°C (35°F), just barely above freezing. At Ulan Bator in Mongolia's Gobi Desert, the average *high* temperature in January is only –19°C (–2°F)! Dry climates are found from the tropics to the high middle latitudes. Consequently, although tropical deserts lack a cold season, deserts in the middle latitudes do experience seasonal temperature changes.

The last two commonly held misconceptions about the world's deserts are more geologic than climatic. One mistaken assumption is that they consist of mile after mile of drifting sand. It is true that sand accumulations do exist in some areas and may be

striking features, but they represent only a small percentage of the total desert area. For example, in the Sahara, the world's largest desert, accumulations of sand cover only one-tenth of its area. The sandiest of all deserts is the Arabian, one-third of which consists of sand.

The final mistaken assumption is the seemingly logical idea that wind is the most important agent of erosion in deserts. Although wind is relatively more significant in dry areas than anywhere else, most erosional landforms in deserts are created by running water. When the rains come, they usually take the form of thunderstorms. Because the heavy rain associated with these storms cannot all soak in, rapid runoff results. Without a thick vegetative cover to protect the ground, erosion is great.

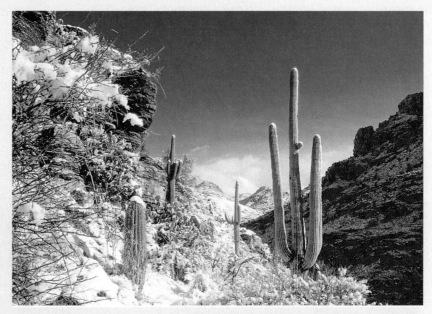

Figure 5.C Snow blankets the rocky ground of the Sonoran Desert near Tuscon, Arizona. As this scene demonstrates, deserts are not necessarily always hot, lifeless, dune-covered expanses. (Photo by T. Wiewandt/DRK Photo)

Weathering

In humid regions, relatively well-developed soils support an almost continuous cover of vegetation. Here the slopes and rock edges are rounded. Such a landscape reflects the strong influence of chemical weathering in a humid climate. By contrast, much of the weathered debris in deserts consists of unaltered rock and mineral fragments—the result of mechanical

weathering processes. In dry lands rock weathering of any type is greatly reduced because of the lack of moisture and the scarcity of organic acids from decaying plants. Chemical weathering, however, is not completely lacking in deserts. Over long spans of time clays and thin soils do form and many iron-bearing silicate minerals oxidize, producing the rust-colored stain found tinting some desert landscapes.

A.   **B.**

Figure 5.23 A. Most of the time, desert stream channels are dry. **B.** An ephemeral stream shortly after a heavy shower. Although such floods are short-lived, large amounts of erosion occur. (Photos by E. J. Tarbuck)

The Role of Water

Permanent streams are normal in humid regions but practically all desert streams are dry most of the time (Figure 5.23A). Desert streams are said to be **ephemeral,** which means that they carry water only in response to specific episodes of rainfall. A typical ephemeral stream might flow only a few days or perhaps just a few hours during the year. In some years the channel may carry no water at all.

This fact is obvious even to the casual observer who, while traveling in a dry region, notices the number of bridges with no streams beneath them or the number of dips in the road where dry channels cross. However, when the rare heavy showers do come, so much rain falls in such a short time that all of it cannot soak in. Because the vegetative cover is sparse, runoff is largely unhindered and consequently rapid, often creating flash floods along valley floors (Figure 5.23B). Such floods, however, are quite unlike floods in humid regions. A flood on a river such as the Mississippi may take several days to reach its crest and then subside. But desert floods arrive suddenly and subside quickly. Because much of the surface material is not anchored by vegetation, the amount of erosional work that occurs during a single short-lived rain event is impressive.

In the dry western United States a number of different names are used for ephemeral streams. Two of the most common are *wash* and *arroyo*. In other parts of the world, a dry desert stream may be called a *wadi* (Arabia and North Africa), a *donga* (South America), or a *nullah* (India).

Humid regions are notable for their integrated drainage systems. But in arid regions streams usually lack an extensive system of tributaries. In fact, a basic characteristic of desert streams are that they are small and die out before reaching the sea. Because the water table is usually far below the surface, few desert streams can draw upon it as streams do in humid regions.

Without a steady supply of water, the combination of evaporation and infiltration soon depletes the stream.

The few permanent streams that do cross arid regions, such as the Colorado and Nile rivers, originate *outside* the desert, often in well-watered mountains. Here the water supply must be great to compensate for the losses occurring as the stream crosses the desert. For example, after the Nile leaves the lakes and mountains of central Africa that are its source, it traverses almost 3000 kilometers of the Sahara *without a single tributary*. By contrast, in humid regions, the discharge of a river increases as it flows downstream because tributaries and groundwater contribute additional water along the way.

It should be emphasized that *running water, although infrequent, nevertheless does most of the erosional work in deserts.* This is contrary to a common belief that wind is the most important erosional agent sculpturing desert landscapes. Although wind erosion is indeed more significant in dry areas than elsewhere, most desert landforms are nevertheless carved by running water. As you will see shortly, the main role of wind is in the transportation and deposition of sediment, which creates and shapes the ridges and mounds we call dunes.

Basin and Range: The Evolution of a Desert Landscape

Because arid regions typically lack permanent streams, they are characterized as having **interior drainage.** This means that they have a discontinuous pattern of intermittent streams that do not flow out of the desert to the ocean. In the United States, the dry Basin and Range region provides an excellent example. The region includes southern Oregon, all of Nevada, western Utah, southeastern California, southern Arizona, and southern New Mexico. The name Basin and Range is an apt description for this almost 800,000-square-kilometer region, because it is characterized by more than 200 relatively small mountain ranges that rise 900 to 1500 meters above the basins that separate them.

In this region, as in others like it around the world, most erosion occurs without reference to the ocean (ultimate base level), because the interior drainage never reaches the sea. Even where permanent streams flow to the ocean, few tributaries exist, and thus only a narrow strip of land adjacent to the stream has sea level as its ultimate level of land reduction.

The block models in Figure 5.24 depict how the landscape has evolved in the Basin and Range region. During and following uplift of the mountains, running water begins carving the elevated mass and depositing large quantities of debris in the basin. In this early stage, relief is greatest, and as erosion lowers the mountains and sediment fills the basins, elevation differences diminish.

When the occasional torrents of water produced by sporadic rains move down the mountain canyons, they are heavily loaded with sediment. Emerging from the confines of the canyon, the runoff spreads over the gentler slopes at the base of the mountains and quickly loses velocity. Consequently, most of its load is dumped within a short distance. The result is a cone of debris known as an **alluvial fan** at the mouth of a canyon (Figure 5.25). Over the years, a fan enlarges, eventually coalescing with fans from adjacent canyons to produce an apron of sediment (*bajada*) along the mountain front.

On the rare occasions of abundant rainfall, or snowmelt in the mountains, streams may flow across the alluvial fans to the center of the basin, converting the basin floor into a shallow **playa lake.** Playa lakes last only a few days or weeks, before evaporation and

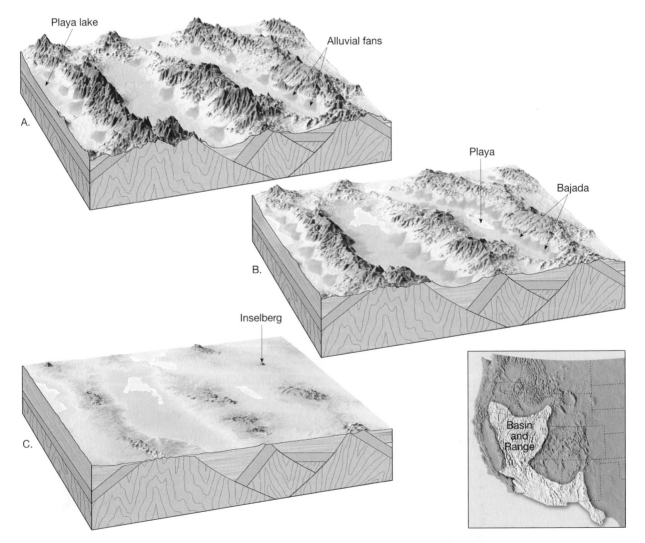

Figure 5.24 Stages of landscape evolution in a mountainous desert such as the Basin and Range region of the West. As erosion of the mountains and deposition in the basins continue, relief diminishes. **A.** Early stage. **B.** Middle stage. **C.** Late stage.

Figure 5.25 Alluvial fans develop where the gradient of a stream changes abruptly from steep to flat. Such a situation exists in Death Valley, California, where streams emerge from the mountains into a flat basin. As a result, Death Valley has many large alluvial fans. (Photo by Michael Collier)

infiltration remove the water. The dry, flat lake bed that remains is termed a *playa*.

Playas occasionally become encrusted with salts left behind by evaporation. These precipitated salts may be unusual. A case in point is the sodium borate (better known as borax) mined from ancient playa lake deposits in Death Valley, California.

With the ongoing erosion of the mountain mass and the accompanying sedimentation, the local relief continues to diminish. Eventually nearly the entire mountain mass is gone. Thus, by the late stages of erosion, the mountain areas are reduced to a few large bedrock knobs (called *inselbergs*) projecting above the sediment-filled basin.

Each of the stages of landscape evolution in an arid climate depicted in Figure 5.24 can be observed in the Basin and Range region. Recently uplifted mountains in an early stage of erosion are found in southern Oregon and northern Nevada. Death Valley, California, and southern Nevada fit into the more advanced middle stage, whereas the late stage, with its inselbergs, can be seen in southern Arizona.

Wind Erosion

Compared to running water and glaciers, wind is a relatively insignificant erosional agent. Recall that even in deserts, most erosion is performed by intermittent running water, not by the wind. Wind erosion is more effective in arid lands than in humid areas because in humid places moisture binds particles together and vegetation anchors the soil. For wind to be an effective erosional force, dryness and scanty vegetation are essential. When such circumstances exist, wind may pick up, transport, and deposit great quantities of fine sediment.

During the 1930s, parts of the Great Plains experienced vast dust storms. The plowing under of the natural vegetative cover for farming, followed by severe drought, exposed the land to wind erosion and led to the area being labeled the Dust Bowl (see Box 5.4).

One way that wind erodes is by **deflation,** the lifting and removal of loose material. Because the competence (ability to transport different-size particles) of moving air is low, it can suspend only fine sediment such as clay and silt. Larger grains of sand are rolled or skipped along the surface (a process called *saltation*) and comprise the bed load. Particles larger than sand are usually not transported by wind. Deflation sometimes is difficult to notice because the entire surface is being lowered at the same time, but it can be significant. In portions of the 1930s Dust Bowl, the land was lowered by a meter or more in only a few years (Figure 5.26).

The most noticeable results of deflation in some places are shallow depressions called **blowouts.** In the Great Plains region, from Texas north to Montana, thousands of blowouts can be seen. They range from small dimples less than 1 meter deep and 3 meters wide to depressions that are over 45 meters deep and several kilometers across.

In portions of many deserts, the surface is characterized by a layer of coarse pebbles and cobbles that are too large to be moved by the wind. This stony veneer, called **desert pavement,** is created as deflation lowers the surface by removing sand and silt until eventually only a continuous cover of coarse particles remains (Figure 5.27). Once desert pavement becomes established, a process that might take hundreds of years, the surface is effectively protected from further deflation if left undisturbed. However, as the layer is only one or two stones thick, the passage of vehicles or

Box 5.4 People and the Environment: Dust Bowl: Soil Erosion in the Great Plains

During a span of dry years in the 1930s, large dust storms plagued the Great Plains. Because of the size and severity of these storms, the region came to be called the "Dust Bowl," and the time period, the "dirty thirties." The heart of the Dust Bowl consisted of nearly 100 million acres in the panhandles of Texas and Oklahoma, as well as adjacent parts of Colorado, New Mexico, and Kansas (Figure 5.D). To a lesser extent, dust storms were also a problem over much of the Great Plains, from North Dakota to west-central Texas.

At times dust storms were so severe that they were called "black blizzards" and "black rollers" because visibility was reduced to only a few feet. Numerous storms lasted for hours and stripped huge volumes of topsoil from the land.

In the spring of 1934, a wind storm that lasted for a day and a half created a dust cloud that extended for 2000 kilometers (1200 miles). As the sediment moved east, "muddy rains" were experienced in New York, and "black snows" in Vermont. Less than a year later, another storm carried dust more than 3 kilometers (2 miles) into the atmosphere and transported it 3000 kilometers from its source in Colorado to create twilight conditions in the middle of the day in parts of New England and New York.

Figure 5.D Dust blackens the sky on May 21, 1937, near Elkhart, Kansas. It was because of storms like this that portions of the Great Plains were called the "Dust Bowl" in the 1930s. (Photo reproduced from the collection of the Library of Congress)

What caused the Dust Bowl? Clearly, the fact that portions of the Great Plains experience some of North America's strongest winds is important. However, it was the expansion of agriculture that set the stage for the disastrous period of soil erosion. Mechanization allowed the rapid transformation of the grass-covered prairies of this semiarid region into farms. Between the 1870s and 1930, the area of cultivation in the region expanded nearly tenfold, from about 10 million acres to more than 100 million acres. As long as precipitation was adequate, the soil remained in place.

However, when a prolonged drought struck in the 1930s, the unprotected fields were vulnerable to the wind. The results were severe soil loss, crop failures, and economic hardship.

Beginning in 1939, a return to rainier conditions brought relief. Moreover, farming practices that were designed to reduce soil loss by wind had also been instituted. Although dust storms are less numerous and not as severe as in the "dirty thirties," soil erosion by strong winds still occurs periodically whenever the combination of drought and unprotected soil exists.

Figure 5.26 This photo was taken north of Granville, North Dakota, in July 1936, during a prolonged drought. Strong winds removed the soil that was not anchored by vegetation. The mounds are 1.2 meters (4 feet) high and show the level of the land prior to deflation. (Photo courtesy of the State Historical Society of North Dakota)

animals can dislodge the pavement and expose the fine-grained material below. If this happens, the surface is once again subject to deflation.

Like glaciers and streams, wind erodes in part by *abrasion*. In dry regions as well as along some beaches, windblown sand will cut and polish exposed rock surfaces. However, abrasion is often given too much credit. Such features as balanced rocks that stand high atop narrow pedestals and intricate detailing on tall pinnacles are not the results of abrasion. Sand seldom travels more than a meter above the surface, so the wind's sandblasting effect is obviously limited in vertical extent. However, in areas prone to such activity, telephone poles have actually been cut through near their bases. For this reason, collars are often fitted on the poles to protect them from being "sawed" down.

Figure 5.27 A. Formation of desert pavement. As these cross-sections illustrate, coarse particles gradually become concentrated into a tightly packed layer as deflation lowers the surface by removing sand and silt. **B.** If left undisturbed, desert pavement such as this in Arizona's Sonoran Desert will protect the surface from further deflation. (Photo by David Muench)

Wind Deposits

Although wind is relatively unimportant in producing *erosional* landforms, significant *depositional* landforms are created by the wind in some regions. Accumulations of windblown sediment are particularly conspicuous in the world's dry lands and along many sandy coasts. Wind deposits are of two distinctive types: (1) extensive blankets of silt, called *loess*, which once were carried in suspension, and (2) mounds and ridges of sand from the wind's bed load, which we call *dunes*.

Loess

In some parts of the world the surface topography is mantled with deposits of windblown silt termed **loess**. Over thousands of years dust storms deposited this material. When loess is breached by streams or road cuts, it tends to maintain vertical cliffs and lacks any visible layers, as you can see in Figure 5.28.

The distribution of loess worldwide indicates that there are two primary sources for this sediment: deserts and glacial deposits of stratified drift. The thickest and most extensive deposits of loess on Earth occur in western and northern China. They were blown there from the extensive desert basins of central Asia. Accumulations of 30 meters are not uncommon, and thicknesses of more than 100 meters have been measured. It is this fine, buff-colored sediment that gives the Yellow River (Hwang Ho) its name.

In the United States, deposits of loess are significant in many areas, including South Dakota, Nebraska, Iowa, Missouri, and Illinois as well as portions of the Columbia Plateau in the Pacific Northwest. Unlike the deposits in China, the loess in the United States, as well as in Europe, is an indirect product of glaciation. Its source is deposits of stratified drift. During the retreat of the ice sheets, many river valleys were choked with sediment deposited by meltwater. Strong westerly winds sweeping across the barren floodplains picked up the finer sediment and dropped it as a blanket on the eastern sides of the valleys.

Sand Dunes

Like running water, wind releases its load of sediment when its velocity falls and the energy available for transport diminishes. Thus, sand begins to accumulate wherever an obstruction across the path of the wind slows its movement. Unlike deposits of loess, which form blanketlike layers over broad areas, winds commonly deposit sand in mounds or ridges called *dunes* (Figure 5.29).

As moving air encounters an object, such as a clump of vegetation or a rock, the wind sweeps around and over it, leaving a shadow of more slowly moving air behind the obstacle as well as a smaller zone of quieter air just in front of the obstacle. Some

Figure 5.28 This vertical loess bluff near the Mississippi River in southern Illinois is about 3 meters high. (Photo by James E. Patterson)

Figure 5.29 Sand sliding down the steep slip face of a dune in White Sands National Monument, New Mexico. (Photo by Michael Collier)

of the saltating sand grains moving with the wind come to rest in these "wind shadows." As the accumulation of sand continues, it forms an increasingly efficient wind barrier to trap even more sand. If there is a sufficient supply of sand and the wind blows steadily long enough, the mound of sand grows into a dune.

Many dunes have an asymmetrical profile, with the leeward (sheltered) slope being steep and the windward slope more gently inclined (Figure 5.30). Sand moves up the gentler slope on the windward side by saltation. Just beyond the crest of the dune, where wind velocity is reduced, the sand accumulates. As more sand collects, the slope steepens and eventually some of it slides or slumps under the pull of gravity (Figure 5.29). In this way, the leeward slope of the dune, called the **slip face,** maintains an angle of about 34 degrees. Continued sand accumulation, coupled with periodic slides down the slip face, results in the slow migration of the dune in the direction of air movement.

As sand is deposited on the slip face, it forms layers inclined in the direction the wind is blowing. These sloping layers are called **cross beds** (Figure 5.30). When the dunes are eventually buried under other layers of sediment and become part of the sedimentary rock record, their asymmetrical shape is destroyed, but the cross beds remain as a testimony to their origin. Nowhere is cross-bedding more prominent than in the sandstone walls of Zion Canyon in Utah (Figure 5.30 photo).

Types of Sand Dunes

Dunes are not just random heaps of wind-blown sediment. Rather, they are accumulations that usually assume patterns that are surprisingly consistent (Figure 5.31). A broad assortment of dune forms exist; generally simplified to a few major types for discussion. Of course, gradations exist among different forms as well as irregularly shaped dunes that do not fit easily into any category. Several factors influence the form and size that dunes ultimately assume. These include wind direction and velocity, availability of sand, and the amount of vegetation present. Six basic dune types are shown in Figure 5.31, with arrows indicating wind directions.

Barchan Dunes. Solitary sand dunes shaped like crescents and with their tips pointing downwind are called **barchan dunes** (Figure 5.31A). These dunes form where supplies of sand are limited and the surface is relatively flat, hard, and lacking vegetation. They migrate slowly with the wind at a rate of up to 15 meters annually. Their size is usually modest, with the largest barchans reaching heights of about 30 meters while the maximum spread between their horns approaches 300 meters. When the wind direction is nearly constant, the crescent form of these dunes is nearly symmetrical. However, when the wind direction is not perfectly fixed, one tip becomes larger than the other.

Transverse Dunes. In regions where the prevailing winds are steady, sand is plentiful, and vegetation is sparse or absent, the dunes form a series of long ridges

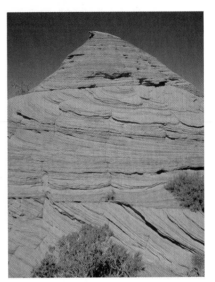

Figure 5.30 Dunes commonly have an asymmetrical shape. The steeper leeward side is called the slip face. Sand grains deposited on the slip face create the cross-bedding of the dunes. A complex pattern develops in response to changes in prevailing winds. The sandstone walls in Zion Canyon, Utah, exhibit excellent cross-bedding. (Photo by Tom Bean)

that are separated by troughs and oriented at right angles to the prevailing wind. Because of this orientation, they are termed **transverse dunes** (Figure 5.31B). Typically, many coastal dunes are of this type. In addition, transverse dunes are common in many arid regions where the extensive surface of wavy sand is sometimes called a *sand sea*. In some parts of the Sahara and Arabian deserts, transverse dunes reach heights of 200 meters, are 1 to 3 kilometers across, and can extend for distances of 100 kilometers or more.

There is a relatively common dune form that is intermediate between isolated barchans and extensive waves of transverse dunes. Such dunes, called **barchanoid dunes,** form scalloped rows of sand oriented at right angles to the wind (Figure 5.31C). The rows resemble a series of barchans that have been positioned side by side. Visitors exploring the gypsum dunes at White Sands National Monument, New Mexico, will recognize this form.

Longitudinal Dunes. **Longitudinal dunes** are long ridges of sand that form more or less parallel to the prevailing wind and where sand supplies are limited (Figure 5.31D). Apparently the prevailing wind direction must vary somewhat, but still remain in the same quadrant of the compass. Although the smaller types are only 3 or 4 meters high and several tens of

meters long, in some large deserts longitudinal dunes can reach great size. For example, in portions of North Africa, Arabia, and central Australia, these dunes may approach a height of 100 meters and extend for distances of more than 100 kilometers (62 miles).

Parabolic Dunes. Unlike the other dunes that have been described thus far, **parabolic dunes** form where vegetation partially covers the sand. The shape of these dunes resembles the shape of barchans except that their tips point into the wind rather than downwind (Figure 5.31E). Parabolic dunes often form along coasts where there are strong onshore winds and abundant sand. If the sand's sparse vegetative cover is disturbed at some spot, deflation creates a blowout. Sand is then transported out of the depression and deposited as a curved rim that grows higher as deflation enlarges the blowout.

Star Dunes. Confined largely to parts of the Sahara and Arabian deserts, **star dunes** are isolated hills of sand that exhibit a complex form (Figure 5.31F). Their name is derived from the fact that the bases of these dunes resemble multipointed stars. Usually three or four sharp-crested ridges diverge from a central high point that in some cases may approach a height of 90 meters. As their form suggests, star dunes develop where wind directions are variable.

Wind

A. Barchan

Wind

B. Transverse

Wind

C. Barchanoid

Wind

D. Longitudinal

Wind

E. Parabolic

Wind

F. Star

Figure 5.31 Sand dune types.

Chapter Summary

- A *glacier* is a thick mass of ice originating on land from the compaction and recrystallization of snow, and it shows evidence of past or present flow. Today, *valley* or *alpine glaciers* are found in mountain areas where they usually follow valleys that were originally occupied by streams. *Ice sheets* exist on a much larger scale, covering most of Greenland and Antarctica.

- On the surface of a glacier, ice is brittle. However, below about 50 meters, pressure is great, causing ice to *flow* like a *plastic material*. A second important mechanism of glacial movement consists of the whole ice mass *slipping* along the ground.

- Glaciers erode land by *plucking* (lifting pieces of bedrock

out of place) and *abrasion* (grinding and scraping of a rock surface). Erosional features produced by valley glaciers include *glacial troughs, hanging valleys, cirques, arêtes, horns,* and *fiords*.

- Any sediment of glacial origin is called *drift*. The two distinct types of glacial drift are (1) *till*, which is unsorted sediment deposited directly by the ice; and (2) *stratified drift*, which is relatively well-sorted sediment laid down by glacial meltwater.

- The most widespread features created by glacial deposition are layers or ridges of till, called *moraines*. Associated with valley glaciers are *lateral moraines*, formed along the sides of the valley, and *medial moraines*, formed between two valley glaciers that have joined. *End moraines*, which mark the former position of the front of a glacier, and *ground moraine*, an undulating layer of till deposited as the ice front retreats, are common to both valley glaciers and ice sheets.

- Perhaps the most convincing evidence for the occurrence of several glacial advances during the *Ice Age* is the widespread existence of *multiple layers of drift* and an uninterrupted record of climate cycles preserved in *seafloor sediments*. In addition to massive erosional and depositional work, other effects of Ice Age glaciers included the *migration of organisms, changes in stream courses, adjustment of the crust* by rebounding after the removal of the immense load of ice, and *climate changes* caused by the existence of the glaciers themselves. In the sea, the most far-reaching effect of the Ice Age was the *worldwide change in sea level* that accompanied each advance and retreat of the ice sheets.

- Any theory that attempts to explain the causes of glacial ages must answer the two basic questions: (1) What causes the onset of glacial conditions? and (2) What caused the alternating glacial and interglacial stages that have been documented for the Pleistocene epoch? Two of the many hypotheses for the cause of glacial ages involve (1) plate tectonics and (2) variations in Earth's orbit.

- Practically all desert streams are dry most of the time and are said to be *ephemeral*. Nevertheless, *running water*

is responsible for most of the erosional work in a desert. Although wind erosion is more significant in dry areas than elsewhere, the main role of wind in a desert is in the transportation and deposition of sediment.

- Many of the landscapes of the Basin and Range region of the western and southwestern United States are the result of streams eroding uplifted mountain blocks and depositing the sediment in interior basins. *Alluvial fans, playas,* and *playa lakes* are features often associated with these landscapes.

- For wind erosion to be effective, dryness and scant vegetation are essential. *Deflation*, the lifting and removal of loose material, often produces shallow depressions called *blowouts* and can also lower the surface by removing sand and silt, leaving behind a stony veneer called *desert pavement. Abrasion*, the "sandblasting" effect of wind, is often given too much credit for producing desert features. However, abrasion does cut and polish rock near the surface.

- Wind deposits are of two distinct types: (1) extensive *blankets of silt*, called *loess*, carried by wind in *suspension*, and (2) *mounds and ridges of sand*, called *dunes*, which are formed from sediment that is carried as part of the wind's *bed load*.

Key Terms

abrasion (p. 129)
alluvial fan (p. 144)
alpine glacier (p. 125)
arête (p. 130)
barchan dune (p. 148)
barchanoid dune (p. 149)
blowout (p. 145)
cirque (p. 130)
crevasse (p. 126)
cross beds (p. 148)
deflation (p. 145)
desert pavement (p. 145)
drift (p. 132)
drumlin (p. 135)
end moraine (p. 134)
ephemeral (p. 143)
esker (p. 136)
fiord (p. 131)

glacial erratic (p. 132)
glacial striations (p. 129)
glacial trough (p. 129)
glacier (p. 125)
ground moraine (p. 134)
hanging valley (p. 129)
horn (p. 130)
ice cap (p. 125)
ice sheet (p. 125)
interior drainage (p. 143)
kame (p. 136)
kettle (p. 134)
lateral moraine (p. 134)
loess (p. 147)
longitudinal dune (p. 149)
medial moraine (p. 134)
outwash plain (p. 134)
parabolic dune (p. 149)

piedmont glacier (p. 125)
playa lake (p. 144)
Pleistocene epoch (p. 137)
plucking (p. 129)
pluvial lake (p. 138)
rock flour (p. 129)
slip face (p. 148)
star dune (p. 149)
stratified drift (p. 132)
till (p. 132)
transverse dune (p. 149)
valley glacier (p. 125)
valley train (p. 134)
zone of accumulation (p. 126)
zone of wastage (p. 126)

Review Questions

1. What is a glacier? What percentage of Earth's land area do glaciers cover?

2. Each of the following statements refers to a particular type of glacier. Name the type of glacier.

    (a) The term *continental* is often used to describe this type of glacier.

    (b) This type of glacier is also called an *alpine glacier*.

    (c) This is a glacier formed when one or more valley glaciers spreads out at the base of a steep mountain front.

    (d) Greenland is the only example in the Northern Hemisphere.

3. Describe how glaciers fit into the hydrologic cycle. What role do they play in the rock cycle?

4. Describe the two components of glacial flow. At what rates do glaciers move? In a valley glacier does all of the ice move at the same rate? Explain.

5. Why do crevasses form in the upper portion of a glacier but not below a depth of about 50 meters?

6. Under what circumstances will the front of a glacier advance? Retreat? Remain stationary?

7. Describe two basic processes of glacial erosion.

8. How does a glaciated mountain valley differ from a mountain valley that was not glaciated?

9. List and describe the erosional features you might expect to see in an area where alpine glaciers exist or have recently existed.

10. What is glacial drift? What is the difference between till and stratified drift? What general effect do glacial deposits have on the landscape?

11. List the four basic moraine types. What do all moraines have in common? Distinguish between terminal and recessional moraines.

12. List and briefly describe four depositional features other than moraines.

13. About what percentage of Earth's land surface was covered at some time by Pleistocene glaciers? How does this compare to the area presently covered by ice sheets and glaciers? (Check your answer with Question 1).

14. List three indirect effects of Ice Age glaciers.

15. How might plate tectonics help us understand the cause of ice ages? Can plate tectonics explain the alternation between glacial and interglacial climates during the Pleistocene?

16. List four common misconceptions about deserts (see Box 5.3).

17. How extensive are the desert and steppe regions of Earth?

18. Describe the characteristics associated with each of the stages in the evolution of a mountainous desert.

19. Why is wind erosion relatively more important in arid regions than in humid areas?

20. Although sand dunes are the best-known wind deposits, accumulations of loess are very significant in some parts of the world. What is loess? Where are such deposits found? What are the origins of this sediment?

21. How do sand dunes migrate?

22. Identify each of the dunes described in the following statements.

    (a) Long ridges of sand oriented parallel to the prevailing wind.

    (b) Solitary, crescent-shaped dunes oriented with their tips pointing downwind.

    (c) Ridges of sand oriented at right angles to the prevailing wind.

    (d) Dunes whose tips point into the wind.

    (e) Scalloped rows of sand oriented at right angles to the wind.

    (f) Isolated dunes consisting of three or four sharp-crested ridges diverging from a central high point.

Examining the Earth System

1. Glaciers are solid, but are a basic part of the hydrologic cycle. Should glaciers be considered a part of the sphere we call the solid Earth or do they belong to the hydrosphere?

2. Some scientists think that ice should be a separate sphere of the Earth system, called the *cryosphere*. Does such an idea have merit?

3. Wind erosion occurs at the interface of the atmosphere, solid Earth, and biosphere and is influenced by the hydrosphere and human activity. With this in mind, describe how human activity contributed to the "Dust Bowl," the period of intense wind erosion in the Great Plains in the 1930s. You might find it helpful to research "Dust Bowl" on the Internet using a search engine such as HotBot (**http://www.hotbot.com**/) or Yahoo (**http://www.yahoo.com/**). You may also find the Smithsonian Institution's Drylands Exhibit site at **http://drylands.nasm.edu:1995/** informative.

Web Resources

The following are informative and interesting World Wide Web sites that address topics related to those presented in the chapter:

Ice Age Exhibit (Illinois State Museum):

- **http://www.museum.state.il.us/exhibits/ice_ages/**

World's Drylands Electronic Exhibit:

- **http://drylands.nasm.edu:1995/**

For direct links to these sites and other appropriate Earth science Web locations, chapter review exercises, and relevant, topical investigations that utilize Web resources, contact the *Earth Science*, 9th Edition Home Page at:

- **http://www.prenhall.com/tarbuck.**

Chapter 6

Earthquakes and Earth's Interior

In January 1995, a strong earthquake toppled this elevated expressway near Kobe, Japan. (Photo by Noboru Hashimoto/Sygma)

What is an earthquake? How does a seismograph record the location of a "quake"? Can earthquakes ever be predicted? If we could view the Earth's interior, what would it look like? In this chapter we shall answer these and other questions. The study of earthquakes is important, not only because of the devastating effect that some earthquakes have on us, but also because they furnish clues about the structure of Earth's interior.

On October 17, 1989, at 5:04 P.M. Pacific Daylight Time, millions of television viewers around the world were settling in to watch the third game of the World Series. Instead, they saw their television sets go black as tremors hit San Francisco's Candlestick Park. Although the earthquake was centered in a remote section of the Santa Cruz Mountains, 100 kilometers to the south, major damage occurred in the Marina District of San Francisco.

The most tragic result of the violent shaking was the collapse of some double-decked sections of Interstate 880, also known as the Nimitz Freeway. The ground motions caused the upper deck to sway, shattering the concrete support columns along a mile-long section of the freeway. The upper deck then collapsed onto the lower roadway, flattening cars as if they were aluminum beverage cans. This earthquake, named the Loma Prieta quake for its point of origin, claimed 67 lives.

In mid-January 1994, less than 5 years after the Loma Prieta earthquake devastated portions of the San Francisco Bay area, a major earthquake struck the Northridge area of Los Angeles. Although not the fabled "Big One," this moderate 6.6- to 6.9-magnitude earthquake left 51 dead, over 5000 injured, and tens of thousands of households without water and electricity. The damage exceeded $15 billion and was attributed to an apparently unknown fault that ruptured 14 kilometers (9 miles) beneath Northridge (Figure 6.1).

The Northridge earthquake began at 4:31 A.M. and lasted roughly 40 seconds. During this brief period, the quake terrorized the entire Los Angeles area. In the three-story Northridge Meadows apartment complex, 16 people died when sections of the upper floors collapsed onto the first-floor units. Nearly 300 schools were seriously damaged and a dozen major roadways buckled. Among these were two of California's major arteries—the Golden State Freeway (Interstate 5), where an overpass collapsed completely and blocked the roadway, and the Santa Monica Freeway. Fortunately, these roadways had practically no traffic at this early morning hour.

In nearby Granada Hills, broken gas lines were set ablaze while the streets flooded from broken water mains. Seventy homes burned in the Sylmar area. A 64-car freight train derailed, including some cars carrying hazardous cargo. But it is remarkable that the destruction was not greater. Unquestionably, the

Figure 6.1 This parking deck in Northridge, California, collapsed during an earthquake in January 1994. (Photo by Spencer Grant/Liaison International)

upgrading of structures to meet the requirements of building codes developed for this earthquake-prone area helped minimize what could have been a much greater human tragedy.

Across the Pacific from California lies Japan, no stranger to earthquakes, and among the most "quake-proofed" countries in the world. Yet at 5:46 A.M. on January 24, 1995, much of the "quake-proofing" proved futile as more than 5000 people perished in a 7.2-magnitude tremor centered near Kobe, the country's sixth-largest city (see chapter-opening photo).

Over 30,000 earthquakes that are strong enough to be felt occur worldwide annually. Fortunately, most are minor tremors and do very little damage. Generally, only about 75 significant earthquakes take place each year, and many of these occur in remote regions. However, occasionally a large earthquake occurs near a large population center. Under these conditions, an earthquake is among the most destructive natural forces on Earth (see Box 6.1).

The shaking of the ground, coupled with the liquefaction of some soils, wreaks havoc on buildings and other structures. In addition, when a quake occurs in a populated area, power and gas lines are often ruptured, causing numerous fires. In the famous 1906 San Francisco earthquake, much of the damage was caused by fires (Figure 6.2), which quickly became uncontrollable when broken water mains left firefighters with only trickles of water.

What Is an Earthquake?

Internal Processes
Earthquakes

An **earthquake** is the vibration of Earth produced by the rapid release of energy. Most often earthquakes are caused by slippage along a fault in Earth's crust. The energy released radiates in all directions from its source, the **focus**, in the form of waves. These waves are analogous to those produced when a stone is dropped into a calm pond (Figure 6.3). Just as the impact of the stone sets water waves in motion, an earthquake generates seismic waves that radiate throughout the Earth. Even though the energy dissipates rapidly with increasing distance from the focus, sensitive instruments located throughout the world record the event.

Earthquakes and Faults

The tremendous energy released by atomic explosions or by volcanic eruptions can produce an earthquake, but these events are relatively weak and infrequent. What mechanism produces a destructive earthquake? Ample evidence exists that Earth is not a static planet. We know that Earth's crust has been uplifted at times, because we have found numerous ancient wave-cut benches many meters above the level of the highest tides. Other regions exhibit evidence of extensive subsidence. In addition to these

Figure 6.2 San Francisco in flames after the 1906 earthquake. (Reproduced from the collection of the Library of Congress)

Box 6.1 People and the Environment: Damaging Earthquakes East of the Rockies

When you think "earthquake," you probably think of California and Japan. However, six major earthquakes have occurred in the central and eastern United States since colonial times. Three of these had estimated Richter magnitudes of 7.5, 7.3, and 7.8, and they were centered near the Mississippi River Valley in southeastern Missouri. Occurring on December 16, 1811, January 23, 1812, and February 7, 1812, these earthquakes, plus numerous smaller tremors, destroyed the town of New Madrid, Missouri, triggered massive landslides, and caused damage over a six-state area. The course of the Mississippi River was altered, and Tennessee's Reelfoot Lake was enlarged. The distances over which these earthquakes were felt are truly remarkable. Chimneys were reported downed in Cincinnati, Ohio, and Richmond, Virginia, while Boston residents, located 1770 kilometers (1100 miles) to the northeast, felt the tremor.

Despite the history of the New Madrid earthquake, Memphis, Tennessee, the largest population center in the area today, does not have adequate earthquake provisions in its building code. Further, because Memphis is located on unconsolidated floodplain deposits, buildings are more susceptible to damage than similar structures built on bedrock. It has been estimated that if an earthquake the size of 1811–1812 New Madrid event were to strike in the next decade, it would result in casualties in the thousands and damages in tens of billions of dollars.

Damaging earthquakes that occurred in Aurora, Illinois (1909), and Valentine, Texas (1931), remind us that other areas in the central United States are vulnerable.

The greatest historical earthquake in the eastern states occurred August 31, 1886, in Charleston, South Carolina. The event, which spanned 1 minute, caused 60 deaths, numerous injuries, and great economic loss within a radius of 200 kilometers (120 miles) of Charleston. Within 8 minutes, effects were felt as far away as Chicago and St. Louis, where strong vibrations shook the upper floors of buildings, causing people to rush outdoors. In Charleston alone, over 100 buildings were destroyed and 90 percent of the remaining structures were damaged. It was difficult to find a chimney still standing (Figure 6.A).

Numerous other strong earthquakes have been recorded in the eastern United States. New England and adjacent areas have experienced sizable shocks since colonial times. The first reported earthquake in the Northeast took place in Plymouth, Massachusetts, in 1683, and was followed in 1755 by the destructive Cambridge, Massachusetts, earthquake. Moreover, ever since records have been kept, New York State alone has experienced over 300 earthquakes large enough to be felt.

Earthquakes in the central and eastern United States occur far less frequently than in California. Yet history indicates that the East is vulnerable. Further, these shocks east of the Rockies have generally produced structural damage over a larger area than counterparts of similar magnitude in California. The reason is that the underlying bedrock in the central and eastern United States is older and more rigid. As a result, seismic waves are able to travel greater distances with less weakening than in the western United States. It is estimated that for earthquakes of similar magnitude, the region of maximum ground motion in the East may be up to 10 times larger than in the West. Consequently, the higher rate of earthquake occurrence in the western United States is balanced somewhat by the fact that central and eastern U.S. quakes can damage larger areas.

Figure 6.A Damage to Charleston, South Carolina, caused by the August 31, 1886, earthquake. Damage ranged from toppled chimneys and broken plaster to total collapse. (Photo courtesy of U.S. Geological Survey)

vertical displacements, offsets in fence lines, roads, and other structures indicate that horizontal movement is common (Figure 6.4). These movements are usually associated with large fractures in Earth's crust called **faults**.

Most of the motion along faults can be satisfactorily explained by the plate tectonics theory. This theory states that large slabs of Earth's crust are in continual slow motion. These mobile plates interact with neighboring plates, straining and deforming the rocks at

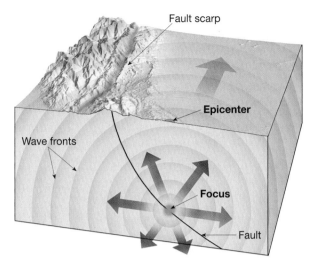

Figure 6.3 The focus of all earthquakes is located at depth. The surface location directly above it is called the epicenter.

their edges. In fact, it is along faults associated with plate boundaries that most earthquakes occur. Furthermore, earthquakes are repetitive—as soon as one is over, the continuous motion of the plates resumes, adding strain to the rocks until they fail again.

Elastic Rebound

The actual mechanism of earthquake generation eluded geologists until H. F. Reid of Johns Hopkins University conducted a study following the great 1906 San Francisco earthquake. The earthquake was accompanied by horizontal surface displacements of several meters along the northern portion of the San Andreas fault. This 1300-kilometer (780-mile) fracture runs north-south through southern California. It is a large fault zone that separates two great sections of Earth's crust, the North American plate and the Pacific plate. Field investigations determined that, during this single

Figure 6.4 Slippage along a fault produced an offset in this orange grove east of Calexico, California. (Photo by John S. Shelton)

earthquake, the Pacific plate lurched as much as 4.7 meters (15 feet) northward past the adjacent North American plate.

The mechanism for earthquake formation that Reid deduced from this information is illustrated in Figure 6.5 A, which shows an existing fault, or break in the rock. In part B, tectonic forces ever so slowly deform the crustal rocks on both sides of the fault, as demonstrated by the bent features. Under these conditions, rocks are bending and storing elastic energy, much like a wooden stick does if bent. Eventually, the frictional resistance holding the rocks together is overcome. As slippage occurs at the weakest point (the focus), displacement will exert stress farther along the fault, where additional slippage will occur until most of the built-up strain is released (Figure 6.5C). This slippage allows the deformed rock to "snap back." The vibrations we know as an earthquake occur as the rock elastically returns to its original shape. The "springing back" of the rock was termed **elastic rebound** by Reid, because the rock behaves elastically, much like a stretched rubber band does when it is released.

In summary, most earthquakes are produced by the rapid release of elastic energy stored in rock that has been subjected to great stress. Once the strength of the rock is exceeded, it suddenly ruptures, causing the vibrations of an earthquake. Earthquakes also occur along existing fault surfaces whenever the frictional forces on the fault surfaces are overcome.

The San Andreas is undoubtedly the most studied fault system in the world. Over the years, investigations have shown that displacement occurs along discrete segments that are 100 to 200 kilometers long. Further, each fault segment behaves somewhat differently from the others. Some portions of the San Andreas exhibit a slow, gradual displacement known as **fault creep**, which occurs relatively smoothly, and therefore with little noticeable seismic activity. Other segments regularly slip, producing small earthquakes.

Still other segments remain locked and store elastic energy for hundreds of years before rupturing in great earthquakes. The latter process is described as *stick-slip* motion, because the fault exhibits alternating periods of locked behavior followed by sudden slippage. It is estimated that great earthquakes should occur about every 50 to 200 years along those sections of the San Andreas fault that exhibit stick-slip motion. This knowledge is useful when assigning a potential earthquake risk to a given segment of the fault zone.

Not all movement along faults is horizontal. Vertical displacement, in which one side is lifted higher in relation to the other, is also common. Figure 6.6 shows a *fault scarp* (cliff) produced by such vertical movement. Further, many earthquakes occur at such great depths that no displacement is evident at the surface.

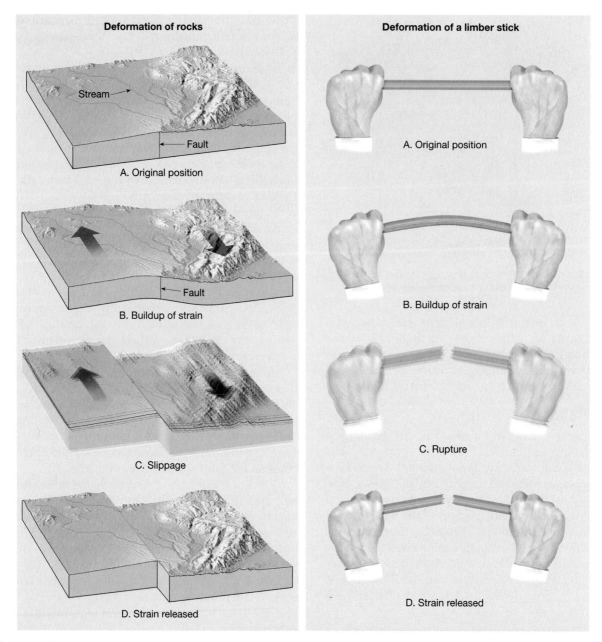

Deformation of rocks

Stream

Fault

A. Original position

Fault

B. Buildup of strain

C. Slippage

D. Strain released

Deformation of a limber stick

A. Original position

B. Buildup of strain

C. Rupture

D. Strain released

Figure 6.5 Elastic rebound. As rock is deformed it bends, storing elastic energy. Once the rock is strained beyond its breaking point it ruptures, releasing the stored-up energy in the form of earthquake waves.

uplift

Figure 6.6 Fault scarp produced from vertical movement during the 1964 Alaskan earthquake. (Courtesy of U.S. Geological Survey)

Foreshocks and Aftershocks

The intense vibrations of the 1906 San Francisco earthquake lasted about 40 seconds. Although most of the displacement along the fault occurred in this rather short period, additional movements along this and other nearby faults continued for several days following the main quake. The adjustments that follow a major earthquake often generate smaller earthquakes called **aftershocks**. Although these aftershocks are usually much weaker than the main earthquake, they can sometimes destroy already badly weakened structures. This occurred, for example, during a 1988 earthquake in Armenia. A large aftershock of magnitude 5.8 collapsed many structures that had been weakened by the main tremor.

In addition, small earthquakes called **foreshocks** often precede a major earthquake by days or, in some cases, by as much as several years. Monitoring of these foreshocks has been used as a means of predicting forthcoming major earthquakes, with mixed success. We will consider the topic of earthquake prediction in a later section of this chapter.

Tectonic Forces and Earthquakes

It is important to understand that the tectonic forces creating the strain that was eventually released during the 1906 San Francisco earthquake are still active.

Currently, laser beams are used to measure the relative motion between the opposite sides of this fault. These measurements reveal a displacement of 2 to 5 centimeters (1 to 2 inches) per year. Although this seems slow, it produces substantial movement over millions of years.

To illustrate, in 30 million years, this rate of displacement would slide the western portion of California northward so that Los Angeles, on the Pacific plate, would be adjacent to San Francisco on the North American plate! More important in the short term, a displacement of just 2 centimeters each year produces 2 meters of offset every 100 years. Consequently, the 4 meters of displacement produced during the 1906 San Francisco earthquake should occur at least every 200 years along this segment of the fault zone. This fact lies behind California's concern for making buildings earthquake-resistant, in anticipation of the inevitable "big one."

Earthquake Waves

Internal Processes
Earthquakes

The study of earthquake waves, **seismology**, dates back to attempts by the Chinese almost 2000 years ago to determine the direction to the source of each earthquake. Modern **seismographs** are instruments that record earthquake waves. Their principle is simple. A

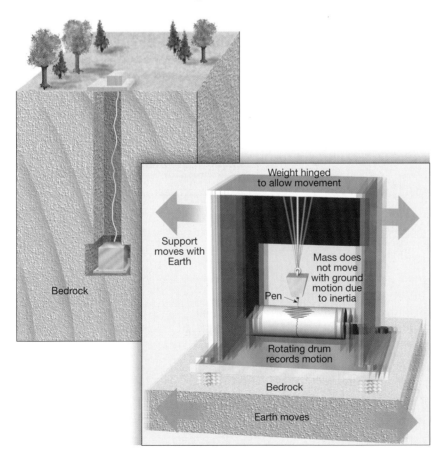

Figure 6.7 Principle of the seismograph. The inertia of the suspended mass tends to keep it motionless, while the recording drum, which is anchored to bedrock, vibrates in response to seismic waves. Thus, the stationary mass provides a reference point from which to measure the amount of displacement occurring as the seismic wave passes through the ground below.

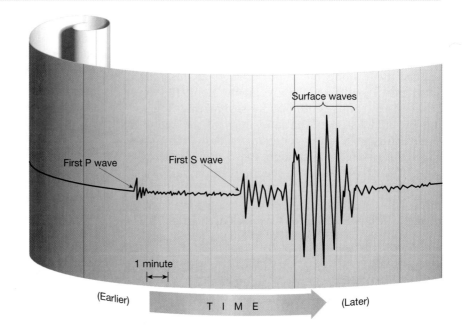

Surface waves

First P wave

First S wave

1 minute

(Earlier) T I M E (Later)

Figure 6.8 Typical seismic record. Note the time interval (about 5 minutes) between the arrival of the first P waves and the arrival of the first S waves.

weight is freely suspended from a support that is attached to bedrock (Figure 6.7). When waves from an earthquake reach the instrument, the inertia of the weight keeps it stationary, while Earth and the support vibrate. The movement of Earth in relation to the stationary weight is recorded on a rotating drum. (*Inertia*

is the tendency of a stationary object to remain still, or a moving object to stay in motion.)

Modern seismographs amplify and record ground motion, producing a trace as shown in Figure 6.8. These records, called **seismograms**, reveal that seismic waves are elastic energy. This energy radiates outward in all

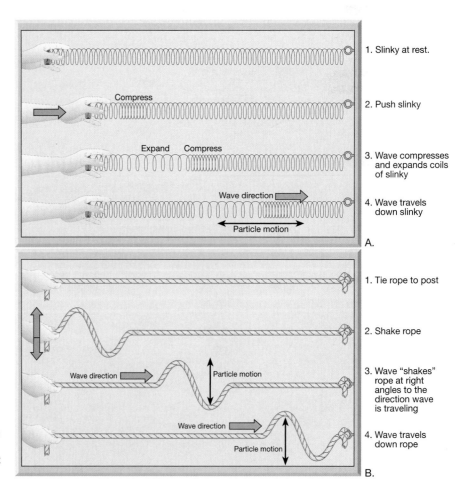

1. Slinky at rest.

2. Push slinky

Compress

3. Wave compresses and expands coils of slinky

Expand   Compress

4. Wave travels down slinky

Wave direction

Particle motion

A.

1. Tie rope to post

2. Shake rope

3. Wave "shakes" rope at right angles to the direction wave is traveling

Wave direction   Particle motion

4. Wave travels down rope

Wave direction

Particle motion

B.

Figure 6.9 Types of seismic waves and their characteristic motion. **A.** P waves compress, causing the particles in the material to vibrate back and forth in the same direction as the waves move. **B.** S waves cause particles to oscillate at right angles to the direction of wave motion.

directions from the focus, as you saw in Figure 6.3. The transmission of this energy can be compared to the shaking of gelatin in a bowl that is jarred. Seismograms reveal that two main types of seismic waves are generated by the slippage of a rock mass. Some travel along Earth's outer layer and are called **surface waves**. Others travel through Earth's interior and are called **body waves**. Body waves are further divided into **primary waves (P waves)** and **secondary waves (S waves)**.

Body waves are divided into P and S waves by their mode of travel through intervening materials. P waves are "push-pull" waves—they push (compress) and pull (expand) rocks in the direction the wave is traveling (Figure 6.9A). Imagine holding someone by the shoulders and shaking the person. This push-pull movement is how P waves move through the Earth. This wave motion is analogous to that generated by human vocal cords as they move air to create sound. Solids, liquids, and gases resist a change in volume when compressed and will elastically spring back once the force is removed. Therefore, P waves, which are compressional waves, can travel through all these materials.

In contrast, S waves "shake" the particles at right angles to their direction of travel. This can be illustrated by fastening one end of a rope and shaking the other end, as shown in Figure 6.9B. Unlike P waves, which temporarily change the *volume* of the intervening material by alternately compressing and expanding it, S waves temporarily change the *shape* of the material that transmits them. Because fluids (gases and liquids) do not respond elastically to changes in shape, they will not transmit S waves.

The motion of surface waves is somewhat more complex. As surface waves travel along the ground, they cause the ground and anything resting upon it to move, much like ocean swells toss a ship. In addition to their up-and-down motion, surface waves have a side-to-side motion similar to an S wave oriented in a horizontal plane. This latter motion is particularly damaging to the foundations of structures.

By observing a "typical" seismic record, as shown in Figure 6.8, you can see a major difference among these seismic waves: P waves arrive at the recording station first; then S waves; and then surface waves. This is a consequence of their speeds. To illustrate, the velocity of P waves through granite within the crust is about 6 kilometers per second. Under the same conditions, S waves travel at 3.5 kilometers per second. Differences in density and elastic properties of the rock greatly influence the velocities of these waves. Generally, in any solid material, P waves travel about 1.7 times faster than S waves, and surface waves can be expected to travel at 90 percent of the velocity of the S waves.

As you shall see, seismic waves allow us to determine the location and magnitude of earthquakes. In addition, seismic waves provide us with a tool for probing Earth's interior.

Finding Earthquake Epicenters

**Internal Processes
Earthquakes**

Recall that the *focus* is the place within Earth where earthquake waves originate. The **epicenter** is the location on the surface directly above the focus (see Figure 6.3).

The difference in velocities of P and S waves provides a method for locating the epicenter. The principle used is analogous to a race between two autos, one faster than the other. The P wave always wins the race, arriving ahead of the S wave. But, the greater the length of the race, the greater will be the difference in the arrival times at the finish line (the seismic station). Therefore, the greater the interval measured on a seismogram between the arrival of the first P wave and the first S wave, the greater the distance to the earthquake source.

A system for locating earthquake epicenters was developed by using seismograms from earthquakes whose epicenters could be easily pinpointed from physical evidence. From these seismograms, travel-time graphs were constructed (Figure 6.10). The first travel-time graphs were greatly improved when seismograms became available from nuclear explosions, because the precise location and time of detonation were known.

Using the sample seismogram in Figure 6.8 and the travel-time curves in Figure 6.10, we can determine the

Figure 6.10 A travel-time graph is used to determine the distance to the epicenter. The difference in arrival times of the first P wave and the first S wave in the example is 5 minutes. Thus, the epicenter is roughly 3800 kilometers (2350 miles) away.

Figure 6.11 Earthquake epicenter is located using the distance obtained from three seismic stations.

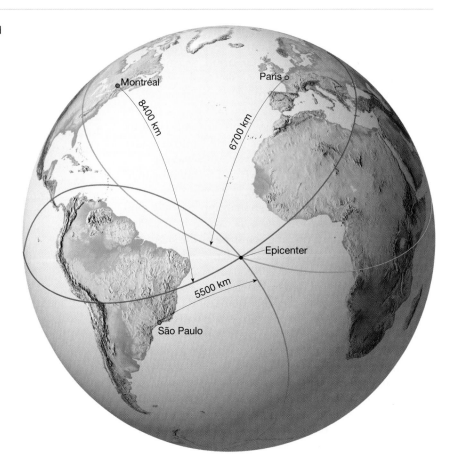

distance separating the recording station from the earthquake in two steps: (1) determine the time interval between the arrival of the first P wave and the first S wave, and (2) find on the travel-time graph the equivalent time spread between the P and S wave curves. From this information, we can determine that this earthquake occurred 3800 kilometers (2350 miles) from the recording instrument.

Now we know the *distance*, but what *direction*? The epicenter could be in any direction from the seismic station. As shown in Figure 6.11, the precise location can be found when the distance is known from three

Figure 6.12 Distribution of the 14,229 earthquakes with magnitudes equal to or greater than 5 for the years 1980 to 1990. (Data from National Geophysical Data Center/NOAA)

or more different seismic stations, On a globe, we draw a circle around each seismic station, Each circle represents the epicenter distance for each station. The point where the three circles intersect is the epicenter of the quake. This method is called "triangulation."

About 95 percent of the energy released by earthquakes originates in a few relatively narrow zones (Figure 6.12). The greatest energy is released along a path around the outer edge of the Pacific Ocean known as the *circum-Pacific belt*. Included in this zone are regions of great seismic activity, such as Japan, the Philippines, Chile, and numerous volcanic island chains, as exemplified by Alaska's Aleutian Islands.

Figure 6.12 reveals another continuous belt that extends for thousands of kilometers through the world's oceans. This zone coincides with the oceanic ridge system, an area of frequent but low-intensity seismic activity. By comparing this figure with Figure 7.8 (pp. 182–183), you can see a close correlation between the location of earthquake epicenters and plate boundaries.

Earthquake Intensity and Magnitude

Until a century ago, earthquake size and strength were described subjectively, making accurate classification of earthquake intensity difficult. Then, in 1902, Giuseppe Mercalli developed a fairly reliable intensity scale based on damage to various types of structures. The U.S. Coast and Geodetic Survey uses a modification of this scale today (Table 6.1).

The **Mercalli intensity scale** assesses the damage from a quake at a specific location. Please note that earthquake intensity depends not only on the strength of the earthquake but also on other factors, such as distance from the epicenter, the nature of surface materials, and building design. A modest 6.9-Richter-magnitude earthquake in Armenia in 1988 was very destructive, mainly because of inferior construction. A 1985 Mexico City quake was deadly because of the soft sediment upon which the city rests. Thus, the destruction wrought by earthquakes is very meaningful to people living there, but it is not a true measure of the earthquake's actual strength. Further, many earthquakes occur beneath the sea or at great depths in the crust and are not felt.

In 1935, Charles Richter of the California Institute of Technology introduced the concept of earthquake **magnitude**. Today, a refined **Richter scale** is used worldwide to describe earthquake magnitude. Richter magnitude is determined by measuring the amplitude (maximum displacement) of the largest wave recorded on the seismogram (see Figure 6.8). For seismic stations worldwide to obtain the same magnitude for a given earthquake, adjustments are made for the weakening of seismic waves with distance as they move away from the focus and for the sensitivity of the recording instrument.

Table 6.1 Modified Mercalli intensity scale.

I	Not felt except by a very few under especially favorable circumstances.
II	Felt only by a few persons at rest, especially on upper floors of buildings.
III	Felt quite noticeably indoors, especially on upper floors of buildings, but many people do not recognize it as an earthquake.
IV	During the day felt indoors by many, outdoors by few. Sensation like heavy truck striking building.
V	Felt by nearly everyone, many awakened. Disturbances of trees, poles, and other tall objects sometimes noticed.
VI	Felt by all; many frightened and run outdoors. Some heavy furniture moved; few instances of fallen plaster or damaged chimneys. Damage slight.
VII	Everybody runs outdoors. Damage negligible in buildings of good design and construction; slight to moderate in well-built ordinary structures; considerable in poorly built or badly designed structures.
VIII	Damage slight in specially designed sructures; considerable in ordinary substantial buildings with partial collapse; great in poorly built structures. (Fall of chimneys, factory stacks, columns, monuments, walls.)
IX	Damage considerable in specially designed structures. Buildings shifted off foundations. Ground cracked conspicuously.
X	Some well-built wooden structures destroyed. Most masonry and frame structures destroyed. Ground badly cracked.
XI	Few, if any (masonry) structures remain standing. Bridges destroyed. Broad fissures in ground.
XII	Damage total. Waves seen on ground surfaces. Objects thrown upward into air.

The largest earthquakes ever recorded had Richter magnitudes of 8.9. These great shocks released energy roughly equivalent to the detonation of a billion tons of TNT. Conversely, earthquakes with a Richter magnitude of less than 2.0 are usually not felt by humans. Table 6.2 shows how earthquake magnitudes and their effects are related.

Earthquakes vary enormously in strength, and great earthquakes produce traces having wave amplitudes that are thousands of times larger than those generated by weak tremors. To accommodate this wide variation, Richter could not use a linear scale, but instead used a *logarithmic scale* to express

Table 6.2 Earthquake magnitudes and expected world incidence.

Richter Magnitudes	Effects Near Epicenter	Estimated Number per Year
<2.0	Generally not felt, but recorded.	600,000
2.0–2.9	Potentially perceptible.	300,000
3.0–3.9	Felt by some.	49,000
4.0–4.9	Felt by most.	6200
5.0–5.9	Damaging shocks.	800
6.0–6.9	Destructive in populous regions.	266
7.0–7.9	Major earthquakes. Inflict serious damage.	18
≥8.0	Great earthquakes. Destroy communities near epicenter.	1.4

Source: *Earthquake Information Bulletin* and others.

magnitude. On this scale, an increase of one magnitude means a *tenfold* increase in wave amplitude. Thus, the amplitude of the largest surface wave for a 5.3-magnitude earthquake is 10 times greater than the wave amplitude produce by an earthquake having a magnitude of 4.3.

More important, each unit of Richter magnitude equates to roughly a *30-fold energy increase*. Thus, an earthquake with a magnitude of 6.5 releases 30 times more energy than one with a magnitude of 5.5, and roughly 900 times (30 × 30) more energy than a 4.5-magnitude quake. A great earthquake with a magnitude of 8.5 releases millions of times more energy than the smallest earthquake felt by humans.

Knowing this should dispel the notion that a moderate earthquake acts as a "pressure relief valve," decreasing the chances for a major quake in the same region. A moderate quake certainly relieves some strain, but thousands of moderate tremors would be needed to release the equivalent energy of one "great" earthquake.

Some of the world's major earthquakes and their corresponding Richter magnitudes are listed in Table 6.3.

Destruction from Earthquakes

The most violent earthquake to jar North America this century—the Good Friday Alaskan Earthquake—occurred in 1964. Felt throughout that state, the earthquake had a Richter magnitude of 8.3–8.4 and reportedly lasted 3 to 4 minutes. This event left 131 people dead, thousands homeless, and the economy of the state badly disrupted because it occurred near major towns and seaports (Figure 6.13). Had the schools and business districts been open on this holiday, the toll surely would have been higher. Within 24 hours of the initial shock, 28 aftershocks were recorded, 10 of which exceeded a Richter magnitude of 6.

Destruction from Seismic Vibrations

The 1964 Alaskan earthquake provided geologists with new insights into the role of ground shaking as a destructive force. Recall that the structural damage from earthquake waves depends on several factors, including (1) the intensity, (2) duration of the vibrations, (3) the nature of the material upon which the structure rests, and (4) the design of the structure.

All multistory structures in Anchorage were damaged by the vibrations, but the more flexible wood-frame residential buildings fared best. Figure 6.14 offers a striking example of how construction variations affect earthquake damage; you can see that the steel-frame building on the left withstood the vibrations, whereas the poorly designed J.C. Penney building was badly damaged. Engineers have learned that unreinforced masonry buildings are the most serious safety threat in earthquakes.

Most large structures in Anchorage were damaged even though they were built to conform to the earthquake provisions of the Uniform Building Code. Perhaps some of that destruction can be attributed to the unusually long duration of this earthquake (estimated at 3 to 4 minutes). (The San Francisco earthquake of 1906 was felt for about 40 seconds; the strong vibrations of the 1989 Loma Prieta earthquake lasted under 15 seconds.)

Table 6.3 Some notable earthquakes.

Year	Location	Deaths (est.)	Magnitude	Comments
1290	Chihli (Hopei), China	100,000		
1556	Shensi, China	830,000		Possibly the greatest natural disaster.
1737	Calcutta, India	300,000		
1755	Lisbon, Portugal	70,000		Tsunami damage extensive.
*1811–1812	New Madrid, Missouri	Few		Three major earthquakes.
*1886	Charleston, South Carolina	60		Greatest historical earthquake in the eastern United States.
*1906	San Francisco, California	1500	8.1–8.2	Fires caused extensive damage.
1908	Messina, Italy	120,000		
1920	Kansu, China	180,000		
1923	Tokyo, Japan	143,000	7.9	Fire caused extensive destruction.
1960	Southern Chile	5700	8.5–8.6	Possibly the largest-magnitude earthquake ever recorded.
*1964	Alaska	131	8.3–8.4	
1970	Peru	66,000	7.8	Great rockslide.
*1971	San Fernando, California	65	6.5	Damage exceeded $1 billion.
1975	Liaoning Province, China	Few	7.5	First major earthquake to be predicted.
1976	Tangshan, China	240,000	7.6	Not predicted.
1985	Mexico City	9500	8.1	Major damage occurred 400 km from epicenter.
1988	Armenia	25,000	6.9	Poor construction practices contributed to destruction.
*1989	San Francisco Bay area	62	7.1	Damages exceeded $6 billion.
1990	Northwestern Iran	50,000	7.3	Landslides and poor construction practices caused great damage.
*1994	Northridge, California	61	6.7	Damages in excess of $15 billion.
1995	Kobe, Japan	5472	6.9	Damage estimated to exceed $100 billion.

*U.S. earthquakes.
Source: U.S. National Oceanic and Atmospheric Administration.

The buildings in Anchorage are situated on soft, unconsolidated sediments, which amplify vibrations more than solid bedrock. Thus, damage in Anchorage was more severe than in Whittier, which rests on a firm foundation of granite. However, Whittier was damaged by a seismic sea wave (described in the next section).

The 1985 Mexican earthquake gave seismologists and engineers a vivid reminder of what had been learned from the 1964 Alaskan earthquake. In central Mexico City, nearly 400 kilometers (250 miles) from the epicenter, the vibrations intensified to five times that experienced in outlying districts, owing to the city's construction on soft sediments, remnants of an ancient lake bed (Figure 6.15).

Where unconsolidated materials are saturated with water, earthquakes can generate a phenomenon known as **liquefaction**. Under these conditions, what

Figure 6.13 Region most affected by the Good Friday earthquake of 1964. Note the epicenter (red dot). (After U.S. Geological Survey)

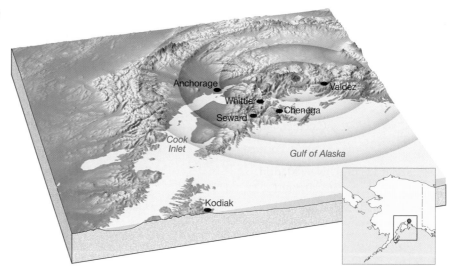

had been a stable soil turns into a fluid that is no longer capable of supporting buildings or other structures (Figure 6.16). As a result, underground objects such as storage tanks and sewer lines may float toward the surface of their newly liquefied environment. Buildings and other structures may settle and collapse. During the Loma Prieta earthquake, in San Francisco's Marina District, foundations failed and geysers of sand and water shot from the ground, indicating that liquefaction had occurred (Figure 6.17).

Tsunami

Most deaths associated with the 1964 Alaskan quake were caused by **seismic sea waves**, or **tsunami**. These destructive waves often are called "tidal waves" by the media. However, this name is inappropriate, for these waves are generated by earthquakes, not by the tidal effect of the Moon or Sun. The name *tsunami* is Japanese for "harbor wave," for Japanese harbors have suffered from many of them.

Most tsunami result from displacement along a fault located on the ocean floor or from a large underwater landslide triggered by an earthquake (Figure 6.18). Once formed, a tsunami resembles the ripples created when a pebble is dropped into a pond. In contrast to ripples, tsunami advance across the ocean at speeds between 500 and 950 kilometers (300 and 600 miles) per hour. Despite this, a tsunami in the open ocean can pass undetected because its height is usually less than 1 meter and the distance between wave crests is great, ranging from 100 to 700 kilometers. However, upon entering shallower coastal water, these destructive waves are slowed and the water begins to pile up

Figure 6.14 Damage to the five-story JC Penney Co. building, Anchorage, Alaska. Very little structural damage was incurred by the adjacent building. (Courtesy of NOAA)

Figure 6.15 During the 1985 Mexican earthquake, multistory buildings swayed back and forth as much as 1 meter. (Photo by James L. Beck)

to heights that occasionally exceed 30 meters (100 feet), as shown in Figure 6.18. As the crest of a tsunami approaches shore, it appears as a rapid rise in sea level with a turbulent and chaotic surface.

Usually the first warning of an approaching tsunami is a rather rapid withdrawal of water from beaches. Coastal residents have learned to heed this warning and move to higher ground, for about 5 to 30 minutes later the retreat of water is followed by a surge capable of extending hundreds of meters inland. In a successive fashion, each surge is followed by a rapid oceanward retreat of the water. These waves are separated by intervals of between 10 and 60 minutes. They are able to traverse thousands of kilometers of the ocean before their energy is dissipated (see Box 6.2).

The tsunami generated in the 1964 Alaskan earthquake heavily damaged communities along the Gulf of Alaskae and killed 107 people. By contrast, only 9 people died in Anchorage as a direct result of the vibrations.

Tsunami damage following the Alaskan earthquake extended along much of the west coast of North America. Despite a 1-hour warning, 12 people perished in Crescent City, California, from the fifth wave. The first wave crested about 4 meters (13 feet) above low tide and was followed by three progressively smaller waves. Believing that the tsunami had ceased, people returned

Figure 6.16 Effects of liquefaction. This tilted building rests on unconsolidated sediment that imitated quicksand during the 1985 Mexican earthquake. (Photo by James L. Beck)

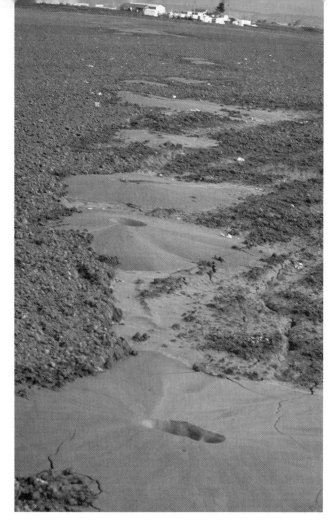

Figure 6.17 These "mud volcanoes" were produced by the Loma Prieta earthquake of 1989. They formed when geysers of sand and water shot from the ground, an indication that liquefaction occurred. (Photo by Richard Hilton, courtesy of Dennis Fox)

to the shore, only to be met by the fifth and most devastating wave. Superimposed upon high tide, it crested about 6 meters higher than the level of low tide.

On July 17, 1998, four villages on New Guinea's north coast were almost entirely swept away (Figure 6.19). Here an otherwise ordinary earthquake with a magnitude of 7.1 is thought to have triggered a massive underwater landslide. (Tremors of at least this size strike the globe every 3 weeks.) Within 5 to 10 minutes a plateau of water averaging 10 meters (30 feet) in height and perhaps 1 or 2 kilometers in width swept over that shore for more than a minute. Two similar waves followed at intervals of several minutes. Officially, the tsunami, the worst in more than two decades, killed 2134 people, with many still unaccounted for.

Landslides and Ground Subsidence

In the 1964 Alaskan earthquake, the greatest damage to structures was from landslides and ground subsidence triggered by the vibrations. At Valdez and Seward, the violent shaking caused river-delta materials to experience liquefaction; the subsequent slumping carried both waterfronts away. Because the disaster could happen again, the entire town of Valdez was relocated about 7 kilometers away on more stable ground. In Valdez, 31 people on a dock died when it slid into the sea.

Most of the damage in Anchorage was attributed to landslides. Many homes were destroyed in Turnagain Heights when a layer of clay lost its strength and over 200 acres of land slid toward the ocean (Figure 6.20). A portion of this landslide has been left in its natural condition as a reminder of this destructive event. The site was named "Earthquake Park." Downtown Anchorage was also disrupted as sections of the main business district dropped by as much as 3 meters (10 feet).

Fire

The 1906 San Francisco earthquake reminds us of the formidable threat of fire. The central city contained

Figure 6.18 Schematic drawing of a tsunami generated by displacement of the ocean floor. The speed of a wave moving across the ocean correlates with ocean depth. As shown, waves moving in deep water advance at speeds in excess of 800 kilometers per hour. Speed gradually slows to 50 kilometers per hour at depths of 20 meters. Decreasing depth slows the movement of the wave column. As waves slow in shallow water, they grow in height until they topple and rush onto shore with tremendous force. The size and spacing of these swells are not to scale.

Box 6.2   People and the Environment: Tsunami Warning System

Tsunamis traverse large stretches of ocean before their energy is fully dissipated. The tsunami generated by a 1960 Chilean earthquake, in addition to destroying villages along an 800-kilometer stretch of coastal South America, traveled 17,000 kilometers across the Pacific to Japan. Here, about 22 hours after the quake, considerable damage occurred in southern coastal villages. For several days afterward, tidal gauges in Hilo, Hawaii, detected these diminishing waves as they reverberated like echoes about the Pacific.

In 1946, a large tsunami struck the Hawaiian Islands without warning. A wave more than 15 meters (50 feet) high left several coastal villages in shambles. This destruction motivated the U.S. Coast and Geodetic Survey to establish a tsunami warning system for coastal areas of the Pacific. From seismic observatories throughout the region, large earthquakes are reported to the Tsunami Warning Center in Honolulu. Scientists at the Center use tidal gauges to determine whether a tsunami has been formed. Within an hour a warning is issued. Although tsunamis travel very rapidly, there is sufficient time to evacuate all but the region nearest the

Figure 6.B Tsunami travel times to Honolulu, Hawaii, from selected locations throughout the Pacific. (Data from NOAA)

epicenter (Figure 6.B). For example, a tsunami generated near the Aleutian islands would take 5 hours to reach Hawaii, and one generated near the coast of Chile would travel 15 hours before reaching Hawaii.

Fortunately, most earthquakes do not generate tsunamis. On the average, only about 1.5 destructive tsunamis are generated worldwide annually. Of these, only about one every 10 years is catastrophic.

Figure 6.19 A man stands before a wall of water about to engulf him at Hilo, Hawaii, on April 1, 1946. This tsunami, which originated in the Aleutian Islands near Alaska, was still powerful enough when it hit Hawaii to rise 9 to 16 meters (33-55 feet). The S.S. *Brigham Victory*, from which this photograph was taken, managed to survive the onslaught, but 159 people in Hawaii, including the man seen here, were killed. (Photo courtesy of Water Resources Center Archives, University of California, Berkeley)

Figure 6.20 Photo of a small portion of the Turnagain Heights slide. (Photo courtesy of U.S. Geological Survey)

mostly large, older wooden structures and brick buildings. The greatest destruction was caused by fires that started when gas and electrical lines were severed. The fires raged uncontrolled for 3 days and devastated over 500 city blocks (see Figure 6.2). The problem was compounded by the initial ground shaking, which broke the city's water lines into hundreds of unconnected pieces.

The fire was finally contained when buildings were dynamited along a wide boulevard to create a *fire break*, the same strategy used in fighting a forest fire. Although only a few deaths were attributed to the fires, such is not always the case. A 1923 earthquake in Japan (their worst quake prior to the 1995 Kobe tremor) triggered an estimated 250 fires, which devastated the city of Yokohama and destroyed more than half the homes in Tokyo. Over 100,000 deaths were attributed to the fires, which were driven by unusually high winds.

Can Earthquakes Be Predicted?

The vibrations that shook Northridge, California, in 1994 inflicted 61 deaths and about $15 billion in damage (Figure 6.21). This was from a brief earthquake (about 40 seconds) of moderate rating (6.6–6.9 on the Richter scale). Seismologists warn that earthquakes of comparable or greater strength will occur along the San Andreas fault, which cuts a 1300-kilometer (800-mile) path through the state. Following the 1995 earthquake in Kobe, Japan, a U.S. Geologist Survey physicist cautioned: "Kobe is almost a dress rehearsal for an earthquake on the Hayward fault (a fault parallel to the San Andreas, near San Francisco)." The obvious question is, can earthquakes be predicted?

Short-Range Predictions

The goal of short-range earthquake prediction is to provide a warning of the location and magnitude of a large earthquake within a narrow time frame. Substantial efforts to achieve this objective are being put forth in Japan, the United States, China, and Russia—countries where earthquake risks are high. This research has concentrated on monitoring possible *precursors*—phenomena that precede and thus provide a warning of a forthcoming earthquake. In California, for example, some seismologists are measuring uplift, subsidence, and strain in the rocks near active faults. Some Japanese scientists are studying peculiar anomalous behavior that may precede a quake.

One claim of a successful short-range prediction was made by Chinese seismologists after the February 4, 1975, earthquake in Liaoning Province. According to reports, very few people were killed, although more than 1 million lived near the epicenter, because the earthquake was predicted and the population was evacuated. Recently, some Western seismologists have questioned this claim and suggest instead that an intense swarm of foreshocks that began 24 hours before the main earthquake may have caused many people to evacuate spontaneously. Further, an official Chinese government report issued 10 years later stated that 1328 people died and 16,980 injuries resulted from this earthquake.

One year after the Liaoning earthquake, at least 240,000 people died in the Tangshan, China, earthquake, which was not predicted. The Chinese have also issued false alarms. In a province near Hong Kong, people reportedly left their dwellings for over a month, but no earthquake followed. Clearly, whatever method the Chinese employ for short-range predictions, it is *not* reliable.

Figure 6.21 Damage to Interstate 5 during the January 17, 1994 Northridge earthquake. (Photo by John T. Barr/Liaison Agency, Inc.)

For a prediction scheme to warrant general acceptance, it must be both accurate and reliable. Thus, *it must have a small range of uncertainty as regards to location and timing, and it must produce few failures, or false alarms.* Can you imagine the debate that would precede an order to evacuate a large city in the United States, such as Los Angeles or San Francisco? The cost of evacuating millions of people, arranging for living accomodations, and providing for their lost work time and wages, would be staggering.

Long-Range Forecasts

In contrast to short-range predictions, which aim to predict earthquakes within a time frame of hours or at most days, long-range forecasts give the probability of a certain magnitude earthquake occurring on a time scale of 30 to 100 years, or more. Stated another way, these forecasts give statistical estimates of the expected intensity of ground motion for a given area over a specified time frame. Although long-range forecasts may not be as informative as we might like, the data is important for updating the Uniform Building Code, which contains nationwide standards for designing earthquake-resistant structures.

Long-range forecasts are based on the premise that earthquakes are repetitive or cyclical, like the weather. In other words, as soon as one earthquake is over, the continuing motions of Earth's plates begin to build strain in the rocks again, until they fail once more. This has led seismologists to study historical records of earthquakes, for patterns, so their probability of recurrence might be established.

One study conducted by the U.S. Geological Survey gives the probability of a rupture occurring along various segments of the San Andreas fault for the 30 years between 1988 and 2018 (Figure 6.22). From this investigation, the Santa Cruz Mountains area was given a 30 percent probability of producing a 6.5-magnitude earthquake during this time period. In fact, it produced the Loma Prieta quake in 1989, of 7.1 magnitude.

The region along the San Andreas Fault given the highest probability (90 percent) of generating a quake is the Parkfield section. This area has been called the "Old Faithful" of earthquake zones because activity here has been very regular since record keeping began in 1857. Another section between Parkfield and the Santa Cruz Mountains is given a very low probability of generating an earthquake. This area has experienced very little seismic activity in historical times; rather, it exhibits a slow, continual movement known as *fault creep*. Such movement is beneficial because it prevents strain from building to high levels in the rocks.

In summary, it appears that the best prospects for making useful earthquake predictions involve forecasting magnitudes and locations in time scales of

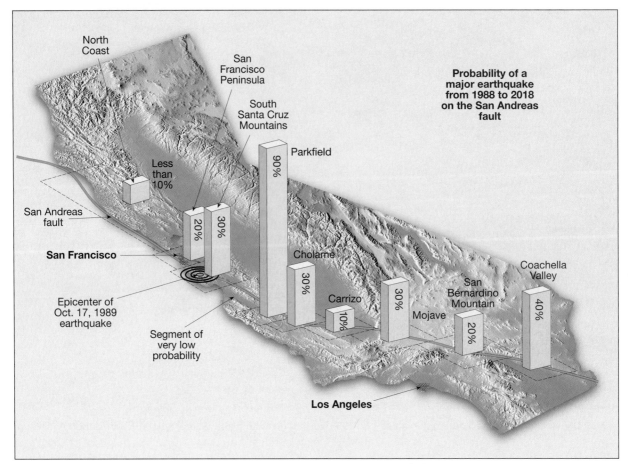

Figure 6.22 Probability of a major earthquake from 1988 to 2018 on the San Andreas fault.

years, or perhaps even decades. These forecasts are important because they provide information used to develop the Uniform Building Code and assist in land-use planning.

Earthquakes and Earth's Interior

Earth's interior lies just below us; however, access is very limited. The deepest well yet drilled has penetrated Earth's crust only 13 kilometers (8 miles), less than 0.2 percent of the distance to the planet's center. Consequently, most knowledge of our planet's interior comes from the study of earthquake waves that travel through Earth and vibrate the surface at some distant point.

From seismograms of earthquakes, seismologists accurately *measure the time* required for seismic waves to travel from the focus of an earthquake to a seismographic station. We know that P waves travel faster, and use this fact to determine distances to earthquake epicenters. But the time required for P and S waves to travel through Earth also depends on the properties of the materials encountered, so seismologists search for variations in travel times that cannot be accounted for simply by differences in the distance traveled. These

variations correspond to changes in properties of the materials encountered.

Changes in rock properties indicate that Earth has four major layers: (1) the **crust**, a very thin outer layer; (2) the **mantle**, a rocky layer beneath the crust (thickness: 2885 kilometers or 1789 miles); (3) the **outer core**, a layer that exhibits the characteristics of a mobile liquid (thickness: about 2270 kilometers or 1407 miles); and (4) the **inner core**, a solid metallic sphere (radius: about 1216 kilometers or 754 miles). Figure 6.23 shows these four layers.

Discovering Earth's Major Layers

The story of how seismologists discovered Earth's core-and-layers is interesting. In 1909, a pioneering Yugoslavian seismologist, Andrija Mohorovičić, presented the first convincing evidence for layering within Earth. By studying seismic records, he found that the velocity of seismic waves increases abruptly below about 50 kilometers of depth. This boundary separates the crust from the underlying mantle and is known as the **Mohorovičić discontinuity** in his honor. For reasons that are obvious, the name for this boundary was quickly shortened to **Moho**.

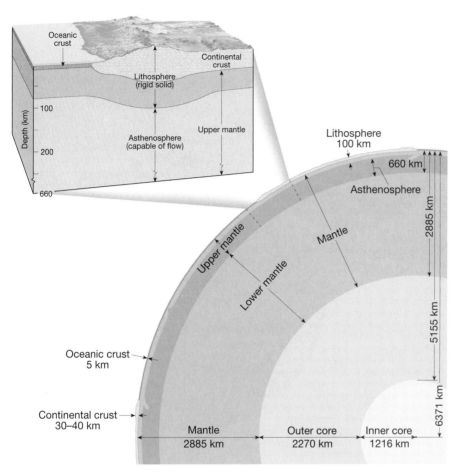

Figure 6.23 Cross-sectional view of Earth showing internal structure.

A few years later, another boundary was discovered by the German seismologist Beno Gutenberg. Generally, seismic waves from even small earthquakes are strong enough to travel around the world. This is why a seismograph in Antarctica can record earthquakes in California or Italy. But Gutenberg observed that P waves diminish and eventually die out about 105 degrees around the globe from an earthquake. Then, about 140 degrees away, the P waves reappear, but about 2 minutes later than would be expected, based on the distance traveled. This belt, where direct P waves are absent, is about 35 degrees wide and has been named the **shadow zone**. Figure 6.24 illustrates how this works.

Gutenberg realized that the shadow zone could be explained if Earth contained a core composed of material unlike the overlying mantle. The core must somehow hinder the transmission of P waves in a manner similar to the light rays being blocked by an opaque object that casts a shadow. However, rather than actually stopping the P waves, the shadow zone bends them, as shown in Figure 6.24. It was further learned that S waves could not travel through the core. Therefore, geologists concluded that at least a portion of this region is liquid.

In 1936, the last major subdivision of Earth's interior was predicted by Inge Lehmann, a Danish seismologist.

Lehmann discovered a new region of seismic reflection within the core. Hence, Earth has a core-within-a-core. The size of the inner core was not accurately calculated until the early 1960s, when underground nuclear tests were conducted in Nevada. Because the precise locations and times of the explosions were known, echoes from seismic waves that bounced off the inner core provided an accurate means of determining its size.

Over the past few decades, advances in seismology have allowed for much refinement of the gross view of Earth's interior presented so far. Of major importance was the discovery of the lithosphere and asthenosphere.

The Lithosphere and Asthenosphere. Earth's outer layer, consisting of the uppermost mantle and crust, forms a relatively cool, rigid shell. Although this layer consists of materials with markedly different chemical compositions, it tends to act as a unit that behaves similarly to mechanical deformation. This outermost rigid unit of Earth is called the **lithosphere** (*sphere of rock*). Averaging about 100 kilometers in thickness, the lithosphere may be 250 kilometers or more in thickness below the older portions of the continents (Figure 6.25). Within the ocean basins the lithosphere is only a few kilometers thick along the oceanic ridges and increases to perhaps 100 kilometers in regions of older and cooler oceanic rocks.

Figure 6.24 The abrupt change in physical properties at the mantle–core boundary causes the wave paths to bend sharply. This abrupt change in wave direction results in a shadow zone for P waves between about 105 and 140 degrees.

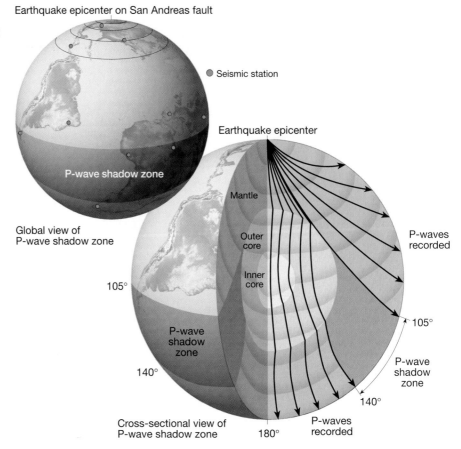

Beneath the lithosphere (to a depth of about 660 kilometers) lies a soft, relatively weak layer located in the upper mantle known as the **asthenosphere** (*weak spere*) (Figure 6.25). The upper 150 kilometers or so of the asthenosphere has a temperature/pressure regime in which a small amount of melting takes place (perhaps 1 to 5 percent). Within this very weak zone, the lithosphere is effectively detached from the asthenosphere located below. The result is that the lithosphere is able to move independently of the asthenosphere.

The discovery of the asthenosphere was an important contribution to an important theory known as *plate tectonics*. According to this model, Earth's rigid outer shell is broken into plates that move relative to each other. Thus, it is the weak nature of the asthenosphere that facilitates movement of the overriding lithospheric plates. This important concept is the topic of the next chapter.

Discovering Earth's Composition

We have examined Earth's structure, so let us now look at the composition of each layer. Composition tells us much about how our planet has developed over its estimated age of 4.6 billion years.

Earth's crust varies in thickness, exceeding 70 kilometers in some mountainous regions and being thinner than 5 kilometers in some oceanic areas (see

Figure 6.23). Early seismic data indicated that the continental crust, which is mostly made of lighter, granitic rocks, is quite different in composition from denser oceanic crust. Until the late 1960s, however, scientists had only seismic evidence from which to determine the composition of oceanic crust, because it lies beneath an average of 3 kilometers of water as well as hundreds of meters of sediment. With the development of deep-sea drilling technology the recovery of ocean floor samples was made possible. The samples were of basaltic composition—very different from the rocks that compose the continents.

Our knowledge of the rocks of the mantle and core is much more speculative. However, we do have some clues. Recall that some of the lava that reaches Earth's surface originates in the partially melted asthenosphere, within the mantle. In the laboratory, experiments have shown that partial melting of a rock called *peridotite* results in a melt that has a basaltic composition similar to lava that emerges during volcanic activity of oceanic islands. Denser rocks like peridotite are thought to make up the mantle and provide the lava for oceanic eruptions.

Surprisingly, meteorites or "shooting stars" that collide with Earth provide evidence of Earth's inner composition. Because meteorites are part of the solar system, they are assumed to be representative samples. Their composition ranges from metallic meteorites made of iron and nickel to stony meteorites composed of dense rock similar to peridotite.

Because Earth's crust contains a much smaller percentage of iron than do meteorites, geologists believe that the dense iron, and other dense metals, literally sank toward Earth's center during the planet's early history. By the same token, lighter substances may have floated to the surface, creating the less-dense crust. Thus, Earth's core is thought to be mainly dense iron and nickel, similar to metallic meteorites, whereas the surrounding mantle is believed to be composed of rocks similar to stony meteorites.

The concept of a molten iron outer core is further supported by Earth's magnetic field. Our planet acts as a large magnet. The most widely accepted mechanism explaining why Earth has a magnetic field requires that the core be made of a material that conducts electricity, such as iron, and which is mobile enough to allow circulation. Both of these conditions are met by the model of Earth's core that was established on the basis of seismic data.

Not only does an iron core explain Earth's magnetic field, but it also explains the high density of inner earth, about 14 times that of water at Earth's center. Even under the extreme pressure at those depths, average crustal rocks with densities 2.8 times that of water would not have the density calculated for the core. But iron, which is 3 times more dense than crustal rocks, has the required density.

In summary, although earthquakes can be very destructive, much of our knowledge about Earth's interior comes from the study of these phenomena. As our knowledge of earthquakes and their causes improves, we learn more of our planet's internal workings and possibly how to reduce the consequences of tremors. In the next chapter, you will see that most earthquakes originate at the boundaries of Earth's great lithospheric plates.

Figure 6.25 Relative positions of the asthenosphere and lithosphere.

Chapter Summary

- *Earthquakes* are vibrations of Earth produced by the rapid release of energy from rocks that rupture because they have been subjected to stresses beyond their limit. This energy, which takes the form of waves, radiates in all directions from the earthquake's source, called the *focus*. The movements that produce most earthquakes occur along large fractures, called *faults*, that are associated with plate boundaries.

- Two main groups of *seismic waves* are generated during an earthquake: (1) *surface waves*, which travel along the outer layer of Earth; and (2) *body waves*, which travel through Earth's interior. Body waves are further divided into *primary*, or *P*, *waves*, which push (compress) and pull (expand) rocks in the direction the wave is traveling, and *secondary*, or *S*, *waves*, which "shake" the particles in rock

at right angles to their direction of travel. P waves can travel through solids, liquids, and gases. Fluids (gases and liquids) will not transmit S waves. In any solid material, P waves travel about 1.7 times faster than do S waves.

- The location on Earth's surface directly above the focus of an earthquake is the *epicenter*. An epicenter is determined using the difference in velocities of P and S waves.

- *There is a close correlation between earthquake epicenters and plate boundaries.* The principal earthquake epicenter zones are along the outer margin of the Pacific Ocean, known as the *circum-Pacific belt*, and through the world's oceans along the *oceanic ridge system*.

- Earthquake *intensity* depends not only on the strength of the earthquake, but on other factors, such as distance from the epicenter, the nature of surface materials, and building design. The *Mercalli intensity scale* assesses the damage from a quake at a specific location. Using the *Richter scale*, the *magnitude* (a measure of the total amount of energy released) of an earthquake is determined. This is done by measuring the *amplitude* (maximum displacement) of the largest seismic wave recorded, with adjustments of the amplitude made for the weakening of seismic waves as they move from the focus, as well as for the sensitivity of the recording instrument. A logarithmic scale is used to express magnitude, in which a tenfold increase in recorded wave amplitude corresponds to an increase of 1 on the magnitude scale. Each unit of Richter magnitude equates to roughly a 32-fold energy increase.

- The most obvious factors that determine the amount of destruction accompanying an earthquake are the *magnitude* of the earthquake and the *proximity* of the quake to a populated area. *Structural damage* attributable to earthquake vibrations depends on several factors, including (1) *intensity*, (2) *duration* of the vibrations, (3) *nature of the material* upon which the structure rests, and (4) the *design* of the structure. Secondary effects of earthquakes include *tsunamis, landslides, ground subsidence*, and *fire*.

- Substantial research to predict earthquakes is underway in Japan, the United States, China, and Russia—countries where earthquake risk is high. No consistent method of short-range prediction has yet been devised. Long-range forecasts are based on the premise that earthquakes are repetitive or cyclical. Seismologists study the history of earthquakes for patterns, so their occurrences might be predicted.

- As indicated by the behavior of P and S waves as they travel through Earth, the four major zones of Earth's interior are the (1) *crust* (the very thin outer layer), (2) *mantle* (a rocky layer located below the crust with a thickness of 2885 kilometers), (3) *outer core* (a layer about 2270 kilometers thick, which exhibits the characteristics of a mobile liquid), and (4) *inner core* (a solid metallic sphere with a radius of about 1216 kilometers).

- The *continental crust* is primarily made of *granitic* rocks, while the *oceanic crust* is of *basaltic* composition. *Ultramafic rocks* such as *peridotite* are thought to make up the *mantle*. The *core* is composed mainly of *iron and nickel*.

- The crust and uppermost mantle form Earth's cool rigid outer shell called the *lithosphere*. Beneath the lithosphere lies a soft, relatively weak layer of the mantle known as the *asthenosphere*.

Key Terms

aftershock (p. 161)
asthenosphere (p. 176)
body wave (p. 163)
crust (p. 174)
earthquake (p. 157)
elastic rebound (p. 159)
epicenter (p. 163)
fault (p. 158)
fault creep (p. 159)
focus (p. 157)

foreshock (p. 161)
inner core (p. 174)
liquefaction (p. 167)
lithosphere (p. 175)
magnitude (p. 165)
mantle (p. 174)
Mercalli intensity scale (p. 165)
Mohorovičić discontinuity (Moho)
 (p. 174)
outer core (p. 174)

primary (P) wave (p. 163)
Richter scale (p. 165)
secondary (S) wave (p. 163)
seismic sea wave (tsunami) (p. 168)
seismogram (p. 162)
seismograph (p. 161)
seismology (p. 161)
shadow zone (p. 175)
surface wave (p. 163)

Review Questions

1. What is an earthquake? Under what circumstances do earthquakes occur?

2. How are faults, foci, and epicenters related?

3. Who was first to explain the actual mechanism by which earthquakes are generated?

4. Explain what is meant by elastic rebound.

5. Faults that are experiencing no active creep may be considered "safe." Rebut or defend this statement.

6. Describe the principle of a seismograph.

7. Using Figure 6.10, determine the distance between an earthquake and a seismic station if the first S wave arrives 3 minutes after the first P wave.

8. List the major differences between P and S waves.

9. Which type of seismic wave causes the greatest destruction to buildings?

10. Most strong earthquakes occur in a zone on the globe known as the _____

11. What factor contributed most to the extensive damage that occurred in the central portion of Mexico City during the 1985 earthquake?

12. The 1988 Armenian earthquake had a Richter magnitude of 6.9, far less than the great quakes in Alaska in 1964 and San Francisco in 1906. Nevertheless, the loss of life was far greater in the Armenian event. Why?

13. An earthquake measuring 7 on the Richter scale releases about _____ times more energy than an earthquake with a magnitude of 6.

14. List four factors that affect the amount of destruction caused by seismic vibrations.

15. In addition to the destruction created directly by seismic vibrations, list three other types of destruction associated with earthquakes.

16. Distinguish between the Mercalli scale and the Richter scale.

17. What is a tsunami? How is one generated?

18. Cite some reasons why an earthquake with a moderate magnitude might cause more extensive damage than a quake with a high magnitude.

19. What evidence do we have that Earth's outer core is molten?

20. Contrast the physical makeup of the asthenosphere and the lithosphere.

21. Why are meteorites considered important clues to the composition of Earth's interior?

22. Describe the composition of the following:

    (a) continental crust   (c) mantle

    (b) oceanic crust   (d) core

Examining the Earth System

1. What potentially disastrous phenomenon often occurs when the energy of an earthquake is tranferred from the solid Earth to the hydrosphere (ocean) at their interface on the floor of the ocean? When the energy from this event is expended along a coast, how might coastal lands and the biosphere be altered?

2. Several research projects are considering ways that humans could interact with the earthquake-generating process and prevent, or significantly reduce, the release of destructive energy. Begin your research of earthquake prevention at this University of Colorado at Boulder Web site, **http://www.Colorado.EDU/hazards/sites/earthquakes.html**. Following your investigation, describe some of the current earthquake prevention proposals being considered by researchers. Do you believe that the proposals are feasible? What are some other ways to lessen the devastating impact of earthquakes?

Web Resources

The following are informative and interesting World Wide Web sites that address topics related to those presented in the chapter:

Earthquake Links (United States Geological Survey):

- **http://www.usgs.gov/network/science/earth/earthquake.html**

Realtime Earthquake Locator:

- **http://quake.wr.usgs. gov/cgi-bin/quake/gldfs.cr.usgs.gov**

For direct links to thse sites and other appropriate Earth science Web locations, chapter review exercises, and relevant, topical investigations that utilize Web resources, contact the *Earth Science*, 9th Edition Home Page at:

- **http://www.prenhall.com/tarbuck.**

Chapter 7

Plate Tectonics

Composite satellite image of a part of North Africa, Arabia, and the Persian Gulf. The movements of giant slabs called plates cause climates to gradually change and ocean basins and mountains to form. (© Worldsat International, Inc./Jim Knighton, 1995. Mississauga, Ontario, Canada. All rights reserved.)

Will California eventually "slide" into the ocean, as some predict? Have continents really "drifted" apart over the centuries? Answers to these questions and many others that have intrigued geologists for decades are now being provided by an exciting theory on large-scale movements taking place within Earth. This theory, called plate tectonics, represents the real frontier of the Earth sciences, and its implications are so far-reaching that it can be considered the framework from which most other geological processes should be viewed.

Early in this century, most geologists thought that the geographic positions of the ocean basins and continents were fixed. During the last few decades, however, vast amounts of new data have dramatically changed our understanding of the nature and workings of our planet. Earth scientists now realize that the continents gradually migrate across the globe. Where landmasses split apart, new ocean basins are created between the diverging blocks. Meanwhile, older portions of the seafloor are carried back into the mantle in regions where trenches occur in the deep ocean floor. Because of these movements, blocks of continental material eventually collide and form Earth's great mountain ranges (Figure 7.1). In short, a revolutionary new model of Earth's tectonic* processes has emerged.

This profound reversal of scientific understanding has been appropriately described as a scientific revolution. Like other scientific revolutions, considerable time elapsed between the idea's inception and its general acceptance. The revolution began early in the twentieth century as a relatively straightforward proposal that the continents drift about the face of Earth. After many years of heated debate, the idea of drifting continents was rejected by the vast majority of Earth scientists. However, during the 1950s and 1960s, new evidence rekindled interest in this proposal. By 1968, these new developments led to the unfolding of a far more encompassing theory than continental drift—a theory known as *plate tectonics*.

---

*Tectonics refers to the deformation of Earth's crust and results in the formation of structural features such as mountains.

Figure 7.1 Mt. McKinley in Alaska's Denali National Park is the highest peak in North America. (Photo by Carr Clifton)

Continental Drift: An Idea Before Its Time

The idea that continents, particularly South America and Africa, fit together like pieces of a jigsaw puzzle originated with improved world maps. However, little significance was given this idea until 1915, when Alfred Wegener, a German meteorologist and geophysicist, published *The Origin of Continents and Oceans*. In this book, Wegener set forth his radical hypothesis of **continental drift**.[†]

Wegener suggested that a supercontinent he called **Pangaea** (meaning "all land") once existed (Figure 7.2). He further hypothesized that, about 200 million years ago, this supercontinent began breaking into smaller continents, which then "drifted" to their present positions (see Figure 7.26).

Wegener and others collected substantial evidence to support these claims. The fit of South America and Africa, and the geographic distribution of fossils, rock structures, and ancient climates all seemed to support the idea that these now-separate landmasses were once joined. Let us examine their evidence.

[†]Wegener's ideas were actually preceded by those of an American geologist, F. B. Taylor, who in 1910 published a paper on continental drift. Taylor's paper provided little supporting evidence for continental drift, which may have been the reason that it had a relatively small impact on the scientific community.

Evidence: The Continental Jigsaw Puzzle

Like a few others before him, Wegener first suspected that the continents might have been joined when he noticed the remarkable similarity between the coastlines on opposite sides of the South Atlantic. However, his use of present-day shorelines to make a fit of the continents was challenged immediately by other Earth scientists. These opponents correctly argued that shorelines are continually modified by erosional processes, and even if continental displacement had taken place, a good fit today would be unlikely. Wegener appeared to be aware of this problem, and, in fact, his original jigsaw fit of the continents was only very crude.

A much better approximation of the true outer boundary of the continents is the continental shelf. Today, the seaward edge of the continental shelf lies submerged, several hundred meters below sea level. In the early 1960s, scientists produced a map that attempted to fit the edges of the continental shelves at a depth of 900 meters. The remarkable fit that was obtained is shown in Figure 7.3. Although the continents overlap in a few places, these are regions where streams have deposited large quantities of sediment, thus enlarging the continental shelves. The overall fit was even better than the supporters of continental drift suspected it would be.

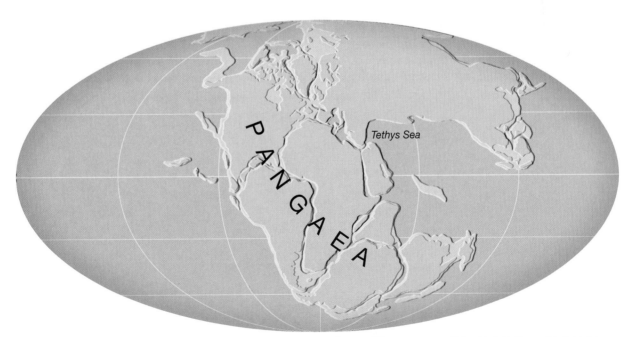

Figure 7.2 Reconstruction of Pangaea as it is thought to have appeared 200 million years ago. (After R. S. Deitz and J. C. Holden. *Journal of Geophysical Research* 75: 4943.)

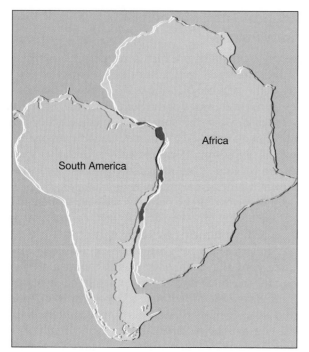

Figure 7.3 This shows the best fit of South America and Africa along the continental slope at a depth of 500 fathoms (about 900 meters). The areas where continental blocks overlap appear in brown. (After A. G. Smith. "Continental Drift." In *Understanding the Earth*, edited by I. G. Gass.)

Evidence: Fossils Match Across the Seas

Although Wegener was intrigued by the jigsaw fit of the continental margins that lie on opposite sides of the Atlantic, he at first thought the idea of a mobile Earth improbable. Not until he came across an article citing fossil evidence for the existence of a land bridge connecting South America and Africa did he begin to take his own idea seriously. Through a search of the literature, Wegener learned that most paleontologists were in agreement that some type of land connection was needed to explain the existence of identical fossils on the widely separated landmasses.

To add credibility to his argument for the existence of the supercontinent of Pangaea, Wegener cited documented cases of several fossil organisms that had been found on different landmasses but that could not have crossed the vast oceans presently separating the continents. The classic example is *Mesosaurus*, a presumably aquatic, snaggle-toothed reptile whose fossil remains are limited to eastern South America and southern Africa (Figure 7.4). If *Mesosaurus* had been able to swim well enough to cross the vast South Atlantic Ocean, its remains should be more widely distributed. As this is not the case, Wegener argued that South America and Africa must have been joined—somehow.

How did scientists explain the discovery of identical fossil organisms separated by thousands of kilometers of open ocean? The idea of land bridges was the most widely accepted solution to the problem of migration (Figure 7.5). We know, for example, that during the most recent glacial period, the lowering of sea level allowed animals to cross the narrow Bering Strait between Asia and North America. Was it possible, then, that one or more land bridges once connected Africa and South America? We are now quite certain that land bridges of this magnitude did not exist, for their remnants should still lie below sea level. But they are nowhere to be found.

Present-Day Organisms. In his book, Wegener also cited the distribution of present-day organisms as evidence to support the concept of drifting continents. For example, modern organisms with similar ancestries clearly had to evolve in isolation during the last few tens of millions of years. Most obvious of these are the Australian marsupials, which have a direct fossil link to the marsupial opossums found in the Americas.

Figure 7.4 Fossils of *Mesosaurus* have been found on both sides of the South Atlantic and nowhere else in the world. Fossil remains of this and other organisms on the continents of Africa and South America appear to link these landmasses during the late Paleozoic and early Mesozoic eras.

Figure 7.5 These sketches by John Holden illustrate various explanations for the occurrence of similar species on landmasses that are presently separated by vast oceans. (Reprinted with permission of John Holden)

Evidence: Rock Types and Structures Match

Anyone who has worked a picture puzzle knows that, in addition to the pieces fitting together, the picture must also be continuous. The "picture" that must match in the "Continental Drift Puzzle" is one of rock types and mountain belts. If the continents were once together to form Pangaea, the rocks found in a particular region on one continent should closely match in age and type those in adjacent positions on the adjoining continent.

Such evidence exists in the form of several mountain belts that terminate at one coastline, only to reappear on a landmass across the ocean. For instance, the mountain belt that includes the Appalachians trends northeastward through the eastern United States and disappears off the coast of Newfoundland (Figure 7.6A). Mountains of comparable age and structure are found in the British Isles and Scandinavia. When these landmasses are reassembled as in Figure 7.6B, the mountain chains form a nearly continuous belt. Numerous other rock structures exist that appear to have formed at the same time and were subsequently split apart.

A.

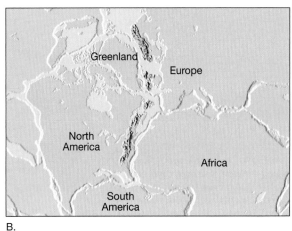

B.

Figure 7.6 Matching mountain ranges across the North Atlantic. **A.** The Appalachian Mountains trend along the eastern flank of North America and disappear off the coast of Newfoundland. Mountains of comparable age and structure are found in the British Isles and Scandanavia. **B.** When these landmasses are placed in their predrift locations, these ancient mountain chains form a nearly continuous belt. These folded mountain belts formed roughly 300 million years ago as the landmasses collided during the formation of the supercontinent of Pangaea.

Wegener was very satisfied that the similarities in rock structure on both sides of the Atlantic linked these landmasses. In his own words, "It is just as if we were to refit the torn pieces of a newspaper by matching their edges and then check whether the lines of print run smoothly across. If they do, there is nothing left but to conclude that the pieces were in fact joined in this way.‡

Evidence: Ancient Climates

Because Alfred Wegener was a meteorologist by training, he was keenly interested in obtaining paleoclimatic (ancient climatic) data in support of continental drift. His efforts were rewarded when he found evidence for dramatic global climatic changes. For instance, glacial deposits indicate that, near the end of the Paleozoic era (between 220 million and 300 million years ago), ice sheets covered extensive areas of the Southern Hemisphere. Layers of glacial till were found in southern Africa and South America, as well as in India and Australia. Below these beds of glacial debris lay striated and grooved bedrock. In some locations,

‡Alfred Wegener, *The Origin of Continents and Oceans.* Translated from the 4th revised German edition of 1929 by J. Birman (London: Methuen, 1966).

the striations and grooves indicated that the ice had moved from what is now the sea onto land. Much of the land area containing evidence of this late Paleozoic glaciation presently lies within 30 degrees of the equator in a subtropical or tropical climate.

Could Earth have gone through a period sufficiently cold to have generated extensive continental glaciers in what is presently a tropical region? Wegener rejected this explanation because, during the late Paleozoic, large tropical swamps existed in the Northern Hemisphere. The lush vegetation of these swamps eventually became the major coal fields of the eastern United States, Europe, and Siberia.

Fossils from these coal fields indicate that the tree ferns which produced the coal deposits had large fronds. This indicates a tropical setting. Furthermore, unlike trees in colder climates, the tree trunks lacked growth rings. Growth rings do not form in tropical plants because there are minimal seasonal fluctuations in temperature.

Wegener believed that a better explanation for the paleoclimatic regimes he observed is provided by fitting together the landmasses as a supercontinent, with South Africa centered over the South Pole (Figure 7.7). This would account for the conditions

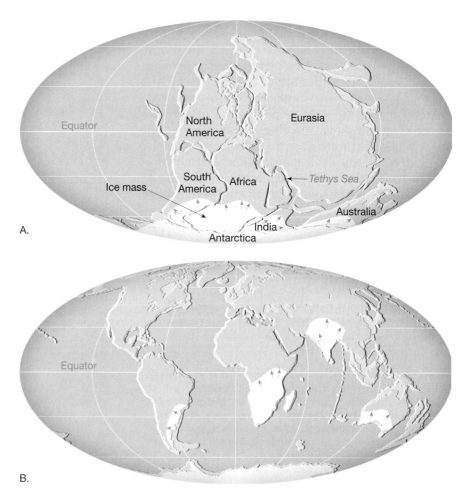

Figure 7.7 A. The supercontinent Pangaea showing the area covered by glacial ice 300 million years ago.
B. The continents as they are today. The shading outlines areas where evidence of the old ice sheets exists.

necessary to generate extensive expanses of glacial ice over much of the Southern Hemisphere. At the same time, this geography would place the northern landmasses nearer the tropics and account for their vast coal deposits.

Wegener was so convinced that his explanation was correct that he wrote, "This evidence is so compelling that by comparison all other criteria must take a back seat."

How does a glacier develop in hot, arid Australia? How do land animals migrate across wide expanses of open water? As compelling as this evidence may have been, 50 years passed before most of the scientific community would accept it and the logical conclusions to which it led.

The Great Debate

Wegener's proposal did not attract much open criticism until 1924 when his book was translated into English. From this time on, until his death in 1930, his drift hypothesis encountered a great deal of hostile criticism. To quote the respected American geologist T. C. Chamberlin, "Wegener's hypothesis …takes considerable liberty with our globe, and is less bound by restrictions or tied down by awkward, ugly facts than most of its rival theories. Its appeal seems to lie in the fact that it plays a game in which there are few restrictive rules and no sharply drawn code of conduct."

One of the main objections to Wegener's hypothesis stemmed from his inability to provide a mechanism that was capable of moving the continents across the globe. Wegener proposed that the tidal influence of the Moon was strong enough to give the continents a westward motion. However, the prominent physicist Harold Jeffreys quickly countered with the argument that tidal friction of the magnitude needed to displace the continents would bring Earth's rotation to a halt in a matter of a few years.

Wegener also proposed that the larger and sturdier continents broke through the oceanic crust, much like ice breakers cut through ice. However, no evidence existed to suggest that the ocean floor was weak enough to permit passage of the continents without themselves being appreciably deformed in the process.

Although most of Wegener's contemporaries opposed his views, even to the point of open ridicule, a few considered his ideas plausible. For these few geologists who continued the search for additional evidence, the exciting concept of continents in motion held their interest. Others viewed continental drift as a solution to previously unexplainable observations.

Plate Tectonics: A Modern Version of an Old Idea

Internal Processes
Plate Tectonics

During the years that followed Wegener's proposal, major strides in technology permitted mapping of the ocean floor. Moreover, extensive data on seismic activity and Earth's magnetic field became available. By 1968, these developments led to the unfolding of a far more encompassing theory than continental drift, known as **plate tectonics**. The implications of plate tectonics are so far-reaching that this theory is today the framework within which to view most geologic processes.

According to the plate tectonics model, the uppermost mantle, along with the overlying crust, behaves as a strong, rigid layer, known as the **lithosphere**. This outermost shell overlies a weaker region in the mantle known as the **asthenosphere**. Further, the lithosphere is broken into numerous segments called **plates**, which are in motion and are continually changing in shape and size. As shown in Figure 7.8, seven major plates are recognized. They are the North American, South American, Pacific, African, Eurasian, Australian, and Antarctic plates. The largest is the Pacific plate, which is located mostly within the ocean. Notice from Figure 7.8 that several of the large plates include an entire continent plus a large area of seafloor (for example, the South American plate). This is a major departure from Wegener's continental drift hypothesis, which proposed that the continents moved through the ocean floor, not with it. Note also that none of the plates are defined entirely by the margins of a continent.

Intermediate-sized plates include the Caribbean, Nazca, Philippine, Arabian, Cocos, and Scotia plates. In addition, there are over a dozen smaller plates that have been identified but are not shown in Figure 7.8.

We know that lithospheric plates move at very slow, but continuous, rates of a few centimeters a year. This movement is ultimately driven by the unequal distribution of heat within Earth. The titanic grinding movements of Earth's lithospheric plates generate earthquakes, create volcanoes, and deform large masses of rock into mountains.

Plate Boundaries

Plates move as coherent units relative to all other plates. Although the interiors of plates may deform, all major interactions among individual plates (and therefore most deformation) occurs along their *boundaries*. In fact, the first attempts to outline plate boundaries were made using locations of earthquakes. Later work showed that plates are bounded by three distinct types of boundaries, which are differentiated by

Figure 7.8 A mosaic of rigid plates constitutes Earth's outer shell.

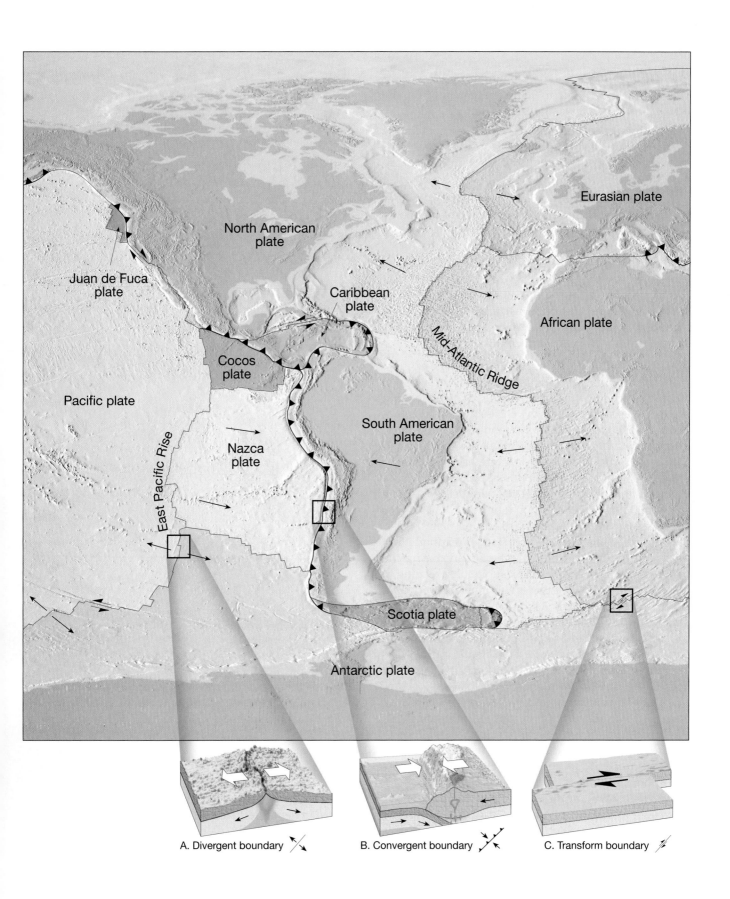

Juan de Fuca plate

North American plate

Caribbean plate

Cocos plate

Pacific plate

East Pacific Rise

Nazca plate

South American plate

Mid-Atlantic Ridge

Eurasian plate

African plate

Scotia plate

Antarctic plate

A. Divergent boundary

B. Convergent boundary

C. Transform boundary

the type of movement they exhibit. These boundaries are depicted in Figure 7.9 and are briefly described here:

1. **Divergent boundaries**—where plates move apart, resulting in upwelling of material from the mantle to create new seafloor (Figure 7.9A).

2. **Convergent boundaries**—where plates move together, resulting in the subduction (consumption) of oceanic lithosphere into the mantle (Figure 7.9B).

3. **Transform fault boundaries**—where plates grind past each other without the production or destruction of lithosphere (Figure 7.9C).

Each plate is bounded by a combination of these zones, as you can see in Figure 7.8. For example, the Nazca plate has a divergent boundary on the west, a convergent boundary on the east, and numerous small transform faults that offset segments of divergent boundaries on the north and south.

Divergent Boundaries

Most divergent boundaries, where plate spreading occurs, are situated along the crests of oceanic ridges (Figure 7.10). Here, as the plates move away from the ridge axis, the fractures created are immediately filled with molten rock that oozes up from the hot asthenosphere. This material cools slowly to produce new slivers of seafloor. In a continuous manner, successive plate spreading and upwelling of magma add new oceanic crust (lithosphere) between the diverging plates.

This mechanism is called **seafloor spreading**. It has produced the floor of the Atlantic Ocean during the past 165 million years. The typical rate of spreading at these ridges is 5 centimeters (2 inches) per year, although it varies from 2 to 20 centimeters per year. This extremely slow rate of lithosphere production is nevertheless rapid enough so that all of Earth's ocean basins could have been generated within the last 200 million years. In fact, none of the ocean floor that has been dated exceeds 180 million years in age.

Along divergent boundaries where molten rock emerges, the ocean floor is elevated (Figure 7.10). Worldwide, this ridge extends for over 70,000 kilometers through all major ocean basins. You can see parts of the **mid-ocean ridge** system in Figure 7.8. As new lithosphere is formed along the oceanic ridge it is slowly yet continually displaced away from the zone of upwelling found along the ridge axis. Thus, it begins to cool and contract, thereby increasing in density. This partially accounts for the greater depth of the older and cooler oceanic crust found in the deep ocean basins (Figure 7.10). In addition, the mantle rocks located below the oceanic crust cool and strengthen with

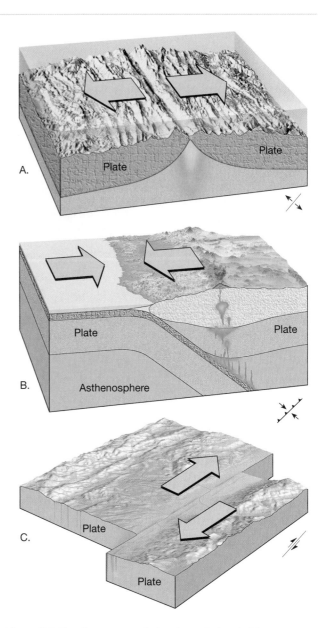

Figure 7.9 The three types of plate boundaries. **A.** Divergent boundary. **B.** Convergent boundary. **C.** Transform fault boundary.

increased distance from the ridge axis, thereby adding to a plate's thickness. Stated another way, both the thickness and the density of oceanic lithosphere are age dependent. The older (cooler) it is, the greater its thickness and density.

Not all spreading centers are found in the middle of large oceans. The Red Sea is the site of a recently formed divergent boundary. Here the Arabian Peninsula separated from Africa and began to move toward the northeast. Consequently, the Red Sea is providing oceanographers with a view of how the Atlantic Ocean may have looked in its infancy. Another narrow, linear sea produced by seafloor spreading in the recent geologic past is the Gulf of

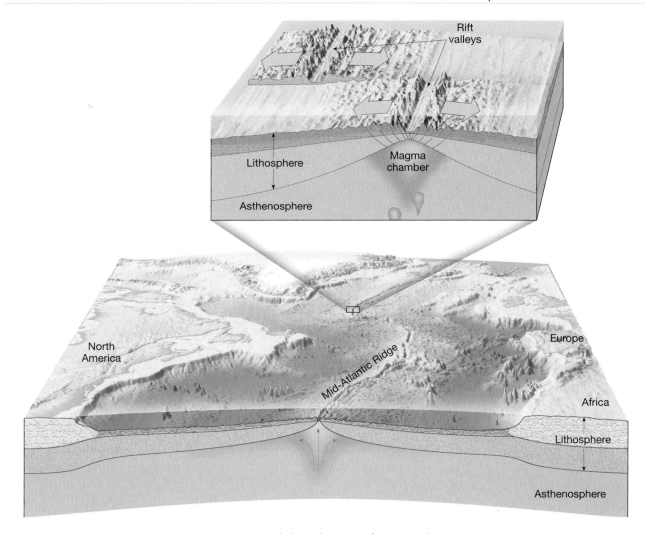

Figure 7.10 Most divergent plate boundaries are situated along the crests of oceanic ridges.

California, which separates the Baja Peninsula from the rest of Mexico.

When a spreading center develops within a continent, the landmass may split into two or more smaller segments as Wegener had proposed for the breakup of Pangaea. The fragmentation of a continent is thought to be associated with the upward movement of hot rock from the mantle. The effect of this activity is doming in the crust directly above the hot rising plume. This uplifting produces extensional forces that stretch and thin the crust as shown in Figure 7.11A. Extension of the crust is accompanied by alternate episodes of faulting and volcanism. Adjacent to the spreading axis, faulted crustal blocks create down-faulted valleys called **rifts** or **rift valleys** (Figure 7.11B). As the spreading continues, the rift valley will lengthen and deepen, eventually extending out into the ocean. At this point the valley will become a narrow linear sea with an outlet to the ocean, similar to the Red Sea today (Figure 7.11C). The zone of rifting

will remain the site of igneous activity, continually generating new seafloor in an ever-expanding ocean basin (Figure 7.11D).

The East African rift valleys represent the initial stage in the breakup of a continent as just described (Figure 7.12). The extensive volcanic activity that accompanies continental rifting is exemplified by large volcanic mountains such as Kilimanjaro and Mount Kenya. If the rift valleys in Africa remain active, East Africa will eventually separate from the mainland in much the same way the Arabian Peninsula split just a few million years ago.

Not all rift valleys develop into full-fledged spreading centers. Running through the central United States is an aborted rift zone extending from Lake Superior to Kansas. This once-active rift valley is filled with rock that was extruded onto the crust more than a billion years ago. Why one rift valley develops into a full-fledged oceanic spreading center while others are abandoned is not yet known.

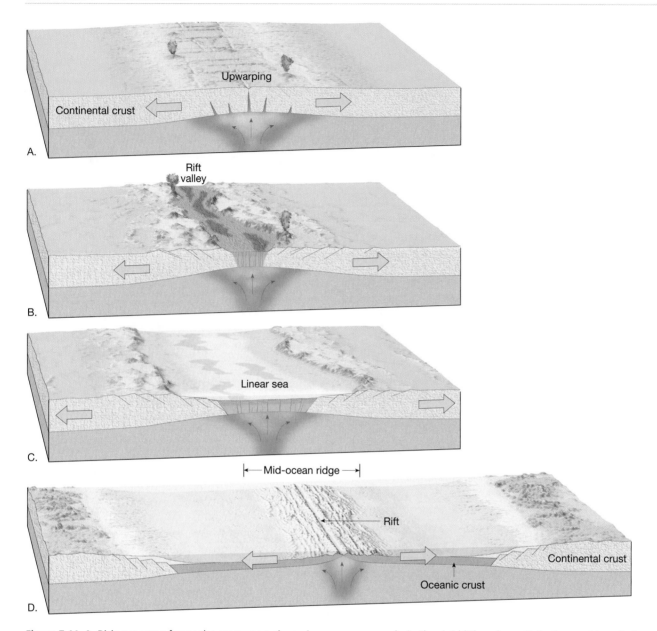

Figure 7.11 A. Rising magma forces the crust upward, causing numerous cracks in the rigid lithosphere. **B.** As the crust is pulled apart, large slabs of rock sink, generating a rift zone. **C.** Further spreading generates a narrow sea. **D.** Eventually, an expansive ocean basin and ridge system are created.

Convergent Boundaries

At spreading centers, new oceanic lithosphere is continually being generated. However, because the total surface area of Earth remains constant, lithosphere must also be consumed. Zones of plate convergence are the sites where oceanic lithosphere is subducted and absorbed into the mantle. When two plates converge, the leading edge of one is bent downward, allowing it to descend beneath the other (Figure 7.13A). The region where an oceanic plate descends into the asthenosphere is called a **subduction zone**. As the oceanic plate slides beneath the overriding plate, the oceanic plate bends, thereby producing a **deep-ocean trench**, like the Peru–Chile trench (Figure 7.14). Trenches formed in this manner can be thousands of kilometers long, 8 to 12 kilometers deep, and about 100 kilometers wide.

The average angle at which oceanic lithosphere descends into the asthenosphere is about 45 degrees. However, depending on its buoyancy, a plate might descend at an angle as small as a few degrees or it might plunge vertically (90 degrees) into the mantle. When a spreading axis is located near a subduction zone, the lithosphere is young and, therefore, warm and buoyant. Thus, the angle of descent is small. This

Figure 7.12 East African rift valleys and associated features.

is the situation along parts of the Peru–Chile trench. Low dip angles are usually associated with a strong coupling between the descending slab and the overriding plate. Consequently, these regions experience great earthquakes. By contrast, some subduction zones, such as the Mariana trench, have steep angles of descent and few strong earthquakes.

Although all convergent zones have the same basic characteristics, they are highly variable features. Each is controlled by the type of crustal material involved and the tectonic setting. Convergent boundaries can form between two oceanic plates, one oceanic and one continental plate, or two continental plates. All these situations are illustrated in Figure 7.13.

Oceanic-Continental Convergence. Whenever the leading edge of a plate capped with continental crust converges with a plate capped with oceanic crust, the plate with the less dense continental material remains "floating," whereas the denser oceanic slab sinks into the asthenosphere (Figure 7.13A). When a descending plate reaches a depth of about 100 to 150 kilometers,

heat drives water and other volatile components from the subducted sediments into the overlying mantle. These substances act as a flux does at a foundry, inducing partial melting of mantle rocks at reduced temperatures. The partial melting of mantle rock generates magmas having a basaltic or possibly andesitic composition. The newly formed magma, being less dense than the rocks of the mantle, will buoyantly rise. Often the magma will pond (accumulate) beneath the overlying continental crust where it may melt some of the silica-enriched crustal rocks. Eventually some of this silica-rich magma may migrate to the surface, where it can give rise to volcanic eruptions, some of which are explosive.

The Andean arc that runs along the western flank of South America is the product of magma generated as the Nazca plate descends beneath the continent (see Figure 7.8). In the central section of the southern Andes the subduction angle is very shallow, which probably accounts for the lack of volcanism in that area. As the South American plate moves westward, it overruns the Nazca plate. The result is a seaward

Figure 7.13 Three types of convergent plate boundaries. **A.** Oceanic-continental. **B.** Oceanic-oceanic. **C.** Continental-continental.

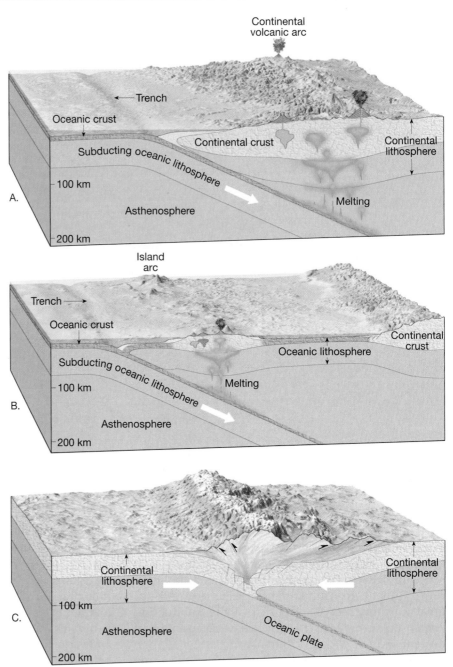

migration of the Peru–Chile trench and a reduction in the size of the Nazca plate.

Mountains such as the Andes, which are produced in part by volcanic activity associated with the subduction of oceanic lithosphere, are called **continental volcanic arcs**. Another active continental volcanic arc is located in the western United States. The Cascade Range of Washington, Oregon, and California consists of several well-known volcanic mountains, including Mount Rainier, Mount Shasta, and Mount St. Helens. As the continuing activity of Mount St. Helens testifies,

the Cascade Range is still active. The magma here arises from melting triggered by the subduction of the Juan de Fuca plate—a small remaining remnant of a once large plate.

A remnant of a formerly extensive continental volcanic arc is California's Sierra Nevada, in which Yosemite National Park is located. The Sierra Nevada is much older than the Cascade Range, and has been inactive for several million years as evidenced by the absence of volcanic cones. Here erosion has stripped away most of the obvious traces of volcanic activity and left exposed the large, crystallized magma chambers that once fed lofty volcanoes.

Oceanic-Oceanic Convergence. When two oceanic slabs converge, one descends beneath the other, initiating volcanic activity in a manner similar to that which occurs at an oceanic-continental convergent boundary. In this case, however, the volcanoes form on the ocean floor rather than on a continent (Figure 7.13B). If this activity is sustained, it will eventually build volcanic structures that emerge as islands. The volcanic islands are spaced about 80 kilometers apart and are built upon submerged ridges a few hundred kilometers wide. This newly formed land consisting of an arc-shaped chain of small volcanic islands is called a **volcanic island arc**. The Aleutian, Mariana, and Tonga islands are examples of volcanic island arcs. Island arcs such as these are generally located 200 to 300 kilometers from the trench axis. Located adjacent to the island arcs just mentioned are the Aleutian trench, the Mariana trench, and the Tonga trench (Figure 7.14).

Only two volcanic island arcs are located in the Atlantic, the Lesser Antilles arc adjacent to the Caribbean Sea, and the Sandwich Islands in the South Atlantic. The Lesser Antilles are a product of the subduction of the Atlantic beneath the Caribbean plate. Located within this arc is the island of Martinique where Mount Pelée erupted in 1902 destroying the town of St. Pierre and killing an estimated 28,000 people, and the island of Montserrat, where volcanic activity has occurred very recently.

In a few places, volcanic arcs are built upon both oceanic and continental crust. For example, the western section of the Aleutian arc consists of numerous islands built on oceanic crust, whereas the volcanoes at the eastern end of the chain are located on the Alaska Peninsula. Further, some volcanic island arcs are built on fragments of continental crust that have been rifted from the mainland. This type of volcanic island arc is exemplified by the Philippines and Japan.

Continental-Continental Convergence. When two plates carrying continental crust converge, neither plate will subduct beneath the other because of the low density and thus the buoyant nature of continental rocks. The result is a collision between the two continental blocks (Figure 7.13C). Such a collision occurred when the subcontinent of India "rammed" into Asia and produced the Himalayas—the most spectacular mountain range on Earth (Figure 7.15). During this collision, the continental crust buckled, fractured, and was generally shortened and thickened. In addition to the Himalayas, several other major mountain systems, including the Alps, Appalachians, and Urals, formed during continental collisions.

Prior to a continental collision, the landmasses involved are separated by an ocean basin. As the continental blocks converge, the intervening seafloor is subducted beneath one of the plates. Subduction initiates partial melting in the overlying mantle rocks, which, in turn, results in the growth of a volcanic arc. Depending on the location of the subduction zone, the volcanic arc could develop on either of the converging landmasses, or if the subduction zone developed several hundred kilometers seaward from the coast, a volcanic island arc would form. Eventually, as the intervening seafloor is consumed, these continental masses collide. This folds and deforms the accumulation of sediments along the continental margin as if they were placed in a gigantic vise. The result is the formation of a new mountain range, composed of deformed and metamorphosed sedimentary rocks, fragments of the volcanic arc, and possibly slivers of oceanic crust.

After continents collide, the subducted oceanic plate may separate from the continental block and continue its downward movement. However, because of its buoyancy, continental lithosphere cannot be carried very far into the mantle. In the case of the Himalayas, the leading edge of the Indian plate was forced partially under Asia, generating an unusually great thickness of continental lithosphere. This accumulation accounts, in part, for the high elevation of the Himalayas and helps explain the elevated Tibetan Plateau to the north.

Transform Fault Boundaries

The third type of plate boundary is the transform fault, where plates grind past one another without the production or destruction of lithosphere. Transform faults were first identified where they join offset segments of an ocean ridge. At first it was erroneously assumed that the ridge system originally formed a long and continuous chain that was later offset by horizontal displacement along these large faults. However, the displacement along these faults was found to be in the exact opposite direction required to produce the offset ridge segments.

The true nature of transform faults was discovered in 1965 by a Canadian researcher who suggested that these large faults connect the global active belts (convergent boundaries, divergent boundaries, and other transform faults) into a continuous network that divides Earth's outer shell into several rigid plates. Thus, he became the first to suggest that Earth was made of individual plates, while at the same time identifying the faults along which relative motion between the plates is made possible.

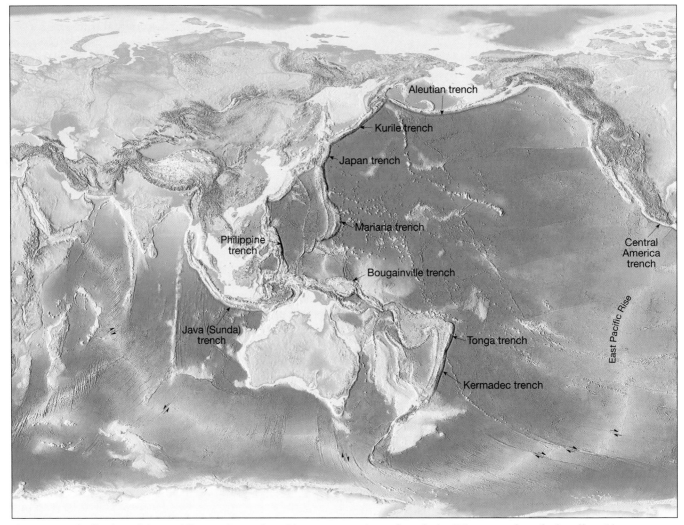

Figure 7.14 Distribution of the world's oceanic trenches, ridge system, and transform faults. Where transform faults offset ridge segments, they permit the ridge to change direction (curve) as can be seen in the Atlantic Ocean.

Most transform faults join two segments of a mid-ocean ridge (Figure 7.16). Here, they are part of prominent linear breaks in the oceanic crust known as *fracture zones* that include both the transform fault as well as their inactive extension into the plate interior. These fracture zones are present approximately every 100 kilometers along the trend of a ridge axis. As shown in Figure 7.16, active transform faults lie only *between* the two offset ridge segments. Here seafloor produced at one ridge axis moves in the opposite direction as seafloor produced at an opposing ridge segment. Thus, between the ridge segments these adjacent slabs of oceanic crust are grinding past each other along a transform fault. Beyond the ridge crests are the inactive zones, which are preserved as linear topographic scars. These fracture zones tend to curve such that small segments roughly parallel the direction of plate motion at the time of their formation.

In another role, transform faults provide the means by which the oceanic crust created at ridge crests can be transported to a site of destruction, the deep-ocean trenches. Figure 7.17 illustrates this situation. Notice that the Juan de Fuca plate moves in a southeasterly direction, eventually being subducted under the west coast of the United States. The southern end of this relatively small plate is bounded by the Mendocino transform fault. This transform fault boundary connects the Juan de Fuca ridge to the Cascade subduction zone (Figure 7.17). Therefore, it facilitates the movement of the crustal material created at the ridge crest to its destination beneath the North American continent (Figure 7.17). Another example of a *ridge-trench* transform fault is found southeast of the tip of South America. Here transform faults on the north and south margins of the Scotia plate connect the trench to a short spreading axis (see Figure 7.8).

Although most transform faults are located within the ocean basins, a few, including California's famous San Andreas fault, cut through continental crust. Notice in Figure 7.17 that the San Andreas fault

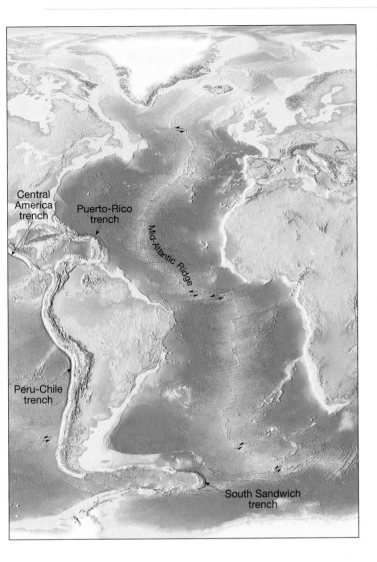

Central America trench

Puerto-Rico trench

Mid-Atlantic Ridge

Peru-Chile trench

South Sandwich trench

been presented. Some of the evidence that was instrumental in solidifying the support for this new concept follows. Note that some of the evidence was not new; rather, it was a new interpretation of old data that swayed the tide of opinion.

Evidence: Paleomagnetism

Probably the most persuasive evidence to the geologic community for the acceptance of the plate tectonics theory comes from the study of Earth's magnetic field. Anyone who has used a compass to find direction knows that the magnetic field has a north pole and a south pole. These magnetic poles align closely, but not exactly, with the geographic poles. (The geographic poles are simply the top and bottom of the spinning sphere we live on, the points through which passes the imaginary axis of rotation.)

In many respects the magnetic field is very much like that produced by a simple bar magnet. Invisible lines of force pass through Earth and extend from one pole to the other. A compass needle, itself a small magnet free to move about, becomes aligned with these lines of force and thus points toward the magnetic poles.

The technique used to study ancient magnetic fields relies on the fact that certain rocks contain minerals that serve as fossil compasses. These iron-rich minerals, such as magnetite, are abundant in lava flows of basaltic composition. When heated above a certain temperature called the *Curie point*, these magnetic minerals lose their magnetism. However, when these iron-rich grains cool below their Curie point (about 580°C) they become magnetized in the direction parallel to the existing magnetic field. Once the minerals solidify, the magnetism they possess will remain "frozen" in this position. In this regard, they behave much like a compass needle inasmuch as they "point" toward the existing magnetic poles. Then, if the rock is moved, or if the magnetic pole changes position, the rock magnetism will, in most instances, retain its original alignment. Rocks formed thousands of millions of years ago thus "remember" the location of the magnetic poles at the time of their formation and are said to possess fossil magnetism, or **paleomagnetism**.

Polar Wandering. A study of lava flows conducted in Europe in the 1950s led to an amazing discovery. The magnetic alignment in the iron-rich minerals in lava flows of different ages was found to vary widely. A plot of the apparent positions of the magnetic north pole revealed that, during the past 500 million years, the location of the pole had gradually wandered from a spot near Hawaii northward through eastern Siberia and finally to its present site (Figure 7.18A). This was clear evidence that either the magnetic poles had migrated through time, an idea known as **polar wandering**, or

connects a spreading center located in the Gulf of California to the Cascade subduction zone and the Mendocino transform fault located along the northwest coast of the United States. Along the San Andreas fault, the Pacific plate is moving toward the northwest, past the North American plate. If this movement continues, that part of California west of the fault zone, including the Baja Peninsula, will become an island off the west coast of the United States and Canada. It could eventually reach Alaska. However, a more immediate concern is the earthquake activity triggered by movements along this fault system.

Testing the Plate Tectonics Model

With the birth of the plate tectonics model, researchers from all of the Earth sciences began testing it (see Box 7.1). Some of the evidence supporting continental drift and seafloor spreading has already

Figure 7.15 The ongoing collision of India and Asia, starting about 45 million years ago, produced the majestic Himalayas.

that the lava flows had moved—in other words, the continents had drifted.

Although the magnetic poles are known to move, studies of the magnetic field indicated that the average positions of the magnetic poles correspond closely to the positions of the geographic poles. This is consistent with our knowledge of Earth's magnetic field, which is generated in part by the rotation of Earth about its axis. If the geographic poles do not wander appreciably, which we believe is true, neither can the

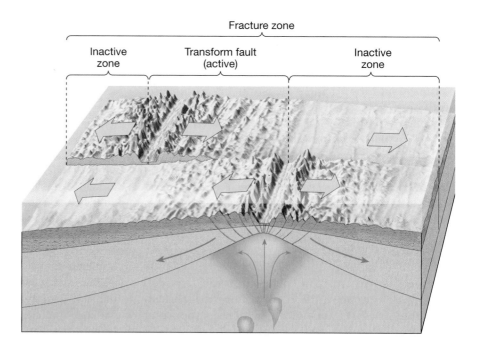

Figure 7.16 Diagram illustrating a transform fault boundary offsetting segments of a divergent boundary (oceanic ridge).

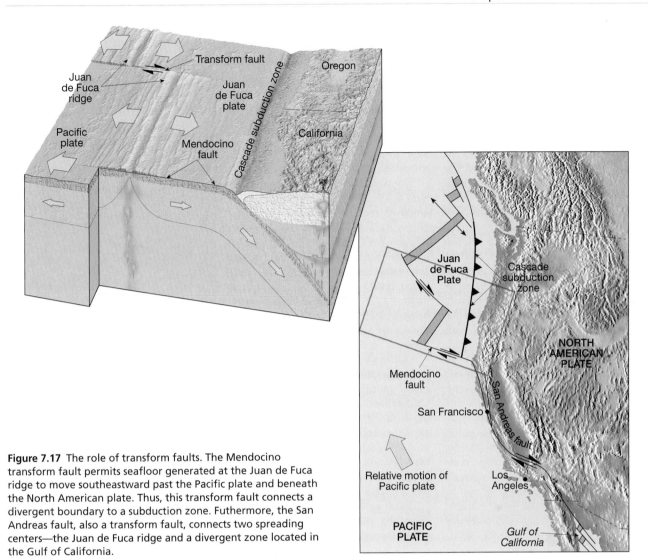

Figure 7.17 The role of transform faults. The Mendocino transform fault permits seafloor generated at the Juan de Fuca ridge to move southeastward past the Pacific plate and beneath the North American plate. Thus, this transform fault connects a divergent boundary to a subduction zone. Futhermore, the San Andreas fault, also a transform fault, connects two spreading centers—the Juan de Fuca ridge and a divergent zone located in the Gulf of California.

magnetic poles. Therefore, a more acceptable explanation for the apparent polar wandering is provided by the plate tectonics theory. *If the magnetic poles remain stationary, their apparent movement was produced by the drifting of the continents.*

Further evidence for plate tectonics came a few years later when polar wandering curves were constructed for North America and Europe (Figure 7.18A). To nearly everyone's surprise, the curves for North America and Europe had similar paths, except that they were separated by about 24 degrees of longitude. When these rocks solidified, could there have been two magnetic north poles that migrated parallel to each other? This is very unlikely. The differences in these migration paths, however, can be reconciled if the two presently separated continents are placed next to one another, as we now believe they were prior to the opening of the Atlantic Ocean (Figure 7.18B).

Magnetic Reversals and Seafloor Spreading. Another discovery came when geophysicists learned that Earth's magnetic field periodically reverses polarity; that is, the north magnetic pole becomes the south magnetic pole, and vice versa. A rock solidifying during one of the periods of reverse polarity will be magnetized with the polarity opposite that of rocks being formed today.

When rocks exhibit the same magnetism as the present magnetic field, they are said to possess **normal polarity**, whereas those rocks exhibiting the opposite magnetism are said to have **reverse polarity**. Evidence for magnetic reversals was obtained from lavas and sediments from around the world. Once the concept of magnetic reversals was confirmed, researchers set out to establish a time scale for polarity reversals. Many areas exist where volcanic activity has occurred sporadically for periods of millions of years (Figure 7.19). The task was to measure the directions of paleomagnetism in numerous lava flows of various ages. These

Box 7.1   A New Test for Plate Tectonics

Until the late 1980s, evidence supporting the theory of plate tectonics was indirectly acquired from the study of geologic phenomena such as volcanoes, earthquakes, and seafloor sediments. With satellite technology, it now is possible to test the theory directly. Scientists have confirmed the fact that Earth's plates shift in relation to one another in the way that the plate tectonics theory predicts.

The new evidence comes from two different techniques that allow unprecedented accuracy of distance measurement between widely separated points. Called *Satellite Laser Ranging* and *Very Long Baseline Interferometry*, these methods can detect the motion of any one site with respect to another at a level of better than 1 centimeter per year. Thus, for the first time, scientists can directly measure the relative motions of Earth's plates. Further, because these techniques are quite different, scientists use them to cross-check one against the other by comparing measurements made for the same sites.

The *Satellite Laser Ranging System* (SLR) employs ground-based stations that bounce laser pulses off satellites whose orbital positions are well-

established. Precise timing of the round-trip travel of these pulses allows scientists to calculate the precise locations of the ground-based stations. By monitoring these stations over time, researchers can establish the relative motions of the sites.

The *Very Long Baseline Interferometry System* (VLBI) uses large radio telescopes to record signals from very distant quasars (quasi-stellar objects) (Figure 7.A). Quasars lie billions of light-years from Earth, so they act as stationary reference points. The millisecond differences in the arrival time of the same signal at different earthbound observational sites provide a means of establishing the distance between receivers.

Confirming data from these two techniques leave little doubt that real plate motion has been detected. Calculations show that Hawaii is moving in a northwesterly direction and approaching Japan at a rate of 8.3 centimeters each year. A site located in Maryland is retreating from one in England at a rate of about 1.7 centimeters per year. This rate is close to the 2.2 centimeters per year of seafloor spreading that was established from paleomagnetic evidence.

Figure 7.A Radio telescopes like these located at Socorro, New Mexico, are used to accurately determine the distance between two distant sites. Data collected by repeated measurements have detected relative plate motions of 1 to 12 centimeters per year between various locations worldwide. (Photo by Geoff Chester/Smithsonian Institution)

---

data were collected from several places and were used to determine the dates when the polarity of Earth's magnetic field changed. Figure 7.20 shows the time scale of the polarity reversals established for the last few million years.

A significant relationship was uncovered between the magnetic reversals and the seafloor spreading hypothesis. Very sensitive instruments called *magnetometers* were towed by research vessels across a segment of the ocean floor located off the west coast of the United States. Here workers from the Scripps Institute of Oceanography discovered alternating strips of high- and low-intensity magnetism that trended in roughly a north-south direction. This relatively simple pattern of magnetic variation defied explanation until 1963, when it was tied to the concept of seafloor spreading. The strips of high-intensity magnetism are regions where the paleomagnetism of the ocean crust is of the normal type. Consequently, these positively magnetized rocks *enhance* the existing

magnetic field. Conversely, the low-intensity strips represent regions where the ocean crust is polarized in the reverse direction and, therefore, *weaken* the existing magnetic field. But how do parallel strips of normally and reversely magnetized rock become distributed across the ocean floor?

As new basalt is added to the ocean floor at the oceanic ridges, it becomes magnetized according to the existing magnetic field (Figure 7.21). Because new rock is added in approximately equal amounts to the trailing edges of both plates, we should expect strips of equal size and polarity to parallel both sides of the ocean ridges, as shown in Figure 7.21C. This explanation of the alternating strips of normal and reverse polarity, which lay as mirror images across the ocean ridges, was the strongest evidence so far presented in support of the concept of seafloor spreading.

Now that the dates of the most recent magnetic reversals have been established, the rate at which spreading occurs at the various ridges can be determined accurately.

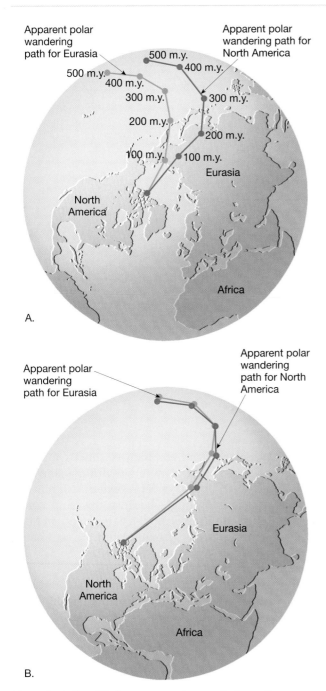

A.

B.

Figure 7.18 Simplified apparent polar wandering paths as established from North American and Eurasian paleomagnetic data. **A.** The more westerly path determined from North American data is thought to have been caused by the westward drift of North America by about 24 degrees from Eurasia. **B.** The positions of the wandering paths when the landmasses are reassembled in their predrift locations.

In the Pacific Ocean, for example, the magnetic strips are much wider for corresponding time intervals than those of the Atlantic Ocean. Hence, we conclude that a faster spreading rate exists for the spreading center of the Pacific as compared to the Atlantic. When we apply absolute dates to these magnetic events, we find that the spreading rate for the North Atlantic Ridge is only 2 centimeters per year. The rate is somewhat faster for the South Atlantic. The spreading rates for the East Pacific Rise generally range between 6 and 12 centimeters per year, with a maximum rate of about 20 centimeters per year in one segment. Thus, we have a magnetic tape recorder that records changes in Earth's magnetic field. This recorder also permits us to determine the rate of seafloor spreading.

Evidence: Earthquake Patterns

By 1968, the basic outline of global tectonics was firmly established. In that same year, three seismologists published papers demonstrating how successfully the new plate tectonics model accounted for the global distribution of earthquakes (Figure 7.22). In particular, these scientists were able to account for the close association between deep-focus earthquakes and ocean trenches. Furthermore, the absence of deep-focus earthquakes along the oceanic ridge system was also shown to be consistent with the new theory.

The close association between plate boundaries and earthquakes can be seen by comparing the distribution of earthquakes shown in Figure 7.22 with the map of plate boundaries in Figure 7.8 and trenches in Figure 7.14. In trench regions where dense slabs of lithosphere plunge into the mantle, this association is especially striking. When the depths of earthquake foci and their locations within the trench systems are plotted, an interesting pattern emerges. Figure 7.23, which shows the distribution of earthquakes in the vicinity of the Japan trench, is an example. Here most shallow-focus earthquakes occur within, or adjacent to, the trench, whereas intermediate- and deep-focus earthquakes occur toward the mainland.

In the plate tectonics model, deep-ocean trenches are produced where cool, dense slabs of oceanic lithosphere plunge into the mantle. Shallow-focus earthquakes are produced as the descending plate interacts with the overriding lithosphere. As the slab descends farther into the asthenosphere, deeper-focus earthquakes are generated (Figure 7.23). Because the earthquakes occur within the rigid subducting plate rather than in the "plastic" mantle, they provide a method for tracking the plate's descent. Very few earthquakes have been recorded below 700 kilometers (435 miles), possibly because the slab has been heated sufficiently to lose its rigidity.

Evidence: Ocean Drilling

Some of the most convincing evidence confirming the plate tectonics theory has come from drilling directly into ocean-floor sediment. From 1968 until 1983, the source of these important data was the Deep Sea Drilling Project, an international program sponsored

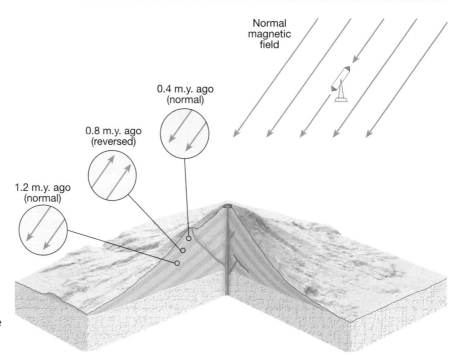

Figure 7.19 Schematic illustration of paleomagnetism preserved in lava flows of various ages. Data such as these from various locales were used to establish the time scale of polarity reversals shown in Figure 7.20.

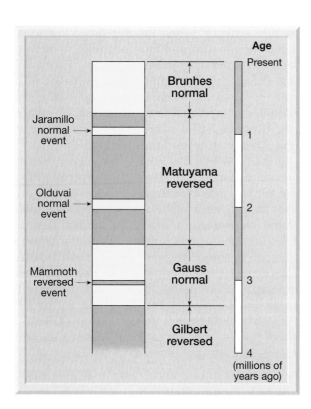

Figure 7.20 Time scale of Earth's magnetic field in the recent past. This time scale was developed by establishing the magnetic polarity for lava flows of known age. (Data from Allen Cox and G. B. Dalrymple)

by several major oceanographic institutions and the National Science Foundation. A new drilling ship was built. The *Glomar Challenger* represented a significant technological breakthrough, because this ship could lower drill pipe thousands of meters to the ocean floor and then drill hundreds of meters into the sediments and underlying basaltic crust.

Operations began in August 1968, in the South Atlantic. At several sites, holes were drilled through the entire thickness of sediments to the basaltic rock below. An important objective was to gather samples of sediment from just above the igneous crust as a means of dating the seafloor at each site. (Radiometric dates of the ocean crust itself are unreliable because seawater alters basalt.)

When the oldest sediment from each drill site was plotted against its distance from the ridge crest, it was revealed that the age of the sediment increased with increasing distance from the ridge. This finding agreed with the seafloor-spreading hypothesis, which predicted that the youngest oceanic crust would be found at the ridge crest, and that the oldest oceanic crust would be at the continental margins.

The data from the Deep Sea Drilling Project also reinforced the idea that the ocean basins are geologically youthful, because no sediment with an age in excess of 180 million years was found. By comparison, some continental crust has been dated at 3.9 billion years.

Figure 7.21 As new basalt is added to the ocean floor at the mid-ocean ridges, it is magnetized according to Earth's existing magnetic field. Hence, it behaves much like a tape recorder as it records each reversal of the planet's magnetic field.

A. Period of normal magnetism

B. Period of reverse magnetism

C. Period of normal magnetism

During its 15 years of operation, the *Glomar Challenger* drilled 1092 holes and obtained more than 96 kilometers (60 miles) of invaluable core samples. The Ocean Drilling Program has succeeded the Deep Sea Drilling Project and, like its predecessor, it is a major international program. A more technologically advanced drilling ship, the *JOIDES Resolution*, now continues the work of the *Glomar Challenger* (Figure 7.24).

Key
● Shallow
○ Intermediate
● Deep

Figure 7.22 Distribution of shallow-, intermediate-, and deep-focus earthquakes. Note that deep-focus earthquakes only occur in association with subduction zones. (Data from NOAA)

Figure 7.23 Distribution of earthquake foci in the vicinity of the Japan trench. Note that intermediate- and deep-focus earthquakes occur only within the sinking slab of oceanic lithosphere. (Data from NOAA)

Key
- ● Shallow
- ○ Intermediate
- ● Deep

Figure 7.24 The *JOIDES Resolution*, the drilling ship of the Ocean Drilling Program. This modern drilling ship has replaced the *Glomar Challenger* in the important work of sampling the floors of the world's oceans. JOIDES is an acronym for Joint Oceanographic Institutions for Deep Earth Sampling. (Photo courtesy of Ocean Drilling Program)

Evidence: Hot Spots

Mapping of seafloor volcanoes (called "seamounts") in the Pacific revealed a chain of volcanic structures extending from the Hawaiian Islands to Midway Island and then continuing northward toward the Aleutian trench (Figure 7.25). Radiometric dates of volcanoes in this chain showed that the volcanoes increase in age with increasing distance from Hawaii. Suiko Seamount, which is located near the Aleutian trench, is 65 million years old, Midway Island is 27 million years old, and the island of Hawaii built up from the seafloor less than a million years ago (Figure 7.25).

Researchers have proposed that a rising plume of mantle material is located below the island of Hawaii. Partial melting of this hot rock as it enters the low-pressure environment near the surface generates a volcanic area or **hot spot**. Presumably, as the Pacific plate moved over the hot spot, successive volcanic mountains have been built. The age of each volcano indicates the time when it was situated over the relatively stationary mantle plume.

This pattern is shown in Figure 7.25. Kauai is the oldest of the large islands in the Hawaiian chain. Five million years ago, when it was positioned over the hot spot, Kauai was the only Hawaiian Island in existence (Figure 7.25). Visible evidence of the age of Kauai can be seen by examining its extinct volcanoes, which have been eroded into jagged peaks and vast canyons. By contrast, the south slopes of the relatively youthful island of Hawaii consist of fresh lava flows, and two of Hawaii's volcanoes, Mauna Loa and Kilauea, remain active.

This evidence supports the fact that the plates do indeed move relative to Earth's interior. The hot spot "tracks" also trace the direction of plate motion. Notice, for example, in Figure 7.25 that the Hawaiian Island–Emperor Seamount chain bends. This particular bend in the trace occurred about 40 million years ago when the motion of the Pacific plate changed from nearly due north to its present northwesterly path.

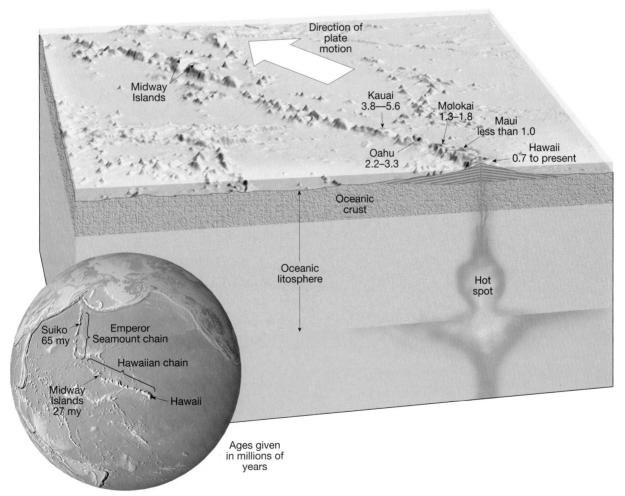

Figure 7.25 The chain of islands and seamounts that extends from Hawaii to the Aleutian trench results from the movement of the Pacific plate over an apparently stationary hot spot. Radiometric dating of the Hawaiian Islands shows that the volcanic activity decreases in age toward the island of Hawaii.

Pangaea: Before and After

Robert Dietz and John Holden have projected the gross details of the migrations of individual continents over the past 500 million years. By extrapolating plate motion back in time using such evidence as the orientation of volcanic structures left behind on moving plates, the distribution and movements of transform faults, and paleomagnetism, Dietz and Holden were able to reconstruct Pangaea (Figure 7.26A). The use of radiometric dating helped them establish the time frame for the formation and eventual breakup of Pangaea, and the

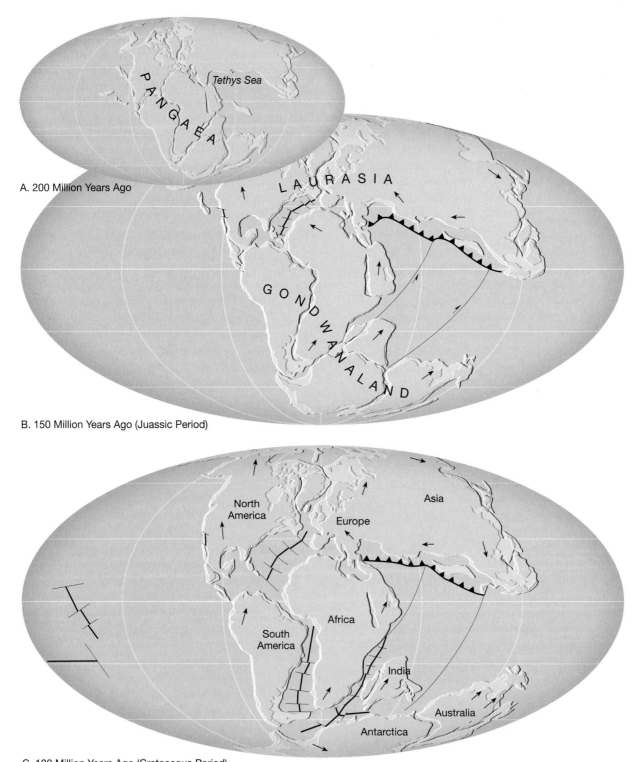

A. 200 Million Years Ago

B. 150 Million Years Ago (Juassic Period)

C. 100 Million Years Ago (Cretaceous Period)

Figure 7.26 Several views of the breakup of Pangaea over a period of 200 million years. (After D. Walsh and C. Scotese) relatively stationary positions of hot spots through time helped to fix the locations of the continents.

Breakup of Pangaea

The fragmentation of Pangaea began about 200 million years ago. Figure 7.26 illustrates the breakup and subsequent paths taken by the landmasses involved. Two major rifts initiated the breakup. The rift zone between North America and Africa generated numerous outpourings of Jurassic-age basalts, which are presently visible along the eastern seaboard of the United States. Radiometric dating of these basalts indicates that rifting

occurred between 200 million and 165 million years ago. This date can be used as the birth date of this section of the North Atlantic. The rift that formed in the southern landmass of Gondwanaland developed a Y-shaped fracture that sent India on a northward journey and simultaneously separated South America-Africa from Australia-Antarctica.

Figure 7.26C illustrates the position of the continents 100 million years ago, after Africa and South America began splitting apart to form the South Atlantic. India can be seen beginning its journey to Asia, while the southern portion of the North Atlantic has widened greatly. By the early Cenozoic, about 50 million years

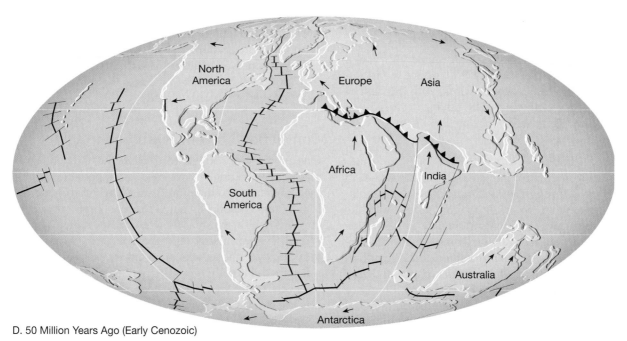

D. 50 Million Years Ago (Early Cenozoic)

D. Present

ago, Madagascar had separated from Africa, and the South Atlantic had emerged as a full-fledged ocean (Figure 7.26D). At this juncture, India had drifted over a hot spot that generated numerous fluid basalt flows across a region in western India now called the Deccan Plateau. These lava flows are very similar to those that make up the Columbia Plateau in the Pacific Northwest.

The current map (Figure 7.26E) shows India in contact with Asia, an event that began about 45 million years ago and created the highest mountains on Earth, the Himalayas, along with the Tibetan Highlands. It is interesting to note that the *average* height of Tibet is 5000 meters, higher than any spot in the contiguous United States. India's continued northward migration is believed to cause the numerous and often destructive earthquakes plaguing that part of the world.

By comparing Figures 7.26D and 7.26E, we can see that the separation of Greenland from Eurasia was a recent event in geologic history. Also notice the recent formation of the Baja Peninsula along with the Gulf of California. This event is thought to have occurred less than 10 million years ago.

Before Pangaea

Prior to the formation of Pangaea, the landmasses had probably gone through several episodes of fragmentation similar to what we see happening today. Also like today, these ancient continents moved away from each other only to collide again at some other location. During the period between 500 million and 225 million years ago, the fragments of an earlier dispersal began collecting to form the continent of Pangaea. Evidence of these earlier continental collisions include the Ural Mountains of Russia and the Appalachian Mountains, which flank the east coast of North America.

Available evidence indicates that about 500 million years ago the northern continent of Laurasia was fragmented into three major sections—North America, northern Europe (southern Europe was part of Africa), and Siberia—with each section separated by a sizable ocean. The southern continent of Gondwanaland probably was intact and lay near the South Pole.

The first collision is believed to have occurred as North America and Europe closed the pre-North Atlantic. This activity resulted in the formation of the northern Appalachians. Parts of the floor of the former ocean can be seen today high above sea level in Nova Scotia. The sliver of eastern Canada and the United States that lies seaward of the zone of this collision is truly a gift from Europe. It is also thought that before North America and Europe collided, part of Scotland, Ireland, and Norway were attached to the North American plate.

While North America and Europe were joining, Siberia was closing the gap between itself and Europe, which lay farther to the west. This closing

culminated about 300 million to 350 million years ago in the formation of the Ural Mountains. The consolidation of these landmasses completed the northern continent of Laurasia.

During the next 50 million years the northern and southern landmasses converged, producing the supercontinent of Pangaea. At this time (about 250 million to 300 million years ago), Africa and North America collided to produce the southern Appalachians (see Box 7.2).

The Driving Mechanism

The plate tectonics theory *describes* plate motion and the *effects* of this motion. Because we accept the theory does not mean that we *understand the forces* that move the plates. In fact, none of the driving mechanisms yet proposed can account for all of the major facets of plate motion. Nevertheless, it is clear that the unequal distribution of heat within Earth is the underlying driving force for plate movement.

Convection Current Hypothesis

One of the first models to explain the movements of plates was originally proposed as a possible driving mechanism for continental drift. Adapted to plate tectonics, this hypothesis suggests that large convection currents within the mantle drive plate motion (Figure 7.27A). The warm, less dense material of the mantle rises very slowly in the regions of oceanic ridges. As the material spreads laterally, it might drag the lithosphere along. Eventually, the material cools and begins to sink back into the mantle, where it is reheated. Partly because of its simplicity, this proposal had wide appeal. However, modern research techniques reveal that the flow of material in the mantle is far more complex than such simple convection cells. Further, it is now known that the lithospheric plates are not passengers carried along by convective flow, but rather are part of the circulation.

Slab-Pull and Slab-Push Hypotheses.

Many other mechanisms that might contribute to plate motion have been suggested. One relies on the fact that, as a newly formed slab of oceanic crust moves away from the ridge crest, it gradually cools and becomes denser. Eventually, the cold oceanic slab becomes denser than the asthenosphere and begins to sink. When this occurs, the dense sinking slab pulls the trailing lithosphere along. This so-called *slab-pull hypothesis* is thought to be an important mechanism for transporting cold material back into the mantle.

Another mechanism proposes that the elevated position of an oceanic ridge could cause the lithosphere to slide under the influence of gravity. However, some

Box 7.2   Earth as a System: Plate Tectonics Into the Future

Two geologists, Robert Dietz and John Holden, extrapolated present-day plate movements into the future. Figure 7.B illustrates where they envision Earth's landmasses will be 50 million years from now if present plate movements persist for this time span.

In North America we see that the Baja Peninsula and the portion of southern California that lies west of the San Andreas fault will have slid past the North American plate. If this northward migration takes place, Los Angeles and San Francisco will pass each other in about 10 million years, and in about 60 million years, Los

Angeles will begin to descend into the Aleutian trench.

Significant changes are seen in Africa, where a new sea emerges as East Africa parts company with the mainland. In addition, Africa will have moved slowly into Europe, perhaps initiating the next major mountain-building stage on our dynamic planet. Meanwhile, the Arabian Peninsula continues to diverge from Africa, allowing the Red Sea to widen and closing the Persian Gulf.

In other parts of the world, Australia is now astride the equator and, along with New Guinea, is on a

collision course with Asia. Meanwhile, North and South America are beginning to separate, while the Atlantic and Indian oceans continue to grow at the expense of the Pacific Ocean.

These projections into the future, although interesting, must be viewed with caution because many assumptions must be correct for these events to unfold as just described. Nevertheless, changes in the shapes and positions of continents that are equally profound will undoubtedly occur for millions of years to come.

Figure 7.B The world as it may look 50 million years from now. (After "The Breakup of Pangaea," Robert S. Dietz and John C. Holden.)

ridge systems are subdued, which would reduce the effectiveness of the *slab-push* model. Nevertheless, the slab-push phenomenon appears to be active in some ridge systems but is less effective than slab-pull (Figure 7.27B).

Hot Plumes Hypothesis

One version of the thermal convection model suggests that hot, buoyant plumes of rock contribute to plate motion (Figure 7.27C). These hot plumes are presumed to extend upward from the vicinity of the mantle–core boundary. Upon reaching the lithosphere, they spread

laterally and facilitate plate motion away from the zone of upwelling. These mantle plumes reveal themselves as long-lived volcanic areas (hot spots) in such places as Iceland. A dozen or so hot spots have been identified along ridge systems where they may contribute to plate divergence. Recall, however, that many hot spots, including the one that generated the Hawaiian Islands, are not located in ridge areas.

The downward limbs of these convection cells are the cold, dense subducting lithospheric plates (Figure 7.27C). Advocates of this view suggest that subducting slabs may descend all the way to the core–mantle boundary.

Figure 7.27 Proposed models of the driving force for plate tectonics. **A.** Large convection cells in the mantle may carry the lithosphere in a conveyor-belt fashion. **B.** Slab-pull results because a subducting slab is more dense than the underlying material. Slab-push is a form of gravity sliding caused by the elevated position of lithosphere at a ridge crest. **C.** The hot plume model suggests that all upward convection is confined to a few plumes, while the downward limbs of these convection cells are the cold, dense subducting oceanic plates.

A.

B.

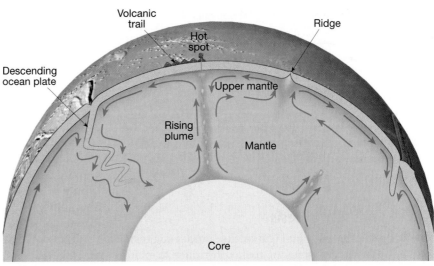

C.

Although there is still much to be learned about the mechanisms that cause plates to move, some facts are clear. The unequal distribution of heat in Earth generates some type of thermal convection in the mantle, which ultimately drives plate motion. Furthermore, the descending lithospheric plates are active components of downwelling, and they serve to transport cold material into the mantle.

Box 7.3   Earth as a System: Plate Tectonics and the Rock Cycle

When the rock cycle was originally presented in Chapter 2, it allowed you to view many of the interrelationships among different parts of the Earth System. Now a basic knowledge of plate tectonics allows you to develop a more complete understanding of the processes that produce and alter Earth's materials.

Figure 7.C presents the rock cycle in terms of the plate tectonics model. Here, weathered material from elevated landmasses is transported to the continental margins, where it is deposited in layers that collectively are thousands of meters thick. Once lithified, these sediments create a thick wedge of sedimentary rocks flanking the continents.

Eventually the relatively quiet activity of sedimentation along a continental margin is interrupted when the region becomes a convergent plate boundary. When this occurs, oceanic lithosphere adjacent to the continent begins to inch downward into the asthenosphere beneath the continent. Along active continental margins such as this, convergence deforms the strata into linear belts of metamorphic rocks.

As the oceanic plate descends, some of the overlying sediments and sedimentary rocks that were not crumpled into mountains are carried downward into the hot asthenosphere, where they too undergo metamorphism. Eventually some of these metamorphic rocks are transported to depths where the conditions are conducive to melting. The newly formed magma then migrates upward through the overlying lithosphere to produce igneous rocks. Some will crystallize prior to reaching the surface and the remainder will erupt and solidify at the surface. After erosion eventually exposes the igneous rocks at the surface, they are immediately attacked by the processes of weathering. Thus, the rock cycle is ready to begin anew.

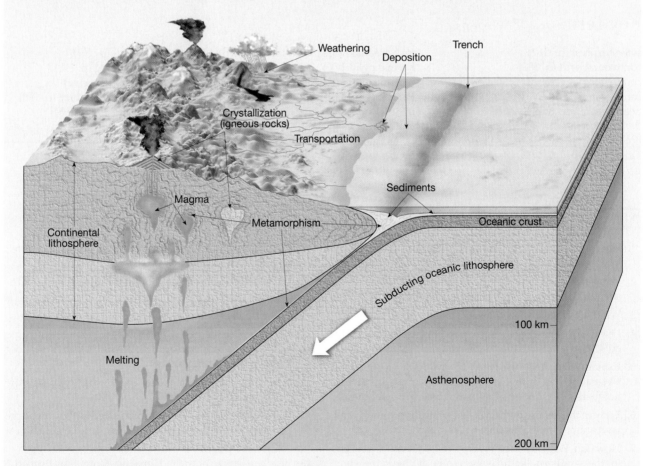

Figure 7.C The rock cycle as it relates to the plate tectonics model.

Chapter Summary

- In the early 1900s *Alfred Wegener* set forth his *continental drift* hypothesis. One of its major tenets was that a supercontinent called *Pangaea* began breaking apart into smaller continents about 200 million years ago. The smaller continental fragments then "drifted" to their present positions. To support the claim that the now-separate continents were once joined, Wegener and others used the *fit of South America and Africa, the distribution of ancient climates, fossil evidence,* and *rock structures.*

- One of the main objections to the continental drift hypothesis was its inability to provide an acceptable mechanism for the movement of continents.

- The theory of *plate tectonics,* a far more encompassing theory than continental drift, holds that Earth's rigid outer shell, called the *lithosphere,* consists of about seven large and numerous smaller segments called *plates* that are in motion relative to each other. Most of Earth's *seismic activity, volcanism,* and *mountain building* occur along the dynamic margins of these plates.

- A major departure of the plate tectonics theory from the continental drift hypothesis is that large plates contain both continental and ocean crust and the entire plate moves. By contrast, in continental drift, Wegener proposed

that the sturdier continents "drifted" by breaking through the oceanic crust, much like ice breakers cut through ice.

- The three distinct types of plate boundaries are (1) *divergent boundaries*—where plates move apart; (2) *convergent boundaries*—where plates move together as in *oceanic-continental convergence, oceanic-oceanic convergence,* or *continental-continental* convergence; and (3) *transform fault boundaries*—where plates slide past each other.

- The theory of plate tectonics is supported by (1) *paleomagnetism,* the direction and intensity of Earth's magnetism in the geologic past; (2) the global distribution of *earthquakes* and their close association with plate boundaries; (3) the ages of *sediments* from the floors of the deep-ocean basins; and (4) the existence of island groups that formed over *hot spots* and that provide a frame of reference for tracing the direction of plate motion.

- Several models for the driving mechanism of plates have been proposed. One model involves large *convection cells* within the mantle carrying the overlying plates. Another proposes that dense *oceanic material descends and pulls* the lithosphere along. No single driving mechanism can account for all of the major facets of plate motion.

Key Terms

asthenosphere (p. 187)
continental drift (p. 183)
continental volcanic arc (p. 194)
convergent boundary (p. 190)
deep-ocean trench (p. 192)
divergent boundary (p. 190)
hot spot (p. 205)
lithosphere (p. 187)

mid-oceanic ridge (p. 190)
normal polarity (p. 199)
paleomagnetism (p. 197)
Pangaea (p. 183)
plate (p. 187)
plate tectonics (p. 187)
polar wandering (p. 197)
reverse polarity (p. 199)

rift (rift valley) (p. 191)
seafloor spreading (p. 190)
subduction zone (p. 192)
transform fault boundary (p. 190)
volcanic island arc (p. 195)

Review Questions

1. Who is credited with developing the continental drift hypothesis?

2. What was probably the first evidence that led some to suspect the continents were once connected?

3. What was Pangaea?

4. List the evidence that Wegener and his supporters gathered to substantiate the continental drift hypothesis.

5. Explain why the discovery of the fossil remains of *Mesosaurus* in both South America and Africa, but nowhere else, supports the continental drift hypothesis.

6. Early in this century, what was the prevailing view of how land animals migrated across vast expanses of ocean?

7. How did Wegener account for the existence of glaciers in the southern landmasses, while at the same time

areas in North America, Europe, and Siberia supported lush tropical swamps?

8. On what basis were plate boundaries first established?

9. What are the three major types of plate boundaries? Describe the relative plate motion at each of these boundaries.

10. What is seafloor spreading? Where is active seafloor spreading occurring today?

11. What is a subduction zone? With what type of plate boundary is it associated?

12. Where is lithosphere being consumed? Why must the production and destruction of lithosphere be going on at approximately the same rate?

13. Briefly describe how the Himalaya Mountains formed.

14. Differentiate between transform faults and the other two types of plate boundaries.

15. Some predict that California will sink into the ocean. Is this idea consistent with the theory of plate tectonics?

16. Define the term *paleomagnetism*.

17. How does the continental drift hypothesis account for the apparent wandering of Earth's magnetic poles?

18. Describe the distribution of earthquake epicenters and foci depths as they relate to oceanic trench systems.

19. What is the age of the oldest sediments recovered by deep-ocean drilling? How do the ages of these sediments compare to the ages of the oldest continental rocks?

20. How do hot spots and the plate tectonics theory account for the fact that the Hawaiian Islands vary in age?

21. With what type of plate boundary are the following places or features associated (be as specific as possible): Himalayas, Aleutian Islands, Red Sea, Andes Mountains, San Andreas fault, Iceland, Japan, Mount St. Helens?

Examining the Earth System

1. As an integral subsystem of the Earth system, plate tectonics has played a major role in determining the events that have taken place on Earth since its formation about 4.5 billion years ago. Briefly comment on the general effect that the changing positions of the continents and the redistribution of land and water over Earth's surface have had on the atmosphere, hydrosphere, and biosphere through time. You may want to review plate tectonics by investigating the topic on the Internet using a search engine such as HotBot (**http://www.hotbot.com/**) or Yahoo (**http://www.yahoo.com/**) and visiting the United States Geological Survey's (USGS), This Dynamic Earth: The Story of Plate Tectonics Web site at **http://pubs.usgs.gov/publications/text/dynamic.html**.

2. Assume that plate tectonics did not cause the breakup of the supercontinent Pangaea. Using Figure 7.2, describe how the climate (atmosphere), vegetation and animal life (biosphere), and geological features (solid Earth) of those locations currently occupied by the cities of Seattle, WA, Chicago, IL, New York, NY, and your college campus location would be different from the conditions that exist today (see Figure 7.2).

Web Resources

The following are informative and interesting World Wide Web sites that address topics related to those presented in the chapter:

Dynamic Earth Tutorial (United States Geological Survey):

• **http://pubs.usgs.gov/publications/text/dynamic.html**

Plate Tectonics and Sea-Floor Spreading (United States Geological Survey):

• **http://vulcan.wr.usgs.gov/Glossary/Plate Tectonics/framework.html**

For direct links to these sites and other appropriate Earth science Web locations, chapter review exercises, and relevant, topical investigations that utilize Web resources, contact the *Earth Science*, 9th Edition Home Page at:

• **http://www.prenhall.com/tarbuck.**

Chapter 8
Igneous Activity

Mount St. Helens emitting volcanic ash on July 22, 1980, two months after the huge May eruption. (Photo by David Weintraub/Photo Researchers, Inc.)

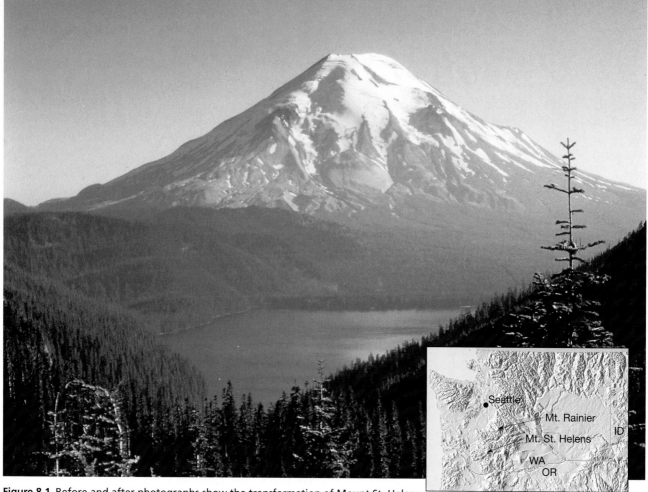

Figure 8.1 Before and after photographs show the transformation of Mount St. Helens caused by the May 18, 1980, eruption. The dark area in the "after" photo is debris-filled Spirit Lake, partially visible in the "before" photo. (Photos courtesy of U.S. Geological Survey)

The significance of igneous activity may not be obvious at first glance. However, because volcanoes extrude molten rock that formed at great depth, they provide the only windows we have for direct observation of processes that occur many kilometers below Earth's surface. Furthermore, the gases emitted during volcanic eruptions are thought to be the material from which the atmosphere and oceans evolved. Either of these facts is reason enough for igneous activity to warrant our attention. All volcanic eruptions are spectacular, but why are some destructive and others quiescent? Is the entire island of Hawaii a volcano as high as Mount Everest resting on the ocean floor? Do the large volcanoes of Washington and Oregon present a threat to human lives? This chapter considers these and other questions as we explore the formation and movement of magma.

On Sunday, May 18, 1980, the largest volcanic eruption to occur in North America in historic times transformed a picturesque volcano into a decapitated remnant (Figure 8.1). On this date in southwestern Washington State, Mount St. Helens erupted with tremendous force. The blast blew out the entire north flank of the volcano, leaving a gaping hole. In one brief moment, a prominent volcano whose summit had been more than 2900 meters (9500 feet) above sea level was lowered by more than 400 meters (1350 feet).

The event devastated a wide swath of timber-rich land on the north side of the mountain (Figure 8.2). Trees within a 400-square-kilometer area lay intertwined and flattened, stripped of their branches and appearing from the air like toothpicks strewn about. The accompanying mudflows carried ash, trees, and water-saturated rock debris 29 kilometers down the Toutle River. The eruption claimed 59 lives, some dying from the intense heat and the suffocating cloud of ash and gases, others from being hurled by the blast, and still others from entrapment in the mudflows.

The eruption ejected nearly a cubic kilometer of ash and rock debris. Following the devastating explosion, Mount St. Helens continued to emit great quantities of hot gases and ash. The force of the blast was so strong that some ash was propelled more than 18,000 meters (over 11 miles) into the stratosphere. During the next few days, this very fine-grained material was carried around Earth by strong upper-air winds. Measurable deposits were reported in Oklahoma and Minnesota, with crop damage into central Montana.

Meanwhile, ash fallout in the immediate vicinity exceeded 2 meters in depth. The air over Yakima, Washington (130 kilometers to the east), was so filled with ash that residents experienced midnight-like darkness at noon. (See Box 8.1.)

Not all volcanic eruptions are as violent as the 1980 Mount St. Helens event. Some volcanoes, such as Hawaii's Kilauea volcano, generate relatively quiet outpourings of fluid lavas. These "gentle" eruptions are not without some fiery displays; occasionally fountains of incandescent lava spray hundreds of meters into the air. Such events, however, are typically short-lived and harmless, and the lava generally falls back into a lava pool.

Testimony to the quiet nature of Kilauea's eruptions is the fact that the Hawaiian Volcanoes Observatory has operated on its summit since 1912. This, despite the fact that Kilhauea has had more than

Figure 8.2 Douglas fir trees were snapped off or uprooted by the lateral blast of Mount St. Helens on May 18, 1980. (Photo by Jeffrey Hutcherson/DRK Photo)

50 eruptive phases since record keeping began in 1823. Further, the longest and largest of Kilauea's eruptions began in 1983 and remains active, although it has received only modest media attention.

Why do volcanoes like Mount St. Helens erupt explosively, whereas others like Kilauea are relatively quiet? Why do volcanoes occur in chains like the Aleutian Islands or the Cascade Range? Why do some volcanoes form on the ocean floor, while others occur on the continents? This chapter will deal with these and other questions as we explore the nature and movement of magma and lava.

The Nature of Volcanic Eruptions

Internal Processes
Igneous Activity

Volcanic activity is commonly perceived as a process that produces a picturesque, cone-shaped structure which periodically erupts in a violent manner. Although some eruptions are very explosive, many others are not. What determines whether a volcano extrudes magma violently or "gently"? The primary factors include the magma's *composition*, its *temperature*, and the amount of *dissolved gases* it contains. To varying degrees these factors affect the magma's **viscosity**. The more viscous ("thicker") the material, the greater its resistance to flow. For example, syrup is more viscous

than water. Magma associated with an explosive eruption may be five times more viscous than magma that is extruded in a quiescent manner.

Factors Affecting Viscosity

The effect of temperature on viscosity is easily seen. Just as heating syrup makes it more fluid (less viscous), the mobility of lava is strongly influenced by temperature. As a lava flow cools and begins to congeal, its mobility decreases and eventually the flow halts.

A more significant influencing volcanic behavior is the chemical composition of magmas. This was discussed in Chapter 2 along with the classification of igneous rocks. Recall that a major difference among various igneous rocks is their silica (SiO_2) content. The same is true of the magmas from which rocks form. Magmas that produce basaltic rocks contain about 50 percent silica, whereas magmas that produce granitic rocks contain over 70 percent silica (Table 8.1).

A magma's viscosity is directly related to its silica content. In general, the more silica in magma, the greater is its viscosity. The flow of magma is impeded because silicate structures link together into long chains, even before crystallization begins. Consequently, because of high silica content, rhyolitic lavas are very viscous and tend to form comparatively short, thick flows. By contrast, basaltic lavas, which contain less silica, tend to be more fluid and

---

Box 8.1   Mount St. Helens: Anatomy of the Eruption

The events leading to the May 18, 1980, eruption of Mount St. Helens began about 2 months earlier as a series of minor Earth tremors centered beneath the awakening mountain (Figure 8.A, part A). The tremors were caused by the upward movement of magma within the mountain. The first volcanic activity took place a week later, when a small amount of ash and steam rose from the summit. Over the next several weeks, sporadic eruptions of varied intensity occurred. Prior to the main eruption, the primary concern had been the potential hazard of mudflows. These moving lobes of saturated soil and rock are created as ice and snow melt from the heat emitted from magma within the volcano.

The only warning of a potential eruption was a bulge on the volcano's north flank (Figure 8.A, part B). Careful monitoring of this dome-shaped structure indicated a very slow but steady growth rate of a few meters per day. If the growth rate of the bulge changed appreciably, an eruption might quickly follow. Unfortunately, no such variation was detected prior to the explosion. In fact, the seismic activity decreased during the 2 days preceding the huge blast.

Dozens of scientists were monitoring the mountain when it exploded. "Vancouver, Vancouver, this is it!" was the only warning—and last words from one scientist—that preceded the unleashing of tremendous quantities of pent-up gases. The trigger was a medium-sized earthquake. Its vibrations sent the north slope of the cone plummeting into the Toutle River, removing the overburden that had trapped the magma below (Figure 8.A, part C). With the pressure reduced, the water in the magma

vaporized and expanded, causing the mountainside to rupture like an overheated steam boiler. Because the eruption originated around the bulge, several hundred meters below the summit, the initial blast was directed laterally rather than vertically. Had the full force of the eruption been upward, far less destruction would have occurred.

Mount St. Helens is one of 15 large volcanoes and innumerable smaller ones that comprise the Cascade Range, which extends from British Columbia to northern California. Eight of the largest volcanoes have been active in the past few hundred years. Of the remaining seven "active" volcanoes, the most likely to erupt again are Mount Baker and Mount Rainier in Washington, Mount Shasta and Lassen Peak in California, and Mount Hood in Oregon.

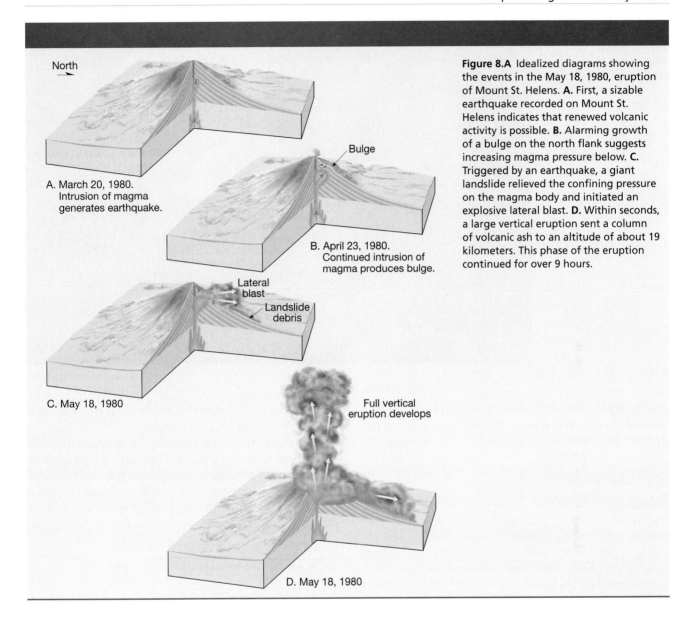

North

A. March 20, 1980.
Intrusion of magma
generates earthquake.

Bulge

B. April 23, 1980.
Continued intrusion of
magma produces bulge.

Lateral
blast
Landslide
debris

C. May 18, 1980

Full vertical
eruption develops

D. May 18, 1980

Figure 8.A Idealized diagrams showing the events in the May 18, 1980, eruption of Mount St. Helens. **A.** First, a sizable earthquake recorded on Mount St. Helens indicates that renewed volcanic activity is possible. **B.** Alarming growth of a bulge on the north flank suggests increasing magma pressure below. **C.** Triggered by an earthquake, a giant landslide relieved the confining pressure on the magma body and initiated an explosive lateral blast. **D.** Within seconds, a large vertical eruption sent a column of volcanic ash to an altitude of about 19 kilometers. This phase of the eruption continued for over 9 hours.

have been known to travel distances of 150 kilometers (90 miles) or more before congealing (Figure 8.3).

In Hawaiian eruptions, the magmas are hot and basaltic, so they are extruded with ease. By contrast, highly viscous magmas are more difficult to force through a vent. On occasion, the vent may become plugged with viscous magma, which results in a buildup of gases and a great pressure increase, so a

Table 8.1 Magmas Have Different Compositions, Which Cause Their Properties to Vary

Composition	Silica Content	Viscosity	Gas Content	Tendency to Form Pyroclastics	Volcanic Landform
Basaltic (Mafic)	Least (~50%)	Least	Least (1–2%)	Least	Shield Volcanoes Basalt Plateaus Cinder Cones
Andesitic (Intermediate)	Intermediate (~60%)	Intermediate	Intermediate (3–4%)	Intermediate	Composite Cones
Granitic (Felsic)	Most (~70%)	Greatest	Most (4–6%)	Greatest	Pyroclastic Flows Volcanic Domes

Figure 8.3 Fluid basaltic lava from the flank of Kilauea Volcano, Hawaii. (Photo by G. Brad Lewis/Liaison Agency, Inc.)

potentially explosive eruption may result. However, a viscous magma is not explosive by itself. It is the gas content that puts the "bang" into a violent eruption.

Importance of Dissolved Gases in Magma

Dissolved gases in magma provide the force that extrudes molten rock from the vent. These gases are mostly water vapor and carbon dioxide. As magma moves into a near-surface environment, such as within a volcano, the confining pressure in the uppermost portion of the magma body is greatly reduced. The reduction in confining pressure allows dissolved gases to be released suddenly, just as opening a soda bottle allows dissolved carbon dioxide gas bubbles to escape.

At high temperatures and low near-surface pressures, these gases will expand to occupy hundreds of times their original volume. Very fluid basaltic magmas allow the expanding gases to bubble upward and escape from the vent with relative ease. As they escape, the gases will often propel incandescent lava hundreds of meters into the air, producing lava fountains. Although spectacular, such fountains are mostly harmless and not generally associated with major explosive events that cause great loss of life and property. Rather, eruptions of fluid basaltic lavas, such as those that occur in Hawaii, are relatively quiescent.

At the other extreme, highly viscous magmas impede the upward migration of expanding gases. The gases collect in bubbles and pockets that increase in size and pressure until they explosively eject the semimolten rock from the volcano. The result is a Mount St. Helens or Mt. Pinatubo.

To summarize, the viscosity of magma, plus the quantity of dissolved gases and the ease with which they can escape, determines the nature of a volcanic eruption. We can now understand the "gentle" volcanic eruptions of hot, fluid lavas in Hawaii and the explosive, violent, dangerous eruptions of viscous lavas in some volcanoes.

What Is Extruded During Eruptions?

 Internal Processes Igneous Activity

Lava may appear to be the primary material extruded from a volcano, but this is not always the case. Just as often, explosive eruptions eject huge quantities of broken rock, lava "bombs," fine ash, and dust. Moreover, all volcanic eruptions emit large amounts of gas. In this section we will examine each of these materials associated with a volcanic eruption.

Lava Flows

Because of their low silica content, basaltic lavas are usually very fluid. They flow in thin, broad sheets or streamlike ribbons. On the island of Hawaii, such lavas have been clocked at speeds of 30 kilometers (20 miles) per hour down steep slopes. These velocities are rare, however, and flow rates of 10 to 300 meters per hour are more common. In contrast, the movement of silica-rich lava is on occasion too slow to be perceptible.

When fluid basaltic lavas of the Hawaiian type congeal, they commonly form a relatively smooth skin that wrinkles as the still-molten subsurface lava continues to advance (Figure 8.4A). These are known as **pahoehoe flows** (pronounced *pah-hoy-hoy*) and resemble the twisted braids in ropes.

Another common type of basaltic lava has a surface of rough, jagged blocks with dangerously sharp edges and spiny projections (Figure 8.4B). The name **aa** (pronounced *ah-ah*) is given to these flows. Active aa flows are relatively cool and thick and, depending on the slope, advance at rates from 5 to 50 meters per hour. Further, escaping gases fragment the cool surface and produce numerous voids and sharp spines in the congealing lava. As the molten interior advances, the outer crust is broken further, giving the flow the appearance of an advancing mass of lava rubble.

Gases

Magmas contain varied amounts of dissolved gases held in the molten rock by confining pressure, just as carbon dioxide is held in soft drinks. As with soft drinks, as soon as the pressure is reduced, the gases begin to escape. Obtaining gas samples from an erupting volcano is often difficult and dangerous, so geologists can often only estimate the amount of gas originally contained within the magma.

The gaseous portion of most magmas is believed to make up 1 to 5 percent of the total weight, with most of this being water vapor. Although the percentage may be small, the actual quantity of emitted gas can exceed thousands of tons each day.

The composition of volcanic gases is important to scientists because these gases are thought to be the original source of the water for the oceans. Further, volcanic eruptions have contributed significantly to gases that make up the atmosphere. Analysis of samples taken during Hawaiian eruptions indicates that the gases are about 70 percent water vapor, 15 percent carbon dioxide, 5 percent nitrogen, 5 percent sulfur, and lesser amounts of chlorine, hydrogen, and argon. Sulfur compounds are easily recognized by their pungent odor and because they readily form sulfuric acid. Volcanoes are a natural source of air pollution, and they can cause serious problems.

Pyroclastic Materials

When basaltic lava is extruded, the dissolved gases escape quite freely and continually. As stated earlier, these gases propel incandescent blobs of lava to great heights, thereby producing spectacular lava fountains (Figure 8.5). Some ejected material may land near the vent and build a cone structure, whereas smaller particles will be carried great distances by the wind.

The gases in highly viscous magmas, on the other hand, become superheated, and upon release they expand a thousandfold as they blow pulverized rock, lava, and glass fragments from the vent. The particles produced by these processes are called **pyroclasts** (meaning "fire fragments"). These ejected lava fragments range in size from very fine dust and sand-sized volcanic ash, to large pieces.

The *ash* particles are produced when the extruded lava contains so many gas bubbles that it resembles the froth flowing from a newly opened bottle of champagne. As the hot gases expand explosively, the lava is

A.

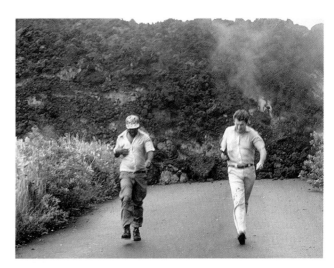

B.

Figure 8.4 A. Typical pahoehoe (ropy) lava flow, Kilauea Hawaii. **B.** Typical slow-moving aa flow. (Photos by J. D. Griggs/U.S. Geological Survey)

Figure 8.5 Fluid basaltic lava erupting from Kilauea Volcano, Hawaii. (Photo by Douglas Peebles)

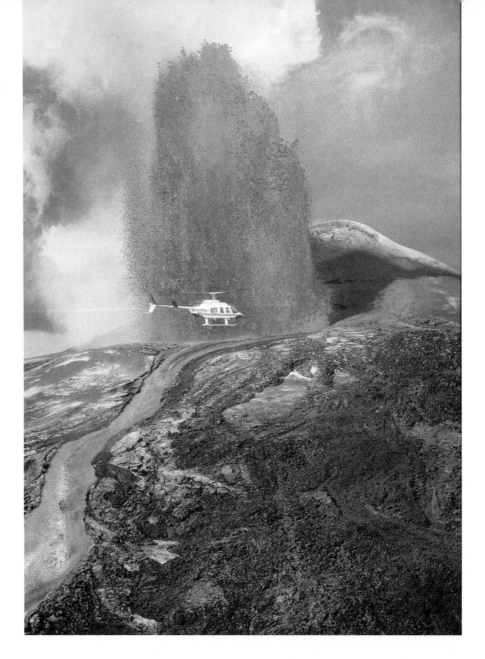

disseminated into very fine glassy fragments. When the hot ash falls, the glassy shards often fuse to form *welded tuff*. Sheets of this material, as well as ash deposits that later consolidated, cover vast portions of the western United States. Sometimes the frothlike lava is ejected in larger pieces called *pumice*. This material has so many voids (air spaces) that it is often light enough to float in water.

Pyroclasts the size of walnuts, called *lapilli* ("little stones"), and pea-sized particles called *cinders* are also very common. Cinders contain numerous voids and form when ejected lava blocks are pulverized by escaping gases. Particles larger than lapilli are called *blocks* when they are made of hardened lava and *bombs* when they are ejected as incandescent lava. Because volcanic bombs are semimolten upon ejection, they often take on a streamlined shape, as shown in Figure 8.6.

Figure 8.6 Volcanic bomb. Ejected lava fragments take on a streamlined shape as they sail through the air. (Photo by GeoScience/PH)

Volcano Types

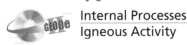

Internal Processes
Igneous Activity

Successive eruptions from a central vent result in a mountainous accumulation of material known as a **volcano**. Located at the summit of many volcanoes is a steep-walled depression called a **crater**. The crater is connected to a magma chamber via a pipelike conduit, or **vent**. Some volcanoes have unusually large summit depressions that exceed 1 kilometer in diameter and are known as **calderas**.

When fluid lava leaves a conduit, it is often stored in the crater or caldera until it overflows. Conversely, viscous lava forms a plug in the pipe. It rises slowly or is blown out, often enlarging the crater. However, lava does not always issue from a central crater. Sometimes it is easier for the magma or escaping gases to push through fissures on the volcano's flanks. Mount Etna in Italy, for example, has more than 200 secondary vents. Some of these emit only gases and are appropriately called *fumaroles*.

The eruptive history of each volcano is unique. Consequently, all volcanoes are somewhat different in form and size. Nevertheless, volcanologists recognize three general eruptive patterns and characteristic forms: shield volcanoes, cinder cones, and composite cones (see Box 8.2).

Shield Volcanoes

When fluid lava is extruded, the volcano takes the shape of a broad, slightly domed structure called a **shield volcano** (Figure 8.7). They are so called because they roughly resemble the shape of a warrior's shield. Shield volcanoes are built primarily of basaltic lava flows and contain only a small percentage of pyroclastic material.

Mauna Loa, probably the largest volcano on Earth, is one of the five shield volcanoes that together make up the island of Hawaii. Its base rests on the ocean floor 5000 meters below sea level, and its summit is 4170 meters

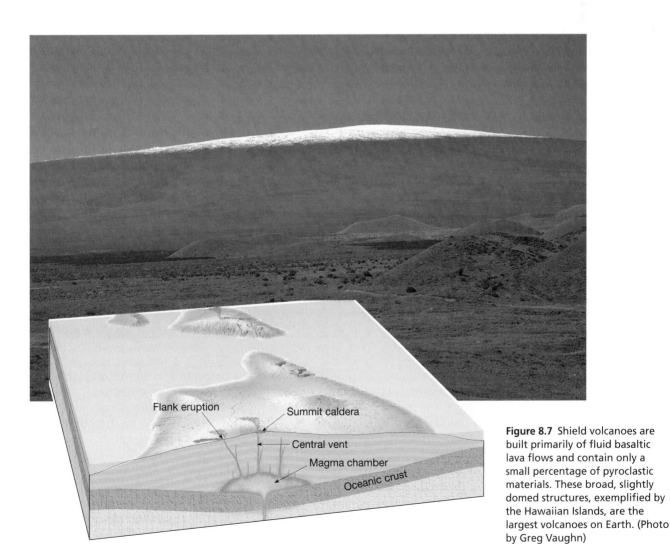

Figure 8.7 Shield volcanoes are built primarily of fluid basaltic lava flows and contain only a small percentage of pyroclastic materials. These broad, slightly domed structures, exemplified by the Hawaiian Islands, are the largest volcanoes on Earth. (Photo by Greg Vaughn)

Box 8.2 Volcano Sizes

In the section on the three basic volcano types (shield, cinder cone, composite cone), we often mentioned their sizes. It is useful to compare the three types side-by-side, as in Figure 8.B.

Smallest are the cinder cones (Figure 8.B). Recall that Parícutin rose from a Mexican cornfield during a period of just 9 years. Because their eruption history is short, most cinder cones do not exceed 300 meters (1000 feet) in height.

In contrast to cinder cones, composite cones and shield volcanoes have recurring eruptions that can span a million years or longer. As shown in Figure 8.B, Mount Rainier, one of the largest composite cones in the Cascade Range, dwarfs a relatively large cinder cone, Sunset Crater in Arizona. Sunset Crater formed over a few years, but Mount Rainier gradually formed over the past 700,000 years.

Mauna Loa, one of five shield volcanoes that make up the island of Hawaii, is considered the world's largest active volcano (Figure 8.B).

This massive pile of basaltic lava has an estimated volume of 40,000 cubic kilometers that was extruded over approximately a million years. From its base on the floor of the Pacific Ocean to its summit, Mauna Loa approaches 9 kilometers in height.

Despite its enormous size, Mauna Loa is not the largest known volcano in the solar system. Olympus Mons, a huge shield volcano on Mars, is 25 kilometers high and 600 kilometers wide (see Chapter 20).

Figure 8.B Profiles of volcanic landforms. **A.** Profile of Mauna Loa, Hawaii, the largest shield volcano in the Hawaiian chain. Note size comparison with Mt. Rainier, Washington, a large composite cone.. **B.** Profile of Mt. Rainier. Note how it dwarfs a typical cinder cone. **C.** Profile of Sunset Crater, Arizona, a typical steep-sided cinder-cone.

(13,680 feet) above the water. Nearly a million years and numerous eruptive cycles built this truly gigantic pile of volcanic rock. Many other volcanic structures, including Midway Island and the Galapagos Islands, have been built in a similar manner from the ocean's depths.

Perhaps the most active and intensively studied shield volcano is Kilauea, located on the island of Hawaii on the southeastern flank of the larger volcano Mauna Loa. Kilauea has erupted more than 50 times in recorded history and is still active today. Several months before an eruptive phase, Kilauea's summit

inflates as magma rises from its source 60 kilometers or more below the surface. This molten rock gradually works its way upward and accumulates in smaller reservoirs 3 to 5 kilometers below the summit. For up to 24 hours in advance of each eruption, swarms of small earthquakes warn of the impending activity.

The longest and largest rift eruption ever recorded at Kilauea began in 1983 and continues intermittently to the present day. The eruption began along a 6.5-kilometer (4-mile) fissure in an inaccessible forested area east of the summit caldera (Figure 8.8). Between 1983 and

1986, nearly 50 eruptive phases were observed. Each consisted of a period of high lava fountaining that lasted a few hours to a few days, followed by an inactive period that averaged one month. After a few months, the eruptions localized in one spot along the fissure, building a major spatter cone 250 meters (820 feet) high.

Lava falling back from these fountains formed aa flows that migrated a few hundred meters per hour. Although inhabitants in a nearby settlement had time to be evacuated, many of their homes were burned and buried in the aa.

By the summer of 1986, the eruptions shifted 3 kilometers downslope from the cone. Here, smooth-surfaced pahoehoe lava formed a lava lake. Eventually, it overflowed, and the fast-moving pahoehoe destroyed nearly a hundred rural homes, covered a major roadway, and eventually flowed into the sea.

Eruptions of the type just described are typical of shield volcanoes and have been occurring sporadically on the Island of Hawaii for nearly a million years. The result is the formation of mountains with summits over 9000 meters (30,000 feet) above the seafloor, the tallest mountains on Earth (see Box 8.3).

Cinder Cones

As the name suggests, **cinder cones** are built from ejected lava fragments. Loose pyroclastic material has a high angle of repose (between 30 and 40 degrees), the steepest angle at which the material remains stable.

Thus, volcanoes of this type have very steep slopes (Figure 8.9). Cinder cones are rather small, usually less than 300 meters (1000 feet) high, and often form as parasitic cones on or near larger volcanoes. In addition, they frequently occur in groups.

One of the very few volcanoes whose formation has been observed by geologists from beginning to end is a cinder cone called Parícutin. In 1943, about 200 miles west of Mexico City, the volcano Parícutin was born (Figure 8.10). It erupted in a cornfield owned by Dionisio Pulido, who witnessed the event as he prepared the field for planting. For 2 weeks prior to the first eruption, numerous earth tremors caused apprehension in the village of Parícutin about 3.5 kilometers away. Then sulfurous smoke began billowing from a small hole that had been in the cornfield for as long as Señor Pulido could remember. During the night, hot, glowing rock fragments thrown into the air from the hole produced a spectacular fireworks display. In one day the cone grew to 40 meters (130 feet) and by the fifth day it was over 100 meters (330 feet) high. Explosive eruptions threw hot fragments 1000 meters (3300 feet) above the crater rim. Larger fragments fell near the crater, some remaining incandescent as they rolled down the slope. These built an aesthetically pleasing cone, while finer ash fell over a much larger area, burning and eventually covering the village of Parícutin. Within 2 years the cone attained its final height of about 400 meters (1300 feet).

Figure 8.8 Lava extruded along the East Rift Zone, Kilauea, Hawaii. (Photo by Greg Vaughn)

Pyroclastic material

Crater

Central vent filled
with rock fragments

Figure 8.9 SP Crater, a cinder cone north of
Flagstaff, Arizona. Note the road across the
lava flow for scale. (Photo by Michael Collier)

Figure 8.10 The village of San
Juan Parangaricutiro engulfed
by lava from Parícutin, shown in
the background. Only the church
towers remain. (Photo by Tad
Nichols)

Box 8.3   Loihi: The Next Hawaiian Island?

Loihi seamount is an undersea volcano (seamount) rising some 3500 meters (11,500 feet) above the floor of the Pacific Ocean. Although hidden beneath the waves, it is taller than Mount St. Helens prior to its explosive 1980 eruption. Situated 32 kilometers (20 miles) off the southeast coast of Hawaii's Big Island, Loihi was found to be active in the 1970s (Figure 8.C). Along with Kilauea, Loihi sits on the flank of Mauna Loa, a huge and still active volcano that makes up a substantial portion of the Big Island. All three derive lava from the same hot-spot source.

Loihi's most recent period of activity dates from July 1996 when the largest swarm of earthquakes ever recorded at any Hawaiian volcano

Figure 8.C Location map of Loihi seamount and other active volcanoes in the Hawaiian chain.

(more than 4000) shook the seamount. Subsequent evidence of ongoing activity at Loihi was gathered in August and September 1997 during dives by the submersible vessel *Pisces*. It was at this time that scientists discovered high-temperature (200°C) hydrothermal (hot water) venting at the seamount's summit.

Will Loihi become the next Hawaiian Island? Any answer, of course, is speculative. If it is going to become an island, Loihi must build upward at least another 930 meters (3100 feet). Given the current rate at which the seamount is growing, it is likely to break the surface of the Pacific Ocean at some time in the distant future, most likely tens of thousands of years from now.

The first lava flow came from a fissure that opened just north of the cone, but after a few months flows began to emerge from the base of the cone itself. In June 1944, a clinkery aa flow 10 meters thick moved over much of the village of San Juan Parangaricutiro, leaving only the church steeple exposed (Figure 8.10). After 9 years, the activity ceased almost as quickly as it had begun. Today, Parícutin is just another one of the numerous quiet cinder cones dotting the landscape in this region of Mexico. Like the others, it will probably not erupt again.

Composite Cones

Earth's most picturesque volcanoes are composite cones. Most active composite cones are in a narrow zone that encircles the Pacific Ocean, appropriately named the *Ring of Fire*. Found in this region are Fujiyama (Mt. Fuji) in Japan, Mount Mayon in the Philippines, and the picturesque volcanoes of the Cascade Range in the northwestern United States, including Mount St. Helens, Mount Rainier, and Mount Shasta (Figure 8.11).

A **composite cone** or **stratovolcano** is a large, nearly symmetrical structure composed of interbedded lava flows and pyroclastic deposits, emitted mainly from a central vent. Just as shield volcanoes owe their shape to the highly fluid nature of the extruded lavas, so too do composite cones reflect the nature of the erupted material.

Composite cones are produced when relatively viscous lavas of andesitic composition are extruded. A composite cone might extrude viscous lava for long periods. Then, suddenly, the eruptive style changes and the volcano violently ejects pyroclastic material. Occasionally, both activities occur simultaneously, and the resulting structure consists of alternating layers of lava and pyroclasts. Two of the most perfect cones, Mount Mayon in the Philippines and Fujiyama in Japan, exhibit the classic form of the stratovolcano with its steep summit area and more gently sloping flanks.

Composite cones represent the most violent type of volcanic activity. An eruption can be unexpected and devastating, as was the A.D. 79 eruption of the Italian volcano we now call Vesuvius. Prior to this eruption, Vesuvius had been dormant for centuries. Although minor earthquakes probably warned of the events to follow, Vesuvius was covered with dense vegetation and hardly looked threatening. On August 24, however, the tranquility ended, and in the next 3 days the city of Pompeii (near Naples) and more than 2000 of its 20,000 residents were buried. They remained so for nearly 17 centuries, until the city was rediscovered and excavated.

Nuée Ardente: A Deadly Pyroclastic Flow. Although the destruction of Pompeii was catastrophic, even more devastating eruptions may occur when hot gases infused with incandescent ash are ejected. Such events produce a fiery pyroclastic flow called a **nuée ardente**. Also referred to as *glowing avalanches*, these turbulent steam clouds and companion ash flows race down steep volcanic slopes at speeds that can approach 200 kilometers (125 miles) per hour (Figure 8.12). The ground-hugging portions of glowing avalanches are

Crater **Pyroclastic layers** **Lava flows**

Figure 8.11 Mount Shasta, California, one of the largest composite cones in the Cascade Range. (Photo by John S. Shelton)

Figure 8.12 Nuée ardente races down the slope of Mount St. Helens on August 7, 1980, at speeds in excess of 100 kilometers (60 miles) per hour. (Photo by Peter W. Lipman, U.S. Geological Survey)

rich in particulate matter that is suspended by hot, buoyant gases. Thus, these flows, which can include larger rock fragments in addition to ash, travel downslope in a nearly frictionless environment cushioned by expanding volcanic gases. This fact explains why some nuée ardente deposits extend more than 100 kilometers (60 miles) from their source.

In 1902, a nuée ardente from Mount Pelée, a small volcano on the Caribbean island of Martinique, destroyed the port town of St. Pierre. The destruction happened in moments and was so devastating that almost all of St. Pierre's 28,000 inhabitants were killed. Only a prisoner protected in a dungeon, a shoemaker, and a few people on ships in the harbor were spared (Figure 8.13).

When we compare the destruction of St. Pierre with that of Pompeii, several differences emerge. Pompeii was buried by an event lasting 3 days, whereas St. Pierre was destroyed in moments and its remains were mantled by only a thin layer of volcanic debris. Also, the structures of Pompeii remained intact, except for roofs that collapsed under the weight of the ash. In St. Pierre, masonry walls nearly 1 meter thick were knocked over like dominoes; large trees were uprooted and cannons were torn from their mounts. Clearly, volcanic hazards vary from one volcano to another.

Lahars. In addition to their violent eruptions, large composite cones often generate a type of debris flow known as a **lahar**. These destructive events occur when

Figure 8.13 St. Pierre as it appeared shortly after the eruption of Mount Pelée, 1902. (Reproduced from the collection of the Library of Congress)

deposits of volcanic ash and debris become saturated with water and flow down steep volcanic slopes, generally following stream valleys. Some lahars are produced when rainfall saturates volcanic deposits, whereas others are triggered as large volumes of ice and snow melt during an eruption.

When Mount St. Helens erupted in May 1980, several lahars formed. The flows and accompanying floods raced down the valleys of the north and south forks of the Toutle River at speeds in excess of 30 kilometers an hour. Water levels reached 4 meters above flood stage, and nearly all the homes and bridges along the river were destroyed or severely damaged (Figure 8.14). Fortunately, the affected area was not densely populated. This was not the case in November 1985, when Nevado del Ruiz, a volcano in the Andes, erupted and

generated a lahar that killed nearly 20,000 people (see the section entitled "Debris Flow" in Chapter 3).

Other Volcanic Landforms

The most obvious volcanic landform is a cone. But other distinctive landforms are associated with volcanic activity.

Calderas and Pyroclastic Flows

Most volcanoes have a steep-walled *crater*. A crater is called a *caldera* when it exceeds 1 kilometer in diameter. Calderas are roughly circular and most form when the summit of a volcanic structure collapses into a partially emptied magma chamber below (Figure 8.15).

Figure 8.14 A house damaged by a lahar along the Toutle River, west-northwest of Mount St. Helens. The end section of the house was torn free and lodged against trees. (Photo by D. R. Crandell, U.S. Geological Survey)

Eruption of
Mount Mazama

Partialy emptied
magma chamber

Collapse of
Mount Mazama

Formation of Crater Lake and Wizard Island

Figure 8.15 Sequence of events that formed Crater Lake, Oregon. About 7000 years ago, the summit of former Mount Mazama collapsed following a violent eruption that partly emptied the magma chamber. Subsequent eruptions produced the cinder cone called Wizard Island. Rainfall and groundwater contributed to form the lake. (After H. Williams, *The Ancient Volcanoes of Oregon*, p. 47. Courtesy of the University of Oregon)

Crater Lake in Oregon, located in such a structure, is 10 kilometers (6 miles) at its widest and 1175 meters (over 3800 feet) deep. The formation of Crater Lake began about 7000 years ago when the volcano, later to be named Mount Mazama, violently extruded 50 to 70 cubic kilometers of pyroclastic material. With the loss of support, 1500 meters (nearly a mile) of this

once-prominent 3600-meter cone collapsed. After the collapse, rainwater filled the caldera. Later volcanic activity built a small cinder cone in the lake called Wizard Island, which today provides a mute reminder of past activity (Figure 8.16).

Although most calderas are produced by *collapse following an explosive eruption*, some are not. For example, Hawaii's active shield volcanoes, Mauna Loa and Kilauea, have large calderas (3 to 5 kilometers, or 2 to 3 miles across, and nearly 200 meters, or 650 feet deep). They formed by gradual subsidence as magma slowly drained from the summit magma chambers during flank eruptions.

At least 138 calderas that exceed 5 kilometers in diameter are known. Unlike Crater Lake, these calderas are so large and irregular that many remained undetected until high-quality aerial, or satellite, images became available. One of these, LaGarita, located in the San Juan Mountains of southern Colorado, is about 32 kilometers wide and 80 kilometers long. Even today, with modern mapping techniques, the entire outline of this structure is still not known with certainty. The formation of a large caldera begins when a granitic (felsic) magma body is emplaced near the surface, upwarping the overlying rocks. Next, fracturing of the roof allows the highly viscous, gas-rich magma to reach the surface, where it explosively ejects huge volumes of pyroclastic materials, mainly ash and pumice fragments. Typically, this material, called a **pyroclastic flow**, avalanches across the landscape at speeds that may exceed 100 kilometers (60 miles) per hour, destroying most living things in its path. After coming to rest, these hot fragments often fuse to form a rock called *welded tuff* that closely resembles a solidified lava flow. Finally, with the loss of support the roof collapses, generating a caldera. The cycle of caldera formation can repeat itself numerous times in the same location.

Perhaps the best-known caldera in the United States is located in the Yellowstone Plateau of northwestern Wyoming. Here a large viscous magma body exists a few kilometers below the surface. Several times over the past 2 million years, fracturing of the rocks overlying the magma chamber has resulted in the ejection of huge quantities of pyroclastic material. One of the eruptions ejected 2500 cubic kilometers of magma in a devastating pyroclastic flow. (Fortunately, no eruption of this type has occurred in historic times.) In Yellowstone National Park, several fossil forests have been discovered, one above another. Following volcanic activity, a forest would develop on a newly formed volcanic surface, only to be covered by ash from the next eruptive phase. Other examples of large calderas that have formed in a similar manner are California's Long Valley Caldera and Valles Caldera located west of Los Alamos, New Mexico.

Figure 8.16 Crater Lake in Oregon occupies a caldera about 10 kilometers (6 miles) in diameter. Wizard Island is a cinder cone located in the caldera. (Photo by Greg Vaughn/Tom Stack and Associates)

Fissure Eruptions and Lava Plateaus

We think of volcanic eruptions as building a cone or mountain from a central vent. But by far the greatest volume of volcanic material is extruded from fractures in the crust called **fissures**. Rather than building a cone, these long, narrow cracks pour forth a low-viscosity Hawaiian-type lava, blanketing a wide area.

The extensive Columbia Plateau in the northwestern United States was formed this way (Figure 8.17). Here, numerous **fissure eruptions** extruded very fluid basaltic lava. Successive flows, some 50 meters thick, buried the existing landscape as they

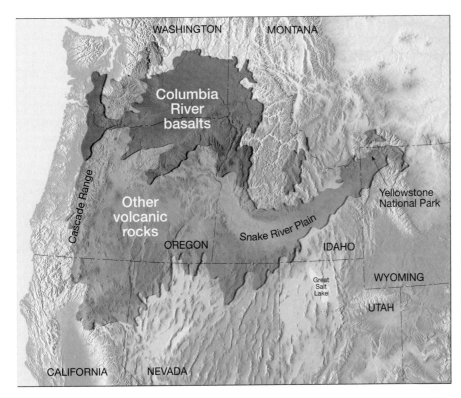

Figure 8.17 Volcanic areas in the northwestern United States. The Columbia River basalts cover an area of nearly 200,000 square kilometers (80,000 square miles). Activity here began about 17 million years ago as lava began to pour out of large fissures, eventually producing a basalt plateau with an average thickness of more than 1 kilometer. (After U.S. Geological Survey)

Figure 8.18 Basalt flows near Idaho Falls. (Photo by John S. Shelton)

built a lava plateau that is nearly a mile thick in places (Figure 8.18). The fluidity is evident, because some lava remained molten long enough to flow 150 kilometers (90 miles) from its source. The term **flood basalts** appropriately describes these flows.

Although a few large continental areas are covered with basalt flows, the greatest activity of this type occurs hidden from view on the ocean floor.

Along oceanic ridges, where seafloor spreading is active, fissure eruptions generate new seafloor. Iceland, which is located astride the Mid-Atlantic Ridge, has experienced numerous fissure eruptions of this type. The largest Icelandic eruptions in historic times occurred in 1783 and were named the Laki eruptions. A rift 25 kilometers long generated over 20 separate vents, which initially extruded sulfurous gases and ash deposits that built several small cinder cones. This activity was followed by huge outpourings of very fluid basaltic lava. The total volume of lava extruded by the Laki eruptions was in excess of 12 cubic kilometers. Volcanic gases stunted grasslands and directly killed most of Iceland's livestock. The ensuing famine caused 10,000 deaths.

Volcanic Necks

Volcanoes, like all land areas, are continually being lowered by weathering and erosion. Cinder cones are easily eroded, because they are composed of unconsolidated materials. As erosion progresses, the rock occupying the vent is often more resistant and may remain standing above the surrounding terrain long after most of the cone has vanished. Ship Rock, New Mexico, is such a feature, called a **volcanic neck** (Figure 8.19). This structure, higher than many skyscrapers, is but one of many such landforms that protrude conspicuously from the red desert landscapes of the Southwest. Ultimately, even these resistant necks will succumb to relentless erosion over geologic time.

Intrusive Igneous Activity

 Internal Processes
Igneous Activity

Volcanic eruptions can be among the most violent and spectacular events in nature and thus receive much scientific attention. Yet most magma is emplaced at depth, intruding into existing rocks. An understanding of this intrusive igneous activity is therefore as important to

Figure 8.19 Ship Rock, New Mexico, is a volcanic neck. This structure, which stands over 420 meters (1380 feet) high, consists of igneous rock that crystallized in the vent of a volcano that has long since been eroded away. The tabular structure in the background is a dike that served to feed lava flows along the flanks of the once active volcano. (Photo by Michael Collier)

geologists as is the study of volcanic events. In mythology, Pluto was lord of the underworld. Thus, underground igneous rock bodies are called *plutons*.

Figure 8.20 shows intrusive igneous bodies (plutons) that form when magma crystallizes within Earth's crust. Notice that some of these structures have a tabular or sheetlike shape, whereas others are bulky masses. Some cut across existing structures, such as layers of sedimentary rocks; others form when magma is injected between sedimentary layers. Because of these differences, plutons are generally classified according to their shape as either *tabular* (sheetlike) or

A. Implacement of igneous structures

B. Crystallization and erosion

C. Extensive uplift and erosion

Figure 8.20 Illustrations showing basic igneous structures. **A.** This cross-sectional view shows the relationship between volcanism and intrusive igneous activity. **B.** This view illustrates the basic intrusive igneous structures, some of which have been exposed by erosion long after their formation. **C.** After millions of years of uplifting and erosion a batholith is exposed at the surface.

massive, and by their orientation with respect to the host rock. Intrusive igneous bodies are said to be *discordant* if they cut across existing sedimentary beds, and *concordant* if they form parallel to the existing sedimentary beds.

Dikes and Sills

Dikes are sheetlike bodies produced when magma is injected into fractures that cut across rock layers. They range in thickness from less than 1 centimeter to more than 1 kilometer. The largest are hundreds of kilometers long. Dikes are often oriented vertically, forming in pathways followed by molten rock that fed ancient lava flows. Frequently dikes are more resistant to weathering than the surrounding rock. When exposed, such dikes have the appearance of a wall.

Sills are tabular plutons formed when magma is injected along sedimentary bedding surfaces (Figures 8.20 and 8.21). Horizontal sills are the most common, although all orientations, even vertical, are known to exist where the strata have been tilted. Because of their relatively uniform thickness and large extent, sills must form from very fluid basaltic magma. The emplacement of a sill requires that the overlying sedimentary rock be lifted to a height equal to the thickness of the sill. Consequently, sills form only at relatively shallow depths where the pressure exerted by the weight of overlying strata is relatively low.

One of the largest and best-known sills in the United States is the Palisades Sill, which is exposed along the Hudson River near New York City. This resistant, 300-meter-thick sill has formed an imposing cliff easily seen from across the river.

In many respects, sills closely resemble buried lava flows. Both are tabular and often exhibit **columnar jointing** (Figure 8.22). Further, because sills generally form in near-surface environments and may be only a few meters thick, the emplaced magma is often chilled quickly enough to generate a fine-grained texture.

When attempts are made to reconstruct the geologic history of a region, it becomes important to differentiate between sills and buried lava flows. Fortunately, under close examination these two structures can be readily distinguished. The upper portion of a buried lava flow usually contains voids produced by entrapped gas bubbles, and only the rocks beneath a lava flow show evidence of metamorphic alteration. Sills, in contrast, form when magma has been forcefully intruded between sedimentary layers. Thus, fragments of the overlying wall rock occur only in sills, as lava flows are extruded before the overlying strata are deposited. Further "baked" zones above and below are trademarks of a sill.

Laccoliths and Batholiths

Laccoliths form the same way as sills, but from more viscous magma, which collects as a lens-shaped mass that arches the overlying strata upward. Consequently, a laccolith can occasionally be detected because of the dome it creates at the surface (Figure 8.20).

Batholiths are by far the largest intrusive igneous bodies. The Idaho batholith, for example, encompasses more than 40,000 square kilometers. Indirect evidence from gravitational studies indicates that batholiths are also very thick, possibly extending tens of kilometers into the crust. Geologists define a batholith as an intrusive body with a surface exposure of more than 100 square kilometers (40 square miles). Similar but smaller intrusive bodies are termed *stocks*. Many stocks appear to be portions of batholiths that are not yet fully exposed.

Batholiths frequently form the cores of mountain systems. Here, uplift and erosion have removed the surrounding rock to expose the resistant igneous body (Figure 8.20C). Some of the highest mountain peaks, such as Mount Whitney in the Sierra Nevada, are carved from such a granitic mass. Large batholiths form over millions of years. The intrusive activity that created the Sierra Nevada batholith, for example, occurred nearly continuously over a 130-million-year span (Figure 8.23).

Large expanses of granitic rock are also exposed in the stable interiors of the continents, such as the Canadian Shield of North America. These relatively flat outcrops are believed to be the remnants of ancient mountains that erosion has long since leveled.

Figure 8.21 Salt River Canyon, Arizona. The dark, essentially horizontal band is a sill of basaltic composition that intruded into horizontal layers of sedimentary rock. (Photo by E. J. Tarbuck)

Figure 8.22 Columnar jointing in basalt, Giants Causeway National Park, Northern Ireland. These five-to-seven-sided columns are produced by contraction and fracturing that results as a lava flow or sill gradually cools. (Photo by Tom Till)

Equally spaced centers of contraction

Igneous Activity and Plate Tectonics

The origin of magma has been controversial in geology, almost from the beginning of the science. How do magmas of different compositions form? Why do volcanoes in the deep-ocean basins primarily extrude basaltic lava, whereas those adjacent to oceanic trenches extrude mainly andesitic lava? Why does an area of igneous activity, commonly called the *Ring of Fire*, surround the Pacific Ocean? New insights from the theory of plate tectonics are providing answers. We will first examine the origin of magma and then look at the global distribution of volcanic activity as viewed by the plate tectonics model.

Origin of Magma

Based on available scientific evidence, *Earth's crust and mantle are composed primarily of solid, not molten, rock.* Although the outer core is a fluid, its iron-rich material is very dense and remains deep within Earth. So, what is the source of magma that produces igneous activity?

Geologists conclude that magma originates when essentially solid rock, located in the crust and upper mantle, melts. The most obvious way to generate magma from solid rock is to raise the temperature above the rock's melting point. In a near-surface environment, silica-rich granitic rocks begin to melt at temperatures around 750°C, whereas basaltic rocks must be heated to temperatures above 1000°C before melting commences. In addition to raising a rock's temperature, rock that is near its melting point may begin to melt if the confining pressure drops, or if fluids (volatiles) are introduced. We will consider the role that heat, pressure, and volatiles play in generating magma.

Role of Heat. What source of heat is sufficient to melt rock? Workers in underground mines know that temperatures get higher as they go deeper. Although the rate of temperature change varies from place to place, it *averages* between 20°C and 30°C per kilometer in the *upper* crust. The change in temperature with depth is known as the *geothermal gradient*. From estimates of the geothermal gradient, the temperature at 100 kilometers ranges between 1200°C and 1400°C, whereas the temperature at the core–mantle boundary is calculated to be about 4500°C and it may exceed 6700°C at Earth's center.

Role of Pressure. If temperature were the only factor that determined whether or not rock melts, our planet would be a molten ball covered with a thin, solid outer shell. This, of course, is not the case. The reason is that pressure also increases with depth.

Figure 8.23 Half Dome in Yosemite National Park, California. This feature is just a tiny portion of the Sierra Nevada batholith, a huge structure that extends for approximately 400 kilometers. (Photo by Galen Rowell/Mountain Light Photography)

Melting, which is accompanied by an increase in volume, *occurs at higher temperatures at depth* because of greater confining pressure. Consequently, an increase in confining pressure causes an increase in the rock's melting temperature. Conversely, reducing confining pressure lowers a rock's melting temperature. When confining pressure drops, melting is triggered. This may occur when rock *ascends* as a result of convective upwelling, thereby moving into zones of lower pressure. (Recall that even though the mantle is a *solid, it does flow* at very slow rates over time scales of millions of years.) This process is responsible for generating magma along ocean ridges where plates are rifting apart (Figure 8.24).

Role of Volatiles. Another important factor affecting the melting temperature of rock is its water content. Water and other volatiles act as salt does to melt ice. That is, volatiles cause rock to melt at lower temperatures. Further, the effect of volatiles is magnified by increased pressure. Consequently, "wet" rock buried at depth has a much lower melting temperature than does "dry" rock of the same composition and under the same confining pressure (Figure 8.25). Therefore, in addition to a rock's

composition, its temperature, depth (confining pressure), and water content determine whether it exists as a solid or liquid.

Volatiles play an important role in generating magma in regions where cool slabs of oceanic lithosphere descend into the mantle (Figure 8.25). As an oceanic plate sinks, both heat and pressure drive water from the subducting crustal rocks. These volatiles, which are very mobile, migrate into the wedge of hot mantle that lies above. This process is believed to lower the melting temperature of mantle rock sufficiently to generate some magma.

Once enough molten rock forms, it will buoyantly rise toward the surface. In a continental setting, the magma body may "pond" beneath crustal rocks, which are already near their melting temperature. This results in the generation of a secondary silica-rich magma.

In summary, magma can be generated under three sets of conditions: (1) *heat* may be added; for example, a magma body from a deeper source intrudes and melts crustal rock; (2) a *decrease in pressure* (without the addition of heat) can cause melting; and (3) the *introduction of volatiles* (principally water) can lower the melting temperature of mantle rock sufficiently to generate magma.

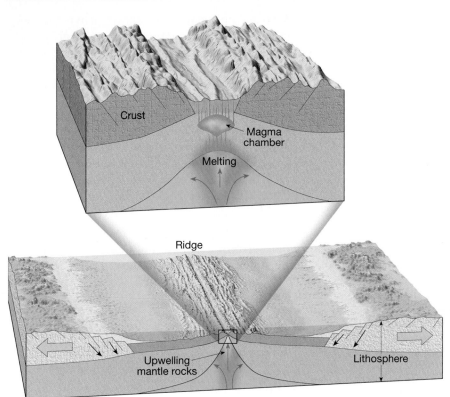

Figure 8.24 As hot mantle rock ascends, it continually moves into zones of lower pressure. This drop in confining pressure can trigger melting, even without additional heat.

Partial Melting and Magma Compositions. An important difference exists between the melting of a substance that consists of a single compound, such as ice, and melting igneous rocks, which are mixtures of several different minerals. Ice melts at a specific temperature, whereas igneous rocks melt over a temperature range of about 200°C. As rock is heated, the minerals with the lowest melting points are the first to melt. Should melting continue, minerals with higher melting points begin to melt and the composition of the magma steadily approaches the overall composition of the rock from which it was derived. Most often, melting is not complete. This process, known as **partial melting**, produces most, if not all, magma.

An important consequence of partial melting is the *production of a magma with a higher silica content than the original rock.* Recall from the discussion of Bowen's reaction series that basaltic rocks contain mostly high-melting-temperature minerals that are comparatively low in silica, whereas granitic rocks are composed primarily of low-melting-temperature silicates that are enriched in silica. Because silica-rich minerals melt first, magmas generated by partial melting are nearer to the granitic end of the

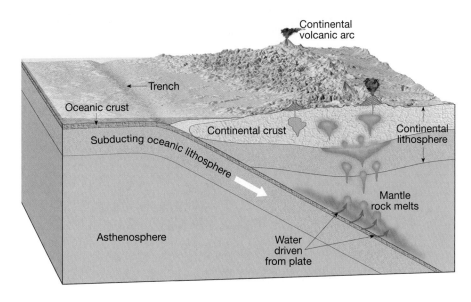

Figure 8.25 As an oceanic plate descends into the mantle, water and other volatiles are driven from the subducting crustal rocks. These volatiles lower the melting temperature of mantle rock sufficiently to generate magma.

compositional spectrum than are the rocks from which they formed (see Table 2.1).

Distribution of Igneous Activity

For many years geologists have realized that the global distribution of igneous activity is not random, but exhibits a definite pattern. In particular, igneous rocks having a granitic or andesitic composition (those rich in silica) are confined largely to the continents and continental margins. By contrast, volcanoes located within the deep-ocean basins, such as those in Hawaii and Iceland, extrude lavas that are basaltic in composition (low in silica). Moreover, basaltic rocks are also commonly found in continental settings. This pattern puzzled geologists until the development of the plate tectonics theory, which greatly clarified the picture.

Most of the more than 800 known active volcanoes are located along those continental margins adjacent to oceanic trenches. Further, volcanic activity occurs along spreading centers of the oceanic ridge system. The latter activity, although extensive, is hidden from view by the world's oceans.

In this section we examine three zones of igneous activity and relate them to global tectonics. These active areas are found along the oceanic ridges (spreading centers), adjacent to ocean trenches (subduction zones), and within the plates themselves (intraplate volcanism). These three settings are shown in Figure 8.26, which highlights two examples of each.

Igneous Activity at Spreading Centers. The greatest volume of volcanic rock is produced along the oceanic

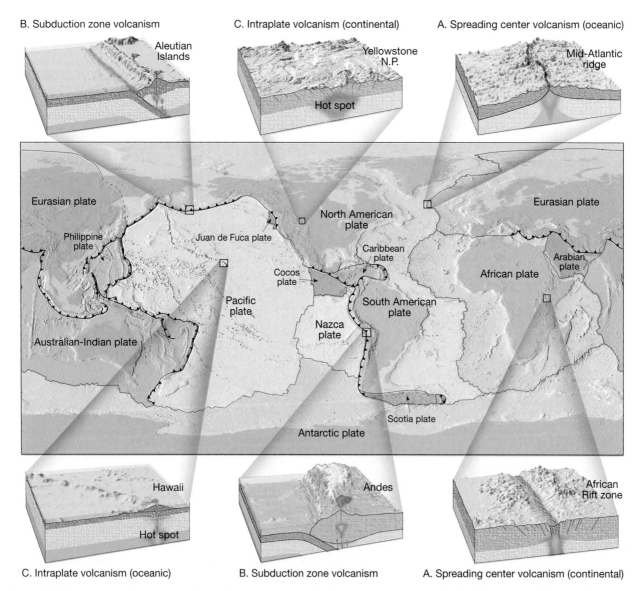

Figure 8.26 Three zones of volcanism. Two of these zones are plate boundaries, and the third includes areas within the plates.

ridge system, where seafloor spreading is active (see Figure 8.24). As plates of rigid lithosphere pull apart, the pressure on the underlying rocks is lessened. This reduced pressure, in turn, lowers the melting temperature of the mantle rocks. Partial melting of these materials (primarily peridotite), produces large quantities of basaltic magma that move upward to fill the newly formed cracks between the diverging plates. In this way new slivers of oceanic crust are generated.

Some of the molten basalt reaches the ocean floor, where it produces extensive lava flows or occasionally grows into a volcanic cone. Sometimes this activity produces a volcanic structure large enough to rise above sea level such as the islands of Vestmann off the southern coast of Iceland (Figure 8.27). Numerous submerged volcanoes also dot the flanks of the ridge system and the adjacent deep-ocean floor. Many of these formed along the ridge crest and gradually moved away as new oceanic crust was created by the seemingly unending process that generates new seafloor.

Igneous Activity at Subduction Zones. Recall that deep-ocean trenches are sites where slabs of water-rich oceanic crust are bent and descend into the mantle. As the slab sinks, volatiles (mainly water and carbon dioxide) are driven from the oceanic crust and migrate upward into the hot, wedge-shaped piece of mantle located directly above. At a depth of 100 to 150 kilometers, these water-rich fluids reduce the melting point of mantle rocks

sufficiently to promote partial melting. In some environments, silica and other components of the subducted sediments may become incorporated into the magma body. This process generates basaltic magma, and may produce some andesitic magma.

After a sufficient quantity of magma has accumulated, it slowly migrates upward because it is less dense than the surrounding rock. When igneous activity occurs along a subduction zone in the ocean, a chain of volcanoes called a *volcanic island arc* is produced. These structures, which usually form 200 to 300 kilometers from an oceanic trench, include such island chains as the Aleutians, the Tongas, and the Marianas.

When subduction of oceanic crust occurs beneath a continent, the magma that forms may become contaminated with silica-rich rocks as it moves through the crust. The assimilation of crustal fragments into the ascending basaltic magma will change it into one exhibiting an andesitic to granitic composition. The *continental volcanic arc* consisting of South America's Andes Mountains is one place where andesitic magma having such an origin is extruded.

Many subduction-zone volcanoes border the Pacific basin. Because of this pattern, the region has come to be called the *Ring of Fire*. Here volcanism is associated with subduction of the Pacific seafloor. As oceanic plates sink, they carry sediments and oceanic crust containing abundant water to great depths. The presence of water contributes to the high gas content and explosive nature of volcanoes that make up the

Figure 8.27 Eruption near the fishing port of Heimaey, off the south coast of Iceland, 1973. (Photo courtesy of Vulcain-Explorer/Photo Researchers, Inc.)

Box 8.4 Earth as a System: Volcanic Activity and Climate

The idea that explosive volcanic eruptions change Earth's climate was first proposed many years ago and still appears to explain some aspects of climatic variability. Explosive eruptions emit huge quantities of gases and fine-grained debris into the atmosphere (Figure 8.D). The greatest eruptions are sufficiently powerful to inject material high into the stratosphere, where it spreads around the globe and remains for many months or even years. The basic premise is that this suspended volcanic material (most importantly, droplets of sulfuric acid) will filter out a portion of the incoming solar radiation, and this, in turn, will lower air temperatures worldwide.

Perhaps the most notable cool period linked to a volcanic event is the "year without a summer" that followed the 1815 eruption of Tambora in Indonesia. The abnormally cold spring and summer of 1816 in many parts of the Northern Hemisphere, including New England, are believed to have been caused by the cloud of volcanic debris and gases ejected from Tambora (Figure 8.C).

When Mount St. Helens erupted in 1980, there was almost immediate speculation: Can an eruption such as

this change our climate? Although spectacular, a single explosive volcanic eruption of the magnitude of Mount St. Helens occurs somewhere in the world every 2 to 3 years. Studies of these events indicate that a very slight cooling of the lower atmosphere does occur. However, the cooling is so slight, less than one-tenth of one degree Celsius, as to be inconsequential.

The huge eruption of the Philippines' Mount Pinatubo in 1991 gave scientists another opportunity to test the connection between volcanism and climatic change. In 1991, this volcano emitted huge quantities of ash and 25 million to 30 million tons of sulfur dioxide gas. The gaseous component turned into a haze of tiny sulfuric acid droplets. Because this material was superhot, it rose to heights of 20 to 30 kilometers and quickly encircled the entire globe. This sun-blocking haze was probably the most massive since the eruption of the Indonesian volcano Krakatau in 1883. Studies indicated that the average global surface temperature declined by about 0.6°C (1°F) when compared to the average of the year before. Scientists hope to learn more about the volcano/climate connection as subsequent data are analyzed.

Figure 8.D Approximate volume of volcanic debris emitted during some well-known eruptions. The 1815 eruption of Tambora, the largest known eruption in historic time, ejected over 100 times more ash than did Mount St. Helens in 1980.

Ring of Fire (see Box 8.4). The volcanoes of the Cascade Range in the northwestern United States, including Mount St. Helens, Mount Rainier, and Mount Shasta, are of this type (Figure 8.28).

Intraplate Igneous Activity. We know why igneous activity occurs along plate boundaries. But why do eruptions occur in the middle of plates?

Because basalts having relatively similar compositions are found in the ocean basins as well as on the continents, partial melting of mantle rocks is the most probable source for these rocks. The sources of some intraplate basaltic magma are hot **mantle plumes** that originate at great depth. Upon reaching the crust, these structures begin to spread laterally. The result is localized volcanic regions a few hundred kilometers across called **hot spots**. More than a hundred hot spots have been identified, and most appear to have persisted for a few tens of millions of years. One hot

spot is situated beneath the island of Hawaii. Another is responsible for the large outpourings of lava that make up Iceland.

With few exceptions, lavas and ash of rhyolitic-composition are restricted to vents located landward of the continental margins. This suggests that remelting of the continental crust may be the mechanism responsible for the formation of these silica-rich magmas. But what mechanism causes large quantities of continental material to be melted? One explanation is that a thick unit of continental crust occasionally becomes situated over a rising plume of hot mantle material. Rather than producing vast outpourings of basaltic lava as occurs at oceanic sites such as Hawaii, the magma from the rising plume is emplaced at depth. Here melting and assimilation of the surrounding host rock result in the formation of a highly altered *secondary magma*. As this buoyant silica-rich magma slowly migrates upward, continued hot-spot activity

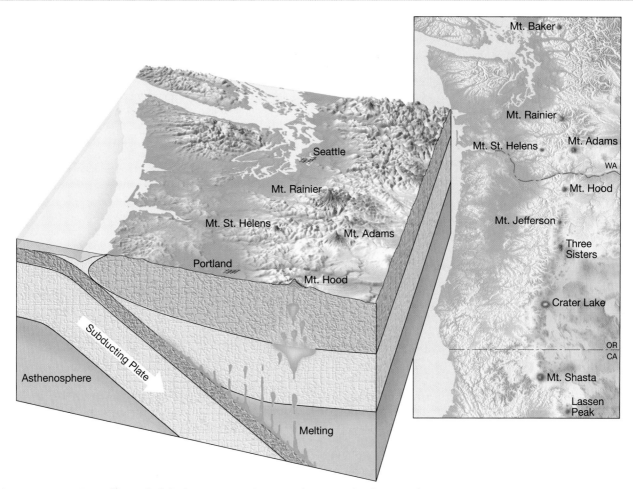

Figure 8.28 Locations of several of the larger composite cones that comprise the Cascade Range.

supplies heat to the rising mass, thereby aiding its ascent. Volcanism in the Yellowstone region may have resulted from this process.

Although the plate tectonics theory has answered many questions regarding the distribution of igneous activity, many new questions have arisen: Why does seafloor spreading occur in some areas but not others? How do mantle plumes and associated hot spots originate? These and other questions are the subject of continuing geologic research.

Chapter Summary

- The primary factors that determine the nature of volcanic eruptions include the magma's *temperature*, its *composition*, and the *amount of dissolved gases* it contains. As lava cools, it begins to congeal, and as *viscosity* increases, its mobility decreases. *The viscosity of magma is directly related to its silica content. Rhyolitic* lava, with its high silica content, is very viscous and forms short, thick flows. *Basaltic* lava, with a lower silica content, is more fluid and may travel a long distance before congealing. Dissolved gases provide the force that propels molten rock from the vent of a volcano.

- The materials associated with a volcanic eruption include *lava flows* (*pahoehoe* and *aa* flows for basaltic lavas), *gases* (primarily in the form of *water vapor*), and *pyroclastic material* (pulverized rock and lava fragments blown from the volcano's vent, which include *ash, pumice, lapilli, cinders, blocks*, and *bombs*).

- *Shield volcanoes* are broad, slightly domed structures built primarily of fluid, basaltic lava. *Cinder cones* have very steep slopes composed of pyroclastic material. *Composite cones*, or *stratovolcanoes*, are large, nearly symmetrical structures built of interbedded lavas and pyroclastic deposits. Composite cones represent the most violent type of volcanic activity.

- Other than volcanoes, regions of volcanic activity may contain *volcanic necks* (rocks that once occupied the vents of volcanoes but are now exposed because of erosion), *craters* (steep walled depressions at the summit of most volcanoes), *calderas* (craters that exceed 1 kilometer in diameter), *fissure eruptions* (volcanic material extruded from fractures in the crust), and *pyroclastic flows*.

- Igneous intrusive bodies are classified according to their *shape* and by their *orientation with respect to the host rock*, generally sedimentary rock. The two general shapes are *tabular* (sheetlike) and *massive*. Intrusive igneous bodies that cut across existing sedimentary beds are said to be *discordant*, whereas those that form parallel to existing sedimentary beds are *concordant*.

- *Dikes* are tabular, discordant igneous bodies produced when magma is injected into fractures that cut across rock layers. Tabular, concordant bodies called *sills* form when magma is injected along the bedding surfaces of sedimentary rocks. *Laccoliths* are similar to sills but form from less-fluid magma that collects as a lens-shaped mass that arches the overlying strata upward. *Batholiths*, the largest intrusive igneous bodies with surface exposures of more than 100 square kilometers (40 square miles), frequently make up the cores of mountains.

- Magma can be generated under three sets of conditions: (1) *heat* may be added; for example, a magma body from a deeper source intrudes and melts crustal rock; (2) a *decrease in pressure* (without the addition of heat) can cause melting; and (3) the *introduction of volatiles* can lower the melting temperature of mantle rock sufficiently to generate magma. Igneous rock, which contains several different minerals, melts over a range of temperatures, with the lower temperature minerals melting first. This process, known as *partial melting*, produces most, if not all, magma. Partial melting often results in the production of a melt with a higher silica content than the parent rock.

- Active areas of *volcanism* are found *along the oceanic ridges* (spreading center volcanism), *adjacent to ocean trenches (subduction zone volcanism)*, as well as *the interiors of plates (intraplate volcanism)*. Most active volcanoes are associated with plate boundaries.

Key Terms

aa flow (p. 221)
batholith (p. 234)
caldera (p. 223)
cinder cone (p. 225)
columnar joints (p. 234)
composite cone (stratovolcano)
    (p. 227)
crater (p. 223)
dike (p. 234)

fissure (p. 231)
fissure eruption (p. 231)
flood basalt (p. 232)
hot spot (p. 240)
laccolith (p. 234)
lahar (p. 228)
mantle plumes (p. 240)
nuée ardente (p. 227)
pahoehoe flow (p. 221)

partial melting (p. 237)
pyroclastic flow (p. 230)
pyroclasts (p. 221)
shield volcano (p. 223)
sill (p. 234)
vent (p. 223)
viscosity (p. 218)
volcanic neck (p. 232)
volcano (p. 223)

Review Questions

1. What triggered the May 18, 1980, eruption of Mount St. Helens?

2. Name five volcanoes in the western United States that geologists believe will erupt again.

3. What is the difference between magma and lava?

4. What three factors determine the nature of a volcanic eruption? What role does each play?

5. Why is a volcano fed by highly viscous magma likely to be a greater threat than a volcano supplied with very fluid magma?

6. Describe pahoehoe and aa lava.

7. List the main gases released during a volcanic eruption.

8. Analysis of samples taken during Hawaiian eruptions on the island of Hawaii indicate that _____ was the most abundant gas released.

9. Describe each type of pyroclastic material.

10. Compare and contrast the main types of volcanoes (size, shape, eruptive style, and so forth).

11. Name one example of each of the three types of volcanoes.

12. Compare the formation of Hawaii with that of Parícutin.

13. How is a caldera different from a crater?

14. Describe the formation of Crater Lake. Compare it to the caldera formed during the eruption of Kilauea.

15. What is Ship Rock, New Mexico, and how did it form?

16. Describe each of the four intrusive features discussed in the text.

17. Why might a laccolith be detected at Earth's surface before being exposed by erosion?

18. What is the largest of all intrusive igneous bodies? Is it tabular or massive? Concordant or discordant (see Figure 8.20)?

19. Explain how most magma is thought to originate.

20. Spreading center volcanism is associated with which rock type? What causes rocks to melt in regions of spreading center volcanism?

21. What is the Ring of Fire?

22. Are volcanic eruptions in the Ring of Fire generally quiet or violent? Name a volcano that would support your answer.

23. The Hawaiian Islands and Yellowstone Park are thought to be associated with which of the three zones of volcanism?

24. Volcanic islands in the deep ocean are composed primarily of what igneous rock type?

25. Explain why andesitic and granitic rocks are confined to the continents and oceanic margins but are absent from deep-ocean basins.

26. Briefly describe the mechanism by which explosive volcanic eruptions are thought to influence Earth's climate (See Box 8.4).

Examining the Earth System

1. Describe some of the consequences that a great and prolonged increase in explosive volcanic activity might have on each of Earth's four spheres. An excellent Web site to begin volcano research is the University of California at Santa Barbara Volcano Information Center at **http://magic.geol.ucsb.edu/~fisher/**

2. Despite the potential for devastating destruction, humans live, work, and play on or near many active volcanoes. What are some of the benefits that a volcano or volcanic region might offer? (List some volcanoes and the assets they provide.)

Web Resources

The following are informative and interesting World Wide Web sites that address topics related to those presented in the chapter:

VolcanoWorld:

• **http://volcano.und.edu/**

Worldwide Volcanism (National Aeronautics and Space Administration):

• **http://skye.gsfc.nasa.gov/wwvolcano.html**

For direct links to these sites and other appropriate Earth science Web locations, chapter review exercises, and relevant, topical investigations that utilize Web resources, contact the *Earth Science*, 9th Edition Home Page at:

• **http://www.prenhall.com/tarbuck.**

Chapter 9
Mountain Building

Crustal Uplift

Rock Deformation

Mountain Types

Mountain Building

Mount Williamson, East Face of the Sierra Nevada, California. The Sierra Nevada play a major role in creating the arid climate to the east of the mountains. (Photo by Carr Clifton)

Mountains provide some of the most spectacular scenery on our planet. This splendor has been captured by poets, painters, and songwriters alike. Geologists believe that at some time all continental regions were mountainous masses, and they have concluded that the continents grow by the addition of mountains to their flanks. Consequently, by unraveling the secrets of mountain formation, geologists will have taken a major step in understanding the evolution of Earth. If continents do indeed grow by adding mountains to their flanks, how do geologists explain the existence of mountains (the Urals, for example) that are located in the interior of a landmass? To answer this and other related questions, this chapter attempts to piece together the sequence of events that is believed to generate these lofty structures.

Mountains often are spectacular features that rise several hundred meters or more above the surrounding terrain (Figure 9.1). Some occur as isolated masses; the volcanic cone Kilimanjaro, for example, stands almost 6000 meters (20,000 feet) above sea level, overlooking the grasslands of East Africa. Other peaks are parts of extensive mountain belts, such as the American Cordillera, which runs continuously from the tip of South America through Alaska, including the Andes and the Rocky Mountains. Chains such as the Himalayas consist of youthful, towering peaks that are still rising, whereas others, including the Appalachian Mountains in the eastern United States, are much older and have been eroded much lower than their original lofty height.

The name for the processes that collectively produce a mountain belt is **orogenesis,** from the Greek *oros* ("mountain") and *genesis* ("to come into being"). Mountain systems show evidence of enormous forces that have bent (folded), broken (faulted), and generally deformed large sections of Earth's crust.

The first encompassing explanation of orogenesis came a little over three decades ago as part of the plate tectonics theory. The idea of crustal plates in slow motion, sometimes converging with great force, has opened many new and exciting avenues to geologists. Before examining mountain building according to the plate tectonics model, however, it will be advantageous to view some basic principles: the processes of crustal uplifting and rock deformation.

Crustal Uplift

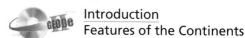

Introduction
Features of the Continents

The fossilized shells of marine invertebrates are often found at high elevations in mountains, an indication that the sedimentary rock in the mountain was once below sea level. This is convincing evidence that some dramatic changes have occurred between the time these animals died and the discovery of their fossilized remains!

Evidence for crustal uplift such as this is common in rocks, which form the "geologic record." Many examples exist along the coast of the western United States. When a coastal area remains undisturbed for an

Figure 9.1 Banner Peak in Ansel Adams Wilderness, California. (Photo by Carr Clifton)

Figure 9.2 Former wave-cut platforms now exist as a series of elevated terraces on the west side of San Clemente Island off the southern California coast. Once at sea level, the highest terraces are now about 400 meters (1320 feet) above it. (Photo by John S. Shelton)

extended period, wave action cuts a gently sloping platform. In parts of California, ancient wave-cut platforms can now be found as terraces, hundreds of meters above sea level (Figure 9.2). Each terrace represents a period when that area was at sea level. Such evidence of crustal uplift is easy to find; unfortunately, the *reasons* for uplift are not so easily determined.

Isostasy

We know that the force of gravity must play an important role in determining the elevation of the land. In particular, the less dense crust is believed to float on top of the denser and more easily deformed rocks of the mantle. This concept—a floating crust in gravitational balance—is called **isostasy.**

Perhaps the easiest way to grasp the concept of isostasy is to envision a series of wooden blocks of different heights floating in water, as shown in Figure 9.3. Note that the taller wooden blocks float higher in water than do the shorter blocks. In a similar manner, mountain belts stand higher above the surface and have "roots" that extend deeper into the supporting material below (Figure 9.4A). The average thickness of the continental crust is about 35 kilometers, but crustal thicknesses for some mountain chains exceeds 70 kilometers. Thus, mountains exist where unusually thick blocks of light crustal material are supported from below by the dense rocks of the mantle.

Carrying this idea one step further, the crust beneath the oceans must be thinner than that of the continents, because its elevation is lower. Not only is this true, but oceanic rocks also have a greater density than do continental rocks, another factor contributing to their lower position.

Isostatic Adjustment

If the concept of isostasy is correct, we should expect that when weight is added to the crust, the crust will respond by subsiding, and when that weight is

removed, the crust will rebound. (Visualize what happens to a ship as cargo is being loaded and unloaded.) This process of establishing a new level of equilibrium is called **isostatic adjustment.**

Evidence for crustal subsidence and crustal rebound is well documented. A classic example is provided by Ice Age glaciers. When continental ice sheets occupied portions of North America during the Pleistocene epoch, the added weight of the 3-kilometer-thick mass of ice caused downwarping of Earth's crust by hundreds of meters. In the 8000 to 10,000 years since the last ice sheet melted, uplifting of as much as 330 meters has occurred in Canada's Hudson Bay region, where the thickest ice had accumulated.

As this example illustrates, isostatic adjustment can account for considerable *vertical* crustal movement. Thus, we can understand why, as erosion slowly lowers the summits of mountains, the crust gradually rises in response to the reduced load. The processes of uplifting and erosion will continue together until the mountain block reaches "normal" crustal thickness (Figure 9.4C). When this occurs, the mountains will have been eroded to near sea level and the deeply buried portions of the mountains will become exposed at the surface. In addition, as the mountains succumb

Figure 9.3 This drawing illustrates how wooden blocks of different thicknesses float in water. In a similar manner, thick sections of crustal material float higher than thinner crustal slabs.

to erosion, sediment derived from the shrinking masses is deposited on the adjacent continental margin, causing the margin to subside (Figure 9.4B,C).

To summarize, we have learned that Earth's major mountain belts consist of unusually thick sections of deformed crustal material that have been elevated above the surrounding terrain. Support for these massive features comes from the buoyancy of light crustal material supported by denser mantle rocks. As erosion lowers the peaks, isostatic adjustment gradually raises the mountains in response. Eventually, the deepest portions of the mountains rise to the shallower depths of the surrounding crust. The question still to be answered is: How do these thickened sections of Earth's crust come into existence? We attempt to answer this question later in the chapter.

Rock Deformation

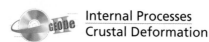

Internal Processes
Crustal Deformation

When rocks are subjected to forces (*stress*) greater than their own strength, they begin to deform, usually by folding or fracturing (Figure 9.5). It is easy to visualize how we could apply a hammer blow to a rock and cause it to break. But how are large rock units, many meters thick and many kilometers long, bent into intricate folds without being appreciably broken during the process? In an attempt to answer this, geologists used the laboratory. Here rocks were subjected to stresses under conditions that simulated those existing at various depths within the crust.

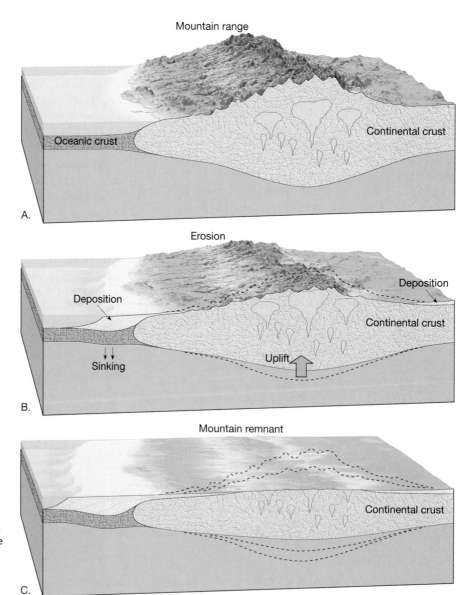

Figure 9.4 This sequence illustrates how the combined effect of erosion and isostatic adjustment results in a thinning of the crust in mountainous regions. **A.** When mountains are young, the continental crust is thickest. **B.** As erosion lowers the mountains, the crust rises in response to the reduced load. **C.** Erosion and uplift continue until the mountains reach "normal" crustal thickness.

Figure 9.5 These Mesozoic-age rocks near Lulworth, Dorset, England, were originally deposited in horizontal beds. They were folded and uplifted as a result of a collision between the African and European crustal plates. (Photo by Tom Bean)

Types of Deformation

Although each rock type deforms somewhat differently, the general characteristics of rock deformation were determined from laboratory experiments. Geologists discovered that when stress is applied, rocks first respond by deforming elastically. Changes resulting from *elastic deformation* are reversible; that is, like a rubber band, the rock will return to nearly its original size and shape when the stress is removed.

However, once the *elastic limit* is surpassed, rocks either deform plastically or they fracture. *Plastic deformation* results in permanent changes; that is, both the size and the shape of a rock unit are altered through folding and flowing. Laboratory experiments confirmed that, at the high temperatures and pressures simulating those deep in the crust, most rocks deform plastically once their elastic limit is surpassed.

Rocks tested under surface conditions also deform elastically, but once their elastic limit is exceeded, most behave like a brittle solid and fracture. Recall that the energy for most earthquakes comes from stored elastic energy that is released as rock fractures and snaps back to its original shape.

One key factor that researchers cannot duplicate in the laboratory is *geologic time*. We know that if stress is applied quickly, as with a hammer, rocks tend to fracture. Conversely, identical rocks may deform plastically if stress is applied over an extended period. For example, marble benches have been known to sag under their own weight over a period of a hundred years or so. In nature, even modest forces applied over long periods surely play an important role in the deformation of rock.

Folds

During mountain building, flat-lying sedimentary and volcanic rocks are often bent into a series of wavelike undulations called **folds** (Figure 9.6). Folds in sedimentary strata are much like those that would form if you were to hold the ends of a sheet of paper and push them together. In nature, folds vary widely in size and shape. Some folds are broad features in which rock units hundreds of meters thick are slightly warped. Others are very tight, microscopic structures found in metamorphic rocks. Size differences notwithstanding, most folds are the result of *compressional stresses* that result in the shortening and thickening of the crust. Occasionally, folds are found singly, but most often they occur as a series of undulations.

Anticlines and Synclines. The two most common types of folds are anticlines and synclines (Figure 9.6). An **anticline** is most commonly formed by the upfolding, or arching, of rock layers. Anticlines are sometimes spectacularly displayed where highways have been cut through deformed strata. Often found in association with anticlines are downfolds, or troughs, called **synclines.** Anticlines and synclines vary in symmetry. They may be *symmetrical, asymmetrical,* or, if one limb has been tilted beyond the vertical, *overturned* (Figure 9.6).

Folds do not continue on forever; rather, their ends die out, much like wrinkles in cloth. These folds are said to be *plunging,* as the axis of the fold plunges to the ground (Figure 9.7). Figure 9.8A shows an example of a plunging anticline and the pattern produced when erosion removes the upper layers of this structure and exposes its interior. Note that the outcrop pattern of an

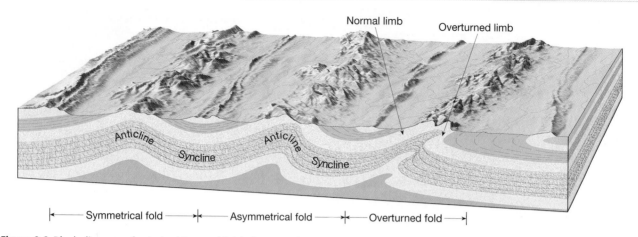

Figure 9.6 Block diagram of principal types of folded strata. The upfolded or arched structures are anticlines. The downfolds or troughs are synclines. Notice that the limb of an anticline is also the limb of the adjacent syncline.

anticline points in the direction it is plunging, whereas the opposite is true for synclines.

A good example of all this—the topography that results when erosional forces wear down folded sedimentary strata for millions of years—is the Valley and Ridge Province of the Appalachians. Here, resistant sandstone beds remain as imposing ridges, separated by valleys that have been cut into more easily eroded shale or limestone beds.

Domes and Basins. Although most folds are caused by horizontal compression, which squeezes strata, some are a consequence of *vertical* displacement. When *upwarping* produces a circular or somewhat elongated structure, the feature is called a **dome** (Figure 9.9A). The Black Hills of western South Dakota are a domal structure. Erosion has stripped away the upwarped sedimentary beds, exposing older igneous and metamorphic rocks in the center (see Figure 9.18).

Structures of similar shape, except *downwarped,* are termed **basins** (Figure 9.9B). Several large basins exist in the United States. The basins of Michigan and Illinois, each roughly the size of its state, have very gently sloping beds, similar to vast saucers.

Because large basins contain sedimentary beds sloping at such low angles, they are usually identified

Figure 9.7 Sheep Mountain, a doubly plunging anticline. Note that erosion has cut the flanking sedimentary beds into low ridges that make a "V" pointing in the direction of plunge. (Photo by John S. Shelton)

Figure 9.8 Plunging folds. **A.** Idealized view of plunging folds in which a horizontal surface has been added. **B.** View of plunging folds as they might appear after extensive erosion. Notice that in a plunging anticline the outcrop pattern "points" in the direction of the plunge, while the opposite is true of plunging synclines.

by the age of the rocks that make up the basins. The youngest rocks are found near the center and the oldest rocks are at the flanks. This is just the opposite order of a domal structure such as the Black Hills, where the oldest rocks form the core.

Faults

Faults are fractures in the crust along which appreciable displacement has occurred, on a scale from centimeters to kilometers. Occasionally, small faults can be recognized in road cuts where sedimentary beds have been

Figure 9.9 Gentle upwarping and downwarping of crustal rocks produce domes **A.** and basins **B.** Erosion of these structures results in an outcrop pattern that is roughly circular or elongated.

Figure 9.10 Faulting caused the vertical displacement of these beds near Kanab, Utah. Arrows show relative movement. (Photo by Tom Bean/DRK Photo)

offset a few meters, as shown in Figure 9.10. Large faults, like the infamous San Andreas fault in California, have displacements of hundreds of kilometers. These large faults consist of multiple interconnecting fault surfaces, forming *fault zones* that can be several kilometers wide. Often they are easier to identify from high-altitude images than at ground level (see Figure 9.B in Box 9.1).

Dip-Slip Faults. Movement along a fault can be horizontal, vertical, or a combination. Faults in which the movement is primarily *vertical* are called **dip-slip faults,** because the displacement is along the inclination (*dip*) of the fault plane. Movement along dip-slip faults can be either up or down the fault plane—that is, rocks on one side of the fault can be either pushed up or dropped down with respect to the other side. Thus, two types of dip-slip faults are recognized. To distinguish between the two, we call the rock that is higher than the fault surface the *hanging wall,* and the rock below the fault surface the *footwall.* This nomenclature arose from prospectors and miners who excavated shafts along fault zones because such zones are frequently sites of mineral deposition. During these operations, the miners would walk on the rocks below the fault plane (hence, footwall), and hang their lanterns on the rocks above (hence, hanging wall).

Dip-slip faults are classified as **normal faults** when the hanging wall moves downward relative to the footwall (Figure 9.11A). Conversely, **reverse faults**

A. Normal fault (tensional)

B. Reverse fault (compressional)

C. Thrust fault (compressional)

D. Strike-slip fault (shear)

Figure 9.11 Block diagrams of four types of faults. **A.** Normal fault. **B.** Reverse fault. **C.** Thrust fault. **D.** Strike-slip fault.

occur when the hanging wall moves upward relative to the footwall (Figure 9.11B). In mountainous regions such as the Alps and the Appalachians, reverse faults with dips less than 45°, called **thrust faults,** have displaced rock as far as 100 kilometers over adjacent strata (Figure 9.11C). Thrust faults of this type result from strong compressional stresses.

Strike-Slip Faults. Faults in which the dominant displacement is along the *strike,* or trend, of the fault are called **strike-slip faults** (Figure 9.11D). Many large strike-slip faults are associated with plate boundaries and are called *transform faults.*

Transform faults have nearly vertical dips and connect large structures, such as segments of an ocean ridge. The San Andreas fault in California is a well-known transform fault in which the displacement exceeds 560 kilometers (see Box 9.1). When faults have both vertical and horizontal movement, they are called **oblique-slip faults.**

Tensional and Compressional Forces. Fault motion provides geologists with a method of determining the nature of the forces at work within Earth. Normal faults indicate *tensional* stresses that pull the crust apart (Figure 9.11A). This "pulling apart" can be accomplished either by uplifting, which causes the surface to stretch and break, or by horizontal forces that actually rip the crust apart.

Normal faulting is known to occur at spreading centers, because plate divergence (spreading apart) is prevalent. Here a central block called a **graben** is bounded by normal faults and drops as the plates separate (Figure 9.12). These grabens produce an elongated valley bounded by upfaulted structures called **horsts.** The Great Rift Valley of East Africa consists of several large grabens, above which tilted horsts produce a linear mountainous topography (see Figure 7.11). This Great Rift Valley, nearly 6000 kilometers (3700 miles) long, contains the excavation sites of some of the earliest human fossils. Other rift valleys include the Rhine Valley in Germany and the valley of the Dead Sea in the Middle East.

Because the blocks involved in reverse and thrust faulting are displaced *toward* one another, geologists conclude that *compressional forces* are at work (Figure 9.11B,C). The primary regions of this activity are convergent zones, where plates are colliding. Compressional forces generally produce folds as well as faults, and they result in a general thickening and shortening of the crust. As you shall see, most mountain building occurs in compressional situations.

Joints

Among the most common rock structures are fractures called **joints.** Unlike faults, joints are fractures along which *no appreciable displacement* has occurred. Although some joints have random orientation, most occur in roughly parallel groups.

We have already considered two types of joints. Earlier we learned that *columnar joints* form when igneous rocks cool and develop shrinkage fractures, producing elongated, pillarlike columns (see Figure 8.23). Also recall that *sheeting* produces a pattern of gently curved joints that develop more or less parallel to the surface of large exposed igneous bodies such as batholiths. Here the jointing is thought to result from the gradual expansion that occurs as erosion removes the pressure of the overlying load (see Figure 3.4).

In contrast to the situations just described, most joints are produced when rocks are deformed by the stresses associated with crustal movements during mountain building. Extensive joint patterns can also develop in response to relatively subtle and often barely perceptible regional upwarping and downwarping of the crust.

Many rock units are broken by two or even three sets of intersecting joints that slice the rock into numerous regularly shaped blocks. These joint sets often exert a strong influence on other geologic processes. For example, chemical weathering tends to concentrate along joints, and in many areas, groundwater movement and the resulting solution

Figure 9.12 Diagrammatic sketch of downfaulted block (graben) and upfaulted block (horst).

Box 9.1 The San Andreas Fault System

The San Andreas, the best-known and largest fault system in North America, first attracted wide attention after the great 1906 San Francisco earthquake and fire. Following this devastating event, geologic studies demonstrated that a displacement of as much as 5 meters along the fault had been responsible for the earthquake. It is now known that this dramatic event is just one of many thousands of earthquakes that have resulted from repeated movements along the San Andreas throughout its 29-million-year history.

Where is the San Andreas fault system located? As shown in Figure 9.A, it trends in a northwesterly direction for

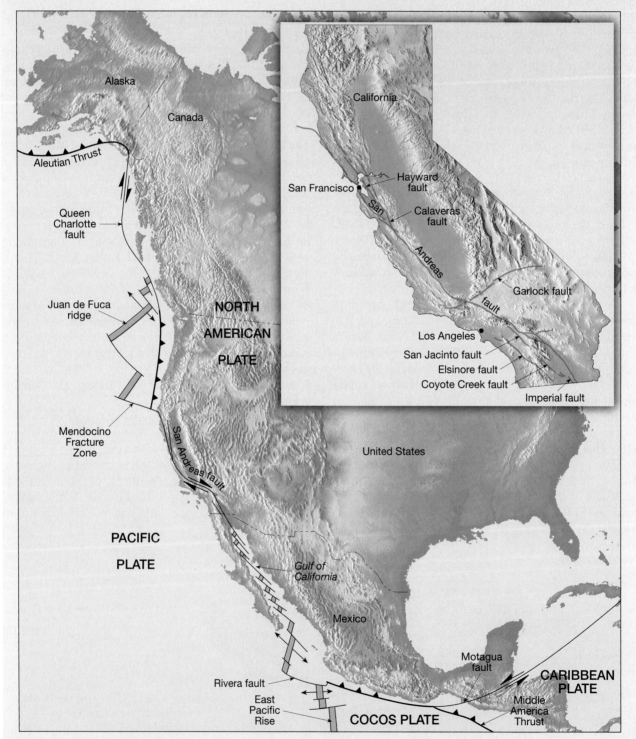

Figure 9.A Map showing the extent of the San Andreas fault system. Inset shows only a few of the many splinter faults that are part of this great fault system.

nearly 1300 kilometers (780 miles) through much of western California. At its southern end, the San Andreas connects with a spreading center located in the Gulf of California. In the north, the fault enters the Pacific Ocean at Point Arena, where it is thought to continue its northwesterly trend, eventually joining the Mendocino fracture zone. In the central section, the San Andreas is relatively simple and straight. However, at its two extremities, several branches spread from the main trace, so that in some areas the fault zone exceeds 100 kilometers (60 miles) in width.

Over much of its extent, a linear trough reveals the presence of the San Andreas fault. When the system is viewed from the air, linear scars, offset stream channels, and elongated ponds mark the trace in a striking manner. On the ground, however, surface expressions of the faults are much more difficult to detect. Some of the most distinctive landforms include long, straight escarpments, narrow ridges, and sag ponds formed by settling of blocks within the fault zone. Further, many stream channels characteristically bend sharply to the right where they cross the fault (Figure 9.B).

With the development of the theory of plate tectonics, geologists began to realize the significance of this great fault system. The San Andreas fault is a transform boundary separating two crustal plates that move very slowly. The Pacific plate, located to the west, moves northwestward relative to the North American plate, causing earthquakes along the fault (Table 9.A).

The San Andreas is undoubtedly the most studied of any fault system in the world. Although many questions remain unanswered, geologists have learned that each fault segment exhibits somewhat different behavior. Some portions of the San Andreas exhibit a slow creep with little noticeable seismic activity. Other segments regularly slip, producing small earthquakes, whereas still other segments seem to store elastic energy for hundreds of years and rupture in great earthquakes. This knowledge is useful when assigning earthquake-hazard potential to a given segment of the fault zone.

Because of the great length and complexity of the San Andreas fault, it is more appropriately referred to as a "fault system." This major fault system consists primarily of the San Andreas fault and several major branches including the Hayward and Calaveras faults of central California and the San Jacinto and Elsinore faults of southern California (see Figure 9.A). These major segments, plus a vast number of smaller faults that include the Imperial fault, San Fernando fault, and the Santa Monica fault, collectively accommodate the relative motion between the North American and Pacific plates.

Ever since the great San Francisco

Figure 9.B Aerial view showing offset stream channel across the San Andreas fault on the Carrizo Plain west of Taft, California. (Photo by Michael Collier/DRK Photo)

earthquake of 1906, when as much as 5 meters of displacement occurred, geologists have attempted to establish the cumulative displacement along this fault over its 29-million-year history. By matching rock units across the fault, geologists have determined that the total accumulated displacement from earthquakes and creep exceeds 560 kilometers (340 miles).

Table 9.A Major earthquakes on the San Andreas fault system.

Date	Location	Magnitude	Remarks
1812	Wrightwood, CA	7	Church at San Juan Capistrano collapsed, killing 40 worshipers.
1812	Santa Barbara channel	7	Churches and other buildings wrecked in and around Santa Barbara.
1838	San Francisco peninsula	7	At one time thought to have been comparable to the great earthquake of 1906.
1857	Fort Tejon, CA	8.25	One of the greatest U.S. earthquakes. Occurred near Los Angeles, then a city of 4000.
1868	Hayward, CA	7	Rupture of the Hayward fault caused extensive damage in San Francisco Bay area.
1906	San Francisco, CA	8.25	The great San Francisco earthquake. As much as 80 percent of the damage caused by fire.
1940	Imperial Valley	7.1	Displacement on the newly discovered Imperial fault.
1952	Kern County	7.7	Rupture of the White Wolf fault. Largest earthquake in California since 1906. Sixty million dollars in damages and 12 people killed.
1971	San Fernando Valley	6.5	One-half billion dollars in damages and 58 lives claimed.
1989	Santa Cruz Mountains	7.1	Loma Prieta earthquake. Six billion dollars in damages, 62 lives lost, and 3757 people injured.
1994	Northridge (Los Angeles area)	6.9	Over 15 billion dollars in damages, 51 lives lost, and over 5000 injured.

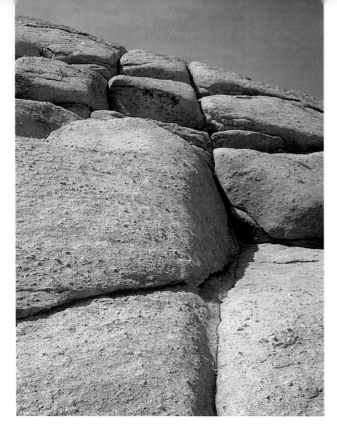

Figure 9.13 Chemical weathering is enhanced along joints in granitic rocks near the top of Lembert Dome, Yosemite National Park. (Photo by E. J. Tarbuck)

activity in soluble rocks is controlled by the joint pattern (Figure 9.13). Moreover, a system of joints can influence the direction that stream courses follow. The rectangular drainage pattern shown in Figure 4.20C is such a case.

Now that you have some knowledge of crustal uplift and rock deformation, we will examine different types of mountains and how they come to be.

Mountain Types

No two mountain ranges are exactly alike, but they can be classified according to their structural characteristics. Using this approach, four main categories result: (1) fault-block mountains, (2) folded mountains (complex mountains), (3) upwarped mountains, and (4) volcanic mountains.

Mountain ranges of the same structural type commonly occur in close proximity, forming a *mountain system*. For example, nearly the entire state of Nevada is composed of numerous elongated fault-block mountains that are separated by faulted basins (Figure 9.14). Further, within any mountainous belt, such as that portion of the American Cordillera in the western United States, mountain ranges representing each of the four groups can be found.

In addition to these basic varieties, some regions exhibit mountainous topography that was produced *without appreciable crustal deformation*. For example,

plateaus (areas of high-standing rocks that are essentially horizontal) can become deeply dissected by erosion into rugged, mountainlike terrains. Although these highlands have the topographical expression of mountains, they lack the structures associated with orogenesis.

In the following sections we will examine three of the four basic mountain types: fault-block mountains, folded mountains (the most common), and upwarped mountains. (Volcanic mountains are treated in detail in Chapter 8.)

Fault-Block Mountains

Most mountain belts, including the Alps, Himalayas, and Appalachians, form in compressional environments, as evidenced by the predominance of large thrust faults and folded strata. However, some mountain building is associated with tensional stresses (Figure 9.15). The mountains that form in such circumstances, termed **fault-block mountains,** are bounded on at least one side by high- to moderate-angle normal faults. Most fault-block mountains form in response to broad uplifting, which causes elongation and faulting. Such a situation is exemplified by the fault blocks that rise above the rift valleys of East Africa.

The Basin and Range province, a region that encompasses Nevada and portions of Utah, New Mexico, Arizona, and California, contains excellent examples of fault-block mountains (see Figure 9.14). Here, the brittle upper crust has literally been broken into hundreds of pieces, giving rise to nearly parallel mountain ranges, averaging about 80 kilometers in length, which rise precipitously above the adjacent sediment-laden basins.

A debate about the mechanisms that created the Basin and Range has gone on for 30 years. It is clear that important geological events preceded the period of extension and faulting. During this earlier time span it is thought that an oceanic plate (Farallon plate) was subducted eastward nearly horizontally under western North America (Figure 9.16, top). This caused compressional stresses in the overlying continental crust, which became tectonically active as far inland as the Rocky Mountains in Colorado.

Then, about 50 million years ago, the once compressional environment became extensional. Some geologists suggest that rifting in the Basin and Range was triggered when the oceanic slab that had been thrust (subducted) nearly horizontally under the continent began to sink (Figure 9.16, bottom). Sinking of the oceanic slab produced an upwelling of hot material from the asthenosphere. The buoyancy of the warm material caused upwarping and rifting that stretched the overlying crust by 200 to 300 kilometers. The lower crust is ductile and easily stretched. The upper crust, on the other hand, is brittle and deformed by faulting. The extensional faulting

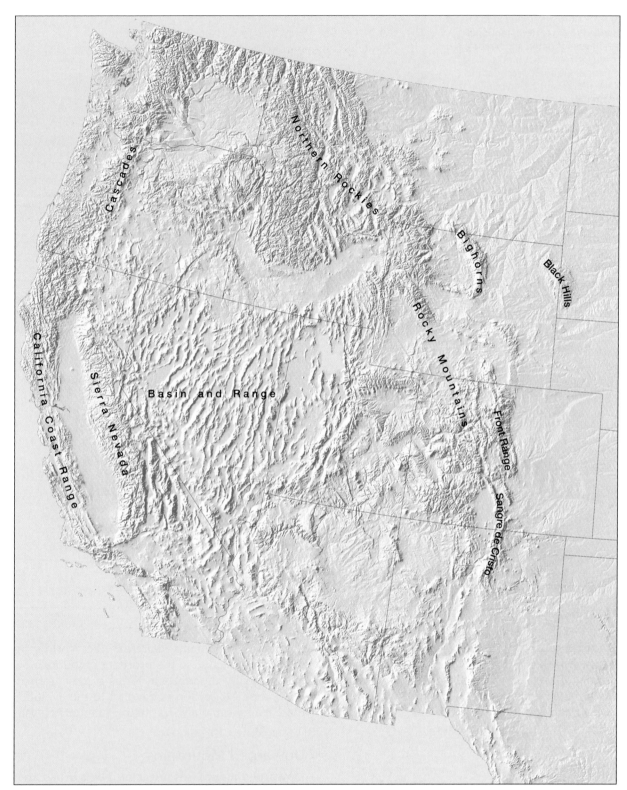

Figure 9.14 Map of major mountainous landforms of the western United States. (After Thelin and Pike, USGS)

rifted the brittle upper crust, causing individual blocks to slump. The high parts of the tilted blocks make up the mountain ranges, while their low parts form the basins, now partially filled with sediment.

Other fault-block mountains in the United States are the Teton Range of Wyoming (see Figure 9.17) and the Sierra Nevada of California. Both are faulted along their eastern flanks, which were uplifted as the blocks

Figure 9.15 Normal faulting in the Basin and Range Province. Here, tensional stresses have elongated and fractured the crust into numerous blocks. Movement along these fractures has tilted the blocks, producing parallel mountain ranges called fault-block mountains. (Photo by Michael Collier)

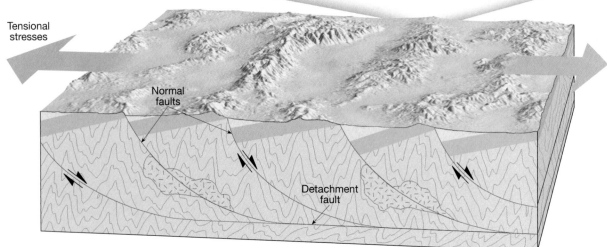

tilted downward to the west. Looking west from Jackson Hole, Wyoming, and Owens Valley, California, the eastern fronts of these ranges (the Tetons and the Sierra Nevada, respectively) rise over 2 kilometers, making them two of the most precipitous mountain fronts in the United States (Figure 9.17).

Folded Mountains

The largest and most complex mountain systems are **folded mountains** (also called **complex mountains**). Although folding is often more conspicuous, faulting, metamorphism, and igneous activity are always present in varying degrees. Most major mountain belts, including the Alps, Urals, Himalayas, Appalachians, and northern Rockies, are of this type (see Box 9.2).

Folded mountains represent the world's major mountain systems, so the process of mountain building is usually described in terms of their formation. Thus, the next section is devoted to the evolution of these majestic and always complex mountain systems.

Upwarped Mountains

Upwarped mountains are caused by a broad arching of the crust or, in some instances, because of great vertical displacement along a high-angle fault. The Black Hills in western South Dakota and the Adirondack Mountains in upstate New York are examples of arching. They consist of older igneous and metamorphic bedrock that was once eroded flat and subsequently mantled with sediment. As these

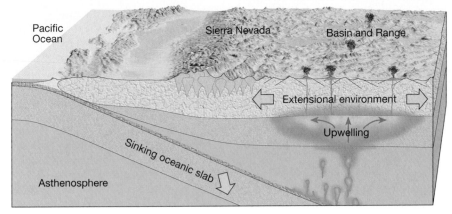

Figure 9.16 One proposal for the formation of the Basin and Range Province. **Top:** Nearly horizontal subduction of an oceanic plate produced compressional stresses that generally thickened the crust in the Basin and Range. **Bottom:** Sinking of this oceanic slab allowed for the upwelling of hot material from the asthenosphere. The buoyancy of the warm material caused upwarping and tensional fracturing in the crust above. This event was associated with volcanism and east-west extension of the crust by nearly 150 kilometers.

Figure 9.17 The Teton Range in Grand Teton National Park, Wyoming, is an example of fault-block mountains. (Photo by Carr Clifton)

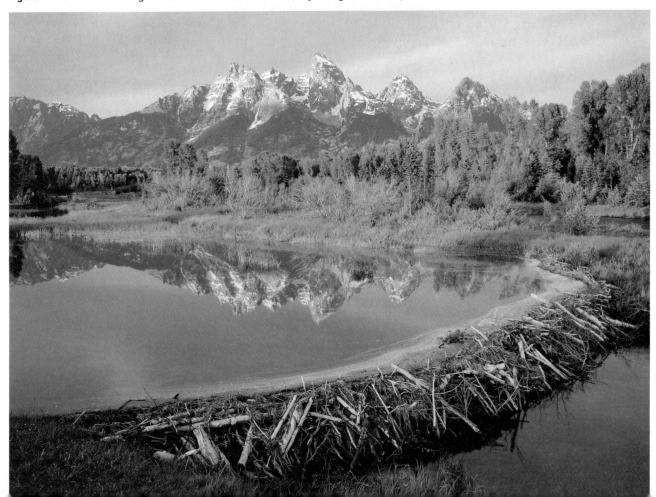

Box 9.2 The Rocky Mountains

The portion of the Rocky Mountains that extends from southern Montana to New Mexico was produced by a period of deformation known as the *Laramide Orogeny*. This event, which created some of the most picturesque scenery in the United States, peaked about 60 million years ago (Figure 9.C). The mountain ranges generated during the Laramide Orogeny include the Front Range of Colorado, the Sangre de Cristo of New Mexico and Colorado, and the Bighorns of Wyoming.

These mountains are structurally much different from the northern Rockies, which include the Canadian Rockies and those portions of the Rockies found in Idaho, western Wyoming, and western Montana. Whereas the latter ranges are folded mountains, composed of thick sequences of sedimentary rocks that were deformed by folding and low-angle thrust faulting, the middle and southern Rockies are upwarped mountains pushed almost vertically upward as part of broad upwarping of the crust.

In general, the middle and southern Rockies consist of ancient basement rocks overlain by relatively thin layers of younger strata. Ever since the time of their formation, much of the sedimentary cover has been eroded from the highest portions of the uplifted blocks, exposing their igneous and metamorphic cores. Examples include a number of granitic outcrops that project as steep summits, such as Pikes Peak and Longs Peak in Colorado's Front Range. In many areas, remnants of the sedimentary strata that once covered this region are visible as prominent angular ridges called *hogbacks* flanking the crystalline cores of the mountain ranges.

Although the Rockies have been extensively studied for over a century, there is still a good deal of debate regarding the mechanisms that led to uplift. According to the plate tectonics model, most mountain belts are produced along continental margins in association with convergent plate boundaries. Here, crustal buckling forces generate thick sequences of folded, faulted, and metamorphosed strata that are often intruded by massive igneous bodies.

However, this model cannot be easily adapted to the middle and southern Rocky Mountains, which consist of elongated, uplifted blocks of Precambrian basement rocks separated by sediment-filled basins.

One hypothesis for the formation of the middle and southern Rocky Mountains proposes that about 65 million years ago a subducting plate, called the Farallon plate, remained nearly horizontal as it pushed eastward under North America as far inland as the Black Hills of South Dakota (see Figure 9.16, top). As the subducted slab scraped beneath North America, compressional stresses shortened and thickened the rocks in the lower crust. Furthermore, this event locally uplifted blocks of ancient basement rocks along high-angle thrust faults to produce the Rockies and intermountain basins. Thus, the Laramide Orogeny may be associated with a special type of convergent plate boundary. Whereas plate subduction along a typical convergent plate boundary occurs at an angle of about 45 degrees, the middle and southern Rocky Mountains may have been produced by the nearly horizontal subduction of an oceanic plate.

Other researchers suggest that the near horizontal subduction of the Farallon plate may have altered the underside of the continents, but not enough to uplift the Rockies. Rather, once the Farallon plate finally sank, hot underlying mantle rocks ascended to replace it. Thus, it was hot, buoyant mantle rocks that provided the force to raise the Rockies. Upwelling in the mantle would also explain the elevated positions of the adjacent Colorado Plateau and the Basin and Range Province.

Figure 9.C The spectacular Maroon Bells are part of the Colorado Rockies. (Photo by Peter Saloutos/The Stock Market)

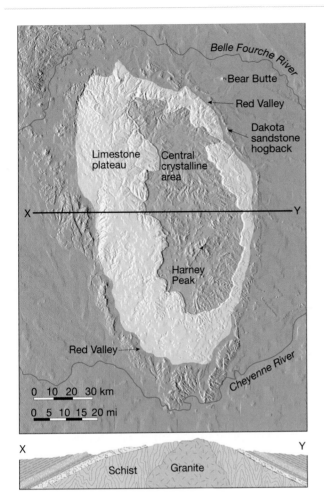

X ———————————————————— Y

Belle Fourche River

Bear Butte

Red Valley

Dakota sandstone hogback

Limestone plateau

Central crystalline area

Harney Peak

Red Valley →

0 10 20 30 km

0 5 10 15 20 mi

X   Y

Schist   Granite

Cheyenne River

Figure 9.18 The Black Hills of South Dakota, a large domal structure with resistant igneous and metamorphic rocks exposed in the core. (After Arthur N. Strahler, *Introduction to Physical Geography,* 3rd ed., New York: John Wiley & Sons, 1973.)

regions were upwarped, erosion removed the veneer of sedimentary strata, leaving a core of igneous and metamorphic rocks standing above the surrounding terrain (Figure 9.18).

Another example of upwarped mountains is that portion of the Rockies extending from southern Montana through Colorado and into New Mexico.

Mountain Building

Mountains are spectacular features that often rise several hundred meters or more above the surrounding terrain. All mountain systems show evidence of enormous forces that have folded, faulted, and generally deformed large sections of Earth's crust (Figure 9.19). Although this folding and faulting have contributed to the majestic structure of mountains, much of the credit for their beauty must be given to erosion by running water and glacial ice,

which sculpture these uplifted masses in an unending effort to lower them to sea level. **Mountain building,** however, refers to the processes that deform and uplift the mountains.

Mountain building has operated during the recent geologic past in several locations around the world. These relatively young mountainous belts include the American Cordillera, which runs along the western margin of the Americas from Cape Horn to Alaska (Rockies, Andes); the Alpine-Himalayan chain, which extends from the Mediterranean through Iran to northern India and into Indochina; and the mountainous terrains of the western Pacific, which include volcanic island arcs such as Japan, the Philippines, and Sumatra. Most of these young mountain belts have come into existence within the last 100 million years. Some, including the Himalayas, began their growth as recently as 45 million years ago.

In addition to these recently formed mountains, several chains of much older mountains exist. Although these structures are deeply eroded and topographically less prominent, they clearly possess the same structural features found in younger mountains. Typical of this older group are the Appalachians in the eastern United States and the Urals in Russia.

The first encompassing explanation of mountain building came a little over three decades ago as part of plate tectonic theory. It is now known that Earth's major mountain systems formed along convergent plate boundaries. Recall that convergence can occur between one oceanic and one continental plate, between two oceanic plates, or between two continental plates. We will consider these sites of mountain building in the following sections.

Mountain Building at Convergent Boundaries

To unravel the events that produce mountains, many studies have been conducted along active subduction zones where plates are converging. At most modern-day subduction zones, volcanic arcs are forming. This situation is typified by Alaska's Aleutian Islands and by the Andes Mountains of western South America. Although all volcanic arcs are similar, *Aleutian-type subduction zones* occur where *two oceanic plates* converge (Figure 9.20). *Andean-type subduction zones* are situated where *oceanic crust is being thrust beneath a continental mass.* Consequently, the events that are generating these volcanic arcs followed different evolutionary paths. Further, although the development of a volcanic arc does result in the formation of mountainous topography, this activity is viewed as just one of the phases in the development of a major mountain belt.

Figure 9.19 Highly deformed sedimentary strata in the Rocky Mountains of British Columbia. These sedimentary rocks are continental shelf deposits that were displaced toward the interior of Canada by low-angle thrust faults. (Photo by John Montagne)

Mountain Building Where Oceanic and Continental Crust Converge. Mountain building along continental margins involves the convergence of an oceanic plate and a plate whose leading edge contains continental crust. Exemplified by the Andes Mountains, this *Andean*

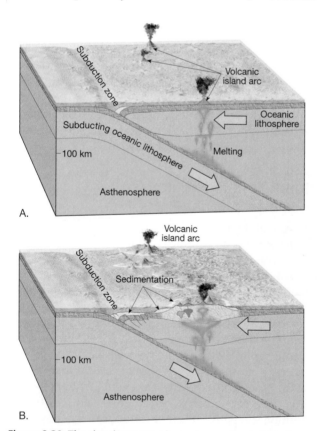

Figure 9.20 The development of a mature volcanic island arc by the convergence of two oceanic plates. These Aleutian-type subduction zones produce a mountainous topography that is similar to the volcanic arc that comprises the Aleutian islands of Alaska.

type of convergence generates structures resembling those of a developing volcanic island arc.

The first stage in the development of an Andean-type mountain belt occurs prior to the formation of the subduction zone. During this period the continental margin is *passive*. It is not a plate boundary but a part of the same plate as the adjoining oceanic crust. The East Coast of the United States is a present-day example of a passive continental margin. Here, as at other passive continental margins surrounding the Atlantic, deposition of sediment along the continental margin is producing a thick wedge of sandstones, limestones, and shales (Figure 9.21A).

At some point the continental margin becomes active. A subduction zone forms and the deformation process begins (Figure 9.21B). A good place to examine such an active continental margin is the western coast of South America. Here, an oceanic plate is being subducted eastward beneath the South American plate along the Peru-Chile trench (see Figure 7.8).

In an idealized Andean-type subduction, convergence of the continental block and the subducting oceanic plate leads to deformation and metamorphism of the continental margin. Once the oceanic plate descends to about 100 kilometers, fluids driven from the plate trigger partial melting in the mantle rock directly above the subducting slab. This process generates magma that slowly migrates upward, intruding and further deforming the overlying strata (Figure 9.21B). During the development of the volcanic arc, sediment derived from the land as well as that scraped from the subducting plate becomes plastered against the landward side of the trench. This chaotic accumulation of sedimentary and metamorphic rocks, including occasional scraps of ocean crust, is called an **accretionary wedge** (Figure 9.21B).

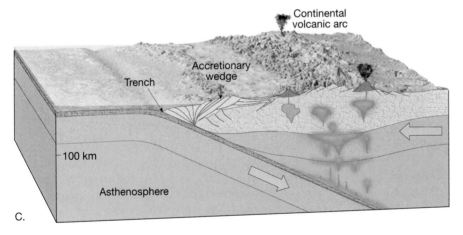

Figure 9.21 Mountain building along an Andean-type subduction zone. **A.** Passive continental margin with extensive wedge of sediments. **B.** Plate convergence generates a subduction zone, and partial melting produces a developing continental volcanic arc. **C.** Continued convergence and igneous activity further deform and thicken the crust, elevating the mountain belt, while the accretionary wedge grows.

Prolonged subduction can build an accretionary wedge that is large enough to stand above sea level (Figure 9.21C).

Andean-type mountain belts, like island arcs, are composed of two roughly parallel zones. The landward segment is the volcanic arc, made up of volcanoes and large bodies of intrusive igneous rock, intermixed with high-temperature metamorphic rocks. The seaward segment is the accretionary wedge. It consists of folded, faulted, and metamorphosed sediments and volcanic debris (Figure 9.21C).

One of the best examples of an inactive Andean-type mountain belt is found in the western United States. It includes the Sierra Nevada and the Coast Ranges in California (Figure 9.22). These parallel mountain belts were produced by the subduction of a portion of the Pacific basin under the western edge of the North American plate. The Sierra Nevada batholith is a remnant of a portion of the continental volcanic arc that was produced by several surges of magma over tens of millions of years. Subsequent uplifting and erosion have removed most evidence of past volcanic activity and exposed a core of crystalline metamorphic and igneous rocks.

In the trench region, sediments scraped from the subducting plate, plus those provided by the eroding

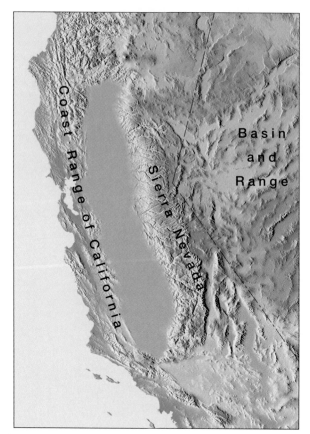

Figure 9.22 Map of the California Coast Ranges and the Sierra Nevada.

continental volcanic arc, were intensely folded and faulted into an accretionary wedge. This chaotic mixture of rock presently constitutes part of California's Coast Ranges. Uplifting of the Coast Ranges took place only recently, as evidenced by the young unconsolidated sediments that still cover portions of these highlands.

Mountain Building Where Continents Converge. So far we have discussed the formation of mountain belts where the leading edge of just one of the two converging lithospheric plates was capped with continental crust. However, it is possible for two converging plates to be carrying continental crust. Because continental crust is too buoyant to undergo any appreciable subduction, a collision between the continental fragments results.

An example of such a collision began about 45 million years ago when India collided with the Eurasian plate (Figure 9.23). India was once part of Antarctica but split from that continent and moved a few thousand kilometers due north. The result of this collision was the formation of the spectacular Himalaya Mountains and the Tibetan Plateau.

Although most of the oceanic crust that separated these landmasses prior to the collision was subducted, some was caught up in the squeeze, along with sediment that lay offshore (Figure 9.23). Today

these sedimentary rocks and slivers of oceanic crust are elevated high above sea level. Geologists believe that, following such a collision, the subducted oceanic plate decouples from the rigid continental plate and continues its descent into the mantle.

A similar but much older collision is believed to have taken place when the European continent collided with the Asian continent to produce the Ural Mountains, which extend north-south through Russia. Prior to the discovery of plate tectonics, geologists had difficulty explaining the existence of mountain ranges, like the Urals, which are located deep within continental interiors. How could thousands of meters of marine sediment be deposited and then become highly deformed while situated in the middle of a large landmass?

Other mountain ranges showing evidence of continental collisions are the Alps and the Appalachians. The Appalachians resulted from collisions among North America, Europe, and northern Africa. Although they have since separated, these landmasses were juxtaposed as part of the supercontinent Pangaea less than 200 million years ago. Detailed studies in the southern Appalachians indicate that the formation of this mountain belt was more complex than once thought. Rather than forming during a single continental collision, the Appalachians resulted from several distinct episodes of mountain building that occurred over a period of nearly 300 million years.

Mountain Building and Continental Accretion

Plate tectonics theory originally suggested two mechanisms for mountain building: (1) Continental collisions were proposed to explain the formation of such mountainous terrains as the Appalachians, Himalayas, and Urals; and (2) subduction of oceanic lithosphere was thought to be the underlying tectonic process for many circum-Pacific mountain chains, as typified by the Andes. Recent investigations, however, indicate a third mechanism of mountain building: *Smaller crustal fragments collide and accrete to continental margins.* Through this process of collision and accretion, many of the mountainous regions rimming the Pacific have been generated.

What is the nature of these small crustal fragments, and where do they come from? Researchers suggest that, prior to their accretion to a continental block, some of the fragments may have been microcontinents similar to the present-day island of Madagascar located in the Indian Ocean just east of Africa. Others were island arcs like Japan, the Philippines, and the Aleutian Islands. Still others may have been submerged crustal fragments that extended high above the ocean floor. It is believed that these fragments originated as submerged continental

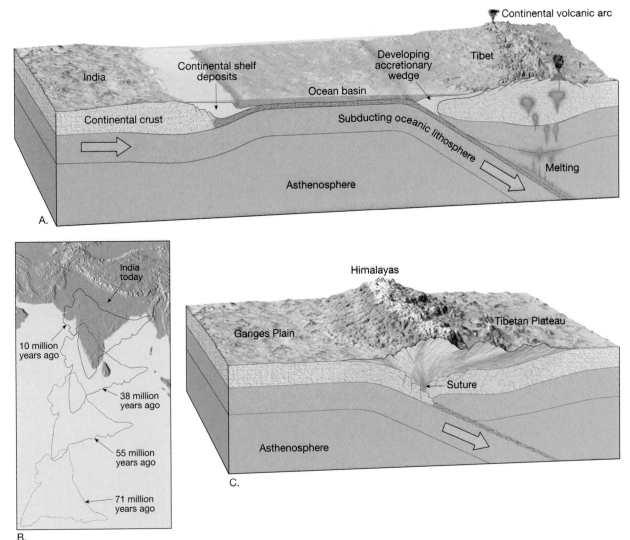

Figure 9.23 Simplified diagrams showing the northward migration and collision of India with the Eurasian plate. **A.** Converging plates generated a subduction zone, while partial melting triggered by the subducting oceanic slab produced a continental volcanic arc. Sediments scraped from the subducting plate were added to the accretionary wedge. **B.** Position of India in relation to Eurasia at various times. (Modified after Peter Molnar) **C.** Eventually the two landmasses collided, deforming and elevating the accretionary wedge and continental shelf deposits. In addition, slices of the Indian crust were thrust up onto the Indian plate.

fragments, extinct volcanic island arcs, or submerged volcanic plateaus associated with hot spot activity.

Accretion of Foreign Terranes. The widely accepted view today is that, as oceanic plates move, they carry the embedded oceanic plateaus or microcontinents to a subduction zone. Here the upper portions of these thickened zones are peeled from the descending plate and thrust in relatively thin sheets onto the adjacent continental block. This newly added material increases the width of the continent. The material may later be overridden and displaced further inland by the addition of other fragments.

Geologists refer to these accreted crustal blocks as terranes. Simply, the term **terrane** designates any crustal fragment whose geologic history is distinct from that of the adjoining terranes. (Do not confuse the

term *terrane* with the word *terrain,* which indicates the topography or lay of the land.) Terranes come in varied shapes and sizes; some are no larger than volcanic islands. Others, such as the one making up the entire Indian subcontinent, are very large.

Accretion and Mountain Building. The idea that mountain building occurs in association with the accretion of small crustal fragments to a continental mass arose principally from studies conducted in the North American Cordillera (Figure 9.24). Some mountainous areas, principally those of Alaska and British Columbia, contain fossil and magnetic evidence that these strata once lay much nearer the equator.

It is now assumed that many other terranes found in the North American Cordillera were once scattered throughout the eastern Pacific, much as we find island

Accreted Terranes

- Island arc
- Submarine deposits
- Ancient ocean floor
- Displaced continental fragments

Wrangellia Terrane

Canada
United States

Sonoma Terrane

Craton

United States
Mexico

0   600 km

Figure 9.24 Map showing terranes thought to have been added to western North America during the past 200 million years. (Redrawn after D. R. Hutchinson and others)

arcs and oceanic plateaus distributed in the western Pacific today. Ever since the breakup of Pangaea began, nearly 200 million years ago, North America has been migrating westward and overriding the Pacific basin (Figure 9.24). Apparently, this activity resulted in the piecemeal addition of fragments to the entire Pacific Coast, from the Baja Peninsula to northern Alaska. In a like manner, many modern microcontinents will eventually be accreted to active continental margins, thus resulting in the formation of new mountainous belts.

Chapter Summary

- The name for the processes that collectively produce a mountain system is *orogenesis*. Earth's less dense crust floats on top of the denser rocks in the mantle, much like wooden blocks floating in water. This concept of a floating crust in gravitational balance is called *isostasy*. As erosion lowers the peaks of mountains, *isostatic adjustment* gradually raises the mountains in response.

- Rocks that are subjected to stresses greater than their own strength begin to deform, usually by *folding* or *fracturing*. During *elastic deformation* a rock behaves like a rubber band and will return to nearly its original size and shape when the stress is removed. Once the *elastic limit* is surpassed, rocks either deform plastically or fracture. *Plastic deformation* results in permanent changes in the size and shape of a rock through folding and flowing.

- The most basic geologic structures associated with rock deformation are *folds* (flat-lying sedimentary and volcanic rocks bent into a series of wavelike undulations) and *faults*. The two most common types of folds are *anticlines*, formed by the upfolding, or arching, of rock layers, and *synclines*, which are downfolds, or troughs. Most folds are the result of horizontal *compressional stresses*. Folds may be *symmetrical*, *asymmetrical*, or, if one limb has been tilted beyond the vertical, *overturned*. *Domes* (upwarped structures) and *basins* (downwarped structures) are circular or somewhat elongated folds formed from vertical displacements of rocks.

- Faults are fractures in the crust along which appreciable displacement has occurred. Faults in which the movement is primarily vertical are called *dip-slip faults*. Dip-slip faults include both *normal* and *reverse faults*. Low-angle reverse faults are also called *thrust faults*. In *strike-slip faults*, horizontal movement causes displacement along the trend, or *strike*, of the fault. Normal faults indicate *tensional stresses* that pull the crust apart. At spreading centers, diverging plates can cause a central block called a *graben*, bounded by normal faults, to drop as the plates separate. Reverse and thrust faulting indicate that *compressional forces* are at work. A *joint* is a fracture along which no appreciable displacement has occurred.

- Mountains can be classified according to their structural characteristics. The four main mountain types are (1) *fault block mountains* (e.g., Basin and Range Province and Sierra Nevada of California), associated with tensional stresses and bounded at least on one side by normal faults; (2) *folded mountains* (e.g., Urals, Himalayas, Appalachians), which are the most complex and form most of the major mountain belts; (3) *upwarped mountains* (e.g., Black Hills, Adirondack Mountains), caused by broad arching; and (4) *volcanic mountains*. Some regions, such as plateaus deeply dissected by erosion, exhibit mountainous topography without appreciable crustal deformation.

- *Major mountain systems form along convergent plate boundaries. Andean-type mountain building* along continental margins involves the convergence of an oceanic plate and a plate whose leading edge contains continental crust. At some point in the formation of Andean-type mountains a subduction zone forms along with a continental volcanic arc. Sediment from the land as well as that scraped from the subducting plate becomes plastered against the landward side of the trench, forming what is called an *accretionary wedge*. A good example of an Andean-type mountain belt is found in the western United States, including the Sierra Nevada and the Coast Ranges in California. *Continental collisions,* in which both plates may be carrying continental crust, have most recently resulted in the formation of the Himalaya Mountains and Tibetan Highlands. Recent investigations indicate that *accretion,* a third mechanism of orogenesis, takes place where *smaller crustal fragments collide and merge with continental margins* along some plate boundaries. Many of the mountainous regions rimming the Pacific have formed in this manner. The accreted crustal blocks are referred to as *terranes*.

Key Terms

accretionary wedge (p. 262)
anticline (p. 249)
basin (p. 250)
dip-slip fault (p. 252)
dome (p. 250)
fault (p. 251)
fault-block mountains (p. 256)
fold (p. 249)
folded mountains (complex mountains) (p. 258)

graben (p. 253)
horst (p. 253)
isostasy (p. 247)
isostatic adjustment (p. 247)
joint (p. 253)
mountain building (p. 261)
normal fault (p. 252)
oblique-slip fault (p. 253)
orogenesis (p. 246)
reverse fault (p. 252)

strike-slip fault (p. 253)
syncline (p. 249)
terrane (p. 265)
thrust fault (p. 253)
upwarped mountains (p. 258)

Review Questions

1. State the two lines of evidence that support the concept of crustal uplift. Can you think of another?

2. What happens to a floating object when weight is added? Subtracted? How do these principles apply to changes in the elevations of mountains? What term is applied to the adjustment that causes crustal uplift of this type?

3. List one line of evidence that supports the idea of a floating crust.

4. What conditions favor rock deformation by folding? By faulting?

5. The San Andreas fault is an excellent example of a _____ fault.

6. Compare the movement of normal and reverse faults. What type of force produces each?

7. At which of the three types of plate boundaries does normal faulting predominate? Reverse faulting? Strike-slip faulting?

8. Describe a horst and a graben. Explain how a graben valley forms and name one.

9. Compare and contrast anticlines and synclines. Domes and basins. Anticlines and domes.

10. Although we classify many mountains as folded, why might this description be misleading?

11. What type of faults are associated with fault-block mountains?

12. During the formation of fault-block mountains, are the forces that act upon the region compressional or tensional?

13. There are four types of mountain ranges. Name the type of mountain range exemplified by each of the following:
   a. Black Hills   b. Basin and Range   c. Adirondacks
   d. Cascades   e. Appalachians   f. Tetons
   g. Bighorns   h. Himalayas   i. Front Range

14. What do we call the site where sediments are deposited along the margin of the continent where they have a good chance of being squeezed into a mountain range?

15. Which type of plate boundary is most directly associated with mountain building?

16. Describe an accretionary wedge and explain its formation.

17. Suppose a sliver of oceanic crust was discovered 1000 kilometers into the interior of a continent. Would this support or refute the theory of plate tectonics? Why?

18. Why might it have been difficult for geologists to conclude that the Appalachian Mountains were formed by plate collision if examples like the Himalayas did not exist?

19. How does the plate tectonics theory explain the existence of fossil marine life on top of the Ural Mountains?

Examining the Earth System

1. A good example of the interactions between Earth's spheres is the influence of mountains on climate. Examine the temperature graphs and annual precipitation amounts for the cities of Seattle and Spokane, WA (Figure 9.25). Notice on the inset map that mountains (the Cascades) separate the two cities and that the general wind direction in the region is from west to east (the prevailing westerlies of the mid-latitudes). Which city receives the greatest annual precipitation? Contrast the summer and winter temperatures that occur at each city? What influence have the mountains apparently had on Spokane's rainfall and temperature?

2. The Cascades have had a profound effect on the amount and type of plant and animal life (biosphere) that inhabit the region around Spokane and Seattle. Provide several specific examples to verify this statement.

Web Resources

The following is an informative and interesting World Wide Web site that addresses topics related to those presented in the chapter:

Structural Geology On The Web:

• **http://home.earthlink.net/~schimmrick/structure/ structure.html**

For direct links to this site and other appropriate Earth science Web locations, chapter review exercises, and relevant, topical investigations that utilize Web resources, contact the *Earth Science*, 9th Edition Home Page at:

• **http://www.prenhall.com/tarbuck.**

Figure 9.25 Comparison of the average monthly temperatures and precipitation for Seattle and Spokane. Note that Seattle receives an average of 38.4 inches of precipitation per year while Spokane receives only 17 inches of precipitation on average.

Chapter 10
Geologic Time

Sedimentary strata representing millions of years of Earth history exposed at Junction Butte, Canyonlands National Park, Utah. (Photo by Carr Clifton)

In the late 1700s, James Hutton recognized that Earth is very old. But how old? For many years there was no reliable method to determine the age of Earth or the dates of various events in the geologic past. Rather, a geologic time scale was developed that showed the sequence of events based on relative dating principles. What are these principles? What part do fossils play? With the discovery of radioactivity and radiometric dating techniques, geologists now can assign fairly accurate dates to many of the events in Earth history. What is radioactivity? Why is it a good "clock" for dating the geologic past? In this chapter we shall answer these questions.

Geology Needs a Time Scale

In 1869, John Wesley Powell, who was later to head the U.S. Geological Survey, led a pioneering expedition down the Colorado River and through the Grand Canyon. Writing about the rock layers that were exposed by the downcutting of the river, Powell observed that "the canyons of this region would be a Book of Revelations in the rock-leaved Bible of geology."

He was undoubtedly impressed with the millions of years of Earth history exposed along the walls of the Grand Canyon (Figure 10.1).

Powell realized that the evidence for an ancient Earth is concealed in its rocks. Like the pages in a long and complicated history book, rocks record the geological events and changing life-forms of the past. The book, however, is not complete. Many pages, especially in the early chapters, are missing. Others are tattered, torn, or smudged. Yet enough of the book remains to allow much of the story to be deciphered.

Interpreting Earth history is a prime goal of the science of geology. Like a modern-day sleuth, the geologist must interpret clues found preserved in the rocks. By studying rocks, especially sedimentary rocks, and the features they contain, geologists can unravel the complexities of the past.

Geological events by themselves, however, have little meaning until they are put into a time perspective. Studying history, whether it be the Civil War or the Age of Dinosaurs, requires a calendar. Among geology's major contributions to human knowledge is the *geologic time scale* and the discovery that Earth history is exceedingly long.

Figure 10.1 The Grand Canyon of the Colorado River as viewed from West Rim Drive. (Photo by Carr Clifton)

The geologists who developed the geologic time scale revolutionized the way people think about time and perceive our planet. They learned that Earth is much older than anyone had previously imagined and that its surface and interior have been changed over and over again by the same geological processes that operate today.

A Short History of Geology

In the mid-1600s, James Ussher, Anglican Archbishop of Armagh, Primate of all Ireland, published a work that had immediate and profound influence on people's view of Earth's age. A respected scholar of the Bible, Ussher constructed a chronology of human and Earth history in which he determined that Earth was only a few thousands of years old, having been created in 4004 B.C. Ussher's treatise earned widespread acceptance among scientific and religious leaders of his day, and his chronology was soon printed in the margins of the Bible itself.

During the seventeenth and eighteenth centuries the doctrine of **catastrophism** strongly influenced people's thinking about Earth. Briefly stated, catastrophists believed that Earth's landscape had been developed primarily by great catastrophes. Features such as mountains and canyons, which today we know take great periods of time to form, were explained as having been produced by sudden and often worldwide disasters triggered by unknowable causes that no longer operate. This philosophy was an attempt to fit the rate of Earth processes to the prevailing ideas on the age of Earth.

Birth of Modern Geology

Modern geology began in the late 1700s when James Hutton, a Scottish physician and gentleman farmer, published his *Theory of the Earth* (Figure 10.2). In this work, Hutton put forth a fundamental principle that is a pillar of geology today: **uniformitarianism**. It simply states that *the physical, chemical, and biological laws that operate today have also operated in the geologic past*. This means that the forces and processes that we observe presently shaping our planet have been at work for a very long time. Thus, to understand ancient rocks, we must first understand present-day processes and their results. This idea is commonly expressed by saying "the present is the key to the past."

Prior to Hutton's *Theory of the Earth*, no one had effectively demonstrated that geological processes occur over extremely long periods of time. However, Hutton persuasively argued that weak, slow-acting processes could, over long spans of time, produce effects just as great as those resulting from sudden catastrophic events. Unlike his predecessors, Hutton carefully cited verifiable observations to support his ideas.

For example, when he argued that mountains are sculpted and ultimately destroyed by weathering and the work of running water, and that their wastes are carried to the oceans by processes that can be observed, Hutton said, "We have a chain of facts which clearly demonstrates...that the materials of the wasted mountains have traveled through the rivers"; and further, "There is not one step in all this progress...that is not to be actually perceived." He then went on to summarize this thought by asking a question and immediately providing the answer: "What more can we require? Nothing but time."

Hutton's literary style was cumbersome and difficult, so his work was not widely read nor easily understood. It is the English geologist Sir Charles Lyell who is given the most credit for advancing the basic principles of modern geology. Between 1830 and 1872, Lyell produced 11 editions of his great work, *Principles of Geology*. As was customary in those days, Lyell's book had a lengthy subtitle that outlined the main theme of the work: *Being an Attempt to Explain the Former Changes of the Earth's Surface, by Reference to Causes Now in Operation*.

Lyell painstakingly illustrated the uniformity of nature through time. He was able to show more convincingly than his predecessors that geologic processes observed today can be assumed to have operated in the past. Although the doctrine of uniformitarianism did not originate with Lyell, he was most successful in interpreting and publicizing it for society at large.

Figure 10.2 James Hutton, the eighteenth-century Scottish geologist who is often called the "founder of modern geology." (The Natural History Museum, London)

Geology Today

Today the basic tenets of uniformitarianism are just as viable as in Lyell's day. Indeed, we realize more strongly than ever that the present gives us insight into the past and that the physical, chemical, and biological laws that govern geological processes remain unchanging through time. However, we also understand that the doctrine should not be taken too literally. To say that geological processes in the past were the same as those occurring today is not to suggest that they always had the same relative importance or that they operated at precisely the same rate. Although the same processes have prevailed through time, their rates have undoubtedly varied.

The acceptance of uniformitarianism meant the acceptance of a very long history for Earth. Although Earth's processes vary in intensity, they still take a very long time to create or destroy major landscape features.

For example, geologists have established that mountains once existed in portions of present-day Minnesota, Wisconsin, and Michigan. Today the region consists of low hills and plains. Erosion gradually destroyed these peaks. Estimates indicate that the North American continent is being lowered at a rate of about 3 centimeters per 1000 years. At this rate, it would take 100 million years for water, wind, and ice to lower mountains that were 300 meters (10,000 feet) high.

But even this time span is relatively short on the time scale of Earth history, for the rock record contains evidence that shows Earth has experienced many cycles of mountain building and erosion. Concerning the ever-changing nature of Earth through great expanses of geologic time, Hutton stated: "We find no vestige of a beginning, no prospect of an end."

It is important to remember that, although many features of our physical landscape may seem to be unchanging over our lifetimes, they are nevertheless changing, but on time scales of hundreds, thousands, or even many millions of years.

Relative Dating—Key Principles

   Geologic Time
Relative Dating

During the late 1800s and early 1900s, various attempts were made to determine the age of Earth. Although some of the methods appeared promising at the time, none proved to be reliable. What these scientists were seeking was a **numerical date**. Such dates specify the actual number of years that have passed since an event occurred—for example, the extinction of the dinosaurs about 65 million years ago. Today our understanding of radioactivity allows us to accurately determine numerical dates for rocks that represent important events in Earth's distant past. We will study radioactivity later in this chapter. Prior to the discovery of radioactivity, geologists had no accurate and dependable method of numerical dating and had to rely solely on relative dating.

Relative dating means placing rocks in their proper *sequence of formation*—which ones formed first, second, third, and so on. Relative dating cannot tell us how long ago something took place, only that it followed one event and preceded another. The relative dating techniques that were developed are valuable and still widely used. Numerical dating methods did not replace these techniques; they simply supplemented them. To establish a relative time scale, a few basic principles or rules had to be discovered and applied. Although they may seem obvious to us today, they were major breakthroughs in thinking at the time, and their discovery and acceptance was an important scientific achievement.

Law of Superposition

Nicolaus Steno, a Danish anatomist, geologist, and priest (1636–1686), is credited with being the first to recognize a sequence of historical events in an outcrop of sedimentary rock layers. Working in the mountains of western Italy, Steno applied a very simple rule that has come to be the most basic principle of relative dating—the **law of superposition**. The law simply states that in an undeformed sequence of sedimentary rocks, each bed is older than the one above it and younger than the one below. Although it may seem obvious that a rock layer could not be deposited unless it had something older beneath it for support, it was not until 1669 that Steno clearly stated the principle.

This rule also applies to other surface-deposited materials such as lava flows and beds of ash from volcanic eruptions. Applying the law of superposition to the beds exposed in the upper portion of the Grand Canyon (Figure 10.3), you can easily place the layers in their proper order. Among those that are shown, the sedimentary rocks in the Supai Group must be the oldest, followed in order by the Hermit Shale, Coconino Sandstone, Toroweap Formation, and Kaibab Limestone.

Principle of Original Horizontality

Steno is also credited with recognizing the importance of another basic principle, called the **principle of original horizontality**. Simply stated it means that layers of sediment are generally deposited in a horizontal position. Thus, if we observe rock layers that are flat, it means they have not been disturbed, and still have their *original* horizontality. The layers in the Grand Canyon illustrate this in Figure 10.1. But if they are folded or inclined at a steep angle, they must have been moved into that position by crustal disturbances sometime after their deposition (Figure 10.4).

A.

B.

Figure 10.3 Applying the law of superposition to these layers exposed in the upper portion of the Grand Canyon, the Supai Group is oldest and the Kaibab Limestone is youngest. (Photo by E. J. Tarbuck)

Principle of Cross-Cutting Relationships

When a fault cuts through other rocks, or when magma intrudes and crystallizes, we can assume that the fault or intrusion is younger than the rocks affected. For example, in Figure 10.5, the faults and dikes clearly must have occurred *after* the sedimentary layers were deposited.

This is the **principle of cross-cutting relationships**. By applying the cross-cutting principle you can see that fault *A* occurred *after* the sandstone layer was deposited, because it "broke" the layer. However, fault *A* occurred *before* the conglomerate was laid down, because that layer is unbroken.

We can also state that dike *B* and its associated sill are older than dike *A*, because dike *A* cuts the sill. In the

Figure 10.4 Most layers of sediment are deposited in a nearly horizontal position. Thus, when we see rock layers that are inclined, we can assume that they must have been moved into that position by crustal disturbances *after* their deposition. Hartland Quay, Devon, England. (Photo by Tom Bean/DRK Photo)

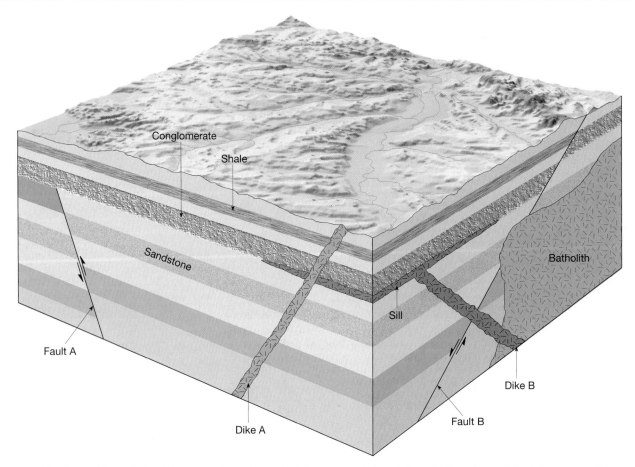

Figure 10.5 Cross-cutting relationships are an important principle used in relative dating. An intrusive rock body is younger than the rocks it intrudes. A fault is younger than the rock layers it cuts.

same manner, we know that the batholith was emplaced after movement occurred along fault *B*, but before dike *B* was formed. This is true because the batholith cuts across fault *B* and dike *B* cuts across the batholith.

Inclusions

Sometimes inclusions can aid the relative dating process. **Inclusions** are pieces of one rock unit that are contained within another. The basic principle is logical and straightforward. The rock mass adjacent to the one containing the inclusions must have been there first in order to provide the rock fragments. Therefore, the rock mass containing inclusions is the younger of the two. Figure 10.6 provides an example. Here the inclusions of granite in the adjacent sedimentary layer indicate the sedimentary layer was deposited on top of a weathered mass of granite rather than being intruded from below by a mass of magma that later crystallized.

Unconformities

When we observe layers of rock that have been deposited essentially without interruption, we call them **conformable**. Particular sites exhibit conformable beds

representing certain spans of geologic time. However, no place on Earth has a complete set of conformable strata.

Throughout Earth history, the deposition of sediment has been interrupted again and again. All such breaks in the rock record are termed "unconformities."

An **unconformity** represents a long period during which deposition ceased, erosion removed previously formed rocks, and then deposition resumed. In each case uplift and erosion are followed by subsidence and renewed sedimentation. Unconformities are important features because they represent significant geologic events in Earth history. Moreover, their recognition helps us identify what intervals of time are not represented by strata and thus are missing from the geologic record.

The rocks exposed in the Grand Canyon of the Colorado River represent a tremendous span of geologic history. It is a wonderful place to take a trip through time. The canyon's colorful strata record a long history of sedimentation in a variety of environments—advancing seas, rivers and deltas, tidal flats, and sand dunes. But the record is not continuous. Unconformities represent vast amounts of time that

Figure 10.6 Because pieces of granite (*inclusions*) are contained within the overlying sedimentary bed, we know the granite must be older. When older intrusive igneous rocks are overlain by younger sedimentary layers, a type of unconformity termed a *nonconformity* is said to exist.

have not been recorded in the canyon's layers. Figure 10.7 is a geologic cross section of the Grand Canyon. Refer to it as you read about the three basic types of unconformities: angular unconformities, disconformities, and nonconformities.

Angular Unconformity. Perhaps the most easily recognized unconformity is an **angular unconformity**. It consists of tilted or folded sedimentary rocks that are overlain by younger, more flat-lying strata. An angular unconformity indicates that during the pause in deposition, a period of deformation (folding or tilting) and erosion occurred (Figure 10.8).

Disconformity. When contrasted with angular unconformities, **disconformities** are more common, but usually far less conspicuous because the strata on either side are essentially parallel. Many disconformities are difficult to identify because the rocks above and below are similar and there is little evidence of erosion. Such a break often resembles an ordinary bedding plane. Other disconformities are easier to identify because the ancient erosion surface is cut deeply into the older rocks below.

Nonconformity. The third basic type of unconformity is a **nonconformity**. Here the break separates older metamorphic or intrusive igneous rocks from younger sedimentary strata (Figure 10.6). Just as angular unconformities and disconformities imply crustal movements, so too do nonconformities. Intrusive igneous masses and metamorphic rocks originate far below the surface. Thus, for a nonconformity to develop, there must be a period of uplift and the erosion of overlying rocks. Once exposed at the surface, the igneous or metamorphic rocks are subjected to weathering and erosion prior to subsidence and the renewal of sedimentation.

Using Relative Dating Principles

If you apply the principles of relative dating to the hypothetical geologic cross section in Figure 10.9, you can place in proper sequence the rocks and the events they represent. The statements within the figure summarize the logic used to interpret the cross section.

In this example, we establish a relative time scale for the rocks and events in the area of the cross section. Remember that this method gives us no idea of how many years of Earth history are represented, for we have no numerical dates. Nor do we know how this area compares to any other.

Correlation of Rock Layers

To develop a geologic time scale that is applicable to the entire Earth, rocks of similar age in different regions must be matched up. Such a task is referred to as **correlation**.

Within a limited area, correlating the rocks of one locality with those of another may be done simply by walking along the outcropping edges. However, this might not be possible when the rocks are mostly concealed by soil and vegetation. Correlation over short distances is often achieved by noting the position of a distinctive rock layer in a sequence of strata. Or, a layer may be identified in another location if it is composed of very distinctive or uncommon minerals.

By correlating the rocks from one place to another, a more comprehensive view of the geologic history of

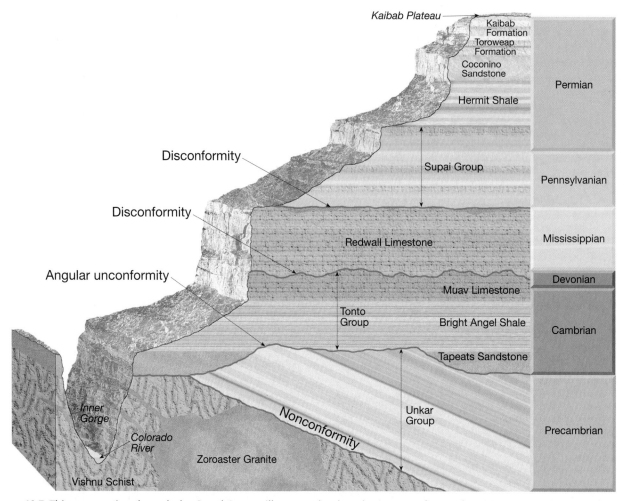

Figure 10.7 This cross section through the Grand Canyon illustrates the three basic types of unconformities. An angular unconformity can be seen between the tilted Precambrian Unkar Group and the Cambrian Tapeats Sandstone. Two disconformities are marked, above and below the Redwall Limestone. A nonconformity occurs between the igneous and metamorphic rocks of the Inner Gorge and the sedimentary strata of the Unkar Group.

a region is possible. Figure 10.10, for example, shows the correlation of strata at three sites on the Colorado Plateau. No single locale exhibits the entire sequence, but correlation reveals a more complete picture of the sedimentary rock record.

Many geologic studies involve relatively small areas. Such studies are important in their own right, but their full value is realized only when the rocks are correlated with those of other regions. Although the methods just described are sufficient to trace a rock formation over relatively short distances, they are not adequate for matching rocks that are separated by great distances. When correlation between widely separated areas or between continents is the objective, the geologist must rely on fossils.

Fossils: Evidence of Past Life

Fossils, the remains or traces of prehistoric life, are important inclusions in sediment and sedimentary rocks. They are important tools for interpreting the geologic past. Knowing the nature of the life-forms that existed at a particular time helps researchers understand past environmental conditions. Further, fossils are important time indicators and play a key role in correlating rocks of similar ages that are from different places.

Types of Fossils

Fossils are of many types. The remains of relatively recent organisms may not have been altered at all. Such objects as teeth, bones, and shells are common examples. Far less common are entire animals, flesh included, that have been preserved because of rather unusual circumstances. Remains of prehistoric elephants called mammoths that were frozen in the Arctic tundra of Siberia and Alaska are examples, as are the mummified remains of sloths preserved in a dry cave in Nevada.

Given enough time, the remains of an organism are likely to be modified. Often fossils become *petrified*

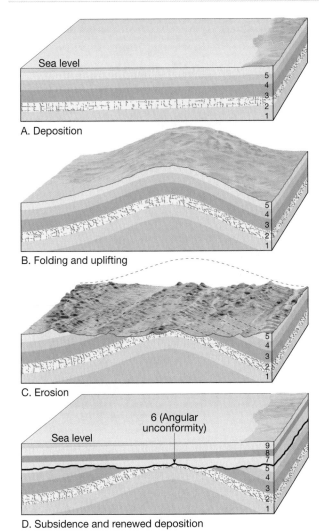

A. Deposition

B. Folding and uplifting

C. Erosion

6 (Angular unconformity)

Sea level

D. Subsidence and renewed deposition

Figure 10.8 Formation of an angular unconformity. An angular unconformity represents a period during which deformation and erosion occurred.

(literally, "turned into stone"), meaning that the small internal cavities and pores of the original structure are filled with precipitated mineral matter (Figure 10.11A). In other instances *replacement* may occur. Here the cell walls and other solid material are removed and replaced with mineral matter. Sometimes the microscopic details of the replaced structure are faithfully retained.

Molds and casts constitute another common class of fossils. When a shell or other structure is buried in sediment and then dissolved by underground water, a *mold* is created. The mold faithfully reflects only the shape and surface marking of the organism; it does not reveal any information concerning its internal structure. If these hollow spaces are subsequently filled with mineral matter, *casts* are created (Figure 10.11B).

A type of fossilization called *carbonization* is particularly effective in preserving leaves and delicate animal forms. It occurs when fine sediment encases the remains of an organism. As time passes, pressure squeezes out the liquid and gaseous components and leaves behind a thin residue of carbon (Figure 10.11C). Black shales deposited as organic-rich mud in oxygen-poor environments often contain abundant carbonized remains. If the film of carbon is lost from a fossil preserved in fine-grained sediment, a replica of the surface, called an *impression*, may still show considerable detail (Figure 10.11D).

Delicate organisms, such as insects, are difficult to preserve, and consequently they are relatively rare in the fossil record. Not only must they be protected from decay but they must not be subjected to any pressure that would crush them. One way in which some insects have been preserved is in *amber*, the hardened resin of ancient trees. The fly in Figure 10.11E was preserved after being trapped in a drop of sticky resin. Resin sealed off the insect from the atmosphere and protected the remains from damage by water and air. As the resin hardened, a protective, pressure-resistant case was formed.

In addition to the fossils already mentioned, there are numerous other types, many of them only traces of prehistoric life. Examples of such indirect evidence include:

1. Tracks—animal footprints made in soft sediment that was later lithified (Figure 10.11F).
2. Burrows—tubes in sediment, wood, or rock made by an animal. These holes may later become filled with mineral matter and preserved. Some of the oldest-known fossils are believed to be worm burrows.
3. Coprolites—fossil dung and stomach contents that can provide useful information pertaining to food habits of organisms.
4. Gastroliths—highly polished stomach stones that were used in the grinding of food by some extinct reptiles.

Conditions Favoring Preservation

Only a tiny fraction of the organisms that have lived during the geologic past have been preserved as fossils. Normally the remains of an animal or plant are destroyed. Under what circumstances are they preserved? Two special conditions appear to be necessary: rapid burial and the possession of hard parts.

When an organism perishes, its soft parts usually are quickly eaten by scavengers or decomposed by bacteria. Occasionally, however, the remains are buried by sediment. When this occurs, the remains are protected from the environment where destructive processes operate. Rapid burial therefore is an important condition favoring preservation.

Figure 10.9 Geologic cross section of a hypothetical region.

Interpretation:

Ocean

1. Applying the *law of superposition*, beds *A*, *B*, *C*, and *E* were deposited in that order.

2. Bed *D* is a sill (a concordant igneous intrusion). Applying the principle of cross-cutting relationships, sill *D* must be younger than the rocks that were intruded. Further evidence that sill *D* is younger than beds *C* and *E* are the *inclusions* in the sill of fragments from these beds. If this igneous mass contains pieces of adjacent strata, then the adjacent strata must have been there first.

3. Following the intrusion of sill *D*, the intrusion of dike *F* occurred. Because the dike cuts through beds *A* through *E*, it must be younger than all of them (principle of cross-cutting relationships).

4. Next, the rocks were tilted and eroded. The tilting happened first because the upturned ends of the strata have been eroded. The tilting and erosion, followed by further deposition, produced an *angular unconformity*.

5. Beds, *G*, *H*, *I*, *J*, and *K* were deposited in that order, again using the *law of superposition*. Although the lava flow (bed *H*) is not a sedimentary rock layer, it is a surface-deposited layer, and thus superposition may be applied.

6. Finally, the irregular surface and the stream valley indicate that another gap in the rock record is being produced by erosion.

In addition, animals and plants have a much better chance of being preserved as part of the fossil record if they have hard parts. Although traces and imprints of soft-bodied animals such as jellyfish, worms, and insects exist, they are not common. Flesh usually decays so rapidly that preservation is exceedingly unlikely. Hard parts such as shells, bones, and teeth predominate in the record of past life.

Because preservation is contingent on special conditions, the record of life in the geologic past is biased. The fossil record of those organisms with hard parts that lived in areas of sedimentation is quite abundant. However, we get only an occasional glimpse of the vast array of other life-forms that did not meet the special conditions favoring preservation.

Fossils and Correlation

The existence of fossils had been known for centuries, yet it was not until the late 1700s and early 1800s that their significance as geologic tools was made evident. During this period an English engineer and canal builder, William Smith, discovered that each rock formation in the canals he worked on contained fossils unlike those in the beds either above or below. Further, he noted that sedimentary strata in widely separated

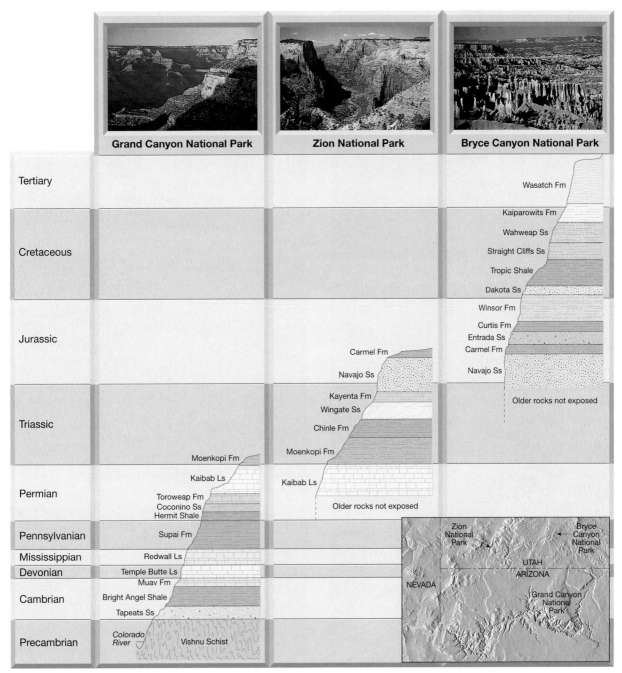

Grand Canyon National Park

Zion National Park

Bryce Canyon National Park

Figure 10.10 Correlation of strata at three locations on the Colorado Plateau reveals a more complete view of the extent of sedimentary rocks in the region. (After U.S. Geological Survey)

areas could be identified—and correlated—by their distinctive fossil content.

Based on Smith's classic observations and the findings of many geologists who followed, one of the most important and basic principles in historical geology was formulated: *Fossil organisms succeed one another in a definite and determinable order, and therefore any time period can be recognized by its fossil content.* This has come to be known as the **principle of fossil succession**. In other words, when fossils are arranged

according to their age by applying the law of superposition to the rocks in which they are found, they do not present a random or haphazard picture. To the contrary, fossils show changes that document the evolution of life through time.

For example, an Age of Trilobites is recognized quite early in the fossil record. Then, in succession, paleontologists recognize an Age of Fishes, an Age of Coal Swamps, an Age of Reptiles, and an Age of Mammals. These "ages" pertain to groups that were

Figure 10.11 There are many types of fossilization. Six examples are shown here. **A.** Petrified wood in Petrified Forest National Park, Arizona. **B.** Natural casts of shelled invertebrates. **C.** A fossil bee preserved as a thin carbon film. **D.** Impressions are common fossils and often show considerable detail. **E.** Insect in amber. **F.** Dinosaur footprint in fine-grained limestone near Tuba City, Arizona. (Photo A by David Muench; Photos B, D, and F by E. J. Tarbuck; Photo C courtesy of the National Park Service; Photo E by Breck P. Kent)

especially plentiful and characteristic during particular time periods. Within each of the "ages," there are many subdivisions based, for example, on certain species of trilobites, and certain types of fish, reptiles, and so on. This same succession of dominant organisms, never out of order, is found on every continent.

Once fossils were recognized as time indicators, they became the most useful means of correlating rocks of similar age in different regions. Geologists pay particular attention to certain fossils called **index fossils**. These fossils are widespread geographically and are limited to a short span of geologic time, so their presence provides an important method of matching rocks of the same age. Rock formations, however, do not always contain a specific index fossil. In such situations, groups of fossils are used to establish the age of the bed. Figure 10.12 illustrates how a group of fossils can be used to date rocks more precisely than could be accomplished by the use of only one of the fossils.

In addition to being important and often essential tools for correlation, fossils are important environmental indicators. Although much can be deduced about past environments by studying the nature and characteristics of sedimentary rocks, a close examination of any fossils present can usually provide a great deal more information.

For example, when the remains of certain clam shells are found in limestone, the geologist can assume that the region was once covered by a shallow sea, because that is where clams live today. Also, by using what we know of living organisms, we can conclude that fossil animals with thick shells capable of withstanding pounding and surging waves must have inhabited shorelines. Conversely, animals with thin, delicate shells probably indicate deep, calm offshore

Figure 10.12 Overlapping ranges of fossils help date rocks more exactly than using a single fossil.

waters. Hence, by looking closely at the types of fossils, the approximate position of an ancient shoreline may be identified.

Further, fossils can indicate the former temperature of the water. Certain present-day corals require warm and shallow tropical seas like those around Florida and the Bahamas. When similar corals are found in ancient limestones, they indicate that a Florida-like marine environment must have existed when the corals were alive. These examples illustrate how fossils can help unravel the complex story of Earth history.

Dating With Radioactivity

 Geologic Time
Radiometric Dating

In addition to establishing relative dates by using the principles described in the preceding sections, it is also possible to obtain reliable numerical dates for events in the geologic past. For example, we know that Earth is about 4.5 billion years old and that the dinosaurs became extinct about 65 million years ago. Dates that are expressed in millions and billions of years truly stretch our imagination because our personal calendars involve time measured in hours, weeks, and years. Nevertheless, the vast expanse of geologic time is a reality and it is radiometric dating that allows us to measure it. In this section you will learn about radioactivity and its application in radiometric dating.

Reviewing Basic Atomic Structure

Recall from Chapter 1 that each atom has a *nucleus* containing protons and neutrons, and that the nucleus is orbited by electrons. *Electrons* have a negative electrical charge and *protons* have a positive charge. A *neutron* is actually a proton and an electron combined, so it has no charge (it is neutral).

The *atomic number* (each element's identifying number) is the number of protons in the nucleus. Every element has a different number of protons, and thus a different atomic number (hydrogen = 1, carbon = 6, oxygen = 8, uranium = 92, etc.). Atoms of the same element always have the same number of protons, so the atomic number stays constant.

Practically all of an atom's mass (99.9%) is in the nucleus, indicating that electrons have virtually no mass at all. So, by adding the protons and neutrons in an atom's nucleus, we derive the atom's *mass number*. The number of neutrons can vary, and these variants, or *isotopes*, have different mass numbers.

To summarize with an example, uranium's nucleus always has 92 protons, so its atomic number always is 92. But its neutron population varies, so uranium has three isotopes: uranium-234 (mass of protons

+ neutrons = 234), uranium-235, and uranium-238. All three isotopes are mixed in nature. They look the same and behave the same in chemical reactions.

Radioactivity

The forces that bind protons and neutrons together in the nucleus usually are strong. However, in some isotopes, the nuclei are unstable because the forces binding protons and neutrons together are not strong enough. As a result, the nuclei spontaneously break apart (decay), a process called **radioactivity**.

What happens when unstable nuclei break apart? Three common types of radioactive decay are illustrated in Figure 10.13 and are summarized as follows:

1. Alpha particles (α particles) may be emitted from the nucleus. An alpha particle consists of 2 protons and 2 neutrons. Consequently, the emission of an alpha particle means that the mass number of the isotope is reduced by 4 and the atomic number is decreased by 2.

2. When a beta particle (β particle), or electron, is given off from a nucleus, the mass number remains unchanged, because electrons have practically no mass. However, because the electron has come from a neutron (remember, a neutron is a combination of a proton and an electron), the nucleus contains one more proton than before. Therefore, the atomic number increases by 1.

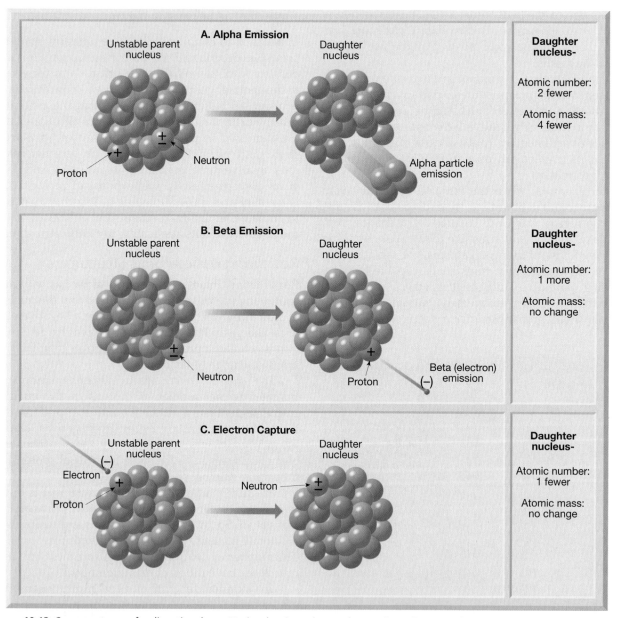

Figure 10.13 Common types of radioactive decay. Notice that in each case the number of protons (atomic number) in the nucleus changes, thus producing a different element.

3. Sometimes an electron is captured by the nucleus. The electron combines with a proton and forms an additional neutron. As in the last example, the mass number remains unchanged. However, as the nucleus now contains one less proton, the atomic number decreases by 1.

An unstable (radioactive) isotope of an element is called the *parent*. The isotopes resulting from the decay of the parent are the *daughter products*. Figure 10.14 provides an example of radioactive decay. Here it can be seen that when the radioactive parent, uranium-238 (atomic number 92, mass number 238), decays, it follows a number of steps, emitting 8 alpha particles and 6 beta particles before finally becoming the stable daughter product lead-206 (atomic number 82, mass number 206). One of the unstable daughter products produced during this decay series is radon. Box 10.1 examines the hazards associated with this radioactive gas.

Certainly among the most important results of the discovery of radioactivity is that it provided a reliable means of calculating the ages of rocks and minerals that contain particular radioactive isotopes. The procedure

is called **radiometric dating**. Why is radiometric dating reliable? Because the rates of decay for many isotopes have been precisely measured and do not vary under the physical conditions that exist in Earth's outer layers. Therefore, each radioactive isotope used for dating has been decaying at a fixed rate since the formation of the rocks in which it occurs, and the products of decay have been accumulating at a corresponding rate. For example, when uranium is incorporated into a mineral that crystallizes from magma, there is no lead (the stable daughter product) from previous decay. The radiometric "clock" starts at this point. As the uranium in this newly formed mineral disintegrates, atoms of the daughter product are trapped and measurable amounts of lead eventually accumulate.

Half-Life

The time required for one-half of the nuclei in a sample to decay is called the **half-life** of the isotope. Half-life is a common way of expressing the rate of radioactive disintegration. Figure 10.15 illustrates what occurs when a radioactive parent decays directly into its stable daughter product. When the quantities of parent and daughter are equal (ratio 1:1), we know that one half-life has transpired. When one-quarter of the original parent atoms remain and three-quarters have decayed to the daughter product, the parent/daughter ratio is 1:3 and we know that two half-lives have passed. After three half-lives, the ratio of parent atoms to daughter atoms is 1:7 (one parent for every seven daughter atoms).

If the half-life of a radioactive isotope is known and the parent/daughter ratio can be measured, the age of the sample can be calculated. For example, assume that the half-life of a hypothetical unstable isotope is 1 million years and the parent/daughter ratio in a sample is 1:15. Such a ratio indicates that four half-lives have passed and that the sample must be 4 million years old.

Radiometric Dating

Notice that the *percentage* of radioactive atoms that decay during one half-life is always the same: 50 percent. However, the *actual number* of atoms that decay with the passing of each half-life continually decreases. Thus, as the percentage of radioactive parent atoms declines, the proportion of stable daughter atoms rises, with the increase in daughter atoms just matching the drop in parent atoms. This fact is the key to radiometric dating.

Of the many radioactive isotopes that exist in nature, five have proven particularly useful in providing radiometric ages for ancient rocks (Table 10.1). Rubidium-87, thorium-232, and the two isotopes of uranium are used only for dating rocks that are millions of years old, but potassium-40 is more

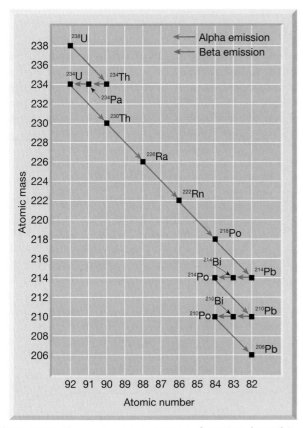

Figure 10.14 The most common isotope of uranium (U-238) is an example of a radioactive decay series. Before the stable end product (Pb-206) is reached, many different isotopes are produced as intermediate steps.

Figure 10.15 The radioactive decay curve shows change that is exponential. Half of the radioactive parent remains after one half-life. After a second half-life one-quarter of the parent remains, and so forth.

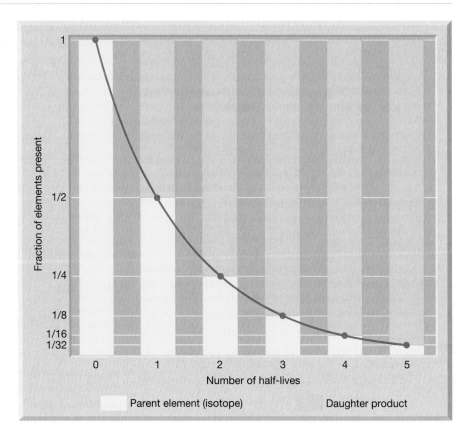

versatile. Although the half-life of potassium-40 is 1.3 billion years, analytical techniques make possible the detection of tiny amounts of its stable daughter product, argon-40, in some rocks that are younger than 100,000 years.

It is important to realize that an accurate radiometric date can be obtained only if the mineral remained a closed system during the entire period since its formation. A correct date is not possible unless there was neither the addition nor loss of parent or daughter isotopes. This is not always the case. In fact, an important limitation of the potassium-argon method arises from the fact that argon is a gas, and it may leak from minerals, throwing off measurements. Cross-checking of samples, using two different radiometric methods, is done where possible to ensure accurate age determinations.

Dating With Carbon-14

To date very recent events, carbon-14 is used. Carbon-14 is the radioactive isotope of carbon. Because the half-life of carbon-14 is only 5730 years, it can be used for dating events from the historic past as well as those from very recent geologic history. In some cases carbon-14 can be used to date events as far back as 75,000 years.

Carbon-14 is continuously produced in the upper atmosphere as a consequence of cosmic ray bombardment. Cosmic rays, which are high-energy particles, shatter the nuclei of gas atoms, releasing neutrons. Some of the neutrons are absorbed by nitrogen atoms (atomic number 7) causing their nuclei to emit a proton. As a result, the atomic number decreases by 1 (to 6), and a different element, carbon-14, is created (Figure 10.16A). This isotope of carbon quickly

Table 10.1 Radioactive isotopes frequently used in radiometric dating.

Radioactive Parent	Stable Daughter Product	Currently Accepted Half-Life Values
Uranium-238	Lead-206	4.5 billion years
Uranium-235	Lead-207	713 million years
Thorium-232	Lead-208	14.1 billion years
Rubidium-87	Strontium-87	47.0 billion years
Potassium-40	Argon-40	1.3 billion years

Box 10.1 People and the Environment: Radon—A Threat to Human Health

Richard L. Hoffman*

Radioactivity is defined as the spontaneous emission of atomic particles and/or electromagnetic waves from unstable atomic nuclei. For example, in a sample of uranium-238, unstable nuclei decay and produce a variety of radioactive progeny or "daughter" products as well as energetic forms of radiation (Table 10.A). One of its radioactive decay products is radon—a colorless, odorless, invisible gas.

Radon gained public attention in 1984 when a worker in a Pennsylvania nuclear power plant set off radiation alarms—not when he left work, but as he entered. His clothing and hair were contaminated with radon decay products. Investigation revealed that his basement at home had a radon level 2800 times the average level in indoor air. The home was located along a geological formation known as the Reading Prong—a mass of uranium-bearing black rock that runs from near Reading, Pennsylvania, to near Trenton, New Jersey.

Originating in the radio-decay of traces of uranium and thorium found in almost all soils, radon isotopes (Rn-222 and Rn-220) are continually renewed in an ongoing, natural process. Geologists estimate that the top six feet of soil from an average acre of land contains about 50 pounds of uranium (about 2 to 3 parts per million); some types of rocks contain more. Radon is continually generated

*Dr. Hoffman is Professor of Chemistry, Illinois Central College.

Table 10.A Decay Products of Uranium-238

Some Decay Products of Uranium-238	Decay Particle Produced	Half-Life
Uranium-238	alpha	4.5 billion years
Radium-226	alpha	1600 years
Radon-222	**alpha**	**3.82 days**
Polonium-218	alpha	3.1 minutes
Lead-214	beta	26.8 minutes
Bismuth-214	beta	19.7 minutes
Polonium-214	alpha	1.6×10^{-4} second
Lead-210	beta	20.4 years
Bismuth-210	beta	5.0 days
Polonium-210	alpha	138 days
Lead-206	none	stable

by the gradual decay of this uranium. Because uranium has a half-life of about 4.5 billion years, radon will be with us forever.

Radon itself decays, having a half-life of only about four days. Its decay products (except lead-206) are all radioactive solids that adhere to dust particles, many of which we inhale. During prolonged exposure to a radon-contaminated environment, some decay will occur while the gas is in the lungs, thereby placing the radioactive radon progeny in direct contact with delicate lung tissue. Steadily accumulating evidence indicates radon to be a significant cause of lung cancer second only to smoking.

A house with a radon level of 4.0 picocuries per liter of air has about 8 to 9 atoms of radon decaying every minute in every liter of air. The EPA suggests indoor radon levels be kept below this level. EPA risk estimates are conservative—they are based on an assumption that one would spend 75 percent of a 70-year time span (about 52 years) in the contaminated space, which most people would not.

Once radon is produced in the soil, it diffuses throughout the tiny spaces between soil particles. Some radon ultimately reaches the soil surface where it dissipates into the air. Radon enters buildings and homes through holes and cracks in basement floors and walls. Radon's density is greater than air, so it tends to remain in basements during its short decay cycle.

The source of radon is as enduring as its generation mechanism within Earth; radon will never go away. However, cost-effective mitigation strategies are available to reduce radon to acceptable levels, generally without great expense.

becomes incorporated into carbon dioxide, which circulates in the atmosphere and is absorbed by living matter. As a result, all organisms contain a small amount of carbon-14, including yourself.

While an organism is alive, the decaying radiocarbon is continually replaced, and the proportions of carbon-14 and carbon-12 remain constant. Carbon-12 is the stable and most common isotope of carbon. However, when any plant or animal dies, the amount of carbon-14 gradually decreases as it decays to nitrogen-14 by beta emission (Figure 10.16B). By comparing the proportions of carbon-14 and carbon-12 in a sample, radiocarbon dates can be determined.

Although carbon-14 is useful in dating only the last small fraction of geologic time, it has become a very valuable tool for anthropologists, archaeologists, and historians, as well as for geologists who study very recent Earth history. In fact, the development of radiocarbon dating was considered so important that the chemist who discovered this application, Willard F. Libby, received a Nobel prize.

Importance of Radiometric Dating

Bear in mind that, although the basic principle of radiometric dating is simple, the actual procedure is quite complex. The analysis that determines the

Figure 10.16 A. Production and **B.** decay of carbon-14. These sketches represent the nuclei of the respective atoms.

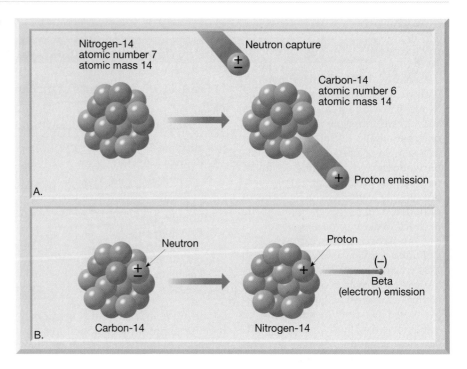

quantities of parent and daughter must be painstakingly precise. In addition, some radioactive materials do not decay directly into the stable daughter product. As you saw in Figure 10.14, uranium-238 produces 13 intermediate unstable daughter products before the 14th and final daughter product, the stable isotope lead-206, is produced.

Radiometric dating methods have produced literally thousands of dates for events in Earth history. Rocks from several localities have been dated at more than 3 billion years, and geologists realize that still-older rocks exist. For example, a granite from South Africa has been dated at 3.2 billion years—and it contains inclusions of quartzite that must be older. Quartzite itself is a metamorphic rock that originally was the sedimentary rock sandstone. Sandstone, in turn, is the product of the lithification of sediments produced by the weathering of existing rocks. Thus, we have a positive indication that much older rocks existed.

Radiometric dating has vindicated the ideas of James Hutton, Charles Darwin, and others who inferred that geologic time must be immense. Indeed, modern dating methods have proven that there has been enough time for the processes we observe to have accomplished tremendous tasks.

The Geologic Time Scale

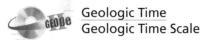

Geologic Time
Geologic Time Scale

Geologists have divided the whole of geologic history into units of varying magnitude. Together they comprise the **geologic time scale** of Earth history (Figure 10.17).

The major units of the time scale were delineated during the 1800s, principally by workers in Western Europe and Great Britain. Because radiometric dating was unavailable at that time, the entire time scale was created using methods of relative dating. It was only in the twentieth century that radiometric dating permitted numerical dates to be added.

The geologic time scale subdivides the 4.5-billion-year history of Earth into many different units and provides a meaningful time frame within which the events of the geologic past are arranged. As shown in Figure 10.17, **eons** represent the greatest expanses of time. The eon that began about 540 million years ago is the **Phanerozoic**, meaning *visible life*. It is an appropriate description because the rocks and deposits of the Phanerozoic eon contain abundant fossils that document major evolutionary trends.

Another glance at the time scale reveals that the Phanerozoic eon is divided into **eras**. The three eras within the Phanerozoic are the **Paleozoic** ("ancient life"), the **Mesozoic** ("middle life"), and the **Cenozoic** ("recent life"). As the names imply, the eras are bounded by profound worldwide changes in life-forms. Each era is subdivided into **periods**. The Paleozoic has seven, the Mesozoic three, and the Cenozoic two. Each of these 12 periods is characterized by a somewhat less profound change in life-forms as compared with the eras.

Finally, periods are divided into still smaller units called **epochs**. As you can see in Figure 10.17, seven epochs have been named for the periods of the Cenozoic. The epochs of other periods, however, are not usually referred to by specific names. Instead, the terms *early, middle,* and *late* are generally applied to the epochs of these earlier periods.

Figure 10.17 The geologic time scale. The numerical dates were added long after the time scale had been established using relative dating techniques. (Data from Geological Society of America)

Notice that the detail of the geologic time scale does not begin until about 540 million years ago, the date for the beginning of the Cambrian period. The more than 4 billion years prior to the Cambrian is divided into three eons, the *Hadean*, the *Archean*, and the *Proterozoic*. It is also common for this vast expanse of time to simply be referred to as the **Precambrian**.

Although it represents about 88 percent of Earth history, the Precambrian is not divided into nearly as many smaller time units as is the Phanerozoic eon.

The quantity of information geologists have deciphered about Earth's past is somewhat analogous to the detail of human history. The further back we go, the less we know. Certainly more data and information exist about the past 10 years than for the first decade of the twentieth century; the events of the nineteenth century have been documented much better than the events of the first century A.D., and so on. Thus it is with Earth history. The more recent past has the freshest, least disturbed, and most observable record. The further back in time the geologist goes, the more fragmented the record and clues become.

Difficulties in Dating the Geologic Time Scale

Although reasonably accurate numerical dates have been worked out for the periods of the geologic time scale (see Figure 10.17), the task is not without difficulty. The primary problem in assigning numerical dates to units of time is the fact that not all rocks can be dated radiometrically. Recall that for a radiometric date to be useful, all minerals in the rock must have formed at approximately the same time. For this reason, radioactive isotopes can be used to determine when minerals in an igneous rock crystallized and when pressure and heat created new minerals in a metamorphic rock.

However, samples of sedimentary rock can only rarely be dated directly by radiometric means. A sedimentary rock may include particles that contain radioactive isotopes, but the rock's age cannot be accurately determined because the grains making up the rock are not the same age as the rock in which they occur. Rather, the sediments have been weathered from rocks of diverse ages.

Radiometric dates obtained from metamorphic rocks may also be difficult to interpret, because the age of a particular mineral in a metamorphic rock does not necessarily represent the time when the rock initially formed. Instead, the date may indicate any one of a number of subsequent metamorphic phases.

If samples of sedimentary rocks rarely yield reliable radiometric ages, how can numerical dates be assigned to sedimentary layers? Usually the geologist must relate them to datable igneous masses, as in Figure 10.18. In this example, radiometric dating has determined the ages of the volcanic ash bed within the Morrison Formation and the dike cutting the Mancos Shale and Mesaverde Formation. The sedimentary beds below the ash are obviously older than the ash, and all the layers above the ash are younger (principle of superposition). The dike is younger than the Mancos Shale and the Mesaverde Formation but older than the Wasatch Formation because the dike does not intrude the Tertiary rocks (cross-cutting relationships).

From this kind of evidence, geologists estimate that a part of the Morrison Formation was deposited about 160 million years ago, as indicated by the ash bed. Further, they conclude that the Tertiary period began after the intrusion of the dike, 66 million years ago. This is one example of literally thousands that illustrates how datable materials are used to *bracket* the various episodes in Earth history within specific time periods. It shows the necessity of combining laboratory dating methods with field observations of rocks.

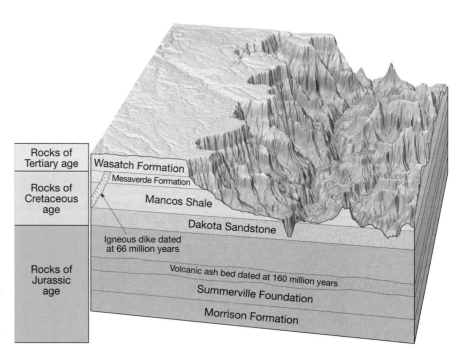

Figure 10.18 Absolute dates for sedimentary layers are usually determined by examining their relationship to igneous rocks. (After U.S. Geological Survey)

Chapter Summary

- *Uniformitarianism*, one of the fundamental principles of modern geology, was formulated by James Hutton in the late 1700s. It states that the physical, chemical, and biological laws that operate today have also operated in the geologic past. The idea is often summarized as "the present is the key to the past." Hutton argued that processes that appear to be slow-acting could, over long spans of time, produce effects that were just as great as those resulting from sudden catastrophic events. In contrast, *catastrophism* states that Earth's landscapes have been developed primarily by great catastrophes.

- The two types of dates used by geologists to interpret Earth history are (1) *relative dates*, which put events in their *proper sequence of formation*, and (2) *numerical dates*, which pinpoint the *time in years* when an event took place.

- Relative dates can be established using the *law of superposition, principle of original horizontality, principle of cross-cutting relationships, inclusions,* and *unconformities.*

- *Correlation*, the matching up of two or more geologic phenomena in different areas, is used to develop a geologic time scale that applies to the entire Earth.

- *Fossils* are the remains or traces of prehistoric life. The special conditions that favor preservation are *rapid burial* and the possession of *hard parts* such as shells, bones, or teeth.

- Fossils are used to correlate sedimentary rocks from different regions by using the rocks' distinctive fossil content and applying the *principle of fossil succession*. It states that fossil organisms succeed one another in a definite and determinable order, and therefore any time period can be recognized by its fossil content.

- *Radioactivity* is the spontaneous breaking apart (decay) of certain unstable atomic nuclei. Three common types of radioactive decay are (1) emission of alpha particles from the nucleus, (2) emission of beta particles (electrons) from the nucleus, and (3) capture of electrons by the nucleus.

- An unstable *radioactive isotope*, called the *parent*, will decay and form stable *daughter products*. The length of time for one-half of the nuclei of a radioactive isotope to decay is called the *half-life* of the isotope. If the half-life of the isotope is known, and the parent/daugher ratio can be measured, the age of a sample can be calculated.

- The *geologic time scale* divides Earth's history into units of varying magnitude. It is commonly presented in chart form, with the oldest time and event at the bottom and the youngest at the top. The principal subdivisions of the geologic time scale, called *eons*, include the *Hadean, Archean, Proterozoic* (together, these three eons are commonly referred to as the *Precambrian*), and, beginning about 540 million years ago, the *Phanerozoic*. The Phanerozoic (meaning "visible life") eon is divided into the following *eras*: *Paleozoic* ("ancient life"), *Mesozoic* ("middle life"), and *Cenozoic* ("recent life").

- A significant problem in assigning numerical dates to units of time is that *not all rocks can be dated radiometrically*. A sedimentary rock may contain particles of many ages that have been weathered from different rocks that formed at various times. One way geologists assign numerical dates to sedimentary rocks is to relate them to datable igneous masses, such as volcanic ash beds.

Key Terms

angular unconformity (p. 277)
catastrophism (p. 273)
Cenozoic era (p. 288)
conformable (p. 276)
correlation (p. 277)
cross-cutting relationships,
    principle of (p. 275)
disconformity (p. 277)
eon (p. 288)
epoch (p. 288)
era (p. 288)

fossil (p. 278)
fossil succession, principle of (p. 281)
geologic time scale (p. 288)
half-life (p. 285)
inclusions (p. 276)
index fossil (p. 283)
Mesozoic era (p. 288)
nonconformity (p. 277)
numerical date (p. 274)
original horizontality, principle of
    (p. 274)

Paleozoic era (p. 288)
period (p. 288)
Phanerozoic eon (p. 288)
Precambrian (p. 289)
radioactivity (p. 284)
radiometric dating (p. 285)
relative dating (p. 274)
superposition, law of (p. 274)
unconformity (p. 276)
uniformitarianism (p. 273)

Review Questions

1. Contrast catastrophism and uniformitarianism. How did the proponents of each perceive the age of Earth?

2. Distinguish between numerical and relative dating.

3. What is the law of superposition? How are cross-cutting relationships used in relative dating?

4. When you observe an outcrop of steeply inclined sedimentary layers, what principle allows you to assume that the beds became tilted *after* they were deposited?

5. Refer to Figure 10.5 and answer the following questions:

   (a) Is fault *A* older or younger than the sandstone layer?

   (b) Is dike *A* older or younger than the sandstone layer?

   (c) Was the conglomerate deposited before or after fault *A*?

   (d) Was the conglomerate deposited before or after fault *B*?

   (e) Which fault is older, *A* or *B*?

   (f) Is dike *A* older or younger than the batholith?

6. A mass of granite is in contact with a layer of sandstone. Using a principle described in this chapter, explain how you might determine whether the sandstone was deposited on top of the granite or the granite was intruded from below after the sandstone was deposited.

7. Distinguish among angular unconformity, disconformity, and nonconformity.

8. What is meant by the term *correlation*?

9. List and briefly describe at least five different types of fossils.

10. List two conditions that improve an organism's chances of being preserved as a fossil.

11. Why are fossils such useful tools in correlation?

12. In addition to being important aids in dating and correlating rocks, how else are fossils helpful in geologic investigations?

13. If a radioactive isotope of thorium (atomic number 90, mass number 232) emits 6 alpha particles and 4 beta particles during the course of radioactive decay, what are the atomic number and mass number of the stable daughter product?

14. Why is radiometric dating the most reliable method of dating the geologic past?

15. Assume that a hypothetical radioactive isotope has a half-life of 10,000 years. If the ratio of radioactive parent to stable daughter product is 1:3, how old is the rock containing the radioactive material? What if the ratio were 1:15?

16. To make calculations easier, let us round the age of Earth to 5 billion years.

    (a) What fraction of geologic time is represented by recorded history (assume 5000 years for the length of recorded history)?

    (b) The first abundant fossil evidence does not appear until the beginning of the Cambrian period (approximately 550 million years ago). What percentage of geologic time is represented by abundant fossil evidence?

17. What subdivisions make up the geologic time scale? What is the primary basis for differentiating the eras?

18. Briefly describe the difficulties in assigning numerical dates to layers of sedimentary rock.

Examining the Earth System

1. Figure 10.11A is a large petrified log in Arizona's Petrified Forest National Park. Describe the transition of this tree from being part of the biosphere to being a component of the solid Earth. How might the hydrosphere and/or atmosphere have played a role in the transition?

2. This famous angular unconformity at Scotland's Siccar Point was originally studied by James Hutton in the late 1700s (Figure 10.19). Can you describe in a general way what occurred to produce this feature? Could all of the spheres of the Earth system have been involved? The Earth system is powered by energy from two sources. How are both sources represented here?

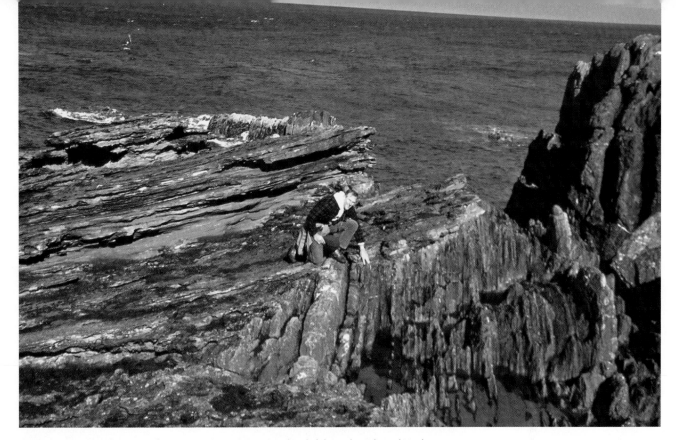

Figure 10.19 Angular unconformity at Siccar Point, Scotland. (Photo by Edward Hay)

Web Resources

The following are informative and interesting World Wide Web sites that address topics related to those presented in the chapter:

Earthtrips: A Virtual Journey Into Earth History:

- **http://www.englib.cornell.edu/pri/earthtrips/fETrips.html**

Time Machine (University of California):

- **http://www.ucmp.berkeley.edu/help/timeform.html**

For direct links to these sites and other appropriate Earth science Web locations, chapter review exercises, and relevant, topical investigations that utilize Web resources, contact the *Earth Science*, 9th Edition Home Page at:

- **http://www.prenhall.com/tarbuck.**

Chapter 11

Earth's History:
A Brief Summary

Fossils of extinct coiled cephalopods. Like their modern descendants, these were highly developed marine organisms. Fossils are crucial to the study of Earth history. (Photo by E.J. Tarbuck)

Earth has a long and complex history. The splitting and colliding of continents has resulted in the formation of new ocean basins and the creation of Earth's great mountain ranges. Furthermore, biological evolution has spurred a constant changing of Earth's life-forms.

Changes on planet Earth occur at a "snail's pace," generally too slowly for people to perceive. Thus, human awareness of evolutionary change is fairly recent. Evolution is not confined to life-forms, for all of Earth's "spheres" have evolved together: the atmosphere, hydrosphere, lithosphere, and biosphere. Examples are evolutionary changes in the air we breath, evolution of the world oceans, the rise of mountains, ponderous movements of crustal plates, the comings and goings of vast ice sheets, and the evolution of a vast array of life-forms. As each facet of Earth has evolved, it has powerfully influenced the others.

As we saw in Chapter 10, geologists have many tools at their disposal for interpreting the clues about Earth's past. Using these tools, and clues that are contained in the rock record, geologists have been able to unravel many of the complex events of the geological past. The goal of this chapter is to provide a brief overview of the history of our planet and its life-forms (Figure 11.1). We will begin nearly 5 billion years ago with the formation of Earth and its atmosphere. Then we will describe how our physical world assumed its present form and how Earth's inhabitants changed through time.

Origin of Earth

At the center of the solar system is the Sun. Around it revolve nine planets, and around seven of the planets revolve one or more moons. This orderly nature of our solar system led most astronomers to conclude that its members formed at essentially the same time and from the same primordial material. This proposal, known as the **nebular hypothesis,** suggests that the bodies of our solar system condensed from an enormous cloud. It was composed

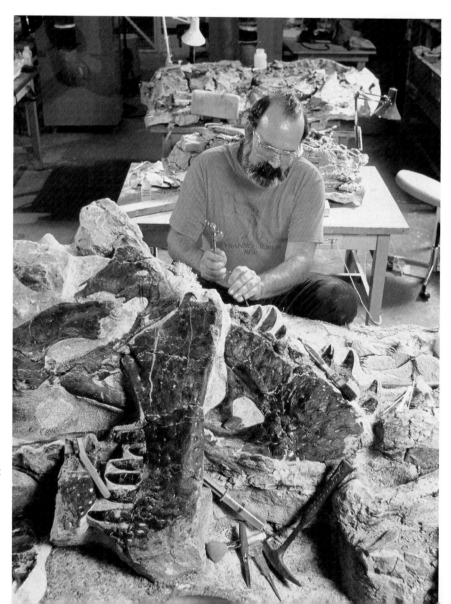

Figure 11.1 Paleontologists study ancient life. This scientist is working with the remains of a dinosaur known as *Tyrannosaurus Rex.* (Photo by Rich Frishman/Tony Stone Images)

mostly of hydrogen and helium, with only a small percentage of all other heavier elements.

Nearly 5 billion years ago, this huge cloud of minute rocky fragments and gases began to contract under its own gravitational influence (Figure 11.2A). The contracting material somehow began to rotate. Like a spinning ice skater pulling in her arms, the cloud rotated faster and faster as it contracted. This rotation in turn caused the nebular cloud to flatten into a disk (Figure 11.2B). Within the rotating disk, smaller condensations formed nuclei from which the planets eventually coalesced. However, the great concentration of material was pulled toward the center of this rotating mass. As it packed inward upon itself, it gravitationally heated, forming the hot *protosun* (sun in the making).

After the protosun formed, the temperature out in the rotating disk dropped significantly. This cooling caused substances with high melting points to condense into small particles, perhaps the size of sand grains. Iron and nickel solidified first. Next to condense were the elements of which rocky substances are composed. As these fragments collided over a few tens of millions of years, they accreted into the planets (Figure 11.2C, D). In the same manner, but on a lesser scale, the processes of condensation and accretion acted to form the moons and other small bodies of the solar system.

As the *protoplanets* (planets in the making) accumulated more and more debris, the space within the solar system began to clear. This removal of debris allowed sunlight to reach planetary surfaces unimpeded and to heat them. The resulting high surface temperatures of the inner planets (Mercury, Venus, Earth, Mars), coupled with their comparatively weak gravitational fields, meant that Earth and its neighbors were unable to retain appreciable amounts of the lighter components of the primordial cloud. These light materials, which included hydrogen, helium, ammonia, methane, and water, vaporized from their surfaces and were eventually whisked from the inner solar system by streams of solar particles called the *solar winds*.

At distances beyond Mars, temperatures were much cooler. Consequently, the large outer planets (Jupiter, Saturn, Uranus, and Neptune) accumulated huge amounts of hydrogen and other light materials from the primordial cloud. The accumulation of these gaseous substances is thought to account for the comparatively large sizes and low densities of the outer planets.

Shortly after Earth formed, the decay of radioactive elements, coupled with heat released by colliding particles, produced at least some melting of the interior. Melting allowed the denser elements, principally iron and nickel, to sink to Earth's center, while the lighter rocky components floated outward, toward the surface. The sorting of material by density, which began early in Earth's history, is believed to be occurring still, but on a much smaller scale. As a result of

Figure 11.2 Nebular hypothesis. **A.** The nebula, a huge rotating cloud of dust and gases, began to contract. **B.** Most of the material was gravitationally swept toward the center, producing the Sun. However, because of rotational motion, some dust and gases remained orbiting the central body as a flattened disk. **C.** The planets began to accrete from the material that was orbiting within the flattened disk. **D.** In time, most of the remaining debris either coalesced to form the nine planets and their moons, or was swept into space by the solar wind.

this differentiation, Earth's interior is not homogeneous. Rather, it consists of shells or spheres of materials that have different properties.

An important consequence of this period of differentiation is that gaseous materials were allowed to escape from Earth's interior, just as gases are emitted today during volcanic eruptions. By this process, an atmosphere gradually evolved, composed chiefly of gasses expelled from within the planet.

Earth's Atmosphere Evolves

Today, the air you breathe is a stable mixture of 79 percent nitrogen, 20 percent oxygen, about 1 percent argon (an inert gas), and trace gases like carbon dioxide and water vapor. But our planet's original atmosphere, several billion years ago, was far different.

Earth's very earliest atmosphere probably was swept into space by the *solar wind*, a vast stream of particles emitted by the Sun. As Earth slowly cooled, a more enduring atmosphere formed. The molten surface solidified into a crust, and gases that had been dissolved in the molten rock were gradually released, a process called **outgassing.** Outgassing continues today from hundreds of active volcanoes worldwide. Thus, geologists hypothesize that Earth's original atmosphere was made up of gases similar to those released in volcanic emissions today: water vapor, carbon dioxide, nitrogen, and several trace gases.

As the planet continued to cool, the water vapor condensed to form clouds, and great rains commenced. At first, the water evaporated in the hot air before reaching the ground, or quickly boiled away upon contacting the surface, just like water sprayed on a hot grill. This accelerated the cooling of Earth's crust. When the surface had cooled below water's boiling point (100°C or 212°F), torrential rains slowly filled low areas, forming the oceans. This reduced not only the water vapor in the air but also the amount of carbon dioxide, for it became dissolved in the water. What remained was a nitrogen-rich atmosphere.

If Earth's primitive atmosphere resulted from volcanic outgassing, we have a problem, because volcanoes do not emit free oxygen. Where did the very significant percentage of oxygen in our present atmosphere (20 percent) come from?

The major source of oxygen is green plants. Put another way, *life itself* has strongly influenced the composition of our present atmosphere. Plants did not just adapt to their environment; they actually influenced it, dramatically altering the composition of the entire planet's atmosphere by using carbon dioxide and releasing oxygen. This is a good example of how Earth operates as a giant system in which living things interact with their environment.

How did plants come to alter the atmosphere? The key is the way in which plants create their own food. They employ *photosynthesis,* in which they use light energy to synthesize food sugars from carbon dioxide and water. The process releases a waste gas, oxygen. Those of us in the animal kingdom rely on oxygen to metabolize our food, and we in turn exhale carbon dioxide as a waste gas. The plants use this carbon dioxide for more photosynthesis, and so on, in a continuing system.

The first life-forms on Earth, probably bacteria, did not need oxygen. Their life processes were geared to the earlier, oxygenless atmosphere. Even today, many *anaerobic* bacteria thrive in environments that lack free oxygen. Later, primitive plants evolved that used photosynthesis and released oxygen. Slowly, the oxygen content of Earth's atmosphere increased. The Precambrian rock record suggests that much of the first free oxygen did not remain free because it combined with (oxidized) other substances dissolved in water, especially iron. Iron has tremendous affinity for oxygen, and the two elements combine to form iron oxides (rust) at any opportunity.

Then, once the available iron satisfied its need for oxygen, substantial quantities of oxygen accumulated in the atmosphere. By the beginning of the Paleozoic era, about 4 billion years into Earth's existence (after seven-eighths of Earth's history had transpired), the fossil record reveals abundant ocean-dwelling organisms that require oxygen to live. Hence, the composition of Earth's atmosphere has evolved together with its life-forms, from an oxygenless envelope to today's oxygen-rich environment.

With this background on the evolution of the solar system, Sun, Earth, and atmosphere, we will look at Earth's history.

Precambrian Time: Vast and Enigmatic

The Precambrian encompasses immense geological time, from Earth's distant beginnings 4.5 billion years ago until the start of the Cambrian period, over 4 billion years later. Thus, the Precambrian spans about 88 percent of Earth's history. Our knowledge of this ancient time is sketchy, for much of the early rock record has been obscured by the very Earth processes you have been studying, especially plate tectonics, erosion, and deposition.

Untangling the long, complex Precambrian rock record is a formidable task, and we are far from done. Most Precambrian rocks are devoid of fossils, which hinders correlation of rocks. Rocks of this great age are metamorphosed and deformed, extensively eroded, and obscured by overlying strata. Consequently, this least-understood span of Earth's history has not been successfully divided into briefer time units, as have later intervals. Indeed, Precambrian history is written in scattered, speculative episodes, like a long book with many missing chapters.

Precambrian Rocks

Looking upon Earth from the Space Shuttle, astronauts see plenty of ocean (71 percent) and much less land area (29 percent). Over large expanses of the continents, the orbiting space scientists gaze upon many Paleozoic, Mesozoic, and Cenozoic rock surfaces, but fewer Precambrian surfaces. This demonstrates the law of superposition: Precambrian rocks in these regions are buried from view beneath varying thicknesses of more recent rocks. Here, Precambrian rocks peek through the surface where younger strata are extensively eroded, as in the Grand Canyon and in some mountain ranges. However, on each continent, large "core areas" of Precambrian rocks dominate the surface, mostly as deformed metamorphic rocks. These areas are called **shields** because they roughly resemble a warrior's shield in shape.

Figure 11.3 shows these shield areas of Precambrian rocks worldwide. In North America (including Greenland), the Canadian Shield encompasses 7.2 million square kilometers (2.8 million square miles), the equivalent of about 10 states of Texas put together.

Much of what we know about Precambrian rocks comes from mining the ores contained in such rocks. The mining of iron, nickel, gold, silver, copper, chromium, uranium, and diamonds has provided Precambrian rock samples for study, and surveys to locate valuable ore deposits have revealed much about the rocks.

Noteworthy are extensive iron ore deposits. Rocks from the middle Precambrian (1.2 billion to 2.5 billion years ago) contain most of Earth's iron ore, mainly as the mineral hematite (Fe_2O_3). These iron-rich sedimentary rocks probably represent the time when oxygen became sufficiently abundant to react with iron dissolved in shallow lakes and seas. Later, after much of the iron was oxidized and deposited on lake and sea bottoms, formation of these iron-rich deposits declined and oxygen levels in the ocean and atmosphere began to increase. Because most of Earth's free oxygen results from plant photosynthesis, the formation of extensive Precambrian iron ore deposits is linked to life in the sea.

Notably absent in the Precambrian are fossil fuels (coal, oil, natural gas). The reason is clear—a virtual absence of land plants to form coal swamps and of certain animals to form petroleum. Fossil fuels are from a later time.

Precambrian Fossils

A century ago, the earliest fossils known dated from the Cambrian period, about 540 million years ago. None were known from the Precambrian. This created a major problem for science at that time: How could complex organisms like trilobites abruptly appear in the geologic record? The answer, or course, was that there were fossils in Precambrian rocks, but they were rare, small, obscure, and simply had not been discovered yet. Today, our knowledge of Precambrian life, although far from complete, is quite extensive.

Precambrian fossils are disappointing if you are expecting to see fascinating plants and large animals, for these had not yet evolved. Instead, the most common Precambrian fossils are **stromatolites.** These are distinctively layered mounds or columns of calcium carbonate (Figure 11.4). Stromatolites are not the remains of actual organisms, but are material deposited by algae. They are

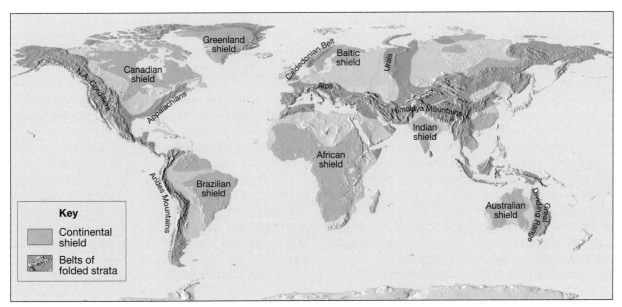

Figure 11.3 Today's continents look very different from the Precambrian. Remnants of Precambrian rocks are the continental shields, composed largely of metamorphic rocks.

A.

B.

Figure 11.4 Stromatolites are among the most common Precambrian fossils. **A.** Precambrian fossil stromatolites composed of calcium carbonate deposited by algae in the Helena formation, Glacier National Park. (Photo by Ken M. Johns/Photo Researchers, Inc.) **B.** Modern stromatolites growing in saline area, Western Australia. (Photo by Bill Bachman/Photo Researchers, Inc.)

indirect evidence of algae because they closely resemble similar deposits made by modern algae.

Stromatolites did not become common until the middle Precambrian, around 2 billion years ago. Stromatolites are large, but most actual organisms preserved in Precambrian rocks are microscopic. Well-preserved remains of many tiny organisms have been discovered, extending the record of life back beyond 3.5 billion years.

Many of these most ancient fossils are preserved in *chert,* a hard, dense chemical sedimentary rock. Chert must be very thinly sliced and studied under powerful microscopes to observe bacteria and algae fossils within it.

Microfossils have been found at several locations worldwide. Two notable areas are in southern Africa, where the rocks date to more than 3.1 billion years, and in the Gunflint Chert (named for its use in flint-lock rifles) of Lake Superior, which dates to 1.7 billion years. In both places, bacteria and blue-green algae have been discovered. The fossils are of the most primitive organisms, *prokaryotes.* Their cells lack organized nuclei, and they reproduce asexually.

More advanced organisms, *eukaryotes,* have cells that contain nuclei. Eukaryotes are among billion-year-old fossils discovered at Bitter Springs in Australia, such as green algae. Unlike prokaryotes, eukaryotes reproduce sexually, which means that genetic material is exchanged between organisms. This reproductive mode permits greatly increased genetic variation. Thus, development of eukaryotes may have dramatically increased the rate of evolutionary change.

Plant fossils date from the middle Precambrian, but animal fossils came a bit later, in the late Precambrian. Many of these fossils are *trace fossils,* meaning that they are not of the animals themselves, but of their activities, such as trails and worm holes. Areas in Australia and Newfoundland have yielded hundreds of fossil impressions of soft-bodied creatures. Most, if not all, of the Precambrian fauna lacked shells, which would develop as protective armor during the Paleozoic.

As the Precambrian came to a close, the fossil record disclosed diverse and complete multicelled organisms. This set the stage for more complex plants and animals to evolve at the dawn of the Paleozoic era.

Paleozoic Era: Life Explodes

Following the long Precambrian, the most recent 540 million years of Earth history are divided into three eras: Paleozoic, Mesozoic, and Cenozoic. The Paleozoic era encompasses about 292 million years and is by far the longest of the three. Seven periods make up the Paleozoic era (Figure 11.5).

Before the Paleozoic, life-forms possessed no hard parts—shells, scales, bones, or teeth. The beginning of the Paleozoic is marked by the appearance of the first life-forms with hard parts (Figure 11.6). Hard parts greatly enhanced their chance of being preserved as part of the fossil record. Therefore, our knowledge of life's diversification improves greatly from the Paleozoic onward. This diversity is demonstrated in Figure 11.7.

Abundant Paleozoic fossils have allowed geologists to construct a far more detailed time scale for the last one-eighth of geologic time than for the preceding seven-eighths, the Precambrian. Moreover, because every organism is associated with a particular environment, the greatly improved fossil record provided invaluable information for deciphering ancient environments. To facilitate our brief tour of the Paleozoic, we divide it into Early Paleozoic (Cambrian, Ordovician, Silurian periods) and Late Paleozoic (Devonian, Mississippian, Pennsylvanian, Permian periods).

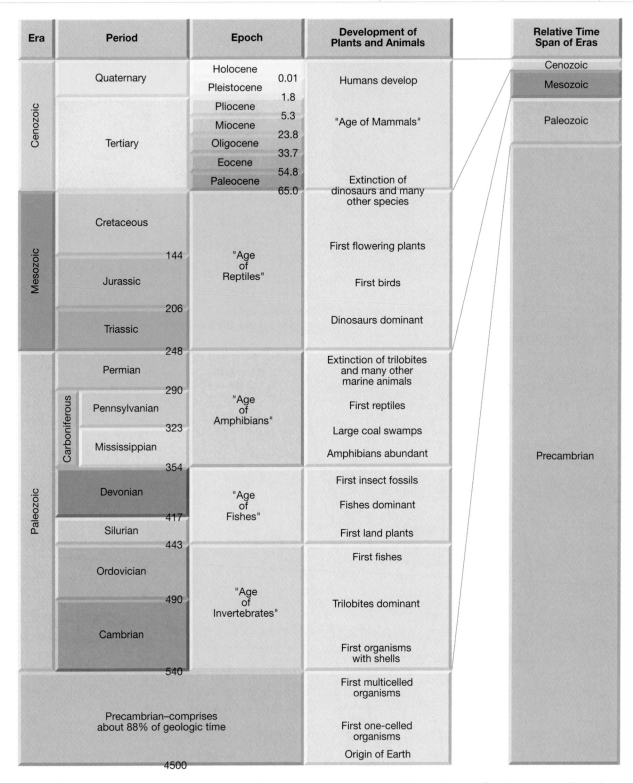

Era	Period	Epoch	Development of Plants and Animals	Relative Time Span of Eras
Cenozoic	Quaternary	Holocene 0.01 / Pleistocene 1.8	Humans develop	Cenozoic
	Tertiary	Pliocene 5.3 / Miocene 23.8 / Oligocene 33.7 / Eocene 54.8 / Paleocene 65.0	"Age of Mammals"	Mesozoic
			Extinction of dinosaurs and many other species	Paleozoic
Mesozoic	Cretaceous	"Age of Reptiles"	First flowering plants	
		144	First birds	
	Jurassic	206		
	Triassic	248	Dinosaurs dominant	
Paleozoic	Permian		Extinction of trilobites and many other marine animals	Precambrian
		290	First reptiles	
	Carboniferous: Pennsylvanian	"Age of Amphibians" 323	Large coal swamps	
	Mississippian	354	Amphibians abundant	
	Devonian	"Age of Fishes" 417	First insect fossils / Fishes dominant	
	Silurian	443	First land plants	
	Ordovician	"Age of Invertebrates" 490	First fishes / Trilobites dominant	
	Cambrian	540	First organisms with shells	
	Precambrian—comprises about 88% of geologic time 4500		First multicelled organisms / First one-celled organisms / Origin of Earth	

Figure 11.5 The geologic time scale. Numbers on the time scale represent time in millions of years before the present. These dates were added long after the time scale had been established using relative dating techniques. The Precambrian accounts for about 88 percent of geologic time. (Data from Geological Society of America)

Early Paleozoic History

The early Paleozoic consists of a 123-million-year span that embraces the Cambrian, Ordovician, and Silurian periods. Anyone approaching Earth from space at this time would have seen the familiar blue planet with plentiful white clouds, but the arrangement of continents would have looked very different from today (Figure 11.8). At this time, the vast southern continent

A.

B.

Figure 11.6 Fossils of common Paleozoic life-forms. **A.** Natural cast of a trilobite. Trilobites dominated the Paleozoic ocean, scavenging food from the bottom. **B.** Extinct coiled cephalopods. Like their modern descendants, these were highly developed marine organisms. (Photos by Edward J. Tarbuck)

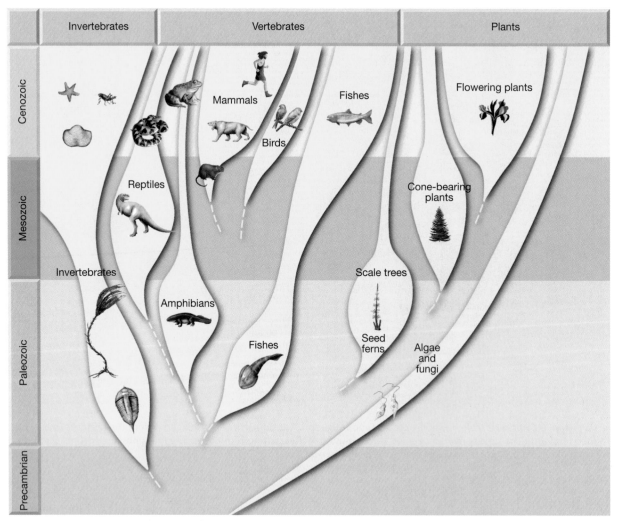

Figure 11.7 This chart indicates the times of appearance and relative abundance of major groups of organisms. The wider the band, the more dominant the group.

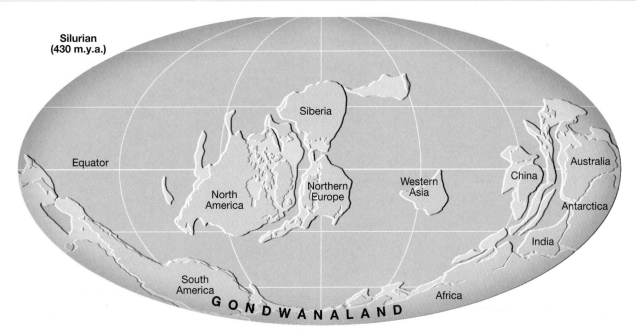

Figure 11.8 Reconstruction of Earth as it may have appeared in early Paleozoic time. The southern continents were joined into a single landmass called Gondwanaland. Four of the five landmasses that would later join to form the northern continent of Laurasia lay scattered roughly along the equator. (After C. Scotese, R. K. Bambach, C. Barton, R. VanderVoo, and A. Ziegler)

of Gondwanaland encompassed five continents (South America, Africa, Australia, Antarctica, India, and perhaps China). Evidence of an extensive continental glaciation places western Africa near the South Pole!

Landmasses that were not part of Gondwanaland existed as five separate units and some scattered fragments. Although the exact position of these "northern" continents is uncertain, ancestral North America and Europe are thought to have been near the equator and separated by a narrow sea, as shown on the map in Figure 11.8.

As the Paleozoic opened, North America was a land with no living things, plant or animal. There were no Appalachian or Rocky Mountains; rather, the continent was largely a barren lowland. Several times during the Cambrian and Ordovician periods, shallow seas moved inland and then receded from the interior of the continent. Deposits of clean sandstones, used today to make glass, mark the edge of these shallow seas in the mid-continent.

Early in the Paleozoic, a mountain-building event affected eastern North America from the present-day central Appalachians to Newfoundland. The mountains produced during this event, the Taconic Orogeny, have since eroded away, leaving behind deformed strata and a large volume of detrital sedimentary rocks that were derived from the weathering of these mountains.

During the Silurian period, much of North America was once again inundated by shallow seas. This time large barrier reefs restricted circulation between shallow marine basins and the open ocean. Water in these basins evaporated, causing deposition of large quantities of rock salt and gypsum. Today these thick *evaporite beds* are important resources for the chemical, rubber, plasterboard, and photographic industries in Ohio, Michigan, and western New York State.

Early Paleozoic Life

Life in early Paleozoic time was restricted to the seas. Vertebrates had not yet evolved, so life consisted of several invertebrate groups (shown in Figure 11.9). The Cambrian period was the golden age of *trilobites.* More than 600 genera of these mud-burrowing scavengers flourished worldwide. By Ordovician times, *brachiopods* outnumbered the trilobites. Brachiopods are among the most widespread Paleozoic fossils and, except for one modern group, are now extinct. Although the adults lived attached to the seafloor, the young larvae were free-swimming. This mobility accounts for the group's wide geographic distribution.

The Ordovician also marked the appearance of abundant *cephalopods*, mobile, highly developed mollusks that became the major predators of their time. The descedents of cephalopods include the modern squid, octopus, and nautilus. Cephalopods were the first truly large organisms on Earth (Figure 11.9). Whereas the largest trilobites seldom exceeded 30 centimeters (12 inches) in length and the biggest brachiopods were no more than about 20 centimeters (8 inches) across, one species of cephalopod reached a length of nearly 10 meters (30 feet).

Figure 11.9 During the Ordovician period (490–443 million years ago), the shallow waters of an inland sea over central North America contained an abundance of marine invertebrates. Shown in this reconstruction are straight-shelled cephalopods, trilobites, brachiopods, snails, and corals. (© The Field Museum, Neg. #GEO80820c, Chicago)

The beginning of the Cambrian period marks an important event in animal evolution. For the first time, organisms appeared that secreted material which formed *hard parts,* such as shells. Why several diverse life-forms began to develop hard parts about the same time remains unanswered. One proposal suggests that, because an external skeleton provides protection from predators, hard parts evolved for survival. Yet, the fossil record does not seem to support this hypothesis. Organisms with hard parts were plentiful in the Cambrian period, whereas predators such as cephalopods were not abundant until the Ordovician period, some 70 million years later.

Whatever the answer, hard parts clearly served many useful purposes and aided adaptations to new ways of life. Sponges, for example, developed a network of fine interwoven silica spicules that allowed them to grow larger and more erect, capable of extending above the surface in search for food. Mollusks (clams and snails) secreted external shells of calcium carbonate that protected them and allowed body organs to function in a more controlled environment. The successful trilobites developed an exoskeleton of a protein called *chitin,* which permitted them to burrow through soft sediment in search of food (Figure 11.10).

Figure 11.10 These trilobites date from the Devonian period, 417 million to 354 million years ago. (Photo by Phil Degginger/Tony Stone Images)

Late Paleozoic History

The late Paleozoic consists of four periods—the Devonian, Mississippian, Pennsylvanian, and Permian—that span about 160 million years. Tectonic forces reorganized Earth's landmasses during this time, culminating with the formation of the supercontinent *Pangaea* (Figure 11.11).

Forming Pangaea, the Supercontinent. As ancestral North America collided with Africa, the narrow sea that separated these landmasses began to close slowly (compare Figure 11.11B and 11.11C). Strong compressional forces from this collision deformed the rocks to produce the original northern Appalachian Mountains of eastern North America.

During the fusion of North America and Africa, the other northern continents began to converge (Figure 11.11).

By the Permian period, this newly formed landmass had collided with western Asia and the Siberian landmass along the line of the Ural Mountains. Through this union, the northern continent of *Laurasia* was born, encompassing present-day North America, Europe, western Asia, Siberia, and perhaps China.

As Laurasia was forming, Gondwanaland migrated northward. By the Pennsylvanian period, Gondwanaland collided with Laurasia, forming a mountainous belt through central Europe. Simultaneously, a collision between the African fragment of Gondwanaland and the southeastern edge of North America rumpled up the southern Appalachian Mountains.

By the close of the Paleozoic, all the continents had fused into the supercontinent of Pangaea (Figure 11.11). With only a single vast continent, the world's climate became very seasonal, having extremes far greater than those we experience today. As illustrated in Box 11.1, these altered climatic conditions caused one of the most dramatic biological declines in all of Earth history.

Late Paleozoic Life

During most of the late Paleozoic, organisms diversified dramatically. Some 400 million years ago, plants that had adapted to survive at the water's edge began to move inland, becoming *land plants.* These earliest land plants were leafless, vertical spikes about the size of your index finger. However, by the end of the Devonian, 40 million years later, the fossil record indicates the existence of forests with trees tens of meters high.

In the oceans, armor-plated fishes that had evolved during the Ordovician continued to adapt. Their armor plates thinned to lightweight scales that increased their speed and mobility (Figure 11.12). Other fishes evolved during the Devonian, including primitive sharks that had a skeleton made of cartilage and bony fishes, the groups to which virtually all modern fishes belong. Because of this, the Devonian period is often called the "age of fishes."

By late Devonian time, two groups of bony fishes, the lung fish and the lobe-finned fish, became adapted to land environments. Not unlike their modern relatives, these fishes had primitive lungs that supplemented their breathing through gills. It is believed that the lobe-finned fish occupied tidal flats or small ponds

---

Box 11.1   Earth as a System: The Great Paleozoic Extinction

The Paleozoic ended with the Permian period, a time when Earth's major landmasses joined to form the supercontinent Pangaea. This redistribution of land and water and changes in the elevations of landmasses brought pronounced changes in world climates. Broad areas of the northern continents became elevated above sea level and the climate grew drier. These changes apparently triggered extinctions of many species on land and sea.

By the close of the Permian, 75 percent of the amphibian families had disappeared, and plants had declined in number and variety. Although many amphibian groups became extinct, their descendants, the reptiles, would become the most successful and advanced animals on Earth. Marine life was not spared. At least 80 percent, and perhaps as much as 95 percent, of marine life disappeared. Many marine invertebrates that had been dominant during the Paleozoic, including all the remaining trilobites as well as some types of corals and brachiopods, failed to adapt to the widespread environmental changes.

The late Paleozoic extinction was the greatest of at least five mass extinctions to occur over the past 500 million years. Each extinction wreaked havoc with the existing biosphere, wiping out large numbers of species. In each case, however, the survivors formed new biological communities that were more diverse than their predecessors. Thus, mass extinctions actually invigo-rated life on Earth, as the few hardy survivors eventually filled more niches than the ones left by the victims.

The cause of the great Paleozoic extinction is uncertain. The climate changes from the formation of Pangaea and the associated drop in sea level undoubtedly stressed many species. In addition, at least 2 million cubic kilometers of lava flowed across Siberia to produce what is called the Siberian Traps. Perhaps debris from these eruptions blocked incoming sunlight, or perhaps enough sulfuric acid was emitted to make the seas virtually uninhabitable. Whatever caused the late Paleozoic extinction, it is clear that, without it, a very different population of organisms would today inhabit this planet.

Figure 11.11 During the late Paleozoic, plate movements were joining together the major landmasses to produce the supercontinent of Pangaea. (After C. Scotese, R. K. Bambach, C. Barton, R. VanderVoo, and A. Ziegler)

Figure 11.12 These placoderms or "plate-skinned" fish were abundant during the Devonian (417–354 million years ago). (Drawing after A. S. Romer)

and that in times of drought they may have used their bone fins to "walk" from dried-up pools in search of other ponds. Through time, the lobe-finned fish began to rely more on their lungs and less on their gills. By late Devonian time, they had evolved into true air-breathing amphibians with fishlike heads and tails. It should be noted that insects had already invaded the land.

Modern amphibians, like frogs, toads, and salamanders, are small and occupy limited biological niches. But conditions during the remainder of the Paleozoic were ideal for these newcomers to the land. Plants and insects, which were their main diet, already were very abundant and large. Having only minimal competition from other land dwellers, the amphibians rapidly diversified. Some groups took on roles and forms that were more similar to modern reptiles, such as crocodiles, than to modern amphibians (Figure 11.13).

By the Pennsylvanian period, large tropical swamps extended across North America, Europe, and Siberia (Figure 11.14). Trees approached 30 meters (100 feet),

Figure 11.13 Permian scene (290-248 million years ago) shows the large amphibian *Eryops*, which exceeded 2 meters (6 feet) in length. *Eryops* was a carnivore, as you can see from the teeth.

Figure 11.14 Restoration of a Pennsylvania coal swamp (323–290 million years ago). Shown are scale trees (left), seed ferns (lower left), and scouring rushes (right). Also note the large dragonfly. (© The Field Museum, Neg. #GEO85637, Chicago. Photographer: John Weinstein.)

with trunks over a meter across. The coal deposits that fueled the Industrial Revolution, and which provide a substantial portion of our electric power today, originated in these vast swamps. Further, it was in the lush coal swamp environment of the late Paleozoic that the amphibians evolved quickly into a variety of species.

Mesozoic Era: Age of the Dinosaurs

Spanning about 183 million years, the Mesozoic era is divided into three periods: the Triassic, Jurassic, and Cretaceous. The Mesozoic era witnessed the beginning of the breakup of the supercontinent Pangaea. Also in this era, organisms that had survived the great Permian extinction began to diversify in spectacular ways (see Box 11.1). On land, dinosaurs became dominant and remained unchallenged for over 100 million years. Because of this, the Mesozoic era is often called the "age of dinosaurs."

Early geologists recognized a profound difference between the fossils in Permian strata and those in younger Triassic rocks, as if someone had drawn a bold line to separate the two time periods. Clearly one-half of the fossil groups that occurred in late Paleozoic rocks were missing in Mesozoic rocks. On this basis, it

was decided to separate the Paleozoic and Mesozoic at the Permian–Triassic boundary.

Mesozoic History

The Mesozoic era began with much of the world's land above sea level. In fact, in North America no period exhibits a more meager marine sedimentary record than does the Triassic period. Of the exposed Triassic strata, most are red sandstones and mudstones that lack fossils and contain features indicating a terrestrial environment.

As the second period opened, the Jurassic, the sea invaded western North America. Adjacent to this shallow sea, extensive continental sediments were deposited on what is now the Colorado Plateau. The most prominent is the Navajo Sandstone, a wind-blown, white quartz sandstone that in places approaches a thickness of 300 meters (1000 feet). These massive dunes indicate that a major desert occupied much of the American Southwest during early Jurassic times.

A well-known Jurassic deposit is the Morrison Formation, within which is preserved the world's richest storehouse of dinosaur fossils. Included are fossilized bones of huge dinosaurs such as *Apatosaurus* (formerly *Brontosaurus*), *Brachiosaurus*, and *Stegosaurus*.

As the Jurassic period gave way to the Cretaceous, shallow seas once again invaded much of western North America, the Atlantic, and Gulf coastal regions. This created great swamps like those of the Paleozoic era, forming Cretaceous coal deposits that are very important economically in the western United States and Canada. For example, on the Crow Indian reservation in Montana, there exists nearly 20 billion tons of high-quality coal of Cretaceous age.

A major event of the Mesozoic era was the breakup of Pangaea (Figure 11.15). A rift developed between what is now the eastern United States and western Africa, marking the birth of the Atlantic Ocean. It also represents the beginning of the breakup of Pangaea, a process that continued for 200 million years, through the Mesozoic and into the Cenozoic.

As Pangaea fragmented, the westward-moving North American plate began to override the Pacific plate. This tectonic event began a continuous wave of deformation that moved inland along the entire western margin of the continent. By Jurassic times, subduction of the Pacific plate had begun to produce the chaotic mixture of rocks that exists today in the Coast Ranges of California. Further inland, igneous activity was widespread, and for nearly 60 million years, huge masses of magma rose to within a few miles of the surface, where they cooled and solidified. The remnants of this intrusive activity included the granitic rocks of the Sierra Nevada, the Idaho batholith, and the Coast Range batholith of British Columbia.

Tectonic activity that began in the Jurassic continued throughout the Cretaceous, ultimately forming the vast mountains of western North America (Figure 11.16). Compressional forces moved huge rock units in a shingle-like fashion toward the east. Throughout much of the western margin of North America, older rocks were thrust eastward over younger strata, for a distance exceeding 150 kilometers (90 miles).

Toward the end of the Mesozoic, the middle and southern ranges of the Rocky Mountains formed. This mountain-building event, called the Laramide Orogeny, resulted when tectonic forces uplifted large blocks of Precambrian rocks in what is today Colorado and Wyoming.

Mesozoic Life

As the Mesozoic era dawned, its life-forms were survivors of the great Paleozoic extinction. These survivors diversified in many new ways to fill the biological voids created at the close of the Paleozoic. On land, conditions favored those that could adapt to drier climates. Among plants, the gymnosperms were one such group. Unlike the first plants to invade the land, the seed-bearing gymnosperms did not depend on free-standing water for fertilization.

Consequently, these plants were not restricted to a life near the water's edge.

The gymnosperms quickly became the dominant trees of the Mesozoic. They included the cycads, the conifers, and the ginkgoes. The cycads resembled a large pineapple plant. The ginkgoes had pan-shaped leaves, much like their modern relatives. Largest were the conifers, whose modern descendants include the pines, firs, and junipers. The best-known fossil occurrence of these ancient trees is in northern Arizona's Petrified Forest National Park. Here, huge petrified logs lie exposed at the surface, having been weathered from rocks of the Triassic Chinle Formation (Figure 11.17).

The Shelled Egg. Among the animals, reptiles readily adapted to the drier Mesozoic environment. They were the first true terrestrial animals. Unlike amphibians, reptiles have shell-covered eggs that can be laid on land. The elimination of a water-dwelling stage (like the tadpole stage in frogs) was an important evolutionary step. Of interest is the fact that the watery fluid within the reptilian egg closely resembles seawater in chemical composition. Because the reptile embryo develops in this watery environment, the shelled egg has been characterized as a "private aquarium" in which the embryos of these land vertebrates spend their water-dwelling stage of life.

Dinosaurs Dominate. With the perfection of the shelled egg, reptiles quickly became the dominant land animals. They continued this dominance for more than 160 million years. Most awesome of the Mesozoic reptiles were the dinosaurs. Some of the huge dinosaurs were carnivorous (*Tyrannosaurus*), whereas others were herbivorous (like the ponderous *Apatosaurus* formerly *Brontosaurus*). The extremely long neck of *Apatosaurus* may have been an adaptation for feeding on tall conifer trees. However, not all dinosaurs were large. In fact, certain small forms closely resembled modern fleet-footed lizards. Further, evidence indicates that some dinosaurs, unlike their present-day reptile relatives, were warm blooded.

The reptiles made one of the most spectacular adaptive radiations in all of Earth history (Figure 11.18). One group, the pterosaurs, took to the air. These "dragons of the sky" possessed huge membranous wings that allowed them rudimentary flight (Figure 11.19). Another group of reptiles, exemplified by the fossil *Archaeopteryx*, led to more successful flyers: the birds. Whereas some reptiles took to the skies, others returned to the sea, including the fish-eating plesiosaurs and ichthyosaurs. These reptiles became proficient swimmers, but retained their reptilian teeth and breathed by means of lungs.

At the close of the Mesozoic, many reptile groups became extinct. Only a few types survived to recent

A. 200 Million Years Ago (Early Jurassic)

B. 150 Million Years Ago (Late Jurassic)

C. 100 Million Years Ago (Cretaceous)

D. 50 Million Years Ago (Early Cenozoic)

Figure 11.15 The breakup of the supercontinent of Pangaea began and continued throughout the Mesozoic era. (After D. Walsh and C. Scotese)

Figure 11.16 Uplifted and deformed sedimentary strata in the Canadian Rockies near Lake Louise, Alberta, Canada. (Photo by Carr Clifton/Minden Pictures)

times, including the turtles, snakes, crocodiles, and lizards. The huge land-dwelling dinosaurs, the marine plesiosaurs, and the flying pterosaurs all are known only through the fossil record. What caused this great extinction? (See Box 11.2.)

Cenozoic Era: Age of Mammals

The Cenozoic era, or "era of recent life," encompasses the past 65 million years of Earth history. It is the "postdinosaur" era, the time of mammals, including

Figure 11.17 Petrified logs that have been weathered out of the Triassic Chinle Formation in the Petrified Forest of Arizona. (Photo by David Muench)

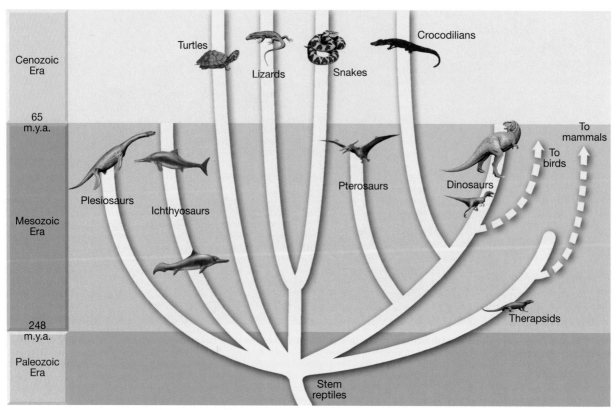

Figure 11.18 Origin and development of the major reptile groups and their descendants. Dinosaurs, pterosaurs (flying reptiles), and large marine reptiles all became extinct by the end of the Mesozoic.

humans. It is during this span that the physical landscapes and life-forms of our modern world came into being. The Cenozoic era represents a much smaller fraction of geologic time than either the Paleozoic or the Mesozoic. Although shorter, it nevertheless possesses a rich history, because the completeness of the geologic record improves as time approaches the present. The rock formations of this time span are more widespread and less disturbed than those of any preceding time.

The Cenozoic era is divided into two periods of very unequal duration, the Tertiary period and the Quaternary period. The Tertiary period includes five epochs and embraces about 63 million years, practically all of the Cenozoic era. The Quaternary period consists of two epochs that represent only the last 2 million years of geologic time.

Cenozoic North America

Most of North America was above sea level throughout the Cenozoic era. However, the eastern and western margins of the continent experienced markedly contrasting events, because of their different relationships with plate boundaries. The Atlantic and Gulf coastal regions, far removed from an active plate boundary, were tectonically stable. In contrast, western North America was the leading edge of the North American plate. As a result, plate interactions during the Cenozoic gave rise to many events of mountain building, volcanism, and earthquakes in the West.

Figure 11.19 Fossils of the great flying reptile Pteranodon have been recovered from Cretaceaous chalk deposits located in Kansas. Although Pteranodon had a wingspan of 7 meters (22 feet), flying reptiles with twice this wingspan have been discovered in strata of west Texas.

Box 11.2   Earth as a System: Demise of the Dinosaurs

Review for a moment the geologic time scale, Figure 11.5. You can see that its divisions represent times of significant geological and/or biological change. Of special interest is the boundary between the Mesozoic era ("middle life") and Cenozoic era ("recent life"), about 65 million years ago. Around this time, more than half of all plant and animal species died out in a *mass extinction*. This boundary marks the end of the era in which dinosaurs and other reptiles dominated the landscape and the beginning of the era when mammals become very important (Figure 11.A). Because the last period of the Mesozoic is the Cretaceous (abbreviated K to avoid confusion with other "C" periods), and the first period of the Cenozoic is the Tertiary (abbreviated T), the time of this mass extinction is called the *Cretaceous-Tertiary* or *KT boundary*.

The extinction of the dinosaurs is generally attributed to this group's inability to adapt to some radical change in environmental conditions. What event could have triggered the rapid extinction of the dinosaurs—one of the most successful groups of land animals ever to have lived?

The most strongly supported hypothesis proposes that about 65 million years ago, a large meteorite about 10 kilometers in diameter collided with Earth (Figure 11.B). The speed at impact was about 70,000 kilometers per hour! The force of the impact vaporized the meteorite and trillions of tons of Earth's crust. Huge quantities of dust and other metamorphosed debris was blasted high into the atmosphere. For months the encircling dust cloud would have greatly restricted the sunlight reaching Earth's surface. With insufficient sunlight for photosynthesis, delicate food chains would have collapsed. Large plant-eating dinosaurs would be affected more adversely than smaller life-forms because of the tremendous volume of vegetation they consumed. Next would come the demise of large carnivores. When the sunlight returned, more than half of the species on Earth, including numerous marine organisms, had become extinct.

This period of mass extinction appears to have affected all land animals larger than dogs. Supporters of

Figure 11.A Dinosaurs dominated the Mesozoic landscape until their extinction at the close of the Cretaceous period. This skeleton of *Tyrannosaurus* stands on display in Chicago's Field Musuem of Natural History. (Photo by Don and Pat Valenti/DRK Photo)

this catastrophic-event scenario suggest that small ratlike mammals were able to survive a breakdown in their food chains lasting perhaps several months. Large dinosaurs, they contend, could not survive this event. The loss of these large reptiles opened habitat for the small mammals that remained. This new habitat, along with evolutionary forces, led to the development of the large mammals that occupy our modern world.

What evidence points to such a catastrophic collision 65 million years ago? First, a thin layer of sediment nearly 1 centimeter thick has been discovered at the KT boundary, worldwide. This sediment contains a high level of the element *iridium*, rare in Earth's crust but found in high proportions in stony meteorites. Could this layer be the scattered remains of the meteorite that was responsible for the environmental changes that led to the demise of many reptile groups?

Despite its growing support, some scientists disagree with the impact hypothesis. They suggest instead that huge volcanic eruptions led to the breakdown in the food chain. To support their hypothesis they cite enormous outpourings of lavas in the Deccan Plateau of northern India about 65 million years ago.

Whatever caused the KT extinction, we now have a greater appreciation of the role of catastrophic events in shaping the history of our planet and the life that occupies it. Could a catastrophic event having similar results occur today? This possibility may explain why an event that occurred 65 million years ago has captured the interest of so many.

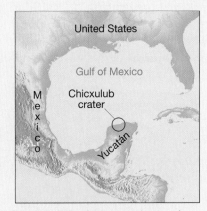

Figure 11.B Employing gravitational measurements, scientists have located a giant impact crater that formed about 65 million years ago and has since been filled with sediment. About 180 kilometers in diameter, Chicxulub crater is regarded by some researchers as the impact site that resulted in the demise of the dinosaurs.

Eastern North America. The stable continental margin of eastern North America was the site of abundant marine sedimentation. The most extensive deposition surrounded the Gulf of Mexico, from the Yucatan Peninsula to Florida. Here, the great buildup of sediment caused the crust to downwarp and produced numerous faults. In many instances, the faults created traps in which oil and natural gas accumulated. Today, these and other petroleum traps are the most economically important resource in Cenozoic strata of the Gulf Coast, as evidenced by the Gulf's well-known off-shore drilling platforms.

By early Cenozoic time, most of the original Appalachians had been eroded to a low plain. Then, by the mid-Cenozoic, isostatic adjustments raised the region once again, rejuvenating its rivers. Streams eroded with renewed vigor, gradually sculpturing the surface into its present-day topography (Figure 11.20). Sediments from all of this erosion were deposited along the eastern margin of the continent, where they attained a thickness of many kilometers. Today, portions of the strata deposited during the Cenozoic are exposed as the gently sloping Atlantic and Gulf coastal plains.

Western North America. In the West, the Laramide Orogeny, which built the Rocky Mountains, was coming to an end. As erosion lowered the mountains, the basins between uplifted ranges filled with sediments. Eastward, a great wedge of sediment from the eroding Rockies was building, creating the Great Plains.

Beginning in the Miocene epoch, a broad region from northern Nevada into Mexico experienced crustal movements that formed more than 150 fault-block mountain ranges. They rise abruptly above the adjacent basins, creating the Basin and Range Province (see Chapter 9).

As the Basin and Range Province was forming, the entire western interior of the continent was gradually uplifted. This uplift reelevated the Rockies and rejuvenated many of the West's major rivers. As the rivers became entrenched, many spectacular gorges were formed, including the Grand Canyon of the Colorado River, the Grand Canyon of the Snake River, and the Black Canyon of the Gunnison. The present topography of the Rocky Mountains is in large measure the result of this late Tertiary uplift and the subsequent

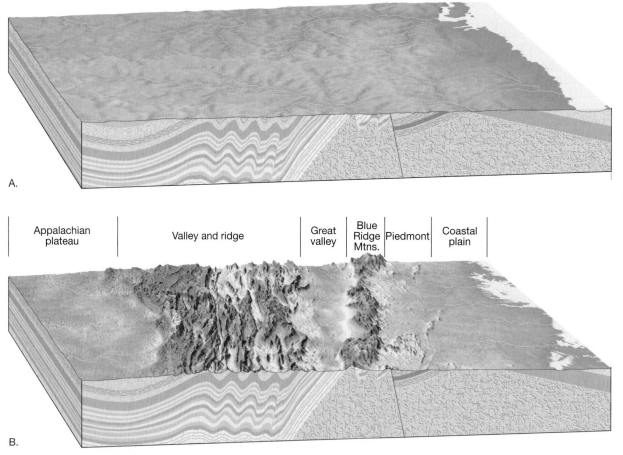

Figure 11.20 The formation of the modern Appalachian Mountains. **A.** The original Appalachians eroded to a low plain. **B.** The recent upwarping and erosion of the Appalachians began nearly 30 million years ago to produce the present topography.

Figure 11.21 Wyoming's spectacular Teton Range is part of the Rocky Mountains. They were uplifted during the late Tertiary, about 10 million years ago. (Photo by Tom Till)

excavation of the early Tertiary basin deposits by rejuvenated streams (Figure 11.21).

Volcanic activity was common in the West during much of the Cenozoic. Beginning in the Miocene epoch, great volumes of fluid basaltic lava flowed from fissures in portions of present-day Washington, Oregon, and Idaho. These eruptions built the extensive (1.3 million square kilometers) Columbia Plateau. Immediately west of the Columbia Plateau, volcanic activity was quite different. Here, thicker magmas with a higher silica content erupted explosively, creating a chain of stratovolcanoes from northern California to the Canadian border, some of which remain active—like Mount St. Helens (Figure 11.22).

A final episode of folding occurred in the West in late Tertiary time, creating the Coast Ranges, which stretch along the Pacific Coast. Meanwhile, the Sierra Nevada became fault-block mountains as they were uplifted along their eastern flank, creating the imposing mountain front we know today.

As the Tertiary period drew to a close, the effects of mountain building, volcanic activity, isostatic adjustments, and extensive erosion and sedimentation had created a physical landscape very similar to the configuration of today. All that remained of Cenozoic time was the final 2-million-year episode called the Quaternary period. During this most recent (and current) phase of Earth history, in which humans evolved, the action of glacial ice and other erosional agents added the finishing touches.

Cenozoic Life

Mammals replaced reptiles as the dominant land animals in the Cenozoic. Angiosperms (flowering plants with covered seeds) replaced gymnosperms as the dominant land plants. Marine invertebrates took on a modern look. Microscopic animals called *foraminifera* became especially important. Today foraminifera are among the most intensely studied of all fossils because their widespread occurrence makes them invaluable in correlating Tertiary sediments. Tertiary strata are very important to the modern world, for they yield more oil than do rocks of any other age.

The Cenozoic is often called the "age of mammals," because these animals came to dominate land life. It could also be called the "age of flowering plants," for the angiosperms enjoy a similar status in the plant world. As a result of advances in seed fertilization and dispersal, angiosperms experienced rapid development and expansion as the Mesozoic drew to a close. Thus, as the Cenozoic era began, angiosperms were already the dominant land plants.

Development of the flowering plants, in turn, strongly influenced the evolution of both birds and

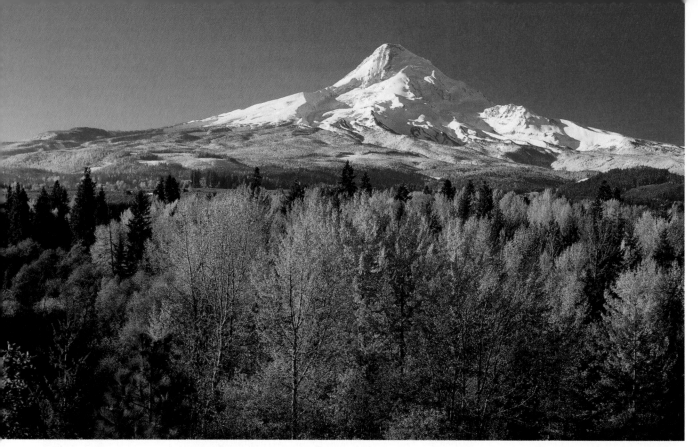

Figure 11.22 Mount Hood, Oregon. This dormant volcano is one of several large composite cones that comprise the Cascade Range. (Photo by John M. Roberts/The Stock Market)

mammals. Birds that feed on seeds and fruits, for example, evolved rapidly during the Cenozoic in close association with the flowering plants. During the middle Tertiary, grasses developed rapidly and spread over the plains. This fostered the emergence of herbivorous (plant-eating) mammals that were mainly grazers. In turn, the development and spread of grazing animals established the setting for the evolution of the carnivorous mammals that preyed upon them.

Mammals Replace Reptiles. Back in the Mesozoic, an important evolutionary event was the appearance of primitive mammals in the late Triassic, about the same time that the dinosaurs emerged. Yet throughout the period of dinosaur dominance, mammals remained in the background as small and inconspicuous animals. By the close of the Mesozoic era, dinosaurs and other reptiles no longer dominated the land. It was only after these large reptiles became extinct that mammals came into their own as the dominant land animals. The transition is a major example in the fossil record of the replacement of one large group by another.

Mammals are distinct from reptiles in important respects. Mammalian young are born live, and mammals maintain a steady body temperature—that is, they are "warm-blooded." This latter adaptation allowed mammals to lead more active and diversified

lives than reptiles because they could survive in cold regions and search for food during any season or time of day. Other mammalian adaptations included the development of insulating body hair and more efficient heart and lungs.

It is worth noting that perfect boundaries cannot be drawn between mammalian and reptilian traits. One minor group of animals, the monotremes, still lay eggs. The two species in this group, the duck-billed platypus and the spiny anteater, are found only in Australia. Moreover, although modern reptiles are "cold-blooded," some paleontologists believe that dinosaurs may have been "warm-blooded."

With the demise of most Mesozoic reptiles, Cenozoic mammals diversified rapidly. The many forms that exist today evolved from small primitive mammals that were characterized by short legs, flat five-toed feet, and small brains. Their development and specialization took four principal directions: (1) increase in size, (2) increase in brain capacity, (3) specialization of teeth to better accommodate a particular diet, and (4) specialization of limbs to better equip the animal for life in a particular environment.

Marsupials, Placentals, and Diversity. Following the reptilian extinctions at the close of the Mesozoic, two groups of mammals, the marsupials and the placentals, evolved and expanded to dominate the Cenozoic. The groups differ principally in their modes of reproduction. Young marsupials are born live but at a very early stage of development. After birth, the tiny and immature young crawl into the mother's external stomach pouch to complete their development. Examples are kangaroos, opossums, and koala bears.

Placental mammals, in contrast, develop within the mother's body for a much longer period, so that birth occurs after the young are relatively mature and independent. Most mammals are placental, including humans.

Today marsupials are found primarily in Australia, where they went through a separate evolutionary expansion during the Cenozoic, largely isolated from placental mammals.

In South America, both primitive marsupials and placentals coexisted before that landmass became completely isolated during the breakup of Pangaea. Evolution and specialization of both groups continued undisturbed for approximately 40 million years until the close of the Pliocene epoch, when the Central American land bridge emerged, connecting the two American continents. Then an invasion of advanced carnivores from North America brought the extinction of many hoofed mammals that had persisted in South America for millions of years. The marsupials, except for opossums, also could not compete and became extinct. Both Australia and South America provide excellent examples of how isolation caused by the separation of continents increased the diversity of animals in the world.

Large Mammals and Extinction. As we have seen, mammals diversified quite rapidly during the Cenozoic era. One tendency was for some groups to became very large. For example, by the Oligocene epoch a hornless rhinoceros that stood nearly 5 meters high (16 feet) had evolved. It is the largest land mammal known to have existed. As time approached the present, many other types evolved to a large size as well; more, in fact, than now exist. Many of these large forms were common as recently as 11,000 years ago. However, a wave of late Pleistocene extinctions rapidly eliminated these animals from the landscape.

In North America, the mastodon and mammoth, both huge relatives of the elephant, became extinct. In addition, saber-toothed cats, giant beavers, large ground sloths, horses, camels, giant bison, and others died out (Figure 11.23). In Europe, late Pleistocene extinctions included woolly rhinos, large cave bears, and the Irish elk. The reason for this recent wave of large animal extinctions puzzles scientists. These animals had survived several major glacial advances and interglacial periods, so it is difficult to ascribe these extinctions to climatic change. Some scientists believe that early humans hastened the decline of these mammals by selectively hunting large forms. Although this hypothesis is preferred by many, it is not yet accepted by all.

Figure 11.23 Partially exposed skeleton of an extinct saber-toothed cat. This lion-sized cat roamed the White River Badlands of South Dakota during the Oligocene epoch. (Photo by T.A. Wiewandt/DRK Photo)

Chapter Summary

- The *nebular hypothesis* describes the formation of the solar system. The planets and Sun began forming about 5 billion years ago from a large cloud of dust and gases composed of hydrogen and helium, with only a small percentage of all the other heavier elements. As the cloud contracted, it began to rotate and assume a disk shape. Material that was gravitationally pulled toward the center became the *protosun*. Within the rotating disk, small centers, called *protoplanets*, swept up more and more of the cloud's debris. Because of their high temperatures and weak gravitational fields, the inner planets (Mercury, Venus, Earth, and Mars) were unable to accumulate and retain many of the lighter components (hydrogen, helium, ammonia, methane, and water). However, because of the very cold temperatures existing far from the Sun, the fragments from which the large outer planets (Jupiter, Saturn, Uranus, and Neptune) formed consisted of huge amounts of hydrogen and other light materials. These gaseous substances account for the comparatively large sizes and low densities of the outer planets.

- The decay of radioactive elements and heat released by colliding particles aided the melting of Earth's interior, allowing the denser elements, principally iron and nickel, to sink to its center. As a result of this *differentiation,* Earth's interior consists of shells or spheres of materials, each having rather distinct properties.

- Earth's primitive atmosphere probably consisted of such gases as water vapor, carbon dioxide, nitrogen, and several trace gases that were released in volcanic emissions, a process called *outgassing*. The first life-forms on Earth, probably *anaerobic bacteria*, did not require oxygen. As life evolved, plants, through the process of *photosynthesis*, used carbon dioxide and water and released oxygen into the atmosphere. Once the available iron on Earth was oxidized (combined with oxygen), substantial quantities of oxygen accumulated in the atmosphere. About 4 billion years into Earth's existence, the fossil record reveals abundant ocean-dwelling organisms that require oxygen to live.

- The *Precambrian* spans about 88% of Earth's history, beginning with the formation of Earth about 4.5 billion years ago and ending 544 million years ago with the diversification of life that marks the start of the Paleozoic era. It is the least understood span of Earth's history because most Precambrian rocks are buried from view. However, on each continent there is a "core area" of Precambrian rocks called the *shield.* The iron ore deposits of Precambrian age represent the time when oxygen became abundant in the atmosphere and combined with iron to form iron oxide. The most common middle Precambrian fossils are *stromatolites.* Microfossils of bacteria and blue-green algae, both primitive *prokaryotes* whose cells lack organized nuclei, have been found in chert, a hard, dense, chemical sedimentary rock in southern Africa (3.1 billion years of age) and near Lake Superior (1.7 billion years of age). *Eukaryotes,* with cells containing organized nuclei, are among billion-year-old fossils discovered in Australia. Plant fossils date from the middle Precambrian, but animal fossils came a bit later, in the late Precambrian. Many of these fossils are *trace fossils,* and not of the animals themselves.

- The *Paleozoic era* extends from 544 million years ago to about 245 million years ago. The beginning of the Paleozoic is marked by the *appearance of the first life-forms with hard parts* such as shells. Therefore, abundant Paleozoic fossils occur and a far more detailed record of Paleozoic events can be constructed. During the early Paleozoic (the Cambrian, Ordovician, and Silurian periods) the vast southern continent of *Gondwanaland* existed. Seas inundated and receded from North America several times, leaving thick evaporite beds of rock salt and gypsum. Life in the early Paleozoic was restricted to the seas and consisted of several invertebrate groups. During the late Paleozoic (the Devonian, Mississippian, Pennsylvanian, and Permian periods), ancestral North America collided with Africa to produce the original northern Appalachian Mountains, and the northern continent of *Laurasia* formed. By the close of the Paleozoic, all the continents had fused into the supercontinent of *Pangaea.* During most of the Paleozoic, organisms diversified dramatically. Insects and plants moved onto the land, and amphibians evolved and diversified quickly. By the Pennsylvanian period, large tropical swamps, which became the major coal deposits of today, extended across North America, Europe, and Siberia. At the close of the Paleozoic, altered climatic conditions caused one of the most dramatic biological declines in all of Earth's history.

- The *Mesozoic era,* often called the *"age of dinosaurs,"* began about 245 million years ago and ended approximately 65 million years ago. Early in the Mesozoic much of the land was above sea level. However, by the middle Mesozoic, seas invaded western North America. As Pangaea began to break up, the westward-moving North American plate began to override the Pacific plate, causing crustal deformation along the entire western margin of the continent. Organisms that had survived extinction at the end of the Paleozoic began to diversify in spectacular ways. *Gymnosperms* (cycads, conifers, and ginkgoes) quickly became the dominant trees of the Mesozoic because they could adapt to the drier climates. Reptiles quickly became the dominant land animals, with one group eventually becoming the birds. The most awesome of the Mesozoic reptiles were the *dinosaurs.* At the close of the Mesozoic, many reptile groups, including the dinosaurs, became extinct.

- The *Cenozoic era,* or *"era of recent life,"* began approximately 65 million years ago and continues today. It is *the time of mammals,* including humans. The widespread, less disturbed rock formations of the Cenozoic provide a rich geologic record. Most of North America was above sea level throughout the Cenozoic. Because of their different relations with tectonic plate boundaries, the eastern and western margins of the North American continent experienced contrasting events. The stable eastern margin was the site of abundant sedimentation as isostatic adjustment raised the Appalachians, causing the streams to erode with renewed vigor and deposit their sediment along the continental margin. In the west, building of the Rocky Mountains (the *Laramide Orogeny*) was coming to an end, the Basin and Range Province was forming, and volcanic activity was extensive. The Cenozoic is often called *"the age of mammals"* because these animals replaced the reptiles as the dominant land life. Two

groups of mammals, the *marsupials* and the *placentals,* evolved and expanded to dominate the era. One tendency was for some mammal groups to become very large. However, a wave of late *Pleistocene* extinctions rapidly eliminated these animals from the landscape. Some scientists believe that early humans hastened the decline of these animals by selectively hunting the larger species. The Cenozoic could also be called the *"age of flowering plants."* As a source of food, flowering plants strongly influenced the evolution of both birds and herbivores (plant-eating) mammals throughout the Cenozoic era.

Key Terms

nebular hypothesis (p. 296)
outgassing (p. 298)

shields (p. 299)
stromatolites (p. 299)

Review Questions

1. Briefly describe the events believed to have led to the formation of the solar system.
2. What is the major source of free oxygen in Earth's atmosphere?
3. Explain why Precambrian history is more difficult to decipher than more recent geological history.
4. Match the following words and phrases to the most appropriate time span. Select among the following: Precambrian, early Paleozoic, late Paleozoic, Mesozoic, Cenozoic.

   (a) Pangaea came into existence.
   (b) Bacteria and blue-green algae preserved in chert.
   (c) The era that encompasses the least amount of time.
   (d) Shields.
   (e) "Age of dinosaurs."
   (f) Formation of the original northern Appalachian mountains.
   (g) Mastodons and mammoths.
   (h) Extensive deposits of rock salt.
   (i) Triassic, Jurassic, and Cretaceous.
   (j) Coal swamps extended across North America, Europe, and Siberia.
   (k) Gulf Coast oil deposits formed.
   (l) Formation of most of the world's major iron ore deposits.
   (m) Massive sand dunes covered large portions of the Colorado Plateau region.

   (n) The "age of fishes" occurred during this span.
   (o) Cambrian, Ordovician, and Silurian.
   (p) Pangaea began to break apart.
   (q) "Age of mammals."
   (r) Animals with hard parts first appeared in abundance.
   (s) Gymnosperms were the dominant trees.
   (t) Columbia Plateau formed.
   (u) Stromatolites are among its more common fossils.
   (v) "Golden age of trilobites" occurred during this span.
   (w) Fault-block mountains form in the Basin and Range Province.

5. Briefly discuss two proposals that attempt to explain why several groups developed hard parts at the beginning of the Cambrian period. Do these proposals appear to provide a satisfactory explanation? Why or why not?
6. List some differences between amphibians and reptiles. List differences between reptiles and mammals.
7. Describe one hypothesis that attempts to explain the extinction of the dinosaurs. Cite a major objection to this proposal.
8. Contrast the eastern and western margins of North America during the Cenozoic era in terms of their relationships to plate boundaries.

Examining the Earth System

1. The Earth system has been responsible for both the conditions that favored the evolution of life on this planet and for the mass extinctions that have occurred throughout geological time. Describe the role of the biosphere, hydrosphere, and solid Earth in forming the current level of atmospheric oxygen. How did Earth's near-space environment interact with the atmosphere and biosphere to contribute to the great mass extinction that marked the end of the dinosaurs?

2. Most of the vast North American coal resources located from Pennsylvania to Illinois began forming during the Pennsylvanian and Mississippian periods of Earth history. (This time period is also referred to as the Carboniferous period.) Using Figure 11.14, a restoration of a Pennsylvanian period coal swamp, describe the climatic and biological conditions associated with this unique environment. Next, examine Figure 11.11B, C.

The maps show the formation of the supercontinent of Pangaea and illustrate the geographic position of North America during the period of coal formation. Where, relative to the equator, was North America located during the time of coal formation? What role did plate tectonics play in determining the conditions that eventually produced North America's eastern coal reserves? Why is it unlikely that the coal-forming environment will repeat itself in North America in the near future? (You may find it helpful to visit the Illinois State Museum Mazon Creek Fossils exhibit at **http://www.museum.state.il.us/exhibits/mazon_creek**, the University of California Time Machine Exhibit at **http://www.ucmp.berkeley.edu/help/timeform.html**, and/or the Kentucky Coal Council's Coal Education Web site at **http://www.coaleducation.org/**)

Web Resources

The following are informative and interesting World Wide Web sites that address topics related to those presented in the chapter:

Extinctions Fossil Company:
- **http://www.extinctions.com/**

Royal Tyrrell Museum, Canada:
- **http://tyrrell.magtech.ab.ca/home.html**

For direct links to these sites and other appropriate Earth science Web locations, chapter review exercises, and relevant, topical investigations that utilize Web resources, contact the *Earth Science*, 9th Edition Home Page at:
- **http://www.prenhall.com/tarbuck.**

Part Two

The Oceans

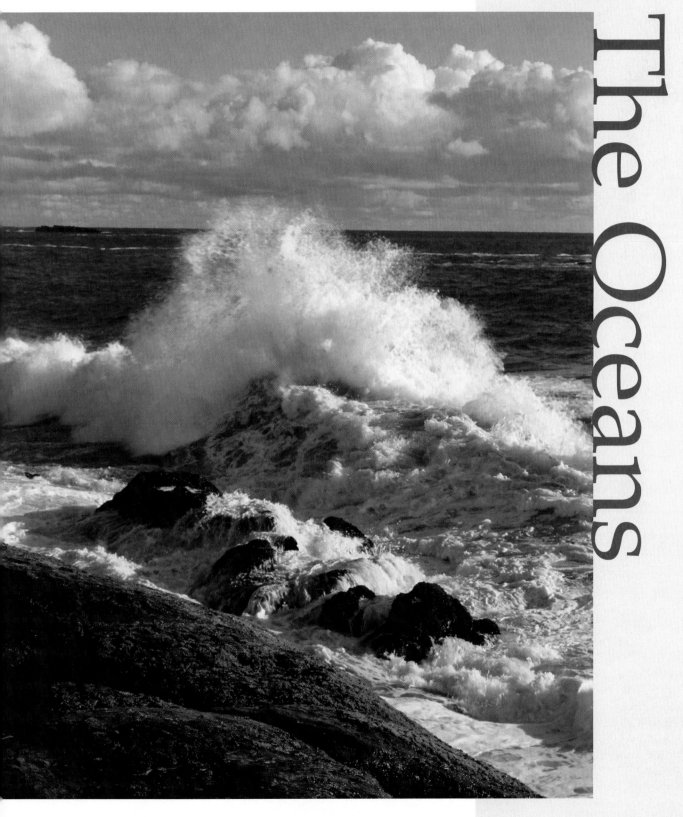

Waves breaking along the Oregon coast.
(Photo by Craig Tuttle/The Stock Market)

Chapter 12

The Ocean Environment

This reef is in the shallow tropical waters off the coast of Papua, New Guinea. (Photo by Steve Wolper/DRK Photo)

How deep is the ocean? How much of Earth is covered by the global sea? These are basic questions, but they went unanswered for thousands of years. People's interest in the oceans undoubtedly dates back to ancient times, but it was not until rather recently that these seemingly simple questions began to be answered. The beginning of this chapter deals with these answers. Then, after examining the composition of seawater and the resources extracted from it, the remainder of the chapter provides a look at Earth beneath the sea. Suppose that all of the water were drained from the ocean. What would we see? Plains? Mountains? Canyons? Plateaus? Indeed, the ocean conceals all of these features, and more. In fact, the topography of the ocean floor is nearly as varied as that of the continents. And what about the carpet of sediment that covers much of the seafloor? Where did it come from and what can we learn by examining it?

Calling Earth the "water planet" is certainly appropriate, because nearly 71 percent of its surface is covered by the global ocean. Although the ocean comprises a much greater percentage of Earth's surface than the continents, it was only in the relatively recent past that the ocean became an important focus of study. Recently there has been a virtual explosion of data about the oceans, and with it, oceanography has grown dramatically. **Oceanography** is a composite science that draws on the methods and knowledge of biology, chemistry, physics, and geology to study all aspects of the world ocean.

The Vast World Ocean

A glance at a globe or a view of Earth from space reveals a planet dominated by the world ocean (Figure 12.1). Indeed, it is for this reason that Earth is often referred to as the *blue planet*. The area of Earth is about 510 million square kilometers (197 million square miles). Of this total, approximately 360 million square kilometers (140 million square miles), or 71 percent, is represented by oceans and marginal seas (meaning seas around the ocean's margin, like the Mediterranean Sea and Caribbean Sea). Continents and islands comprise the remaining 29 percent, or 150 million square kilometers (58 million square miles).

By studying a globe or world map, it is readily apparent that the continents and oceans are not evenly divided between the Northern and Southern hemispheres (Figure 12.1). When we compute the percentages of land and water in the Northern Hemisphere, we find that nearly 61 percent of the surface is water, whereas about 39 percent is land. In the Southern Hemisphere, on the other hand, almost 81 percent of the surface is water, and only 19 percent is land. It is no

Figure 12.1 These views of Earth show the uneven distribution of land and water between the Northern and Southern hemispheres. Almost 81 percent of the Southern Hemisphere is covered by the oceans—20 percent more than the Northern Hemisphere.

wonder then that the Northern Hemisphere is called the *land hemisphere*, and the Southern Hemisphere the *water hemisphere*.

Figure 12.2 shows the distribution of land and water in the Northern and Southern hemispheres. Between latitudes 45 degrees north and 70 degrees north, there is actually more land than water, whereas between 40 degrees south and 65 degrees south there is almost no land to interrupt the oceanic and atmospheric circulation.

The volume of the ocean basins is many times greater than the volume of the continents above sea level. In fact, the volume of all land above sea level is only 1/18 that of the ocean.

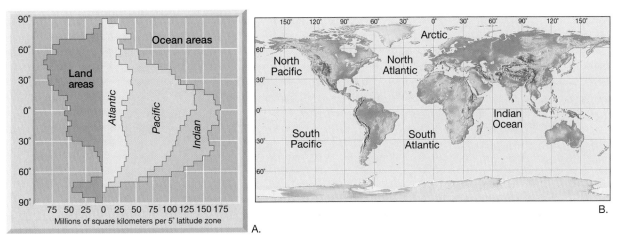

Figure 12.2 Distribution of land and water. **A.** The graph shows the amount of land and water in each 5-degree latitude belt. **B.** The world map provides a more familiar view.

An obvious difference between continents and the ocean basins is their relative levels. The average elevation of the continents above sea level is about 840 meters (2750 feet), whereas the average depth of the oceans is more than 4.5 times this figure—3800 meters (12,500 feet). If Earth's solid mass were perfectly smooth (level) and spherical, the oceans would cover it to a uniform depth of more than 2000 meters (1.2 miles).

A comparison of the three major oceans shows that the Pacific is by far the largest; it is nearly as large as the Atlantic and Indian oceans combined (Figure 12.2). The Pacific contains slightly more than half of the water in the world ocean, and because it includes few shallow seas along its margins, it has the greatest average depth—3940 meters (12,900 feet).

The Atlantic, bounded by almost parallel continental margins, is a relatively narrow ocean when compared to the Pacific. When the Arctic Ocean is included, the Atlantic has the greatest north–south extent and connects the two polar regions. Because the Atlantic has many shallow adjacent seas, including the Caribbean, Gulf of Mexico, Baltic, and Mediterranean, as well as wide continental shelves along its borders, it is the shallowest of the three oceans, with an average depth of 3310 meters (10,860 feet).

As Figure 12.2 shows, the Indian Ocean covers less area than the Pacific or the Atlantic. Unlike the others, it is largely a Southern Hemisphere water body.

Composition of Seawater

Seawater is a complex solution of salts, consisting of about 3.5 percent (by weight) dissolved mineral substances. Although the percentage of salts may seem small, the actual quantity is huge. If all of the water were evaporated from the oceans, a layer of salt approaching 60 meters (200 feet) thick would cover the entire ocean floor.

Salinity and Its Variations

Salinity is the proportion of dissolved salts to pure water. In many things, we express proportion in parts-per-hundred, or percent, %. But because the proportion of salts in seawater is a small number, scientists express salinity in parts-per-thousand, noted ‰. Thus, the average salinity of the ocean is about 35‰.

The principal elements that contribute to the ocean's salinity are shown in Figure 12.3. If we made our own seawater, we could come reasonably close by following the recipe shown in Table 12.1. From this table it is evident that most of the salt is sodium chloride (common table salt). Sodium chloride together with the next four most abundant salts comprise 99 percent of the salt in the sea. Although only eight elements make up these five most abundant salts, seawater contains more than 70 of Earth's other naturally occurring elements. Despite their presence in minute quantities, many of these elements are very important in maintaining the necessary chemical environment for life in the sea.

The relative abundances of the major components in sea salt are essentially constant, no matter where the ocean is sampled. Variations in salinity, therefore, are primarily a consequence of changes in the water content of the solution. As a result, high salinities are found where evaporation is high, as is the case in the dry subtropics. Conversely, where heavy precipitation dilutes ocean waters, as in the mid-latitudes and near the equator, lower salinities prevail (Figure 12.4).

Whereas salinity variations in the open ocean normally range from 33‰ to 37‰, some marginal seas demonstrate extraordinary extremes. For example, in

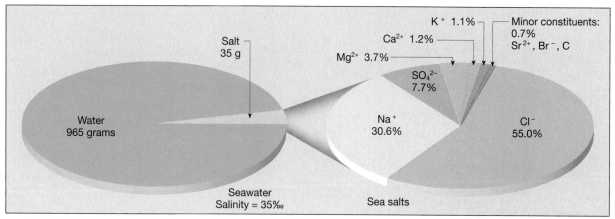

Figure 12.3 Relative proportions of water and dissolved salts in seawater. Ions shown by chemical symbol are chlorine (Cl), sodium (Na), sulfate (SO₄), magnesium (Mg), calcium (Ca), potassium (K), strontium (Sr), bromine (Br), and carbon (C).

the restricted waters of the Persian Gulf and the Red Sea, where evaporation far exceeds precipitation, salinity may exceed 42‰. Conversely, very low salinities occur where large quantities of freshwater are supplied by rivers and precipitation. Such is the case for the Baltic Sea, where salinity is often below 10‰.

Sources of Sea Salts

What are the primary sources for the vast quantities of salts in the ocean? Chemical weathering of rocks on the continents is one source. These soluble materials are delivered to the oceans by streams at an estimated rate of more than 2.5 billion tons annually. The second major source of elements found in ocean water is Earth's interior. Through volcanic eruptions, large quantities of water and dissolved gases have been emitted during much of geologic time. This process, called **outgassing**, is the principal source of water in the oceans as well as in the atmosphere. Certain elements, notably chlorine, bromine, sulfur, and boron, are much more abundant in

the ocean than in Earth's crust. Because they result from outgassing, their abundance in the sea tends to confirm the hypothesis that volcanic action is largely responsible for the present oceans.

Although rivers and volcanic activity continually contribute salts to the oceans, the salinity of seawater is not increasing. In fact, many oceanographers believe that the composition of seawater has been relatively stable for a large span of geologic time. Why doesn't the sea get saltier? Obviously, material is being removed just as rapidly as it is added. Some elements are withdrawn from seawater by plants and animals as they build hard parts. Others are removed when they chemically precipitate out of the water as sediment. The net effect is that the overall makeup of seawater remains relatively constant.

Table 12.1 Recipe for artificial seawater.

MIX:

Sodium chloride (NaCl)	23.48 grams
Magnesium chloride (MgCl₂)	4.98
Sodium sulfate (Na₂SO₄)	3.92
Calcium chloride (CaCl₂)	1.10
Potassium chloride (KCl)	0.66
Sodium bicarbonate (NaHCO₃)	0.192
Potassium bromide (KBr)	0.096
Hydrogen borate (H₃BO₃)	0.026
Strontium chloride (SrCl₂)	0.024
Sodium fluoride (NaF)	0.003

ADD:

Water (H₂O) to form 1000 grams of solution.

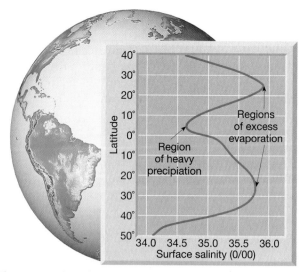

Figure 12.4 The subtropics are dry. On the continents, deserts such as the Sahara occur in this latitude belt. Over the oceans, high evaporation rates remove more water than is replaced by the meager rainfall. Thus, surface salinities are high here. By contrast, near the equator rainfall is heavy; therefore, surface salinities are reduced here.

Resources From Seawater

You have seen that the ocean is a great storehouse of dissolved materials. Some are presently being extracted from seawater, whereas others remain as untapped potential resources. Commercial products obtained from seawater include common salt (sodium chloride), the lightweight metal magnesium, and bromine, which is used primarily in gasoline additives and in the manufacture of fireproofing materials. For many years seawater provided a high percentage of the world's magnesium and bromine. In the late 1960s, 70 percent of the world's bromine and 61 percent of the world's magnesium came from the chemical processing of seawater. Since that time, the production of these two elements has shifted away from seawater for economic reasons. Today the primary sources of magnesium and bromine are water from Utah's Great Salt Lake, and brine wells and evaporite deposits in the United States and other areas. In countries lacking brine and evaporate deposits, such as Japan and the Scandinavian countries, seawater is still a source of these elements.

Ever since ancient times the ocean has been an important source of salt for human consumption, and the sea remains a significant supplier. To obtain pure sodium chloride, evaporating ponds are used. The idea is that different salts precipitate from seawater at different salinities. In the first pond, evaporation proceeds until the salt concentration is such that calcium carbonate and calcium sulfate precipitate. Then the remaining brine is transferred to another pond where evaporation continues. Practically pure sodium chloride precipitates there. When most sodium chloride has precipitated, the remaining liquid (now enriched in potassium, bromine, and magnesium) is drained off and the salt is harvested (Figure 12.5). In this manner about 30 percent of the world's salt is produced.

One of the resources derived from seawater is freshwater. The removal of salts and other chemicals from seawater is termed **desalination**. The purpose of this process is to extract low-salinity ("fresh") water suitable for drinking, industry, and agriculture from water that is so saline that it is not suitable for these purposes. Although hundreds of desalination plants are now operating and others are being planned, the cost of desalinated water is still high, and the total production remains relatively small.

Although freshwater produced by various desalination technologies is becoming more important as a source of water for human and even industrial use, it is unlikely to be an important supply for agricultural purposes. Because human water requirements are quite low, the cost of supplying areas that do not have adequate natural supplies can often be justified. In addition, the cost of water may represent only a small

Figure 12.5 Harvesting salt produced by the evaporation of seawater. (Photo by William E. Townsend, Jr./Photo Researchers, Inc.)

fraction of the ultimate cost of many industrial goods and hence can be absorbed even in competitive markets. However, the cost of water for irrigation represents a large percentage of the total cost of the crop being produced. For this reason, the cost of desalinated water is still much too high for crop production. Consequently, making the deserts "bloom" by irrigating them with freshwater produced by desalination processes is only a dream and, for economic reasons, is likely to remain so for the foreseeable future.

The Ocean's Layered Structure

By sampling ocean waters, oceanographers have found that temperature and salinity vary with depth. Generally, they recognize a three-layered structure in the open ocean: a shallow surface mixed zone, a transition zone, and a deep zone (Figure 12.6).

Because solar energy is received at the ocean surface, it is here that water temperatures are warmest. The mixing of these waters by waves as well as the turbulence from currents creates a rapid vertical heat transfer. Hence, this mixed surface zone has nearly uniform temperatures. The thickness and temperature of this layer vary, depending on latitude and season. The zone may attain a thickness of 450 meters (1500 feet) or more, and temperatures typically range between 21°C and 26°C (70°F and 80°F). In equatorial latitudes, the temperatures may be even higher.

Below the Sun-warmed zone of mixing, the temperature falls abruptly with depth (Figure 12.7). This layer of rapid temperature change, known as the **thermocline**, marks the transition between the warm surface layer and the deep zone of cold water below. Below the thermocline temperatures fall only a few more degrees. At depths greater than about 1500 meters (5000 feet),

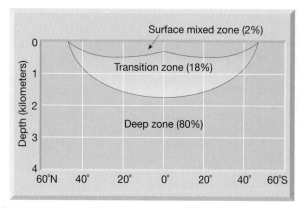

Figure 12.6 Oceanographers recognize three "layers" of the ocean based on temperature and salinity. The thickness of each varies with latitude, as shown here, and with the season of the year. Clearly, most of the ocean's water is in the deep zone.

ocean-water temperatures are consistently below 4°C (39°F). Because the average ocean depth is approximately 3800 meters (12,500 feet), you can see that the temperature of most seawater is not much above freezing. In the high polar latitudes, surface waters are cold and temperature changes with depth are slight. Consequently, the three-layered structure is not present (Figure 12.6).

Salinity variations with depth correspond to the general three-layered system described for temperatures. Generally, in the low and middle latitudes, a surface zone of higher salinity is created when freshwater is removed by evaporation. Below the surface zone, salinity decreases rapidly. This layer of rapid change, the **halocline**, corresponds closely to the thermocline. Below the halocline salinity variations are small.

Figure 12.7 In the low-to-mid latitudes, water temperatures in the surface mixed zone are relatively high and then drop off rapidly in the transition zone. Salinity variations generally show a similar pattern. (Photo by Carr Clifton/Minden Pictures)

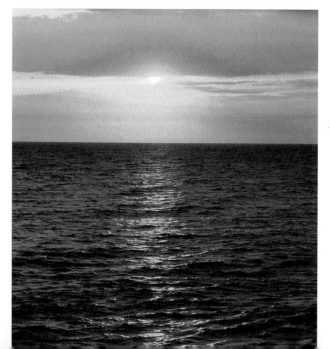

Mapping the Ocean Floor

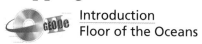
Introduction
Floor of the Oceans

If all water were drained from the ocean basins, a great diversity of features would be seen including linear chains of volcanoes, deep canyons, rift valleys, and large submarine plateaus. The scenery would be nearly as varied as that on the continents.

An understanding of the ocean floor's varied topography did not unfold until the historic $3\frac{1}{2}$-year voyage of the H. M. S. *Challenger* (Figure 12.8). From December 1872 to May 1876, the *Challenger* expedition made the first, and perhaps still most comprehensive, study of the global ocean ever attempted by one agency. The 110,000-kilometer (68,000-mile) trip took the ship and its crew of scientists to every ocean except the Arctic. Throughout the voyage they sampled the depth of water by laboriously lowering a weighted line overboard. Not many years later, the knowledge gained by the *Challenger* of the ocean's great depths and varied topography was further expanded with the laying of transatlantic cables. However, as long as ocean depth had to be measured with weighted line, our knowledge of the sea floor remained slight

In the 1920s a technological breakthrough occurred with the invention of electronic depth-sounding equipment (**echo sounder**). (The echo sounder is also referred to as *sonar*.) The echo sounder works by transmitting sound waves toward the ocean bottom (Figure 12.9A). A delicate receiver intercepts the echo reflected from the bottom, and a clock precisely measures the time interval to fractions of a second. By knowing the velocity of sound waves in water (about 1500 meters per second) and the time required for the energy pulse to reach the ocean floor and return, depth can be established. The depths determined from continuous monitoring of these echoes are plotted so that a profile of the ocean floor is obtained. By laboriously combining profiles from several adjacent traverses, a chart of the seafloor is produced.

In the last few decades, researchers have designed even more sophisticated echo sounders to map the ocean floor. In contrast to simple echo sounders, *multibeam sonar* employs an array of sound sources and listening devices. Thus, rather than obtaining the depth of a single point every few seconds, this technique obtains a profile of a narrow strip of seafloor (Figure 12.9B). These profiles are recorded every few seconds as the research vessel advances, building a continuous swath of relatively detailed coverage. (A ship will map a section of seafloor traveling in a back-and-forth pattern like that used when mowing a lawn.)

Despite their greater efficiency and enhanced detail, research vessels equipped with multibeam sonar travel at a mere 10 to 20 kilometers per hour. It would

Figure 12.8 The H.M.S. *Challenger.* (From C.W. Thomson and Sir John Murray, *Report on the Scientific Results of the Voyage of the H.M.S. Challenger,* Vol. 1. Great Britain: Challenger Office, 1895, Plate 1)

take at least 100 vessels outfitted with this equipment hundreds of years to map the entire seafloor.

Another technological breakthrough that has enhanced our understanding of the seafloor involves measuring the shape of the ocean surface from space. The ocean surface is not perfectly flat. Gravity attracts water toward regions where massive features occur.

Therefore, mountains and ridges produce elevated areas on the ocean surface. Conversely, canyons and trenches cause slight depressions. Satellites equipped with radar altimeters are able to measure these subtle differences by bouncing microwaves off the sea surface. These devices can measure variations as small as 3 to 6 centimeters. Such data have added greatly to our knowledge of

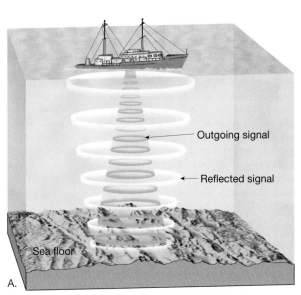

Outgoing signal

Reflected signal

Sea floor

A.

Sea floor

B.

Figure 12.9 Echo sounders. **A.** An echo sounder determines the water depth by measuring the time interval required for an acoustic wave to travel from a ship to the seafloor and back. The speed of sound in water is 1500 m/sec. Therefore, depth = 1/2 (1500 m/sec x echo travel time). **B.** Modern multibeam sonar obtains a profile of a narrow swath of seafloor every few seconds.

Figure 12.10 The topography of Earth's solid surface is shown on these two pages.

ocean floor topography. Coupled with traditional sonic depth measurements, more detailed ocean floor maps are being produced, such as the one in Figure 12.10.

Oceanographers studying the topography of the ocean floor have delineated three major units: *continental margins, deep-ocean basins,* and *mid-ocean ridges.* The map in Figure 12.11 outlines these provinces for the North Atlantic, and the profile at the bottom of the illustration shows the varied topography. Such profiles usually have their vertical dimension exaggerated many times—40 times in this case—to make topographic features more conspicuous. Because of this, the slopes shown in the seafloor profile appear to be *much* steeper than they actually are.

Continental Margins

Introduction
Floor of the Oceans

Two main types of **continental margins,** have been identified—passive and active. Passive margins are found along most of the coastal areas that surround the Atlantic Ocean, including the east coasts of North and South America, as well as the coastal areas of Western Europe and Africa. Passive margins are *not* associated with plate boundaries and therefore experience very little volcanism and few earthquakes. Here, weathered materials eroded from the adjacent landmass accumulate to form a thick, broad wedge of relatively undisturbed sediments.

By contrast, active continental margins occur where oceanic lithosphere is being subducted beneath the edge of a continent. The result is a relatively narrow

Figure 12.11 This map and profile show the major topographic divisions of the North Atlantic.

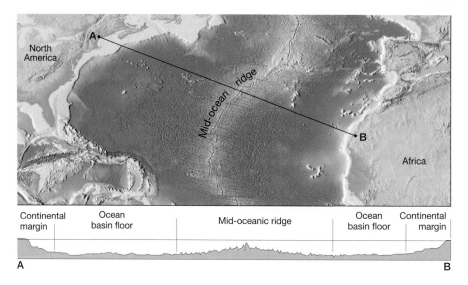

Figure 12.11 This map and profile show the major topographic divisions of the North Atlantic.

margin, consisting of highly deformed sediments that were scraped from the descending lithospheric slab. Active continental margins are common around the Pacific Rim, where they parallel deep oceanic trenches.

Passive Continental Margins

The features comprising a **passive continental margin** include the continental shelf, the continental slope, and the continental rise (Figure 12.12).

Continental Shelf. The **continental shelf** is a gently sloping submerged surface extending from the shoreline toward the deep-ocean basin. Because it is underlain by continental crust, it is clearly a flooded extension of the continents. The continental shelf varies greatly in width. Almost nonexistent along some continents, the shelf may extend seaward as far as 1500 kilometers (930

miles) along others. On the average, the continental shelf is about 80 kilometers (50 miles) wide and 130 meters (423 feet) deep at its seaward edge. The average inclination of the continental shelf is only about one-tenth of 1 degree, a drop of only about 2 meters per kilometer (10 feet per mile). The slope is so slight that it would appear to an observer to be a horizontal surface.

Although continental shelves represent only 7.5 percent of the total ocean area, they have economic and political significance because they contain important mineral deposits, including large reservoirs of petroleum and natural gas, as well as huge sand and gravel deposits. The waters of the continental shelf also contain many important fishing grounds that are significant sources of food.

Although the continental shelf is relatively featureless, some areas are mantled by extensive glacial

Figure 12.12 Schematic view showing the provinces of a passive continental margin. Note that the slopes shown for the continental shelf and continental slope are greatly exaggerated. The continental shelf has an average slope of one-tenth of 1 degree, whereas the continental slope has an average slope of about 5 degrees.

deposits and are thus quite rugged. The most profound features are long valleys running from the coastline into deeper waters. Many of these *shelf valleys* are the seaward extensions of river valleys on the adjacent landmass. Such valleys appear to have been excavated during the Pleistocene epoch (Ice Age).

During this time great quantities of water were stored in vast ice sheets on the continents. This caused sea level to drop by 100 meters or more, exposing large areas of the continental shelves (see Figure 5.B, p. 137). Because of this drop in sea level, rivers extended their courses, and land-dwelling plants and animals inhabited the newly exposed portions of the continents. Dredging off the coast of North America has produced the remains of numerous land dwellers, including mammoths, mastodons, and horses, adding to the evidence that portions of the continental shelves were once above sea level.

Most passive continental shelves, such as those along the east coast of the United States, consist of thick accumulations of shallow-water sediments. These sediments are frequently several kilometers thick and are interbedded with limestones that formed during earlier periods of coral reef building, a process that occurs only in shallow water. Such evidence led researchers to conclude that these thick accumulations of sediment are produced along a gradually subsiding continental margin.

Continental Slope. Marking the seaward edge of the continental shelf is the **continental slope,** a relatively steep structure (as compared with the shelf) that marks the boundary between continental crust and oceanic crust (Figure 12.12). Although the inclination of the continental slope varies greatly from place to place, it averages about 5 degrees and in places may exceed 25 degrees. Further, the continental slope is a relatively narrow feature, averaging only about 20 kilometers in width.

Continental Rise. In regions where trenches do not exist, the steep continental slope merges into a more gradual incline known as the **continental rise.** Here the slope drops to about one-third degree, or about 6 meters per kilometer. Whereas the width of the continental slope averages about 20 kilometers, the continental rise may extend for hundreds of kilometers into the deep-ocean basin.

The continental rise consists of a thick accumulation of sediment that moved downslope from the continental shelf to the deep-ocean floor. The sediments are delivered to the base of the continental slope by *turbidity currents* that follow submarine canyons (see Figure 12.14). (We will discuss these shortly.) When these muddy currents emerge from the mouth of a canyon onto the relatively flat ocean floor, they deposit sediment that forms a **deep-sea fan.** Deep-sea fans have the same basic shape as alluvial fans, which form at the foot of steep mountain slopes on land. As fans from adjacent submarine canyons grow, they coalesce to produce the continuous apron of sediment at the base of the continental slope that we call the continental rise.

Active Continental Margins

Along some coasts the continental slope descends abruptly into a deep-ocean trench located between the continent and ocean basin. Thus, the landward wall of the trench and the continental slope are essentially the same feature. In such locations, the continental shelf is very narrow, if it exists at all.

Active continental margins are located primarily around the Pacific Ocean in areas where the leading edge of a continent is overrunning oceanic lithosphere (Figure 12.13). Here sediments from the ocean floor and pieces of oceanic crust are scraped from the descending oceanic plate and plastered against the edge of the overriding continent. This chaotic accumulation of deformed sediment and scraps of oceanic crust is called an *accretionary wedge.* Prolonged subduction, along with the accretion of sediments on the landward side of the trench, can produce a large accumulation of sediments along a continental margin. For

Figure 12.13 Active continental margin. Here sediments from the ocean floor are scraped from the descending plate and added to the continental crust as an accretionary wedge.

example, a large accretionary wedge is found along the northern coast of Japan's Honshu Island.*

Some subduction zones have little or no accumulation of sediments, indicating that ocean sediments are being carried into the mantle with the subducting plate. Here the continental margin is very narrow as the trench may lie a mere 50 kilometers off-shore.

Submarine Canyons and Turbidity Currents

Deep, steep-sided valleys known as **submarine canyons** are cut into the continental slope and may extend across the entire continental rise to the deep-ocean basin (Figure 12.14). Although some of these canyons appear to be the seaward extensions of river valleys such as the Hudson valley and Amazon River, others are not directly associated with existing river systems. Furthermore, because these canyons extend to depths far below the lowest level of the sea during the Ice Age, we cannot attribute their formation to stream erosion. These features must be created by some process that operates far below the ocean surface.

---

*For more on accretionary wedges, see *Mountain Building at Convergent Boundaries* in Chapter 9.

Most available information favors the view that submarine canyons have been eroded, at least in part, by turbidity currents (Figure 12.14). **Turbidity currents** are downslope movements of dense, sediment-laden water. They are created when sand and mud on the continental shelf and slope are dislodged, perhaps by an earthquake, and are thrown into suspension. Because such mud-choked water is denser than normal sea water, it flows downslope, eroding and accumulating more sediment. The erosional work repeatedly carried on by these muddy torrents is thought to be the major force in the excavation of most submarine canyons.

Narrow continental margins, such as the one located along the California coast, are dissected by numerous submarine canyons. Here, erosion has extended many of these canyons landward into shallow water where longshore currents are active. As a result, sediments carried to the coasts by rivers are transported along the shore until they reach a submarine canyon. This steady supply of sediment collects until it becomes unstable and moves as a massive landslide (turbidity current) to the deep-ocean floor.

Turbidity currents eventually lose momentum and come to rest along the floor of the ocean basin (Figure 12.14). As these currents slow, the suspended sediments begin to settle out. First, the coarser, heavier sand is dropped, followed by successively finer

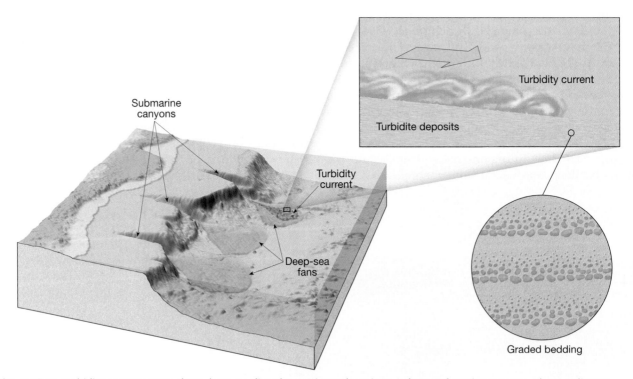

Figure 12.14 Turbidity currents move downslope, eroding the continental marin to enlarge submarine canyons. These sediment-laden density currents eventually lose momentum and deposit their loads of sediment as deep-sea fans. Beds deposited by these currents are called *turbidites.* Each event produces a single bed characterized by a decrease in sediment size from bottom to top, a feature known as a *graded bed.*

deposits of silt and then clay. Consequently, these deposits, called **turbidites,** are characterized by a decrease in sediment grain size from bottom to top, a phenomenon known as **graded bedding** (Figure 12.14).

Although there is still more to be learned about the complex workings of turbidity currents, it has been well established that they are an important mechanism of sediment transport in the ocean. By the action of turbidity currents, submarine canyons are created and sediments are carried to the deep-ocean floor.

The Ocean Basin Floor

Introduction
Floor of the Oceans

Between the continental margin and the mid-ocean ridge lies the ocean basin floor (see Figure 12.10). The size of this region—almost 30 percent of Earth's surface—is roughly comparable to the percentage of the surface that presently projects above sea level as land. Here we find remarkably flat regions known as abyssal plains, broad volcanic peaks called seamounts, and deep-ocean trenches, which are extremely deep linear depressions in the ocean floor.

Deep-Ocean Trenches

Deep-ocean trenches are long, relatively narrow features that form the deepest parts of the ocean. Most trenches are located in the Pacific Ocean where some approach or exceed 10,000 meters in depth, and at least a portion of one, the Challenger Deep in the Mariana trench, is more than 11,000 meters below sea level.

Although deep-ocean trenches represent only a very small portion of the area of the lithospheric ocean floor, they are nevertheless significant geologic features. Trenches are the sites where moving crustal plates plunge back into the mantle. In addition to the earthquakes created as one plate descends beneath another, volcanic activity is also associated with these regions. Thus, trenches are often paralleled by volcanic island arcs. Furthermore, volcanic mountains, such as those making up portions of the Andes, are located parallel to trenches that lie adjacent to continental margins. The release of volatiles (water) from a descending plate triggers melting in the wedge of asthenosphere above it. This buoyant material slowly migrates upward and gives rise to volcanic activity at the surface.

Abyssal Plains

Abyssal plains are incredibly flat features; in fact, these regions are likely the most level places on Earth. The abyssal plain found off the coast of Argentina, for example, has less than 3 meters of

relief over a distance exceeding 1300 kilometers. The monotonous topography of abyssal plains will occasionally be interrupted by the protruding summit of a partially buried volcanic structure.

Using seismic profilers, instruments whose signals penetrate far below the ocean floor, researchers have determined that abyssal plains owe their relatively featureless topography to thick accumulations of sediment that have buried an otherwise rugged ocean floor. The nature of the sediment indicates that these plains consist primarily of sediments transported far out to sea by turbidity currents.

Abyssal plains are found in all of the oceans. However, because the Atlantic Ocean has fewer trenches to act as traps for the sediments carried down the continental slope, it has more extensive abyssal plains than does the Pacific.

Seamounts

Dotting the ocean floors are isolated volcanic peaks called **seamounts.** These features may rise hundreds of meters above the surrounding topography. Although these steep-sided conical peaks are found on the floors of all the oceans, the greatest number have been identified in the Pacific (see Figure 12.10).

Some, like the Emperor Seamount chain that stretches from the Hawaiian Islands to the Aleutian trench, form in association with volcanic hot spots. Others are born near oceanic ridges, divergent plate boundaries where the plates of the lithosphere move apart (see Chapter 7). If the volcano grows large enough before being carried from the zone that nourishes it, the structure emerges as an island. Examples in the Atlantic include the Azores, Ascension, Tristan da Cunha, and St. Helena.

During the time they exist as islands, some of these volcanoes are eroded to near sea level by running water and wave action. Over a span of millions of years the islands gradually sink as the moving plate slowly carries them away from the elevated oceanic ridge or hot spot where they originated. These submerged, flat-topped seamounts are called **guyots.**

Mid-Ocean Ridges

Introduction
Floor of the Oceans

Our knowledge of **mid-ocean ridges** comes from soundings taken of the ocean floor, core samples obtained from deep-sea drilling, visual inspection using deep-diving submersibles (Figure 12.15), and even firsthand inspection of slices of ocean floor that have been shoved up onto dry land (see Box 12.1). The ocean ridge system is characterized by an elevated

Figure 12.15 The deep-diving submersible *Alvin* is 7.6 meters long, weights 16 tons, has a cruising speed of 1 knot, and can reach depths as great as 4000 meters. A pilot and two scientific observers are along during a normal 6- to 10-hour dive. Vessels such as this are important tools for detecting and studying the fine-scale features of the ocean floor. (Photo by Rod Catanach/ © Woods Hole Oceanographic Institution)

position, extensive faulting, and numerous volcanic structures that have developed on the newly formed crust (Figure 12.10).

The interconnected ocean ridge system is the longest topographic feature on Earth's surface, exceeding 70,000 kilometers (43,000 miles) in length. Representing 20 percent of Earth's surface, the ocean ridge system winds through all major oceans in a manner similar to the seam on a baseball (Figure 12.10). The term *ridge* may be misleading as these features are not narrow but have widths of from 3000 to 4000 kilometers and, in places, may occupy as much as one-half the total area of the ocean floor. An examination of Figure 12.10 shows that the ridge system is broken into segments that are offset by large transform faults. Further, along the axis of some segments are deep down-faulted structures called **rift valleys.**

Although ocean ridges stand 2 to 3 kilometers above the adjacent deep-ocean basins, they are much different from mountains found on the continents. Rather than consisting of thick sequences of folded and faulted sedimentary rocks, oceanic ridges consist of layer upon layer of basaltic rocks that have been faulted and uplifted.

The flanks of most ridges are topographically relatively featureless and rise very gradually (slope less than 1 degree) toward the ridge crest. Approaching the ridge crest, the topography begins to exhibit greater local relief as volcanic structures and faulted valleys become more prominent. The most rugged topography is found on those ridges that exhibit large rift valleys.

Partly because of its accessibility to both American and European scientists, the Mid-Atlantic Ridge has been studied more thoroughly than other ridge systems (Figure 12.10). The Mid-Atlantic Ridge is a broad, submerged structure standing 2500 to 3000 meters above the adjacent ocean basin floor. In a few places, such as Iceland, the ridge has actually grown above sea level. Throughout most of its length, however, this divergent plate boundary lies 2500 meters below sea level. Another prominent feature of the Mid-Atlantic Ridge is its deep linear rift valley extending along the ridge axis. In places this rift valley is deeper than the Grand Canyon of the Colorado River and two or three times as wide. The name *rift valley* has been applied to this feature because it is so strikingly similar to continental rift valleys such as the East Africa Rift Valley.

Marine Life Zones

Oceanography includes biology.[*] The presence and distribution of living marine organisms is affected by the chemistry, physics, and geology of the world oceans. Ocean plants and animals are influenced by available sunlight, distance from shore, water depth, salinity, temperature, type of bottom sediments, and dissolved oxygen.

Availability of Light

The upper part of the ocean into which any sunlight penetrates is the **photic zone.** The clarity of seawater is affected by air bubbles, mineral content, decaying organic particles, and microscopic living organisms. Because these factors vary, sunlight penetrates to different depths in different areas.

The **euphotic zone** is the portion of the photic zone near the surface where light is bright enough for photosynthesis to occur. In clear water this zone can reach a depth of 100 meters, but the zone will be much shallower in cloudy water. In the euphotic zone small free-floating algae (*phytoplankton*) use sunlight to produce food molecules and become the basis of most ocean food chains. Phytoplankton are eaten by tiny floating organisms (*zooplankton*) such as protozoans, worms, crustaceans, jellyfishes, and larvae. Zooplankton in turn are eaten by small fish, and so on up the food chain to top predators such as sharks and killer whales.

[*]This section was prepared by David Cox, Professor of Biology at Illinois Central College.

Box 12.1 Sampling the Ocean Floor

The *JOIDES Resolution* is the drilling ship of the Ocean Drilling Program. The "*JOIDES*" in the ship's name stands for Joint Oceanographic Institutions for Deep Earth Sampling and reflects the international commitment from the program's 19 member countries. The "*Resolution*" honors the *HMS Resolution*, commanded more than 200 years ago by the well-known explorer Captain James Cook.

During cruises, holes are drilled deep into the seafloor. The cores of sediment and rock that are recovered represent millions of years of Earth history and are used by scientists to study many aspects of Earth science, including changes in global climate.

The ship is 143 meters long and 21 meters wide with a derrick that towers 62 meters above the waterline. A computer-controlled positioning system maintains the ship over a specific location. The vessel can drill in water depths up to 8200 meters (about 27,000 feet) and can deploy as much as 9100 meters (about 30,000 feet) of drill pipe (Figure 12.A).

The *JOIDES Resolution* has living quarters for 50 scientists and technicians plus a crew of 65. The 12 onboard laboratories are equipped with the largest and most varied array of seagoing research equipment in the world.

Since 1985, the ship has drilled in the Atlantic, Pacific, Indian, and Arctic oceans as well as several seas. The result has been the recovery of more than 168,000 meters (more than 104 miles!) of core samples.

Figure 12.A *JOIDES Resolution* can reenter holes in the seafloor years after initial drilling. The ship's dynamic positioning system consists of powerful thrusters (small propellers) that allow it to remain in place. Previous drill sites are located by bouncing sound waves between the ship's hydrophones and sonar beacons. A remote television camera aids in positioning the drill pipe into the reentry cone.

Thrusters

Thrusters

Hydrophones

Drill pipe

Maximum water depth 8,200 meters (27,000 feet)

Television camera

Reentry cone

Sonar beacon

Sediment layer

Hard rock layer

Sunlight is absorbed as it passes through seawater. The longer wavelength red-orange colors are absorbed first, the greens and yellows next, and the shorter wavelength blue-violets penetrate the farthest (Table 12.2). At clear-water depths of 1000 meters, only blues and violets are still visible.

Although photosynthesis cannot occur much below 100 meters, there is enough light in the lower photic zone for marine animals to find food, avoid predators, recognize their species, and locate mates. But even deeper, in the **aphotic zone,** there is no sunlight. Most of the deep ocean is lacking in light, except for the faint amount produced chemically by some *bioluminescent* bacteria and animals.

Distance From Shore

Another way to subdivide marine life zones uses horizontal geography. The area where the land and ocean meet and overlap is the **intertidal zone.** This narrow strip between high and low tidelines is covered or uncovered by seawater with each tidal change. Here the land and sea appear to be in constant conflict. Only a few species such as rockweeds, barnacles, chitons, and mole crabs can survive the dehydration and temperature changes, or withstand pounding wave action and shore erosion. Other species live in small pools isolated as the tide recedes.

Seaward from the low tideline is the **neritic zone.** This covers the gently sloping continental shelf out to

Table 12.2 Marine life zones and habitats

Life Zone	Subdivision	Characteristics
Based on available light		
Photic		Has light
	Euphotic	Has enough light for photosynthesis
Aphotic		No light
Based on distance from shore		
Intertidal		Between tidelines, waves hit shore
Neritic		Above continental shelf, rich in biodiversity
Oceanic		Open ocean, beyond shelf break
Based on depth		
Pelagic		All water above ocean floor, organisms swim or float
Benthic		Bottom of ocean, organisms attach to, burrow into, or crawl on seafloor
	Abyssal	Deep-sea bottom, dark, cold, high pressure
Other habitats		
Estuaries		Rivers flow into ocean, tidal cycles
Coral Reefs		Coral skeletons in tropical waters
Hydrothermal Vents		Black smokers, chemosynthesis

the *shelf break* (see Figure 12.12). This zone can be very narrow or may extend for hundreds of kilometers. The entire neritic zone is sometimes in the photic zone because sunlight may reach to its bottom.

Although the neritic zone covers only about 5 percent of the world ocean, it is rich in both numbers and diversity of living organisms. Photosynthesis occurs readily, nutrients wash in from land, and its bottom provides nest sites and protection. Ninety percent of commercial fisheries are in this marine zone.

Beyond the continental shelf is the **oceanic zone.** The open ocean reaches great depths, but usually has far smaller populations than the more productive continental shelf areas. Mid-ocean often resembles a biological desert with relatively few organisms present. This is the case because the oceanic zone is constantly mixed by currents and has very low nutrient concentrations. Food and energy in the dead bodies of plankton and fishes instead sink to the ocean floor.

Tropical oceans have a permanent *thermal stratification* (see Figure 12.6) that usually prevents a return of nutrients from great depths to the surface. By contrast, the open ocean in the middle latitudes is more productive than in tropical regions because nutrient overturn occurs in the spring and fall. The spring *upwelling* of nutrients is particularly effective in stimulating phytoplankton growth. Such "blooms" result in rapid population increases for many marine organisms. Another

important effect of upwellings on sea communities is illustrated by *El Niño.* Its ecological chain reaction affects oceanic food webs and weather patterns (see "El Niño and Global Weather" in Chapter 16).

Depth

A third method of organizing marine habitats uses ocean depth. Open ocean of *any* depth is called the **pelagic zone.** Animals in this zone swim or float freely. The photic part of the pelagic zone is home to phytoplankton, zooplankton, fishes such as tuna and swordfish, and also sea turtles and dolphins. The aphotic part has strange species like viperfish and giant squid.

But animals such as sponges, crabs, and tube worms that attach to, crawl upon, or burrow into the seafloor are part of the **benthic zone.** This includes any sea-bottom surface regardless of its distance from shore.

In near-coastal water, benthic communities usually get sunlight and have a bounty of organisms: bacteria, seaweeds, sponges, sea anemones, sea stars, worms, crustaceans, and flatfishes.

In deeper water, nutrients drift down from above and permit food chains in dark depths even kilometers below the ocean surface. The **abyssal zone** is a subdivision of the benthic zone and is characterized by extremely high pressures, consistently low temperatures (3°C), low oxygen, few nutrients, and no sunlight. Although rarely studied, *abyssal plains* are among the most extensive environments on Earth.

Box 12.2   Earth as a System: Hydrothermal Vent Communities

David Cox*

Hot springs called *hydrothermal vents* form along active *rift zones* of the mid-ocean ridge. Here seawater percolates a kilometer or two down into the hot and newly formed oceanic crust. During its journey the water becomes saturated with minerals before boiling out at a hydrothermal vent. At some vents, metal sulfide minerals precipitate as the hot ventwater enters the cold seawater, forming cone-shaped chimneys called *black smokers* (Figure 12.B).

Water temperatures at some vents reach as high as 660°C, far too hot for anything to live. But water temperatures of 100°C or lower at other vents nourish dense, exotic communities of living organisms found nowhere else. Over 400 species *new to science* have been discovered surrounding these deep-sea habitats.

Hydrothermal vent communities were discovered in 1977 by deep-sea

*David Cox is a professor of Biology at Illinois Central College.

submersibles (see Figure 12.15). The best-studied hydrothermal vent regions are along the Galapagos Rift, East Pacific Rise, and Mid-Atlantic Ridge. One large vent area off the Azores covers over 15 acres.

Photosynthesis was previously thought to be the basis of all food chains on Earth. But study of hydrothermal vents reveals that the Sun does not provide energy and nutrients to this dark, hot, sulfur-rich but oxygen-deficient environment. Instead, *archaebacteria* ("ancient bacteria") living in and near the vents do *chemosynthesis.* The hydrothermal vent provides heat energy for these bacteria to oxidize the hydrogen sulfide (H_2S) formed by the reaction of hot water with dissolved sulfate (SO_4^{-2}). The chemosynthesis joins CO_2 molecules to produce sugars and other foods that enable the bacteria as well as other organisms to live in this very unusual and extreme environment.

Some archaebacteria species live inside giant gutless tube-dwelling

Figure 12.C Tube worms are among the organisms found in the extreme environment of mid-ocean ridges where sunlight is nonexistent. These organisms derive nourishment and energy from processes associated with hydrothermal vents. (Photo by Al Giddings Images, Inc.)

worms (Figure 12.C). The bacteria provide food for the tube worms to grow as rapidly as 1 meter each year and over 3 meters long. Other chemosynthetic bacteria are consumed by specialized yellow mussels, giant white clams, and pink sea urchins. These, in turn, are eaten by a variety of unique crabs and fishes. The bacteria are the foundation of a living ecosystem that *does not* require sunlight. Some scientists suggest the chemistry of vent communities may be similar to how life on Earth first began.

Figure 12.B A black smoker spewing hot, mineral-rich water along the East Pacific Rise. As heated solutions meet cold seawater, sulfides of copper, iron, and zinc precipitate immediately, forming mounds of minerals around these vents. (Photo by Dudley Foster, Woods Hole Oceanographic Institution)

Two typical food sources exist at the abyssal depths. Tiny decaying particles steadily "raining" from above provide food for filter-feeders, brittle stars, and burrowing worms. Large fragments or entire dead bodies falling in scattered sites supply meals for actively searching fish, such as the tripodfish and grenadier, that locate food by chemical sensing (see Box 12.2).

Marine biologists often use all three methods of organizing marine life zones simultaneously. Table 12.2 and Figure 12.16 summarize marine life zones.

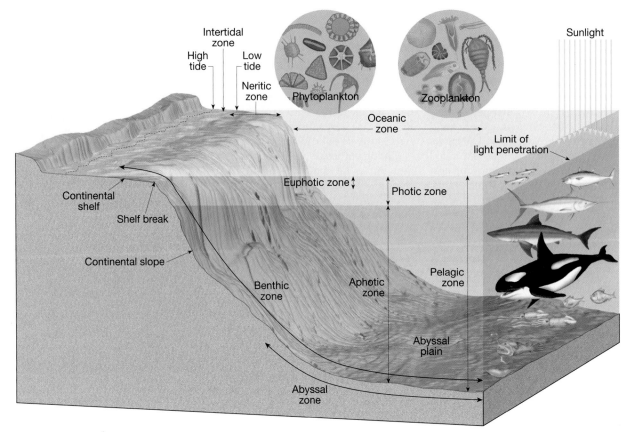

Figure 12.16 Marine life zones.

Other Marine Habitats

Estuaries

Where a river flows into the ocean, an estuary may occur. An **estuary** is a partially enclosed coastal water body that is connected to the ocean (see Figure 13.26, p. 371). The salinity of seawater in the estuary is measurably reduced by the freshwater inflow of the river water. Animals living here need specialized kidneys to adjust their blood chemistry. The daily ebb and flow of the tides alternately expose and flood near-shore areas, so plants must adapt to drying and flooding.

One advantage for life in an estuary is the constant supply of nutrients and organic matter provided by the river. The growth of saltmarsh grasses and phytoplankton can be very lush. Estuaries are nurseries for many sea animals such as commercially important fishes, oysters, and crabs.

Coral Reefs and Atolls

Many neritic zones in tropical oceans have currents supplying nutrients and sunlight penetrating to the ocean floor. These conditions may allow growth of an unusual jellyfish relative called *coral.* Each individual coral animal protects itself by growing a bony external skeleton made of calcium carbonate. Where corals reproduce and grow over many centuries, their skeletons fuse into a structure called a **reef.** Other corals, as well as sponges and algae, begin to attach to the reef, enlarging it further. Eventually fishes, sea slugs, octopus, and other invertebrates are attracted to these diverse and productive habitats.

Coral reefs are confined to the warm, clear waters of the Pacific and Indian oceans, although a few occur elsewhere. Reef-building corals grow best in waters with an average annual temperature of about 24°C (75°F). They can survive neither sudden temperature changes nor prolonged exposure to temperatures below 18°C (65°F). In addition, these reef-builders require clear sunlit water. Consequently, the limiting depth of active reef growth is only about 45 meters (150 feet).

In 1831, the naturalist Charles Darwin set out aboard the British ship H.M.S. *Beagle* on its famous 5-year expedition that circumnavigated the globe. One outcome of Darwin's studies was development of a hypothesis on the formation of coral islands, called **atolls.** As Figure 12.17 illustrates, atolls consist of a continuous or broken ring of coral reef surrounding a central lagoon.

Darwin's hypothesis explained what seemed to be a paradox; that is, how can corals, which require

Figure 12.17 A. An aerial view of Tetiaroa Atoll in the Pacific. The light blue waters of the relatively shallow lagoon contrast with the dark blue color of the deep ocean surrounding the atoll. (Photo by Douglass Peebles Photography) **B.** View from space of a group of atolls in the Pacific Ocean. (Courtesy of NASA)

warm, shallow, sunlit water no deeper than a few dozen meters to live, create structures that reach thousands of meters to the floor of the ocean? The essence of Darwin's hypothesis was that coral reefs form on the flanks of sinking volcanic islands. As an island slowly sinks, the corals continue to build the reef complex upward (Figure 12.18).

Thus, atolls, like guyots, are thought to owe their existence to the gradual sinking of oceanic crust. Drilling on atolls has revealed that volcanic rock does indeed underlie the thick coral reef structure. This finding confirmed Darwin's explanation.

Seafloor Sediments

Except for a few areas, such as near the crests of mid-ocean ridges, the ocean floor is mantled with sediment. Part of this material has been deposited by turbidity currents, and the rest has slowly settled to the bottom from above. The thickness of this carpet of debris varies greatly. In some trenches, which act as traps for sediments originating on the continental margin, accumulations may approach 10 kilometers (6 miles). In general, however, sediment accumulations are considerably less. In the Pacific Ocean, uncompacted sediment measures about 600 meters (2000 feet) or less, whereas on the floor of the Atlantic, the thickness varies from 500 to 1000 meters (1500 to 3000 feet).

Although deposits of sand-sized particles are found on the deep-ocean floor, mud is the most common sediment covering this region. Muds also predominate on the continental shelf and slope, but the sediments in these areas are coarser overall because of greater quantities of sand.

Types of Seafloor Sediments

Seafloor sediments can be classified according to their origin into three broad categories: (1) **terrigenous** ("derived from land"); (2) **biogenous** ("derived from organisms"); and (3) **hydrogenous** ("derived from water"). Although each category is discussed separately, remember that all seafloor sediments are mixtures. No body of sediment comes entirely from a single source.

Terrigenous Sediments. Terrigenous sediment consists primarily of mineral grains that were weathered from continental rocks and transported to the ocean. The sand-sized particles settle near shore. However, because the very smallest particles take years to settle to the ocean floor, they may be carried for thousands of kilometers by ocean currents. As a consequence, virtually every part of the ocean receives some terrigenous sediment. The rate at which this sediment accumulates on the deep-ocean floor, though, is very slow. From 5000 to 50,000 years are necessary for a 1-centimeter layer to form. Conversely, on the continental margins near the mouths of large rivers, terrigenous sediment accumulates rapidly. In the Gulf of Mexico, for example, the sediment has reached a depth of many kilometers.

Because fine particles remain suspended in the water for a very long time, ample opportunity exists for chemical reactions to occur. Because of this, the colors of the deep-sea sediments are often red or brown. This results when iron in the particle or in the water reacts with dissolved oxygen in the water and produces a coating of iron oxide (rust).

Biogenous Sediments. Biogenous sediment consists of shells and skeletons of marine animals and plants (Figure 12.19). This debris is produced mostly by microscopic organisms living in the sunlit waters near the ocean surface. Their remains constantly "rain" down on the seafloor.

The most common biogenous sediments are known as *calcareous* ($CaCO_3$) *oozes,* and, as their name implies, they have the consistency of thick mud. These

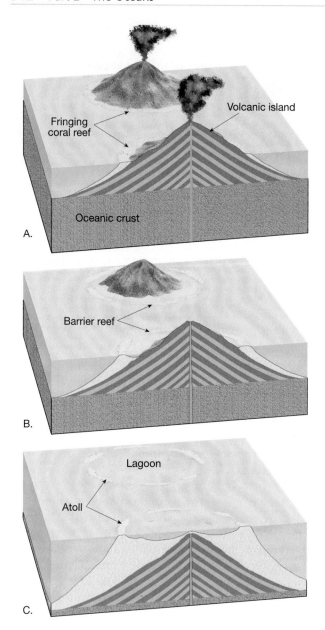

Figure 12.18 Formation of a coral atoll caused by the gradual sinking of oceanic crust.

sediments are produced by organisms that inhabit warm surface waters. When calcareous hard parts slowly sink through a cool layer of water, they begin to dissolve. This results because cold seawater is rich in carbon dioxide and is thus more acidic than is warm water. In seawater deeper than about 4500 meters (15,000 feet), calcareous shells will completely dissolve before they reach bottom. Consequently, calcareous ooze does not accumulate in the deep-ocean basins.

Other biogenous sediments include *siliceous* (SiO_2) *oozes* and phosphate-rich materials. The former are composed primarily of opaline skeletons of diatoms (single-celled algae) and radiolarians (single-celled animals),

whereas the latter are derived from the bones, teeth, and scales of fish and other marine organisms.

Hydrogenous Sediments. Hydrogenous sediment consists of minerals that crystallize directly from seawater through various chemical reactions. For example, some limestones are formed when calcium carbonate precipitates directly from the water; however, most limestone is composed of biogenous sediment.

One type of hydrogenous sediment is significant for its economic potential. **Manganese nodules** are rounded dark lumps of mixed composition (Figure 12.20). About 30 percent of a typical nodule is manganese dioxide (MnO_2). Roughly 20 percent of each nodule is iron oxide (Fe_2O_3). The manganese and iron are important, but the real interest lies in more valuable metals that may be enriched in the nodules: copper, nickel, and cobalt.

Nodules are widely distributed, but not all regions are equally promising for development. Good locations have abundant nodules (more than 5 kilograms per square meter) that contain the economically optimum mix of copper, nickel, and cobalt. Sites meeting these criteria relatively limited. Also, the logistics of extracting nodules from deep-ocean basins must be worked out, in addition to political problems of ownership and access.

Seafloor Sediments and Climate Change

Reliable climate records go back only a couple of hundred years, at best. How do scientists learn about climates and climate changes prior to that time? The obvious answer is that they must reconstruct past climates from *indirect evidence;* that is, they must examine and analyze phenomena that respond to and reflect changing atmospheric conditions. An interesting and important technique for analyzing Earth's climate history is the study of sediments from the ocean floor.

Most seafloor sediments contain the remains of organisms that once lived near the sea surface (the ocean–atmosphere interface). When such near-surface organisms die, their shells slowly settle to the ocean floor where they become part of the sedimentary record. These seafloor sediments are useful recorders of worldwide climate change because the numbers and types of organisms living near the sea surface change with the climate:

...[W]e would expect that in any area of the ocean/atmosphere interface the average annual temperature of the surface water of the ocean would approximate that of the contiguous atmosphere. The temperature equilibrium established between surface seawater and the air above it should mean that...changes in climate should be reflected in changes in organisms living near the surface of the deep sea....

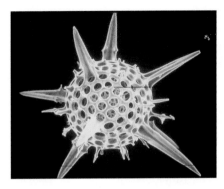

Figure 12.19 Microscopic radiolarian hard parts are examples of biogenous sediments. These photomicrographs have been enlarged *hundreds* of times. (Photo by Manfred Kage/Peter Arnold, Inc.)

When we recall that the seafloor sediments in vast areas of the ocean consist mainly of shells of pelagic foraminifers, and that these animals are sensitive to variations in water temperature, the connection between such sediments and climatic change become obvious.*

Thus, in seeking to understand climate change as well as other environmental transformations, scientists are tapping the huge reservoir of data in seafloor sediments. The sediment cores gathered by drilling ships and other research vessels have provided invaluable data that have significantly expanded our knowledge and understanding of past climates (see Box 12.1).

One notable example of the importance of seafloor sediments to our understanding of climate change relates to unraveling the fluctuating atmospheric conditions of the Ice Age. The record of temperature changes contained in cores of sediment from the ocean floor have proven critical to our present understanding of this recent span of Earth history.+

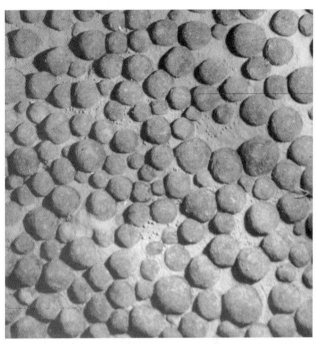

Figure 12.20 Manganese nodules photographed at a depth of 2909 fathoms (5323 meters) beneath the *Robert Conrad* south of Tahiti. (Photo courtesy of Lawrence Sullivan, Lamont-Doherty Earth Observatory/Columbia University)

*Richard F. Flint, *Glacial and Quaternary Geology* (New York: Wiley, 1971), p. 718.
+For more on this topic, see "Causes of Glaciation" in Chapter 5.

Chapter Summary

- *Oceanography* is a composite science that draws on the methods and knowledge of biology, chemistry, physics, and geology to study all aspects of the world ocean.

- *Earth is a planet dominated by oceans.* Seventy-one percent of Earth's area consists of oceans and marginal seas. In the Southern Hemisphere, often called the *water hemisphere*, about 81% of the surface is water. Of the three major oceans, Pacific, Atlantic, and Indian, the *Pacific Ocean* is the *largest*, contains slightly *more than half of the water* in the world ocean, and has the *greatest average depth*—3940 meters (12,900 feet).

- *Salinity* is the proportion of dissolved salts to pure water, usually expressed in parts per thousand (‰). The average salinity in the open ocean ranges from 35‰ to 37‰. The principal elements that contribute to the ocean's salinity are *chlorine* (55%) and *sodium* (31%). The primary *sources of the elements in sea salt* in the ocean are *chemical weathering* of

rocks on the continents and *volcanic outgassing*. Outgassing is also considered to be the principal source of water in the oceans as well as in the atmosphere.

- Commercial products obtained from seawater include common salt, the lightweight metal magnesium, and bromine. Another resource derived from seawater is freshwater. The removal of salts from seawater to produce usable freshwater is termed *desalination*.

- In most regions outside of the polar latitudes, the open ocean exhibits a *three-layered temperature and salinity structure*. Ocean water temperatures are warmest at the surface because of solar energy. The mixing of waves as well as the turbulence from currents can distribute this heat to a depth of about 450 meters or more. Beneath the Sun-warmed zone of mixing, a layer of rapid temperature change, called the *thermocline*, occurs. Below the thermocline, in the deep zone, temperatures fall only a few more degrees. The changes in salinity with increasing depth correspond to the general three-layered temperature structure. In the low and middle latitudes, a surface zone of higher salinity is underlain by a layer of rapidly decreasing salinity, called the *halocline*. Below the halocline, salinity changes are small.

- Ocean depths are determined using *echo sounders* and *multibeam sonars*. These devices bounce sonic signals off the ocean floor. Delicate receivers intercept the reflected echos. The time interval of the signal must also be carefully measured. Ocean depths are plotted to produce ocean floor profiles. Recently, satellite measurements of the shape of the ocean surface have added data for mapping ocean floor features.

- The zones that collectively make up a *passive continental margin* include the *continental shelf* (a gently sloping, submerged surface extending from the shoreline toward the deep-ocean basin); *continental slope* (the true edge of the continent, which has a steep slope that leads from the continental shelf into deep water); and in regions where trenches do not exist, the steep continental slope merges into a more gradual incline known as the *continental rise*. The continental rise consists of sediments that have moved downslope from the continental shelf to the deep-ocean floor.

- *Active continental margins* are located primarily around the Pacific Ocean in areas where the leading edge of a continent is overrunning oceanic lithosphere. Here sediment scraped from the descending oceanic plate is plastered against the continent to form a collection of sediments called an *accretionary wedge*. An active continental margin generally has a narrow continental shelf, which grades into a deep-ocean trench.

- *Submarine canyons* are deep, steep-sided valleys that originate on the continental slope and may extend to depths of 3 kilometers. Many submarine canyons have been excavated by *turbidity currents* (downslope movements of dense, sediment-laden water).

- The *ocean basin floor* lies between the continental margin and the mid-oceanic ridge system. The features of the ocean basin floor include *deep-ocean trenches* (the deep-est parts of the ocean, where moving crustal plates descend into the mantle), *abyssal plains* (the most level places on Earth, consisting of thick accumulations of sediments that were deposited atop the low, rough portions of the ocean floor by turbidity currents), and *seamounts* (isolated volcanic peaks on the ocean floor that originate near oceanic ridges or in association with volcanic hot spots).

- *Mid-ocean ridges,* the sites of seafloor spreading, are found in all major oceans and represent more than 20 percent of Earth's surface. These broad features are characterized by an elevated position, extensive faulting, and volcanic structures that have developed on newly formed oceanic crust. Most of the geologic activity associated with ridges occurs along a narrow region on the ridge crest, called the *rift valley,* where magma from the asthenosphere moves upward to create new slivers of oceanic crust.

- Three criteria are frequently used for organizing the habitats of marine life: availability of light, distance from shore, and water depth.

- *Reefs,* which are confined largely to the warm, sunlit waters of the Pacific and Indian oceans, are constructed over thousands of years primarily from the skeletal remains and secretions of corals and certain algae. Coral islands, called *atolls,* consist of a continuous or broken ring of coral reef surrounding a central lagoon. Atolls form from corals that grow on the flanks of sinking volcanic islands, where the corals continue to build the reef complex upward as the island sinks.

- *There are three broad categories of seafloor sediments. Terrigenous sediment* consists primarily of mineral grains that were weathered from continental rocks and transported to the ocean. *Biogenous sediment* consists of shells and skeletons of marine animals and plants. *Hydrogenous sediment* includes minerals that crystallize directly from seawater through various chemical reactions. Seafloor sediments are helpful when studying worldwide climate changes because they often contain the remains of organisms that once lived near the sea surface. The numbers and types of these organisms change as the climate changes, and their remains in the sediments record these changes.

Key Terms

abyssal plain (p. 335)
abyssal zone (p. 328)
active continental margin (p. 333)
aphotic zone (p. 337)
atoll (p. 340)
benthic zone (p. 338)
biogenous sediment (p. 341)
continental margin (p. 331)
continental rise (p. 333)
continental shelf (p. 332)
continental slope (p. 333)
deep-ocean trench (p. 335)
deep-sea fan (p. 335)
desalination (p. 327)

echo sounder (p. 328)
estuary (p. 340)
euphotic zone (p. 336)
graded bedding (p. 335)
guyot (p. 335)
halocline (p. 328)
hydrogenous sediment (p. 341)
intertidal zone (p. 337)
manganese nodule (p. 342)
mid-ocean ridge (p. 335)
neritic zone (p. 337)
oceanic zone (p. 338)
oceanography (p. 324)

outgassing (p. 326)
passive continental margin (p. 332)
pelagic zone (p. 338)
photic zone (p. 336)
reef (p. 340)
rift valley (p. 336)
salinity (p. 325)
seamount (p. 335)
submarine canyon (p. 334)
terrigenous sediment (p. 341)
thermocline (p. 327)
turbidite (p. 335)
turbidity current (p. 334)

Review Questions

1. How does the area covered by the oceans compare with that of the continents? Describe the distribution of land and water on Earth.

2. Answer the following questions about the Atlantic, Pacific, and Indian oceans:

    (a) Which is the largest in area? Which is smallest?

    (b) Which has the greatest north-south extent?

    (c) Which is shallowest? Explain why its average depth is least.

3. How does the average depth of the ocean compare to the average elevation of the continents?

4. What is meant by *salinity?* What is the average salinity of the ocean?

5. Why do variations in salinity occur? Give some specific examples to illustrate your answer.

6. What is the origin of the elements that make up sea salt?

7. List three commercial products that are extracted from seawater.

8. Why is desalination not likely to be a significant source of water for agriculture in the foreseeable future?

9. Assuming that the average speed of sound waves in water is 1500 meters per second, determine the water depth if the signal sent out by an echo sounder requires 6 seconds to strike bottom and return to the recorder (see Figure 12.9).

10. List the three major subdivisions of a passive continental margin. Which subdivision is considered a flooded extension of the continent? Which has the steepest slope?

11. How does an active continental margin differ from a passive continental margin?

12. Defend or rebut the statement "Most submarine canyons found on the continental slope and rise were formed during the Ice Age when rivers extended their valleys seaward."

13. What are turbidites? What is meant by the term *graded bedding?*

14. Why are abyssal plains more extensive on the floor of the Atlantic than on the floor of the Pacific?

15. How are mid-ocean ridges and deep-ocean trenches related to plate tectonics?

16. List three factors that influence the abundance and distribution of marine life.

17. Why are there greater numbers and types of organisms in the neritic zone than in the oceanic zone?

18. What is an atoll? Describe Darwin's proposal on the origin of atolls. Was it ever confirmed?

19. Distinguish among the three basic types of seafloor sediment.

20. Why are seafloor sediments useful in studying climates of the past?

Examining the Earth System

1. Describe some of the material and energy exchanges that take place at the interface between the (a) ocean surface and atmosphere, (b) ocean water and ocean floor, and (c) ocean biosphere and ocean water.

2. The photo in Figure 12.21 shows Bora-Bora, a volcanic island surrounded by a coral reef. Use this photo as the basis for an essay that describes Earth as a system. Be sure to include several examples of interactions among solid Earth, hydrosphere, atmosphere, and biosphere. Indicate the sources of energy for the processes that produced the features shown in the photo. You may find the informaiton at Indiana University's Reef Resource Page Web Site (http://www.indiana.edu/~reefpage/) interesting, informative, and useful.

Figure 12.21 Bora-Bora is a volcanic island surrounded by a barrier reef. (Photo by Jean-Marc Truchet/Tony Stone Images)

Web Resources

The following are informative and interesting World Wide Web sites that address topics related to those presented in the chapter:

Ocean Planet (Smithsonian Institution):

• **http://seawifs.gsfc.nasa.gov/ocean_planet.html**

Reef Resource Page:

• **http://www.indiana.edu/~reefpage/**

For direct links to these sites and other appropriate Earth science Web locations, chapter review exercises, and relevant, topical investigations that utilize Web resources, contact the *Earth Science*, 9th Edition Home Page at:

• **http://www.prenhall.com/tarbuck.**

Chapter 13

The Restless Ocean

Coastal erosion forced the abandonment of this Long Island, New York shoreline. The shoreline is a dynamic place that can change rapidly in response to natural forces. (Photo by Mark Wexler/Woodfin Camp and Associates)

The restless waters of the ocean are constantly in motion. Winds generate surface currents, the Moon and Sun produce tides, and density differences create deep-ocean circulation. Further, waves carry the energy from storms to distant shores, where their impact erodes the land. This chapter examines the movements of ocean waters and their importance.

Ocean circulation involves surface currents and movements of deep-water masses. Surface waters are set in motion by the wind (Figure 13.1). Some of these currents are short-lived and affect only small areas. Such water movements are responses to local or seasonal influences. Other surface currents are relatively permanent phenomena that extend over large portions of the oceans. These major horizontal movements of surface waters are closely related to the general atmospheric circulation, which in turn is driven by the unequal heating of Earth by the Sun.

In contrast to the largely horizontal movements of surface currents, the deep-ocean circulation has a significant vertical component and accounts for the thorough mixing of deep-water masses. This component of ocean circulation is a response to density differences among water masses and reflects contrasts in temperature and salinity.

Surface Currents

There is a river in the ocean. In the severest droughts it never fails, and in the mightiest floods...it never overflows. Its banks and its bottoms are of cold water, while its current is of warm. The Gulf of Mexico is its fountain, and its mouth is in the Arctic Sea. It is the Gulf Stream. There is in the world no other such majestic flow of waters.

This quote, from Matthew Fontaine Maury's 1855 book, *The Physical Geography of the Sea*, describes perhaps the best-known and most-studied of the surface ocean currents, the Gulf Stream. This current is just a portion of a huge, slowly moving, circular whorl, or **gyre.** Figure 13.2 shows the entire gyre, which fills the North Atlantic, and other major gyres of the world ocean.

Ocean Circulation Patterns

Where the atmosphere and ocean are in contact, energy is passed from moving air to the water through friction. The drag exerted by winds blowing steadily across the ocean causes the surface layer of water to move. Thus, winds are the primary driving force of surface currents. This means there is a clear relationship between the oceanic circulation and the general

Figure 13.1 Wind is not only responsible for creating waves, but it also provides the force that drives the ocean's surface circulation. (Photo by Otto Rogge/The Stock Market)

Figure 13.2 Average position and extent of the principal surface ocean currents.

atmospheric circulation. A comparison of Figures 13.2 and 16.13 (p. 452) illustrates this.

A further clue to the influence of winds on ocean circulation is provided by the currents in the northern Indian Ocean. Here, there are seasonal wind shifts known as the summer and winter monsoons. When the winds change direction, the surface currents also reverse direction.

Although wind is the force that generates surface currents, other factors also influence the movement of ocean waters. The most significant of these is the **Coriolis effect**. Because of Earth's rotation, currents are deflected to the right in the Northern Hemisphere and to the left in the Southern Hemisphere. (The Coriolis effect will be explained in Chapter 16.) As a consequence, the direction of surface currents does not coincide precisely with the wind direction. Rather, they are altered by Earth's rotation.

North and south of the equator are two westward-moving currents, the equatorial currents (Figure 13.2). They derive their energy principally from the trade winds that blow from the northeast and southeast, respectively, toward the equator. The equatorial currents can be thought of as the "backbone" of the system of ocean currents. Because of the Coriolis effect, these currents are deflected poleward to form clockwise gyres in the Northern Hemisphere and counterclockwise gyres in the Southern Hemisphere. These nearly circular patterns are found in each major ocean basin (Figure 13.2).

In the North Atlantic, the equatorial current is deflected northward through the Caribbean, where it becomes the Gulf Stream (Figure 13.2). As the Gulf Stream moves along the East Coast of the United States, it is strengthened by the prevailing westerly winds and is deflected to the east (to the right) between the Carolinas and New England. As it continues northeastward beyond the Grand Banks, it gradually widens and slows until it becomes a vast, slowly moving current known as the North Atlantic Drift (Figure 13.2).

As the North Atlantic Drift approaches Western Europe, it splits, with part moving northward past Great Britain, Norway, and Iceland, carrying heat to these otherwise chilly areas. The other part is deflected southward as the cool Canaries Current. As the Canaries Current moves southward, it eventually merges into the North Equatorial Current, completing the gyre.

The clockwise circulation of the North Atlantic leaves a large central area that has no well-defined currents. This zone of calmer waters is known as the Sargasso Sea, named for the large quantities of *Sargassum*, a type of seaweed, encountered there.

In the South Atlantic, surface ocean circulation is very much the same as in the North Atlantic. And, circulation in the North and South Pacific generally parallels that of the North and South Atlantic (Figure 13.2).

Around the ice-covered continent of Antarctica, with no large landmasses in the way, cold surface waters circulate continuously. They move in response to prevailing westerly winds, giving this circulation the name West Wind Drift.

Ocean Currents and Upwelling

In addition to producing surface currents, winds can also cause *vertical* water movements. **Upwelling**, the rising of cold water from deeper layers to replace warmer surface water, is a common wind-induced vertical movement. It is most characteristic along the eastern shores of the oceans, most notably along the coasts of California, South America, and West Africa.

Upwelling occurs in these areas when winds blow toward the equator and parallel to the coast. Owing to the Coriolis effect, the surface water movement is deflected away from the shore. As the surface layer moves away from the coast, it is replaced by water that "upwells" from below the surface. This slow upward flow from depths of 50 to 300 meters (165 to 1000 feet) brings water that is cooler than the original surface water and creates a characteristic zone of lower temperatures near the shore.

For swimmers who are accustomed to the warm waters along the mid-Atlantic shore of the United States, a swim in the Pacific off the coast of central California can be a chilling surprise. In August, when temperatures in the Atlantic are 21°C (70°F) or higher, central California's surf is only about 15°C (60°F). Coastal upwelling also brings to the ocean surface greater concentrations of dissolved nutrients, such as nitrates and phosphates. These nutrient-enriched waters from below promote the growth of plankton, which in turn supports extensive populations of fish.

The Importance of Ocean Currents

Anyone who navigates the oceans needs to be aware of currents. By understanding the direction and strength of ocean currents, sailors soon realize that their voyage time can be reduced if they travel with a current, or increased if they travel against a current. For example, the first good chart of the Gulf Stream, which was printed by Benjamin Franklin, was developed as a result of this need to know (Figure 13.3). P.L. Richardson describes the situation as follows:

> While he was in London as Deputy Postmaster General for the American colonies, Franklin was consulted on the question of why the mail packets took a fortnight longer to sail to America than the merchant ships. In October 1768 Franklin discussed this problem with his cousin Timothy Folger, a Nantucket ship captain then visiting London. Folger told him the packet captains were ignorant of the Gulf Stream and frequently sailed in this current.... Folger sketched the Gulf Stream on a chart and added written notes on how to avoid the Gulf Stream and Franklin had the chart printed in 1769 or 1770.

Even today, the Franklin-Folger chart remains a good summary of the average path of the Gulf Stream.

Currents also have an important effect on climates. The moderating effect of poleward-moving

warm ocean currents is well known. The North Atlantic Drift, an extension of the warm Gulf Stream, keeps Great Britain and much of northwestern Europe warmer during the winter than one would expect for their latitudes, which are similar to the latitudes of Alaska and Labrador. The prevailing westerly winds carry the moderating effects far inland. For example, Berlin (52 degrees north latitude) has an average January temperature similar to that experienced at New York City, which lies 12 degrees farther south.

In contrast to warm ocean currents whose effects are felt most in the middle latitudes in winter, the influence of cold currents is most pronounced in the tropics or during summer months in the middle latitudes. Cold currents, such as the Benguela Current along Africa and the Peruvian Current (see Figure 13.2), moderate the tropical heat.

In addition to influencing temperatures of adjacent land areas, cold currents have other climatic influences. For example, where tropical deserts exist along the west coasts of continents, cold ocean currents have a dramatic impact. The principal west coast deserts are the Atacama in Peru and Chile, and the Namib in southern Africa. The aridity along these coasts is intensified because the lower air is chilled by cold offshore waters. When this occurs, the air becomes very stable and resists the upward movement necessary to create precipitation-producing clouds. In addition, the presence of cold currents causes temperatures to approach and often reach the dew point, the temperature at which water vapor condenses. As a result, these areas are characterized by high relative humidities and much fog. Thus, not all tropical deserts are hot with low humidities and clear skies. Rather, the presence of cold currents transforms some tropical deserts into relatively cool, damp places that are often shrouded in fog.

Ocean currents also play a major role in maintaining Earth's heat balance. They accomplish this task by transferring heat from the tropics, where there is an excess of heat, to the polar regions, where a deficit exists (Figure 13.3). Ocean water movement accounts for about a quarter of this heat transport, and winds transport the remaining three-quarters.

Deep-Ocean Circulation

Ocean circulation is not confined to surface currents. Significant deep-water movements occur as well. Unlike the wind-induced movements of surface and near-surface waters, deep-ocean circulation is governed by gravity and driven by density differences.

Two factors—temperature and salinity—are most significant in creating a dense mass of water. Water that is cold and salty is denser than warmer, less salty

---

* P. L. Richardson, "Benjamin Franklin and Timothy Folger's First Printed Chart of the Gulf Stream," *Science* 207 (4431): 643.

Figure 13.3 To aid ships crossing the Atlantic, Benjamin Franklin, with the assistance of his cousin, Timothy Folger, produced the first detailed chart of the Gulf Stream about 1769 or 1770. (Courtesy of NOAA) Today satellite images provide views of the Gulf Stream's complexities. Reds and yellows denote warmer waters. The current transports heat from lower latitudes toward the North Pole. Meanders of the Stream pinch off to form eddies that may move about the ocean for up to 2 years before dissipating. (Courtesy of O. Brown, R. Evans, and M. Carle/University of Miami Rosenstiel School of Marine and Atmospheric Science)

water. Consequently, deep-ocean circulation is called **thermohaline circulation** (thermo- "heat," haline- "salt"). After sinking from the surface of the ocean, waters will not reappear at the surface for an average of 500 to 2000 years.

The basic idea of thermohaline circulation is easily understood. Water at the surface of the ocean is made colder (by heat loss to the atmosphere) and/or more salty (by removal of water by evaporation or freezing) and becomes more dense. This denser water then sinks toward the ocean bottom and displaces lighter (less dense) water, which moves back toward the zone where the denser water formed. In this manner, cold, dense waters flow away from their source near the poles and are replaced by warmer waters from lower latitudes.

Arctic and Antarctic waters represent the two major regions where dense water masses are created. Antarctic

waters are chilled during the winter. The temperatures here are low enough to form sea ice, and because sea salts do not get frozen into the ice, the remaining water becomes saltier. The result is the densest water in all of the oceans (Figure 13.4). This cold saline brine slowly sinks to the seafloor, where it moves through the ocean basins in sluggish currents. Centuries later, subsurface waters return to the sea surface.

Sometimes ocean circulation is compared to a conveyor belt that travels from the Atlantic, through the Indian and Pacific oceans and back again (Figure 13.5). In this simplified model, warm water in the ocean's upper layers flows poleward, converts to dense water, and returns equatorward as cold deep water that eventually upwells to complete the circuit. As this "conveyor belt" moves around the globe, it influences global climate by converting warm water to cold and liberating heat to the atmosphere.

Figure 13.4 Sea ice near the Antarctic Peninsula. When seawater freezes, sea salts do not become part of the ice. Consequently, the salt content of the remaining seawater becomes more concentrated, which makes it denser and prone to sink. (Photo by Wolfgang Kaehler)

Tides

Tides are daily changes in the elevation of the ocean surface. Their rhythmic rise and fall along coastlines have been known since antiquity. Other than waves, they are the easiest ocean movements to observe. An exceptional example of extreme daily tides is shown in Figure 13.6.

Although known for centuries, tides were not explained satisfactorily until Sir Isaac Newton applied the law of gravitation to them. Newton showed that there is a mutual attractive force between two bodies, as between Earth and the Moon. Because both the atmosphere and the ocean are fluids, and free to move, both are deformed by this force. Hence, ocean tides result from the gravitational attraction exerted upon Earth by the Moon, and to a lesser extent by the Sun.

Causes of Tides

To illustrate how tides are produced, we will assume Earth is a rotating sphere covered to a uniform depth with water (Figure 13.7). It is easy to see how the Moon's gravitational force can cause the water to bulge on the side of Earth nearest the Moon. In addition, however, an equally large tidal bulge is produced on the side of Earth directly opposite the Moon.

Both tidal bulges are caused, as Newton discovered, by the pull of gravity. Gravity is inversely proportional to the square of the distance between two objects, meaning simply that it quickly weakens with distance. In this case, the two objects are the Moon and Earth. Because the force of gravity decreases with distance, the Moon's gravitational pull on Earth is slightly greater on the near side of Earth than on the far side. The result of this differential pulling is to stretch (elongate) the solid Earth, very slightly. In contrast, the world ocean, which is mobile, is deformed quite dramatically by this effect to produce the two opposing tidal bulges.

Because the position of the Moon changes only moderately in a single day, the tidal bulges remain in place while Earth rotates "through" them. For this reason, if you stand on the seashore for 24 hours, Earth

Figure 13.5 A simple "conveyor belt" model of ocean circulation. As warm surface waters move northward in the Atlantic Ocean, they cool and sink. After this dense subsurface flow rounds Africa, it eventually rises to the surface in the Indian and Pacific oceans. In this model, the Atlantic acts as a "radiator" that converts imported warm water to exported cold water.

Figure 13.6 High tide and low tide on Nova Scotia's Minas Basin in the Bay of Fundy. The areas exposed during low tide and flooded during high tide are called *tidal flats*. Tidal flats here are extensive. (Courtesy of Nova Scotia Deptartment of Tourism and Culture)

will rotate you through alternating areas of deeper and shallower water. As you are carried into regions of deeper water, the tide rises, and as you are carried away, the tide falls. Therefore, during one day you experience two high tides and two low tides.

Further, the tidal bulges migrate as the Moon revolves around Earth every 29 days. As a result, the

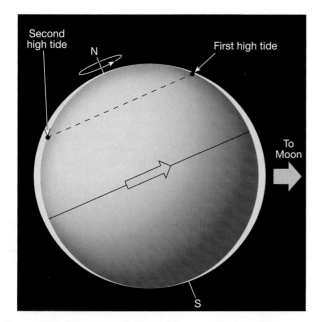

Figure 13.7 Tides on an Earth that is covered to a uniform depth with water. Depending on the Moon's position, tidal bulges may be inclined to the equator. In this situation an observer will experience two unequal high tides.

tides, like the time of moonrise, shift about 50 minutes later each day. After 29 days, the cycle is complete and a new one begins.

There may be an inequality between the high tides during a given day. Depending on the Moon's position, the tidal bulges may be inclined to the equator as in Figure 13.7. This figure illustrates that the first high tide experienced by an observer in the Northern Hemisphere is considerably lower than the high tide half a day later. In contrast, a Southern Hemisphere observer would experience the opposite effect.

Spring and Neap Tides

The Sun also influences the tides. It is far larger than the Moon, but because it is so far away, the effect is considerably less than that of the Moon. In fact, the tide-generating potential of the Sun is slightly less than half that of the Moon.

Near the times of new and full moons, the Sun and Moon are aligned and their forces are added together (Figure 13.8A). Accordingly, the combined gravity of these two tide-producing bodies cause higher tidal bulges (high tides) and lower tidal troughs (low tides). These are called the **spring tides**. Spring tides create the largest daily tidal range, that is, the largest variation between high and low tides. Conversely, at about the time of the first and third quarters of the Moon, the gravitational forces of the Moon and Sun act on Earth at right angles, and each partially offsets the influence of the other (Figure 13.8B). As a result, the daily tidal range is less. These are called **neap tides**.

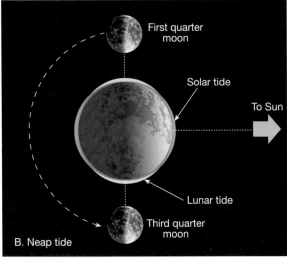

Figure 13.8 Relationship of the Moon and Sun to Earth during **A.** spring tides and **B.** neap tides.

Types of Tides

So far we have explained the basic causes and patterns of tides. But keep in mind these theoretical considerations cannot be used to predict either the height or the time of actual tides at a particular place. The shape of coastlines and the configuration of ocean basins greatly influence the tides. Consequently, tides at various locations respond differently to the tide-producing forces. This being the case, the nature of the tide at any place can be determined most accurately by actual observation. The predictions in tidal tables and the tidal data on nautical charts are based on such observations.

Tides are classified as one of three types according to the tidal pattern occurring at a particular locale. The *semidiurnal type* fits the twice-daily pattern described earlier. There are two high and two low tides each tidal day, with a relatively small difference in the high and low water heights. Tides along the Atlantic coast of the United States are representative of the semidiurnal type, which is illustrated by the tidal curve for Boston in Figure 13.9A. *Diurnal tides* are characterized by a single high and low water height each tidal day (Figure 13.9B). Tides of this type occur along the northern shore of the Gulf of Mexico, among other locations. More common than the diurnal tide, the *mixed* type of

tide is characterized by a large inequality in high water heights, low water heights, or both (Figure 13.9C). In this case, there are usually two high and two low waters each day. Such tides are prevalent along the Pacific coast of the United States and in many other parts of the world.

Tidal Currents

Tidal current is the term used to describe the *horizontal* flow of water accompanying the rise and fall of the tide. These water movements induced by tidal forces can be important in some coastal areas. Tidal currents that advance into the coastal zone as the tide rises are called *flood currents*. As the tide falls, seaward-moving water generates *ebb currents*. Periods of little or no current, called *slack water*, separate flood and ebb. The areas affected by these alternating tidal currents are called **tidal flats** (see Figure 13.6). Depending on the

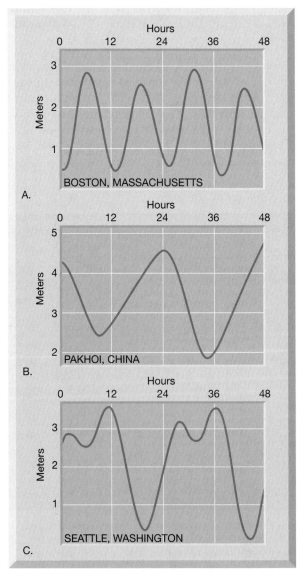

Figure 13.9 Types of tides. **A.** Semidiurnal. **B.** Diurnal. **C.** Mixed.

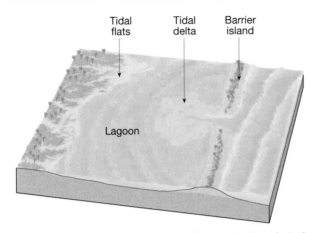

Tidal flats   Tidal delta   Barrier island

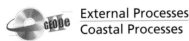

Lagoon

Figure 13.10 Because this tidal delta is forming in the relatively quiet waters on the landward side of a barrier island, it is termed a flood delta. As a rapidly moving tidal current emerges from the inlet, it slows and deposits sediment. The shapes of tidal deltas range from irregular to typical delta shaped.

nature of the coastal zone, tidal flats vary from narrow strips seaward of the beach to zones that may extend for several kilometers.

Although tidal currents are not important in the open sea, they can be rapid in bays, river estuaries, straits, and other narrow places. Off the coast of Brittany in France, for example, tidal currents that accompany a high tide of 12 meters (40 feet) may attain a speed of 20 kilometers (12 miles) per hour. Tidal currents are not generally considered to be major agents of erosion and sediment transport, but notable exceptions occur where tides move through narrow inlets. Here they scour the small entrances to many good harbors that would otherwise be blocked.

Sometimes deposits called **tidal deltas** are created by tidal currents (Figure 13.10). They may develop either as *flood deltas* landward of an inlet or as *ebb deltas* on the seaward side of an inlet. Because wave activity and longshore currents are reduced on the sheltered landward side, flood deltas are more common and actually more prominent (see Figure 13.21, p. 364). They form after the tidal current moves rapidly through an inlet. As the current emerges into more open waters from the narrow passage, it slows and deposits its load of sediment.

Waves Modify the Shoreline

 External Processes
Coastal Processes

The waters of the ocean are constantly in motion. The restless nature of the water is most noticeable along the shore—the dynamic interface between land and sea. Here we can observe the rhythmic rise and fall of tides and see waves constantly rolling in and breaking. Sometimes the waves are low and

gentle. At other times, they pound the shore with an awesome fury.

Although it may not be obvious, the shoreline is constantly being modified by waves. For example, along Cape Cod, Massachusetts, wave activity is eroding cliffs of poorly consolidated glacial sediment so aggressively that the cliffs are retreating inland up to 1 meter per year (Figure 13.11A). By contrast, at Point Reyes, California, the far more durable bedrock cliffs are far less susceptible to wave attack and therefore are retreating much more slowly (Figure 13.11B). Along both coasts, wave activity is moving sediment along the shore and building narrow sand bars that protrude into and across some bays.

The nature of present-day shorelines is not just the result of the relentless attack of the land by the sea. Indeed, the shore has a complex character that results from multiple geologic processes. For example, practically all coastal areas were affected by the worldwide rise in sea level that accompanied the melting of glaciers at the close of the Pleistocene epoch. As the sea encroached landward, the shoreline retreated, becoming superimposed upon existing landscapes that had resulted from such diverse processes as stream erosion, glaciation, volcanic activity, and the forces of mountain building.

Today the coastal zone is experiencing intensive human activity. Unfortunately, people often treat the shoreline as if it were a stable platform on which structures can safely be built. This attitude inevitably leads to conflicts between people and nature. As we shall see, many coastal landforms, especially beaches and barrier islands, are relatively fragile, short-lived features that are inappropriate sites for development.

Waves

 External Processes
Coastal Processes

Wind-generated waves provide most of the energy that shapes and modifies shorelines. Where the land and sea meet, waves that may have traveled unimpeded for hundreds of kilometers suddenly encounter a barrier that will not allow them to advance farther and that must absorb their energy. Stated another way, the shore is where a practically irresistible force confronts an almost immovable object. The conflict that results is never-ending and sometimes dramatic.

Characteristics of Waves

The undulations of the water surface, called waves, derive their energy and motion from the wind. If a breeze of less than 3 kilometers (2 miles) per hour starts to blow across still water, small wavelets appear almost instantly. When the breeze dies, the ripples disappear as suddenly as they formed. However, if the

Figure 13.11 A. This satellite image includes the familiar outline of Cape Cod. Boston is in the upper left corner. The two large islands off the south shore of Cape Cod are Martha's Vineyard (left) and Nantucket (right). Although the work of waves constantly modifies this coastal landscape, shoreline processes are not responsible for creating it. Rather, the present size and shape of Cape Cod result from the positioning of moraines and other glacial materials deposited during the Pleistocene epoch. (Photo courtesy of Earth Satellite Corporation/Science Photo Library/Photo Researchers, Inc.)

wind exceeds 3 kilometers per hour, more stable waves gradually form and progress with the wind.

All waves have the characteristics illustrated in Figure 13.12. The tops of the waves are the *crests*, which are separated by *troughs*. The vertical distance between trough and crest is called the **wave height**, and the horizontal distance separating successive crests is the **wavelength**. The **wave period** is the time interval between the passage of successive crests at a stationary point.

The height, length, and period that are eventually achieved by a wave depend on three factors: (1) wind speed; (2) length of time the wind has blown; and (3) **fetch**, the distance that the wind has traveled across open water. As the quantity of energy transferred from the wind to the water increases, both the height and steepness of the waves also increase. Eventually, a critical point is reached where waves grow so tall that they topple over, forming ocean breakers called *whitecaps*.

When the wind stops or changes direction, or the waves leave the stormy area where they were created, they continue on without relation to local winds. The waves also undergo a gradual change to *swells*, which are lower in height and longer in length and may carry

Figure 13.11 B. High-altitude image of the Point Reyes area north of San Francisco, California. The 5.5-kilometer-long south-facing cliffs at Point Reyes (lower left corner) are exposed to the full force of the waves from the Pacific Ocean. Nevertheless, this promontory retreats slowly because the bedrock from which it formed is very resistant. (Photo courtesy of USDA-ASCS)

a storm's energy to distant shores. Because many independent wave systems exist at the same time, the sea surface acquires a complex and irregular pattern.

Hence, the sea waves we watch from the shore are usually a mixture of swells from faraway storms and waves created by local winds.

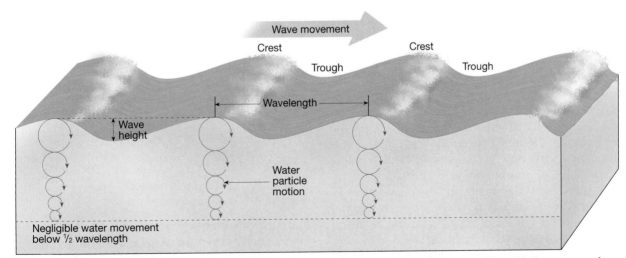

Figure 13.12 This diagram illustrates the basic parts of a wave as well as the movement of water particles with the passage of a wave. Negligible water movement occurs below a depth equal to one-half the wavelength (the level of the dashed line).

Types of Waves

When observing waves, always remember that you are watching *energy* travel through a medium (water). If you make waves by tossing a pebble into a pond, or by splashing in a pool, or by blowing across the surface of an aquarium, you are imparting *energy* to the water, and the waves you see are just the visible evidence of the energy passing through.

In the open sea, it is the wave form that moves forward, not the water itself. Each water particle moves in a circular path during the passage of a wave (Figure 13.12). As a wave passes, a water particle returns almost to its original position. The circular orbits followed by the water particles at the surface have a diameter equal to the wave height. When water is part of the wave crest, it moves in the same direction as the advancing wave form. In the trough, the water moves in the opposite direction. This is demonstrated by observing the behavior of a floating object as a wave passes. The toy boat in Figure 13.13 merely seems to bob up and down and sway slightly to and fro without advancing appreciably from its original position. (The wind does drag the water forward slightly, causing the surface circulation of the oceans.) For this reason, waves in the open sea are called **waves of oscillation**.

The energy contributed by the wind to the water is transmitted not only along the surface of the sea but also downward. However, beneath the surface, the circular motion rapidly diminishes until, at a depth equal to about one-half the wavelength, the movement of water particles becomes negligible. This is shown by the rapidly diminishing diameters of water-particle orbits in Figure 13.13.

As long as a wave is in deep water, it is unaffected by water depth (Figure 13.14, left). However, when a wave approaches the shore, the water becomes shallower and influences wave behavior. The wave begins to "feel bottom" at a water depth equal to about one-half its wavelength. Such depths interfere with water movement at the base of the wave and slow its advance (Figure 13.14, center).

As a wave advances toward the shore, the slightly faster waves farther out to sea catch up, decreasing the wavelength. As the speed and length of the wave diminish, the wave steadily grows higher. Finally, a critical point is reached when the steep wave front collapses or *breaks* (Figure 13.14, right). What had been a wave of oscillation now becomes a **wave of translation** in which the water advances up the shore.

The turbulent water created by breaking waves is called **surf**. On the landward margin of the surf zone, the turbulent sheet of water from collapsing breakers, called *swash*, moves up the slope of the beach. When the energy of the swash has been expended, the water flows back down the beach toward the surf zone as *backwash*.

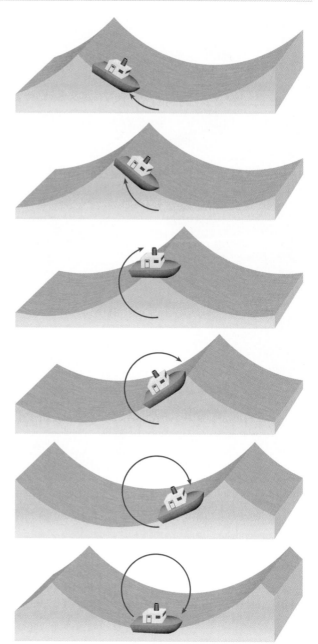

Figure 13.13 The movements of the toy boat show that the wave form advances, but the water does not advance appreciably from its original position. In this sequence, the wave moves from left to right as the boat (and the water in which it is floating) rotates in an imaginary circle. The boat moves slightly to the left up the front of the approaching wave, then after reaching the crest, slides to the right down the back of the wave.

Wave Erosion

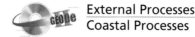

**External Processes
Coastal Processes**

During calm weather wave action is minimal. However, just as streams do most of their work during floods, so too do waves perform most of their work during storms. The impact of high, storm-induced waves against the

Figure 13.14 Changes that occur when a wave moves onto shore.

shore can be awesome in its violence. Each breaking wave may hurl thousands of tons of water against the land, sometimes causing the ground literally to tremble. The pressures exerted by Atlantic waves in wintertime, for example, average nearly 10,000 kilograms per square meter (more than 2000 pounds per square foot). The force during storms is even greater.

It is no wonder that cracks and crevices are quickly opened in cliffs, seawalls, breakwaters, and anything else that is subjected to these enormous shocks. Water is forced into every opening, causing air in the cracks to become highly compressed by the thrust of crashing waves. When the wave subsides, the air expands rapidly, dislodging rock fragments and enlarging and extending preexisting fractures (Figure 13.15).

In addition to the erosion caused by wave impact and pressure, **abrasion**, the sawing and grinding action of the water armed with rock fragments, is also

Figure 13.15 The force of breaking waves can be powerful. Here waves batter a seawall at Sea Bright, New Jersey. (Photo by Rafael Marcia/Photo Researchers, Inc.)

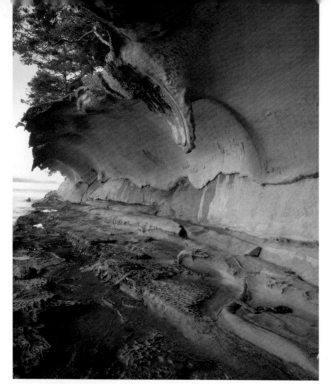

Figure 13.16 Sandstone cliff undercut by wave erosion. Gabriola Island, British Columbia, Canada. (Photo by Fletcher and Baylis/Photo Researchers, Inc.)

important. In fact, abrasion is probably more intense in the surf zone than in any other environment. Smooth, rounded stones and pebbles along the shore are obvious reminders of the relentless grinding action of rock against rock in the surf zone. Further, such fragments are used as "tools" by the waves as they cut horizontally into the land (Figure 13.16).

Wave Refraction

 External Processes
Coastal Processes

The bending of waves, called **wave refraction**, plays an important part in shoreline processes. It affects the distribution of energy along the shore, and thus strongly influences where and to what degree erosion, sediment transport, and deposition will take place.

Waves seldom approach the shore straight on. Rather, most waves move toward the shore at an angle. However, when they reach the shallow water of a smoothly sloping bottom they are bent and tend to become parallel to the shore. Such bending occurs because the part of the wave nearest the shore touches bottom and slows first, whereas the end that is still in deep water continues forward at its full speed. The net result is a wave front that may approach nearly parallel to the shore regardless of the original direction of the wave.

Because of refraction, wave impact is concentrated against the sides and ends of headlands that project into the water, whereas wave attack is weakened in bays. This differential wave attack along irregular coastlines is illustrated in Figure 13.17. Because the waves reach the shallow water in front of the headland sooner than they do in adjacent bays, they are bent more nearly parallel to the protruding land and strike it from all three sides. By contrast, refraction in the bays causes waves to diverge and expend less energy.

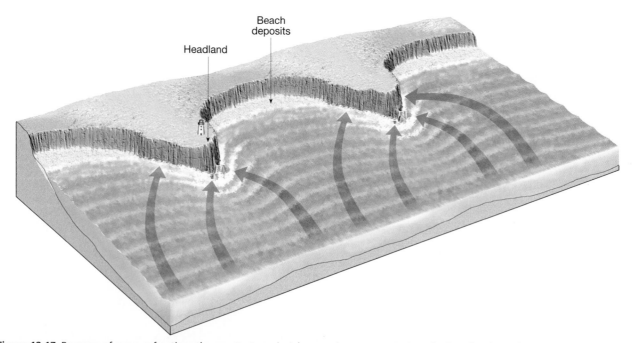

Figure 13.17 Because of wave refraction, the greatest erosional power is concentrated on the headlands. In the bays, the force of the waves is much weaker.

In these zones of weakened wave activity, sediments can accumulate and form sandy beaches. Over a long period, erosion of the headlands and deposition in the bays will straighten an irregular shoreline.

Moving Sand Along the Beach

 External Processes
Coastal Processes

Although waves are refracted, most still reach the shore at an angle, however slight. Consequently, the uprush of water from each breaking wave (the swash) is oblique. However, the backwash is straight down the slope of the beach. The effect of this pattern of water movement is to transport sediment in a zigzag pattern along the beach (Figure 13.18). This movement is called **beach drift**, and it can transport sand and pebbles hundreds or even thousands of meters daily.

Oblique waves also produce currents within the surf zone that flow parallel to the shore and move substantially more sediment than beach drift. Because the water here is turbulent, these **longshore currents** easily move the fine suspended sand, and roll larger sand and gravel along the bottom. When the sediment transported by longshore currents is added to the quantity moved by beach drift, the total amount can be

very large. At Sandy Hook, New Jersey, for example, the quantity of sand transported along the shore over a 48-year period averaged almost 750,000 tons annually. For a 10-year period at Oxnard, California, more than 1.5 million tons of sediment moved along the shore each year.

There should be little wonder that beaches have been characterized as "rivers of sand." At any point along a beach there is likely to be more sediment that was derived elsewhere than material eroded from the land immediately behind it. It is also worth noting that much of the sediment making up beaches is not wave-eroded debris. Rather, in many areas sediment-laden rivers that discharge into the ocean are the major source of material (see Box 13.1). For that reason, if it were not for beach drift and longshore currents, many beaches would be nearly sandless.

Shoreline Features

 External Processes
Coastal Processes

Shoreline features vary, depending on the rocks of the shore, currents, wave intensity, and whether the coast is stable, sinking, or rising. This section summarizes many shoreline features.

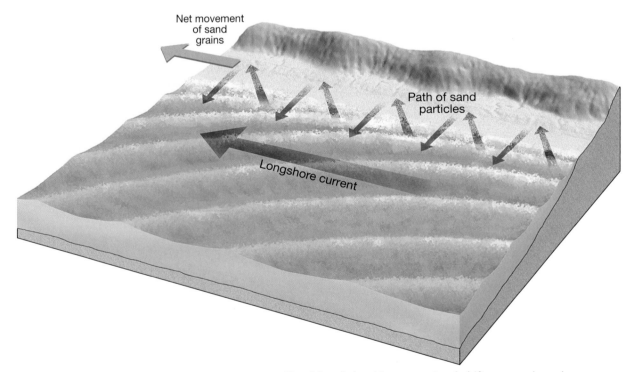

Figure 13.18 Beach drift and longshore currents are created by obliquely breaking waves. Beach drift occurs as incoming waves carry sand obliquely up the beach, while the water from spent waves carries it directly down the slope of the beach. Similar movements occur offshore in the surf zone to create the longshore current. These processes transport large quantities of material along the beach and in the surf zone.

Box 13.1 People and the Environment: Louisiana's Vanishing Coastal Wetlands

Coastal wetlands form in sheltered environments that include swamps, tidal flats, coastal marshes, and bayous. They are rich in wildlife and provide nesting grounds and important stopovers for waterfowl and migratory birds, as well as spawning areas and valuable habitats for fish.

The delta of the Mississippi River in Louisiana contains about 40 percent of all coastal wetlands in the lower 48 states. Louisiana's wetlands are sheltered from the damaging wave action of hurricanes and winter storms by low-lying barrier islands. Both the wetlands and the protecting islands have formed as a result of the shifting of the Mississippi River during the past 5000 to 6000 years.

Because the wetlands depend on the Mississippi River for sediment and freshwater, they are vulnerable when changes to the river system occur. Moreover, the reliance on barrier islands for protection from storm waves leaves coastal wetlands vulnerable when these narrow offshore islands are eroded.

Today, the coastal wetlands of Louisiana are disappearing at an alarming rate. The U.S. Army Corps of Engineers estimates that about 90 square kilometers (35 square miles) of Louisiana's wetlands are lost *annually*. Fifty years from now the shoreline is predicted to be many kilometers from where it is today. As the shoreline retreats, southern Louisiana's wetlands are being submerged beneath the Gulf of Mexico. And if the predicted rise in sea level occurs, it will only add to the coastal land loss. (See Box 13.2 for more on rising sea level.)

Why are Louisiana's wetlands shrinking? The answer is twofold: *natural change* and *human activity*. First, the Mississippi delta and its wetlands are continually changing naturally. As sediment accumulates and builds the delta in one area, erosion and subsidence cause losses elsewhere (Figure 13.A). When the river shifts, the zones of delta growth and destruction also shift. Second, ever since people arrived, the rate at which the delta and its wetlands are destroyed has accelerated.

Before Europeans settled the delta, the Mississippi River regularly overflowed its banks in seasonal floods. The huge quantities of sediment that were deposited renewed the soil and kept the delta from sinking below sea level. However, with settlement came flood-control efforts and the desire to maintain and improve navigation on the river.* Artificial levees were constructed to contain the rising river during flood stage. Over time, the levees were extended all the way to the mouth of the Mississippi to keep the channel open for navigation.

The effects have been straightforward. The levees prevent sediment and freshwater from being dispersed into the wetlands. Instead, the river is forced to carry its load to the deep waters at the mouth. Meanwhile, the processes of compaction, subsidence, and wave erosion continue. Because not enough sediment is added to offset these forces, the size of the delta and

*See the section on "Floods and Flood Control" in Chapter 4.

the extent of its wetlands gradually shrink.

The problem has been aggravated by a decline in the sediment transported by the Mississippi, decreasing by approximately 50 percent over the past century. A substantial portion of the reduction results from the trapping of sediment in large reservoirs created by dams built on tributaries to the Mississippi, especially the Missouri and Arkansas rivers. The diversion of part of the Mississippi's flow to the Atchafalaya distributary also reduces the amount of sediment available to build and maintain the delta.

Understanding and modifying the impact of people is a necessary basis for any plan to reduce the loss of wetlands in the Mississippi delta. Although some projects are underway to divert more sediment and freshwater to the wetlands, plans to cope effectively with the problem have yet to be implemented.

Figure 13.A Satellite image of the Mississippi delta. For the past 600 years or so, the main flow of the river has been along its present course, extending southeast from New Orleans. During that span, the delta advanced into the Gulf of Mexico at a rate of about 10 kilometers (6 miles) per century. (Photo courtesy of NASA)

Figure 13.19 Elevated wave-cut platform along the California coast north of San Francisco. A new platform is being created at the base of the cliff. (Photo by John S. Shelton)

Wave-Cut Cliffs and Platforms

Whether along the rugged and irregular New England coast or along the steep shorelines of the West Coast, the effects of wave erosion are often easy to see. **Wave-cut cliffs**, as the name implies, originate by the cutting action of the surf against the base of coastal land. As erosion progresses, rocks overhanging the notch at the base of the cliff crumble into the surf and the cliff retreats (see Figure 13.16). A relatively flat, benchlike surface, the **wave-cut platform**, is left behind by the receding cliff (Figure 13.19). The platform broadens as wave attack continues. Some debris produced by the breaking waves remains along the water's edge as part of the beach, while the remainder is transported farther seaward.

Arches, Stacks, Spits, and Bars

Headlands that extend into the sea are vigorously attacked by waves because of refraction. The surf erodes the rock selectively, wearing away the softer or more highly fractured rock at the fastest rate. At first, sea caves may form. When two caves on opposite sides of a headland unite, a **sea arch** results (Figure 13.20). Eventually, the arch falls in, leaving an isolated remnant, or **sea stack**, on the wave-cut platform (Figure 13.20). In time, it too will be consumed by the action of the waves.

Where beach drift and longshore currents are active, several features related to the movement of sediment along the shore may develop. **Spits** are elongated ridges of sand that project from the land into the mouth of an

Figure 13.20 Sea arch and sea stack at the tip of Mexico's Baja Peninsula. (Photo by Mark A. Johnson/The Stock Market)

Figure 13.21 High-altitude image of a well-developed spit and baymouth bar along the coast of Martha's Vineyard, Massachusetts. Also notice the tidal delta in the lagoon adjacent to the inlet through the baymouth bar. (Photo courtesy of USDA-ASCS)

adjacent bay. Often the end in the water hooks landward in response to wave-generated currents (Figure 13.21). The term **baymouth bar** is applied to a sand bar that completely crosses a bay, sealing it off from the open ocean (Figure 13.21). Such a feature tends to form across bays where currents are weak, allowing a spit to extend to the other side. A **tombolo**, a ridge of sand that connects an island to the mainland or to another island, forms in much the same manner as a spit.

Barrier Islands

The Atlantic and Gulf Coastal Plains are relatively flat and slope gently seaward. The shore zone is characterized by **barrier islands**. These low ridges of sand parallel the coast at distances from 3 to 30 kilometers offshore. From Cape Cod, Massachusetts, to Padre Island, Texas, nearly 300 barrier islands rim the coast (Figure 13.22).

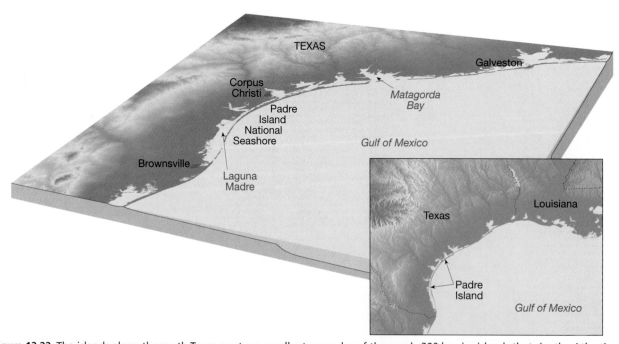

Figure 13.22 The islands along the south Texas coast are excellent examples of the nearly 300 barrier islands that rim the Atlantic and Gulf coasts.

Most barrier islands are from 1 to 5 kilometers wide and between 15 and 30 kilometers long. The tallest features are sand dunes, which usually reach heights of 5 to 10 meters. The lagoons that separate these narrow islands from the shore are zones of relatively quiet water that allow small craft traveling between New York and northern Florida to avoid the rough waters of the North Atlantic.

Barrier islands probably formed in several ways. Some originated as spits that were subsequently severed from the mainland by wave erosion or by the general rise in sea level following the last episode of glaciation. Others were created when turbulent waters in the line of breakers heaped up sand that had been scoured from the bottom. Finally, some barrier islands may be former sand dune ridges that originated along the shore during the last glacial period, when sea level was lower. As the ice sheets melted, sea level rose and flooded the area behind the beach-dune complex.

The Evolving Shore

A shoreline continually undergoes modification regardless of its initial configuration. At first, most coastlines are irregular, although the degree of and reason for the irregularity may differ considerably from place to place. Along a coastline that is characterized by varied geology, the pounding surf may at first increase its irregularity because the waves will erode the weaker rocks more easily than the stronger ones. However, if a shoreline remains stable, marine erosion and deposition will eventually produce a straighter, more regular coast. Figure 13.23 illustrates the evolution of an initially irregular coast. As waves erode the headlands, creating cliffs and a wave-cut platform, sediment is carried along the shore. Some material is deposited in the bays, while other debris is formed into spits and baymouth bars. At the same time rivers fill the bays with sediment. Ultimately, a smooth coast results.

Shoreline Erosion Problems

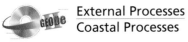

External Processes
Coastal Processes

Today the coastal zone teems with human activity. Unfortunately, people often treat the shoreline as if it were a stable platform on which structures can be built safely. This attitude jeopardizes both people and the shoreline. Many coastal landforms are relatively fragile, short-lived features that are easily damaged by development. As anyone who has endured a tropical storm knows, the shoreline is not always a safe place to live.

Compared with other natural hazards such as earthquakes, volcanic eruptions, and landslides, shoreline erosion appears to be a more continuous and predictable process that causes relatively modest damage to limited areas. In reality, the shoreline is a dynamic place that can change rapidly in response to natural forces. Exceptional storms are capable of eroding beaches and cliffs at rates that far exceed the long-term average. Such bursts of accelerated erosion not only have a significant impact on the natural evolution of a coast, but can also have a profound impact on people who reside in the coastal zone (Figure 13.24). Erosion along our coasts causes significant property damage. Huge sums are spent annually not only to repair damage, but also to prevent or control erosion. Already a problem at many sites, shoreline erosion is certain to become increasingly serious as extensive coastal development continues.

Although the same processes cause change along every coast, not all coasts respond in the same way. Interactions among different processes and the relative importance of each process depend on local factors. The factors include: (1) the proximity of a coast to sediment-laden rivers, (2) the degree of tectonic activity, (3) the topography and composition of the land, (4) prevailing winds and weather patterns, and (5) the configuration of the coastline and near-shore areas.

During the past 100 years, growing affluence and increasing demands for recreation have brought unprecedented development to many coastal areas. As both the number and the value of buildings have increased, so too have efforts to protect property from storm waves. Also, controlling the natural migration of sand is an ongoing struggle in many coastal areas. Such interference can result in unwanted changes that are difficult and expensive to correct.

Groins

To maintain or widen beaches that are losing sand, groins are sometimes constructed. A **groin** is a barrier built at a right angle to the beach to trap sand that is moving parallel to the shore (Figure 13.25). These structures often do their job so effectively that the longshore current beyond the groin becomes sand-starved. As a result, the current erodes sand from the beach on the leeward side of the groin.

To offset this effect, property owners downcurrent from the structure may erect a groin on their property. In this manner, the number of groins multiplies. An example of such proliferation is the shoreline of New Jersey, where hundreds of these structures have been built. Because it has been shown that groins often do not provide a satisfactory solution, they are no longer the preferred method of keeping beach erosion in check.

Breakwaters and Seawalls

In some coastal areas a **breakwater** might be constructed parallel to the shoreline. The purpose of such

Figure 13.23 These diagrams illustrate the changes that can take place through time along an initially irregular coastline that remains relatively stable. The coastline shown in part **A** gradually evolves to **B** and then **C**. The diagrams also serve to illustrate many of the features described in the section on shoreline features. (Photos by E. J. Tarbuck)

a structure is to protect boats from the force of large breaking waves by creating a quiet water zone near the shore. However, when this is done, the reduced wave activity along the shore behind the structure may allow sand to accumulate. If this happens, the marina will eventually fill with sand while the downstream beach erodes and retreats. At Santa Monica, California, where the building of a breakwater created such a problem, the city had to install a dredge to remove sand from the protected quiet water zone and deposit it down the beach where longshore currents and beach drift could recirculate the sand (Figure 13.26).

As development has moved ever closer to the beach, seawalls are sometimes built to defend property from the force of breaking waves. **Seawalls** are simply massive barriers intended to prevent waves from reaching the areas behind the wall (see Figure 13.15). Waves expend much of their energy as they move across an open beach. Seawalls cut this process short by reflecting the force of unspent waves sea-

Figure 13.24 This destruction occurred when Hurricane Emily hit North Carolina's Outer Banks on August 31, 1993. (Photo courtesy of *Raleigh News and Observer*)

ward. As a consequence, the beach to the seaward side of the seawall experiences significant erosion and may, in some instances, be eliminated entirely. Once the width of the beach is reduced, the seawall is subjected to even greater pounding by the waves. Eventually this battering will cause the wall to fail, and a larger, more expensive wall must be built to take its place.

The wisdom of building temporary protective structures along shorelines is increasingly questioned. The feelings of many coastal scientists and engineers are expressed in the following excerpt from a position

Figure 13.25 A series of groins at Ship Bottom, New Jersey. Because groins trap sand on the upcurrent side, the movement of sand along this coast, caused by the longshore current, must be toward the bottom of the photograph. (Photo by John S. Shelton)

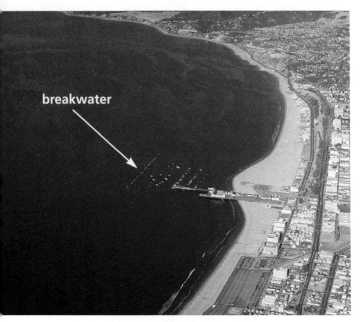

Figure 13.26 Aerial view of a breakwater at Santa Monica, California. The breakwater appears as a faint line in the water behind which many boats are anchored. The construction of the breakwater disrupted longshore transport and caused the seaward growth of the beach. (Photo by John S. Shelton)

paper that grew out of a conference on America's Eroding Shoreline:

> It is now clear that halting the receding shoreline with protective structures benefits only a few and seriously degrades or destroys the natural beach and the value it holds for the majority. Protective structures divert the ocean's energy temporarily from private properties, but usually refocus the energy on the adjacent natural beaches. Many interrupt the natural sand flow in coastal currents, robbing many beaches of vital sand replacement.*

Beach Nourishment

Beach nourishment represents another approach to stabilizing shoreline sands. As the term implies, this practice simply involves the addition of large quantities of sand to the beach system. By building the beaches seaward, beach quality and storm protection are both improved. Beach nourishment, however, is not a permanent solution to the problem of shrinking beaches. The same processes that removed the sand in the first place will eventually remove the replacement sand as well. In addition, beach nourishment is very expensive. When beach nourishment was used to renew 24 kilometers (15 miles) of Miami Beach, the cost was $64 million. Here the restoration must be redone every 10 to 12 years.

In some instances, beach nourishment can lead to unwanted environmental effects. For example, beach replenishment at Waikiki Beach, Hawaii, involved replacing coarse calcareous sand with softer, muddier calcareous sand. Destruction of the soft beach sand by breaking waves increased the water's turbidity and killed offshore coral reefs. At Miami Beach, increased turbidity also damaged local coral communities.

Beach nourishment appears to be an economically viable long-range solution to the beach preservation problem only in areas where there exists dense development, large supplies of sand, relatively low wave energy, and reconcilable environmental issues. Unfortunately, few areas possess all these attributes.

Abandonment and Relocation

So far, two basic responses to shoreline erosion problems have been considered: (1) the building of structures, such as groins and seawalls, to hold the shoreline in place, and (2) the addition of sand to replenish eroding beaches. However, a third option is also available. Many coastal scientists and planners are calling for a policy shift from defending and rebuilding beaches and coastal property in high hazard areas to removing storm-damaged buildings in those places and letting nature reclaim the beach. This approach is similar to that adopted by the federal government for river floodplains following the devastating 1993 Mississippi River floods in which vulnerable structures are abandoned and relocated on higher, safer ground.

Such proposals, of course, are controversial. People with significant near-shore investments shudder at the thought of not rebuilding and defending coastal developments from the erosional wrath of the sea. Others, however, argue that with sea level rising, the impact of coastal storms will only get worse in the decades to come (see Box 13.2). This group advocates that oft-damaged structures be abandoned or relocated to improve personal safety and to reduce costs. Such ideas will no doubt be the focus of much study and debate as states and communities evaluate and revise coastal land-use policies.

Contrasting the Atlantic and Pacific Coasts

The shoreline along the Pacific Coast of the United States is strikingly different from that characterizing the Atlantic and Gulf Coast regions. Some of the differences are related to plate tectonics. The West Coast represents the leading edge of the North American plate, and because of this, it experiences active uplift and deformation. By contrast, the East Coast is a tectonically quiet region that is far from any active plate margin. Because of this basic geological difference, the nature of shoreline erosion problems along America's opposite coasts is different.

---

* "Strategy for Beach Preservation Proposed," Geotimes 30 (No. 12, December 1985): 15.

Box 13.2   Earth as a System: Is Global Warming Causing Sea Level to Rise?

The shifting dynamic nature of barrier islands and the ineffectiveness of most shoreline protection measures are now relatively well-established facts. Unfortunately, recent research, which indicates that sea level is rising, has compounded this already distressing situation. Studies indicate that sea level has risen between 10 and 15 centimeters over the past century. Furthermore, some investigators predict an accelerated sea-level rise in the years to come—30 cm or more by the middle of the next century. Although such a vertical change may seem modest, many coastal geologists believe that any given rise in sea level along the gently sloping Atlantic and Gulf coasts will cause from 10 to 1000 times as much horizontal shoreline retreat (Figure 13.B).

The idea that sea level will continue to rise in the coming decades is linked to the results of climatic studies that predict a global warming trend. Such predictions are based on the now well-established fact that the carbon dioxide (CO_2) content of the atmosphere has been rising at an accelerating rate for more than a century. The CO_2 is added primarily as a by-product of the combustion of ever-increasing quantities of fossil fuels.

If we assume that the use of fossil fuels will continue to rise at projected rates, current estimates indicate that the atmosphere's CO_2 content will grow by an additional 40 percent or more by some time in the second half of the next century.

The importance of CO_2 lies in the fact that it traps a portion of the radiation emitted by Earth and thereby keeps the air near the surface warmer than it would be without CO_2. Because CO_2 is an important heat-absorbing gas, an increase in the air's CO_2 content is believed to contribute to higher atmospheric temperatures. In addition, other trace gases generated by human activities also play a role.

How is a warmer atmosphere related to a global rise in sea level? First, higher temperatures can cause glacial ice to melt. About one-half of the 10- to 15-centimeter rise in sea level over the past century is attributed to the melting of small glaciers and ice caps. Second, a warmer atmosphere causes an increase in ocean volume through thermal expansion. That is, higher air temperatures raise the temperature of the upper layers of the ocean, which, in turn, causes the water to expand and sea level to rise.

Because rising sea level is a gradual phenomenon, it may be overlooked by coastal residents as a significant contributor to shoreline erosion problems. Rather, the blame is assigned to other forces, especially storm activity. Although a given storm may be the immediate cause, the magnitude of its destruction may result from the relatively small sea level rise that allowed the storm's power to cross a much greater land area.

Figure 13.B The slope of a shoreline is critical to determining the degree to which sea-level changes will affect it. **A** When the slope is gentle, small changes in sea level cause a substantial shift. **B** The same sea-level rise along a steep coast results in only a small shoreline shift.

Atlantic and Gulf Coasts. Much of the coastal development along the Atlantic and Gulf coasts has occurred on barrier islands. Typically, barrier islands consist of a wide beach that is backed by dunes and separated from the mainland by marshy lagoons. The broad expanses of sand and exposure to the ocean have made barrier islands exceedingly attractive sites for development. Unfortunately, development has taken place more rapidly than our understanding of barrier island dynamics.

Because barrier islands face the open ocean, they receive the full force of major storms that strike the coast. When a storm occurs, the barriers absorb the energy of the waves primarily through the movement of sand. This process and the dilemma that results have been described as follows:

> Waves may move sand from the beach to offshore areas or, conversely, into the dunes; they may erode the dunes, depositing sand onto the beach or carrying it out to sea; or they may carry sand from the beach and the dunes into the marshes behind the barrier, a process known as overwash. The common factor is movement. Just as a flexible reed may survive a wind that destroys an oak tree, so the barriers survive hurricanes and nor'easters not through unyielding strength but by giving before the storm.

This picture changes when a barrier is developed for homes or a resort. Storm waves that previously rushed harmlessly through gaps between the dunes now encounter buildings and roadways. Moreover, since the dynamic nature of the barriers is readily perceived only during storms, homeowners tend to attribute damage to a particular storm, rather than to the basic mobility of coastal barriers. With their homes or investments at stake, local residents are more likely to seek to hold the sand in place and the waves at bay than to admit that development was improperly placed to begin with.

Pacific Coast. In contrast to the broad, gently sloping coastal plains of the East, much of the Pacific Coast is characterized by relatively narrow beaches that are backed by steep cliffs and mountain ranges. Recall that America's western margin is a more rugged and tectonically active region than the eastern margin. Because uplift continues, the apparent rise in sea level in the West is not so readily apparent. Nevertheless, like the shoreline erosion problems facing the East's barrier islands, West Coast difficulties also stem largely from the alteration of a natural system by people.

A major problem facing the Pacific shoreline, and especially portions of southern California, is a significant narrowing of many beaches. The bulk of the sand on many of these beaches is supplied by rivers that transport it from the mountainous regions to the coast. Over the years this natural flow of material to the coast has been interrupted by dams built for irrigation and flood control. The reservoirs effectively trap the sand that would otherwise nourish the beach environment. When the beaches were wider, they served to protect the cliffs behind them from the force of storm waves. Now, however, the waves move across the narrowed beaches without losing much energy and cause more rapid erosion of the sea cliffs.

Although the retreat of the cliffs provides material to replace some of the sand impounded behind dams, it also endangers homes and roads built on the bluffs. In addition, development atop the cliffs aggravates the problem. Urbanization increases runoff, which, if not carefully controlled, can result in serious bluff erosion. Watering lawns and gardens adds significant quantities of water to the slope. This water percolates downward toward the base of the cliff, where it may emerge in small seeps. This action reduces the slope's stability and facilitates mass wasting.

Shoreline erosion along the Pacific Coast varies considerably from one year to the next, largely because of the sporadic occurrence of storms. As a consequence, when the infrequent but serious episodes of erosion occur, the damage is often blamed on the unusual storms and not on coastal

development or the sediment-trapping dams that may be great distances away. If, as predicted, sea level rises at an increasing rate in the years to come, increased shoreline erosion and sea cliff retreat should be expected along many parts of the Pacific Coast (see Box 13.2).

Emergent and Submergent Coasts

The great variety of shorelines demonstrates their complexity. Indeed, to understand any particular coastal area, many factors must be considered, including rock types, size and direction of waves, frequency of storms, tidal range, and submarine profile. Moreover, recent tectonic events and changes in sea level must also be taken into account. These many variables make shoreline classification difficult.

Many geologists classify coasts based on changes that have occurred with respect to sea level. This commonly used classification divides coasts into two very general categories: emergent and submergent. **Emergent coasts** develop either because an area experiences uplift or as a result of a drop in sea level. Conversely, **submergent coasts** are created when sea level rises or the land adjacent to the sea subsides.

In some areas the coast is clearly emergent because rising land or falling water levels expose wave-cut cliffs and platforms above sea level. Excellent examples include portions of coastal California where uplift has occurred in the recent geological past. The elevated wave-cut platform shown in Figure 13.19 illustrates this. In the case of the Palos Verdes Hills, south of Los Angeles, seven different terrace levels exist, indicating seven episodes of uplift. The ever-persistent sea is now cutting a new platform at the base of the cliff. If uplift follows, it too will become an elevated marine terrace.

Other examples of emergent coasts include regions that were once buried beneath great ice sheets. When glaciers were present, their weight depressed the crust and when the ice melted, the crust began gradually to spring back. Consequently, prehistoric shoreline features today are found high above sea level. The Hudson Bay region of Canada is one such area, portions of which are still rising at a rate of more than a centimeter annually.

In contrast to the preceding examples, other coastal areas show definite signs of submergence. The shoreline of a coast that has been submerged in the relatively recent past is often highly irregular because the sea typically floods the lower reaches of river valleys flowing into the ocean. The ridges separating the valleys, however, remain above sea level and project into the sea as headlands. These drowned river mouths, which are called *estuaries*, characterize many coasts today. Along the Atlantic coastline, the Chesapeake and Delaware bays are examples of

---

* Frank Lowenstein, "Beaches or Bedrooms—The Choice as Sea Level Rises," *Oceanus* 28 (No. 3, Fall 1985): 22.

estuaries created by submergence (Figure 13.27). The picturesque coast of Maine, particularly in the vicinity of Acadia National Park, is another excellent example of an area that was flooded by the post-glacial rise in sea level and transformed into a highly irregular submerged coastline.

Keep in mind that most coasts have a complicated geologic history. With respect to sea level, many have at various times emerged and then submerged again. Each time they retain some of the features created during the previous situation.

Figure 13.27 Estuaries along the East Coast of the United States. The lower portions of many river valleys were submerged by the rise in sea level that followed the end of the Ice Age. Chesapeake Bay and Delaware Bay are especially prominent examples.

Chapter Summary

- *Surface ocean currents* are parts of huge, slowly moving, circular whirls, or *gyres*, that begin near the equator in each ocean. *Wind is the driving force* for the ocean's surface currents. Where wind is in contact with the ocean, it passes energy to the water through friction and causes the surface layer to move. The most significant factor other than wind that influences the movement of ocean waters is the *Coriolis effect*, the deflective force of Earth's rotation that causes free-moving objects to be deflected to the right in the Northern Hemisphere and to the left in the Southern Hemisphere. Because of the Coriolis effect, surface currents form clockwise gyres in the Northern Hemisphere and counterclockwise gyres in the Southern Hemisphere.

- *Upwelling*, the rising of colder water from deeper layers, is a common wind-induced vertical movement that is most characteristic along the eastern shores of the oceans.

- Ocean currents are important in navigation and travel and for the effect that they have on climates. The moderating effect of poleward-moving warm ocean currents during the winter in middle latitudes is well known. Cold currents exert their greatest influence in summer and year-round in the tropics. In addition to cooler temperatures, cold currents are associated with greater fog frequency and drought.

- In contrast to surface currents, *deep-ocean circulation* is governed by gravity and driven by *density* differences. The two factors that are most significant in creating a dense mass of water are *temperature* and *salinity*. The colder and saltier the water, the denser it is.

- *Tides*, the daily rise and fall in the elevation of the ocean surface at a specific location, are caused by the *gravitational attraction* of the Moon and, to a lesser extent, by the Sun. Near the times of new and full moons, the Sun and Moon are aligned and their gravitational forces are added together to produce especially high and low tides. These are called the *spring tides*. Conversely, at about the times of the first and third quarters of the Moon, when the gravitational forces of the Moon and Sun are at right angles, the daily *tidal range* is less. These are called *neap tides*.

- *Tidal currents* are horizontal movements of water that accompany the rise and fall of tides. *Tidal flats* are the areas that are affected by the advancing and retreating tidal currents. When tidal currents slow after emerging from narrow inlets, they deposit sediment that may eventually create *tidal deltas*.

- The three factors that influence the *height, wavelength,* and *period* of a wave are (1) *wind speed*, (2) *length of time the wind has blown*, and (3) *fetch*, the distance that the wind has traveled across the open water.

- The two types of wind-generated waves are (1) *waves of oscillation*, which are waves in the open sea in which the wave form advances as the water particles move in circular orbits, and (2) *waves of translation*, the turbulent advance of water formed near the shore as waves of oscillation collapse, or *break*, and form *surf*.

- Wave erosion is caused by *wave impact pressure* and *abrasion* (the sawing and grinding action of water armed with rock fragments). The bending of waves is called *wave refraction*. Owing to refraction, wave impact is concentrated against the sides and ends of headlands.

- Most waves reach the shore at an angle. The uprush (swash) and backwash of water from each breaking wave moves the sediment in a zigzag pattern along the beach. This movement, called *beach drift*, can transport sand hundreds or even thousands of meters each day. Oblique waves also produce *longshore currents* within the surf zone that flow parallel to the shore.

- Features produced by *shoreline erosion* include *wave-cut cliffs* (which originate from the cutting action of the surf against the base of coastal land), *wave-cut platforms* (relatively flat, benchlike surfaces left behind by receding cliffs), *sea arches* (formed when a headland is eroded and two caves from opposite sides unite), and *sea stacks* (formed when the roof of a sea arch collapses).

- Some of the features formed when sediment is moved by *beach drift* and *longshore currents* are *spits* (elongated ridges of sand that project from the land into the mouth of an adjacent bay), *baymouth bars* (sand bars that completely cross a bay), and *tombolos* (ridges of sand that connect an island to the mainland or to another island). Along the Atlantic and Gulf Coastal Plains, the shore zone is characterized by *barrier islands*, low ridges of sand that parallel the coast at distances from 3 to 30 kilometers offshore.

- Local factors that influence shoreline erosion are (1) the proximity of a coast to sediment-laden rivers, (2) the degree of tectonic activity, (3) the topography and composition of the land, (4) prevailing winds and weather patterns, and (5) the configuration of the coastline and near-shore areas.

- Three basic responses to shoreline erosion problems are (1) building *structures* such as *groins* (short walls built at a right angle to the shore to trap moving sand), *breakwaters* (structures built parallel to the shoreline to protect it from the force of large breaking waves), and *seawalls* (barriers constructed to prevent waves from reaching the area behind the wall) to hold the shoreline in place; (2) *beach nourishment*, which involves the addition of sand to replenish eroding beaches; and (3) *relocating* buildings away from the beach.

- Because of basic geological differences, the *nature of shoreline erosion problems along America's Pacific and Atlantic coasts is very different.* Much of the development along the Atlantic and Gulf coasts has occurred on barrier islands, which receive the full force of major storms. Much of the Pacific Coast is characterized by narrow beaches backed by steep cliffs and mountain ranges. A major problem facing the Pacific shoreline is a narrowing of beaches caused because the natural flow of materials to the coast has been interrupted by dams built for irrigation and flood control.

- One frequently used classification of coasts is based upon changes that have occurred with respect to sea level. *Emergent coasts*, often with wave-cut cliffs and wave-cut platforms above sea level, develop either because an area experiences uplift or as a result of a drop in sea level. Conversely, *submergent coasts*, with their drowned river mouths, called *estuaries*, are created when sea level rises or the land adjacent to the sea subsides.

Key Terms

abrasion (p. 359)
barrier island (p. 364)
baymouth bar (p. 364)
beach drift (p. 361)
beach nourishment (p. 368)
breakwater (p. 364)
Coriolis effect (p. 349)
emergent coast (p. 370)
fetch (p. 356)
groin (p. 365)
gyre (p. 348)
longshore current (p. 361)

neap tide (p. 353)
sea arch (p. 363)
sea stack (p. 363)
seawall (p. 366)
spit (p. 363)
spring tide (p. 353)
submergent coast (p. 370)
surf (p. 358)
thermohaline circulation (p. 351)
tidal current (p. 354)
tidal delta (p. 355)
tidal flat (p. 355)

tide (p. 351)
tombolo (p. 364)
upwelling (p. 349)
wave-cut cliff (p. 362)
wave-cut platform (p. 363)
wave height (p. 356)
wavelength (p. 356)
wave of oscillation (p. 358)
wave of translation (p. 358)
wave period (p. 356)
wave refraction (p. 360)

Review Questions

1. What is the primary driving force of surface ocean currents? How does the Coriolis effect influence these currents?

2. Describe the process of coastal upwelling.

3. How do ocean currents influence climate?

4. What is the driving force of deep-ocean circulation?

5. Discuss the origin of ocean tides.

6. Explain why an observer can experience two unequal high tides during a day (see Figure 13.7).

7. How does the Sun influence tides?

8. How do semidiurnal, diurnal, and mixed tides differ?

9. Distinguish between flood current and ebb current.

10. How has the construction of artificial levees and dams on the Mississippi River and its tributaries contributed to the shrinking of the Mississippi's delta and its extensive wetlands? (See Box 13.1.)

11. List three factors that determine the height, length, and period of a wave.

12. Describe the motion of a water particle as a wave passes (see Figure 13.13).

13. Explain what happens when a wave breaks.

14. How do waves cause erosion?

15. What is wave refraction? What is the effect of this process along irregular coastlines?

16. Why are beaches often called "rivers of sand"?

17. Describe the formation of the following features: wave-cut cliff, wave-cut platform, sea stack, spit, baymouth bar, tombolo.

18. Examine the satellite image of Cape Cod in Figure 13.11A. Locate some examples of spits and baymouth bars.

19. List three ways that barrier islands may form.

20. For what purpose is a groin built? Why might the building of one groin lead to the building of others?

21. How was the beach at Santa Monica, California, affected when a breakwater was constructed (see Figure 13.26)?

22. How might a seawall increase beach erosion?

23. Relate the damming of rivers to the shrinking of beaches at some locations along the West Coast of the United States.

24. What observable features would lead you to classify a coastal area as emergent?

25. Are estuaries associated with submergent or emergent coasts? Why?

Examining the Earth System

1. Describe the exchanges of energy between the ocean and the atmosphere that are responsible for (1) surface ocean currents and (2) the ocean's deep circulation. If the warm North Atlantic Drift were to cease, how might the climate of Western Europe change? Speculate about how such a climate change might influence the biosphere.

2. It is believed that global temperatures will be increasing in the decades to come. How can a warmer atmosphere lead to a rise in sea level? How could geologic events associated with the solid Earth contribute to a rise in sea level? Describe some possible changes in coastal areas if sea level were to rise. You may find some interesting and useful information at this Environmental Protection Agency Web site (**http://www.epa.gov/global warming/climate/future/sealevel/**).

Web Resources

The following are informative and interesting World Wide Web sites that address topics related to those presented in the chapter:

Coastal Programs Division (National Oceanic and Atmospheric Administration):

- **http://www.nos.noaa.gov/ocrm/cpd/**

Marine and Coastal Geology Program Fact Sheets (United States Geological Survey):

- **http://marine.usgs.gov/fact-sheets/**

For direct links to these sites and other appropriate Earth science Web locations, chapter review exercises, and relevant, topical investigations that utilize Web resources, contact the *Earth Science*, 9th Edition Home Page at:

- **http://www.prenhall.com/tarbuck.**

Part Three

The Atmosphere

Cumulonimbus clouds are associated with lightining, thunderstorms, and other forms of severe weather.
(Photo by Richard Kaylin/Tony Stone Images)

Chapter 14

Composition, Structure, and Temperature

As this climber in South America's Andes mountains knows, temperatures decrease with an increase in altitude in the troposphere. (Photo by Ned Gillette/The Stock Market) arth's atmosphere is unique. No other planet in our solar system has an atmosphere with the exact mixture of gases or the heat and moisture conditions necessary to sustain life as we know it. The gases that make up Earth's atmosphere and the controls to which they are subject are vital to our existence. In this chapter we begin our examination of the ocean of air in which we all must live. We will try to answer a number of basic questions. What is the composition of the atmosphere? At what point do we leave the atmosphere and enter outer space? What causes the seasons? How is air heated? What factors control temperature variations over the globe?

Weather influences our everyday activities, our jobs, and our health and comfort (Figure 14.1). Many of us pay little attention to the weather unless we are inconvenienced by it or when it adds to our enjoyment outdoors. Nevertheless, there are few other aspects of our physical environment that affect our lives more than the phenomena we collectively call the weather.

The United States has the greatest variety of weather of any country in the world. Beyond its direct impact on the lives of individuals, the weather has a strong effect on the world economy by influencing agriculture, energy use, water resources, transportation, and industry.

Although the atmosphere influences our lives a great deal, it is also important to realize that the reverse is true: People influence the atmosphere and its behavior. There are, and will continue to be, significant political and scientific decisions that must be made involving these impacts. Important examples are air pollution control and the effects of human industry on global climate and the atmosphere's protective ozone layer (see Box 14.1). We all need increased awareness and understanding of our atmosphere and its behavior.

Weather and Climate

Acted on by the combined effects of Earth's motions and energy from the Sun, our planet's formless and invisible envelope of air reacts by producing an infinite variety of weather, which, in turn, creates the basic pattern of global climates. Although not identical, weather and climate have much in common.

Weather is constantly changing, sometimes from hour to hour and at other times from day to day. It is a term that denotes the state of the atmosphere at a given time and place. Whereas changes in the weather are continuous and sometimes seemingly erratic, it is nevertheless possible to arrive at a generalization of these variations. Such a description of aggregate weather conditions is termed **climate**. It is based on observations that have been accumulated over many years. Climate is often defined simply as "average weather," but this is an inadequate definition. To portray the character of an area more accurately, variations and extremes must also be included as well as the probabilities that such departures will take place. Thus, climate is the sum of all statistical weather information that helps describe a place or region.

Figure 14.1 In January, 1997, many parts of the northern Great Plains experienced paralyzing blizzards. This one struck Bismarck, North Dakota on January 9, 1997. Probably no other aspect of our physical enviroment affects the daily lives of people more than the weather. (Photo by Mike McCleary/*Bismarck Tribune*)

The nature of weather and climate is expressed in terms of the same basic **elements**, those quantities or properties that are measured regularly. The most important are (1) air temperature, (2) humidity, (3) type and amount of cloudiness, (4) type and amount of precipitation, (5) air pressure, and (6) the speed and direction of the wind. These elements are the major variables from which weather patterns and climate types are deciphered. Although we shall study these elements separately at first, keep in mind that they are very much interrelated. A change in any one of the elements will often bring about changes in the others.

Composition of the Atmosphere

Air is *not* a unique element or compound. Rather, **air** is a *mixture* of many discrete gases, each with its own physical properties, in which varying quantities of tiny solid and liquid particles are suspended.

Major Components

The composition of air is not constant; it varies from time to time and from place to place. If the water vapor, dust, and other variable components were removed from the atmosphere, we would find that its makeup is very stable worldwide up to an altitude of about 80 kilometers (50 miles).

As you can see in Figure 14.2, two gases, nitrogen and oxygen, make up 99 percent of the volume of clean, dry air. Although these gases are the most plentiful components of air and are of great significance to life on Earth, they are of minor importance in affecting weather phenomena. The remaining 1 percent of dry air is mostly the inert gas argon (0.93 percent) plus tiny quantities of a number of other gases. Carbon dioxide,

although present in only minute amounts (0.036 percent), is nevertheless an important constituent of air, because it has the ability to absorb heat energy radiated by Earth and thus helps keep the atmosphere warm.

Variable Components

Air includes many gases and particles that vary significantly from time to time and place to place. Important examples include water vapor, dust particles, and ozone. Although usually present in small percentages, they can have significant effects on weather and climate.

Water Vapor. The amount of water vapor in the air varies considerably, from practically none at all up to about 4 percent by volume. Why is such a small fraction of the atmosphere so significant? Certainly the fact that water vapor is the source of all clouds and precipitation would be enough to explain its importance. However, water vapor has other roles. Like carbon dioxide, it has the ability to absorb heat given off by Earth as well as some solar energy. It is therefore important when we examine the heating of the atmosphere.

When water changes from one state to another (see Figure 15.2, p. 407), it absorbs or releases heat. This energy is termed *latent heat*, which means "hidden" heat. As we shall see in later chapters, water vapor in the atmosphere transports this latent heat from one region to another, and it is the energy source that helps drive many storms.

Aerosols. The movements of the atmosphere are sufficient to keep a large quantity of solid and liquid particles suspended within it. Although visible dust sometimes clouds the sky, these relatively large particles are too heavy to stay in the air for very long. Still, many particles are microscopic and remain suspended for considerable periods of time. They may originate from many sources, both natural and human made, and include sea salts from breaking waves, fine soil blown into the air, smoke and soot from fires, pollen and microorganisms lifted by the wind, ash and dust from volcanic eruptions, and more. Collectively, these tiny solid and liquid particles are called **aerosols**.

From a meteorological standpoint, these tiny, often invisible particles can be significant. First, many act as surfaces on which water vapor can condense, an important function in the formation of clouds and fog. Second, aerosols can absorb or reflect incoming solar radiation. Thus, when an air pollution episode is occurring or when ash fills the sky following a volcanic eruption, the amount of sunlight reaching Earth's surface can be measurably reduced. Finally, aerosols contribute to an optical phenomenon we have all observed—the varied hues of red and orange at sunrise and sunset (Figure 14.3).

Figure 14.2 Proportional volume of gases composing dry air. Nitrogen and oxygen clearly dominate.

Box 14.1   People and the Environment: Altering the Atmosphere's Composition: Sources and Types of Air Pollution

Air pollutants are airborne particles and gases that occur in concentrations that endanger the health and well-being of organisms or disrupt the orderly functioning of the environment. One category of pollutants, the *primary pollutants*, are emitted directly from identifiable sources. They pollute the air immediately upon being emitted. Figure 14.A shows the major primary pollutants and the sources that produce them. When the sources are examined, the significance of the transportation category is obvious. It accounts for nearly half of our air pollution (by weight). In addition to highway vehicles, this category includes trains, ships, and airplanes. Still, the tens of millions of cars and trucks on U.S. roads are, without a doubt, the greatest contributors (Figure 14.B).

Sometimes the direct impact of primary pollutants on human health and the environment is less severe than the effects of the secondary pollutants they form. *Secondary pollutants* are not emitted directly into the air, but form in the atmosphere when reactions take place among primary pollutants. The chemicals that make up smog are important examples, as is the sulfuric acid that falls as acid precipitation. After the primary pollutant, sulfur dioxide, is emitted into the atmosphere, it combines with oxygen to produce sulfur trioxide, which then combines with water to create this irritating and corrosive acid (see Box 3.1).

Many reactions that produce secondary pollutants are triggered by strong sunlight and so are called *photochemical reactions*. One common example occurs when nitrogen oxides absorb solar radiation, initiating a chain of complex reactions. When certain volatile organic compounds are present, the result is the formation of a number of undesirable secondary products that are very reactive, irritating, and toxic. Collectively, this noxious mixture of gases and particles is called *photochemical smog*.

Air quality standards have not yet been met in a large number of places.

Figure 14.B This crowded freeway reminds us that the transportation category accounts for nearly half of the air pollution (by weight) in the United States. (Photo by Derek Trask/The Stock Market)

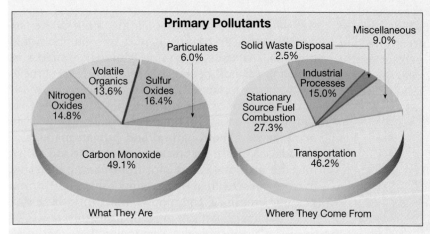

Figure 14.A Major primary pollutants and their sources. Percentages are calculated on the basis of weight. (Data from the U.S. Environmental Protection Agency)

But significant improvements in air quality have occurred ever since the U.S. Environmental Protection Agency initiated air pollution control programs. Between 1970 and 1997 national total emissions of primary pollutants declined 31%. This occurred during a span when the U.S. population increased 31% and vehicle miles traveled increased 127%. Despite continued improvements in air quality, approximately 107 million people lived in counties with unhealthy air in 1997.

Ozone. Another important component of the atmosphere is **ozone**. It is a form of oxygen that combines three oxygen atoms into each molecule (O_3). Ozone is not the same as oxygen we breathe, which has two atoms per molecule (O_2). There is very little of this gas in the atmosphere and its distribution is not uniform. In the lowest portion of the atmosphere, ozone represents less than one part in 100 million. It is concentrated well above the surface in a layer called the *stratosphere*, between 10 and 50 kilometers (6 and 31 miles).

In this altitude range, oxygen molecules (O_2) are split into single atoms of oxygen (O) when they absorb ultraviolet radiation emitted by the Sun. Ozone is then created when a single atom of oxygen

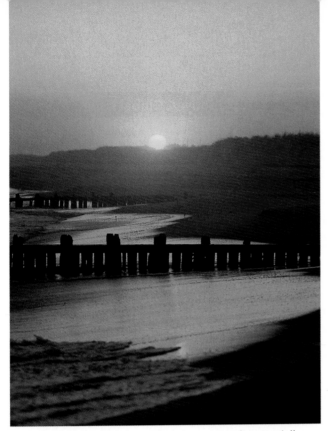

Figure 14.3 Dust in the air can cause sunsets to be especially colorful. (Photo by Steve Elmore/The Stock Market)

(O) and a molecule of oxygen (O_2) collide. This must happen in the presence of a third, neutral molecule that acts as a *catalyst* by allowing the reaction to take place without itself being consumed in the process. Ozone is concentrated in the 10- to 50-kilometer height range because a crucial balance exists there: The ultraviolet radiation from the Sun is sufficient to produce single atoms of oxygen, and there are enough gas molecules to bring about the required collisions.

The presence of the ozone layer in our atmosphere is crucial to those of us who dwell on Earth. The reason is that ozone absorbs the potentially harmful ultraviolet (UV) radiation from the Sun. If ozone did not filter a great deal of the ultraviolet radiation, and if the Sun's UV rays reached the surface of Earth undiminished, our planet would be uninhabitable for most life as we know it. Thus, anything that reduces the amount of ozone in the atmosphere could affect the well-being of life on Earth. Just such a problem exists and is described in Box 14.2.

Height and Structure of the Atmosphere

To say that the atmosphere begins at Earth's surface and extends upward is obvious. However, where does the atmosphere end and outer space begin? There is no sharp boundary; the atmosphere rapidly thins as you travel away from Earth, until there are too few gas molecules to detect.

Pressure Changes

To understand the vertical extent of the atmosphere, let us examine the changes in the atmospheric pressure with height. Atmospheric pressure is simply the weight of the air above. At sea level, the average pressure is slightly more than 1000 millibars. This corresponds to a weight of slightly more than 1 kilogram per square centimeter (14.7 pounds per square inch). Obviously the pressure at higher altitudes is less (Figure 14.4).

One-half of the atmosphere lies below an altitude of 5.6 kilometers (3.5 miles). At about 16 kilometers (10 miles), 90 percent of the atmosphere has been traversed, and above 100 kilometers (62 miles), only 0.00003 percent of all the gases making up the atmosphere remains. Even so, traces of our atmosphere extend far beyond this altitude, gradually merging with the emptiness of space.

Temperature Changes

By the early twentieth century, much had been learned about the lower atmosphere. The upper atmosphere was partly known from indirect methods. Data from balloons and kites had revealed that the air temperature dropped with increasing height above Earth's surface.

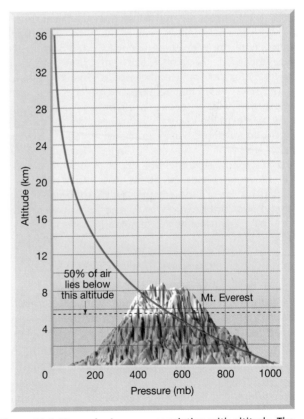

Figure 14.4 Atmospheric pressure variation with altitude. The rate of pressure decrease with an increase in altitude is not constant. Rather, pressure decreases rapidly near Earth's surface and more gradually at greater heights.

Box 14.2 People and the Environment: Ozone Depletion—A Global Issue

The loss of ozone high in the atmosphere as a consequence of human activities is a serious global-scale environmental problem (Figure 14.C). Over the past half century, people have unintentionally placed the ozone layer in jeopardy by polluting the atmosphere. The offending chemicals are known as chlorofluorocarbons (CFCs for short). They are versatile compounds that are chemically stable, odorless, nontoxic, noncorrosive, and inexpensive to produce. Over the decades many uses were developed for CFCs, including as coolants for air conditioning and refrigeration equipment, cleaning solvents for electronic components, propellants for aerosol sprays, and the production of certain plastic foams.

No one worried about how CFCs might affect the atmosphere until three scientists, Paul Crutzen, F. Sherwood Rowland, and Mario Molina, studied the relationship. In 1974, they alerted the world when they reported that CFCs were probably reducing the average concentration of ozone in the stratosphere. In 1995, these scientists were awarded the Nobel Prize in chemistry for their pioneering work.

They discovered that because CFCs are practically inert (that is, not chemically active) in the lower atmosphere, a portion of these gases gradually makes its way to the ozone layer, where sunlight separates the chemicals into their constituent atoms. The chlorine atoms released this way, through a complicated series of reactions, have the net effect of removing some of the ozone.

TOTAL OZONE

150 180 210 240 270 300 330 360 390
Dobson Units

Figure 14.C Ozone distribution of the Southern Hemisphere for the month of October 1995. The image was produced with data acquired by the TIROS Observational Vertical Sounder (TOVS). The area of greatest depletion, called the ozone "hole," forms over Antarctica during the Southern Hemisphere spring. In October 1995, ozone levels reached a minimum of about 120 Dobson units, far below the 220 Dobson units typically seen over Antarctica before the hole forms. (Data courtesy of NOAA)

Because ozone filters out most of the ultraviolet (UV) radiation from the Sun, a decrease in its concentration permits more of these harmful wavelengths to reach Earth's surface. The most serious threat to human health is an increased risk of skin cancer. An increase in damaging UV radiation also can impair the human immune system as well as promote cataracts, a clouding of the eye lens that reduces vision and may cause blindness if not treated.

In response to this problem, an international agreement known as the *Montreal Protocol* was developed under the sponsorship of the United Nations to eliminate the production and use of CFCs.

Although relatively strong action was taken, CFC levels in the atmosphere will not drop rapidly. Once in the atmosphere, CFC molecules can take many years to reach the ozone layer and, once there, they can remain active for decades. This does not promise a near-term reprieve for the ozone layer. Even after CFC levels begin to decline, the ozone layer probably will not cleanse itself of most of the pollutants until sometime in the second half of the twenty-first century.

This phenomenon is felt by anyone who has climbed a high mountain, and is obvious in pictures of snow-capped mountaintops rising above snow-free lowlands (Figure 14.5). We divide the atmosphere vertically into four layers on the basis of temperature (Figure 14.6).

Troposphere. The bottom layer in which we live, where temperature decreases with an increase in altitude, is the **troposphere**. The term literally means the region where air "turns over," a reference to the appreciable vertical mixing of air in this lowermost zone. The troposphere is the chief focus of meteorologists, because it is in this layer that essentially all important weather phenomena occur.

The temperature decrease in the troposphere is called the **environmental lapse rate**. Its average value is 6.5°C per kilometer (3.5°F per 1000 feet), a figure known as the *normal lapse rate*. It should be emphasized, however, that the environmental lapse rate is not a constant, but rather can be highly variable, and must be regularly measured. To determine the actual environmental lapse

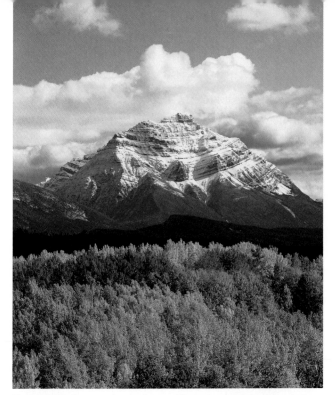

Figure 14.5 Temperatures drop with an increase in altitude in the troposphere. Therefore, it is possible to have snow on a mountaintop and warmer, snow-free lowlands below. Mount Kerkeslin, Jasper National Park, Alberta, Canada. (Photo by Carr Clifton/Minden Pictures)

rate as well as gather information about vertical changes in pressure, wind, and humidity, radiosondes are used. The *radiosonde* is an instrument package that is attached to a balloon and transmits data by radio as it ascends through the atmosphere (Figure 14.7).

The thickness of the troposphere is not the same everywhere; it varies with latitude and the season. On the average, the temperature drop continues to a height of about 12 kilometers (7.4 miles). The outer boundary of the troposphere is the *tropopause*.

Stratosphere. Beyond the tropopause is the **stratosphere**. In the stratosphere, the temperature remains constant to a height of about 20 kilometers and then begins a gradual increase that continues until the *stratopause*, at a height of nearly 50 kilometers above Earth's surface. Below the tropopause, atmospheric properties like temperature and humidity are readily transferred by large-scale turbulence and mixing. Above the tropopause, in the stratosphere, they are not. Temperatures increase in the stratosphere because it is in this layer that the atmosphere's ozone is concentrated. Recall that ozone absorbs ultraviolet radiation from the Sun. As a consequence, the stratosphere is heated.

Mesosphere. In the third layer, the **mesosphere**, temperatures again decrease with height until, at the *mesopause*, more than 80 kilometers (50 miles) above the surface, the temperature approaches -90°C.

383

Thermosphere. The fourth layer extends outward from the mesopause and has no well-defined upper limit. It is the **thermosphere**, a layer that contains only a tiny fraction of the atmosphere's mass. In the extremely rarefied air of this outermost layer, temperatures again increase owing to the absorption of very short-wave, high-energy solar radiation by atoms of oxygen and nitrogen.

Temperatures rise to extremely high values of more than 1000°C in the thermosphere. But such temperatures are not comparable to those experienced near Earth's surface. Temperature is defined in terms of the average speed at which molecules move. Because the gases of the thermosphere are moving at very high speeds, the temperature is very high. But the gases are so sparse that, collectively, they possess only an insignificant quantity of heat. For this reason, the temperature of a satellite orbiting Earth in the thermosphere is determined chiefly by the amount of solar radiation it absorbs and not by the high temperature of the almost nonexistent surrounding air. If an astronaut inside were to expose his or her hand, it would not feel hot.

Earth–Sun Relationships

Always remember that nearly all of the energy that drives Earth's variable weather and climate comes from the Sun. Earth intercepts only a minute percentage of the energy given off by the Sun—less than one two-billionth. This may seem to be an insignificant amount until we realize that it is several hundred thousand times the electrical-generating capacity of the United States.

If solar energy were distributed uniformly over Earth's surface, our planet would have more uniform temperatures. But solar energy is not distributed evenly over Earth's surface. It is this unequal heating that drives the ocean's currents and creates the winds, which, in turn, transport heat from the tropics to the poles in an unending attempt to reach an energy balance. If the Sun were to be "turned off," global winds would quickly subside. Yet as long as the Sun shines, the winds will blow and the phenomena we know as weather will persist.

Energy from the Sun is the most important control of our weather and climate. To understand atmospheric processes, we must understand what causes the variations in the amount of solar energy reaching Earth. The principal causes are variations in Sun angle and exposure time.

Earth has two principal motions—rotation and revolution. **Rotation** is the spinning of Earth about its axis. The axis is an imaginary line running through the poles. Our planet rotates once every 24 hours, producing the daily cycle of daylight and darkness. At any moment, half of Earth is experiencing daylight,

Figure 14.6 Thermal structure of the atmosphere to a height of about 110 kilometers (68 miles).

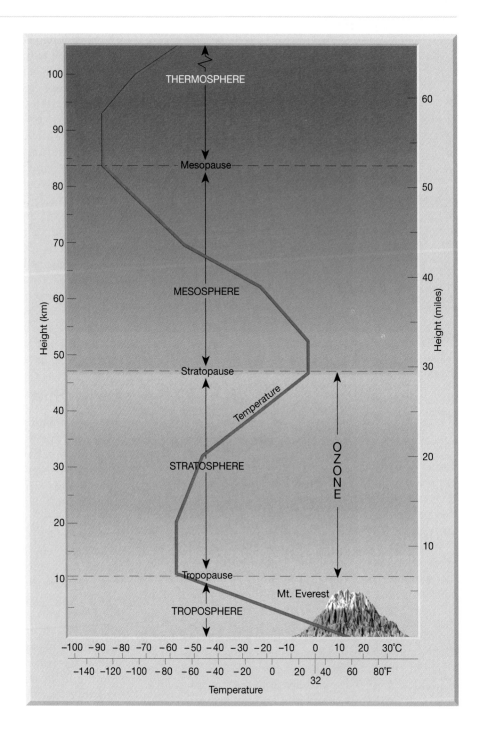

and the other half darkness. The line separating the dark half of Earth from the lighted half is called the **circle of illumination**.

Revolution refers to the movement of Earth in its orbit around the Sun. Hundreds of years ago, most people believed that Earth was stationary in space and that the Sun and stars revolved around our planet. Today, we know that Earth is traveling at nearly 113,000 kilometers per hour in its orbit about the Sun.

Seasons

We know that it is colder in winter than in summer. But why? Length of daylight certainly accounts for some of the difference. Long summer days expose us to more solar radiation, whereas short winter days expose us to less. But an even bigger factor is the angle of the Sun above the horizon.

During mid-summer the Sun is high above the horizon as it makes its daily journey across the sky. But as summer gives way to autumn, the noon Sun

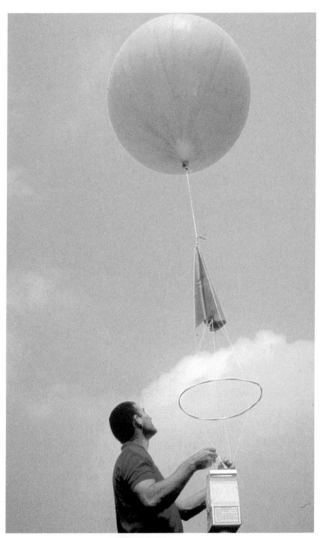

Figure 14.7 Atmospheric soundings using radiosondes supply data on vertical changes in temperature, pressure, and humidity. (Courtesy of NOAA)

appears lower and lower in the sky. What we observe here is the annual shifting of the solar angle or *altitude* of the Sun.

The seasonal variation in the altitude of the Sun affects the amount of energy received at Earth's surface in two ways. First, when the Sun is high in the sky, the solar rays are most concentrated (you can see this in Figure 14.8A). The lower the angle, the more spread out and less intense is the solar radiation reaching the surface (Figure 14.8B, C). To illustrate this principle, hold a flashlight at a right angle to a surface and then change the angle.

Second, and of lesser importance, the angle of the Sun determines the amount of atmosphere the rays must penetrate (Figure 14.9). When the Sun is directly overhead, the rays pass through a thickness of only 1 atmosphere, whereas rays entering at a 30-degree angle travel through twice this amount, and 5-degree rays travel through a thickness roughly equal to 11 atmospheres. The longer the path, the greater are the chances for absorption, reflection, and scattering by the atmosphere, all of which reduce the intensity at the surface. The same effects account for the fact that we cannot look directly at the midday Sun, but we can enjoy gazing at a sunset.

It is also important to remember that Earth has a spherical shape. Hence, on any given day only places located at a particular latitude receive vertical (90-degree) rays from the Sun. As we move either north or south of this location, the Sun's rays strike at an ever-decreasing angle. Thus, the nearer a place is to the latitude receiving vertical rays of the Sun, the higher will be its noon Sun and the more intense will be the radiation it receives.

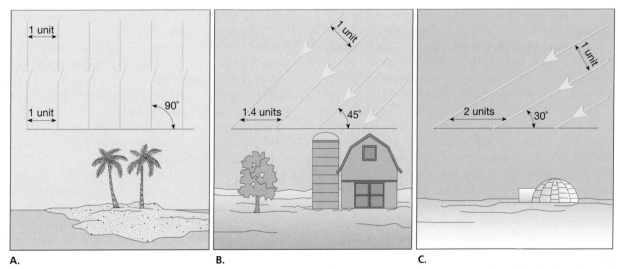

A.   B.   C.

Figure 14.8 Changes in the Sun angle cause variation in the amount of solar energy reaching Earth's surface. The higher the angle, the more intense the solar radiation. Notice that one unit of solar energy striking at a 30° angle (C) spreads over twice as much area as that striking at a 90° angle (A).

Figure 14.9 Rays striking Earth at a low angle (toward the poles) must travel through more of the atmosphere than rays striking at a high angle (around the equator) and thus are subject to greater depletion by reflection and absorption.

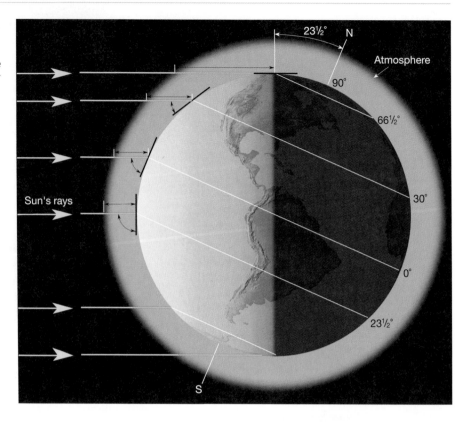

Earth's Orientation

What causes the fluctuations in the Sun angle and length of daylight that occur during the course of a year? They occur because *Earth's orientation to the Sun continually changes as it travels along its orbit*. Earth's axis is not perpendicular to the plane of its orbit around the Sun. Instead it is tilted 23 1/2 degrees from the perpendicular as shown in Figure 14.10. This is called the **inclination of the axis**. As we shall see, if the axis were not inclined, we would have no seasonal changes. In addition, because the axis remains pointed in the same direction (toward the North Star) as Earth journeys around the Sun, the orientation of Earth's axis to the Sun's rays is constantly changing (Figure 14.10).

For example, on one day in June each year the axis is such that the Northern Hemisphere is "leaning" 23 1/2 degrees *toward* the Sun. Six months later, in December, when Earth has moved to the opposite side of its orbit, the Northern Hemisphere "leans" 23 1/2 degrees *away* from the Sun. On days between these extremes, Earth's axis is leaning at amounts less than 23 1/2 degrees to the rays of the Sun. This change in orientation causes the spot where the Sun's rays are vertical to make an annual migration from 23 1/2 degrees north of the equator to 23 1/2 degrees south of the equator.

In turn, this migration causes the altitude of the noon Sun to vary by up to 47 degrees (23 1/2 + 23 1/2) for many locations during the year. For example, a mid-latitude city like New York (about 40 degrees north latitude) has a maximum noon Sun angle of 73 1/2 degrees when the Sun's vertical rays reach their farthest northward location in June and a minimum noon Sun angle of 26 1/2 degrees 6 months later.

Solstices and Equinoxes

Historically, 4 days a year have been given special significance based on the annual migration of the direct rays of the Sun and its importance to the yearly weather cycle. On June 21 or 22, Earth is in a position where the axis in the Northern Hemisphere is tilted 23 1/2 degrees toward the Sun (Figure 14.11A). At this time the vertical rays of the Sun strike 23 1/2 degrees north latitude (23 1/2 degrees north of the equator), a latitude known as the **Tropic of Cancer**. For people in the Northern Hemisphere, June 21 or 22 is known as the **summer solstice**.

Six months later, on about December 21 or 22, Earth is in an opposite position, with the Sun's vertical rays striking at 23 1/2 degrees south latitude (Figure 14.11B). This parallel is known as the **Tropic of Capricorn**. For those in the Northern Hemisphere, December 21 and 22 is the **winter solstice**, the first day of winter. However, at the same time in the Southern Hemisphere, people are experiencing just the opposite—the summer solstice.

The equinoxes occur midway between the solstices. September 22 or 23 is the date of the **autumnal equinox** in the Northern Hemisphere, and March 21 or 22 is the date of the **spring equinox**. On these dates,

Figure 14.10 Earth–Sun relationships.

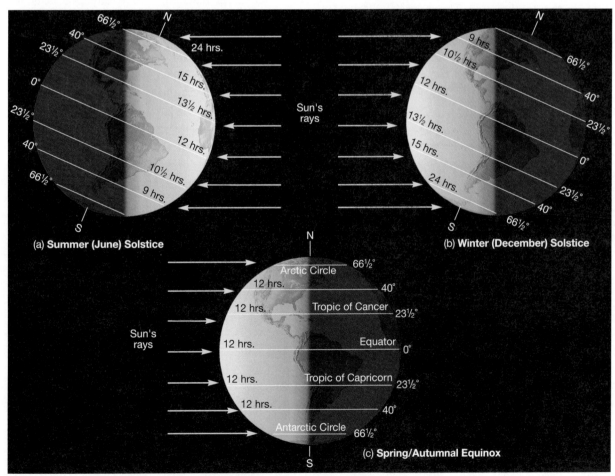

Figure 14.11 Characteristics of the solstices and equinoxes.

Table 14.1 Length of daylight.

Latitude (degrees)	Summer Solstice	Winter Solstice	Equinoxes
0	12 h	12 h	12 h
10	12 h 35 min	11 h 25 min	12
20	13 12	10 48	12
30	13 56	10 04	12
40	14 52	9 08	12
50	16 18	7 42	12
60	18 27	5 33	12
70	24 h (for 2 mo)	0 00	12
80	24 h (for 4 mo)	0 00	12
90	24 h (for 6 mo)	0 00	12

the vertical rays of the Sun strike the equator (0 degrees latitude) because Earth is in such a position in its orbit that the axis is tilted neither toward nor away from the Sun (Figure 14.11C).

The length of daylight versus darkness is also determined by Earth's position in orbit. The length of daylight on June 21, the summer solstice in the Northern Hemisphere, is greater than the length of night. This fact can be established from Figure 14.11A by comparing the fraction of a given latitude that is on the "day" side of the circle of illumination with the fraction on the "night" side. The opposite is true for the winter solstice, when the nights are longer than the days. Again for comparison let us consider New York City, which has 15 hours of daylight June 21 and only 9 hours on December 21 (you can see this in Figure 14.11 and Table 14.1). Also note from Table 14.1 that the

farther you are north of the equator on June 21, the longer the period of daylight. When you reach the Arctic Circle (66 1/2° N), the length of daylight is 24 hours. This is the land of the "midnight Sun," which does not set for about 6 months at the North Pole (Figure 14.12).

During an equinox (meaning "equal night"), the length of daylight is 12 hours everywhere on Earth, because the circle of illumination passes directly through the poles, dividing the latitudes in half.

As a review of the characteristics of the summer solstice for the Northern Hemisphere, examine Figure 14.11A and Table 14.1 and consider the following facts:

1. The solstice occurs on June 21 or 22.
2. The vertical rays of the Sun are striking the Tropic of Cancer (23 1/2 degrees north latitude).
3. Locations in the Northern Hemisphere are experiencing their greatest length of daylight. (Opposite for the Southern Hemisphere.)
4. Locations north of the Tropic of Cancer are experiencing their highest noon Sun angles. (Opposite for places south of the Tropic of Capricorn.)
5. The farther you are north of the equator, the longer the period of daylight, until the Arctic Circle is reached, where daylight lasts for 24 hours. (Opposite for the Southern Hemisphere.)

The facts about the winter solstice are just the opposite. It should now be apparent why a mid-latitude location is warmest in the summer. It is then that the days are longest and the Sun's altitude is highest.

Figure 14.12 Multiple exposures of the midnight Sun in late June or July in high northern latitudes—Alaska, Scandinavia, northern Canada, etc. (Photo by Brian Stablyk/Tony Stone Images)

In summary, seasonal variations in the amount of solar energy reaching places on Earth's surface are caused by the migrating vertical rays of the Sun and the resulting variations in Sun angle and length of daylight.

All places at the same latitude have identical Sun angles and lengths of daylight. If the Earth–Sun relationships just described were the only controls of temperature, we would expect these places to have identical temperatures as well. Obviously this is not the case. Although the altitude of the Sun is an extremely important control of temperature, it is not the only control, as we shall see.

Heat and Temperature

The concepts of *heat* and *temperature* are often confused. The phrase "in the heat of the day" is one common expression in which the word "heat" is misused to describe the concept of temperature. **Heat** is a form of energy. Recall that all matter is composed of atoms or molecules that are constantly vibrating and therefore possess kinetic energy, which is the energy of motion. Heat is defined as the *total* kinetic energy of all the atoms and molecules that make up a substance. By contrast, **temperature** refers to intensity—that is, the degree of "hotness." It is a measure of the *average* kinetic energy of the individual atoms or molecules in a substance. When heat is added to a substance, its atoms move faster and faster and its temperature rises. When heat is removed, its atoms move slower and its temperature drops.

Although the ideas of temperature and heat are distinct, they are nevertheless related. Certainly, the addition or subtraction of heat causes temperatures to increase or decrease. In addition, differences in temperature determine the direction of heat flow. When two bodies having different temperatures are in contact, *heat always moves from the higher-temperature body to the lower-temperature body.* For example, when you touch a hot stove, heat enters your hand because the stove is much warmer than your hand. By contrast, when you hold an ice cube, heat is transferred from your hand to the ice cube. Thus, if two objects of different temperature are in contact, the warmer object will become cooler and the cooler will become warmer until they reach the same temperature.

Mechanisms of Heat Transfer

Three mechanisms of heat transfer are recognized: conduction, convection, and radiation. Although we present them separately, all three processes go on simultaneously in the atmosphere. In addition, these mechanisms operate to transfer heat between Earth's surface (both land and water) and the atmosphere.

Conduction

Conduction is familiar to all of us. Anyone who has touched a metal spoon that was left in a hot pan has discovered that heat was conducted through the spoon. **Conduction** is *the transfer of heat through matter by molecular activity.* The energy of molecules is transferred through collisions from one molecule to another, with the heat flowing from the higher temperature to the lower temperature.

The ability of substances to conduct heat varies considerably. Metals are good conductors, as those of us who have touched hot metal have quickly learned (Figure 14.13). Air, conversely, is a very poor conductor of heat. Consequently, conduction is important only between Earth's surface and the air directly in contact with the surface. As a means of heat transfer for the atmosphere as a whole, conduction is the least significant.

Convection

Much of the heat transport that occurs in the atmosphere is carried on by convection. **Convection** is *the transfer of heat by mass movement or circulation within a substance.* It can only take place in fluids (liquids like the ocean and gases like air) where the atoms and molecules are free to move about. In solids, such movement does not occur, so heat transfer through solids takes place by conduction.

The pan of water in Figure 14.13 illustrates the nature of simple convective circulation. Radiation from the fire warms the bottom of the pan, which conducts heat to the water near the bottom of the container. As the water is heated, it expands and thus becomes less dense than the water above. Because of this new buoyancy, the warmer water begins to rise. At the same time, cooler, denser water near the top of the pan sinks to the bottom, where it becomes heated. As long as the water is heated unequally—that is, from the bottom up—the water will

Figure 14.13 The three mechanisms of heat transfer: conduction, convection, and radiation.

continue to "turn over," producing a *convective circulation*. In a similar manner, most of the heat acquired in the lowest layer of the atmosphere by way of radiation and conduction is transferred by convective flow.

On a global scale, convection in the atmosphere creates a huge, worldwide air circulation. This is responsible for the redistribution of heat between hot equatorial regions and the frigid poles. This important process will be discussed in detail in Chapter 16.

Radiation

The third mechanism of heat transfer is radiation. As shown in Figure 14.13, radiation travels out in all directions from its source. Unlike conduction and convection, which need a medium to travel through, radiant energy readily travels through the vacuum of space. Thus, radiation is the heat-transfer mechanism by which solar energy reaches our planet.

From our everyday experience we know that the Sun emits light and heat as well as the ultraviolet rays that cause suntan. Although these forms of energy comprise a major portion of the total energy that radiates from the Sun, they are only part of a large array of energy called **radiation** or **electromagnetic radiation**. This array, or spectrum, of electromagnetic energy is shown in Figure 14.14. All radiation, whether X rays, radio waves, or heat waves, travel through the vacuum of space at 300,000 kilometers (186,000 miles) per second and only slightly slower through our atmosphere.

It helps to understand radiant energy by picturing ripples made in a pond by tossing in a pebble or by observing ocean waves. Like these waves, *electromagnetic waves*, as waves of radiant energy are called, come in various sizes. For our purpose, the most important difference among electromagnetic waves is their *wavelength*, or the distance from one crest to the next. Radio waves have the longest wavelengths, ranging to tens of kilometers. Gamma waves are the shortest, being less than a billionth of a centimeter long.

Visible light, as the name implies, is the only portion of the spectrum we can see. We often refer to visible light as "white" light as it appears "white" in color. However, it is easy to show that white light is really a mixture of colors, each corresponding to a specific wavelength (Figure 14.15). By using a prism, white light can be divided into the colors of the rainbow, from violet with the shortest wavelength—0.4 micrometer (1 micrometer is 0.0001 centimeter)—to red with the longest wavelength—0.7 micrometer.

Located adjacent to red, and having a longer wavelength, is **infrared** radiation, which we cannot see but which we can detect as heat. The closest invisible waves to violet are called **ultraviolet** rays. They are responsible for the sunburn that can occur after an intense exposure to the Sun. Although we divide radiant energy into groups based on our ability to perceive them, all forms of radiation are basically the same. When any form of radiant energy is absorbed by an object, the result is an increase in molecular motion, which causes a corresponding increase in temperature.

To understand better how the atmosphere is heated, it is useful to have a general understanding of the basic laws governing radiation.

1. *All objects, at whatever temperature, emit radiant energy.* Thus, not only hot objects like the Sun but

Figure 14.14 The electromagnetic spectrum, illustrating the wavelengths and names of various types of radiation.

Figure 14.15 Visible light consists of an array of colors we commonly call the "colors of the rainbow." Rainbows are relatively common optical phenomena produced by the bending and reflection of light by drops of water. This scene is in Alaska's Denali National Park, a subarctic region characterized by coniferous forest *(taiga)*. (Photo by Michael Giannechini/Photo Researchers, Inc.)

also Earth, including its polar ice caps, continually emit energy.

2. *Hotter objects radiate more total energy per unit area than do colder objects.*

3. *The hotter the radiating body, the shorter the wavelength of maximum radiation.* The Sun, with a surface temperature of nearly 6000°C, radiates maximum energy at 0.5 micrometer, which is in the visible range. The maximum radiation for Earth occurs at a wavelength of 10 micrometers, well within the infrared (heat) range. Because the maximum Earth radiation is roughly 20 times longer than the maximum solar radiation, it is often called long-wave radiation, and solar radiation is called short-wave radiation.

4. *Objects that are good absorbers of radiation are good emitters as well.* Earth's surface and the Sun approach being perfect radiators because they absorb and radiate with nearly 100 percent efficiency for their respective temperatures. On the other hand, *gases are selective absorbers and radiators.* Thus, the atmosphere, which is nearly transparent to (does not absorb) certain wavelengths of radiation, is nearly opaque (a good absorber) to others. Our experience tells us that the atmosphere is transparent to visible light; hence, it readily reaches Earth's surface. This is not the case for the longer wavelength radiation emitted by Earth.

Paths Taken by Incoming Solar Radiation

Although the atmosphere is largely transparent to incoming solar radiation, only about 25 percent penetrates directly to Earth's surface without some sort of interference by the atmosphere. The remainder is either *absorbed* by the atmosphere, *scattered* about until it reaches Earth's surface or returns to space, or is *reflected* back to space (Figure 14.16). What determines whether radiation will be absorbed, scattered, or reflected outward? It depends greatly on the wavelength of the energy being transmitted, as well as on the nature of the intervening material.

Scattering

Although solar radiation travels in a straight line, the gases and dust particles in the atmosphere can redirect this energy. This process, called *scattering*, explains how light reaches into a shaded area or a room when direct sunlight is absent. Some of the light is backscattered into space, while the remainder continues Earthward, where it interacts with other molecules that scatter it further by changing the direction of the radiation, but not its wavelength.

Reflection

About 30 percent of the solar energy reaching the outer atmosphere is reflected back to space. Included in this figure is the amount sent skyward by backscattering.

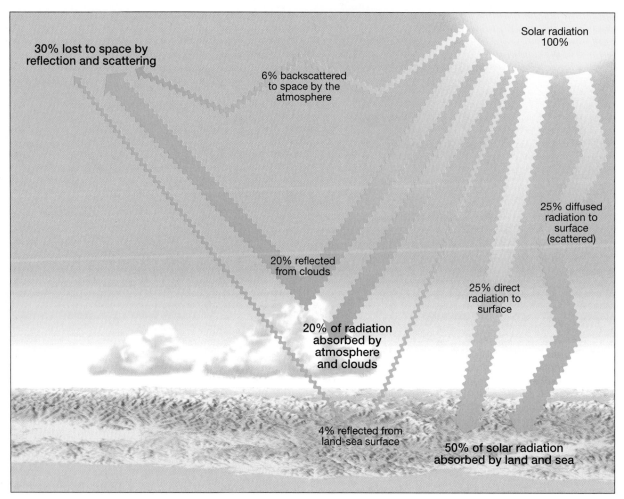

30% lost to space by
reflection and scattering

6% backscattered
to space by the
atmosphere

Solar radiation
100%

25% diffused
radiation to
surface
(scattered)

20% reflected
from clouds

25% direct
radiation to
surface

20% of radiation
absorbed by
atmosphere
and clouds

4% reflected from
land-sea surface

50% of solar radiation
absorbed by land and sea

Figure 14.16 Average distribution of incoming solar radiation by percentage. More solar energy is absorbed by Earth's surface than by the atmosphere. Consequently, the air is not heated directly by the Sun, but is heated indirectly from Earth's surface.

This energy is lost to Earth and does not play a role in heating the atmosphere.

The fraction of the total radiation that is reflected by a surface is called its **albedo**. Thus, the albedo for Earth as a whole (the planetary albedo) is 30 percent. However, the albedo from place to place as well as from time to time in the same locale varies considerably, depending on the amount of cloud cover and particulate matter in the air, as well as on the angle of the Sun's rays and the nature of the surface. A lower Sun angle means that more atmosphere must be penetrated, thus making the "obstacle course" longer, and therefore the loss of solar radiation greater (see Figure 14.9). Table 14.2 gives the albedo for various surfaces. Note that the angle at which the Sun's rays strike a water surface greatly affects its albedo.

Absorption

As stated earlier, gases are selective absorbers, meaning that they absorb strongly in some wavelengths, moderately in others, and only slightly in still others. When a gas molecule absorbs light waves, this energy

is transformed into internal molecular motion, which is detectable as a rise in temperature.

Nitrogen, the most abundant constituent in the atmosphere, is a poor absorber of all types of incoming radiation. Oxygen and ozone are efficient absorbers of ultraviolet radiation. Oxygen removes most of the shorter ultraviolet radiation high in the atmosphere, and ozone absorbs most of the remaining ultraviolet rays in the stratosphere. The absorption of UV radiation in the stratosphere accounts for the high temperatures experienced there. The only other significant absorber of incoming solar radiation is water vapor, which along with oxygen and ozone accounts for most of the solar radiation absorbed within the atmosphere.

For the atmosphere as a whole, none of the gases are effective absorbers of visible radiation. This explains why most visible radiation reaches Earth's surface and why we say that the atmosphere is *transparent* to incoming solar radiation. Thus, the atmosphere does not acquire the bulk of its energy directly from the Sun. Rather, it is heated chiefly by energy that is first absorbed by Earth's surface and then reradiated to the sky.

Table 14.2 Albedo of various surfaces.

Surface	Percent Reflected
Clouds, stratus	
<150 meters thick	25–63
150–300 meters thick	45–75
300–600 meters thick	59–84
Average of all types and	
thicknesses	50–55
Concrete	17–27
Crops, green	5–25
Forest, green	5–10
Meadows, green	5–25
Ploughed field, moist	14–17
Road, blacktop	5–10
Sand, white	30–60
Snow, fresh-fallen	80–90
Snow, old	45–70
Soil, dark	5–15
Soil, light (or desert)	25–30
Water	8*

* Typical albedo value for a water surface. The albedo of a water surface varies greatly depending upon the Sun angle. If the Sun angle is greater than 30 degrees, the albedo is less than 5 percent. When the Sun is near the horizon (Sun angle less than 3 degrees), the albedo is more than 60 percent.

Heating the Atmosphere: The Greenhouse Effect

Approximately 50 percent of the solar energy that strikes the top of the atmosphere reaches Earth's surface and is absorbed. Most of this energy is then reradiated skyward. Because Earth has a much lower surface temperature than the Sun, the radiation that it emits has longer wavelengths than solar radiation.

The atmosphere as a whole is an efficient absorber of the longer wavelengths emitted by Earth (*terrestrial radiation*). Water vapor and carbon dioxide are the principal absorbing gases. Water vapor absorbs roughly five times more terrestrial radiation than do all other gases combined and accounts for the warm temperatures found in the lower troposphere, where it is most highly concentrated. Because the atmosphere is quite transparent to shorter-wavelength solar radiation and more readily absorbs longer-wavelength terrestrial radiation, the atmosphere is heated from the ground up rather than vice versa. This explains the general drop in temperature with increasing altitude experienced in the troposphere. The farther from the "radiator," the colder it becomes.

When the gases in the atmosphere absorb terrestrial radiation, they warm, but they eventually radiate this energy away. Some travels skyward, where it may be reabsorbed by other gas molecules, a possibility less likely with increasing height because the concentration of water vapor decreases with altitude. The remainder travels Earthward and is again absorbed by Earth. For this reason, Earth's surface is continually being supplied with heat from the atmosphere as well as from the Sun. Without these absorptive gases in our atmosphere, Earth would not be a suitable habitat for humans and numerous other life-forms.

This very important phenomenon has been termed the **greenhouse effect** because it was once thought that greenhouses were heated in a similar manner (Figure 14.17). The gases of our atmosphere, especially water vapor and carbon dioxide, act very much like the glass in the greenhouse. They allow shorter-wavelength solar radiation to enter, where it is absorbed by the objects inside. These objects in turn reradiate the heat, but at longer wavelengths, to which glass is nearly opaque. The heat therefore is trapped in the greenhouse. However, a more important factor in keeping a greenhouse warm is the fact that the greenhouse itself prevents mixing of air inside with cooler air outside. Nevertheless, the term *greenhouse effect* is still used.

Temperature Measurement and Data

Temperature is one of the basic elements of weather and climate. When someone asks what it is like outside, air temperature is often the first element we mention. At a weather station, the temperature is read on a regular basis from instruments mounted in an instrument shelter (Figure 14.18). The shelter protects the instruments from direct sunlight and allows a free flow of air. In addition to a standard mercury thermometer, the shelter is likely to contain a thermograph to continuously record temperature and a set of maximum–minimum thermometers. As their name implies, these thermometers record the highest and lowest temperatures during a measurement period, usually 24 hours.

The daily maximum and minimum temperatures are the bases for many of the temperature data compiled by meteorologists:

1. By adding the maximum and minimum temperatures and then dividing by two, the *daily mean temperature* is calculated.

2. The *daily range* of temperature is computed by finding the difference between the maximum and minimum temperatures for a given day.

3. The *monthly mean* is calculated by adding together the daily means for each day of the month and dividing by the number of days in the month.

4. The *annual mean* is an average of the 12 monthly means.

Figure 14.17 The heating of the atmosphere. Most of the solar radiation that is not reflected back to space passes through the atmosphere and is absorbed at Earth's surface. Earth's surface, in turn, emits longer wavelength radiation. A portion of this energy is absorbed by certain gases in the atmosphere.

1. Much of the incoming, short wavelength, solar radiation penetrates the atmosphere and heats Earth's surface.

3. Greenhouse gases absorb outgoing, long wavelength, radiation and reradiate some of this energy earthward, thus trapping heat in the lower atmosphere.

2. Objects on Earth's surface emit long wavelength radiation skyward.

5. The *annual temperature range* is computed by finding the difference between the highest and lowest monthly means.

Mean temperatures are particularly useful for making comparisons, whether on a daily, monthly, or annual basis (see Box 14.3). It is common to hear a weather reporter state, "Last month was the hottest July on record," or "Today Chicago was ten degrees warmer than Miami." Temperature ranges are also useful statistics, because they give an indication of extremes.

Many different applications of temperature data have been devised. One familiar use of temperature data relates to human perception of temperature. Television weather reports use indices that attempt to portray levels of human comfort and discomfort. Such indices are based on the fact that our sensation of temperature is often different from the actual temperature of the air as recorded by a thermometer.

The human body is a heat engine that continually releases energy. Anything that influences the rate of heat loss from the body also influences our sensation of temperature, thereby affecting our feeling of comfort. Several factors control the thermal comfort of the human body, and certainly air temperature is a major factor. Other environmental conditions are also significant, such as relative humidity, wind, and solar radiation. For example, many of us know that a cold, windy winter day can feel much colder than the air temperature would seem to indicate (Table 14.3).

Controls of Temperature

A "temperature control" is any factor that causes temperature to vary from place to place and from time to time. Earlier in this chapter we examined the most important cause for temperature variations—differences in the receipt of solar radiation. Because variations in Sun angle and length of daylight depend on

Figure 14.18 Standard instrument shelter. A shelter protects instruments from direct sunlight and allows for the free flow of air. (Courtesy of Qualimetrics, Inc.)

Box 14.3   Showing Temperature Distribution Using Isotherms

To examine the distribution of air temperatures over large areas, isotherms are commonly used. An *isotherm* is a line that connects points on a map that have the same temperature (*iso* = equal, *therm* = temperature). Therefore, all points through which an isotherm passes have identical temperatures for the time period indicated. Generally, isotherms representing 5° or 10° temperature differences are used, but any interval may be chosen. Figure 14.D illustrates how isotherms are drawn on a map. Notice that most isotherms do not pass directly through the observing stations, because the station readings may not coincide with the values chosen for the isotherms. Only an occasional station temperature will be exactly the same as the value of the isotherm, so it is usually necessary to draw the lines by estimating the proper position between stations.

Isothermal maps are valuable tools because they clearly make temperature distribution visible at a glance. Areas of low and high temperatures are easy to pick out. In addition, the amount of temperature change per unit of distance, called the *temperature gradient*, is easy to visualize. Closely spaced isotherms indicate a rapid rate of temperature change, whereas more widely spaced lines indicate a more gradual rate of change. You can see this in Figure 14.D. The isotherms are closer in Colorado and Utah (steeper temperature gradient), whereas the isotherms are spread farther in Texas (gentler temperature gradient). Without isotherms, a map would be covered with numbers representing temperatures at dozens or hundreds of places, which would make patterns difficult to see.

Figure 14.D Isothermal map. Isotherms are lines that connect points of equal temperature. Showing temperature distribution in this way makes patterns easier to see. On television, and in many newspapers, temperature maps are in color. Rather than labeling isotherms, the area *between* isotherms is labeled. For example, the zone between the 60° and 70° isotherms is labeled "60s".

latitude, they are responsible for warm temperatures in the tropics and colder temperatures at more poleward locations. Of course, seasonal temperature changes at a given latitude occur as the Sun's vertical rays migrate toward and away from a place during the year.

However, latitude is not the only control of temperature. If it were, we would expect all places along the same parallel of latitude to have identical temperatures. This is clearly not the case. For example, Eureka, California, and New York City are both coastal cities at about the same latitude and both have an annual mean temperature of 11°C (52°F). However, New York City is 9°C (16°F) warmer than Eureka in July and 10°C cooler in January. In another example,

Table 14.3 Windchill equivalent temperature (°C).

Actual Temperature (°C)	Wind Speed (km per hr)								
	0	10	20	30	40	50	60	70	80
12	12	9	5	3	1	0	0	-1	-1
8	8	5	0	-3	-5	-6	-7	-7	-8
4	4	0	-5	-8	-11	-12	-13	-14	-14
0	0	-4	-10	-14	-17	-18	-19	-20	-21
-4	-4	-8	-15	-20	-23	-25	-26	-27	-27
-8	-8	-13	-21	-25	-29	-31	-32	-33	-34
-12	-12	-17	-26	-31	-35	-37	-39	-40	-40
-16	-16	-22	-31	-37	-41	-43	-45	-46	-47
-20	-20	-26	-36	-43	-47	-49	-51	-52	-53
-24	-24	-31	-42	-48	-53	-56	-58	-59	-60
-28	-28	-35	-47	-54	-59	-62	-64	-65	-66
-32	-32	-40	-52	-60	-65	-68	-70	-72	-73
-36	-36	-44	-57	-65	-71	-74	-77	-78	-79
-40	-40	-49	-63	-71	-77	-80	-83	-85	-86

Source: NOAA, National Weather Service.

two cities in Ecuador, Quito and Guayaquil, are relatively close to each other, yet the annual mean temperatures at these two cities differ by 12°C (21°F). To explain these situations and countless others, we must realize that factors other than latitude also exert a strong influence on temperature. In the next sections, we will examine these other controls, which include differential heating of land and water, altitude, geographic position, cloud cover and albedo, and ocean currents.*

Land and Water

The heating of Earth's surface controls the heating of the air above it. Therefore, to understand variations in air temperature, we must examine the nature of the surface. Different land surfaces absorb varying amounts of incoming solar energy, which in turn cause variations in the temperature of the air above. The largest contrast, however, is not between different land surfaces, but between land and water. *Land heats more rapidly and to higher temperatures than water and cools more rapidly and to lower temperatures than water.* Temperature variations, therefore, are considerably greater over land than over water.

Among the reasons for the differential heating of land and water are the following:

1. The *specific heat* (amount of energy needed to raise 1 gram of a substance 1°C) is far greater for water than for land. Thus, water requires a great deal more heat to raise its temperature the same amount as an equal quantity of land.

2. Land surfaces are opaque, so heat is absorbed only at the surface. Water, being more transparent, allows some solar radiation to penetrate to a depth of many meters.

3. The water that is heated often mixes with water below, thus distributing the heat through an even larger mass.

4. Evaporation (a cooling process) from water bodies is greater than that from land surfaces.

All these factors collectively cause water to warm more slowly, store greater quantities of heat, and cool more slowly than land.

Monthly temperature data for two cities will demonstrate the moderating influence of a large water body and the extremes associated with land (Figure 14.19). Vancouver, British Columbia, is located along the windward Pacific coast, whereas Winnipeg, Manitoba, is in a continental position far from the influence of water. Both cities are at about the same latitude and thus experience similar Sun angles and lengths of daylight. Winnipeg, however, has a mean January temperature that is 20°C lower than Vancouver's. Conversely, Winnipeg's July mean is 2.6°C higher than Vancouver's. Although their latitudes are nearly the same, Winnipeg, which has no water influence, experiences much greater temperature extremes than does Vancouver, which does. The key to Vancouver's moderate year-round climate is the Pacific Ocean.

On a different scale, the moderating influence of water may also be demonstrated when temperature variations in the Northern and Southern hemispheres are compared. In the Northern Hemisphere, 61 percent

---

* For a discussion of the effects of ocean currents on temperature, see Chapter 13.

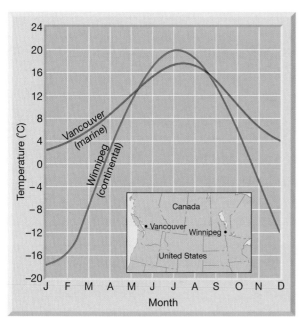

Figure 14.19 Mean monthly temperatures for Vancouver, British Columbia, and Winnipeg, Manitoba. Vancouver has a much smaller annual temperature range owing to the strong marine influence of the Pacific Ocean. Winnipeg illustrates the greater extremes associated with an interior location.

is covered by water, and land accounts for the remaining 39 percent. However, in the Southern Hemisphere, 81 percent is covered by water and 19 percent by land. The Southern Hemisphere is correctly called the *water hemisphere* (see Figure 12.1, p. 324). Table 14.4 portrays the considerably smaller annual temperature variations in the water-dominated Southern Hemisphere as compared with the Northern Hemisphere.

Altitude

The two cities in Ecuador mentioned earlier—Quito and Guayaquil—demonstrate the influence of altitude on mean temperature. Both cities are near the equator and relatively close to one another, but the annual mean temperature at Guayaquil is 25°C (77°F) compared to Quito's mean of 13°C (55°F). The difference

Table 14.4 Variation in annual mean temperature range (°C) with latitude.

Latitude	Northern Hemisphere	Southern Hemisphere
0	0	0
15	3	4
30	13	7
45	23	6
60	30	11
75	32	26
90	40	31

may be understood when the cities' elevations are noted. Guayaquil is only 12 meters (40 feet) above sea level, whereas Quito is high in the Andes Mountains at 2800 meters (9200 feet).

Recall that temperatures drop an average of 6.5°C per kilometer in the troposphere; thus cooler temperatures are to be expected at greater heights (see Figure 14.5). Yet, the magnitude of the difference is not explained completely by the normal lapse rate. If the normal lapse rate is used, we would expect Quito to be about 18°C cooler than Guayaquil, but the difference is only 12°C. The fact that high-altitude places such as Quito are warmer than the value calculated using the normal lapse rate results from the absorption and reradiation of solar energy by the ground surface.

Geographic Position

The geographic setting can greatly influence the temperatures experienced at a specific location. A coastal location where prevailing winds blow from the ocean onto the shore (a *windward* coast) experiences considerably different temperatures than does a coastal location where the prevailing winds blow from the land toward the ocean (a *leeward* coast). In the first situation, the windward coast will experience the full moderating influence of the ocean—cool summers and mild winters, compared to an inland station at the same latitude.

A leeward coast, on the other hand, will have a more continental temperature pattern because the winds do not carry the ocean's influence onshore. Eureka, California, and New York City, the two cities mentioned in the section on temperature controls, illustrate this aspect of geographic position. The annual temperature range at New York City is 19°C (34°F) greater than Eureka's (Figure 14.20).

Seattle and Spokane, both in the state of Washington, illustrate a second aspect of geographic position—mountains that act as barriers. Although Spokane is only about 360 kilometers (220 miles) east of Seattle, the towering Cascade Range separates the cities. Consequently, Seattle's temperatures show a marked marine influence, but Spokane's are more typically continental (Figure 14.21). Spokane is 7°C (13°F) cooler than Seattle in January and 4°C (7°F) warmer than Seattle in July. The annual range at Spokane is 11°C (20°F) greater than at Seattle. The Cascade Range effectively cuts Spokane off from the moderating influence of the Pacific Ocean.

Cloud Cover and Albedo

The extent of cloud cover is a factor that influences temperatures in the lower atmosphere. Cloud cover is important because many clouds have a high albedo and therefore reflect a significant portion of the sunlight that

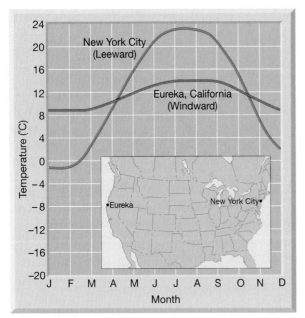

Figure 14.20 Monthly mean temperatures for Eureka, California, and New York City. Both cities are coastal and located at about the same latitude. Because Eureka is strongly influenced by prevailing winds from the ocean and New York City is not, the annual temperatures range at Eureka is much smaller.

A.

B.

Figure 14.22 Clouds reduce the daily temperature range. **A.** During daylight hours, clouds reflect solar radiation back to space. Therefore, the maximum temperature is lower than if the sky were clear. **B.** At night, the minimum temperature will not fall as low because clouds retard the loss of heat.

strikes them back to space. By reducing the amount of incoming solar radiation, daytime temperatures will be lower than if the clouds were absent and the sky were clear (Figure 14.22A).

At night, clouds have the opposite effect as during daylight. They act as a blanket by absorbing outgoing terrestrial radiation and reradiating a portion of it back to the surface (Figure 14.22B). Consequently, some of the heat that otherwise would have been lost remains near the ground. Thus, nighttime air temperatures do not drop as low as they would on a clear night. The effect of cloud cover is to reduce the daily temperature range by lowering the daytime maximum and raising the nighttime minimum.

Clouds are not the only phenomenon that increase albedo and thereby reduce air temperatures. We also recognize that snow- and ice-covered surfaces have high albedos. This is one reason why mountain glaciers do not melt away in the summer and why snow may still be present on a mild spring day. In addition, during the winter, when snow covers the ground, daytime maximums on a sunny day are less than they otherwise would be because energy that the land would have absorbed and used to heat the air is reflected and lost.

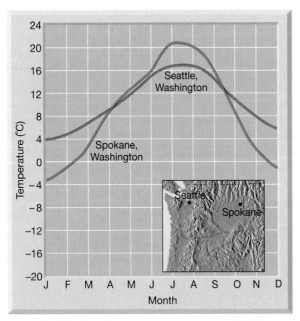

Figure 14.21 Monthly mean temperatures for Seattle and Spokane, Washington. Because the Cascade Mountains cut off Spokane from the moderating influence of the Pacific Ocean, its annual temperature range is greater than Seattle's.

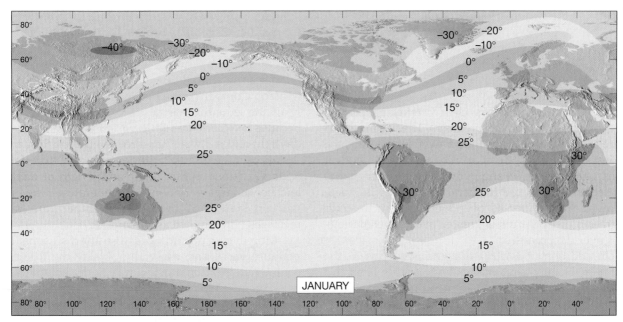

Figure 14.23 World mean sea-level temperatures in January in degrees Celsius.

World Distribution of Temperature

Take a moment to study the two world isothermal maps (Figures 14.23 and 14.24). From hot colors near the equator to cool colors toward the poles, these maps portray sea-level temperatures in the seasonally extreme months of January and July. On these maps you can study global temperature patterns and the effects of the controlling factors of temperature, especially latitude, the distribution of land and water, and ocean currents. Like most isothermal maps of large regions, all temperatures on these world maps have

been reduced to sea level to eliminate the complications caused by differences in altitude.

On both maps, the isotherms generally trend east and west and show a decrease in temperatures poleward from the tropics. They illustrate one of the most fundamental aspects of world temperature distribution: that the effectiveness of incoming solar radiation in heating Earth's surface and the atmosphere above it is largely a function of latitude.

Moreover, there is a latitudinal shifting of temperatures caused by the seasonal migration of the Sun's vertical rays. To see this, compare the color bands by

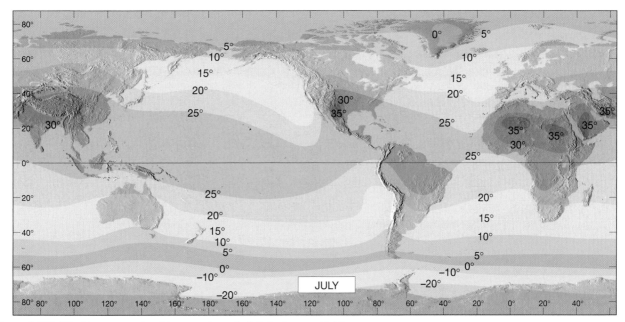

Figure 14.24 World mean sea-level temperatures in July in degrees Celsius.

latitude on the two maps. For example, on the January map, the "hot spots" of 30°C are *south* of the equator, but in July they have shifted *north* of the equator.

If latitude were the only control of temperature distribution, our analysis could end here, but this is not the case. The added effect of the differential heating of land and water is clearly reflected on the January and July temperature maps. The warmest and coldest temperatures are found over land—note the coldest area, a purple oval in Siberia, and the hottest areas, the deep orange ovals—all over land. Consequently, because temperatures do not fluctuate as much over water as over land, the north-south migration of isotherms is greater over the continents than over the oceans.

In addition, it is clear that the isotherms in the Southern Hemisphere, where there is little land and where the oceans predominate, are much more regular than in the Northern Hemisphere, where they bend sharply northward in July and southward in January over the continents.

Isotherms also reveal the presence of ocean currents. Warm currents cause isotherms to be deflected toward the poles, whereas cold currents cause an equatorward bending. The horizontal transport of water poleward warms the overlying air and results in air temperatures that are higher than would otherwise be expected for the latitude. Conversely, currents moving toward the equator produce cooler than expected air temperatures.

Figures 14.23 and 14.24 show the seasonal extremes of temperature, so comparing them enables us to see the annual range of temperature from place to place. Comparing the two maps shows that a station near the equator has a very small annual range because it experiences little variation in the length of daylight and it always has a relatively high Sun angle. A station in the middle latitudes, however, experiences wide variations in Sun angle and length of daylight and hence large variations in temperature. Therefore, we can state that the annual temperature range increases with an increase in latitude.

Moreover, land and water also affect seasonal temperature variations, especially outside the tropics. A continental location must endure hotter summers and colder winters than a coastal location. Consequently, outside the tropics the annual range will increase with an increase in continentality.

A classic example of the effect of latitude and continentality on annual temperature range is Yakutsk, Russia. This city in the heart of Siberia is just a few degrees south of the Arctic Circle and far from the influence of water. As a result, Yakutsk has an average annual temperature range of 62.2°C, among the greatest ranges in the world.

Chapter Summary

- *Weather* is the state of the atmosphere at a particular place for a short period of time. *Climate*, on the other hand, is a generalization of the weather conditions of a place over a long period of time.

- The most important *elements*, those quantities or properties that are measured regularly, of weather and climate are (1) *air temperature*, (2) *humidity*, (3) type and amount of *cloudiness*, (4) type and amount of *precipitation*, (5) *air pressure*, and (6) the speed and direction of the *wind*.

- If water vapor, dust, and other variable components of the atmosphere were removed, clean, dry air would be composed almost entirely of *nitrogen* (N), about 78% of the atmosphere by volume, and *oxygen* (O_2), about 21%. *Carbon dioxide* (CO_2), although present only in minute amounts (0.036%), is important because it has the ability to absorb heat radiated by Earth and thus helps keep the atmosphere warm. Among the variable components of air, *water vapor* is very important because it is the source of all clouds and precipitation and, like carbon dioxide, it is also a heat absorber.

- *Ozone* (O_3), the triatomic form of oxygen, is concentrated in the 10- to 50-kilometer altitude range of the atmosphere, and is important to life because of its ability to absorb potentially harmful ultraviolet radiation from the Sun.

- Because the atmosphere gradually thins with increasing altitude, it has no sharp upper boundary but simply blends into outer space. Based on temperature, the atmosphere is divided vertically into four layers. The *troposphere* is the lowermost layer. In the troposphere, temperature usually decreases with increasing altitude. This *environmental lapse rate* is variable, but averages about 6.5°C per kilometer (3.5°F per 1000 feet). Essentially all important weather phenomena occur in the troposphere. Above the troposphere is the *stratosphere*, which exhibits warming because of absorption of ultraviolet radiation by ozone. In the *mesosphere*, temperatures again decrease. Upward from the mesosphere is the *thermosphere*, a layer with only a minute fraction of the atmosphere's mass and no well-defined upper limit.

- The two principal motions of Earth are (1) *rotation*, the spinning of Earth about its axis, which produces the daily cycle of daylight and darkness; and (2) *revolution*, the movement of Earth in its orbit around the Sun.

- *Several factors act together to cause the seasons.* Earth's axis is inclined 23 1/2° from the perpendicular to the plane of its orbit around the Sun and remains pointed in the same direction (toward the North Star) as Earth journeys around the Sun. As a consequence, Earth's orientation to the Sun continually changes. The yearly fluctuations in the angle of the Sun and length of daylight brought about by Earth's changing orientation to the Sun cause the seasons.

- The three mechanisms of heat transfer are (1) *conduction*, the transfer of heat through matter by molecular activity; (2) *convection*, the transfer of heat by the movement of a mass or substance from one place to another; and (3) *radiation*, the transfer of heat by electromagnetic waves.

- *Electromagnetic radiation* is energy emitted in the form of rays, called electromagnetic waves. All radiation is capable of transmitting energy through the vacuum of space. One of the most important differences between electromagnetic waves are their *wavelengths*, which range from very long *radio waves* to very short *gamma rays*. *Visible light* is the only portion of the electromagnetic spectrum we can see. Some of the basic laws that govern radiation as it heats the atmosphere are (1) all objects emit radiant energy; (2) hotter objects radiate more total energy than do colder objects; (3) the hotter the radiating body, the shorter the wavelengths of maximum radiation; and (4) objects that are good absorbers of radiation are good emitters as well.

- The general drop in temperature with increasing altitude in the troposphere supports the fact that the *atmosphere is heated from the ground up.* Approximately 50 percent of the solar energy that strikes the top of the atmosphere is ultimately absorbed at Earth's surface. Earth releases the absorbed radiation in the form of long-wave radiation. The atmospheric absorption of this long-wave *terrestrial radiation*, primarily by water vapor and carbon dioxide, is responsible for heating the atmosphere.

- The factors that cause temperature to vary from place to place, also called the *controls of temperature*, are (1) *differences in the receipt of solar radiation*— the greatest single cause; (2) the unequal heating and cooling of *land and water*, in which land heats more rapidly and to higher temperatures than water and cools more rapidly and to lower temperatures than water; (3) *altitude*; (4) *geographic position*; (5) *cloud cover and albedo*; and (6) *ocean currents*.

- Temperature distribution is shown on a map by using *isotherms*, which are lines that connect equal temperatures. Differences between January and July temperatures around the world can be explained in terms of the basic controls of temperature.

Key Terms

aerosols (p. 380)
air (p. 380)
albedo (p. 392)
autumnal equinox (p. 386)
circle of illumination (p. 384)
climate (p. 378)
conduction (p. 389)
convection (p. 389)
electromagnetic radiation (p. 390)
element (of weather and climate)
  (p. 379)

environmental lapse rate (p. 382)
greenhouse effect (p. 393)
heat (p. 389)
inclination of the axis (p. 386)
infrared (p. 390)
mesosphere (p. 383)
ozone (p. 380)
radiation (p. 390)
revolution (p. 384)
rotation (p. 383)
spring equinox (p. 386)

stratosphere (p. 383)
summer solstice (p. 386)
temperature (p. 389)
thermosphere (p. 383)
Tropic of Cancer (p. 386)
Tropic of Capricorn (p. 386)
troposphere (p. 382)
ultraviolet (p. 390)
visible light (p. 390)
weather (p. 378)
winter solstice (p. 386)

Review Questions

1. Distinguish between weather and climate.

2. List the basic elements of weather and climate.

3. What are the two major components of clean, dry air? Are they important meteorologically?

4. Why are water vapor and aerosols important constituents of our atmosphere?

5. What source is responsible for the most pollution? (See Figure 14.A.)

6. What is the difference between primary and secondary pollutants? (See Box 14.1)

7. (a) Why is ozone important to life on Earth? (b) What is the most serious threat to human health of a decrease in the stratosphere's ozone? (See Box 14.2)

8. The atmosphere is divided vertically into four layers on the basis of temperature. List the names of these layers in order (from lowest to highest) and describe how temperature changes in each layer.

9. Why do temperatures increase in the stratosphere?

10. Briefly explain the primary cause of the seasons.

11. After examining Table 14.1, write a general statement that relates the season, the latitude, and the length of daylight.

12. Describe the relationship between the temperature of a radiating body and the wavelengths it emits.

13. Distinguish between heat and temperature.

14. Describe the three basic mechanisms of heat transfer.

15. Figure 14.16 illustrates what happens to incoming solar radiation. The percentages shown, however, are only global averages. In particular, the amount of solar radiation reflected (albedo) may vary considerably. What factors might cause variations in albedo?

16. How does Earth's atmosphere act as a "greenhouse"?

17. How are the following temperature data computed: daily mean, daily range, monthly mean, annual mean, annual range?

18. Referring to Table 14.3 determine the wind-chill equivalent temperature if the wind speed were 10 kilometers per hour and the air temperature were -8°C. What if the air temperature remained unchanged but the wind speed were 30 kilometers per hour?

19. Quito, Ecuador, is located on the equator and is not a coastal city. It has an average annual temperature of only 13°C (55°F). What is the likely cause for this low average temperature?

20. In what ways can geographic position be considered a control of temperature?

21. How does cloud cover influence the maximum temperature on an overcast day? How is the nighttime minimum influenced by clouds?

22. Yakutsk is located in Siberia at about 60 degrees north latitude. This Russian city has one of the highest average annual temperature ranges in the world: 62.2°C (112°F). Explain the reasons for the very high annual temperature range.

Examining the Earth System

1. Give and example or two of how the Earth system might be affected if Earth's axis were perpendicular to the plane of its orbit instead of being inclined 23 1/2°.

2. Speculate on the changes in global temperatures that might occur if Earth had substantially more land area and less ocean area than at present. How might such changes influence the biosphere?

Web Resources

The following are informative and interesting World Wide Web sites that address topics related to those presented in the chapter:

National Oceanic and Atmospheric Administration (NOAA):

- **http://www.noaa.gov/**

Weather World 2010 (University of Illinois):

- **http://www2010.atmos.uicu.edu/(Gh)/home.rxml**

For direct links to these sites and other appropriate Earth science Web locations, chapter review exercise, and relevant, topical investigations that utilize Web resources, contact the Earth Science, 9th Edition Home Page at:

- **http://www.prenhall.com/tarbuck.**

Chapter 15

Moisture, Clouds, and Precipitation

People caught in a heavy downpour. This image reminds us once again that no other aspect of our physical environment affects the daily lives of people more than the weather. (Photo by Mary Fulton/Liaison International)

As you observe day-to-day weather changes, you witness the powerful role of water in the air. This raises questions: What is humidity and how is it measured? Why do clouds form on some occasions but not on others? What processes produce clouds and precipitation? Why are some clouds thin, white, and harmless, while others are towering, gray, and ominous? What is the difference between sleet and hail? What is fog? Are all fogs alike? This chapter investigates the remarkable phenomena associated with moisture in the air.

Scientists agree: when it comes to understanding atmospheric processes, *water vapor is the most important gas in the atmosphere.* Water vapor constitutes only a small fraction of the gases in the atmosphere, varying from nearly 0 to about 4 percent by volume. But the importance of water in the air vastly exceeds what these small percentages would indicate.

Recall from Chapter 14 that water vapor is an important heat-absorbing gas. This property is critical to the heating of the atmosphere. Water vapor is also the source of all condensation and precipitation. Clouds and fog, as well as rain, snow, sleet, and hail, are among the most conspicuous and observable of weather phenomena (Figure 15.1).

We all expect rain to fall when certain cloud types are present, yet most of us have no idea of the complex processes that must take place in a cloud to produce rain. The formation of a single average raindrop requires water from nearly a million microscopic cloud droplets. What mechanisms foster the creation of raindrops? The wide variations in the amount of precipitation from place to place, as well as local differences from time to time, have a significant impact on the nature of the physical landscape and on our lifestyles.

Changes of State

Water vapor is an odorless, colorless gas that mixes freely with the other gases of the atmosphere. Oxygen and nitrogen, the two gases that comprise 99 percent of the atmosphere, change state (solid to liquid to gas) at extremely low temperatures (−200°C or so). But water vapor changes state (solid to liquid to gas) at the temperatures and pressures experienced near Earth's surface.

It is because of these changes of state at "ordinary" temperatures that water leaves the oceans as a gas and returns again as a liquid. The processes that involve such changes of state require that *heat* be absorbed or released (Figure 15.2). This heat energy is measured in calories. One **calorie** is the amount of heat required to raise the temperature of 1 gram of water 1°C. Thus, when 10 calories of heat are added to 1 gram of water, a 10°C temperature rise occurs.

Importance of Latent Heat

Under certain conditions, heat may be added to a substance without an accompanying temperature change. This might sound impossible, but consider what

Figure 15.1 Thunderstorm over Eagle Mesa in Arizona's Monument Valley (Photo by Greg Gawlowski/Dembinsky Photo Associates)

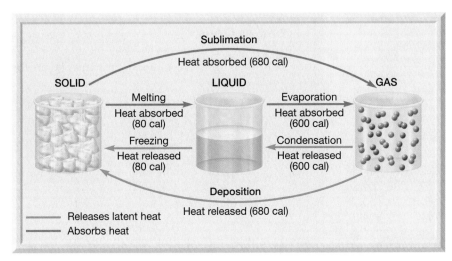

Figure 15.2 Changes of state.

occurs during a change in state. To illustrate, let us apply heat to a container of ice water. Its temperature is 0°C. We keep adding heat, but the temperature remains constant until all the ice is melted. *Then* the temperature begins to rise.

Where did the heat go during melting? In this case, the energy was used to disrupt the internal crystalline structure of the ice cubes, freeing the water molecules to move around—in other words, to melt. *Because this heat energy is not associated with a temperature change*, it is referred to as **latent heat** (latent means hidden). This energy is not available as heat until the liquid returns to the solid state, when the heat is released.

As we shall examine later, latent heat plays a crucial role in many atmospheric processes. In particular, the release of latent heat aids the formation of the towering clouds often seen on warm summer days, and it is the major source of energy for violent thunderstorms, tropical storms, and hurricanes.

Liquid/Gas Change of State. The process of converting a liquid to a gas is termed **evaporation**. It takes approximately 600 calories of energy to convert 1 gram of water to water vapor (Figure 15.2). The energy absorbed by the water molecules during evaporation is used solely to give them the motion needed to escape the surface of the liquid and become a gas. This energy is referred to as *latent heat of vaporization*. During the process of evaporation, it is the higher-temperature (faster-moving) molecules that escape the surface. As a result, the average molecular motion (temperature) of the remaining water is reduced, causing a cooling effect—hence the common expression that, "evaporation is a cooling process." You have undoubtedly experienced this cooling effect when stepping dripping wet from a swimming pool or bathtub.

Condensation is the process whereby water vapor changes to the liquid state. For condensation to occur, the water molecules must release their stored heat energy (*latent heat of condensation*) equal to what was absorbed during evaporation. This released energy

plays an important role in producing violent weather and can transfer great quantities of heat from tropical oceans to more poleward locations. When condensation occurs in the atmosphere, it results in the formation of fog and clouds.

Solid/Liquid Change of State. **Melting** is the process by which a solid is changed to a liquid. It requires absorption of approximately 80 calories of heat per gram of water (*latent heat of melting*). **Freezing**, the reverse process, releases these 80 calories per gram as *latent heat of fusion*.

Solid/Gas Change of State. The final processes illustrated in Figure 15.2 are sublimation and deposition. **Sublimation** is the conversion of a solid directly to a gas, without passing through the liquid state. You may have observed this change in watching the sublimation of "dry ice" (frozen carbon dioxide) into white, wispy vapor, sometimes used to generate "fog" or "smoke" in theatrical productions. Water ice also sublimates, which is why unused ice cubes gradually shrink in a freezer.

The term **deposition** is used to denote the reverse process, the conversion of a vapor directly to a solid. This change occurs, for example, when water vapor is deposited as ice on cold objects such as grass or windows (Figure 15.3). These deposits are termed *white frost* or *hoar frost* but are generally just called *frost*. A household example of deposition is the "frost" that accumulates in a freezer compartment. As shown in Figure 15.2, sublimation and deposition involve an amount of energy equal to the total of the other two processes.

Humidity: Water Vapor in the Air

Humidity is the general term for the amount of water vapor in air. There are different ways of looking at humidity; we will examine two—specific humidity and relative humidity.

Figure 15.3 Frost on a window pane is an example of deposition. (Photo by D. Cavagnaro/DRK Photo)

However, before we consider these humidity measures, you need to understand the concept of **saturation**. Imagine a closed jar half full of water and half full of dry air, both at the same temperature. As the water begins to evaporate from the water surface, a small increase in pressure can be detected in the air above. This increase is the result of the motion of the water vapor molecules that were added to the air through evaporation. In the open atmosphere, this pressure is termed **vapor pressure** and is defined as that part of the total atmospheric pressure that can be attributed to the water vapor content.

In the closed container, as more and more molecules escape from the water surface, the steadily increasing vapor pressure in the air above forces more and more of these molecules to return to the liquid. Eventually the number of vapor molecules returning to the surface will balance the number leaving. At that point, the air is said to be *saturated*, or filled to capacity. However, if we add heat to the container, increasing the temperature of the water and air, more water will evaporate before a balance is reached. Consequently, at higher temperatures, more moisture is required for saturation. Stated another way, the water vapor capacity of air is *temperature dependent*, with warm air having a much greater capacity than cold air. The amount of water vapor required for saturation at various temperatures varies surprisingly, as shown in Table 15.1.

Specific Humidity

Not all air is saturated, of course. Thus, we need ways to express how humid a parcel of air is. One method specifies the amount of water vapor contained in a unit of air. **Specific humidity** is the weight of water vapor per weight of a chosen mass of air, including the water vapor. Because it is measured in units of weight (usually in grams per kilogram), specific humidity is not affected by changes in pressure or temperature. *Specific humidity is a measure of the actual quantity of water vapor in a given mass of air.*

Relative Humidity

The most familiar, and perhaps the most misunderstood, term used to describe the moisture content of air is relative humidity. Stated in an admittedly oversimplified manner, **relative humidity** *is the ratio of the air's actual water vapor content to its potential water vapor capacity at a given temperature.* To illustrate, we see from Table 15.1 that at 25°C air is saturated when it contains 20 grams of water vapor per kilogram of air. Thus, if the air contains only 10 grams per kilo-

Table 15.1 Amount of water vapor (grams) required to saturate a kilogram of air at various temperatures.

Temperature		Grams of water vapor per kg of air
(°C)	(°F)	
-40	-40	0.1
-30	-22	0.3
-20	-4	0.75
-10	14	2
0	32	3.5
5	41	5
10	50	7
15	59	10
20	68	14
25	77	20
30	86	26.5
35	95	35
40	104	47

gram on a 25°C day, the relative humidity is expressed as 10/20 or 50 percent. Further, if air with a temperature of 25°C had a water vapor content of 20 grams per kilogram, the relative humidity would be expressed as 20/20 or 100 percent. On those occasions when the relative humidity reaches 100 percent, the air is saturated.

Because relative humidity is a comparison of the air's water vapor content (specific humidity) and its capacity, relative humidity can be changed in either of two ways. First, if moisture is added to or subtracted from air, its relative humidity will change. This is illustrated in Figure 15.4.

The second condition that affects relative humidity is air temperature. Examine Figure 15.5 carefully, and note in Part A that when air at 20°C contains 7 grams of water vapor per kilogram of air it has a relative humidity of 50 percent. When the air is cooled from 20°C to 10°C as shown in Part B, it still contains 7 grams of water vapor, but the relative humidity increases from 50 percent to 100 percent. We can conclude from this that, when the water vapor content remains constant, *a decrease in temperature results in an increase in relative humidity*.

What happens when the air is cooled further, below the temperature at which saturation occurs? Part C of Figure 15.5 illustrates this situation. Notice from Table 15.1 that, when the flask is cooled to 0°C, the air is saturated at 3.5 grams of water vapor per kilogram of air. Because this flask originally contained 7 grams of water vapor, 3.5 grams of water vapor have to go somewhere, because there is no room for them in the cooler air. They will condense to form liquid droplets that collect on the walls of the container. In the meantime, the relative humidity of the air inside remains at 100 percent. This brings up an important concept. When air aloft is cooled below its saturation point, some of the water vapor condenses to form clouds. Because clouds are made of liquid droplets, this moisture is no longer part of the *water vapor* content of the air.

We can generalize the effects of temperature on relative humidity as follows. When the water vapor content of air remains at a constant level, a decrease in air temperature results in an increase in relative humidity, and an increase in temperature causes a decrease in relative humidity. In Figure 15.6 the variations in temperature and relative humidity during a typical day demonstrate the relationship just described (see Box 15.1).

Figure 15.4 When the temperature remains constant, relative humidity will increase as water vapor is added to the air. Here the water vapor capacity remains constant at 20 grams per kilogram, while the relative humidity rises from 25 percent to 100 percent as the water vapor content (specific humidity) increases.

Figure 15.5 When the water vapor content remains constant, the relative humidity can be changed by increasing or decreasing the air temperature. In this example, when the temperature of the air in the flask was lowered from 20°C to 10°C, the relative humidity increased from 50 percent to 100 percent. Thus, 10°C is the dew point. Further cooling (from 10°C to 0°C) causes one-half of the water vapor to condense because colder air holds less moisture. In nature, cooling of air below its dew point generally causes condensation in the form of clouds, dew, or fog.

Dew Point

Another important idea related to relative humidity is the dew-point temperature. **Dew point** is the *temperature* to which air would have to be cooled to reach saturation. Note that in Figure 15.5 unsaturated air at 20°C is cooled to 10°C before saturation occurs.

Therefore, 10°C would be the dew-point temperature for this air. If this same parcel of air were cooled further, the air's capacity would be exceeded, and the excess vapor would condense, typically as dew, fog, or clouds. The term "dew point" stems from the fact that during nighttime hours, objects near the ground often

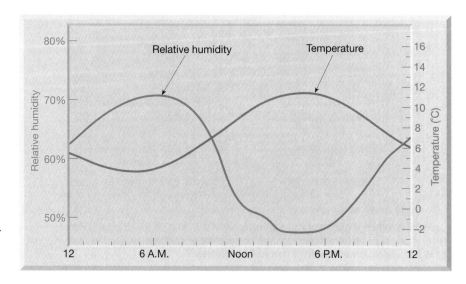

Figure 15.6 Typical daily variations in temperature and relative humidity during a spring day at Washington, D.C. When temperature increases, relative humidity drops (see midafternoon) and vice versa.

Box 15.1   Dry Air at 100 Percent Relative Humidity?

Does air with higher relative humidity always contain more water vapor than air with low relative humidity? At the *same temperature*, yes. However, if temperatures are different, there can be less water in "humid" air than in "dry" air. To illustrate, let us compare a typical January day at Winnipeg, Manitoba, to one at Phoenix, Arizona. On this hypothetical day the temperature in Winnipeg is a frigid –10°C and the relative humidity is 100 percent. By referring to Table 15.1, we can see that saturated –10°C air has a water vapor content of 2 grams per kilogram. By contrast, the desert air at Phoenix on this January day is a warm 25°C and the relative humidity is just 20 percent. A look at Table 15.1 reveals that 25°C air has a capacity of 20 grams per kilogram. Therefore, with a relative humidity of 20 percent, the air at Phoenix has a water vapor content of 4 grams per kilogram (20 grams times 20 percent). Consequently, the "dry" air at Phoenix actually contains twice the water vapor as the "humid" air at Winnipeg.

It now should be clear why places that are very cold are also very dry. The low water vapor content of frigid air helps to explain why many arctic areas receive only meager precipitation and are really polar deserts. This also explains why you may experience dry skin and chapped lips during the winter months. The water vapor content of cold air is low, even when compared to some hot arid regions.

---

become coated with dew (Figure 15.7). Contrary to popular belief, white frost is not frozen dew. Rather, white frost forms on occasions when the dew point of the air is 0°C or below. Thus, frost forms when water vapor changes directly from a gas into a solid (ice) without entering the liquid state. This process, called *deposition*, produces delicate patterns of ice crystals that frequently decorate windows on frigid winter days (see Figure 15.3).

Measuring Humidity

Relative humidity is commonly measured using a **hygrometer**. One type of hygrometer, called a **psychrometer** consists of two identical thermometers mounted side by side (Figure 15.8). One thermometer, the *dry-bulb*, gives the present air temperature. The other, called the *wet-bulb* thermometer, has a thin muslin wick tied around the end (see end of thermometers in the photo).

To use the psychrometer, the cloth sleeve is saturated with water and a continuous current of air is passed over the wick. This is done either by swinging the instrument freely in the air or by fanning air past it. As a consequence, water evaporates from the wick, and the heat absorbed by the evaporating water makes the temperature of the wet bulb drop. The loss of heat that was required to evaporate water from the wet bulb lowers the thermometer reading.

The amount of cooling that takes place is directly proportional to the dryness of the air. The drier the air, the more moisture evaporates. The more heat the evaporating water absorbs, the greater the cooling. Therefore, the larger the difference that is observed between the thermometer readings, the lower the relative humidity; the smaller the difference, the higher the relative humidity. If the air is saturated, no evaporation will occur, and the two thermometers will have identical readings.

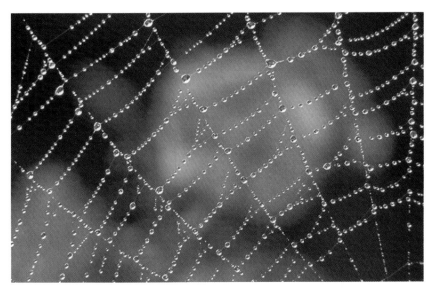

Figure 15.7 Dew on a spider web. (Photo by Wolfgang Kaehler)

Figure 15.8 Sling psychrometer. This instrument is used to determine both relative humidity and dew point. The dry-bulb thermometer gives the current air temperature. The web-bulb thermometer is covered with a cloth wick that is dipped in water. The thermometers are spun until the temperature of the wet-bulb thermometer stops declining. Then the thermometers are read and the data used in conjunction with Tables 15.2 and 15.3. (Photo by E. J. Tarbuck)

To determine the precise relative humidity from the thermometer readings, a standard table is used (Table 15.2). With the same information, but using a different table (Table 15.3), the dew-point temperature can also be calculated.

Another instrument used for measuring relative humidity, the *hair hygrometer*, can be read directly, without the use of tables. The hair hygrometer operates on the principle that hair or certain synthetic fibers change their length in proportion to changes in the relative humidity, lengthening as relative humidity increases and shrinking as the relative humidity drops. The tension of a bundle of hairs is linked mechanically to an indicator that is calibrated between 0 and 100 percent. Thus, we need only glance at the dial to determine the relative humidity. Unfortunately, the hair hygrometer is less accurate than the psychrometer. Further, it

requires frequent calibration and is slow in responding to changes in humidity, especially at low temperatures.

A different type of hygrometer is used in remote-sensing instrument packages such as radiosondes that transmit upper-air observations back to ground stations. The *electric hygrometer* contains an electrical conductor coated with a moisture-absorbing chemical. It works on the principle that the passage of current varies as the relative humidity varies.

The Basis of Cloud Formation: Adiabatic Cooling

Up to this point, we have considered basic properties of water vapor and how its variability is measured. We are now ready to examine some of the important roles that water vapor plays in weather, especially the formation of clouds.

Fog and Dew Versus Cloud Formation

Recall that condensation occurs when water vapor changes to a liquid. This condensation may form dew, fog, or clouds. Although these three forms are different, all require saturated air to develop. As indicated earlier, saturation occurs either when water vapor is added to the air or, more commonly, when the air is cooled to its dew point. Near Earth's surface heat is readily exchanged between the ground and the air above. During evening hours, the surface radiates heat away, and the surface and adjacent air cool rapidly. This "radiation cooling" accounts for the formation of dew and some types of fog. At higher altitudes, however, clouds often form during the warmest part of the day, not during the cool evening or night. Consequently, some other mechanism must operate to form clouds.

Adiabatic Temperature Change

The process that is responsible for most cloud formation is easily visualized if you have ever pumped up a bicycle tire and noticed that the pump barrel became quite warm. The heat you felt was the consequence of the work you did on the air to compress it. When energy is used to compress air, the motion of the gas molecules increases and therefore the temperature of the air rises. Conversely, air that is allowed to escape from a bicycle tire *expands and cools*. This results because the expanding air pushes (does work on) the surrounding air and must cool by an amount equivalent to the energy expended.

You may have experienced the cooling effect of an expanding gas while applying a spray deodorant. As the compressed gas propellant in the aerosol can is released, it quickly expands and cools. This drop in temperature occurs *even though heat is neither added nor subtracted*. Such variations are known as **adiabatic temperature changes** and result when air is compressed or

Table 15.2 Relative humidity (percent)*

Dry bulb (°C)	Depression of Wet-Bulb Temperature (Dry-Bulb Temperature Minus Wet-Bulb Temperature = Depression of the Wet Bulb)																					
	1	2	3	4	5	6	7	8	9	10	11	12	13	14	15	16	17	18	19	20	21	22
-20	28																					
-18	40																					
-16	48	0																				
-14	55	11																				
-12	61	23																				
-10	66	33	0																			
-8	71	41	13																			
-6	73	48	20	0																		
-4	77	54	32	11																		
-2	79	58	37	20	1																	
0	81	63	45	28	11																	
2	83	67	51	36	20	6																
4	85	70	56	42	27	14																
6	86	72	59	46	35	22	10	0														
8	87	74	62	51	39	28	17	6														
10	88	76	65	54	43	33	24	13	4													
12	88	78	67	57	48	38	28	19	10	2												
14	89	79	69	60	50	41	33	25	16	8	1											
16	90	80	71	62	54	45	37	29	21	14	7	1										
18	91	81	72	64	56	48	40	33	26	19	12	6	0									
20	91	82	74	66	58	51	44	36	30	23	17	11	5									
22	92	83	75	68	60	53	46	40	33	27	21	15	10	4	0							
24	92	84	76	69	62	55	49	42	36	30	25	20	14	9	4	0						
26	92	85	77	70	64	57	51	45	39	34	28	23	18	13	9	5						
28	93	86	78	71	65	59	53	45	42	36	31	26	21	17	12	8	4					
30	93	86	79	72	66	61	55	49	44	39	34	29	25	20	16	12	8	4				
32	93	86	80	73	68	62	56	51	46	41	36	32	27	22	19	14	11	8	4			
34	93	86	81	74	69	63	58	52	48	43	38	34	30	26	22	18	14	11	8	5		
36	94	87	81	75	69	64	59	54	50	44	40	36	32	28	24	21	17	13	10	7	4	
38	94	87	82	76	70	66	60	55	51	46	42	38	34	30	26	23	20	16	13	10	7	5
40	94	89	82	76	71	67	61	57	52	48	44	40	36	33	29	25	22	19	16	13	10	7

Dry-Bulb (Air) Temperature (vertical axis label)

Relative Humidity Values (diagonal label)

* To determine the relative humidity, find the air (dry-bulb) temperature on the vertical axis (far left) and the depression of the wet bulb on the horizontal axis (top). Where the two meet, the relative humidity is found. For example, when the dry-bulb temperature is 20°C and a wet-bulb temperature is 14°C, then the depression of the wet-bulb is 6°C (20°C - 14°C). From Table 15.2, the relative humidity is 51 percent and from Table 15.3, the dew point is 10°C.

allowed to expand. In summary, when air is allowed to expand, it cools, and when it is compressed, it warms.

Dry Adiabatic Rate. As you travel from Earth's surface upward through the atmosphere, the atmospheric pressure rapidly diminishes, because there are fewer and fewer gas molecules. Thus, any time a parcel of air moves upward, it passes through regions of successively lower pressure. As a result, the ascending air expands. As it expands, it cools adiabatically. Unsaturated air cools at the rather constant rate of 10°C for every 1000 meters of ascent (1°C per 100 meters).

Conversely, descending air comes under increasingly higher pressures, compresses, and is heated 10°C for every 1000 meters of descent. This rate of cooling or heating applies only to *unsaturated air* and is known as the **dry adiabatic rate**.

Wet Adiabatic Rate. If air rises high enough, it will cool sufficiently to reach the dew point and cause condensation. From this point on along its ascent, *latent heat of condensation* stored in the water vapor will be liberated. Although the air will continue to cool after condensation begins, the released latent heat works against the adiabatic process, thereby reducing the rate at which the air cools. This slower rate of cooling caused by the addition of latent heat is called the **wet adiabatic rate** of cooling. Because the amount of latent heat released depends on the quantity of moisture present in the air, the wet adiabatic rate varies from 5°C per 1000 meters for air with a high moisture content to 9°C per 1000 meters for dry air.

Figure 15.9 illustrates the role of adiabatic cooling in the formation of clouds. Note that from the surface up to the condensation level the air cools at the dry

Table 15.3 Dew-point temperature (°C).*

Dry bulb (°C)	Dry-Bulb Temperature Minus Wet-Bulb Temperature = Depression of the Wet Bulb																					
	1	2	3	4	5	6	7	8	9	10	11	12	13	14	15	16	17	18	19	20	21	22

Dew point temperatures

adiabatic rate. The wet adiabatic rate commences at the condensation level.

Stability of Air

As you have learned, if air rises, it will cool adiabatically and eventually produce clouds. But why does air rise on some occasions and not on others? Why do the size of clouds and the amount of precipitation vary so much when air does rise? Answers to these questions are closely related to the stability of the air.

Imagine, if you will, a large bubble of air with a thin flexible cover that allows it to expand but prevents it from mixing with the surrounding air. If the imaginary bubble were forced to rise, its temperature would *decrease because of expansion*. By comparing the bubble's temperature to that of the surrounding air, we can determine the stability of the bubble.

If the bubble's temperature is lower than that of its environment, it will be denser, and if allowed to move freely, it would sink to its original position. Air of this type is termed *stable air* because it resists vertical displacement.

Conversely, if our imaginary bubble were warmer, and therefore less dense than the surrounding air, it would continue to rise until it reached an altitude having the same temperature, much as a hot-air balloon rises as long as it is less dense than the surrounding air (Figure 15.10). This type of air is called *unstable air*.

Determining Stability

The stability of air is determined by examining the temperature of the atmosphere at various heights. Recall from Chapter 14 that this measure is termed the *environmental lapse rate*. The environmental lapse rate is the temperature of the atmosphere as

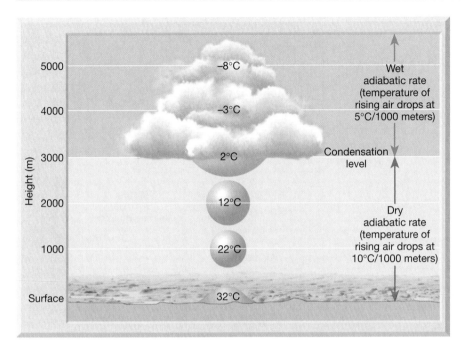

Figure 15.9 Rising air cools at the dry adiabatic rate of 10° per 1000 meters, until the air reaches the dew point and condensation (cloud formation) begins. As air continues to rise, the latent heat released by condensation reduces the rate of cooling. The wet adiabatic rate is therefore always less than the dry adiabatic rate.

determined from observations made by radiosondes and aircraft. It is important not to confuse this with *adiabatic temperature changes*, which are changes in temperature caused by expansion or compression as a parcel of air rises or descends.

Stable Air. For illustration, we will examine a situation where the environmental lapse rate is 5°C per 1000 meters (Figure 15.11). Under this condition, when air at the surface has a temperature of 25°C, the air at 1000

meters will be 5 degrees cooler, or 20°C, the air at 2000 meters will have a temperature of 15°C, and so forth. At first glance it appears that the air at the surface is less dense than the air at 1000 meters, because it is 5 degrees warmer. However, if the air near the surface is unsaturated and were to rise to 1000 meters, it would expand and cool at the dry adiabatic rate of 10°C per 1000 meters. Therefore, upon reaching 1000 meters, its temperature would have dropped 10°C. Being 5 degrees cooler than its environment, it would be denser and

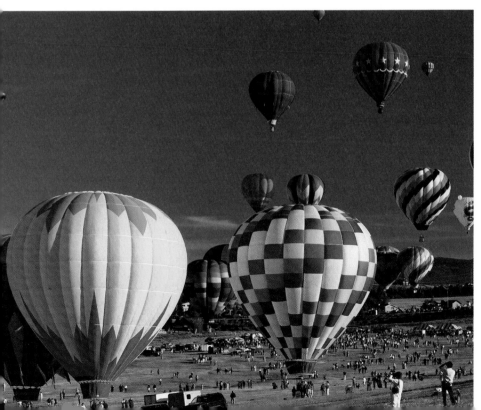

Figure 15.10 As long as air is warmer than its surroundings, it will rise. Hot air balloons rise up through the atmosphere for this reason. (Photo by Jim Steinberg/Photo Researchers, Inc.)

Figure 15.11 In a stable atmosphere, as an unsaturated parcel of air is lifted, it expands and cools at the dry adiabatic rate of 10°C per 1000 meters. Because the temperature of the rising parcel of air is lower than that of the surrounding environment, it will be heavier and, if allowed to do so, will sink to its original position.

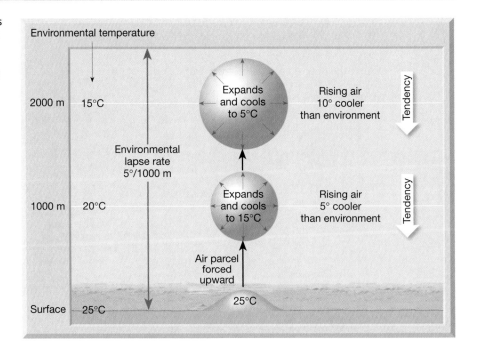

tend to sink to its original position. Hence, we say that the air near the surface is potentially cooler than the air aloft and therefore will not rise on its own. The air just described is *stable* and resists vertical movement.

Stated quantitatively, **absolute stability** prevails when the environmental lapse rate is less than the wet adiabatic rate. Figure 15.12 depicts this situation using an environmental lapse rate of 5°C per 1000

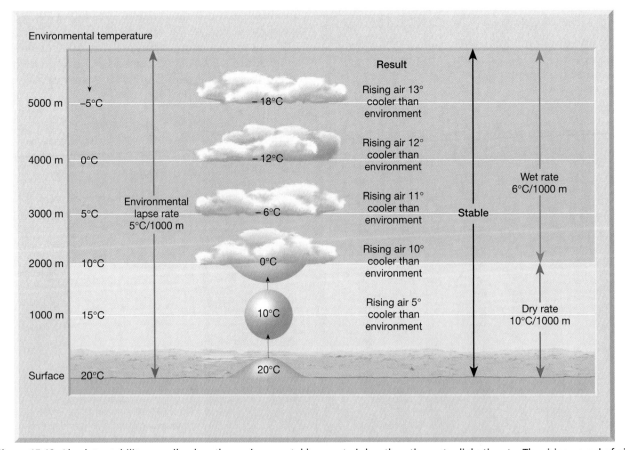

Figure 15.12 Absolute stability prevails when the environmental lapse rate is less than the wet adiabatic rate. The rising parcel of air is therefore always cooler and denser than the surrounding air. When stable air is forced to rise, it spreads out, producing flat, layered clouds.

meters and a wet adiabatic rate of 6°C per 1000 meters. Note that at 1000 meters the temperature of the surrounding air is 15°C, while the rising parcel of air has cooled to 10°C and is therefore the denser air. Even if this stable air were to be forced above the condensation level, it would remain cooler and denser than its environment, and thus it would tend to return to the surface.

The most stable conditions occur when the temperature in a layer of air actually increases with altitude. When such a reversal occurs, a *temperature inversion* is said to exist. Temperature inversions frequently occur on clear nights as a result of radiation cooling of Earth's surface. Under these conditions, an inversion is created because the ground and the air immediately above will cool more rapidly than the air aloft. When warm air overlies cooler air, it acts as a lid and prevents appreciable vertical mixing. Because of this, temperature inversions are responsible for trapping pollutants in a narrow zone near Earth's surface. This idea is explored more fully in Box 15.2.

Unstable Air. At the other extreme, air is said to exhibit **absolute instability** when the environmental lapse rate is greater than the dry adiabatic rate. As shown in Figure 15.13, the ascending parcel of air is always warmer than its environment and will continue to rise because of its own buoyancy. However, absolute instability is generally limited to near Earth's surface. On hot, sunny days the air above some surfaces, like

paved shopping malls, is heated more than the air over adjacent surfaces. These invisible pockets of more intensely heated air, being less dense than the air aloft, will rise like a hot-air balloon. This phenomenon produces the small fluffy clouds we associate with fair weather. Occasionally, when the surface air is considerably warmer than the air aloft, clouds with considerable vertical development can form, as shown in Figure 15.13. Clouds of this type can be associated with rain showers.

Conditional Instability. A more common type of atmospheric instability is called **conditional instability**. This situation prevails when moist air has an environmental lapse rate between the dry and wet adiabatic rates (between 5 and 10°C per 1000 meters). Simply, the atmosphere is said to be conditionally unstable when it is *stable* for an *unsaturated* parcel of air, but *unstable* for a *saturated* parcel of air. Notice in Figure 15.14 that the rising parcel of air is cooler than the surrounding air for the first 4000 meters. With the addition of latent heat above the lifting condensation level, the parcel becomes warmer than the surrounding air. From this point along its ascent the parcel will continue to rise because of its own buoyancy, without an outside force. Thus, conditional instability depends on whether or not the rising air is saturated. The word *conditional* is used because the air must be forced upward, such as over mountainous terrain, before it becomes unstable and rises because of its own buoyancy.

Figure 15.13 Absolute instability illustrated using an environmental lapse rate of 12°C per 1000 meters. The rising air is always warmer and therefore lighter than the surrounding air.

Box 15.2 People and the Environment: Atmospheric Stability Influences Air Pollution

Air quality is not just a function of the quantity and types of pollutants emitted into the air, but it is also closely linked to the atmosphere's ability to disperse these noxious substances. Perhaps you have heard this following well-known phrase: "The solution to pollution is dilution." To a significant degree this is true. If the air into which the pollution is released is not dispersed, the air will become more toxic. Two of the most important atmospheric conditions affecting the dispersion of pollutants are the strength of the wind and the stability of the air. These factors are critical because they determine how rapidly pollutants are diluted by mixing with the surrounding air after leaving the source.

The way in which wind speed influences the concentration of pollutants is straightforward. When winds are weak or calm, the concentration of pollutants is higher than when winds are strong. High wind speeds mix polluted air into a greater volume of surrounding air and therefore cause the pollution to be more diluted. When winds are light, there is less turbulence and mixing so the concentration of pollutants is higher.

Whereas wind speed governs the amount of air into which pollutants are initially mixed, atmospheric stability determines the extent to which vertical motions will mix the pollution with cleaner air above. The distance between Earth's surface and the height to which vertical air movements extend is termed the *mixing depth*. Generally, the greater the mixing depth, the better the air quality. When the mixing depth is several kilometers, pollutants are mixed through a large volume of cleaner air and dilute rapidly. When the mixing depth is shallow, pollutants are confined to a much smaller volume of air and concentrations can reach unhealthy levels. When air is stable, vertical motions are suppressed and mixing depths are small. Conversely, an unstable atmosphere promotes vertical air movements and greater mixing depths. Because heating of Earth's surface by the Sun enhances convectional movements, mixing depths are usually greater during the afternoon hours. For the same reason, mixing depths during the summer months are typically greater than during the winter months.

Temperature inversion represents a situation in which the atmosphere is very stable and the mixing depth is significantly restricted. Warm air overlying cooler air acts as a lid and prevents upward movement, leaving the pollutants trapped in a relatively narrow zone near the ground (Figure 15.A).

Inversions are generally classified into one of two categories—those that form near the ground and those that form aloft. A *surface inversion* develops close to the ground on clear and relatively calm nights. It forms because the ground is a more effective radiator than the air above. This being the case, radiation from the ground to the clear night sky causes more rapid cooling at the surface than higher in the atmosphere. The result is that the air close to

Figure 15.A Air pollution in downtown Los Angeles. Temperature inversions act as lids to trap pollutants below. (Photo by Ted Spiegel/Black Star)

Summary. In summary, the stability of air is determined by examining the temperature of the atmosphere at various heights. In simple terms, a column of air is deemed unstable when the air near the bottom of this layer is significantly warmer (less dense) than the air aloft, indicating a steep environmental lapse rate. Under these conditions the air actually turns over, as the warm air below rises and displaces the colder air aloft. Conversely, the air is considered to be stable when the temperature drops gradually with increasing altitude. The most stable conditions occur during a temperature inversion when the temperature actually increases with height. Under these conditions, there is very little vertical air movement.

Stability and Daily Weather

From the previous discussion we can conclude that stable air resists vertical movement, whereas unstable air ascends freely because of its own buoyancy. But how do these facts manifest themselves in our daily weather?

Because stable air resists upward movement, we might conclude that clouds will not form when stable conditions prevail in the atmosphere. Although this seems reasonable, processes exist that *force* air aloft. These will be discussed in the following section, "Processes That Lift Air." On occasions when stable air is forced aloft, the clouds that form are widespread

Figure 15.B Generalized temperature profile for a surface inversion.

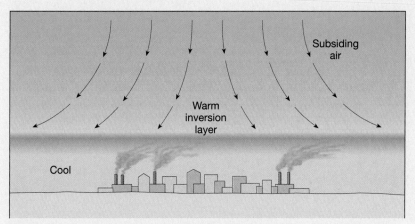

Figure 15.C Inversions aloft frequently develop in association with slow-moving centers of high pressure where the air aloft subsides and warms by compression. The turbulent surface zone does not subside as much. Thus, an inversion often forms between the lower turbulent zone and the subsiding layers above.

the ground is cooled more than the air above, yielding a temperature profile similar to the one shown in Figure 15.B. After sunrise the ground is heated and the inversion disappears.

Although surface inversions are usually shallow, they may be quite thick in regions where the land surface is uneven. Because cold air is denser (heavier) than is warm air, the chilled air near the surface gradually drains from the uplands and slopes into adjacent lowlands and valleys. As might be expected, these thicker surface inversions will not dissipate as quickly after sunrise.

Many extensive and long-lived air pollution episodes are linked to temperature inversions that develop in association with the sinking air that characterizes slow-moving centers of high pressure (Figure 15.C). As the air sinks to lower altitudes, it is compressed and so its temperature rises. Because turbulence is almost always present near the ground, this lowermost portion of the atmosphere is generally prevented from participating in the general subsidence. Thus, an inversion develops aloft between the lower turbulent zone and the subsiding warmed layers above.

and have little vertical thickness when compared to their horizontal dimension, and precipitation, if any, is light to moderate.

By contrast, clouds associated with unstable air are towering and are usually accompanied by heavy precipitation (Figure 15.15). For this reason, we can conclude that on a dreary, overcast day with light drizzle, stable air has been forced aloft. On the other hand, during a day when cauliflower-shaped clouds appear to be growing as if bubbles of hot air are surging upward, we can be fairly certain that the ascending air is unstable.

In summary, the role of stability in determining our daily weather is very important. To a large degree, stability determines the type of clouds that develop and whether precipitation will come as a gentle shower or a heavy downpour.

Processes That Lift Air

Earlier you learned that on some summer days, surface heating can be sufficient to cause pockets of air to be warmed more than the surrounding air. If the temperature of this air is sufficiently high and ample moisture is present, these unstable parcels will rise to produce clouds and occasionally precipitation. By contrast, air that is stable will not rise on its own. It requires some mechanism to initiate the vertical movement. Three such mechanisms are orographic lifting, frontal wedging, and convergence.

Orographic Lifting

Orographic lifting occurs when elevated terrains, such as mountains, act as barriers to flowing air (Figure 15.16). As air is pushed up a mountain slope, adiabatic

Figure 15.14 Conditional instability illustrated using an environmental lapse rate of 8°C per 1000 meters, which lies between the dry adiabatic rate and the wet adiabatic rate. The rising parcel of air is cooler than the surrounding air below 4000 meters and warmer above 4000 meters.

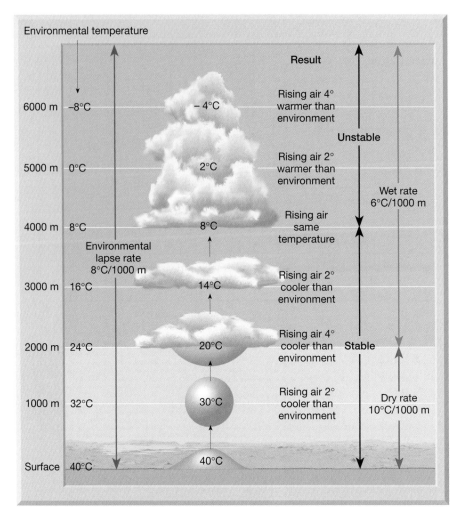

Environmental temperature

			Result
6000 m	−8°C	− 4°C	Rising air 4° warmer than environment
			Unstable
5000 m	0°C	2°C	Rising air 2° warmer than environment
			Wet rate 6°C/1000 m
4000 m	8°C	8°C	Rising air same temperature
	Environmental lapse rate 8°C/1000 m		
3000 m	16°C	14°C	Rising air 2° cooler than environment
2000 m	24°C	20°C	Rising air 4° cooler than environment — **Stable**
1000 m	32°C	30°C	Rising air 2° cooler than environment — Dry rate 10°C/1000 m
Surface	40°C	40°C	

Figure 15.15 Cauliflower-shaped clouds provide evidence of unstable conditions in the atmosphere. (Photo by Henry Lansford)

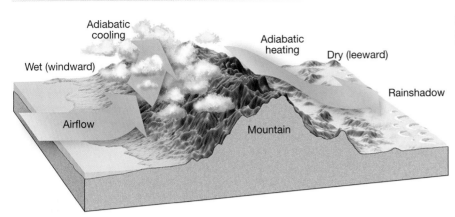

Figure 15.16 Orographic lifting. The name comes from *oro-*, which means "mountain."

cooling often generates clouds and copious precipitation. In fact, many of the rainiest places in the world are located on windward mountain slopes, such as the Olympic Mountains in Washington State.

In addition to providing the lift to render air unstable, mountains remove additional moisture in other ways. By slowing the horizontal flow of air, they retard the passage of storm systems. Moreover, the irregular topography of mountains enhances differential heating and instability. These combined effects account for the generally higher precipitation associated with mountainous regions compared with surrounding lowlands.

By the time air reaches the leeward side of a mountain, much of its moisture has been lost. And, if the air descends, it becomes compressed and warms adiabatically, making condensation and precipitation even less likely. As shown in Figure 15.16, the result can be a **rainshadow desert**. The Great Basin Desert of the western United States lies only a few hundred kilometers from the Pacific Ocean but is effectively cut off from the ocean's moisture by the imposing Sierra Nevada range (Figure 15.17). The Gobi Desert of Mongolia, the Takla Makan of China, and the Patagonia Desert of Argentina are other examples of rainshadow deserts found on the leeward sides of mountains.

Frontal Wedging

If orographic lifting were the only mechanism that forced air aloft, the relatively flat central portion of North America would be an expansive desert. Fortunately, this is not the case. **Frontal wedging** occurs when cool air acts as a barrier over which warmer, less dense air rises. Figure 15.18 illustrates frontal wedging of both stable and unstable air. As this figure shows, forceful lifting is important in producing clouds. The stability of the air, however, determines to a great extent the type of clouds formed and the amount of precipitation that may be expected.

Figure 15.17 Rainshadow desert. The arid conditions in California's Death Valley can be partially attributed to the adjacent mountains, which orographically remove the moisture from air originating over the Pacific. (Photo by James E. Patterson)

Figure 15.18 A. When stable air is lifted, layered clouds usually result. **B.** When warm, unstable air is forced to rise over cooler air, "towering" clouds develop.

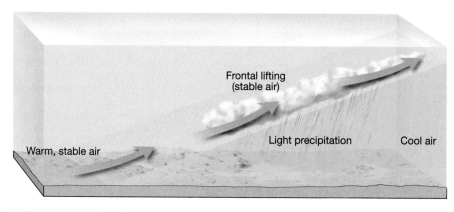

Frontal lifting (stable air)

Light precipitation

Cool air

Warm, stable air

Unstable air

Heavy precipitation

Cool air

Warm, unstable air

These weather-producing fronts are part of the storm systems called *middle-latitude cyclones*. These cyclones, which show up as the "lows" you see on weather maps, are responsible for producing a high proportion of the precipitation in the middle latitudes. We will examine them closely in Chapter 17.

Convergence

Whenever air masses flow together, **convergence** is said to occur. Such flow results in general upward movement, because when air converges, it has to go somewhere, so the height of the air column increases. Consequently, air within the column must move upward.

The Florida peninsula provides an excellent example of the role that convergence can play in initiating cloud development and precipitation. On warm days, the airflow is from the ocean to the land along both coasts of Florida. This leads to general convergence over the peninsula. Such a pattern of air movement and the uplift that results is aided by intense solar heating. Rising parcels of humid air expand and cool adiabatically, the water vapor condenses to form clouds, and the result is the greatest frequency of midafternoon thunderstorms in the United States.

More important, convergence as a mechanism of forceful lifting is a major contributor to the stormy weather associated with middle-latitude cyclones and hurricanes, discussed in Chapter 17.

Condensation and Cloud Formation

To review briefly, condensation occurs when water vapor in the air changes to a liquid. The result of this process may be dew, fog, or clouds. For any of these forms of condensation to occur, the air must be saturated. Saturation occurs most commonly when air is cooled to its dew point, or less often when water vapor is added to the air.

Generally, there must be a surface on which the water vapor can condense. When dew occurs, objects at or near the ground serve this purpose, like grass and car windows. But when condensation occurs in the air above the ground, tiny bits of particulate matter, known as **condensation nuclei**, serve as surfaces for water vapor condensation. These nuclei are very important, for in their absence, a relative humidity well in excess of 100 percent is needed to produce clouds.

Condensation nuclei such as microscopic dust, smoke, and salt particles (from the ocean) are profuse in the lower atmosphere. Because of this abundance of particles, relative humidity rarely exceeds 101 percent. Some particles, such as ocean salt, are particularly good nuclei because they absorb water. These particles are termed **hygroscopic** ("water-seeking") **nuclei**. When condensation takes place, the initial growth rate of cloud droplets is rapid. It diminishes quickly because the excess water vapor is quickly absorbed by the numerous competing particles. This results in the formation of a cloud consisting of millions upon millions of tiny water droplets, all so fine that they remain suspended in air. When cloud formation occurs at below-freezing temperatures, tiny ice crystals form. Thus, a cloud might consist of water droplets, ice crystals, or both.

The slow growth of cloud droplets by additional condensation and the immense size difference between cloud droplets and raindrops suggest that condensation alone is not responsible for the formation of drops large enough to fall as rain. We will first examine clouds and then return to the questions of how precipitation forms.

Types of Clouds

Clouds are among the most conspicuous and observable aspects of the atmosphere and its weather. **Clouds** are a form of condensation best described as *visible aggregates of minute droplets of water or tiny crystals of ice.* In addition to being prominent and sometimes spectacular features in the sky, clouds are of continual interest to meteorologists, because they provide a visible indication of what is going on in the atmosphere. Anyone who observes clouds with the hope of recognizing different types often finds that there is a bewildering variety of these familiar white and gray masses streaming across the sky. Still, once one comes to know the basic classification scheme for clouds, most of the confusion vanishes.

Clouds are classified on the basis of their *form* and *height* (Figure 15.19). Three basic forms are recognized: cirrus, cumulus, and stratus.

- **Cirrus** clouds are high, white, and thin. They can occur as patches or as delicate veil-like sheets or extended wispy fibers that often have a feathery appearance.
- **Cumulus** clouds consist of globular individual cloud masses. Normally they exhibit a flat base and have the appearance of rising domes or towers. Such clouds are frequently described as having a cauliflower structure.
- **Stratus** clouds are best described as sheets or layers that cover much or all of the sky. While there

may be minor breaks, there are no distinct individual cloud units.

All other clouds reflect one of these three basic forms or are combinations or modifications of them.

Three levels of cloud heights are recognized: high, middle, and low (Figure 15.19). **High clouds** normally have bases above 6000 meters (20,000 feet), **middle clouds** generally occupy heights from 2000 to 6000 meters (6500 to 20,000 feet), and **low clouds** form below 2000 meters (6,500 feet). The altitudes listed for each height category are not hard and fast. There is some seasonal as well as latitudinal variation. For example, at high latitudes or during cold winter months in the mid-latitudes, high clouds are often found at lower altitudes.

High Clouds. Three cloud types make up the family of high clouds (above 6000 meters): cirrus, cirrostratus, and cirrocumulus. *Cirrus* clouds are thin and delicate and sometimes appear as hooked filaments called "mares' tails" (Figure 15.20A). As the names suggest, *cirrocumulus* clouds consist of fluffy masses (Figure 15.20B), whereas *cirrostratus* clouds are flat layers (Figure 15.20C). Because of the low temperatures and small quantities of water vapor present at high altitudes, all high clouds are thin and white and are made up of ice crystals. Further, these clouds are not considered precipitation makers. However, when cirrus clouds are followed by cirrocumulus clouds and increased sky coverage, they may warn of impending stormy weather.

Middle Clouds. Clouds that appear in the middle range (2000 to 6000 meters) have the prefix *alto* as part of their name. *Altocumulus* clouds are composed of globular masses that differ from cirrocumulus clouds in that they are larger and denser (Figure 15.20D). *Altostratus* clouds create a uniform white to grayish sheet covering the sky with the Sun or Moon visible as a bright spot (Figure 15.20E). Infrequent light snow or drizzle may accompany these clouds.

Low Clouds. There are three members in the family of low clouds: stratus, stratocumulus, and nimbostratus. *Stratus* are a uniform foglike layer of clouds that frequently covers much of the sky. On occasions these clouds may produce light precipitation. When stratus clouds develop a scalloped bottom that appears as long parallel rolls or broken globular patches, they are called *stratocumulus* clouds.

Nimbostratus clouds derive their name from the Latin *nimbus*, which means "rainy cloud," and *stratus*, which means "to cover with a layer" (Figure 15.20F). As the name suggests, nimbostratus clouds are one of the chief precipitation producers. Nimbostratus clouds form in association with stable conditions. We might not expect clouds to grow or persist in stable air, yet cloud growth of this type is common when air is

Figure 15.19 Classification of clouds according to height and form.

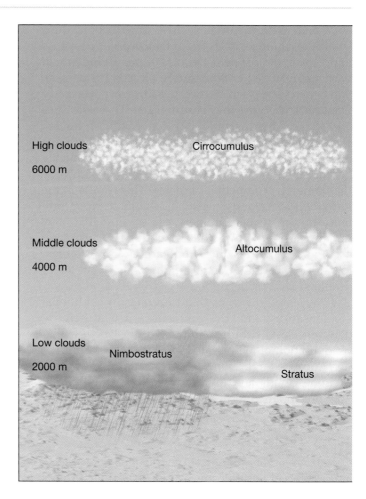

forced to rise, as occurs along a mountain range, a front, or near the center of a cyclone where converging winds cause air to ascend. Such forced ascent of stable air leads to the formation of a stratified cloud layer that is large horizontally compared to its depth.

Clouds of Vertical Development. Some clouds do not fit into any one of the three height categories mentioned. Such clouds have their bases in the low height range but often extend upward into the middle or high altitudes. Consequently, clouds in this category are called **clouds of vertical development**. They are all related to one another and are associated with unstable air. Although *cumulus* clouds are often connected with "fair" weather (Figure 15.20G), they may grow dramatically under the proper circumstances. Once upward movement is triggered, acceleration is powerful, and clouds with great vertical extent form. The end result is often a towering cloud, called a *cumulonimbus*, that may produce rain showers or a thunderstorm (Figure 15.20H).

Definite weather patterns can often be associated with particular clouds or certain combinations of cloud types, so it is important to become familiar with cloud descriptions and characteristics. Table 15.4 lists the 10 basic cloud types that are recognized internationally and gives some characteristics of each.

Fog

Fog is generally considered to be an atmospheric hazard. When it is light, visibility is reduced to 2 or 3 kilometers. However, when it is dense, visibility may be cut to a few dozen meters or less, making travel by any mode not only difficult but often dangerous. Officially, visibility must be reduced to 1 kilometer or less before fog is reported. While this figure is arbitrary, it does provide an objective criterion for comparing fog frequencies at different locations.

Fog is defined as *a cloud with its base at or very near the ground.* Physically, there is no basic difference between a fog and a cloud; their appearance and structure are the same. The essential difference is the method and place of formation. Whereas clouds result when air rises and cools adiabatically, most fogs are the consequence of radiation cooling or the movement of air over a cold surface. (The exception is upslope fog.) In other circumstances, fogs form when enough water vapor is added to the air to bring about saturation (evaporation fogs). We will examine both of these types.

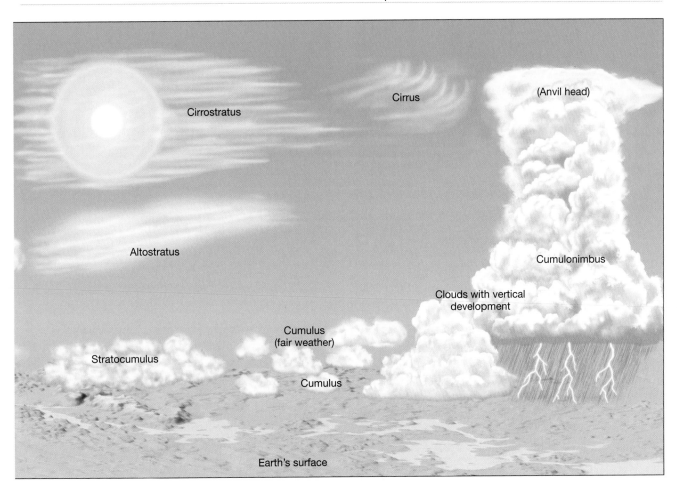

Fogs Caused by Cooling

When warm, moist air moves over a cool surface, the result might be a blanket of fog called **advection fog** (Figure 15.21). Examples of such fogs are very common. The foggiest location in the United States, and perhaps in the world, is Cape Disappointment, Washington. The name is indeed appropriate, because this station averages 2552 hours of fog each year (there are about 8760 hours in a year). The fog experienced at Cape Disappointment, and at other West Coast locations, is produced when warm, moist air from the Pacific Ocean moves over the cold California Current and is then carried onto shore by the prevailing winds. Advection fogs are also quite common in the winter season when warm air from the Gulf of Mexico blows across cold, often snow-covered surfaces of the Midwest and East.

Radiation fog forms on cool, clear, calm nights, when Earth's surface cools rapidly by radiation. As the night progresses, a thin layer of air in contact with the ground is cooled below its dew point. As the air cools and becomes denser, it drains into low areas, resulting in "pockets" of fog. The largest "pockets" are often river valleys, where thick accumulations may occur (Figure 15.22).

As its name implies, **upslope fog** is created when relatively humid air moves up a gradually sloping plain or up the steep slopes of a mountain. As a result of this upward movement, air expands and cools adiabatically. If the dew point is reached, an extensive layer of fog may form. In the United States, the Great Plains offers an excellent example. When humid easterly or southeasterly winds blow westward from the Mississippi River upslope toward the Rocky Mountains, the air gradually rises, resulting in an adiabatic temperature decrease of about 13°C. When the difference between the air temperature and dew point of westward-moving air is less than 13°C, an extensive fog can result in the western plains.

Evaporation Fogs

When the saturation of air occurs primarily because of the addition of water vapor, the resulting fogs are called *evaporation fogs*. Two types of evaporation fogs are recognized: steam fog and frontal or precipitation fog.

When cool air moves over warm water, enough moisture may evaporate from the water surface to produce saturation. As the rising water vapor meets the cold air, it immediately recondenses and rises with the

A. Cirrus

B. Cirrocumulus

C. Cirrostratus

Figure 15.20 These photos depict common forms of several different cloud types. (Photos A, B, D, E, F, and G by E. J. Tarbuck. Photo C by A. and J. Verkaik/The Stock Market. Photo H by Dough Millar/Science Source/Photo Researchers, Inc.)

D. Altocumulus

E. Altostratus

F. Nimbostratus

G. Cumulus

H. Cumulonimbus

Table 15.4 Cloud types and characteristics.

Cloud Family and Height	Cloud Type	Characteristics
High clouds—above 6000 meters (20,000 feet)	Cirrus	Thin, delicate, fibrous, ice-crystal clouds. Sometimes appear as hooked filaments called "mares' tails." (Figure 15.20A)
	Cirrocumulus	Thin, white, ice-crystal clouds in the form of ripples, waves, or globular masses all in a row. May produce a "mackerel sky." Least common of the high clouds. (Figure 15.20B)
	Cirrostratus	Thin sheet of white, ice-crystal clouds that may give the sky a milky look. Sometimes produce halos around the Sun or Moon. (Figure 15.20C).
Middle clouds—2000–6000 meters (6500–20,000 feet)	Altocumulus	White to gray clouds often composed of separate globules; "sheep-back" clouds. (Figure 15.20D)
	Altostraatus	Stratified veil of clouds that are generally thin and may produce very light precipitation. When thin, the Sun or Moon may be visible as a "bright spot," but no halos are produced. (Figure 15.20E)
Low clouds—below 2000 meters (6500 feet)	Stratocumulus	Soft, gray clouds in globular patches or rolls. Rolls may join together to make a continuous cloud.
	Stratus	Low uniform layer resembling fog but not resting on the ground. May produce drizzle.
	Nimbostratus	Amorphous layer of dark gray clouds. One of the chief precipitation-producing clouds. (Figure 15.20F)
Clouds of vertical development—500–18,000 meters (1600–60,000 feet)	Cumulus	Dense, billowy clouds often characterized by flat bases. May occur as isolated clouds or closely packed. (Figure 15.20G)
	Cumulonimbus	Towering cloud sometimes spreading out on top to form an "anvil head." Associated with heavy rainfall, thunder, lightning, hail, and tornadoes. (Figure 15.20H)

air that is being warmed from below. Because the water has a steaming appearance, the phenomenon is called **steam fog** (Figure 15.23). Steam fog is fairly common over lakes and rivers in the fall and early winter, when the water may still be relatively warm and the air is rather crisp. Steam fog is often shallow because as the steam rises it reevaporates in the unsaturated air above.

When frontal wedging occurs, warm air is lifted over colder air. If the resulting clouds yield rain, and the cold air below is near the dew point, enough rain will evaporate to produce fog. A fog formed in this manner is called **frontal fog**, or **precipitation fog**. The result is a more or less continuous zone of condensed water droplets reaching from the ground up through the clouds.

In summary, both steam fog and frontal fog result from the addition of moisture to a layer of air. As you saw, the air is usually cool or cold and already near saturation. Because air's capacity to hold water vapor at low temperatures is small, only a relatively modest amount of evaporation is necessary to produce saturated conditions and fog.

Figure 15.21 Advection fog rolling into San Francisco Bay. (Photo by Ed Pritchard/Tony Stone Worldwide)

Figure 15.22 Radiation fog in the Illinois River valley at Peoria, Illinois. (Photo by David Zalaznik/Peoria Journal Star)

Figure 15.23 Early-morning steam fog on pond, Acadia, Maine. (Photo by Nancy Rotenberg/*Animals Animals/Earth Scenes*)

How Precipitation Forms

All clouds contain water, but why do some produce precipitation while others drift placidly overhead? This simple question perplexed meteorologists for many years. First, cloud droplets are very small, averaging less than 10 micrometers in diameter (for comparison, a human hair is about 75 micrometers in diameter). Because of their small size, cloud droplets fall incredibly slowly. Theoretically, an average cloud droplet falling from a cloud base at 1000 meters would require about 48 hours to reach the ground. Of course, it would never complete its journey. Even falling through humid air, a cloud droplet would evaporate before it fell a few meters below the cloud base. In addition, clouds are made up of many billions of these droplets, all competing for the available water vapor; thus, their continued growth via condensation is very slow.

A raindrop large enough to reach the ground without evaporating contains roughly a million times more water than a cloud droplet. Therefore, for precipitation to form, millions of cloud droplets must somehow coalesce (join together) into drops large enough to sustain themselves during their descent. Two mechanisms have been proposed to explain this phenomenon: the Bergeron process and the collision-coalescence process.

Bergeron Process

The **Bergeron process** is named after its discoverer, Swedish meteorologist Tor Bergeron. The process relies on two interesting properties of water: supercooling and supersaturating.

Supercooling. Cloud droplets do not freeze at 0°C as expected. In fact, pure water suspended in air does not freeze until it reaches a temperature of nearly –40°C (–40°F). Water in the liquid state below 0°C is referred to as **supercooled**. Supercooled water will readily freeze if sufficiently agitated. This explains why airplanes collect ice when they pass through a cloud composed of supercooled droplets.

In addition, supercooled droplets will freeze upon contact with solid particles that have a crystal form closely resembling that of ice. Such materials are termed **freezing nuclei**. The need for freezing nuclei to initiate the freezing process is similar to the requirement for condensing nuclei in the process of condensation. However, in contrast to condensation nuclei, freezing nuclei are very sparse in the atmosphere and do not generally become active until the temperature reaches –10°C or less. Only at temperatures well below freezing will ice crystals begin to form in clouds, and even at that, they will be few and far between. Once ice crystals form, they are in direct competition with the supercooled droplets for the available water vapor.

Supersaturation. When air is saturated (100 percent relative humidity) with respect to water, it is supersaturated (relative humidity is greater than 100 percent) with respect to ice. Table 15.5 shows that at –10°C, when the relative humidity is 100 percent with respect to water, the relative humidity with respect to ice is nearly 110 percent. Thus, ice crystals cannot coexist with water droplets, because the air always "appears" supersaturated to the ice crystals. Hence, the ice crystals

Table 15.5 Relative humidity with respect to ice when relative humidity with respect to water is 100 percent.

Temperature (°C)	Relative Humidity with Respect to:	
	Water (%)	Ice (%)
0	100	100
-5	100	105
-10	100	110
-15	100	116
-20	100	121

begin to consume the "excess" water vapor, which lowers the relative humidity near the surrounding droplets. In turn, the water droplets evaporate to replenish the diminishing water vapor, thereby providing a continual source of vapor for the growth of the ice crystals (Figure 15.24).

Because the level of supersaturation with respect to ice can be quite great, the growth of ice crystals is generally rapid enough to generate crystals large enough to fall. During their descent, these ice crystals enlarge as they intercept cloud drops, which freeze upon them. Air movement will sometimes break up these delicate crystals and the fragments will serve as

freezing nuclei. A chain reaction develops, producing many ice crystals, which by accretion form into large crystals called snowflakes. When the surface temperature is above 4°C (39°F), snowflakes usually melt before they reach the ground and continue their descent as rain. Even a summer rain may have begun as a snowstorm in the clouds overhead.

Collision-Coalescence Process

A few decades ago, meteorologists believed that the Bergeron process was responsible for the formation of most precipitation. However, it was discovered that copious rainfall can be associated with clouds located well below the freezing level (*warm clouds*), particularly in the tropics. This led to the proposal of a second mechanism thought to produce precipitation—the **collision-coalescence process**.

Research has shown that clouds composed entirely of liquid droplets must contain droplets larger than 20 micrometers if precipitation is to form. These large droplets form when "giant" condensation nuclei are present, or when hygroscopic particles such as sea salt exist. Hygroscopic particles begin to remove water vapor from the air at relative humidities under 100 percent and can grow quite large. Because the rate at which drops fall is size-dependent, these "giant"

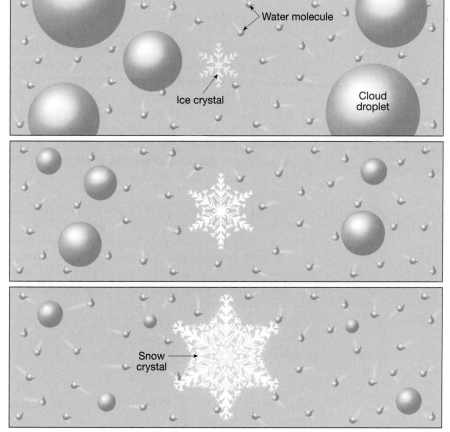

Water molecule

Ice crystal

Cloud droplet

Snow crystal

Figure 15.24 The Bergeron process. Ice crystals grow at the expense of cloud droplets until they are large enough to fall. The size of these particles has been greatly exaggerated.

droplets fall most rapidly. As they collide with smaller, slower droplets and coalesce. Growing larger in the process, they fall even more rapidly (or in an updraft, rise more slowly), increasing their chances of collision and rate of growth (Figure 15.25). After a great many such collisions they are large enough to fall to the surface without completely evaporating. Because of the number of collisions required for growth to raindrop size, droplets in clouds that have great vertical thickness and abundant moisture have a better chance of reaching the required size. Updrafts also aid this process, for they allow the droplets to traverse the cloud repeatedly.

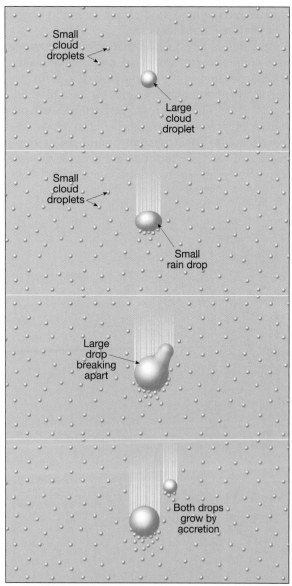

Figure 15.25 The collision-coalescence process. Because large cloud droplets fall more rapidly than smaller droplets, they are able to sweep up the smaller ones in their path and grow. Most cloud droplets are so small that the motion of the air keeps them suspended. Even if these small cloud droplets were to fall, they would evaporate long before reaching the surface.

Raindrops can grow to a maximum size of 5 millimeters, at which point they fall at the rate of 33 kilometers (20 miles) per hour. At this size and speed, the water's surface tension, which holds the drop together, is overcome by the drag imposed by the air, which in turn pulls the drops apart. The resulting breakup of a large raindrop produces numerous smaller drops that begin anew the task of sweeping up cloud droplets. Drops that are less than 0.5 millimeter upon reaching the ground are termed *drizzle* and require about 10 minutes to fall from a cloud 1000 meters (3300 feet) overhead.

The collision-coalescence process is not quite as simple as described. First, as the large droplets descend, they produce an air stream around them similar to that produced by a fast-moving automobile. If an automobile is driven at night and we use the bugs that are often out to be analogous to the cloud droplets, it is easy to visualize how most cloud droplets are swept aside. The larger the cloud droplet (or bug), the better its chance of colliding with the giant droplet (or car). Second, collision does not guarantee coalescence. Experiments indicate that atmospheric electricity may hold these droplets together once they collide. If a droplet with a negative charge should collide with a positively charged droplet, electrical attraction of opposite charges may bind them together.

Forms of Precipitation

Much of the world's precipitation begins as snow crystals or other solid forms such as hail or graupel (Table 15.6). Entering the warmer air below the cloud, these ice particles often melt and reach the ground as raindrops. In some parts of the world, particularly the subtropics, precipitation often forms in clouds that are warmer than 0°C (32°F). These rains frequently occur over the ocean where cloud condensation nuclei are not plentiful and those that exist vary in size. Under such conditions, cloud droplets can grow rapidly by the collision-coalescence process to produce copious amounts of rain.

Because atmospheric conditions vary greatly from place to place as well as seasonally, several different forms of precipitation are possible. Rain and snow are the most common and familiar, but the other forms of precipitation listed in Table 15.6 are important as well. The occurrence of sleet, glaze, or hail is often associated with important weather events. Although limited in occurrence and sporadic in both time and space, these forms, especially glaze and hail, can cause considerable damage.

Rain and Drizzle

We are familiar with rain. In meteorology, the term **rain** is restricted to drops of water that fall from a cloud and have a diameter of at least 0.5 millimeter.

Table 15.6 Forms of precipitation.

Type	Appropriate Size	State of Matter	Description
Mist	0.005 to 0.05 mm	Liquid	Droplets large enough to be felt on the face when air is moving 1 meter/second. Associated with stratus clouds.
Drizzle	Less than 0.5 mm	Liquid	Small uniform drops that fall from stratus clouds, generally for several hours.
Rain	0.5 to 5 mm	Liquid	Generally produced by nimbostratus or cumulonimbus clouds. When heavy, it can show high variability from one place to another.
Sleet	0.5 to 5 mm	Solid	Small, spherical to lumpy ice particles that form when raindrops freeze while falling through a layer of sub-freezing air. Because the ice particles are small, damage, if any, is generally minor. Sleet can make travel hazardous.
Glaze	Layers 1 mm to 2 cm thick	Solid	Produced when supercooled raindrops freeze on contact with solid objects. Glaze can form a thick coating of ice having sufficient weight to seriously damage trees and power lines
Rime	Variable accumulations	Solid	Deposits usually consisting of ice feathers that point into the wind. These delicate, frostlike accumulations form as supercooled cloud or fog droplets encounter objects and freeze on contact.
Snow	1mm to 2 cm	Solid	The crystalline nature of snow allows it to assume many shapes including six-sided crystals, plates, and needles. Produced in supercooled clouds where water vapor is deposited as ice crystals that remain frozen during their descent.
Hail	5 mm to 10 cm or larger	Solid	Precipitation in the form of hard, rounded pellets or irregular lumps of ice. Produced in large convective, cumulonimbus clouds, where frozen ice particles and supercooled water coexist.
Gaupel	2 to 5 mm	Solid	Sometimes called soft hail, graupel forms as rime collects on snow crystals to produce irregular masses of "soft" ice. Because these particles are softer than hailstones, they normally flatten out upon impact.

Most rain originates either in nimbostratus clouds or in towering cumulonimbus clouds that are capable of producing unusually heavy rainfalls known as *cloudbursts*. Raindrops rarely exceed about 5 millimeters in diameter. Larger drops do not survive because surface tension, which holds the drops together, is exceeded by the frictional drag of the air. Consequently, large raindrops regularly break apart into smaller ones.

Fine, uniform drops of water having a diameter less than 0.5 millimeter are called *drizzle*. Drizzle can be so fine that the tiny drops appear to float and their impact is almost imperceptible. Drizzle and small raindrops generally are produced in stratus or nimbostratus clouds where precipitation can be continuous for several hours, or on rare occasions for days.

Snow

Snow is precipitation in the form of ice crystals (snowflakes) or, more often, aggregates of crystals. The size, shape, and concentration of snowflakes depend to a great extent on the temperature at which they form.

Recall that at very low temperatures, the moisture content of air is small. The result is the generation of very light, fluffy snow made up of individual six-sided ice crystals. This is the "powder" that downhill skiers talk so much about. By contrast, at temperatures warmer than about –5°C, the ice crystals join together into larger clumps consisting of tangled aggregates of crystals. Snowfalls composed of these composite snowflakes are generally heavy and have a high moisture content, which makes them ideal for making snowballs.

Sleet and Glaze

Sleet is a wintertime phenomenon and refers to the fall of small particles of ice that are clear to translucent. For sleet to be produced, a layer of air with temperatures above freezing must overlie a subfreezing layer near the ground. When raindrops, which are often melted snow, leave the warmer air and encounter the colder air below, they freeze and reach the ground as small pellets of ice the size of the raindrops from which they formed.

On some occasions, when the vertical distribution of temperatures is similar to that associated with the formation of sleet, freezing rain or **glaze** results instead. In such situations, the subfreezing air near the ground is not thick enough to allow the raindrops to freeze. The raindrops, however, do become super-cooled as they fall through the cold air and turn to ice upon colliding with solid objects. The result can be a thick coating of ice having sufficient weight to break tree limbs, down power lines, and make walking or driving extremely hazardous (Figure 15.26).

Figure 15.26 Glaze forms when supercooled raindrops freeze on contact with objects. (Photo by Annie Griffiths Belt/DRK Photo)

Figure 15.27 A cross section of the Coffeyville hailstone. This largest recorded hailstone fell over Kansas in 1970 and weighed 766 grams (1.67 pounds). (University Corporation for Atmospheric Research/National Center for Atmospheric Research/National Science Foundation)

Hail

Hail is precipitation in the form of hard, rounded pellets or irregular lumps of ice. Moreover, large hailstones often consist of a series of nearly concentric shells of differing densities and degrees of opaqueness (Figure 15.27). Most hailstones have diameters between 1 centimeter (pea size) and 5 centimeters (golf ball size), although some can be as big as an orange or larger. Occasionally, hailstones weighing a pound or more have been reported. Many of these were probably composites of several stones frozen together.

The heaviest authenticated hailstone on record fell on Coffeyville, Kansas, September 3, 1970 (Figure 15.27). With a 14-centimeter (5.5-inch) diameter, this "giant" weighed 766 grams (1.67 pounds). It is estimated that this stone hit the ground at a speed in excess of 160 kilometers (100 miles) per hour.

The destructive effects of large hailstones are well known, especially to farmers whose crops have been devastated in a few minutes and to people whose windows and roofs have been damaged (Figure 15.28). In the United States, hail damage each year is significant. In 1994 alone, there was more than $450 million in property damage attributed to hail.

Hail is produced only in large cumulonimbus clouds where updrafts can sometimes reach speeds approaching 160 kilometers (100 miles) per hour, and where there is an abundant supply of supercooled water. Hailstones begin as small embryonic ice pellets that grow by collecting supercooled water droplets as they fall through the cloud. If they encounter a strong updraft, they may be carried upward again and begin the downward journey anew. Each trip through the supercooled portion of the cloud may be represented by an additional layer of ice. Hailstones can also form

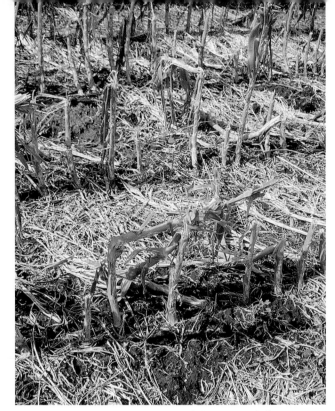

Figure 15.28 Hail damage to a corn field in central Illinois. (University Corporation for Atmospheric Research/National Center for Atmospheric Research/National Science Foundation)

from a single descent through an updraft. Either way, the process continues until the hailstone encounters a downdraft or grows too heavy to remain suspended by the thunderstorm's updraft.

Figure 15.29 Rime consists of delicate ice crystals that form when supercooled fog or cloud droplets freeze on contact with objects. Here, rime has formed on the needles of a pine tree. (Photo by D. Cavagnaro/DRK Photo)

Rime

Rime is a deposit of ice crystals formed by the freezing of supercooled fog or cloud droplets on objects whose surface temperature is below freezing. When rime forms on trees, it adorns them with its characteristic ice feathers, which can be spectacular to behold (Figure 15.29). In these situations, objects such as pine needles act as freezing nuclei, causing the supercooled droplets to freeze on contact. On occasions when the wind is blowing, only the windward surfaces of objects will accumulate the layer of rime.

Measuring Precipitation

The most common form of precipitation, rain, is the easiest to measure. Any open container having a consistent cross section throughout can be a rain gauge. In general practice, however, more sophisticated devices are used so that small amounts of rainfall can be measured more accurately, and to reduce losses from evaporation. The *standard rain gauge* (Figure 15.30) has a diameter of about 20 centimeters (8 inches) at the top. Once the water is caught, a funnel conducts the rain into a cylindrical measuring tube that has a cross-sectional area only one-tenth as large as the receiver. Consequently, rainfall depth is magnified 10 times, which allows for accurate measurements to the nearest 0.025 centimeter (0.01 inch). The narrow opening also minimizes evaporation. When the amount of rain is less than 0.025 centimeter, it is reported as a *trace* of precipitation.

Figure 15.30 Precipitation measurement. The standard rain gauge allows for accurate rainfall measurement to the nearest 0.025 centimeter (0.01 inch). Because the cross-sectional area of the measuring tube is only one-tenth as large as the collector, rainfall is magnified 10 times.

In addition to the standard rain gauge, several types of recording gauges are routinely used. These instruments record not only the amount of rain but also its time of occurrence and intensity (amount per unit of time).

No matter which rain gauge is used, proper exposure is critical. Errors arise when the gauge is shielded from obliquely falling rain by buildings, trees, or other high objects. Hence, the instrument should be at least as far away from such obstructions as the objects are high. Another cause of error is the wind. It has been shown that with increasing wind and turbulence, it becomes more difficult to collect a representative quantity of rain.

When snow records are kept, two measurements are normally taken: depth and water equivalent.

Usually the depth of snow is measured with a calibrated stick. The actual measurement is simple, but choosing a *representative* spot often poses a dilemma. Even when winds are light or moderate, snow drifts freely. As a rule, it is best to take several measurements in an open place away from trees and obstructions and then average them. To obtain the water equivalent, samples can be melted and then weighed or measured as rain.

The quantity of water in a given volume of snow is not constant. A general ratio of 10 units of snow to 1 unit of water is often used when exact information is not available, but the actual water content of snow may deviate widely from this figure. It may take as much as 30 centimeters of light, fluffy dry snow or as little as 4 centimeters of wet snow to produce 1 centimeter of water.

Chapter Summary

- *Water vapor*, an odorless, colorless gas, changes from one state of matter (solid, liquid, or gas) to another at the temperatures and pressures experienced near Earth's surface. The processes involved in *changing the state of matter of water* are *evaporation, condensation, melting, freezing, sublimation*, and *deposition*.

- *Humidity* is the general term to describe the amount of water vapor in the air. *Relative humidity*, the *ratio* (expressed as a percent) of the air's water vapor *content* to its water vapor *capacity* at a given temperature, is the most familiar term used to describe humidity. The water vapor capacity of air is temperature dependent, with warm air having a much greater capacity than cold air.

- *Relative humidity can be changed in two ways.* One is by *adding or subtracting water vapor*. The second is by *changing the air's temperature*. When air is cooled, its relative humidity increases. Air is said to be *saturated* when it contains the maximum quantity of water vapor that it can hold at any given temperature and pressure. *Dew point* is the temperature to which air would have to be cooled in order to reach saturation.

- The cooling of air as it rises and expands owing to successively lower air pressure is the basic cloud-forming process. Temperature changes in air brought about by compressing or expanding the air are called *adiabatic temperature changes*. Unsaturated air warms by compression and cools by expansion at the rather constant rate of 10°C per 1000 meters of altitude change, a figure called the *dry adiabatic rate*. If air rises high enough it will cool sufficiently to cause condensation and form a cloud. From this point on, air that continues to rise will cool at the *wet adiabatic rate*, which varies from 5°C to 9°C per 1000 meters of ascent. The difference in the wet and dry adiabatic rates is caused by the condensing water vapor releasing *latent heat*, thereby reducing the rate at which the air cools.

- The *stability of air* is determined by examining the temperature of the atmosphere at various altitudes. Air is said to be *unstable* when the *environmental lapse rate* (the rate of temperature decrease with increasing altitude in the troposphere) is greater than the *dry adiabatic rate*. Stated differently, a column of air is unstable when the air near the bottom is significantly warmer (less dense) than the air aloft.

- Three mechanisms that can initiate the vertical movement of air are (1) *orographic lifting*, which occurs when elevated terrains, such as mountains, act as barriers to the flow of air; (2) *frontal wedging*, when cool air acts as a barrier over which warmer, less dense air rises; and (3) *convergence*, which happens when air flows together and a general upward movement of air occurs.

- *For condensation to occur, air must be saturated.* Saturation takes place either when air is cooled to its dew point, which most commonly happens, or when water vapor is added to the air. There must also be a surface on which the water vapor can condense. In cloud and fog formation, tiny particles called *condensation nuclei* serve this purpose.

- *Clouds* are classified on the basis of their *appearance* and *height*. The three basic forms are *cirrus* (high, white, thin, wispy fibers), *cumulus* (globular, individual cloud masses), and *stratus* (sheets or layers that cover much or all of the sky). The four categories based on height are *high clouds* (bases normally above 6000 meters), *middle clouds* (from 2000 to 6000 meters), *low clouds* (below 2000 meters), and *clouds of vertical development*.

- *Fog* is defined as a cloud with its base at or very near the ground. Fogs from when air is cooled below its dew point or when enough water vapor is added to the air to bring about saturation. Various types of fog include *advection fog*, *radiation fog*, *upslope fog*, *steam fog*, and *frontal* (or *precipitation*) *fog*.

- For *precipitation* to form, millions of cloud droplets must somehow join together into large drops. Two mechanisms for the formation of precipitation have been pro-

posed. (1) In clouds where the temperatures are below freezing, ice crystals form and fall as snowflakes. At lower altitudes the snowflakes melt and become raindrops before they reach the ground. (2) *large* droplets form in warm clouds that contain large *hygroscopic* ("water-seeking") *nuclei*, such as salt particles. As these big droplets descend, they collide and join with smaller water droplets. After many collisions the droplets are large enough to fall to the ground as rain.

- The forms of precipitation include *rain, snow, sleet, hail,* and *rime*.

Key Terms

absolute instability (p. 417)
absolute stability (p. 406)
adiabatic temperature change (p. 412)
advection fog (p. 425)
Bergeron process (p. 430)
calorie (p. 406)
cirrus (p. 423)
cloud (p. 423)
cloud of vertical development (p. 424)
collision-coalescence process (p. 429)
condensation (p. 407)
condensation nuclei (p. 422)
conditional instability (p. 417)
convergence (p. 422)
cumulus (p. 423)
deposition (p. 407)

dew point (p. 410)
dry adiabatic rate (p. 413)
evaporation (p. 407)
fog (p. 424)
freezing (p. 407)
freezing nuclei (p. 430)
frontal fog (p. 428)
frontal wedging (p. 421)
glaze (p. 431)
hail (p. 434)
high cloud (p. 423)
humidity (p. 407)
hygrometer (p. 411)
hygroscopic nuclei (p. 423)
latent heat (p. 407)
low cloud (p. 423)
melting (p. 407)
middle cloud (p. 423)

orographic lifting (p. 419)
precipitation fog (p. 428)
radiation fog (p. 425)
rain (p. 432)
rainshadow desert (p. 421)
relative humidity (p. 408)
rime (p. 435)
saturation (p. 408)
sleet (p. 434)
snow (p. 433)
specific humidity (p. 408)
steam fog (p. 428)
stratus (p. 423)
sublimation (p. 407)
supercooled (p. 430)
upslope fog (p. 425)
vapor pressure (p. 408)
wet adiabatic rate (p. 413)

Review Questions

1. Summarize the processes by which water changes from one state to another. Indicate whether heat energy is absorbed or liberated.

2. After studying Table 15.1, write a generalization relating temperature and the capacity of air to hold water vapor.

3. How do relative and specific humidity differ?

4. Referring to Figure 15.6, answer the following questions and then write a generalization relating changes in air temperature to changes in relative humidity.

   (a) During a typical day, when is the relative humidity highest? Lowest?

   (b) At what time of day would dew most likely form?

5. If the temperature remains unchanged and the specific

humidity decreases, how will relative humidity change?

6. On a cold winter day when the temperature is −10°C and the relative humidity is 50 percent, what is the specific humidity (refer to Table 15.1)? What is the specific humidity for a day when the temperature is 20°C and the relative humidity is 50 percent?

7. Explain the principle of the sling psychrometer and the hair hygrometer.

8. Using the standard tables (Tables 15.2 and 15.3), determine the relative humidity and dew-point temperature if the dry-bulb thermometer reads 16°C and the wet-bulb thermometer reads 12°C. How would the relative humidity and dew point change if the wet-bulb thermometer read 8°C?

9. On a warm summer day when the relative humidity is high, it may seem even warmer than the thermometer indicates. Why do we feel so uncomfortable on a "muggy" day?

10. Why does air cool when it rises through the atmosphere?

11. Explain the difference between environmental lapse rate and adiabatic cooling.

12. If unsaturated air at 23°C were to rise, what would its temperature be at 500 meters? If the dew-point temperature at the condensation level were 13°C, at what altitude would clouds begin to form?

13. Why does the adiabatic rate of cooling change when condensation begins? Why is the wet adiabatic rate not a constant figure?

14. The contents of an aerosol can are under very high pressure. When you push the nozzle on such a can, the spray feels cold. Explain.

15. How do orographic lifting and frontal wedging act to force air to rise?

16. Explain why the Great Basin area of the western United States is so dry. What term is applied to such a situation?

17. How does stable air differ from unstable air? Describe the general nature of the clouds and precipitation expected with each.

18. How do temperature inversions influence air pollution? (See Box 15.2.)

19. What is the function of condensation nuclei in cloud formation? The function of the dew point?

20. As you drink an ice-cold beverage on a warm day, the outside of the glass or bottle becomes wet. Explain.

21. What is the basis for the classification of clouds?

22. Why are high clouds always thin?

23. Which cloud types are associated with the following characteristics: thunder, halos, precipitation, hail, mackerel sky, lightning, mares' tails?

24. List five types of fog and discuss the details of their formation.

25. What is the difference between precipitation and condensation?

26. List the forms of precipitation and the circumstances of their formation.

27. Sometimes, when rainfall is light, the amount is reported as a trace. When this occurs, how much (or little) rain has fallen?

Examining the Earth System

1. The interrelationships among Earth's spheres have produced the Great Basin area of the western United States, which includes some of the driest areas in the world. Examine the map of the region in Figure 9.14 on page 251. Although the area is only a few hundred miles from the Pacific Ocean, it is a desert. Why? Did any geologic factor(s) contribute to the formation of this desert environment? Do any major rivers have their source in the Great Basin? Explain. (For information about deserts in the United States, visit the United States Geological Survey (USGS) Deserts: Geology and Resources Web site at **http://pubs.usgs.gov/gip/deserts/**)

2. Phoenix and Flagstaff, Arizona, are located nearby each other in the southwestern United States. Using the climate diagrams for the two cities found in Figure 18.12, describe the impact that elevation has on the precipitation and temperature of each city. Compare the natural vegetation of these nearby locations. (To examine the current weather conditions and forecasts for Phoeniz and Flagstaff, Arizona, contact *USA Today* at **http://www. usatoday.com/weather/** and/or *The Weather Channel* at **http://www.weather.com/twc/homepage.twc**)

3. The amount of precipitation that falls at any particular place and time is controlled by the quantity of moisture in the air and many other factors. How might each of the following alter the precipitation at a particular locale? (1) An increase in the elevation of the land. (b) A decrease in the area covered by forests and other types of vegetation. (c) Lowering of average ocean-surface temperatures. (d) An increase in the percentage of time that the winds blow from an adjacent body of water. (e) A major episode of global volcanism lasting for decades.

Web Resources

The following are informative and interesting World Wide Web sites that address topics related to those presented in the chapter:

Cloud Types (Plymouth State College):

- **http://vortex.plymouth.edu/clouds.html**

Clouds and Precipitation Online Tutorial (University of Illinois):

- **http://www2010.atmos.uiuc.edu/(Gh)/guides/mtr /cld/home.rxml**

For direct links to these sites and other appropriate Earth science Web locations, chapter review exercises, and relevant, topical investigations that utilize Web resources, contact the *Earth Science*, 9th Edition Home Page at:

- **http://www.prenhall.com/tarbuck.**

Chapter 16
Air Pressure and Wind

View of a wind farm near Palm Springs, California. Some estimate that in the next 50 to 60 years wind power could meet 5 percent or more of the country's demand for electrical energy. (Photo by John Mead/Photo Researchers, Inc.)

Of the various elements of weather and climate, changes in air pressure are the least noticeable. In listening to a weather report, generally we are interested in moisture conditions (humidity and precipitation), temperature, and perhaps wind. It is the rare person, however, who wonders about air pressure. Although the hour-to-hour and day-to-day variations in air pressure are not perceptible to human beings, they are very important in producing changes in our weather. For example, it is variations in air pressure from place to place that generate winds that in turn can bring changes in temperature and humidity (Figure 16.1). Air pressure is one of the basic weather elements and is a significant factor in weather forecasting. As we shall see, air pressure is closely tied to the other elements of weather in a cause-and-effect relationship.

Measuring Air Pressure

In Chapter 14 you saw that air has weight; at sea level, it exerts a pressure of 1 kilogram per square centimeter (14.7 pounds per square inch). The **air pressure** at a particular place is simply the force exerted by the weight of the air above. With an increase in altitude, there is less air, so the weight of the air, and thus the pressure, decreases. It decreases rapidly at first, then much more slowly (see Figure 14.4, p. 381, and Table 16.1).

When meteorologists measure atmospheric pressure, they employ a unit called the *millibar*. Standard sea level pressure is 1013.2 millibars. Although millibars are used almost exclusively by meteorologists, you might be better acquainted with the expression "inches of mercury" that is used by the media to describe atmospheric pressure.

Table 16.1 Pressure changes with altitude.

Altitude (kilometers)	Altitude (miles)	Pressure (millibars)
0	0	1013
1.0	0.6	899
2.0	1.2	795
3.0	1.9	701
4.0	2.5	617
5.0	3.1	540
10.0	6.2	265
20.0	12.4	55
30.0	18.6	12
40.0	24.8	3

This expression dates from 1643, when Torricelli, a student of the famous Italian scientist Galileo, invented the **mercury barometer**. Torricelli correctly described the atmosphere as a vast ocean of air that exerts pressure on us and all objects about us. To measure this force he filled a glass tube, which was closed at one end, with mercury. He then inverted the tube into a dish of mercury (Figure 16.2A). Torricelli found that the mercury flowed out of the tube until the weight of the column was balanced by the pressure that the atmosphere exerted on the surface of the mercury in the dish. In other words, the weight of mercury in the column equaled the weight of the same diameter column of air that extended from the ground to the top of the atmosphere.

When air pressure increases, the mercury in the tube rises. Conversely, when air pressure decreases, so does the height of the mercury column. With some refinements the mercurial barometer invented by Torricelli is still the standard pressure-measuring instrument used today. Standard atmospheric pressure at sea level equals 29.92 inches of mercury.

Figure 16.1 Windsurfers at the Cable Beach Regatta, Nassau, Bahamas. Wind, the horizontal movement of air, is created by differences in air pressure at Earth's surface. (Photo by Darrel Jones/The Stock Market)

Figure 16.2 **A.** Simple mercury barometer. The weight of the column of mercury is balanced by the pressure exerted on the dish of mercury by the air above. If the pressure decreases, the column of mercury falls; if the pressure increases, the column rises. **B.** An aneroid barograph makes a continuous record of pressure changes. One important advantage of the aneroid barometer is that it is easily adapted to a recording mechanism.

The need for a smaller and more portable instrument for measuring air pressure led to the development of the **aneroid** ("without liquid") **barometer**. Instead of having a mercury column held up by air pressure, the aneroid barometer uses a partially evacuated metal chamber. The chamber, being very sensitive to variations in air pressure, changes shape,

compressing as the pressure increases and expanding as the pressure decreases. An advantage of the aneroid barometer is that it can be easily connected to a recording mechanism. The resulting instrument is a **barograph**, which provides a continuous record of pressure changes with the passage of time (Figure 16.2B).

Factors Affecting Wind

In Chapter 15 we examined the upward movement of air and its role in cloud formation. As important as vertical motion is, far more air moves horizontally, the phenomenon we call **wind**. What causes wind?

Simply stated, wind is the result of horizontal differences in air pressure. *Air flows from areas of higher pressure to areas of lower pressure.* You may have experienced this when opening a vacuum-packed can of coffee. The noise you hear is caused by air rushing from the higher pressure outside the can to the lower pressure inside. Wind is nature's attempt to balance such inequalities in air pressure. Because unequal heating of Earth's surface generates these pressure differences, *solar energy is the ultimate driving force of wind.*

If Earth did not rotate, and if there were no friction between moving air and Earth's surface, air would flow in a straight line from areas of higher pressure to areas of lower pressure. But because both factors exist, wind is controlled by a combination of forces: (1) the pressure-gradient force, (2) Coriolis effect, and (3) friction. We will now examine each of these factors.

Pressure-Gradient Force

Pressure differences create wind, and the greater these differences, the greater the wind speed (see Box 16.1). Over Earth's surface, variations in air pressure are determined from barometric readings taken at hundreds of weather stations. These pressure data are shown on a weather map using **isobars**, lines that connect places of equal air pressure (Figure 16.3). The spacing of isobars indicates the amount of pressure change occurring over a given distance and is expressed as the **pressure gradient**.

You might find it easier to visualize a pressure gradient if you think of it as being similar to the slope of a hill. A steep pressure gradient, like a steep hill, causes greater acceleration of an air parcel than does a weak pressure gradient (a gentle hill). Thus, the relationship between wind speed and the pressure gradient is straightforward: *Closely spaced isobars indicate a steep pressure gradient and high winds, whereas widely spaced isobars indicate a weak pressure gradient and light winds.* Figure 16.3 illustrates the relationship between the spacing of isobars and wind speed. Notice that wind speeds are greater in Ohio, Kentucky, Michigan, and Illinois, where isobars are

Box 16.1 People and the Environment: Wind Energy: An Alternative With Potential

Approximately 0.25 percent (one-quarter of 1 percent) of the solar energy that reaches the lower atmosphere is transformed into wind. Although it is just a minuscule percentage, the absolute amount of energy is enormous. According to one estimate, if the winds of North and South Dakota could be harnessed, they would provide 80 percent of the electrical energy used in the United States.

Wind has been used for centuries as an almost free and nonpolluting source of energy. Sailing ships and wind-powered grist mills represent two of the early ways that this renewable resource was harnessed. Further, as rural America was settled, there was a strong reliance on wind power to pump water and later to generate small amounts of electricity.

Following the "energy crisis" that was brought about by an oil embargo in the 1970s, interest in wind power as well as other alternative forms of energy increased dramatically. In 1980, the federal government initiated a program to develop wind-power systems. A project sponsored by the U.S. Department of Energy involved setting up experimental wind farms in mountain passes known to have strong persistent winds. One of these facilities, located at Altamont Pass near San Francisco, now employs more than 7000 wind turbines. In 1995, wind supplied about 1 percent of California's electricity (Figure 16.A).

In September 1998, an array of 143 20-story-high wind turbines went on line near Lake Benton in southwestern Minnesota. The 107-megawatt project is the largest single wind farm in the world. When the wind is blowing, the Lake Benton farm can power 40,000 homes. When the project is completed in 2002, the facility's output will be 425 megawatts, nearly four times the 1998 output.

As technology has improved, efficiency has increased and the costs of wind-generated electricity have become more competitive. Between 1983 and 1999 technological advances cut the cost of wind power by about 85 percent. Some estimate that in the next 50 to 60 years, wind power could meet between 5 and 10 percent of the country's demand for electrical energy. One area for the expansion of wind energy will likely be islands and other regions far from electrical grids that must import fuel for generating power.

Although the future for wind power appears promising, it is not without difficulties. In addition to technical advances that must continue to be made, noise pollution and the costs of large tracts of land in populated areas present significant obstacles to development. In the Netherlands, where windmills have a long history, proposals to locate wind turbines atop the country's dike system have been met with strong opposition. Part of the problem is that the Netherlands, like much of Europe, is densely populated and, hence, has few remote locations.

Another significant limitation of wind energy is that it is intermittent. If wind were to constitute a high proportion of the total energy supply, any shortfall could cause severe economic

Figure 16.A Wind power old and new. Altamont Pass California. (Photo by Joe Sohm/Chromosohm/Photo Researchers, Inc.)

disruptions. Better means of storage would allow wind to contribute a significantly higher percentage of our energy needs. One proposal would use wind energy to produce hydrogen by electrolysis of water. This combustible gas would then be distributed and stored in a manner similar to natural gas.

more closely spaced, than in the western states where isobars are more widely spaced.

The pressure gradient is the driving force of wind, and it has both magnitude and direction. Its magnitude is determined from the spacing of isobars. The direction of force is always from areas of higher pressure to areas of lower pressure and at right angles to the isobars. Once the air starts to move, the Coriolis effect and friction come into play, but then only to modify the movement, not to produce it.

Coriolis Effect

The weather map in Figure 16.3 shows the typical air movements associated with high- and low-pressure systems. As expected, the air moves out of the regions of higher pressure and into the regions of lower pressure. However, the wind does not cross the isobars at right angles as the pressure-gradient force directs it. This deviation is the result of Earth's rotation and has been named the **Coriolis effect** after the French scientist who first thoroughly described it.

ff	Miles per hour
◎	Calm
——	1–2
⌐	3–8
⌐	9–14
⌐⌐	15–20
⌐⌐	21–25
⌐⌐⌐	26–31
⌐⌐⌐	32–37
⌐⌐⌐⌐	38–43
⌐⌐⌐⌐	44–49
⌐⌐⌐⌐⌐	50–54
◤	55–60
◤⌐	61–66
◤⌐	67–71
◤⌐⌐	72–77
◤⌐⌐	78–83
◤⌐⌐⌐	84–89
◤◤⌐	119–123

Figure 16.3 The black lines are isobars that connect places of equal barometric pressure. They show the distribution of pressure on weather maps. The lines usually curve and often join around cells of high and low pressure. The flags indicate the airflow surrounding cells of high and low pressure and are plotted like flags flying with the wind. Wind speed is indicated by flags and "feathers" as shown along the right-hand side of this drawing.

All free-moving objects or fluids, including the wind, are deflected to the *right* of their path of motion in the Northern Hemisphere and to the *left* in the Southern Hemisphere. The reason for this deflection can be illustrated by imagining the path of a rocket launched from the North Pole toward a target located on the equator (Figure 16.4). If the rocket took an hour to reach its target, Earth would have rotated 15 degrees to the east during its flight. To someone standing on Earth it would look as if the rocket veered off its path and hit Earth 15 degrees west of its target. The true path of the rocket is straight and would appear so to someone out in space looking down at Earth. It was Earth turning under the rocket that gave it its *apparent* deflection.

Note that the rocket was deflected to the right of its path of motion because of the counterclockwise rotation of the Northern Hemisphere. In the Southern Hemisphere, the effect is reversed. Clockwise rotation produces a similar deflection, but to the left of the path of motion. The same deflection is experienced by wind regardless of the direction it is moving.

We attribute the apparent shift in wind direction to the Coriolis effect. This deflection (1) is always directed at right angles to the direction of airflow; (2) affects only wind direction, not wind speed; (3) is affected by wind speed (the stronger the wind, the greater the deflection); and (4) is strongest at the poles and weakens equatorward, becoming nonexistent at the equator.

Friction With Earth's Surface

The effect of friction on wind is important only within a few kilometers of Earth's surface. Friction acts to slow air movement and, as a consequence, alters wind direction. To illustrate friction's effect on wind direction, let us look at a situation in which it has no role. Above the friction layer, the pressure-gradient force and Coriolis effect work together to direct the flow of air. Under these conditions, the pressure-gradient force causes air to start moving across the isobars. As soon as the air starts to move, the Coriolis effect acts at right angles to this motion. The faster the wind speed, the greater the deflection.

Eventually, the Coriolis effect will balance the pressure-gradient force and the wind will blow parallel to the isobars (Figure 16.5). Upper-air winds generally take this path and are called **geostrophic winds**. Because of the lack of friction with Earth's surface, geostrophic winds travel at higher speeds than do surface winds. This can be observed in Figure 16.6 by noting the wind flags, many of which indicate winds of 50 to 100 miles per hour.

The most prominent features of upper-level flow are the **jet streams**. First encountered by high-flying bombers during World War II, these fast-moving "rivers" of air travel between 120 and 240 kilometers (75 and 150 miles) per hour in a west-to-east direction. One such stream is situated over the polar front, which

Figure 16.4 The Coriolis effect illustrated using a 1-hour flight of a rocket traveling from the North Pole to a location on the equator. **A.** On a nonrotating Earth, the rocket would travel straight to its target. **B.** However, Earth rotates 15° each hour. Thus, although the rocket travels in a straight line, when we plot the path of the rocket on Earth's surface, it follows a curved path that veers to the right of the target.

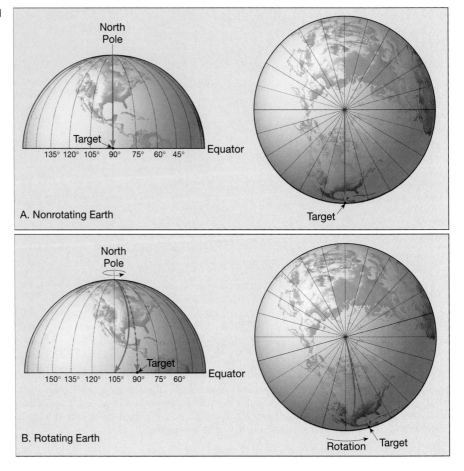

is the zone separating cool polar air from warm subtropical air.

Below 600 meters (2000 feet), friction complicates the airflow just described. Recall that the Coriolis effect is proportional to wind speed. Friction lowers the wind speed, so it reduces the Coriolis effect. Because the pressure-gradient force is not affected by wind speed, it wins the tug of war shown in Figure 16.7. The result is a movement of air at an angle across the isobars toward the area of lower pressure.

The roughness of the terrain determines the angle of airflow across the isobars. Over the smooth ocean surface, friction is low and the angle is small. Over rugged terrain, where friction is higher, the angle that air makes as it flows across the isobars can be as great as 45 degrees.

In summary, upper airflow is nearly parallel to the isobars, whereas the effect of friction causes the surface winds to move more slowly and cross the isobars at an angle.

Figure 16.5 The geostrophic wind. Upper-level winds are deflected by the Coriolis effect until the Coriolis effect just balances the pressure-gradient force. Above 600 meters (1970 feet), where friction is negligible, these winds will flow nearly parallel to the isobars and are called geostrophic winds.

ff	Miles per hour
◎	Calm
—	1–2
⌐	3–8
⌐	9–14
⌐	15–20
⌐	21–25
⌐	26–31
⌐	32–37
⌐	38–43
⌐	44–49
⌐	50–54
▌	55–60
▌	61–66
▌	67–71
▌	72–77
▌	78–83
▌	84–89
▌	119–123

A. Upper-level weather chart

B. Representation of upper-level chart

Figure 16.6 Upper-air winds. This map shows the direction and speed for the upper-air wind for a particular day. Note that the airflow is nearly parallel to the contours. These isolines are height contours for the 500-millibar level.

Lows and Highs

Among the most common features on any weather map are areas designated as pressure centers. **Lows**, or **cyclones**, are centers of low pressure, and **highs**, or **anticyclones**, are high-pressure centers. As Figure 16.8 illustrates, the pressure decreases from the outer isobars toward the center in a low. In a high, just the opposite is the case—the values of the isobars increase from the outside toward the center. By knowing just a few basic facts about centers of high and low pressure, you can greatly increase your understanding of current and forthcoming weather.

Cyclonic and Anticyclonic Winds

From the preceding section, you learned that the two most significant factors that affect wind are the pressure-gradient force and the Coriolis effect. Winds move from higher pressure to lower pressure and are deflected to the right or left by Earth's rotation. When these controls of airflow are applied to pressure centers in the Northern Hemisphere, the result is that

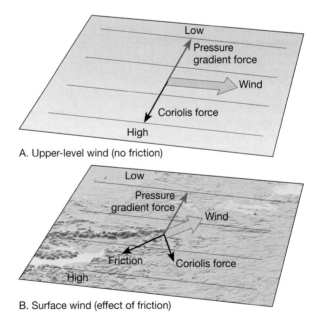

A. Upper-level wind (no friction)

B. Surface wind (effect of friction)

Figure 16.7 Comparison between upper-level winds and surface winds showing the effects of friction on airflow. Friction slows surface wind speed, which weakens the Coriolis effect, causing the winds to cross the isobars and move toward the lower pressure.

Figure 16.8 Cyclonic and anticyclonic winds in the Northern Hemisphere. Arrows show that winds blow into and counterclockwise around a low. By contrast, around a high, winds blow outward and clockwise.

winds blow inward and counterclockwise around a low. Around a high, they blow outward and clockwise (Figure 16.8).

In the Southern Hemisphere the Coriolis effect deflects the winds to the left, and therefore winds around a low blow clockwise, and winds around a high move counterclockwise. In either hemisphere, friction causes a net inflow (**convergence**) around a cyclone and a net outflow (**divergence**) around an anticyclone.

Weather Generalizations About Highs and Lows

Rising air is associated with cloud formation and precipitation, whereas subsidence produces clear skies. In this section we will learn how the movement of air can itself create pressure change and hence generate winds. After doing so, we will examine the relation between horizontal and vertical flow, and their effects on the weather.

Let us first consider the situation around a surface low-pressure system where the air is spiraling inward. Here the net inward transport of air causes a shrinking of the area occupied by the air mass, a process that is termed *horizontal convergence*. Whenever air converges horizontally, it must pile up, that is, increase in height to allow for the decreased area it now occupies. This generates a "taller" and therefore heavier air column. Yet a surface low can exist only as long as the column of air above exerts less pressure than that occurring in surrounding regions. We seem to have encountered a paradox—a low-pressure center causes a net accumulation of air, which increases its pressure. Consequently, a surface cyclone should quickly

eradicate itself in a manner not unlike what happens when a vacuum-packed can is opened.

You can see that for a surface low to exist for very long, compensation must occur aloft. For example, surface convergence could be maintained if divergence (spreading out) aloft occurred at a rate equal to the inflow below. Figure 16.9 shows the relationship between surface convergence (inflow) and divergence (outflow) aloft that is needed to maintain a low-pressure center.

Note that surface convergence about a cyclone causes a net upward movement. The rate of this vertical movement is slow, generally less than 1 kilometer per day. Nevertheless, because rising air often results in cloud formation and precipitation, a low-pressure center is generally related to unstable conditions and stormy weather (Figure 16.10A).

As often as not, it is divergence aloft that creates a surface low. Spreading out aloft initiates upflow in the atmosphere directly below, eventually working its way to the surface, where inflow is encouraged.

Like their cyclonic counterparts, anticyclones must be maintained from above. Outflow near the surface is accompanied by convergence aloft and general subsidence of the air column (Figure 16.9). Because descending air is compressed and warmed, cloud formation and precipitation are unlikely in an anticyclone. Thus, "fair" weather can usually be expected with the approach of a high (Figure 16.10B).

For reasons that should now be obvious, it has been common practice to print on household barometers the words "stormy" at the low-pressure end and "fair" on the high-pressure end. By noting whether the pressure is rising, falling, or steady, we have a good indication of what the forthcoming weather will be. Such a determination,

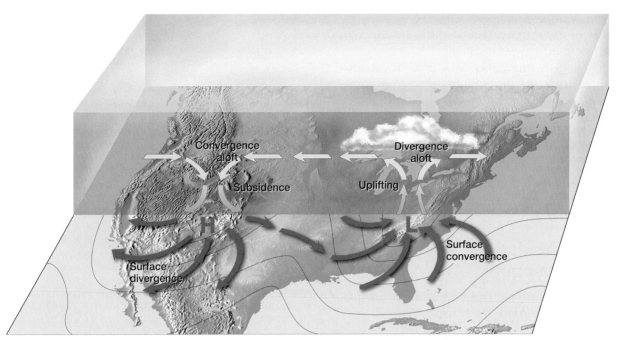

Figure 16.9 Airflow associated with surface cyclones and anticyclones. A low, or cyclone, has converging surface winds and rising air causing cloudy conditions. A high, or anticyclone, has diverging surface winds and descending air, which leads to clear skies and fair weather.

called the **pressure**, or **barometric tendency**, is a very useful aid in short-range weather prediction.

In conclusion, you should now be better able to understand why television weather reporters emphasize the locations and projected paths of cyclones and anticyclones. The "villain" on these weather programs is always the low-pressure center, which produces "bad" weather in any season. Lows move in roughly a west-to-east direction across the United States and require a few days to more than a week for the journey.

Figure 16.10 These two photographs illustrate the basic weather generalizations associated with pressure centers. **A.** New Yorkers struggle during a storm. Centers of low pressure are frequently associated with cloudy conditions and precipitation. (Photo by Alex Brandon/AP/Wide World Photos) **B.** By contrast, clear skies and "fair" weather may be expected when an area is under the influence of high pressure. Sunbathers on a beach in Hawaii. (Photo by E.J. Tarbuck)

A.

B.

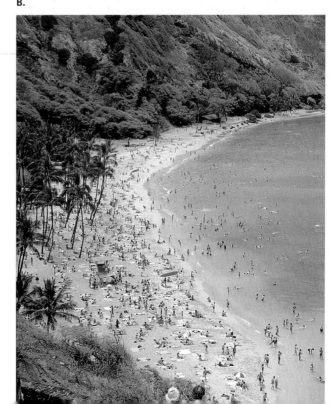

Their paths can be somewhat erratic; thus, accurate prediction of their movement is difficult, although essential, for short-range forecasting.

Meteorologists must also determine if the flow aloft will intensify an embryo storm or act to suppress its development. Because of the close tie between conditions at the surface and those aloft, a great deal of emphasis has been placed on the importance and understanding of the total atmospheric circulation, particularly in the mid-latitudes. We will now examine the workings of Earth's general atmospheric circulation, and then again consider the structure of the cyclone in light of this knowledge.

General Circulation of the Atmosphere

As noted, the underlying cause of wind is unequal heating of Earth's surface. In tropical regions, more solar radiation is received than is radiated back to space. In polar regions the opposite is true—less solar energy is received than is lost. Attempting to balance these differences, the atmosphere acts as a giant heat transfer system, moving warm air poleward and cool air equatorward. On a smaller scale, but for the same reason, ocean currents also contribute to this global heat transfer. The general circulation is very complex, and there is a great deal that has yet to be explained. We can, however, develop a general understanding by first considering the circulation that would occur on a nonrotating Earth having a uniform surface. We will then modify this system to fit observed patterns.

Circulation on a Nonrotating Earth

On a hypothetical nonrotating planet with a smooth surface of either all land or all water, two large thermally produced cells would form (Figure 16.11). The heated equatorial air would rise until it reached the tropopause, which, acting like a lid, would deflect the air poleward. Eventually, this upper-level airflow would reach the poles, sink, spread out in all directions at the surface, and move back toward the equator. Once there, it would be reheated and start its journey over again. This hypothetical circulation system has upper-level air flowing poleward and surface air flowing equatorward.

If we add the effect of rotation, this simple convection system will break down into smaller cells. Figure 16.12 illustrates the three pairs of cells proposed to carry on the task of heat redistribution on a rotating planet. The polar and tropical cells retain the characteristics of the thermally generated convection described earlier. The nature of the mid-latitude circulation is more complex and will be discussed in more detail in a later section.

Figure 16.11 Global circulation on a nonrotating Earth. A simple convection system is produced by unequal heating of the atmosphere on a nonrotating Earth.

Idealized Global Circulation

Near the equator, the rising air is associated with the pressure zone known as the **equatorial low**— a region marked by abundant precipitation. As the upper-level flow from the equatorial low reaches 20 to 30 degrees latitude, north or south, it sinks back toward the surface. This subsidence and associated adiabatic heating produce hot, arid conditions. The center of this zone of subsiding dry air is the **subtropical high**, which encircles the globe near 30 degrees latitude, north and south (Figure 16.12). The great deserts of Australia, Arabia, and North Africa exist because of the stable dry conditions associated with the subtropical highs.

At the surface, airflow is outward from the center of the subtropical high. Some of the air travels equatorward and is deflected by the Coriolis effect, producing the reliable **trade winds**. The remainder travels poleward and is also deflected, generating the prevailing **westerlies** of the mid-latitudes. As the westerlies move poleward, they encounter the cool **polar easterlies** in the region of the **subpolar low.** The interaction of these warm and cool winds produces the stormy belt known as the **polar front.** The source region for the variable polar easterlies is the **polar high.** Here, cold polar air is subsiding and spreading equatorward.

In summary, this simplified global circulation is dominated by four pressure zones. The subtropical and polar highs are areas of dry subsiding air that flows outward at the surface, producing the prevailing winds. The low-pressure zones of the equatorial and subpolar regions are associated with inward and upward airflow accompanied by clouds and precipitation.

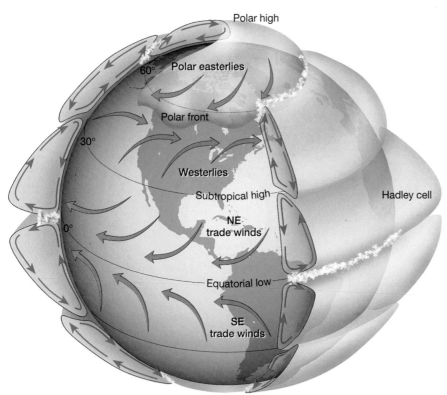

Figure 16.12 Idealized global circulation.

Influence of Continents

Up to this point, we have described the surface pressure and associated winds as continuous belts around Earth. However, the only truly continuous pressure belt is the subpolar low in the Southern Hemisphere. Here the ocean is uninterrupted by landmasses. At other latitudes, particularly in the Northern Hemisphere where landmasses break up the ocean surface, large seasonal temperature differences disrupt the pattern. Figure 16.13 shows the resulting pressure and wind patterns for January and July. The circulation over the oceans is dominated by semipermanent cells of high pressure in the subtropics and cells of low pressure over the subpolar regions. The subtropical highs are responsible for the trade winds and westerlies, as mentioned earlier.

The large landmasses, on the other hand, particularly Asia, become cold in the winter and develop a seasonal high-pressure system from which surface flow is directed off the land (Figure 16.13). In the summer, the opposite occurs; the landmasses are heated and develop a low-pressure cell, which permits air to flow onto the land. These seasonal changes in wind direction are known as the **monsoons**. During warm months, areas such as India experience a flow of warm, water-laden air from the Indian Ocean, which produces the rainy summer monsoon. The winter monsoon is dominated by dry continental air. A similar situation exists, but to a lesser extent, over North America.

In summary, the general circulation is produced by semipermanent cells of high and low pressure over the oceans and is complicated by seasonal pressure changes over land.

Circulation in the Mid-Latitudes

The circulation in the mid-latitudes, the zone of the westerlies, is complex and does not fit the convection system proposed for the tropics. Between about 30 and 60 degrees latitude, the general west-to-east flow is interrupted by migrating cyclones and anticyclones. In the Northern Hemisphere these cells move from west to east around the globe, creating an anticyclonic (clockwise) flow or a cyclonic (counterclockwise) flow in their area of influence. A close correlation exists between the paths taken by these surface pressure systems and the position of the upper-level airflow, indicating that the upper air strongly influences the movement of cyclonic and anticyclonic systems.

Among the most obvious features of the flow aloft are the seasonal changes. The steep temperature gradient across the middle latitudes in the winter months corresponds to a stronger flow aloft. In addition, the polar jet stream fluctuates seasonally such that its average position migrates southward with the approach of winter and northward as summer nears. By midwinter, the jet core may penetrate as far south as central Florida.

Because the paths of low-pressure centers are guided by the flow aloft, we can expect the southern

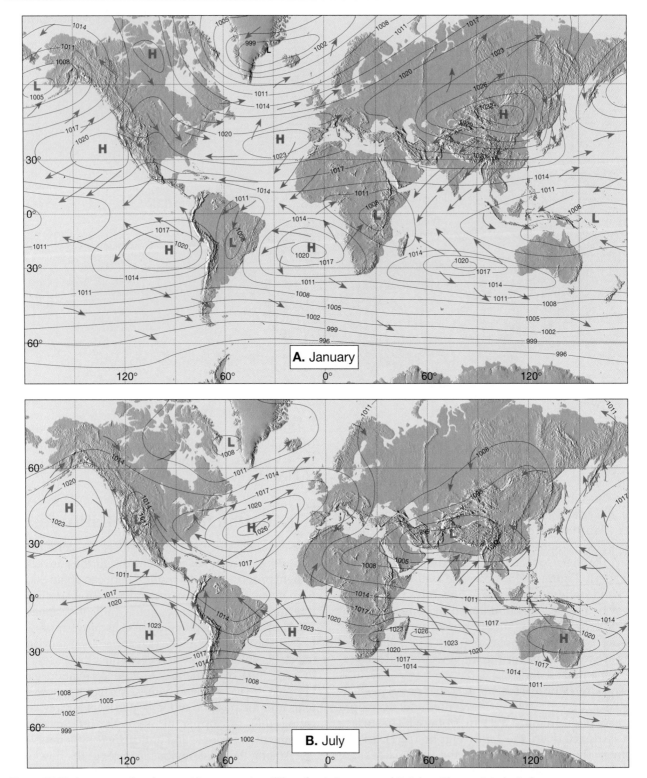

Figure 16.13 Average surface barometric pressure in millibars for **A.** January and **B.** July, with associated winds.

tier of states to experience more of their stormy weather in the winter season. During the hot summer months, the storm track is across the northern states, and some cyclones never leave Canada. The northerly storm track associated with summer applies also to Pacific storms, which move toward Alaska during the warm months, thus producing an extended dry season for much of our western coast. The number of cyclones generated is seasonal as well, with the largest number occurring in the cooler months when the temperature gradients are greatest. This fact is in agreement with the role of cyclonic storms in the distribution of heat across the mid-latitudes.

El Niño and Global Weather

As can be seen in Figure 16.14, the cold Peruvian current flows equatorward along the coast of Ecuador and Peru. This flow encourages upwelling of deep, nutrient-filled waters that serve as the primary food source for millions of fish, particularly anchovies.

Near the end of each year, a weak, warm countercurrent flows southward along the coasts of Ecuador and Peru replacing the cold Peruvian current. Centuries ago the local residents named this annual event El Niño ("the child") after the Christ child

because it usually appeared during the Christmas season. Normally, these warm countercurrents last for at most a few weeks when they again give way to the cold Peruvian flow. However, every 3 to 7 years this countercurrent is unusually warm and strong. Accompanying this event is a pool of warm, ocean surface water in the central and eastern Pacific. Today, scientists use the term **El Niño** for these episodes of ocean warming that greatly exceed the relatively weak event that originally bore the name.

The onset of El Niño is marked by abnormal weather patterns that drastically affect the economies

A. Normal conditions

B. El Niño

Figure 16.14 The relationship between the Southern Oscillation and El Niño is illustrated on these simplified maps. **A.** Normally, the trade winds and strong equatorial currents flow toward the west. At the same time, the strong Peruvian current causes upwelling of cold water along the west coast of South America. **B.** When the Southern Oscillation occurs, the pressure over the eastern and western Pacific flip-flops. This causes the trade winds to diminish, leading to an eastward movement of warm water along the equator. As a result, the surface waters of the central and eastern Pacific warm, with far-reaching consequences to weather patterns.

of Ecuador and Peru. As shown in Figure 16.14B, these unusually warm currents amass large quantities of warm water in the eastern Pacific. This warm pool blocks the upwelling of colder, nutrient-filled water, and the anchovies starve from lack of food, devastating the local economies. At the same time, however, some inland areas that are normally arid receive an uncommon abundance of rain. These climatic fluctuations have been known for years but were always considered local phenomena. Today, we know that El Niño is part of the global circulation and affects the weather at great distances from Peru and Ecuador.

The strongest El Niño on record occurred in 1997 and 1998, and was responsible for weather extremes of a variety of types in many parts of the world. Heavy rains and flooding during this event plagued normally dry portions of Ecuador and Peru. Some locations that usually receive only 10 to 13 centimeters of rain each year had as much as 350 centimeters of precipitation. At the same time, severe drought beset Indonesia, Australia, and the Philippines. Huge crop losses, property damage, and much human suffering were recorded.

Farther north, ferocious storms that struck the California coast brought unprecedented beach erosion, landslides, and floods. In the southern United States, heavy rains also brought floods to Texas and the Gulf states. The same energized jet stream that produced storms in the South, upon reaching the Atlantic, sheared off the northern portions of hurricanes, destroying the storms. It was one of the quietest Atlantic hurricane seasons in years.

Major El Niño events, such as the one in 1997 and 1998, are intimately related to the large-scale atmospheric circulation. Each time an El Niño occurs, the barometric pressure drops over large portions of the southeastern Pacific, whereas in the western Pacific, near Indonesia and northern Australia, the pressure rises (Figure 16.14A, B). Then, as a major El Niño event comes to an end, the pressure difference between these two regions swings back in the opposite direction. This see-saw pattern of atmospheric pressure between the eastern and western Pacific is called the **Southern Oscillation**. It is an inseparable part of the El Niño warmings that occur in the central and eastern Pacific every 3 to 7 years. Therefore, this phenomenon is often termed El Niño/Southern Oscillation, or ENSO for short.

Winds in the lower atmosphere are the link between the pressure change associated with the Southern Oscillation and the extensive ocean warming associated with El Niño. During a typical year, the trade winds converge near the equator and flow westward toward Indonesia (Figure 16.14A). This steady westward flow creates a warm surface current that moves from east to west along the equator. The result is a "piling up" of a thick layer of warm surface water that produces higher sea levels (by 30 centimeters) in the western Pacific. Meanwhile, the eastern Pacific is characterized by a strong Peruvian current, upwelling of cold water, and lower sea levels.

Then when the Southern Oscillation occurs, the normal situation just described changes dramatically. Barometric pressure rises in the Indonesian region, causing the pressure gradient along the equator to weaken or even to reverse. As a consequence, the once-steady trade winds diminish and may even change direction. This reversal creates a major change in the equatorial current system, with warm water flowing eastward. With time, water temperatures in the central and eastern Pacific increase and sea level in the region rises (Figure 16.14B). This eastward shift of the warmest surface water marks the onset of El Niño and sets up changes in atmospheric circulation that affect areas far outside the tropical Pacific (see Box 16.2).

When an El Niño began in the summer of 1997, forecasters predicted that the pool of warm water over the Pacific would displace the paths of both the subtropical and polar jet streams, which steer weather systems across North America (Figure 16.14B). As predicted, the subtropical jet brought rain to the Gulf Coast, where Tampa, Florida, received more than three times its normal winter precipitation. Furthermore, the polar jet pumped warm air far north into the continent. As a result, winter temperatures west of the Rockies were significantly above normal.

The effects of El Niño are highly variable depending in part on the temperatures and size of the warm pools. During one El Niño, an area may experience flooding, only to be hit by drought during the next event. Nevertheless, some locales appear to be affected more consistently. In particular, during most El Niños, warmer-than-normal winters occur in the northern United States and Canada. In addition, normally arid portions of Peru and Ecuador, as well as the eastern United States, experience wet conditions. By contrast, drought conditions are generally observed in Indonesia, Australia, and the Philippines. Furthermore, during El Niño years, fewer Atlantic hurricanes are recorded than during non-El Niño years.

In summary, the effects of El Niño on world climate are widespread and variable. Following an event in 1982–83, one researcher stated that

> there is no place on Earth where the weather is indifferent to air and ocean conditions in the tropical Pacific: the Southern Oscillation and El Niño events associated with it are now understood to have a significant influence on the state of weather and climate almost everywhere.[*]

---

[*] Murray Mitchell, "El Niño: The Global Weather Connection," *Weatherwise*, 36, no. 4 (August 1983), 168.

Box 16.2   Tracking El Niño From Space

The images in Figure 16.B show the progression of the 1997–98 El Niño as derived from the TOPEX/Poseidon satellite.* This satellite uses radar altimeters to bounce radar signals off the ocean surface to obtain precise measurements of the distance between the satellite and the sea surface. This information is combined with high-precision orbital data from the Global Positioning System (GPS) of satellites to produce maps of sea-surface height. Such maps show the topography of the sea surface. Elevated topography ("hills") indicates warmer-than-average water, whereas areas of low topography ("valleys") indicate cooler-than-normal water. With this detailed information, scientists can determine the speed and direction of surface ocean currents.

The colors represented in these images show sea-level height relative to average. Remember, "hills" are warm and "valleys" are cool. The white and red areas indicate places of higher-than-normal sea-surface heights ("hills"). In the white areas, the sea surface is between 14 and 32 centimeters above normal; in the red areas, sea level is elevated by about 10 centimeters. Green areas indicate average conditions, whereas purple shows zones that are at least 18 centimeters below average sea level.

The images depict the progression of the large warm water mass from

Figure 16.B The image for April 25, 1997, shows the Pacific on the eve of the 1997–98 El Niño event. By examining the images for July 25 and November 10, you can see the dramatic buildup of warm water (white and red areas) in the eastern Pacific and the enlarging region of cool water (purple) in the western Pacific. By the time of the March 14, 1998 image, the area of warm water in the eastern Pacific was *much* smaller. (Courtesy of NASA Goddard Space Flight Center).

*This box is based on material provided by NASA's Goddard Space Flight Center.

west to east across the equatorial Pacific. At its peak in November 1997, the surface area covered by the warm water mass was about one and a half times the size of the contiguous 48 states. The added amount of warm water in the eastern Pacific with a temperature between 21° and 30°C was about 30 times the combined volume of the water in all of the U.S. Great Lakes.

Local Winds

Having examined Earth's large-scale circulation, let us turn briefly to winds that influence much smaller areas. Remember that all winds are produced for the same reason: temperature differences that arise because of unequal heating of Earth's surface. In turn, these differences generate pressure variations. *Local winds* are simply small-scale winds produced by a locally generated pressure-gradient. Those described here are caused either by topographic effects or variations in surface composition in the immediate area.

Sea and Land Breezes

In coastal areas during the warm summer months, the land surface is heated more intensely during the daylight hours than is the adjacent body of water (see the section on "Land and Water" in Chapter 14). As a result, the air above the land surface heats, expands, and rises, creating an area of lower pressure. A **sea breeze** then develops, because cooler air over the water (higher pressure) moves toward the warmer land (lower pressure) (Figure 16.15, top). The sea breeze begins to develop shortly before noon and generally reaches its greatest intensity during the mid- to

Figure 16.15 Sea and land breezes. During the daytime, land heats more intensely than water. A low is created over the land, and the air moves from the water (higher pressure) to the land (lower pressure). At night, the land cools more rapidly, resulting in higher pressure over land and a reversal of the wind.

late afternoon. These relatively cool winds can be a significant moderating influence on afternoon temperatures in coastal areas.

At night, the reverse may take place. The land cools more rapidly than the sea and the **land breeze** develops (Figure 16.15, bottom). Small-scale sea breezes can also develop along the shores of large lakes. People who live in a city near the Great Lakes, such as Chicago, recognize this lake effect, especially in the summer. They are reminded daily by weather reports of the cool temperatures near the lake as compared to warmer outlying areas.

Valley and Mountain Breezes

A daily wind similar to land and sea breezes occurs in many mountainous regions. During daylight hours, the air along the slopes of the mountains is heated more intensely than the air at the same elevation over the valley floor. Because this warmer air is less dense, it glides up along the slope and generates a **valley breeze** (Figure 16.16, left). The occurrence of these daytime

upslope breezes can often be identified by the cumulus clouds that develop on adjacent mountain peaks.

After sunset, the pattern may reverse. Rapid radiation cooling along the mountain slopes produces a layer of cooler air next to the ground. Because cool air is denser than warm air, it drains downslope into the valley. Such a movement of air is called a **mountain breeze** (Figure 16.16, right). The same type of cool air drainage can occur in places that have very modest slopes. The result is that the coldest pockets of air are usually found in the lowest spots. Like many other winds, mountain and valley breezes have seasonal preferences. Although valley breezes are most common during the warm season when solar heating is most intense, mountain breezes tend to be more dominant in the cold season.

Chinook and Santa Ana Winds

Warm, dry winds are common on the eastern slopes of the Rockies, where they are called **chinooks**. Such winds are created when air descends the leeward (sheltered) side of a mountain and warms by compression. Because condensation may have occurred as the air ascended the windward side, releasing latent heat, the air descending the leeward slope will be warmer and drier than it was at a similar elevation on the windward side. Although the temperature of these winds is generally less than 10°C (50°F), which is not particularly warm, they occur mostly in the winter and spring when the affected areas may be experiencing below-freezing temperatures. Thus, by comparison, these dry, warm winds often bring a drastic change. When the ground has a snow cover, these winds are known to melt it in short order.

A chinooklike wind that occurs in southern California is the **Santa Ana**. These hot, desiccating winds greatly increase the threat of fire in this already dry area.

How Wind Is Measured

Two basic wind measurements, direction and speed, are particularly significant to the weather observer. *Winds are always labeled by the direction from which they blow.* A north wind blows *from* the north *toward* the south, an east wind *from* the east *toward* the west. The instrument most commonly used to determine wind direction is the **wind vane** (Figure 16.17, left). This instrument, a common sight on many buildings, always points *into* the wind. Often the wind direction is shown on a dial that is connected to the wind vane (Figure 16.18). The dial indicates wind direction, either by points of the compass (N, NE, E, SE, etc.) or by a scale of 0° to 360°. On the latter scale, 0 degrees or 360 degrees are both north, 90 degrees is east, 180 degrees

Figure 16.16 Valley and mountain breezes. Heating during the daylight hours warms the air along the mountain slopes. This warm air rises, generating a valley breeze. After sunset, cooling of the air near the mountain can result in cool air drainage into the valley, producing the mountain breeze.

is south, and 270 degrees is west. When the wind consistently blows more often from one direction than from any other, it is termed a **prevailing wind**.

Wind speed is commonly measured using a **cup anemometer** (Figure 16.17, right). The wind speed is read from a dial much like the speedometer of an automobile. Places where winds are steady and speeds are relatively high are potential sites for tapping wind energy.

By knowing the locations of cyclones and anticyclones in relation to where you are, you can predict the changes in wind direction that will be experienced as a pressure center moves past. Because changes in wind direction often bring changes in temperature and moisture conditions, the ability to predict the winds can be very useful. In the Midwest, for example, a north wind may bring cool, dry air from Canada, whereas a south wind may bring warm, humid air from the Gulf of Mexico. Sir Francis Bacon summed it up nicely when he wrote, "Every wind has its weather."

Figure 16.17 Wind vane and cup anemometer. The wind vane shows wind direction and the anemometer measures wind speed. (Photo by Scott T. Smith)

Figure 16.18 Wind direction. Wind direction can be expressed using the points of the compass or a scale of 0–360 degrees. Winds are always labeled according to the direction *from* which they are blowing.

Global Distribution of Precipitation

A casual glance at Figure 16.19 shows a relatively complex pattern for the distribution of precipitation. Although the map appears to be complicated, the general features of the map can be explained by applying our knowledge of global winds and pressure systems.

In general, regions influenced by high pressure, with its associated subsidence and diverging winds, experience relatively dry conditions. On the other hand, regions under the influence of low pressure and its converging winds and ascending air receive ample precipitation. This pattern is illustrated by noting that the tropical region dominated by the equatorial low is the rainiest region on Earth. It includes the rain forests of the Amazon basin in South America and the Congo basin in Africa. Here the warm, humid trade winds converge to yield abundant rainfall throughout the year. By way of contrast, areas dominated by the subtropical high pressure cells clearly receive much smaller amounts of precipitation. These are regions of extensive deserts. In the Northern Hemisphere the largest is the Sahara. Examples in the Southern Hemisphere include the Kalahari in southern Africa and the dry lands of Australia.

If Earth's pressure and wind belts were the only factors controlling precipitation distribution, the pattern shown in Figure 16.19 would be simpler. The inherent nature of the air is also an important factor in determining precipitation potential. Because cold air has a low capacity for moisture compared with warm air, we would expect a latitudinal variation in precipitation, with low latitudes receiving the greatest amounts of precipitation and high latitudes receiving the smallest amounts. Figure 16.19 indeed reveals heavy rainfall in equatorial regions and meager precipitation in high-latitude areas. Recall that the dry region in the warm subtropics is explained by the presence of the subtropical high.

In addition to latitudinal variations in precipitation, the distribution of land and water complicates the precipitation pattern. Large landmasses in the middle latitudes commonly experience decreased precipitation toward their interiors. For example, central North America and central Eurasia receive considerably less precipitation than do coastal regions at the same latitude. Furthermore, the effects of mountain barriers alter the idealized precipitation patterns we would expect solely from global wind and pressure systems. Windward mountain slopes receive abundant rainfall resulting from orographic lifting, whereas leeward slopes and adjacent lowlands are usually deficient in moisture (see Figure 15.16, p. 421).

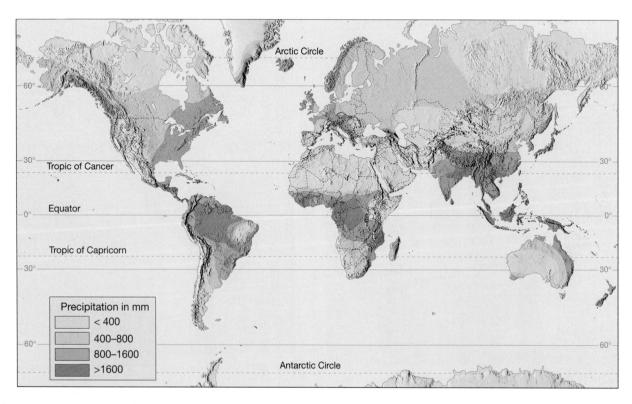

Figure 16.19 Average annual precipitation in millimeters.

Chapter Summary

- *Air has weight*: at sea level it exerts a pressure of 1 kilogram per square centimeter (14.7 pounds per square inch). *Air pressure* is the force exerted by the weight of air above. With increasing altitude there is less air above to exert a force, and thus air pressure decreases with altitude, rapidly at first, then much more slowly. The unit used by meteorologists to measure atmospheric pressure is the *millibar*. *Standard sea level pressure* is expressed as 1013.2 millibars. *Isobars* are lines on a weather map that connect places of equal air pressure.

- A *mercury barometer* measures air pressure using a column of mercury in a glass tube that is sealed at one end and inverted in a dish of mercury. As air pressure increases, the mercury in the tube rises; conversely, when air pressure decreases, so does the height of the column of mercury. A mercury barometer measures atmospheric pressure in *"inches of mercury,"* the height of the column of mercury in the barometer. Standard atmospheric pressure at sea level equals 29.92 inches of mercury. *Aneroid* ("without liquid") *barometers* consist of partially evacuated metal chambers that compress as air pressure increases and expand as pressure decreases.

- *Wind* is the horizontal flow of air from areas of higher pressure toward areas of lower pressure. Winds are controlled by the following combination of forces: (1) the *pressure-gradient force* (amount of pressure change over a given distance); (2) *Coriolis effect* (deflective effect of Earth's rotation—to the right in the Northern Hemisphere and to the left in the Southern Hemisphere); and (3) *friction* with Earth's surface (slows the movement of air and alters wind direction).

- Upper-air winds, called *geostrophic winds*, blow parallel to the isobars and reflect a balance between the pressure-gradient force and the Coriolis effect. Upper-air winds are faster than surface winds because friction is greatly reduced aloft. Friction slows surface winds, which in turn reduces the Coriolis effect. The result is air movement at an angle across the isobars toward the area of lower pressure.

- The two types of pressure centers are (1) *cyclones*, or *lows* (centers of low pressure), and (2) *anticyclones*, or *highs* (high-pressure centers). In the Northern Hemisphere, winds around a low (cyclone) are counterclockwise and inward. Around a high (anticyclone), winds are clockwise and outward. In the Southern Hemisphere, the Coriolis effect causes winds to be clockwise around a low and counterclockwise around a high. Because air rises and cools adiabatically in low-pressure centers, cloudy conditions and precipitation are often associated with their passage. In high-pressure centers, descending air is compressed and warmed; therefore, cloud formation and precipitation are unlikely, and "fair" weather is usually expected.

- Earth's *global pressure zones* include the *equatorial low*, *subtropical high*, *subpolar low*, and *polar high*. The *global surface winds* associated with these pressure zones are the *trade winds*, *westerlies*, and *polar easterlies*.

- Particularly in the Northern Hemisphere, large seasonal temperature differences over continents disrupt the idealized, or zonal, global patterns of pressure and wind. In winter, large, cold landmasses develop a seasonal high-pressure system from which surface air flow is directed off the land. In summer, landmasses are heated and low pressure develops over them, which permits air to flow onto the land. The seasonal changes in wind direction are known as *monsoons*.

- In the middle latitudes, between 30 and 60 degrees latitude, the general west-to-east flow of the westerlies is interrupted by migrating cyclones and anticyclones. The paths taken by these pressure systems are closely related to upper-level air flow and the polar *jet stream*. The average position of the polar jet stream, and hence the paths followed by cyclones, migrates equatorward with the approach of winter and poleward as summer nears.

- *El Niño* is the name given to the periodic warming of the ocean that occurs in the central and eastern Pacific. It is associated with periods when a weakened pressure gradient causes the trade winds to diminish. A major El Niño event triggers extreme weather in many parts of the world.

- *Local winds* are small-scale winds produced by a locally generated pressure gradient. Local winds include *sea* and *land breezes* (formed along a coast because of daily pressure differences caused by the differential heating of land and water); *valley* and *mountain breezes* (daily wind similar to sea and land breezes except in a mountainous area where the air along slopes heats differently from the air at the same elevation over the valley floor); and *chinook* and *Santa Ana winds* (warm, dry winds created when air descends the leeward side of a mountain and warms by compression).

- The two basic wind measurements are *direction* and *speed*. Winds are always labeled by the direction *from* which they blow. Wind direction is measured with a *wind vane*, and wind speed is measured using a *cup anemometer*.

- The global distribution of precipitation is strongly influenced by the global pattern of air pressure and wind, latitude, and distribution of land and water.

Key Terms

air pressure (p. 442)
aneroid barometer (p. 443)
anticyclone (p. 447)
barograph (p. 443)
barometric tendency (p. 449)
chinook (p. 456)
convergence (p. 448)
Coriolis effect (p. 444)
cup anemometer (p. 457)
cyclone (p. 447)
divergence (p. 448)
equatorial low (p. 450)
geostrophic wind (p. 445)

high (p. 447)
isobar (p. 443)
jet stream (p. 445)
land breeze (p. 456)
low (p. 447)
mercury barometer (p. 442)
monsoon (p. 451)
mountain breeze (p. 456)
polar easterlies (p. 450)
polar front (p. 450)
polar high (p. 450)
pressure gradient (p. 443)
pressure tendency (p. 449)

prevailing wind (p. 457)
Santa Ana (p. 456)
sea breeze (p. 455)
Southern Oscillation (p. 454)
subpolar low (p. 450)
subtropical high (p. 450)
trade winds (p. 450)
valley breeze (p. 456)
westerlies (p. 450)
wind (p. 443)
wind vane (p. 456)

Review Questions

1. What is standard sea-level pressure in millibars? In inches of mercury? In pounds per square inch?

2. Mercury is 13 times heavier than water. If you built a barometer using water rather than mercury, how tall would it have to be to record standard sea-level pressure (in centimeters of water)?

3. Describe the principle of the aneroid barometer.

4. What force is responsible for generating wind?

5. Write a generalization relating the spacing of isobars to the speed of wind.

6. How does the Coriolis effect modify air movement?

7. Contrast surface winds and upper-air winds in terms of speed and direction.

8. Describe the weather that usually accompanies a drop in barometric pressure and a rise in barometric pressure.

9. Sketch a diagram (isobars and wind arrows) showing the winds associated with surface cyclones and anticyclones in both the Northern and Southern hemispheres.

10. If you live in the Northern Hemisphere and are directly west of the center of a cyclone, what most probably will be the wind direction? What will the wind direction be if you are west of an anticyclone?

11. The following questions relate to the global pattern of air pressure and winds.

(a) The trade winds diverge from which pressure zone?

(b) Which prevailing wind belts converge in the stormy region known as the polar front?

(c) Which pressure belt is associated with the equator?

12. What influence does upper-level airflow seem to have on surface pressure systems?

13. Describe the monsoon circulation of India.

14. Describe the relationship between the Southern Oscillation and a major El Niño event (see Figure 16.14).

15. What is a local wind? List three examples.

16. A northeast wind is blowing *from* the _____ (direction) *toward* the _____ (direction).

17. The wind direction is 225 degrees. From what compass direction is the wind blowing (Figure 16.18)?

18. Which global pressure system is responsible for deserts such as the Sahara and Kalahari in Africa? With which global pressure belt are the tropical rain forests of the Amazon and Congo basins associated?

19. Other than Earth's pressure and wind belts, list two other factors that exert a significant influence on the global distribution of precipitation.

Examining the Earth System

1. Examine the image of Africa in Figure I.3, p. 3, and pick out the region dominated by the equatorial low and the areas influenced by the subtropical highs in each hemisphere. What clue(s) did you use? Speculate on the differences in the biosphere between the regions dominated by high pressure and the zone influenced by low pressure.

2. How are global winds related to surface ocean currents? (Try comparing Figure 16.13 with Figure 13.2, p. 349) What is the ultimate source of energy that drives both of these circulations?

3. Winds and ocean currents change in the tropical Pacific during an El Niño event. How might this impact the biosphere and solid Earth in Peru and Ecuador? How about in Indonesia? (Two excellent Web Sites that deal with El Niño are the University of Southern California site at **http://www.usc.edu/org/seagrant/elnino/link.html** and NOAA's *El Niño Theme Page* at **http://www.pmel. noaa.gov/toga-tao/el-nino/nino-home.html**)

Web Resources

The following are informative and interesting World Wide Web sites that address topics related to those presented in the chapter:

American Wind Energy Association:

- **http://www.igc. apc.ord/awea/**

Forces and Winds Online Tutorial (University of Illinois):

- **http://www.2010.atmos.uiuc.edu/(Gh)/guides /mtr/fw/home.rxml**

For direct links to these sites and other appropriate Earth science Web locations, chapter review exercises, and relevant, topical investigations that utilize Web resources, contact the *Earth Science*, 9th Edition Home Page, at:

- **http://www.prenhall.com/tarbuck.**

Chapter 17

Weather Patterns and Severe Storms

Satellite image of a storm over the British Isles. Satellites are invaluable tools for tracking storms and gathering atmospheric data. (Courtesy of European Space Agency/Science Photo Library/Photo Researchers, Inc.)

Tornadoes and hurricanes rank among nature's most destructive forces. Each spring, newspapers report the death and destruction left in the wake of a "band" of tornadoes. During late summer and fall we hear occasional reports about hurricanes in the news. Storms with names like Andrew, Fran, Georges, and Mitch have made front-page headlines. Thunderstorms, although less intense and far more common than tornadoes and hurricanes, will also be part of our discussion on severe weather in this chapter. Before looking at violent weather, however, we will study those atmospheric phenomena that most often affect our day-to-day weather: air masses, fronts, and traveling middle-latitude cyclones. Here we will see the interplay of the elements of weather discussed in Chapters 14, 15, and 16.

Air Masses

Any person living in the middle latitudes has experienced hot, "sticky" summer heat waves and frigid winter cold waves. In the first case, after several days of high temperatures and oppressive humidities the spell may come to an abrupt end that is marked by thundershowers and followed by a few days of relatively cool relief. In the second case, the clear skies that often accompany a span of frigid subzero days may be replaced by thick gray clouds and a period of snow as temperatures rise to levels that seem mild when compared to those that existed just a day earlier. In both examples, what was experienced was a period of generally uniform weather conditions followed by a relatively short period of change, and then the subsequent reestablishment of a new set of weather conditions that remained for perhaps several days before changing again.

The weather patterns just described result from movements of large bodies of air, called air masses. An **air mass,** as the term implies, is an immense body of air, usually 1600 kilometers (1000 miles) or more across, and perhaps several kilometers thick, that is characterized by a similarity of temperature and moisture at any given altitude. When this air moves out of its region of origin, it will carry these temperatures and moisture conditions with it, eventually affecting a large portion of a continent (Figure 17.1).

An excellent example of the influence of an air mass is illustrated in Figure 17.2. Here a cold, dry mass from northern Canada moves southward. With a beginning temperature of –46°C, the air mass warms 13 degrees, to –33°C, by the time it reaches Winnipeg. It continues to warm as it moves southward through the Great Plains and into Mexico. Throughout its southward journey, the air mass becomes warmer. But it also brings some of the coldest weather of the winter to the places in its path. Thus, the air mass is modified, but it also modifies the weather in the areas over which it moves.

The horizontal uniformity of an air mass is not complete because it may extend through 20 degrees or more of latitude and cover hundreds of thousands to millions of square kilometers. Consequently, small differences in temperature and humidity from one point to another at the same level are to be expected. Still the

Figure 17.1 When an air mass that forms over a desert in summer moves out from its region of origin, it will bring hot, dry conditions to other regions. (Photo by Warren Faidley/Weatherstock)

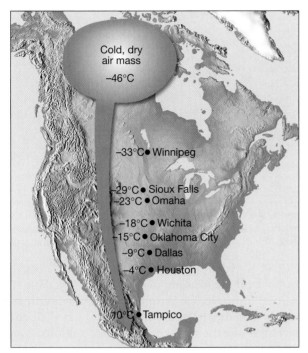

Figure 17.2 As this frigid Canadian air mass moved southward, it brought some of the coldest weather of the winter. (After Tom L. McKnight, *Physical Geography*, 5th ed. Upper Saddle River, NJ: Prentice Hall, 1996, p. 174)

differences observed within an air mass are small in comparison to the rapid changes experienced across air mass boundaries.

Because it may take several days for an air mass to traverse an area, the region under its influence will probably experience fairly constant weather, a situation called **air-mass weather.** Certainly some day-to-day variations may exist, but the events will be very unlike those in an adjacent air mass. For this reason the boundary between two adjoining air masses that have contrasting characteristics, called a *front,* marks a change in weather.

Source Regions

When a portion of the lower atmosphere moves slowly or stagnates over a relatively uniform surface, the air will assume the distinguishing features of that area, particularly with regard to temperature and moisture conditions.

The area where an air mass acquires its characteristic properties of temperature and moisture is called its **source region.** The source regions that produce air masses influencing North America are shown in Figure 17.3.

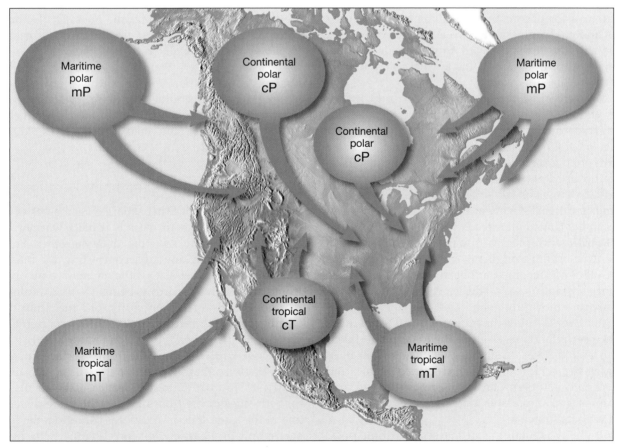

Figure 17.3 Air masses are classified on the basis of their source region. The designation continental (c) or maritime (m) gives an indication of moisture content, whereas polar (P) and tropical (T) indicate temperature conditions.

Air masses are classified according to their source region. **Polar (P) air masses** originate in high latitudes toward Earth's poles, whereas those that form in low latitudes are called **tropical (T) air masses.** The designation *polar* or *tropical* gives an indication of the temperature characteristics of the air masses. *Polar* indicates cold and *tropical* indicates warm.

In addition, air masses are classified according to the nature of the surface in the source region. **Continental (c) air masses** form over land and **maritime (m) air masses** originate over water. The designation *continental* or *maritime* thus suggests the moisture characteristics of the air mass. Continental air is likely to be dry, and maritime air, humid.

The four basic types of air masses according to this scheme of classification are continental polar (cP), continental tropical (cT), maritime polar (mP), and maritime tropical (mT).

Weather Associated With Air Masses

Continental polar and maritime tropical air masses influence the weather of North America most, especially east of the Rocky Mountains. Continental polar air masses originate in northern Canada, interior Alaska, and the Arctic—areas that are uniformly cold and dry in winter and cool and dry in summer. In winter, an invasion of continental polar air brings the clear skies and cold temperatures we associate with a cold wave as it moves southward from Canada into the United States. In summer, this air mass may bring a few days of cooling relief.

Although cP air masses are not, as a rule, associated with heavy precipitation, those that cross the Great Lakes during late autumn and winter sometimes bring snow to the leeward shores. These localized storms often form when the surface weather map indicates no apparent cause for a snowstorm to occur. These are known as **lake-effect snows** and they make Buffalo and Rochester, New York, among the snowiest cities in the United States (see Box 17.1).

Maritime tropical air masses affecting North America most often originate over the warm waters of the Gulf of Mexico, the Caribbean Sea, or the adjacent Atlantic Ocean. As we might expect, these air masses are warm, moisture-laden, and usually unstable. Maritime tropical air is the source of much, if not most, of the precipitation received in the eastern two-thirds of the United States. In summer, when an mT air mass invades the central and eastern United States, and occasionally southern Canada, it brings the high temperatures and oppressive humidity typically associated with its source region.

Of the two remaining air masses, maritime polar and continental tropical, the latter has the least

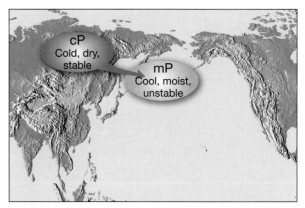

Figure 17.4 During winter, maritime polar (mP) air masses in the North Pacific usually begin as continental polar (cP) air masses in Siberia. The cP air is modified to mP as it slowly crosses the ocean.

influence on the weather of North America. Hot, dry continental tropical air, originating in the Southwest and Mexico during the summer, only occasionally affects the weather outside its source region.

During the winter, maritime polar air masses coming from the North Pacific often originate as continental polar air masses in Siberia. As these air masses move across the Pacific, they become unstable as they warm and accumulate moisture (Figure 17.4). Upon entering North America, mP air masses drop much of their moisture because of orographic lifting in the western mountains. Maritime polar air also originates in the North Atlantic off the coast of eastern Canada and occasionally influences the weather of the northeastern United States. In winter, when New England is on the northern or northwestern side of a passing low, the counterclockwise cyclonic winds draw in maritime polar air. The result is a storm characterized by snow and cold temperatures, known locally as a *nor'easter.*

Fronts

Fronts are boundaries that separate air masses of different densities. One air mass is usually warmer and often contains more moisture than the other. Fronts can form between any two contrasting air masses. Considering the vast size of the air masses involved, fronts are relatively narrow, being 15- to 200-kilometer-wide bands of discontinuity. On the scale of a weather map, they are generally narrow enough to be represented by a broad line (as in Figure 17.10).

Above Earth's surface, the frontal surface slopes at a low angle so that warmer air overlies cooler air, as shown in Figure 17.5. In the ideal case, the air masses on both sides of the front move in the same direction and at the same speed. Under this condition, the front acts simply as a barrier, that travels along with the air masses, and that neither mass can penetrate.

Box 17.1 Earth as a System: Lake-Effect Snow

Every winter a unique and interesting weather phenomenon takes place along the downwind shores of the Great Lakes. Periodically, brief, heavy snow showers issue from dark clouds that move onshore from the lakes. Seldom do the storms move more than about 40 kilometers inland from the shore before the snows come to an end. These highly localized storms occurring along the leeward (downwind) shores of the Great Lakes create what are known as lake-effect snows.

Lake-effect storms account for a high percentage of the snowfall in many areas adjacent to the lakes. The strips of land that are most frequently affected, called *snowbelts*, are shown in Figure 17.A. When snowfall statistics for large cities were examined for a recent 10-year span, the two snowiest metropolitan areas in the United States turned out to be Buffalo and Rochester in upstate New York. Buffalo, on the eastern shore of Lake Erie, had more than 2700 centimeters (nearly 90 feet) during the period, while Rochester, situated on the south shore of Lake Ontario, recorded slightly less (2635 centimeters). A comparison of average snowfalls totals at Thunder Bay in Ontario, Canada, on the north shore of Lake Superior, and Marquette, Michigan, along the southern shore, provides another excellent example of lake-effect snow. Amounts are in centimeters.

Because Marquette is situated on the leeward shore of the lake, it

Figure 17.A The snowbelts of the Great Lakes region are the zones that most frequently experience lake-effect snowstorms. As continental polar air crosses the lakes in winter, it acquires moisture and is made unstable because of warming from below. Snow showers on the lee side of the lakes are often the consequence of this air-mass modification.

receives substantial lake-effect snow and therefore has a much higher snowfall total than does Thunder Bay.

What causes lake-effect snow? The answer is closely linked to the differential heating of water and land (Chapter 14) and to the concept of atmospheric instability (Chapter 15). During the summer months water bodies, including the Great Lakes, absorb huge quantities of energy from the Sun and from the warm air that passes over them. Although these water bodies do not reach particularly high temperatures, they nevertheless represent huge reservoirs of heat. In

contrast, the surrounding land cannot store heat nearly as effectively.

Consequently, during autumn and winter, the temperature of the land drops quickly, whereas water bodies lose their heat more gradually and cool slowly. From late November through late January, the contrasts in average temperatures between water and land range from about 8°C in the southern Great Lakes to 17°C farther north. However, the temperature differences can be much greater (perhaps 25°C) when a very cold cP air mass pushes southward across the lakes. When such a dramatic temperature contrast exists, the lakes interact with the air to produce major lake-effect storms. As a cP air mass moves across one of the Great Lakes, the air acquires large quantities of heat and moisture from the relatively warm lake surface. By the time it reaches the opposite shore, the air mass is humid and unstable and heavy snow showers are likely.

	Oct.	Nov.	Dec.	Jan.
Thunder Bay, Ontario	3.0	14.9	19.0	22.6
Marquette, Michigan	5.3	37.6	56.4	53.1

Generally, however, the distribution of pressure across a front is such that one air mass moves faster than the other. Thus, one air mass actively advances into another and "clashes" with it. In

fact, the boundaries were tagged *fronts* during World War I by Norwegian meteorologists who visualized them as analogous to battle lines between two armies.

Figure 17.5 Warm front produced as warm air glides up over a cold air mass. Precipitation is moderate and occurs within a few hundred kilometers of the surface front.

As one air mass moves into another, some mixing does occur along the frontal surface, but for the most part the air masses retain their distinct identities as one is displaced upward over the other. No matter which air mass is advancing, *it is always the warmer, less dense air that is forced aloft, whereas the cooler, denser air acts as the wedge upon which lifting takes place.*

Warm Fronts

When the surface (ground) position of a front moves so that warm air occupies territory formerly covered by cooler air, it is called a **warm front** (Figure 17.5). On a weather map, the surface position of a warm front is shown by a line with semicircles extending into the cooler air.

East of the Rockies, warm tropical air often enters the United States from the Gulf of Mexico and over-runs receding cool air. As the cold air retreats, friction with the ground greatly slows the advance of the surface position of the front compared to its position aloft. Consequently, the boundary separating these air masses acquires a very gradual slope. The average slope of a warm front is about 1:200, which means that if you are 200 kilometers ahead of the surface location of a warm front, you will find the frontal surface at a height of 1 kilometer.

As warm air ascends the retreating wedge of cold air, it cools adiabatically to produce clouds, and frequently, precipitation. The sequence of clouds shown in Figure 17.5 typically precedes a warm front. The first sign of the approaching warm front is the appearance of cirrus clouds (Figure 17.6). These high clouds form where the overrunning warm air has ascended high up the wedge of cold air, 1000 kilometers or more ahead of the surface front.

Figure 17.6 Cirrus clouds are high, thin clouds that are often the first sign that a warm front is approaching. As warm air is wedged aloft, clouds form at increasingly greater heights. (Photo by E. J. Tarbuck)

As the front nears, cirrus clouds grade into cirrostratus, which blend into denser sheets of altostratus. About 300 kilometers ahead of the front, thicker stratus and nimbostratus clouds appear and rain or snow begins.

Because of their slow rate of advance and very low slope, warm fronts usually produce light-to-moderate precipitation over a large area for an extended period. Warm fronts, however, are occasionally associated with cumulonimbus clouds and thunderstorms. This occurs when the overrunning air is unstable and the temperatures on opposite sides of the front contrast sharply. At the other extreme, a warm front associated with a dry air mass could pass unnoticed at the surface.

A gradual increase in temperature occurs with the passage of a warm front. As you would expect, the increase is most apparent when a large contrast exists between adjacent air masses. Moreover, a wind shift from the east to the southwest is generally noticeable. (The reason for this shift will become evident later.) The moisture content and stability of the encroaching warm air mass largely determine when clear skies will return. During the summer, cumulus, and occasionally cumulonimbus, clouds are embedded in the warm unstable air mass that follows the front. These clouds can produce precipitation, but it is usually restricted in extent and of short duration.

Cold Fronts

When cold air actively advances into a region occupied by warmer air, the boundary is called a **cold front** (Figure 17.7). As with warm fronts, friction tends to slow the surface position of a cold front more so than its position aloft. However, because of the relative positions of the adjacent air masses, the cold front steepens as it moves. On the average, cold fronts are about twice as steep as warm fronts, having a slope of perhaps 1:100. In addition, cold fronts advance more rapidly than do warm fronts. These two differences—rate of movement and steepness of slope—largely account for the more violent nature of cold-front weather compared to the weather generally accompanying a warm front.

The forceful lifting of air along a cold front is often so rapid that the latent heat released when water vapor condenses will increase the air's buoyancy appreciably. The heavy downpours and vigorous wind gusts associated with mature cumulonimbus clouds frequently result. Because a cold front produces roughly the same amount of lifting as a warm front, but over a shorter distance, the precipitation intensity is greater, but of shorter duration.

As a cold front approaches, generally from the west or northwest, towering clouds often can be seen in the distance. Near the front, a dark band of ominous clouds foretells the ensuing weather. A marked temperature drop and a wind shift usually accompany the passage of the front. (Again, the reason for this shift, from the south or southwest to the west or northwest, will be explained later.)

The sometimes violent weather and sharp temperature contrast along the cold front are symbolized on a weather map by a line with triangle-shaped points that extend into the warmer air mass (Figure 17.7).

The weather behind a cold front is dominated by a subsiding and relatively cold air mass. Thus, clearing begins relatively soon after the front passes. Although the compression of air by subsidence causes some adiabatic heating, the effect on surface temperatures is minor. In winter, the clear skies that follow the passing of a cold front further reduce surface temperatures because of the high rate of radiation cooling that occurs during the

Figure 17.7 Fast-moving cold front and cumulonimbus clouds. Often thunderstorms occur if the warm air is unstable.

night. If the continental polar air mass, which most frequently accompanies a cold front, moves into a relatively warm and humid area, surface heating of the air can produce shallow convection. This, in turn, may generate low cumulus or stratocumulus clouds behind the front.

Stationary Fronts and Occluded Fronts

Occasionally the flow on both sides of a front is neither toward the cold air mass nor toward the warm air mass, but almost parallel to the line of the front. Thus, the surface position of the front does not move. This condition is called a **stationary front.** On a weather map, stationary fronts are shown with triangular points on one side of the front and semicircles on the other. At times some overrunning occurs along a stationary front. On these occasions, gentle to moderate precipitation is most likely.

The fourth type of front is the **occluded front.** Here an active cold front overtakes a warm front, as shown in Figure 17.8. As the advancing cold air wedges the warm front upward, a new front emerges between the advancing cold air and the air over which the warm front is gliding. The weather of an occluded front is generally complex. Most precipitation is associated with the warm air being forced aloft. When conditions are suitable, however, the newly formed front is capable of initiating precipitation of its own.

A word of caution is in order concerning the weather associated with various fronts. Although the preceding discussion will help you recognize the weather patterns associated with fronts, remember that these descriptions are generalizations. The weather generated along any individual front may or may not conform fully to this idealized picture. Fronts, like all aspects of nature, never lend themselves to classification as nicely as we would like.

The Middle-Latitude Cyclone

So far we have examined the basic elements of weather as well as the dynamics of atmospheric motions. We are now ready to apply our knowledge of these diverse phenomena to an understanding of day-to-day weather patterns in the middle latitudes. For our purposes, *middle latitudes* refers to the region between southern Florida and Alaska.

The primary weather producers here are **middle-latitude cyclones.** They are large centers of low pressure that generally travel from west to east. Lasting from a few days to more than a week, these weather systems have a counterclockwise circulation with an air flow inward toward their centers. Most middle-latitude cyclones also have a cold front extending from the central area of low pressure, and frequently a warm front as well. Convergence and forceful lifting initiate cloud development and often cause abundant precipitation.

As early as the 1800s it was known that cyclones were the bearers of precipitation and severe weather. But it was not until the early part of the twentieth century that a model was developed to explain how cyclones form. It was formulated by a group of Norwegian scientists and published in 1918. The model was created primarily from near-surface observations.

Years later, as data from the middle and upper troposphere and from satellite images became available, modifications were necessary. Yet this model is still an accepted working tool in interpreting the weather. If you keep this model in mind when you observe changes in the weather, the changes will no longer come as a surprise. Furthermore, you will begin to see some order in what had appeared to be disorder, and you might even correctly forecast the impending weather.

Life Cycle

Middle-latitude cyclones form along fronts where they change in a somewhat predictable way. Figure 17.9 is a schematic representation of the stages in the development of a "typical" middle-latitude cyclone. As the figure shows, cyclones originate along a front where

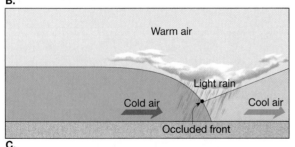

Figure 17.8 Stages in the formation of an occluded front.

A. Front develops   B. Wave develops   C. Cyclonic circulation established

D. Occlusion begins   E. Occluded front developed   F. Cyclone dissipates

Figure 17.9 Stages in the life cycle of a hypothetical middle-latitude cyclone.

air masses of different densities (temperatures) are moving parallel to the front in opposite directions.

In the classic model, continental polar air associated with the polar easterlies is north of the front, and maritime tropical air of the westerlies is south of the front. The result of this opposing airflow is counterclockwise (cyclonic) rotation. (To better visualize this effect, place a pencil between the palms of your hands. Now move your right hand ahead of your left hand and notice that your pencil rotates in a counterclockwise fashion.) Under the correct conditions, the frontal surface will take on a wave shape. The waves are usually several hundred kilometers long. Some dampen out, whereas others become unstable and grow in amplitude. The latter ones change in shape with time much as a gentle ocean swell does as it moves into shallow water and becomes a tall breaking wave.

Once a small wave forms, warm air moves poleward, while the surrounding cold air moves equatorward. This change causes a readjustment in air pressure that results in nearly circular isobars, with the low pressure centered at the apex of the wave. The creation of the low encourages the inflow (convergence) of air and general vertical lifting, particularly where warm air is overrunning colder air.

You can see in Figure 17.10 that the air in the warm sector is flowing from the southwest toward colder air flowing from the southeast. Because the warm air is moving faster than the cold air in a direction perpendicular to this front, we can conclude that warm air is invading a region formerly occupied by cold air;

hence, this must be a warm front. Similar reasoning indicates that in the rear of the cyclonic disturbance, cold air is underrunning the air of the warm section, generating a cold front.

Generally, the position of the cold front advances faster than the warm front and begins to close the warm sector, as shown in Figure 17.9. This process, called occlusion, results in an occluded front with the displaced warm sector located aloft. The cyclone enters maturity (maximum intensity) when it reaches this stage in its development. A steep pressure gradient and strong winds develop as lifting continues. Eventually all of the warm sector is forced aloft and cold air surrounds the cyclone at low levels (Figure 17.9F). Once the sloping discontinuity (front) between the air masses no longer exists, the pressure gradient weakens. At this point, the cyclone has exhausted its source of energy, and the storm comes to an end.

Idealized Weather

The middle-latitude cyclone model provides a useful tool for examining the weather patterns of the middle latitudes. Figure 17.10 illustrates the distribution of clouds and thus the regions of possible precipitation associated with a mature system. Compare this drawing to the satellite image of a cyclone shown in Figure 17.11.

Guided by the westerlies aloft, cyclones generally move eastward across the United States. Therefore, we can expect the first signs of a cyclone's arrival to appear in the western sky. In the region of

Figure 17.10 Cloud patterns typically associated with a mature middle-latitude cyclone. The middle section is a map view. Note the cross-section lines (F–G, A–E). Above the map is a vertical cross section along line F–G. Below the map is a section along A–E. For cloud abbreviations, refer to Figures 17.5 and 17.7.

the Mississippi valley, however, cyclones often begin a more northeasterly path and occasionally move directly northward. Typically, a mid-latitude cyclone requires 2 to 4 days to pass over a region. During that brief period, abrupt changes in atmospheric conditions may occur, particularly in the spring of the year when the greatest temperature contrasts occur across the mid-latitudes.

Using Figure 17.10 as a guide, let us examine these weather producers and the changes we can expect as

Figure 17.11 Satellite view of a mature cyclone over the eastern United States. It is easy to see why we often refer to the cloud pattern of a cyclone as having a "comma" shape. (Courtesy of John Jensenius/National Weather Service)

they pass in the spring of the year. To facilitate our discussion, Figure 17.10 includes two profiles along lines *A–E* and *F–G*.

- Imagine the change in weather as you move from right to left along profile *A–E* (bottom of figure). At point *A*, the sighting of high cirrus clouds is the first sign of the approaching cyclone. These high clouds can precede the surface front by 1000 kilometers or more and they are normally accompanied by falling pressure. As the warm front advances, a lowering and thickening of the cloud deck is noticed.

- Within 12 to 24 hours after the first sighting of cirrus clouds, light precipitation begins (point *B*). As the front nears, the rate of precipitation increases, a rise in temperature is noticed, and winds begin to change from east or southeast to south or southwest.

- With the passage of the warm front, an area is under the influence of a maritime tropical air mass (point *C*). Usually the region affected by this sector of the cyclone experiences warm temperatures, south or southwest winds, and generally clear skies, although fair-weather cumulus or altocumulus clouds are not uncommon here.

- The relatively warm, humid weather of the warm sector passes quickly in the spring of the year and is replaced by gusty winds and precipitation generated along the cold front. The approach of a rapidly advancing cold front is marked by a wall of dark clouds (point D). Severe weather accompanied by heavy precipitation, hail, and an occasional tornado is a definite possibility at this time of year. The passage of the cold front is easily detected by a wind shift; the southwest winds are replaced by winds from the west to northwest and by a pronounced drop in temperature. Also, rising pressure hints of the subsiding cool, dry air behind the front.

- Once the front passes, skies clear as cooler air invades the region (point *E*). Often a day or two of almost cloudless deep-blue skies occurs unless another cyclone is edging into the region.

A very different set of weather conditions will prevail in those regions north of the storm's center along profile *F–G* of Figure 17.10. Often the storm reaches its greatest intensity in this zone and the area along profile *F–G* receives the brunt of the storm's fury. Here temperatures remain cold during the passage of the system, and heavy snow, sleet, and/or freezing rain may develop during the winter months.

The Role of Airflow Aloft

When the earliest studies of cyclones were made, little was known about the nature of the airflow in the middle and upper troposphere. Since then, a close relationship has been established between surface disturbances and the flow aloft. Airflow aloft plays an important role in maintaining cyclonic and anticyclonic circulation. In fact, more often than not, these rotating surface wind systems are actually generated by upper-level flow.

Recall that the airflow around a surface low is inward, a fact that leads to mass convergence, or coming

Figure 17.12 Idealized diagram depicting the support that divergence and convergence aloft provide to cyclonic and anticyclonic circulation at the surface. Divergence aloft initiates upward air movement, reduced surface pressure, and cyclonic flow. In contrast, convergence along the jet stream results in general subsidence of the air column, increased surface pressure, and anticyclonic surface winds.

together (Figure 17.12). The resulting accumulation of air must be accompanied by a corresponding increase in surface pressure. Consequently, we might expect a low-pressure system to "fill" rapidly and be eliminated, just as the vacuum in a coffee can is quickly dissipated when we open it. However, this does not occur.

On the contrary, cyclones often exist for a week or longer. For this to happen, surface convergence must be offset by a mass outflow at some level aloft (Figure 17.12). As long as divergence (spreading out) aloft is equal to, or greater than, the surface inflow, the low pressure and its accompanying convergence can be sustained.

Because cyclones are bearers of stormy weather, they have received far more attention than anticyclones. Nevertheless, a close relation exists, which makes it difficult to separate any discussion of these two types of pressure systems. The surface air that feeds a cyclone, for example, generally originates as air flowing out of an anticyclone. Consequently, cyclones and anticyclones typically are found adjacent to one another. Like the cyclone, an anticyclone depends on the flow far above to maintain its circulation. In this instance, divergence at the surface is balanced by convergence aloft and general subsidence of the air column (Figure 17.12).

What's in a Name?

Up to now we have examined the middle-latitude cyclones, which play such an important role in causing day-to-day weather changes. Yet the use of the term *cyclone* is often confusing. To many people, the term implies only an intense storm, such as a tornado or a hurricane. When a hurricane unleashes its fury on India or Bangladesh, for example, it is usually reported in the media as a cyclone (the term denoting a hurricane in that part of the world).

Similarly, tornadoes are referred to as cyclones in some places. This custom is particularly common in portions of the Great Plains of the United States. Recall that in the *Wizard of Oz*, Dorothy's house was carried from her Kansas farm to the land of Oz by a cyclone. Indeed, the nickname for the athletic teams at Iowa State University is the Cyclones. Although hurricanes and tornadoes are, in fact, cyclones, the vast majority of cyclones are *not* hurricanes or tornadoes. The term *cyclone* simply refers to the circulation around any low-pressure center, no matter how large or intense it is.

Tornadoes and hurricanes are both smaller and more violent than middle-latitude cyclones. Middle-latitude cyclones can have a diameter of 1600 kilometers or more. By contrast, hurricanes average only

600 kilometers across, and tornadoes, with a diameter of just 0.25 kilometer, are much too small to show up on a weather map.

The thunderstorm, a much more familiar weather event, hardly needs to be distinguished from tornadoes, hurricanes, and mid-latitude cyclones. Unlike the flow of air about these latter storms, the circulation associated with thunderstorms is characterized by strong up-and-down movements. Winds in the vicinity of a thunderstorm do not follow the inward spiral of a cyclone, but they are typically variable and gusty.

Although thunderstorms form "on their own" away from cyclonic storms, they also form in conjunction with cyclones. For instance, thunderstorms are frequently spawned along the cold front of a mid-latitude cyclone, where on rare occasions a tornado may descend from the thunderstorm's cumulonimbus tower. Hurricanes also generate widespread thunderstorm activity. Thus, thunderstorms are related in some manner to all three types of cyclones mentioned here.

Thunderstorms

This is the first of three severe weather types we will examine in this chapter. Sections on tornadoes and hurricanes follow this look at thunderstorms.

Severe weather has a fascination that everyday weather phenomena cannot provide. The lightning display and booming thunder generated by a severe thunderstorm can be a spectacular event that elicits both awe and fear (Figure 17.13). Of course, hurricanes and tornadoes also attract a great deal of much-deserved attention. A single tornado outbreak or hurricane can cause billions of dollars in property damage as well as many deaths (see Box 17.2).

Thunderstorm Occurrence

Almost everyone has observed a small-scale phenomenon that is the result of the vertical movements of relatively warm, unstable air. Perhaps you have seen a dust devil form over an open field on a hot day and whirl its dusty load to great heights, or seen a bird

Figure 17.13 Cumulonimbus clouds can produce lightning, thunderstorms, and other forms of severe weather. (Photo by Richard Kaulin/Tony Stone Images)

Box 17.2 People and The Environment: The Deadliest Weather

Tornadoes and hurricanes are nature's most awesome storms. Because of this status, they are logically the focus of much well-deserved attention. Yet, surprisingly, these dreaded events are not responsible for the greatest number of weather-related deaths. That distinction is reserved for lightning and flash floods.

Not everyone who is struck by lightning is killed. In fact, fatalities represent only about 20 percent of lightning casualties. The risks from lightning are actually greater than the official statistics because some deaths and injuries go unreported.

Lightning as a safety threat receives less attention than the other causes of weather-related deaths. Tornadoes, hurricanes, and flash floods usually receive considerably more publicity because a single event often results in multiple deaths, many injuries, and widespread damage. By contrast, lightning usually takes its victims one or two at a time.

For a recent 30-year span, lightning ranked second to flash floods as a cause of weather-related deaths (Figure 17.B). Slow-moving intense thunderstorms are most often responsible for triggering the sudden overflow of stream channels. Dam failures and the sudden breakup of ice jams can also release large volumes of water in a short time. Such floods rise rapidly with little or no warning and diminish almost as quickly.

Flash floods are particularly common in mountainous terrain, where

Figure 17.B Annual average number of weather-related deaths for a recent 30-year period. (Source: National Weather Service)

steep slopes can quickly channel runoff into narrow valleys. The hazard is most acute when the soil is

already nearly saturated from earlier rains or consists of impermeable materials. The disastrous nature of flash floods is illustrated by the Big Thompson River flood of July 31, 1976, in Colorado (Figure 17.C). During a 4-hour period more than 30 centimeters of rain fell on portions of the small area drained by the river. This amounted to nearly three-quarters of the average yearly total. The flash flood in the narrow canyon lasted only a few hours, but cost 139 people their lives. Damages were estimated at more than $35 million. Despite efforts to improve observations and warning procedures, flash floods remain an elusive natural killer.

Figure 17.C Effects of the July 1976 flash flood in Colorado's Big Thompson Canyon. (Photo courtesy of U.S. Geological Survey)

glide effortlessly skyward on an invisible thermal of hot air. These examples illustrate the dynamic thermal instability that occurs during the development of a **thunderstorm.** Thunderstorm activity is associated with cumulonimbus clouds that generate heavy rainfall, thunder, lightning, and occasionally hail and tornadoes.

At any given time there are an estimated 2000 thunderstorms in progress on Earth. As we would expect, the greatest number occur in the tropics where warmth, plentiful moisture, and instability are always present. About 45,000 thunderstorms take place each day and more than 16 million occur annually around

the world. The lightning from these storms strikes Earth 100 times each second!

Annually, the United States experiences about 100,000 thunderstorms and millions of lightning strikes. A glance at Figure 17.14 shows that thunderstorms are most frequent in Florida and the eastern Gulf Coast region, where such activity is recorded between 70 and 100 days each year. The region on the east side of the Rockies in Colorado and New Mexico is next, with thunderstorms occurring on 60 to 70 days each year. Most of the rest of the nation experiences thunderstorms on 30 to 50 days annually. Clearly, the

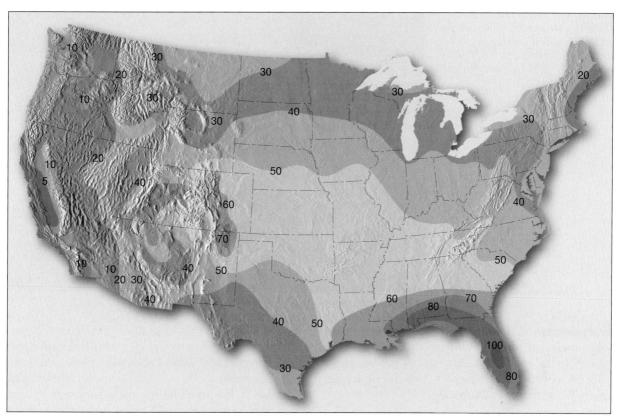

Figure 17.14 Average number of days per year with thunderstorms. Because of its close proximity to the source region for warm, humid, and unstable air masses, the Gulf Coast receives much of its precipitation from thunderstorms. (Source: Environmental Data Service, NOAA)

western margin of the United States has little thunderstorm activity. The same is true for the northern tier of states and for Canada, where warm, moist, unstable mT air seldom penetrates.

Most thunderstorms occur in association with relatively short-lived cumulonimbus clouds that produce local precipitation. In the United States these clouds typically form within warm, humid (maritime tropical) air masses that originate over the Gulf of Mexico and migrate northward. Occasionally, thunderstorms grow very large and remain active for hours. These *severe thunderstorms* produce frequent lightning and are accompanied by locally damaging winds or hail. Most severe thunderstorms in the middle latitudes form along or ahead of cold fronts. Here, forceful lifting of unstable mT air masses triggers thunderstorm development.

Stages of Thunderstorm Development

All thunderstorms require warm, moist air, which, when lifted, will release sufficient latent heat to provide the buoyancy necessary to maintain its upward flight. This instability and associated buoyancy are triggered by a number of different processes, yet all thunderstorms have a similar life history.

Because instability and buoyancy are enhanced by high surface temperatures, thunderstorms are most common in the afternoon and early evening. Surface heating is generally not sufficient in itself to cause the growth of towering cumulonimbus clouds. A solitary cell of rising hot air produced by surface heating alone can, at best, produce a small cumulus cloud, which would evaporate within 10 to 15 minutes.

The development of 12-kilometer (or on rare occasions 18-kilometer) cumulonimbus towers requires a continuous supply of moist air (Figure 17.15). Each new surge of warm air rises higher than the last, adding to the height of the cloud (Figure 17.16). These updrafts must occasionally reach speeds over 100 kilometers (60 miles) per hour to accommodate the size of hailstones they are capable of carrying upward.

Usually within an hour the amount and size of precipitation that has accumulated is too much for the updrafts to support, and in one part of the cloud downdrafts develop, releasing heavy precipitation. This represents the most active stage of the thunderstorm. Gusty winds, lightning, heavy precipitation, and sometimes hail are experienced.

Eventually downdrafts dominate throughout the cloud. The cooling effect of falling precipitation, coupled with the influx of colder air aloft, marks the end

Figure 17.15 This developing cumulonimbus cloud became a towering August thunderstorm over the high plains near Boulder, Colorado. (Photo by Henry Lansford)

of the thunderstorm activity. The life span of a single cumulonimbus cell within a thunderstorm complex is only an hour or two, but as the storm moves, fresh supplies of warm, water-laden air generate new cells to replace those that are dissipating.

Tornadoes

Tornadoes are local storms of short duration that must be ranked high among nature's most destructive forces. Their sporadic occurrence and violent winds cause many deaths each year. Tornadoes are violent windstorms that take the form of a rotating column of air or *vortex* that extends downward from a cumulonimbus cloud (Figure 17.17).

Pressures within some tornadoes have been estimated to be as much as 10 percent lower than immediately outside the storm. Drawn by the much lower pressure in the center of the vortex, air near the ground rushes into the tornado from all directions. As the air streams inward, it is spiraled upward around the core until it eventually merges with the airflow of the parent thunderstorm deep in the cumulonimbus tower. Because of the tremendous pressure gradient associated with a strong tornado, maximum winds can sometimes approach 480 kilometers (300 miles) per hour.

Tornado Occurrence and Development

An average of about 770 tornadoes are reported annually in the United States. Still, the actual number that occur from one year to the next varies greatly. During a recent 34-year span, for example, yearly totals ranged from a low of 421 to a high of 1102.

Tornadoes occur during every month of the year. April through June is the period of greatest tornado frequency in the United States, and December and

Figure 17.16 Stages in the development of a thunderstorm. During the cumulus stage, strong updrafts act to build the storm. The mature stage is marked by heavy precipitation and cool downdrafts in part of the storm. When the warm updrafts disappear completely, precipitation becomes light, and the cloud begins to evaporate.

Figure 17.17 A tornado is a violently rotating column of air in contact with the ground. The air column is visible when it contains condensation or when it contains dust and debris. Often the appearance is the result of both. When the column of air is aloft and does not produce damage, the visible portion is properly called a funnel cloud. This tornado hit Oklahoma May 22,1981. (Photo by Howard Bluestien/Photo Researchers)

January are the months of lowest activity (Figure 17.18). Of the nearly 27,000 confirmed tornadoes reported over the contiguous 48 states during the period, an average of almost five per day occurred during May. At the other extreme, a tornado was reported only every other day in January.

Tornadoes form in association with severe thunderstorms that produce high winds, heavy (sometimes torrential) rainfall, and often damaging hail. Fortunately, less than 1 percent of all thunderstorms produce tornadoes. Nevertheless, a much higher number must be monitored as potential tornado producers.

Although meteorologists are still not sure what triggers tornado formation, it is apparent that they are the product of the interaction between strong updrafts in the thunderstorm and winds in the troposphere. Despite recent advances in modeling the many variables that eventually produce a strong tornado, our knowledge still must be considered limited. However, the general atmospheric conditions that are most likely to develop into tornado activity are known.

General Atmospheric Conditions. Severe thunderstorms —and hence tornadoes—are most often spawned along the cold front of a middle-latitude cyclone. Throughout

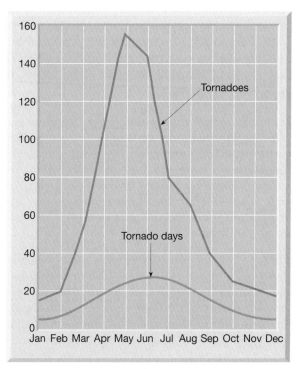

Figure 17.18 Average number of tornadoes and tornado days each month in the United States for a 27-year period. (After NOAA)

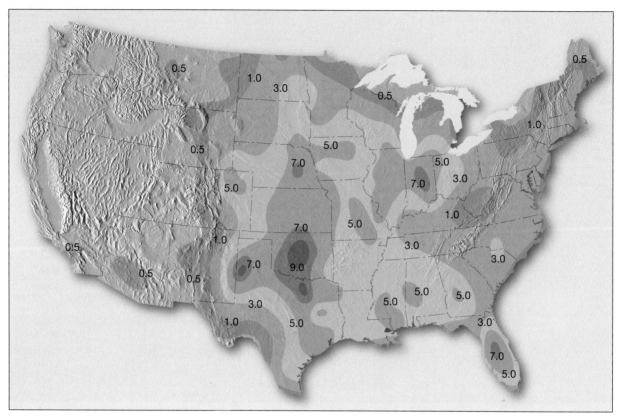

Figure 17.19 Average annual tornado incidence per 10,000 square miles (26,000 square kilometers) for a 27-year period.

spring, air masses associated with middle-latitude cyclones are most likely to have greatly contrasting conditions. Continental polar air from Canada may still be very cold and dry, whereas maritime tropical air from the Gulf of Mexico is warm, humid, and unstable. The greater the contrast when these air masses meet, the more intense the storm.

These two contrasting air masses are most likely to meet in the central United States, because there is no significant natural barrier separating the center of the country from the arctic or the Gulf of Mexico. Consequently, this region generates more tornadoes than any other area of the country or, in fact, the world. Figure 17.19, which depicts tornado incidence in the United States for a 27-year period, readily substantiates this fact.

Profile of a Tornado. An average tornado has a diameter of between 150 and 600 meters (500 and 2000 feet), travels across the landscape at approximately 45 kilometers (30 miles) per hour, and cuts a path about 10 kilometers (6 miles) long.* Because tornadoes usually occur slightly ahead of a cold front, in the zone of

southwest winds, most move toward the northeast. The Illinois example demonstrates this movement (Figure 17.20). Figure 17.20 also shows that many tornadoes do not fit the description of the "average" tornado.

Of the hundreds of tornadoes reported in the United States each year, over half are comparatively weak and short-lived. Most of these small tornadoes have lifetimes of 3 minutes or less and paths that seldom exceed 1 kilometer in length and 100 meters wide. Typical wind speeds are on the order of 150 kilometers per hour or less. On the other end of the tornado spectrum are the infrequent and often long-lived violent tornadoes. Although large tornadoes constitute only a small percentage of the total reported, their effects are often devastating. Such tornadoes may exist for periods in excess of 3 hours and produce an essentially continuous damage path more than 150 kilometers long and perhaps a kilometer or more wide. Maximum winds range upward to 450 kilometers per hour.

Tornado Destruction

The potential for tornado destruction depends largely on the strength of the winds generated by the storm. Because tornadoes generate the strongest winds in nature, they have accomplished many seemingly impossible tasks, such as driving a piece of straw through a thick wooden plank and uprooting huge

---

*The 10-kilometer figure applies to documented tornadoes. Because many small tornadoes go undocumented, the real average path of all tornadoes is unknown, but shorter than 10 kilometers.

Figure 17.20 Paths of Illinois tornadoes (1916–1969). Because most tornadoes occur slightly ahead of a cold front, in the zone of southwest winds, they tend to move toward the northeast. Tornadoes in Illinois verify this. Over 80 percent exhibited directions of movement toward the northeast through east. (After John W. Wilson and Stanley A. Changnon, Jr., *Illinois Tornadoes,* Illinois State Water Survey Circular 103, 1971, pp. 10, 24)

trees (Figure 17.21). Although it may seem impossible for winds to cause some of the fantastic damage attributed to tornadoes, tests in engineering facilities have repeatedly demonstrated that winds in excess of 320 kilometers (200 miles) per hour are capable of incredible feats.

One commonly used guide to tornado intensity is the *Fujita intensity scale,* or simply the *F-scale* (Table 17.1). Because tornado winds cannot be measured directly, a rating on the F-scale is determined by assessing the worst damage produced by a storm.

Tornadoes take many lives each year, sometimes hundreds in a single day. When tornadoes struck an area stretching from Canada to Georgia on April 3, 1974, the death toll exceeded 300, the worst in half a century. Most tornadoes, however, do not result in a loss of life. In one statistical study that examined a 29-year period, there were 689 tornadoes that resulted in deaths. This figure represents slightly less than 4 percent of the total 19,312 reported storms.

Figure 17.21 The force of the wind during a tornado near Wichita, Kansas in April 1991 was enough to drive this piece of metal into a utility pole. (Photo by John Sokich/NOAA)

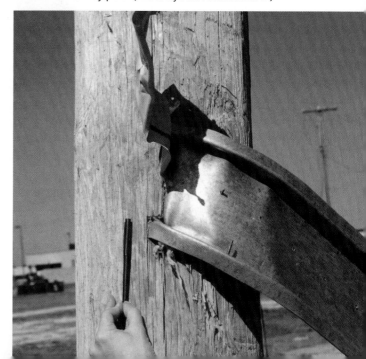

Table 17.1 Fujita Intensity Scale.

Scale	Wind Speed Km/Hr	Wind Speed Mi/Hr	Damage
F0	<116	<72	Light damage
F1	116–180	72–112	Moderate damage
F2	181–253	113–157	Considerable damage
F3	254–332	158–206	Severe damage
F4	333–419	207–260	Devastating damage
F5	>419	>260	Incredible damage

Although the percentage of tornadoes that result in death is small, each tornado is potentially lethal. When tornado fatalities and storm intensities are compared, the results are quite interesting: The majority (63 percent) of tornadoes are weak (F0 and F1), and the number of storms decreases as tornado intensity increases. The distribution of tornado fatalities, however, is just the opposite. Although only 2 percent of tornadoes are classified as violent (F4 and F5), they account for nearly 70 percent of the deaths. If there is some question as to the causes of tornadoes, there certainly is no question about the destructive effects of these violent storms (Figure 17.22).

Predicting Tornadoes

Because severe thunderstorms and tornadoes are small and relatively short-lived phenomena, they are among the most difficult weather features to forecast precisely.

Tornado Watch. When conditions appear favorable for tornado formation, a **tornado watch** is issued for areas covering about 65,000 square kilometers (25,000 square miles). Between 35 and 40 percent of these predictions are correct; that is, one or more tornadoes occur somewhere in the specified region. The incorrect forecasts are about evenly divided between cases in which no tornadoes are sighted and cases when tornadoes occur outside, but near, the watch area.

Tornado Warning. Whereas a tornado watch is designed to alert people to the possibility of tornadoes, a **tornado warning** is issued when a funnel cloud has actually been sighted or is indicated by radar. It warns of high probability of imminent danger. Warnings are issued for much smaller areas than watches, usually covering a portion of a county or counties. In addition, they are in effect for much shorter periods, typically 30 to 60 minutes. Because a tornado warning may be based on an actual sighting, it is not necessarily issued in advance of the tornado.

If the direction and approximate speed of the storm are known, an estimate of the storm's most probable path can be made. Because tornadoes often move erratically, the warning area is fan-shaped downwind from the point where the tornado has been spotted. Since the late 1960s, warnings have been given for most major tornadoes. It is believed that such warnings substantially reduce the number of deaths and serious injuries that might otherwise occur.

Doppler Radar. The tornado warning system that has been used throughout the United States for many years relies heavily on visual sightings by a few trained observers as well as the general public. Unfortunately, such a system is prone to incomplete coverage and mistakes. The errors are most likely to occur at night when tornadoes may go unnoticed or harmless clouds may be mistaken for funnel clouds. Hence, there may be

Figure 17.22 A dozen tornadoes devastated the Kissimee, Florida area during the night of February 22, 1998. Ponderosa Park shown here was totally obliterated. (Photo by Craig Bailey/ *Florida Today*/Liaison Agency, Inc.)

a lack of adequate warning on the one hand, or unnecessary warnings on the other.

Many of the difficulties that once limited the accuracy of tornado warnings have been reduced or eliminated by an advancement in radar technology called **Doppler radar.** Doppler radar not only performs the same tasks as conventional radar, but also has the ability to detect motion directly. Doppler radar can detect the initial formation and subsequent development of a *mesocyclone*, an intense rotating wind system in the lower part of a thunderstorm that frequently precedes tornado development.

Almost all mesocyclones produce damaging hail, severe winds, or tornadoes. Those that produce tornadoes (about 60 percent) can be distinguished by their stronger wind speeds and their sharper gradients of wind speeds.

It should also be pointed out that not all tornado-bearing storms have clear-cut radar signatures and that other storms can give false signatures. Detection, therefore, is sometimes a subjective process and a given display could be interpreted in several ways. Consequently, trained observers will continue to form an important part of the warning system in the foreseeable future.

Although some operational problems exist, the benefits of Doppler radar are many. As a research tool, it is not only providing data on the formation of tornadoes, but is also helping meteorologists gain new insights into thunderstorm development, the structure and dynamics of hurricanes, and air turbulence hazards that plague aircraft. As a practical tool for tornado detection, it has significant advantages over a system that uses conventional radar.

Hurricanes

Most of us view the weather in the tropics with favor. Places like Hawaii and the islands of the Caribbean are known for their lack of significant day-to-day variations. Warm breezes, steady temperatures, and rains that come as heavy but brief tropical showers are expected. It is ironic that these relatively tranquil regions sometimes produce the most violent storms on Earth.

The whirling tropical cyclones that on occasion have wind speeds attaining 300 kilometers (185 miles) per hour are known in the United States as **hurricanes** and are the greatest storms on Earth. Out at sea, they can generate 15-meter (50-foot) waves capable of inflicting destruction hundreds of kilometers from their source. Should a hurricane smash onto land, strong winds coupled with extensive flooding can impose billions of dollars in damage and great loss of life (Figure 17.23). The vast majority of hurricane-related deaths and damage are caused by relatively infrequent, yet powerful, storms.

Hurricanes are becoming a growing threat because more and more people are living and working along and near coasts. At the close of the twentieth century, more than 50 percent of the U.S. population lived within 75 kilometers (45 miles) of a coast. This number is projected to increase substantially in the early decades of the twenty-first century. The concentration of such large numbers of people near the shoreline means that hurricanes and other large storms place millions at risk. Moreover, the potential costs of property damage are incredible.

Figure 17.23 This destruction on North Carolina's Topsail Isle occurred when Hurricane Fran struck on September 5, 1996. The storm caused an estimated $3 billion in total damages. (Photo by Cheryl Hogue/Photo Researchers, Inc.

Profile of a Hurricane

Most hurricanes form between the latitudes of 5 degrees and 20 degrees over all the tropical oceans except those of the South Atlantic and eastern South Pacific. They are known by different names in various parts of the world. In the western Pacific, they are called *typhoons* and in the Indian Ocean, *cyclones.* The North Pacific has the greatest number of storms, averaging 20 per year. Fortunately for those living in the coastal regions of the southern and eastern United States, fewer than five hurricanes, on the average, develop annually in the warm sector of the North Atlantic.

Although many tropical disturbances develop each year, only a few reach hurricane status. By international agreement, a hurricane has wind speeds in excess of 119 kilometers (74 miles) per hour and a rotary circulation. Mature hurricanes average 600 kilometers (375 miles) in diameter and often extend 12,000 meters (40,000 feet) above the ocean surface. From the outer edge to the center, the barometric pressure has on occasion dropped 60 millibars, from 1010 millibars to 950 millibars. The lowest pressures ever recorded in the Western Hemisphere are associated with these storms. A steep pressure gradient generates the rapid, inward-spiraling winds of a hurricane.

As the inward rush of warm, moist surface air approaches the core of the storm, it turns upward and ascends in a ring of cumulonimbus towers. This doughnut-shaped wall of intense convective activity surrounding the center of the storm is called the **eye wall.** It is here that the greatest wind speeds and heaviest rainfall occur. Surrounding the eye wall are curved bands of clouds that trail away in a spiral fashion. Near the top of the hurricane the airflow is outward, carrying the rising air away from the storm center, thereby providing room for more inward flow at the surface.

At the very center of the storm is the **eye** of the hurricane (Figure 17.24). This well-known feature is a zone about 20 kilometers (12.5 miles) in diameter where precipitation ceases and winds subside. It offers a brief but deceptive break from the extreme weather in the enormous curving wall clouds that surround it. The air within the eye gradually descends and heats by compression, making it the warmest part of the storm. Although many people believe that the eye is characterized by clear blue skies, such is usually not the case because the subsidence in the eye is seldom strong enough to produce cloudless conditions. Although the sky appears much brighter in this region, scattered clouds at various levels are common.

Hurricane Formation and Decay

A hurricane is a heat engine that is fueled by the latent heat liberated when huge quantities of water vapor condense. The amount of energy produced by a typical hurricane in just a single day is truly immense—roughly equivalent to the entire electrical energy production of the United States in a year. The release of latent heat warms the air and provides buoyancy for its upward flight. The result is to reduce the pressure near the surface, which encourages a more rapid inward flow of air. To get this engine started, a large quantity of warm, moisture-laden air is required, and a continual supply is needed to keep it going.

Hurricanes develop most often in the late summer when water temperatures have reached 27°C (80°F) or higher and thus are able to provide the necessary heat and moisture to the air. This ocean-water temperature requirement accounts for the fact that hurricanes do not form over the relatively cool waters of the South Atlantic and the eastern South Pacific. For the same reason, few hurricanes form poleward of 20 degrees of

Figure 17.24 A satellite image of Hurricane Fran as the storm approached the Florida coast September 4, 1996. The eye is clearly visible in the center of the storm. The eye wall, the most intense part of the storm, is a doughnut-shaped wall of cumulonimbus development surrounding the eye. Satellites are essential tools for detecting and tracking hurricanes. (Photo by WSI Corp./Liaison Agency, Inc.)

latitude. Although water temperatures are sufficiently high, hurricanes do not form within 5 degrees of the equator, because the Coriolis effect is too weak to initiate the necessary rotary motion.

The initial stage of a tropical cyclone's life cycle is not well understood, attributable in part to the complexity of these storms. Another important factor is the lack of observations in regions where storms form. However, we do know that smaller tropical cyclones initiate the process. These initial disturbances are regions of low-level convergence and lifting. The U.S. National Weather Service labels such incipient storms *tropical disturbances.*

Many tropical disturbances of this type occur each year, but only a few develop into full-fledged hurricanes. By international agreement, lesser tropical cyclones are given different names based on the strength of their winds. When a cyclone's strongest winds do not exceed 61 kilometers (38 miles) per hour, it is called a **tropical depression**. When winds are between 61 and 119 kilometers (38 and 74 miles) per hour, the cyclone is termed a **tropical storm.** It is during this phase that a name is given (Andrew, Mitch, Opal, etc.). Should the tropical storm become a hurricane, the name remains the same. Each year between 80 and 100 tropical storms develop around the world. Of them, usually half or more eventually become hurricanes.

Hurricanes diminish in intensity whenever they (1) move over ocean waters that cannot supply warm, moist tropical air; (2) move onto land; or (3) reach a location where the large-scale flow aloft is unfavorable. Whenever a hurricane moves onto land, it loses its punch rapidly. The most important reason for this rapid demise is the fact that the storm's source of warm, moist air is cut off. When an adequate supply of water vapor does not exist, condensation and the release of latent heat must diminish. In addition, friction from the increased roughness of the land surface rapidly slows surface wind speeds. This factor causes the winds to move more directly into the center of the low, thus helping to eliminate the large pressure differences.

Hurricane Destruction

A location only a few hundred kilometers from a hurricane—just one day's striking distance away—may experience clear skies and virtually no wind. Prior to the age of weather satellites, such a situation made the task of warning people of impending storms very difficult.

The worst natural disaster in U.S. history resulted from a hurricane that struck an unprepared Galveston, Texas, on September 8, 1900. The strength of the storm, coupled with the lack of adequate warning, caught the population by surprise and cost 6000 people in the city their lives. At least 2000 more were killed elsewhere. Fortunately, hurricanes are no longer the unheralded killers they once were.

Once a storm develops cyclonic flow and the spiraling bands of clouds characteristic of a hurricane, it receives continuous monitoring. When hurricanes form, satellites are able to identify and track the storms long before they make landfall (Figure 17.24). In the United States, early warning systems have greatly reduced the number of deaths caused by hurricanes. At the same time, however, there has been an astronomical rise in the amount of property damage. The primary reason, of course, has been the extensive building of homes and businesses in coastal areas.

Although the amount of damage caused by a hurricane depends on several factors, including the size and population density of the area affected and the shape of the ocean bottom near the shore, certainly the most significant factor is the strength of the storm itself. By studying past storms, a scale has been established to rank the relative intensity of hurricanes (Table 17.2). As Table 17.2 indicates, a *category 5* storm is the worst possible, whereas a *category 1* hurricane is least severe.

During the hurricane season it is common to hear scientists and reporters alike use the numbers from the *Saffir-Simpson Hurricane Scale.* The famous Galveston hurricane just mentioned, with winds in excess of 209 kilometers (130 miles) per hour and a pressure of 931 millibars, would be placed in category 4. Storms that fall into category 5 are rare. Hurricane Camille, a 1969 storm that caused catastrophic

Table 17.2 Saffir–Simpson hurricane scale.

Scale Number (category)	Central Pressure (millibars)	Winds (km/hr)	Storm Surge (meters)	Damage
1	≥980	119–153	1.2–1.5	Minimal
2	965–979	154–177	1.6–2.4	Moderate
3	945–964	178–209	2.5–3.6	Extensive
4	920–944	210–250	3.7–5.4	Extreme
5	<920	>250	>5.4	Catastrophic

Figure 17.25 In 1969, Hurricane Camille hit the coast of Mississippi. It was a rare category 5 storm. These classic photos document the devastating force of the hurricane's 25-foot storm surge at Pass Christian, Mississippi. **A.** The Richelieu Apartments before the hurricane. This substantial-looking three-story building was directly across the highway from the beach. **B.** The same apartments after the hurricane. (Photos by Chauncey T. Hinman)

A.

B.

damage along the coast of Mississippi, is one well-known example (Figure 17.25).

Damage caused by hurricanes can be divided into three categories: (1) wind damage, (2) storm surge, and (3) inland freshwater flooding.

Wind Damage. For some structures, the force of the wind is sufficient to cause total destruction. This was demonstrated in southern Florida in 1992. The billions of dollars in property damages from Hurricane Andrew were largely the result of strong winds. However, wind damage is not necessarily responsible for a hurricane's greatest destructiveness.

Storm Surge. The most devastating damage in the coastal zone is caused by the storm surge (Figure 17.25). It not only accounts for a large share of coastal property losses, but is also responsible for 90 percent of all hurricane-caused deaths. A **storm surge** is a dome of water 65 to 80 kilometers (40 to 50 miles) wide that sweeps across the coast near the point where the eye makes landfall. Ignoring wave activity, the storm surge is the height of the water above normal tide level. Thus,

a storm surge commonly adds 2 to 3 meters (6 to 10 feet) to normal tide heights—to say nothing of tremendous wave activity superimposed atop the surge.

We can easily imagine the damage this surge of water can inflict on low-lying coastal areas. In the delta region of Bangladesh, for example, most of the land is less than 2 meters above sea level. When a storm surge superimposed upon normal high tide inundated that area on November 13, 1970, the official death toll was 200,000. This was one of the worst disasters of modern times. In May 1991 a similar event struck Bangladesh again.

Inland Flooding. The torrential rains that accompany most hurricanes represent a third significant threat—flooding. Whereas the effects of storm surge and strong winds are concentrated in coastal areas, heavy rains may affect places hundreds of kilometers from the coast for several days after the storm has lost its hurricane-force winds.

Hurricanes weaken rapidly as they move inland, yet the remnants of the storm can still yield 15 to 30 centimeters (6 to 12 inches) or more of rain as they move inland. A good example of such destruction was

Hurricane Agnes (1972). Although this was just a category 1 storm on the Saffir–Simpson scale, it created more than $2 billion in damage and took 122 lives. Most destruction was attributed to flooding caused by an inordinate amount of rainfall.

To summarize, extensive damage and loss of life in the coastal zone can result from storm surge, torrential rains, and strong winds. When loss of life occurs, it is commonly caused by the storm surge, which can devastate entire barrier islands and low-lying land along the coast. Although wind damage is usually not as catastrophic as the storm surge, it affects a much larger area. Where building codes are inadequate, economic losses can be especially severe. Because hurricanes weaken as they move inland, most wind damage occurs within 200 kilometers of the coast. Far from the coast a weakening storm can produce extensive flooding long after the winds have diminished below hurricane levels. Sometimes the damage from inland flooding exceeds storm-surge destruction.

Chapter Summary

- An *air mass* is a large body of air, usually 1600 kilometers (1000 miles) or more across, which is characterized by a *sameness of temperature and moisture* at any given altitude. When this air moves out of its region of origin, called the *source region*, it will carry these temperatures and moisture conditions elsewhere, perhaps eventually affecting a large portion of a continent.

- Air masses are classified according to (1) the nature of the surface in the source region and (2) the latitude of the source region. *Continental (c)* designates an air mass of land origin, with the air likely to be dry; whereas a *maritime (m)* air mass originates over water, and therefore will be humid. *Polar (P)* air masses originate in high latitudes and are cold. *Tropical (T)* air masses form in low latitudes and are warm. According to this classification scheme, the *four basic types of air masses* are *continental polar* (cP), *continental tropical* (cT), *maritime polar* (mP), and *maritime tropical* (mT). Continental polar (cP) and maritime tropical (mT) air masses influence the weather of North America most, especially east of the Rocky Mountains. Maritime tropical air is the source of much, if not most, of the precipitation received in the eastern two-thirds of the United States.

- *Fronts* are boundaries that separate air masses of different densities, one warmer and often higher in moisture content than the other. A *warm front* occurs when the surface position of the front moves so that warm air occupies territory formerly covered by cooler air. Along a warm front, a warm air mass overrides a retreating mass of cooler air. As the warm air ascends, it cools adiabatically to produce clouds and, frequently, light-to-moderate precipitation over a large area. A *cold front* forms where cold air is actively advancing into a region occupied by warmer air. Cold fronts are about twice as steep and move more rapidly than do warm fronts. Because of these two differences, precipitation along a cold front is more intense and of shorter duration than is precipitation associated with a warm front.

- The primary weather producers in the middle latitudes are *large centers of low pressure* that generally travel from *west to east,* called *middle-latitude cyclones.* These *bearers of stormy weather,* which last from a few days to a week, have a *counterclockwise circulation* pattern in the Northern Hemisphere, with an *inward flow of air* toward their centers. Most middle-latitude cyclones have a *cold front and frequently a warm front* extending from the central area of low pressure. *Convergence and forceful lifting along the fronts* initiate cloud development and frequently cause precipitation. As a middle-latitude cyclone with its associated fronts passes over a region, it often brings with it abrupt changes in the weather. The particular weather experienced by an area depends on the path of the cyclone.

- *Thunderstorms* are caused by the upward movement of warm, moist, unstable air, triggered by a number of different processes. They are associated with cumulonimbus clouds that generate heavy rainfall, thunder, lightning, and occasionally hail and tornadoes.

- *Tornadoes,* destructive, local storms of short duration, are violent windstorms associated with severe thunderstorms that take the form of a rotating column of air that extends downward from a cumulonimbus cloud. Tornadoes are most often spawned along the cold front of a middle-latitude cyclone, most frequently during the spring months.

- *Hurricanes,* the greatest storms on Earth, are tropical cyclones with wind speeds in excess of 119 kilometers (74 miles) per hour. These complex tropical disturbances develop over tropical ocean waters and are fueled by the latent heat liberated when huge quantities of water vapor condense. Hurricanes form most often in late summer when ocean-surface temperatures reach 27°C (80°F) or higher and thus are able to provide the necessary heat and moisture to the air. Hurricanes diminish in intensity whenever they (1) move over cool ocean water that cannot supply adequate heat and moisture, (2) move onto land, or (3) reach a location where large-scale flow aloft is unfavorable.

Key Terms

air mass (p. 464)
air-mass weather (p. 465)
cold front (p. 469)
continental (c) air mass (p. 466)
Doppler radar (p. 483)
eye (p. 484)
eye wall (p. 484)
front (p. 466)
hurricane (p. 483)

lake-effect snow (p. 466)
maritime (m) air mass (p. 466)
middle-latitude cyclone (p. 470)
occluded front (p. 470)
polar (P) air mass (p. 466)
source region (p. 465)
stationary front (p. 470)
storm surge (p. 486)
thunderstorm (p. 476)

tornado (p. 478)
tornado warning (p. 482)
tornado watch (p. 482)
tropical (T) air mass (p. 466)
tropical depression (p. 485)
tropical storm (p. 485)
warm front (p. 468)

Review Questions

1. Describe the weather associated with a continental polar air mass in the winter and in the summer. When would this air mass be most welcome in the United States?

2. What are the characteristics of a maritime tropical air mass? Where are the source regions for the maritime tropical air masses that affect North America? Where are the source regions for the maritime polar air masses?

3. Describe the weather along a cold front where very warm, moist air is being displaced.

4. The formation of an occluded front marks the beginning of the end of a middle-latitude cyclone. Why is this true?

5. For each of the weather elements that follow, describe the changes that an observer experiences when a middle-latitude cyclone passes with its center north of the observer: wind direction, pressure tendency, cloud type, cloud cover, precipitation, temperature.

6. Describe the weather conditions an observer would experience if the center of a middle-latitude cyclone passed to the south.

7. Briefly explain how the flow aloft aids the formation of cyclones at the surface.

8. Compare the wind speeds and sizes of middle-latitude cyclones, tornadoes, and hurricanes. How are thunderstorms related to each?

9. What is the primary requirement for the formation of thunderstorms?

10. Based on your answer to Question 9, where would you expect thunderstorms to be most common on Earth? In the United States?

11. Why do tornadoes have such high wind speeds?

12. What general atmospheric conditions are most conducive to the formation of tornadoes?

13. When is the "tornado season"? That is, during what months is tornado activity most pronounced?

14. Distinguish between a tornado watch and a tornado warning.

15. What advantage does Doppler radar have over conventional radar?

16. During what time of year do most of the hurricanes that affect North America form? Why is hurricane formation favored at this time?

17. Which has stronger winds, a tropical storm or a tropical depression?

18. Why does the intensity of a hurricane diminish rapidly when it moves onto land?

19. Hurricane damage can be divided into three broad categories. Name them. Which category is responsible for the highest percentage of hurricane-related deaths?

20. Great damage and significant loss of life can take place a day or more after a hurricane has moved ashore and weakened. When this occurs, what is the likely cause?

21. A hurricane has slower wind speeds than a tornado, yet it inflicts more total damage. How might this be explained?

22. Although tornadoes and hurricanes are dangerous storms, they are not responsible for the greatest number of weather-related deaths. What weather phenomena cause more deaths than tornadoes and hurricanes? (See Box 17.2.)

Examining the Earth System

1. Which parts of the Earth system interact in the Great Lakes region of North America to produce the high snowfalls on the leeward (downwind) shores of the lakes? What term is applied to these heavy snows? Describe what causes this effect. (A good short review of this process can be found at **http://covis.atmos.uiuc. edu/guide/rsteve/html/snglband.html**)
2. Hurricanes are among the most severe storms experienced on Earth. When a hurricane makes landfall in the southeastern United States what impact might it have on coastal lands, drainage networks (hydrosphere), and natural vegetation (biosphere)? (You may find some helpful information at the *USA Today* hurricane Web site at **http://www.usatoday.com/weather/whuro.htm** and the National Hurricane Center's site at **http://www.nhc. noaa.gov/**)

Web Resources

The following are informative and interesting World Wide Web sites that address topics related to those presented in the chapter:

Middle-latitude Cyclones Online Tutorial (University of Illinois):

• **http://www.2010.atmos.uiuc.edu/(Gh)/guides/mtr/cyc/ home.rxml**

National Weather Service (National Oceanic and Atmospheric Administration):

• **http://www.nws.noaa.gov/**

For direct links to these sites and other appropriate Earth science Web locations, chapter review exercises, and relevant, topical investigations that utilize Web resources, contact the *Earth Science*, 9th Edition Home Page at:

• **http://www.prenhall.com/tarbuck.**

Chapter 18
Climate

Air pollution episode in Los Angeles. Human impact on the atmosphere is not confined to localized areas. Humans may be significantly contributing to global climate change. (Photo by Lester Lefkowitz/The Stock Market)

A nyone who has the opportunity to travel around the world will find such an incredible variety of climates that it is hard to believe they could all occur on the same planet. The broad diversity of climates around the globe is an important focus of this chapter. Climate strongly influences the nature of plant and animal life, the soil, and many external geological processes. Climate influences people as well.

Although climate has a significant impact on people, we are learning that people also have a strong influence on climate. The latter portion of this chapter examines the ways in which humans may be changing global climate.

The Climate System

The focus of this chapter is *climate*. In Chapter 14, where this term was introduced, climate was characterized as being an aggregate of weather. It consists not only of average atmospheric values, but also involves the variability of elements and information on the occurrence of extreme events.

To understand and appreciate climate, it is important to realize that climate involves more than just the atmosphere. Indeed, we must recognize that there is a **climate system.** In addition to the atmosphere, it consists of the hydrosphere, solid Earth, biosphere, and cryosphere (Figure 18.1). The *cryosphere* refers to the ice and snow that exist at Earth's surface. Powered by the energy from the Sun, the climate system involves the exchanges of energy and moisture that occur among the five parts. These exchanges link the atmos-

phere to the other parts of the system to produce an integrated and extremely complex interactive unit.

> The atmosphere is the central component of the complex, connected, and interactive global environmental system upon which all life depends. Climate may be broadly defined as the long-term behavior of this environmental system. To understand fully and to predict changes in the atmospheric component of the climate system, one must understand the Sun, oceans, ice sheets, solid Earth, and all forms of life.*

The climate system provides a framework for the study of climate. The interactions and exchanges among the parts of the climate system create a complex network that links the five spheres. Changes to the system do not occur in isolation. Rather, when one part of this interactive unit changes, the other components also react. This well-established relationship is demonstrated often when we study climate and climate change.

World Climates

Previous chapters have already presented the spatial and seasonal variations of the major elements of weather and climate. Chapter 14 examined the controls of temperature and the world distribution of temperature. In Chapter 16, you studied the general

---

*The American Meterological Society and the University Corporation for Atmospheric Research. "Weather and the Nation's Well-Being." *Bulletin of American Meteorological Society,* 73, no. 12 (December 1991): 2038.

Figure 18.1 The climate system involves the complex interactions that occur among the atmosphere, hydrosphere, solid earth, cryosphere, and biosphere. All of the components are represented in this scene in Colorado's Maroon Bells. (Photo by Pete Saloutos/The Stock Market)

circulation of the atmosphere and the global distribution of precipitation. You are now ready to investigate the *combined* effects of these variations in different parts of the world. The varied nature of Earth's surface and the many interactions that occur between atmospheric processes give every location on our planet a distinctive, even unique, climate. Our intention, however, is not to describe the unique climatic character of countless different locales. Instead, the purpose is to introduce the major climatic regions of the world. The discussion examines large areas and uses particular places only to illustrate the characteristics of these major climatic regions.

Temperature and precipitation are the most important elements in a climatic description because they have the greatest influence on people and their activities and also have an important impact on the distribution of such phenomena as vegetation and soils. Nevertheless, other factors are also important for a complete climatic description. When possible, some of these factors are introduced into our discussion of world climates (see Box 18.1).

Climate Classification

The distribution of the major atmospheric elements is, to say the least, complex. Because of the many differences from place to place as well as from time to time at a particular locale, it is unlikely that any two sites on Earth experience exactly the same weather conditions. The fact that the number of places on Earth is virtually infinite makes it readily apparent that the number of different climates must be extremely large.

Of course, having a great diversity of information to investigate is not unique to the study of the atmosphere; it is a problem that is basic to all science. To cope with such variety, it is not only desirable, but essential to devise some means of classifying the vast array of data to be studied. By establishing groups consisting of items that have certain important characteristics in common, order and simplicity are introduced. Bringing order to large quantities of information not only aids comprehension and understanding, but it also facilitates analysis and explanation.

Over the past 100 years, many climate-classification schemes have been devised. It should be remembered that the classification of climates (or of anything else) is not a natural phenomenon but the product of human ingenuity. The value of any particular classification is determined largely by its intended use. A system designed for one purpose is not necessarily applicable to another.

In this chapter, we use a classification devised by Russian-born German climatologist Wladimir Köppen (1846–1940). As a tool for presenting the general world pattern of climates, the **Köppen classification** has

been the best-known and most used system for more than 70 years. It is widely accepted for many reasons. For one, it uses only easily obtained data: mean monthly and annual values of temperature and precipitation. Furthermore, the criteria are unambiguous, relatively simple to apply, and divide the world into climatic regions in a realistic way.

Köppen believed that the distribution of natural vegetation was an excellent expression of the totality of climate. Consequently, the boundaries he chose were largely based on the limits of certain plant associations. Five principal groups were recognized; each group was designated by a capital letter as follows:

A *Humid tropical.* Winterless climates; all months having a mean temperature above 18°C.

B *Dry.* Climates where evaporation exceeds precipitation; there is a constant water deficiency.

C *Humid middle-latitude.* Mild winters; the average temperature of the coldest month is below 18°C but above –3°C.

D *Humid middle-latitude.* Severe winters; the average temperature of the coldest month is below –3°C and the warmest monthly mean exceeds 10°C.

E *Polar.* Summerless climates; the average temperature of the warmest month is below 10°C.

Notice that four of the major groups (A, C, D, E) are defined on the basis of temperature characteristics, and the fifth, the B group, has precipitation as its primary criterion. Each of the five groups is further subdivided by using the criteria and symbols presented in Table 18.1. In addition, the world distribution of climates according to the Köppen* classification is shown in Figure 18.2. You will be referred to this figure several times as Earth's climates are discussed in the following pages.

Humid Tropical (A) Climates

Within the A group of climates, two main types are recognized—wet tropical climates (Af and Am) and tropical wet and dry (Aw).

The Wet Tropics

The constantly high temperatures and year-round rainfall in the wet tropics combine to produce the most luxuriant vegetation found in any climatic realm: the **tropical rain forest** (Figure 18.3).

---

*When classifying climatic data using Table 18.1, you should first determine whether the data meet the criteria for the E climates. If the station is not a polar climate, proceed to the criteria for B climates. If your data do not fit into either the E or B groups, check the data against the criteria for A, C, and D climates, in that order.

Box 18.1 The World's Hottest and Coldest Places

Maximum Temperature Records

Most people living in the United States have experienced temperatures of 38°C (100°F) or more. When statistics for the 50 states are examined, we find that *every* state has a maximum temperature of 38°C or higher. Even Alaska has recorded a temperature this high. The record was set June 27, 1915, at Fort Yukon, a town along the Arctic Circle in the interior of the state. With one exception, every other state has a maximum temperature record in *excess* of 38°C. Surprisingly, the state that ties Alaska for the "lowest high" is Hawaii. Panaloa, on the south coast of the big island, recorded 38°C on April 27, 1931. Although humid tropical and subtropical places like Hawaii are known for being warm throughout the year, they seldom experience maximum temperatures that surpass the low- to mid-30s Celsius (90s Fahrenheit).

The highest accepted temperature record for the United States as well as the entire Western Hemisphere is 57°C (134°F). This long-standing record was set at Death Valley, California , on July 10, 1913. Summer temperatures at Death Valley are consistently among the highest in the Western Hemisphere. During June, July, and August, temperatures exceeding 49°C (120°F) are to be expected. Fortunately, Death Valley has no summertime residents.

Why are summer temperatures at Death Valley so high? In addition to having the lowest elevation in the Western Hemisphere (53 meters below sea level), Death Valley is a desert. Although it is only about 300 kilometers from the Pacific Ocean, mountains cut off the valley from the ocean's moderating influence and moisture. Clear skies allow a maximum of sunshine to strike the dry, barren surface. Because no energy is used to evaporate moisture as occurs in humid regions, all of the energy is available to heat the ground. In addition, subsiding air that warms by compression as it descends is also common to the region and contributes

to its high maximum temperatures.

When maximum temperature records for other continents are examined (Table 18.A), we find situations similar to that in North America. That is, the highest temperatures have been recorded at locations that are arid or nearly so. The nearly 59°C reading at Azizia, Libya, in North Africa's Sahara Desert, is the world record. Of course, Antarctica's highest recorded temperature of 15°C (59°F) does not come close to the others.

Minimum Temperature Records

Table 18.B presents minimum temperature records for each of the continents plus Greenland. The stations are ranked, with the lowest minimum at the top of the list. Remember that many other places have no doubt experienced equally low or even lower temperatures; they just were not officially recorded.

The temperature controls that make these places so frigid are predictable, and should come as no surprise. We should expect cold temperatures during winter in high-latitude places that lack marine influence. Moreover, stations located on ice sheets and glaciers should be especially cold, as should stations positioned high in the mountains. Clearly *all* of these criteria apply to Vostok in Antarctica and to Greenland's Northice Station. Snag in Canada's Yukon Territory holds the record for North America, so no U.S. locations appear on the list. Nevertheless, Prospect Creek, located north of the Arctic Circle in the Endicott Mountains of Alaska, came close to the North American record on January 23, 1971, when the temperature plunged to –62°C (–80°F). In the lower 48 states, the record of –57°C (70°F) was set in the mountains of Rogers Pass, Montana, on January 20, 1954.

Table 18.A Maximum temperature records.

Area	Highest (°C)	(°F)	Place	Elevation (Meters)	Date
Africa	58	136	Azizia, Libya	114	Sept. 13, 1992
North America	57	134	Death Valley, Calif.	–53	July 10, 1913
Asia	54	129	Tirat Tsvi, Israel	–217	June 21, 1942
Australia	53	128	Cloncurry, Queensland	187	Jan. 16, 1889
Europe	50	122	Seville, Spain	8	Aug. 4, 1881
South America	49	120	Rivadavia, Argentina	203	Dec. 11, 1905
Oceania	42	108	Tuguegarao, Philippines	22	April 29, 1912
Antarctica	15	59	Vanda Station	8	Jan. 5, 1974

Table 18.B Minimum temperature records.

Area	Lowest (°C)	(°F)	Place	Elevation (Meters)	Date
Antarctica	–89	–129	Vostok	3366	Aug. 24, 1960
Asia	–68	–90	Oymykon, Russia	788	Feb. 6, 1933
Greenland	–66	–87	Northice	2307	Jan. 9, 1954
North America	–63	–81	Snag, Yukon, Canada	578	Feb. 3, 1947
Europe	–55	–67	Ust'Shchugor, Russia	84	Jan.*
South America	–33	–27	Sarmiento, Argentina	264	June 1, 1907
Africa	–24	–11	Ifrane, Morocco	1609	Feb. 11, 1935
Australia	–22	–8	Charlotte Pass, NSW	†	July 22, 1947**

*Exact date unknown; lowest in 15-year period.
**And earlier dates.
†elevation unknown.

Table 18.1 Köppen system of climatic classification.

1st	2nd	3rd	
A			Average temperature of the coldest month is 18°C or higher
	f		Every month has 6 cm of precipitation or more
	m		Short dry season; precipitation in driest month less than 6 cm but equal to or greater than 10 − R/25 (R is annual rainfall in cm)
	w		Well-defined winter dry season; precipitation in driest month less than 10 − R/25
	s		Well-defined summer dry season (rare)
B			Potential evaporation exceeds precipitation. The dry-humid boundary is defined by the following formulas: (Note: R is the average annual precipitation in cm and T is the average annual temperature in °C)
			R < 2T + 28 — when 70% or more of rain falls in warmer 6 months
			R < 2T — when 70% or more of rain falls in cooler 6 months
			R < 2T + 14 — when neither half year has 70% or more of rain
	S		Steppe — The BS-BW boundary is 1/2 the dry-humid boundary
	W		Desert — The BS-BW boundary is 1/2 the dry-humid boundary
		h	Average annual temperature is 18°C or greater
		k	Average annual temperature is less than 18°C
C			Average temperature of the coldest month is under 18°C and above −3°C
	w		At least ten times as much precipitation in a summer month as in the driest winter month
	s		At least three times as much precipitation in a winter month as in the driest summer month; precipitation in driest summer month less than 4 cm
	f		Criteria for w and s cannot be met
		a	Warmest month is over 22°C; at least 4 months over 10°C
		b	No month above 22°C; at least 4 months over 10°C
		c	One to 3 months above 10°C
D			Average temperature of coldest month is −3°C or below; average temperature of warmest month is greater than 10°C
	s		Same as under C
	w		Same as under C
	f		Same as under C
		a	Same as under C
		b	Same as under C
		c	Same as under C
		d	Average temperature of the coldest month is −38°C or below
E			Average temperature of the warmest month is below 10°C
	T		Average temperature of the warmest month is greater than 0°C and less than 10°C
	F		Average temperature of the warmest month is 0°C or below

Figure 18.2 Climates of the world based on the Köppen classification.

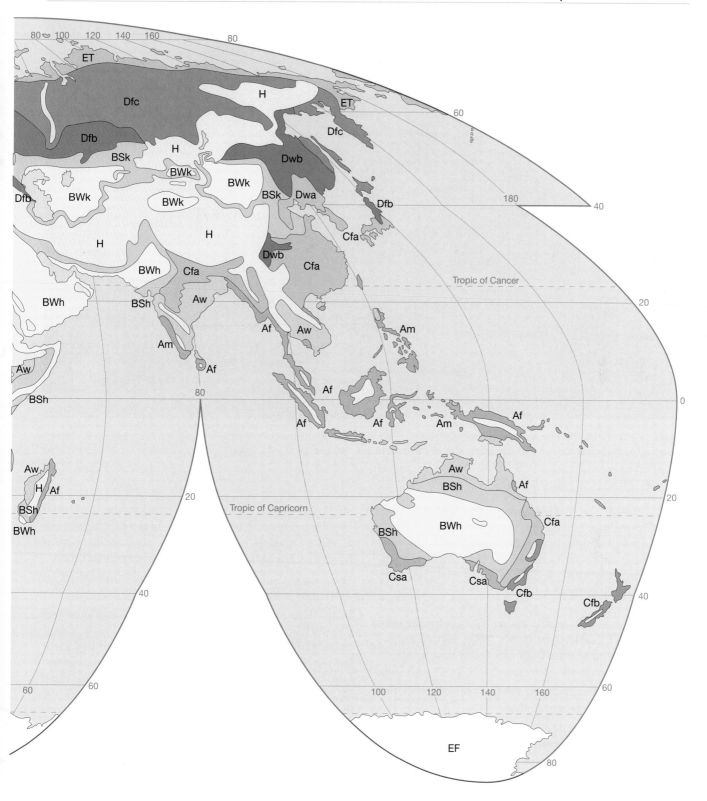

The environment of the wet tropics characterizes almost 10 percent of Earth's land area. An examination of Figure 18.2 shows that Af and Am climates form a discontinuous belt astride the equator that typically extends 5° to 10° into each hemisphere. The poleward margins are most often marked by diminishing rainfall, but occasionally decreasing temperatures mark the boundary. Because of the general decrease in temperature with height in the troposphere, this climate region is restricted to elevations below 1000 meters.

Figure 18.3 Unexcelled in luxuriance and characterized by hundreds of different species per square kilometer, the tropical rain forest is a broadleaf evergreen forest that dominates the wet tropics. Lambir Hills, Sarawak, Malaysia (Photo by Michael J. Doolittle/Rainbow)

Consequently, the major interruptions near the equator are principally cooler highland areas.

Data for some representative stations in the wet tropics are shown in Figure 18.4A, B. A brief examination reveals the most obvious features that characterize the climate in these areas.

1. Temperatures usually average 25°C or more each month. Consequently, not only is the annual mean temperature high, but the annual temperature range is also very small.
2. The total precipitation for the year is high, often exceeding 200 centimeters.
3. Although rainfall is not evenly distributed throughout the year, tropical rain forest stations are generally wet in all months. If a dry season exists, it is a very short one.

Because places with an Af or Am designation lie near the equator, the reason for the uniform temperature rhythm experienced in such locales is clear: The intensity of solar radiation is consistently high. The vertical rays of the Sun are always relatively close, and changes in the length of daylight throughout the year are slight; therefore, seasonal temperature variations are minimal.

The region is strongly influenced by the equatorial low. Its converging trade winds and the accompanying ascent of warm, humid, unstable air produce conditions that are ideal for the formation of precipitation.

Tropical Wet and Dry

In the latitude zone poleward of the wet tropics and equatorward of the subtropical deserts lies a transitional climatic region called **tropical wet and dry.** Here the rain forest gives way to the *savanna*, a tropical grassland with scattered drought-tolerant trees (Figure 18.5). Because temperature characteristics among all A climates are quite similar, the primary factor that distinguishes the Aw climate from Af and Am is precipitation. Although the overall amount of precipitation in the tropical wet and dry realm is often considerably less than in the wet tropics, the most distinctive feature of this climate is not the annual rainfall total but the markedly seasonal character of the rainfall. The climatic diagram for Normanton, Australia (Figure 18.4C), clearly illustrates this trait. As the equatorial low advances poleward in summer, the rainy season commences and features weather patterns typical of the wet tropics. Later, with the retreat of the equatorial low, the subtropical high advances into the region and brings

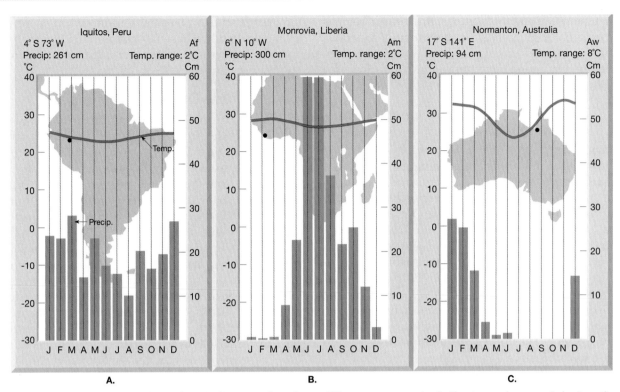

Figure 18.4 By comparing these three climatic diagrams, the primary differences among the A climates can be seen. **A.** Iquitos, the Af station, is wet throughout the year. **B.** Monrovia, the Am station, has a short, dry season. **C.** As is true for all Aw stations, Normanton has an extended dry season and a higher annual temperature range than the others.

Figure 18.5 Tropical savanna grassland in Tanzania's Serengeti National Park. This tropical savanna, with its stunted, drought-resistant trees, probably resulted from seasonal burnings carried out by native human populations. (Photo by Stan Osolinski/Dembinsky/Photo Associates)

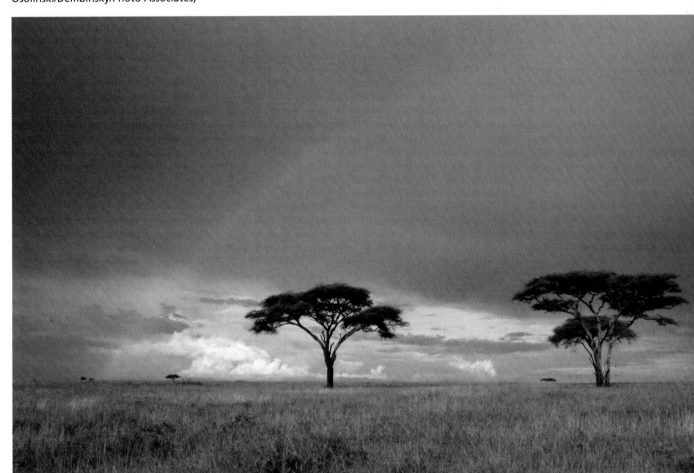

with it intense drought conditions. In some Aw regions such as India, Southeast Asia, and portions of Australia, the alternating periods of rainfall and drought are associated with a pronounced monsoon circulation (see Chapter 16).

Dry (B) Climates

It is important to realize that the concept of dryness is a relative one and refers to any situation in which a water deficiency exists. Climatologists define a dry climate as one in which the yearly precipitation is not as great as the potential loss of water by evaporation. Thus, dryness is not only related to annual rainfall totals, but is also a function of evaporation, which in turn is closely dependent upon temperature.

To establish the boundary between dry and humid climates, the Köppen classification uses formulas that involve three variables: average annual precipitation, average annual temperature, and seasonal distribution of precipitation. The use of average annual temperature reflects its importance as an index of evaporation. The amount of rainfall defining the humid–dry boundary increases as the annual mean temperature increases. The use of seasonal precipitation as a variable is also related to this idea. If rain is concentrated in the warmest months, loss to evaporation is greater than if the precipitation were concentrated in the cooler months.

Within the regions defined by a general water deficiency are two climatic types: **arid** or **desert** (BW) and **semiarid** or **steppe** (BS) (Figure 18.6). These two groups have many features in common; their differences are primarily a matter of degree.

The semiarid is a marginal and more humid variant of the arid and represents a transition zone that surrounds the desert and separates it from the bordering humid climates.

Low-Latitude Deserts and Steppes

The heart of low-latitude dry climates lies in the vicinities of the Tropics of Cancer and Capricorn. A glance at Figure 18.6 shows a virtually unbroken desert environment stretching for more than 9300 kilometers from the Atlantic coast of North Africa to the dry lands of northwestern India. In addition to this single great expanse, the Northern Hemisphere contains another much smaller area of subtropical desert and steppe in northern Mexico and the southwestern United States. In the Southern Hemisphere, dry climates dominate Australia. Almost 40 percent of the continent is desert, and much of the remainder is steppe. In addition, arid and semi-arid areas are found in southern Africa and make a limited appearance in coastal Chile and Peru.

The existence of this dry subtropical realm is primarily the result of the prevailing global distribution of air pressure and winds. Earth's low-latitude deserts and steppes coincide with the subtropical high-pressure belts (see Figures 16.12, p. 451, and 16.13, p. 452). Here, air is subsiding. When air sinks, it is compressed and warmed. Such conditions are just opposite of what is needed for cloud formation and precipitation. Therefore, clear skies, a maximum of sunshine, and drought are to be expected. The climate diagrams for Cairo, Egypt, and Monterrey, Mexico (Figure 18.7A, B), illustrate the characteristics of low-latitude dry climates.

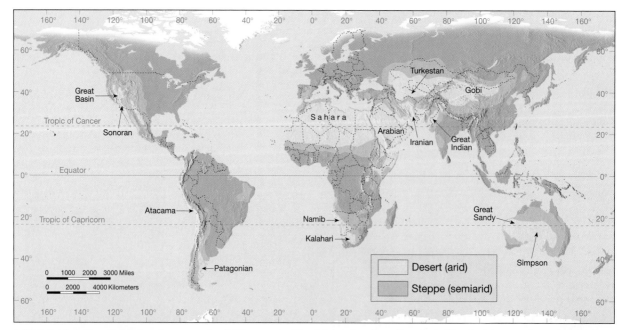

Figure 18.6 Arid and semiarid climates cover about 30 percent of Earth's land surface. No other climatic group covers so large an area.

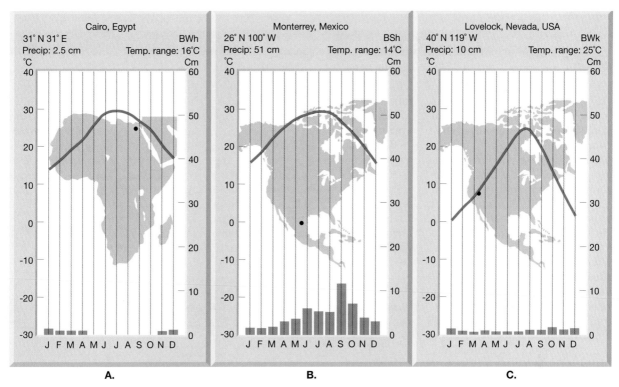

Figure 18.7 Climatic diagrams for representative arid and semiarid stations. Stations **A.** and **B.** are in the subtropics, whereas **C.** is in the middle-latitudes. Cairo and Lovelock are classified as deserts; Monterey is a steppe. Lovelock, Nevada, may also be called a rainshadow desert.

Middle-Latitude Deserts and Steppes

Unlike their low-latitude counterparts, middle-latitude deserts and steppes are not controlled by the subsiding air masses associated with high pressure. Instead, these dry lands exist principally because of their positions in the deep interiors of large landmasses far removed from the oceans, which are the ultimate source of moisture for cloud formation and precipitation. In addition, the presence of high mountains across the paths of prevailing winds further acts to separate these areas from water-bearing, maritime air masses.

Windward sides of mountains are often wet. As prevailing winds meet mountain barriers, the air is forced to ascend, producing clouds and precipitation (see Box 18.2). By contrast, the leeward sides of mountains are usually much drier and are often arid enough to be referred to as **rainshadow deserts** (see Figure 15.16, p. 421). Because many middle-latitude deserts occupy sites on the leeward sides of the mountains, they can also be classified as rainshadow deserts (Figure 18.7C). In North America, the Coast Ranges, Sierra Nevada, and Cascades are the foremost mountain barriers. In Asia, the great Himalayan chain prevents the summertime monsoon flow of moist Indian Ocean air from reaching the interior. Because the Southern Hemisphere lacks extensive land areas in the middle latitudes, only a small area of desert and steppe is found in this latitude range, existing primarily in the rainshadow of the towering Andes.

In the case of middle-latitude deserts, we have an example of the impact of tectonic processes on climate. Rainshadow deserts exist by virtue of the mountains produced when plates collide. Without such mountain-building episodes, wetter climates would prevail where many dry regions exist today.

Humid Middle-Latitude Climates With Mild Winters (C Climates)

Although the term *subtropical* is often used for the C climates, it can be misleading. Although many areas with C climates do indeed possess some near-tropical characteristics, other regions do not. For example, we would be stretching the use of the term *subtropical* to describe the climates of coastal Alaska and Norway, which belong to the C group. Within the C group of climates, several subgroups are recognized.

Humid Subtropics

Located on the eastern sides of the continents, in the 25- to 40-degree latitude range, the **humid subtropical climate** dominates the southeastern United States, as well as other similarly situated areas around the world

Box 18.2 Precipitation Records and Mountainous Terrain

Many of the rainiest places in the world are located on windward mountain slopes. Typically these rainy areas are situated such that the mountains act as a barrier to the general circulation. Thus, the prevailing winds are forced to ascend the sloping terrain, thereby generating clouds and often abundant precipitation. A station at Mt. Waialeale, Hawaii, for example, records the highest average annual rainfall in the world, some 1234 centimeters (486 inches). The station is located on the windward (northeast) coast of the island of Kauai at an elevation of 1569 meters (5148 feet). Incredibly, only 31 kilometers (19 miles) away lies the popular sunspot Barking Sands, with an annual precipitation that averages less than 50 centimeters (20 inches).

The greatest recorded rainfall for a 12-month period occurred at Cherrapunji, India, where an astounding 2647 centimeters (1042 inches), over 86 feet, fell. Cherrapunji, which is located at an elevation of 1293 meters (4309 feet), lies just north of the Bay of Bengal in an ideal location to receive the full effect of India's wet, summer monsoon. Most all of this rainfall occurred in the summer, particularly during the month of July, when a record of 930 centimeters (366 inches) fell. For comparison, 10 times more rain fell in a single month at Cherrapunji than falls in an average year at Chicago, Illinois.

Because mountains can be sites of abundant precipitation, they are frequently very important sources of water. This is particularly true for arid regions such as the southwestern United States. Here the snowpack that accumulates high in the mountains during the winter is a major source of water for the summer season when precipitation is light and demand is high (Figure 18.A). Reservoirs in the Sierra Nevada, for example, accumulate and store spring runoff, which is then delivered to cities such as Los Angeles by way of an extensive network of canals. The record for greatest annual snowfall in the United States goes to Paradise Ranger Station, located on Mount Rainier, where 2850 centimeters (1122 inches) of snow fell during the winter of 1971/72.

Figure 18.A Heavy snowpack along Trail Ridge Road, Rocky Mountain National Park, Colorado. (Photo by Henry Lansford)

(Figures 18.2 and 18.8A). In the summer, the humid subtropics experience hot, sultry weather of the type one expects to find in the rainy tropics. Daytime temperatures are generally high, and because both specific and relative humidities are high, the night brings little relief. An afternoon or evening thunderstorm is also possible, for these areas experience such storms on an average of 40 to 100 days each year, the majority during the summer months.

As summer turns to autumn, the humid subtropics lose their similarity to the rainy tropics. Although winters are mild, frosts are common in the higher-latitude Cfa areas and occasionally plague the tropical margins as well. The winter precipitation is also different in character from the summer. Some is in the form of snow, and most is generated along fronts of the frequent middle-latitude cyclones that sweep over these regions.

Marine West Coast

Situated on the western (windward) side of continents, from about 40 to 65 degrees north and south latitude, is a climatic region dominated by the onshore flow of oceanic air. In North America, the **marine west coast climate** extends from near the U.S.–Canadian border northward as a narrow belt into southern Alaska (Figure 18.8B). The largest area of Cfb climate is found in Europe, for here there is no mountain barrier blocking the movement of cool maritime air from the North Atlantic.

The prevalence of maritime air masses means that mild winters and cool summers are the rule, as is an ample amount of rainfall throughout the year.

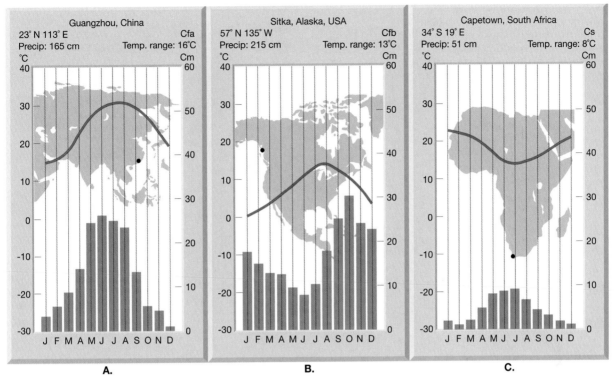

Figure 18.8 Each of these climatic diagrams represents one of the three main types of C climates: **A.** humid subtropical, **B.** marine west coast, and **C.** dry-summer subtropical.

Although there is no pronounced dry period, there is a drop in monthly precipitation totals during the summer. The reason for the reduced summer rainfall is the poleward migration of the oceanic subtropical highs. Although the areas of marine west coast climate are situated too far poleward to be dominated by these dry anticyclones, their influence is sufficient to cause a decrease in warm season rainfall.

Dry-Summer Subtropics

The **dry-summer subtropical climate** is typically located along the west sides of continents between latitudes 30 and 45°. Situated between the marine west coast climate on the poleward side and the subtropical steppes on the equatorward side, this climatic region is best described as transitional in character. It is unique because it is the only humid climate that has a strong winter rainfall maximum, a feature that reflects its intermediate position (Figure 18.8C). In summer, the region is dominated by stable conditions associated with the oceanic subtropical highs. In winter, as the wind and pressure systems follow the Sun equatorward, it is within range of the cyclonic storms of the polar front. Thus, during the course of a year, these areas alternate between becoming a part of the dry tropics and an extension of the humid middle latitudes. Although middle-latitude

changeability characterizes the winter, tropical constancy describes the summer.

As was the case for the marine west coast climate, mountain ranges limit the dry-summer subtropics to a relatively narrow coastal zone in both North and South America. Because Australia and southern Africa barely reach to the latitudes where dry-summer climates exist, the development of this climatic type is limited on these continents as well. Consequently, because of the arrangement of the continents, and of their mountain ranges, inland development occurs only in the Mediterranean basin. Here the zone of subsidence extends far to the east in summer; in winter, the sea is a major route of cyclonic disturbances. Because the dry-summer climate is particularly extensive in this region, the name *Mediterranean climate* is often used as a synonym (Figure 18.9).

Humid Middle-Latitude Climates With Severe Winters (D Climates)

The C climates that were just described characteristically have mild winters. By contrast, D climates experience severe winters. Two types of D climates are recognized, the humid continental and the subarctic. Climatic diagrams of representative locations are shown in Figure 18.9. The D climates are land-controlled climates, the

Figure 18.9 *D* climates are associated with the interiors of large landmasses in the mid-to-high latitudes of the Northern Hemisphere. Although winters can be harsh in Chicago's humid continental (Dfa) climate, the subarctic environment (Dfc) of Moose Factory is more extreme.

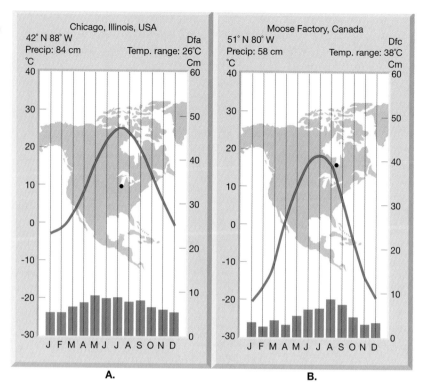

A.

B.

result of broad continents in the middle latitudes. Because continentality is a basic feature, D climates are absent in the Southern Hemisphere where the middle-latitude zone is dominated by the oceans.

Humid Continental

The **humid continental climate** is confined to the central and eastern portions of North America and Eurasia in the latitude range between approximately 40 and 50 degrees north latitude. It may at first seem unusual that a continental climate should extend eastward to the margins of the ocean. However, because the prevailing atmospheric circulation is from the west, deep and persistent incursions of maritime air from the east are not likely to occur.

Both winter and summer temperatures in the humid continental climate can be characterized as relatively severe. Consequently, annual temperature ranges are high throughout the climate.

Precipitation is generally greater in summer than in winter. Precipitation totals generally decrease toward the interior of the continents as well as from south to north, primarily because of increasing distance from the sources of mT air. Furthermore, the more northerly stations are also influenced for a greater part of the year by drier polar air masses.

Wintertime precipitation in humid continental climates is chiefly associated with the passage of fronts connected with traveling middle-latitude cyclones. Part of this precipitation is in the form of snow, the proportion increasing with latitude. Although precipitation is often considerably less during the cold season, it is usually more conspicuous than the greater amounts that fall during summer. An obvious reason is that snow remains on the ground, often for extended periods, and rain, of course, does not.

Subarctic

Situated north of the humid continental climate and south of the polar tundra is an extensive **subarctic climate** region covering broad, uninterrupted expanses from western Alaska to Newfoundland in North America and from Norway to the Pacific coast of Russia in Eurasia (see Figure 14.15 p. 391). It is often referred to as the *taiga* climate, for its extent closely corresponds to the northern coniferous forest region of the same name. Although scrawny, the spruce, fir, larch, and birch trees in the taiga represent the largest stretch of continuous forest on the surface of Earth.

Here in the source regions of continental polar air masses, the outstanding feature is certainly the dominance of winter. Not only is it long, but temperatures are also bitterly cold. Winter minimum temperatures are among the lowest ever recorded outside the ice sheets of Greenland and Antarctica. In fact, for many years, the world's coldest temperature was attributed to Verkhoyansk in east central Siberia, where the temperature dropped to −68°C on February 5 and 7, 1892. Over a 23-year period, this same station had an average monthly minimum of −62°C during January.

Although exceptional temperatures, they illustrate the extreme cold that envelop the taiga in winter.

By contrast, summers in the subarctic are remarkably warm, despite their short duration. However, when compared with regions farther south, this short season must be characterized as cool. The extremely cold winters and relatively warm summers combine to produce the highest annual temperature ranges on Earth. Because these far northerly continental interiors are the source regions for cP air masses, there is very limited moisture available throughout the year. Precipitation totals are therefore small, with a maximum occurring during the warmer summer months.

Polar (E) Climates

Polar climates are those in which the mean temperature of the warmest month is below 10°C. Thus, just as the tropics are defined by their year-round warmth, the polar realm is known for its enduring cold. As winters are periods of perpetual night, or nearly so, temperatures at most polar locations are understandably bitter. During the summer months temperatures remain cool despite the long days, because the Sun is so low in the sky that its oblique rays are not effective in bringing about a genuine warming. Although polar climates are classified as humid, precipitation is generally meager. Evaporation, of course, is also limited. The scanty precipitation totals are easily understood in view of the temperature characteristics of the region. The amount of water vapor in the air is always small because low specific humidities must accompany low temperatures. Usually precipitation is most abundant during the warmer summer months when the moisture content of the air is highest.

Two types of polar climates are recognized. The **tundra climate** (ET) is a treeless climate found almost exclusively in the Northern Hemisphere (Figure 18.10). Because of the combination of high latitude and continentality, winters are severe, summers are cool, and annual temperature ranges are high (Figure 18.11A). Further, yearly precipitation is small, with a modest summer maximum.

The **ice cap climate** (EF) does not have a single monthly mean above 0°C (Figure 18.11B). Consequently, because the average temperature for all months is below freezing, the growth of vegetation is prohibited and the landscape is one of permanent ice and snow. This climate of perpetual frost covers a surprisingly large area—more than 15.5 million square kilometers, or about 9 percent of Earth's land area. Aside from scattered occurrences in high mountain areas, it is confined to the ice sheets of Greenland and Antarctica (see Figure 5.2, p. 125).

Highland Climates

It is a well-known fact that mountains have climate conditions that are distinctively different from those found in adjacent lowlands. Compared to nearby places at lower elevations, sites with **highland climates** are cooler and usually wetter. Unlike the world climate types already discussed, which consist of large, relatively homogeneous regions, the outstanding characteristic of highland climates is the great diversity of climatic conditions that occur.

The best-known climatic effect of increased altitude is lower temperatures. In addition, an increase in precipitation due to orographic lifting usually occurs at higher elevations. Despite the fact that mountain stations are colder and often wetter than locations at

Figure 18.10 The tundra in bloom north of Nome, Alaska. It is a region almost completely devoid of trees. Bogs and marshes are common, and plant life frequently consists of mosses, low shrubs, and flowering herbs. (Photo by Fred Bruemmer/DRK Photo)

Figure 18.11 These climatic diagrams represent the two basic types of polar climates. **A.** Barrow, Alaska, exhibits a tundra (ET) climate. **B.** Eismitte, Greenland, a station located on a massive ice sheet, is classified as an ice cap (EF) climate.

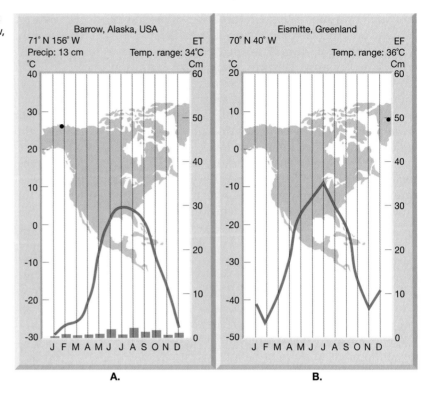

A.   B.

lower elevations, highland climates are often very similar to those in adjacent lowlands in terms of seasonal temperature cycles and precipitation distribution. Figure 18.12 illustrates this relationship.

Phoenix, at an elevation of 338 meters, lies in the desert lowlands of southern Arizona. By contrast, Flagstaff is located at an altitude of 2100 meters on the Colorado Plateau in northern Arizona (Figure 18.13). When summer averages climb to 34°C in Phoenix, Flagstaff is experiencing a pleasant 19°C, which is a full 15°C cooler. Although the temperatures at each city are quite different, the pattern of monthly temperature changes for each place is similar. Both experience their minimum and maximum monthly means in the same months. When precipitation data are examined, both places have a similar seasonal pattern, but the amounts at Flagstaff are higher in every month. In addition, owing to its higher altitude, much of Flagstaff's winter precipitation is in the form of snow. By contrast, all of the precipitation at Phoenix is rain.

Perhaps the terms *variety* and *changeability* best describe mountain climates. Because atmospheric conditions fluctuate rapidly with changes in altitude and exposure, a nearly limitless variety of local climates occur in mountainous regions. The climate in a protected valley is very different from that of an exposed peak. Conditions on windward slopes contrast sharply with those on the leeward sides, whereas slopes facing the Sun are unlike those that lie mainly in the shadows.

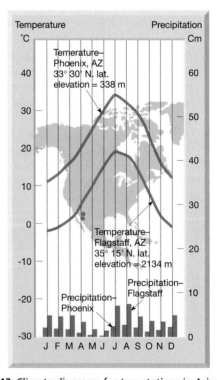

Figure 18.12 Climate diagrams for two stations in Arizona illustrate the general influence of elevation on climate. Flagstaff is cooler and wetter because of its position on the Colorado Plateau, nearly 1800 meters higher than Phoenix.

A.

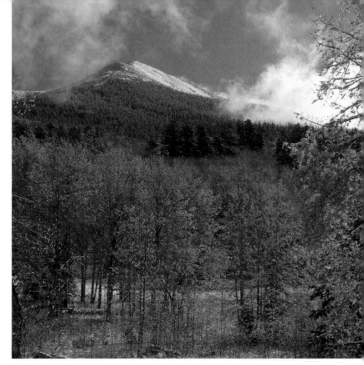

B.

Figure 18.13 A. Only scanty drought-tolerant natural vegetation can survive in the hot, dry climate of southern Arizona, near Phoenix (Photo by Charlie Ott Photography/Photo Researchers, Inc.). **B.** The natural vegetation associated with the cooler, wetter highlands near Flagstaff, Arizona, is much different from the desert lowlands. (Photo by Larry Ulrich/DRK Photo)

Human Impact on Global Climate

The proposals to explain global climatic change are many and varied. In Chapter 5, we examined some possible causes for ice-age climates. These hypotheses, which included the movement of lithosphere plates and variations in Earth's orbit, involved natural forcing mechanisms. Another natural forcing mechanism, discussed in Chapter 8, is the possible role of explosive volcanic eruptions in modifying the atmosphere. It is important to remember that these mechanisms, as well as others, not only have contributed to climatic changes in the geologic past, but will also be responsible for future shifts in climate. However, when relatively recent and future changes in our climate are considered, we must also examine the possible impact of human beings (see Box 18.3). In this section we shall examine the major way in which humans might be contributing to global climatic change. This impact results from the addition of carbon dioxide and other gases to the atmosphere.

Human influence of regional and global climate probably did not just begin with the onset of the modern industrial period. There is good evidence that humans have been modifying the environment over extensive areas for thousands of years. The use of fire as well as the overgrazing of marginal lands by domesticated animals have negatively affected the abundance and distribution of vegetation. By altering ground cover, such important climatological factors as surface albedo, evaporation rates, and surface winds

have been, and continue to be, modified. Commenting on this aspect of human-induced climatic modification, the authors of one study observed:

> In contrast to the prevailing view that only modern humans are able to alter climate, we believe it is more likely that the human species has made a substantial and continuing impact on climate since the invention of fire.*

It should be pointed out that when any hypothesis of climatic change is examined, whether it depends on natural or human-induced causes, a degree of caution must be exercised. It is safe to say that most, if not all, hypotheses are to some degree controversial and speculative. This is to be expected if we consider the fact that at present all of our models of Earth's climate are far from complete. Because atmospheric processes are so large and complex, they cannot be physically reproduced in laboratory experiments. Instead, the climate must be simulated mathematically with the aid of computers. Although such models are sophisticated enough to serve as primary tools for climate research, they cannot yet approach the actual complexity of the atmosphere. Computer models are powerful and essential aids, but climate forecasts based on such simulations still contain many uncertainties.

---

*Carl Sagan et al., "Anthropogenic Albedo Changes and the Earth's Climate," *Science* 206(4425):1367.

Box 18.3 People and the Environment: The Urban Heat Island

The most apparent human impact on climate is the modification of the atmospheric environment by the building of cities. The construction of every factory, road, office building, and house destroys microclimates and creates new ones of great complexity.

The most studied and well-documented urban climatic effect is the *urban heat island*. The term simply refers to the fact that temperatures within cities are generally higher than in rural areas. The heat island is evident when temperature data such as appear in Table 18.C are examined. As is typical, the data for Philadelphia show the heat island is most pronounced when minimum temperatures are examined. The magnitude of the temperature differences shown by this table is probably even greater than the figures indicate, because temperatures observed at suburban airports are usually higher than those in truly rural environments.

Figure 18.B, which shows the distribution of average minimum temperatures in the Washington, D.C., metropolitan area for the 3-month winter period over a 5-year span, also illustrates a well-developed heat island. The warmest winter temperatures occurred in the heart of the city, while the suburbs and surrounding countryside experienced average minimum temperatures that were as much as 3.3°C lower. Remember that these temperatures are averages, because on many clear, calm nights the temperature difference between the city center and the countryside was considerably greater, often 11°C (20°F) or more. Conversely, on many overcast or windy nights the temperature differential approached zero degrees.

Why are cities warmer? The radical change in the surface that results when rural areas are transformed into cities is a significant cause of the urban heat island. First, the tall buildings and the concrete and asphalt of the city absorb and store greater quantities of solar radiation than do the vegetation and soil typical of rural areas. In addition, because the city surface is impermeable, the runoff of water following a rain is rapid, resulting in a severe

Table 18.C Average temperatures (°C) for suburban Philadelphia Airport and downtown Philadelphia (10-year averages).

	Airport	Downtown
Annual mean	12.8	13.6
Mean June max	27.8	28.2
Mean December max	6.4	6.7
Mean June min	16.5	17.7
Mean December min	−2.1	−0.4

Source: After H. Neuberger and J. Cahir, *Principles of Climatology* (New York: Holt, Rinehart and Winston, 1969), 128.

reduction on the evaporation rate. Hence, heat that once would have been used to convert liquid water to a gas now goes to increase further the surface temperature. At night, while both the city and countryside cool by radiative losses, the stone-like surface of the city gradually releases the additional heat accumulated during the day, keeping the urban air warmer than that of the outlying areas.

A portion of the urban temperature rise is also attributed to waste heat from sources such as home heating and air conditioning, power generation, industry, and transportation. In addition, the "blanket" of pollutants over a city contributes to the heat island by absorbing a portion of the upward-directed long-wave radiation emitted by the surface and re-emitting some of it back to the ground.

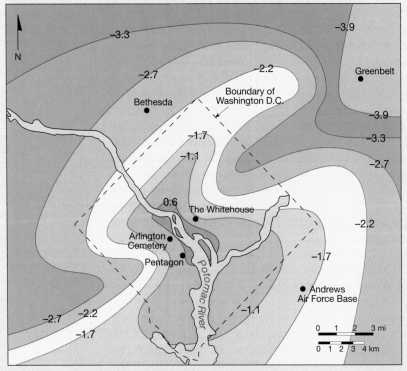

Figure 18.B The heat island of Washington, D.C., as shown by the average minimum temperatures (°C) during the winter season (December through February). The city center had an average minimum that was nearly 4°C higher than some outlying areas. (After Clarence A. Woolum, "Notes from the Study of the Microclimatology of the Washington, D.C. Area for the Winter and Spring Seasons," *Weatherwise*, 17, no. 6 (1964), 264, 267.)

Carbon Dioxide, Trace Gases, and Global Warming

In Chapter 14, we learned that although carbon dioxide (CO_2) represents only about 0.036 percent of the gases that make up clean, dry air, it is nevertheless a meteorologically significant component. The importance of carbon dioxide lies in the fact that it is transparent to incoming short-wavelength solar radiation, but it is not transparent to some of the longer-wavelength outgoing terrestrial radiation. A portion of the energy leaving the ground is absorbed by carbon dioxide and subsequently reemitted, part of it toward the surface, thereby keeping the air near the ground warmer than it would be without carbon dioxide.

Thus, along with water vapor, carbon dioxide is largely responsible for the *greenhouse effect* of the atmosphere. Carbon dioxide is an important heat absorber, and it follows logically that any change in the air's carbon dioxide content could alter temperatures in the lower atmosphere.

Earth's tremendous industrialization of the past two centuries has been fueled—and still is fueled—by burning fossil fuels—coal, natural gas, and petroleum (Figure 18.14). Combustion of these fuels has added great quantities of carbon dioxide to the atmosphere.

The use of coal and other fuels is the most prominent means by which humans add CO_2 to the atmosphere, but it is not the only way. The clearing of forests also contributes substantially because CO_2 is released as vegetation is burned or decays. Deforestation is particularly pronounced in the tropics, where vast tracts are cleared for ranching and agriculture or subjected to inefficient commercial logging operations.

Although some of the excess CO_2 is taken up by plants or is dissolved in the ocean, it is estimated that 45 to 50 percent remains in the atmosphere. Figure 18.15 shows CO_2 concentrations over the past thousand years based on ice-core records and (since 1958) measurements taken at Mauna Loa Observatory, Hawaii. The rapid increase in CO_2 concentration since the onset of industrialization is obvious and has closely followed the increase in CO_2 emissions from burning fossil fuels. Figure 18.15B shows this parallel growth.

We might expect that global temperature has increased as a result of the growing carbon dioxide level. In fact, since the late 1800s the mean global temperature has risen by about 0.3 to 0.6°C and by about 0.2 to 0.3°C over the last 45 years. Moreover, recent years have been among the warmest since the mid-1800s despite the cooling effect of the 1991 Mount Pinatubo eruption. Are these temperature trends caused by human activities or would they have occurred anyway? Scientists are cautious, but now they seem convinced that "the observed trend in global mean temperature over the past 100 years is unlikely to be natural in origin," and that the data "point towards a human influence on global climate."*

---

*Intergovernmental Panel on Climate Change, *Climate Change 1995: The Science of Climate Change*, New York: Cambridge University Press, 1996, p. 4.

A.

B.

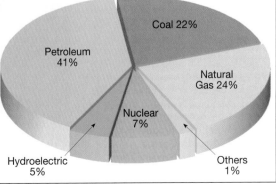

Figure 18.14 A. Paralleling the rapid growth of industrialization, which began in the nineteenth century, has been the combustion of fossil fuels, which has added great quantities of carbon dioxide to the atmosphere. (Photo by Bruce Forster/Tony Stone Images) **B.** Energy consumption in the United States, 1998. Fossil fuels (petroleum, coal, and natural gas) represent about 86 percent of the total. (Data from U.S. Department of Energy)

Coal 22%

Petroleum 41%

Natural Gas 24%

Nuclear 7%

Others 1%

Hydroelectric 5%

Figure 18.15 A. Carbon dioxide (CO_2) concentrations over the past 1000 years. Most of the record is based on data obtained from Antarctic ice cores. Bubbles of air trapped in the glacial ice provide samples of past atmospheres. The record since 1958 comes from direct measurements of atmospheric CO_2 taken at Mauna Loa Observatory, Hawaii. **B.** The rapid increase in CO_2 concentration since the onset of industrialization in the late 1700s is clear and has followed closely the rise in CO_2 emissions from fossil fuels.

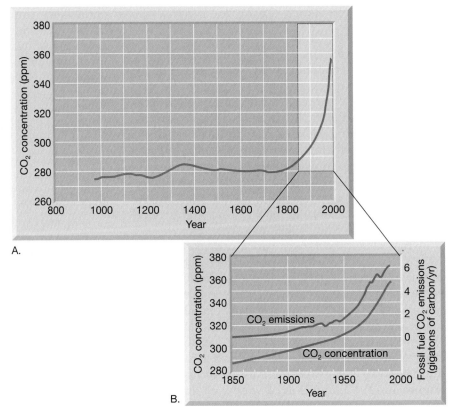

A.

B.

If fossil fuel use continues to increase at projected rates, current estimates are that the atmosphere's CO_2 content will approach 400 ppm by the year 2010 and 600 ppm during the second half of the twenty-first century. With such an increase, the greenhouse effect would become much more dramatic and measurable than in the past. The most realistic models predict an increase in the mean global surface temperature of about 2.5°C. A change of this magnitude would be unprecedented in human history. It would approach the warming that has taken place since the peak of the most recent glacial stage 18,000 years ago, but it would occur *much more rapidly.*

Carbon dioxide is not the only gas contributing to a possible global increase in temperature. In recent years, atmospheric scientists have come to realize that the industrial and agricultural activities of people are causing a buildup of several trace gases that may also play a significant role. The substances are called *trace gases* because their concentrations are so much smaller than that of carbon dioxide. The trace gases that appear to be most important are methane (CH_4) (Figure 18.16), nitrous oxide (N_2O), and certain types of chlorofluorocarbons (CFCs). These gases absorb wavelengths of outgoing Earth radiation that would otherwise escape into space. Although individually

Figure 18.16 Methane is produced by *anaerobic* bacteria in wet places where oxygen is scarce (anaerobic means "without air," specifically oxygen). Such places include swamps, bogs, wetlands, and the guts of termites and grazing animals like cattle and sheep. Methane is also generated in flooded paddy fields ("artificial swamps") used for growing rice. These paddies are in India's Ganges lowlands. Mining of coal and drilling for oil and natural gas are other sources because methane is a product of their formation. (Photo by George Holton/Photo Researchers, Inc.)

their impact is modest, taken together the effects of these trace gases may be nearly as great as CO_2 in warming Earth.

Sophisticated computer models of the atmosphere show that the warming of the lower atmosphere caused by CO_2 and trace gases will not be the same everywhere. Rather, the temperature response in polar regions could be two to three times greater than the global average. Part of the reason for such a response is an expected reduction in sea ice. This topic is explored more fully in Box 18.4.

Some Possible Consequences of Greenhouse Warming

What consequences can be expected if the carbon dioxide content of the atmosphere reaches a level that is twice what it was early in the twentieth century? Because the climate system is so complex, predicting the distribution of particular regional changes is still very speculative. It is not yet possible to pinpoint specifics, such as where or when it will become drier or wetter. Nevertheless, plausible scenarios can be given for larger scales of space and time. As computers become larger and data improve, scientists will gradually develop models that can provide more specific and reliable results.

As noted, the magnitude of the temperature increase will not be the same everywhere. The temperature rise will probably be smallest in the tropics and increase toward the poles. As for precipitation, the models indicate that some regions will experience significantly more precipitation and runoff. However, other regions will experience a decrease in runoff due to reduced precipitation or because of greater evaporation caused by higher temperatures.

Water Resources and Agriculture. Such changes could profoundly alter the distribution of the world's water resources and hence affect the productivity of agricultural regions that depend on rivers for irrigation water. For example, a 2°C warming and 10 percent precipitation decrease in the region drained by the Colorado River could diminish the river's flow by 50 percent or more. Because the present flow of the river barely meets current demand for irrigation agriculture, the negative effect would be serious (Figure 18.17). Many other rivers are the basis for extensive irrigated agriculture, and the projected reduction of their flow could have equally grave consequences. In contrast, large precipitation increases in other areas would increase the flow of some rivers and bring more frequent destructive floods.

Harder to estimate is the effect on nonirrigated crops that depend on direct rainfall for moisture. Some places will no doubt experience productivity loss due to a decrease in rainfall or increase in evaporation. Still, these losses may be offset by gains elsewhere. Warming in the high latitudes could lengthen the growing season, for instance. This, in turn, could allow expansion of agriculture into areas that are presently not suited to crop production.

Sea-Level Rise. Another impact of a human-induced global warming is a probable rise in sea level. How is a warmer atmosphere related to a global rise in sea level? The most obvious connection, the melting of glaciers, is important, but *not* the most significant. Far more significant is that a warmer atmosphere causes an increase in ocean volume due to thermal expansion. Higher air temperatures warm the adjacent upper layers of the ocean, which in turn causes the water to expand and sea level to rise (see Box 13.2, p. 369).

Figure 18.17 It is not yet possible to specify the magnitude and location of particular climate changes that may result from greenhouse warming. Many consequences are possible. Decreased rainfall and increased evaporation rates could diminish the flow of certain rivers and force the abandonment of some presently productive irrigated farmland. (Photo by E. J. Tarbuck)

Box 18.4 Earth as a System: Climatic Feedback Mechanisms

Because climate is a very complex interactive physical system, scientists must consider many possible outcomes when one of the system's elements is altered. These various possibilities, termed *climatic feedback mechanisms*, not only complicate climatic modeling efforts, but they also add greater uncertainty to climatic predictions.

The most important and obvious of the feedback effects related to increases in carbon dioxide arises from the fact that higher surface temperatures produce greater evaporation rates. This, in turn, increases the amount of water vapor in the atmosphere. Remember that water vapor is also a powerful absorber of long-wavelength radiation emitted by Earth. Therefore, with more water vapor in the air, the temperature increase caused by carbon dioxide alone is reinforced.

Recall that the temperature increase at high latitudes is expected to be as much as two or three times greater than the global average. This assumption is based in part on the likelihood that the area covered by sea ice will decrease as surface temperatures rise. Because ice reflects a much larger percentage of incoming solar radiation than does open water, the melting of the pack ice would replace a highly reflective surface with a relatively dark surface (Figure 18.C). The result would be a substantial increase in the amount of solar energy absorbed at the surface. This, in turn, would feed back to the atmosphere and magnify the initial temperature increase created by higher carbon dioxide levels.

So far, the climate feedback mechanisms discussed have magnified the temperature rise caused by the buildup of carbon dioxide. The effects that reinforce an initial change are termed *positive feedback mechanisms*. As we shall see, however, other effects

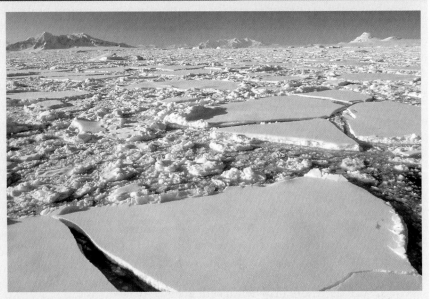

Figure 18.C Sea ice in the Matha Strait, Antarctica. A reduction in sea ice would act as a positive feedback mechanism because surface albedo would decrease and the amount of solar energy absorbed at the surface would increase. (Photo by Kim Heacox/DRK Photo)

must be classified as *negative feedback mechanisms* because they produce results that are just the opposite of the initial change and tend to offset it.

One probable result of a global rise in temperatures would be an accompanying increase in cloud cover resulting from the higher moisture content of the atmosphere. Most clouds are good reflectors of solar radiation. At the same time, however, they are also good absorbers and emitters of terrestrial radiation. Consequently, clouds produce two opposite effects. They act as a negative feedback mechanism because they increase albedo and thus diminish the amount of solar energy available to heat the atmosphere. On the other hand, clouds act as a positive feedback mechanism by absorbing and emitting terrestrial radiation that would otherwise be lost from the troposphere.

Which effect, if either, is stronger? Atmospheric modeling shows that the negative effect of a higher albedo is the more dominant. Therefore, the net

result of an increase in cloudiness should be a decrease in air temperature. The magnitude of this negative feedback, however, is not believed to be as strong as the positive feedback caused by added moisture and decreased sea ice. Thus, although increases in cloud cover may partly offset a global temperature increase, climatic models show that the ultimate effect of the projected increase in CO_2 and trace gases will still be a temperature increase.

The problem of global warming induced by increases in carbon dioxide and trace gases continues to be one of the most studied and discussed aspects of climatic change. Although there are no models that can yet incorporate the full range of potential influences and feedbacks, the consensus in the scientific community is that altering atmospheric composition will eventually lead to a warmer planet with a different distribution of climatic regimes.

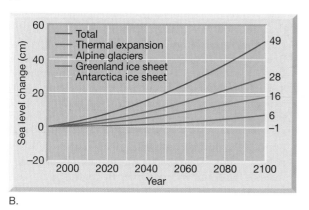

A. High, middle, and low projections of global sea-level rise, 1990–2100. **B.** Projected individual contributions to global sea-level change, 1990–2100, for the middle projection in part A. Thermal expansion is responsible for 28 centimeters of the sea-level rise. Melting alpine glaciers and ice caps contribute another 16 centimeters. The impact shown for Antarctica indicates that this massive ice sheet is projected to grow larger during this period.

Research indicates that sea level has risen between 10 and 25 centimeters over the past century and that the trend will continue at an accelerated rate. Some models indicate that the rise may approach or even exceed 50 centimeters by the end of the twenty-first century (Figure 18.18A).

As mentioned, a warmer climate will cause glaciers to melt. In fact, a portion of the 10- to 25-centimeter rise in sea level over the past century is attributed to the melting of alpine glaciers. This contribution is projected to continue through the twenty-first century, as shown in Figure 18.18B. Of course, if the Greenland and Antarctic ice sheets were to experience a significant increase in melting, it would lead to a much greater rise in sea level and a major encroachment by the sea in coastal zones. It should be emphasized, however, that a significant melting of major ice sheets,

although possible at some future date, is *not* expected during the next century.

New Weather Patterns. Atmospheric scientists also expect that weather patterns will change as a result of the projected global warming. Potential weather changes include the following:

1. A higher frequency and greater intensity of hurricanes because of warmer ocean temperatures.
2. Shifts in the paths of large-scale cyclonic storms, which, in turn, would affect the distribution of precipitation and the occurrence of severe storms, including tornadoes.
3. More intense heat waves and droughts in some regions and fewer such events in other places.

The impact on climate of an increase in atmospheric CO_2 and trace gases is obscured by many unknowns and uncertainties. The changes will probably be gradual environmental shifts, imperceptible from year to year. Nevertheless, the effects, accumulated over decades, will have powerful economic, social, and political consequences.

Policymakers are confronted with responding to the risks posed by emissions of greenhouse gases in the face of significant scientific uncertainties. However, they are also faced with the fact that climate-induced environmental changes cannot be reversed quickly, if at all, owing to the lengthy time scales associated with the climate system. Addressing this issue, the Intergovernmental Panel on Climate Change states:

> Uncertainty does not mean that a nation or the world community cannot position itself better to cope with the broad range of possible climate changes or protect against potentially costly future outcomes. Delaying such measures may leave a nation or the world poorly prepared to deal with adverse changes and may increase the possibility of irreversible or very costly consequences. Options for adapting to change or mitigating change that can be justified for other reasons today (e.g., abatement of air and water pollution) and make society more flexible or resilient to anticipated adverse effects of climate change appear particularly desirable.*

*Intergovernmental Panel on Climate Change, *Climate Change 1995: Impacts, Adaptations and Mitigation of Climate Change: Scientific–Technical Analysis.* New York: Cambridge University Press, 1996, p. 23. *Note:* The Intergovernmental Panel on Climate Change was jointly established by the World Meteorological Organization and the United Nations Environment Program to evaluate and make recommendations regarding the many aspects of global climate change.

Chapter Summary

- *Climate* is the aggregate of weather conditions for a place or region over a long period of time. Earth's *climate system* involves the exchanges of energy and moisture that occur among the atmosphere, hydrosphere, solid Earth, biosphere, and *cryosphere* (the ice and snow that exist at Earth's surface).

- Climate classification brings order to large quantities of information, which aids comprehension and understanding, and facilitates analysis and explanation. *Temperature and precipitation are the most important elements in a climatic description.* Many climate-classification schemes have been devised, with the value of each determined by its intended use. The *Köppen classification,* which uses mean monthly and annual values of temperature and precipitation, has been widely used for more than 70 years. The boundaries Köppen chose were largely based on the limits of certain plant associations. Five principal climate groups, each with subdivisions, were recognized. Each group is designated by a capital letter. Four of the climate groups (A, C, D, and E) are defined on the basis of temperature characteristics, and the fifth, the B group, has precipitation as its primary criterion.

- *Humid tropical (A) climates* are winterless with all months having a mean temperature above 18°C. *Wet tropical climates* (Af and Am), which lie near the equator, have constantly high temperatures, enough rainfall to support the most luxuriant vegetation (tropical rain forest) found in any climatic realm. *Tropical wet and dry climates* (Aw) are found poleward of the wet tropics and equatorward of the tropical deserts where the rain forest gives way to the tropical grasslands and scattered drought-tolerant trees of the savanna. The most distinctive feature of this climate is the seasonal character of the rainfall.

- *Dry (B) climates,* in which the yearly precipitation is less than the potential loss of water by evaporation, are subdivided into two types: *arid* or *desert* (BW) and *semiarid* or *steppe* (BS). Their differences are primarily a matter of degree, with semiarid being a marginal and more humid variant of arid. Low-latitude deserts and steppes coincide with the clear skies caused by subsiding air beneath the subtropical high-pressure belts. Middle-latitude deserts and steppes exist principally because of their position in the deep interiors of large landmasses far removed from the ocean. Because many middle-latitude deserts occupy sites on the leeward sides of mountains, they can also be classified as *rainshadow deserts.*

- *Middle-latitude climates with mild winters* (C climates) occur where the average temperature of the coldest month is below 18°C but above –3°C. Several C climate subgroups exist. *Humid subtropical climates* (Cfa) are located on the eastern sides of the continents, in the 25- to 40-degree latitude range. Summer weather is hot and sultry, and winters are mild. In North America, the *marine west coast climate* (Cfb, Cfc) extends from near the United States–Canadian border northward as a narrow belt into southern Alaska. The prevalence of maritime air masses means that mild winters and cool summers are the rule. *Dry-summer subtropical climates* (Csa, Csb) are typically located along the west sides of continents between latitudes 30 and 45 degrees. In summer, the regions are dominated by stable, dry conditions associated with the oceanic subtropical highs. In winter they are within range of the cyclonic storms of the polar front.

- *Humid middle-latitude climates with severe winters* (D climates) are land-controlled climates that are absent in the Southern Hemisphere. The D climates have severe winters. The average temperature of the coldest month is –3°C or below and the warmest monthly mean exceeds 10°C. *Humid continental climates* (Dfa, Dfb, Dwa, Dwb) are confined to the eastern portions of North America and Eurasia in the latitude range between approximately 40 and 50 degrees north latitude. Both winter and summer temperatures can be characterized as relatively severe. Precipitation is generally greater in summer than in winter. *Subarctic climates* (Dfc, Dfd, Dwc, Dwd) are situated north of the humid continental climates and south of the polar tundras. The outstanding feature of subarctic climates is the dominance of winter. By contrast, summers in the subarctic are remarkably warm, despite their short duration. The highest annual temperature ranges on Earth occur here.

- *Polar (E) climates* are summerless, with the average temperature of the warmest month below 10°C. Two types of polar climates are recognized. The *tundra climate* (ET) is a treeless climate found almost exclusively in the Northern Hemisphere. The *ice cap climate* (EF) does not have a single monthly mean above 0°C. As a consequence, the growth of vegetation is prohibited, and the landscape is one of permanent ice and snow.

- Compared to nearby places of lower elevation, *highland climates* are cooler and usually wetter. Because atmospheric conditions fluctuate rapidly with changes in altitude and exposure, these climates are best described by their variety and changeability.

- For centuries humans have had an impact on global climate by their use of fire as well as the overgrazing of marginal lands by domesticated animals. Since the mid-1800s, the combustion of fossil fuels (coal, natural gas, and petroleum) has added great quantities of carbon dioxide, an important heat absorber, to the atmosphere. If the use of fossil fuels continues to increase at projected rates, models predict an increase in the mean global surface temperature of about 2.5°C. The atmospheric buildup of trace gases such as methane, nitrous oxide, and certain chlorofluorocarbons will also contribute to any future global temperature increase. Possible consequences of a greenhouse warming include significant changes in precipitation patterns, rise in global mean sea level, higher frequency and intensity of hurricanes, shifts in the paths of cyclonic storms, changes in the frequency and intensity of heat waves and droughts, and more.

Key Terms

arid climate (p. 500)
climate system (p. 492)
desert climate (p. 500)
dry-summer subtropical climate
   (p. 503)
highland climate (p. 505)
humid continental climate (p. 504)

humid subtropical climate (p. 501)
ice cap climate (p. 505)
Köppen classification (p. 493)
marine west coast climate (p. 502)
polar climate (p. 505)
rainshadow desert (p. 501)
semiarid climate (p. 500)

steppe climate (p. 500)
subarctic climate (p. 504)
tropical rain forest (p. 493)
tropical wet and dry climate (p. 498)
tundra climate (p. 505)

Review Questions

1. List the five parts of the climate system.

2. Why is classification often a helpful or even necessary task in science?

3. What climatic data are needed in order to classify a climate using the Köppen scheme?

4. (a) What primary factor distinguishes Aw climates from Af and Am?

   (b) How is this difference reflected in the vegetation?

5. Describe the influence of the equatorial low and the subtropical high on the precipitation regime in the Aw climate.

6. Why is the amount of precipitation that defines the humid–dry boundary variable?

7. What is the primary reason (control) for the existence of the dry subtropical realm (BWh and BSh)?

8. What is the primary cause for the existence of the middle-latitude deserts and steppes?

9. Describe and explain the differences between summertime and wintertime precipitation in the humid subtropics (Cfa).

10. Why is the marine west coast climate (Cfb and Cfc) represented by only slender strips of land in North and South America and why is it very extensive in Western Europe?

11. In this chapter the dry–summer subtropics were described as transitional. Explain why this statement is true.

12. Why is the humid continental climate confined to the Northern Hemisphere?

13. Although generally characterized by small precipitation totals, subarctic and polar climates are considered humid. Explain.

14. Describe and explain the annual temperature range one should expect in the realm of the taiga.

15. Although polar regions experience extended periods of almost perpetual sunlight in the summer, temperatures remain cool. Explain.

16. The tundra climate is not confined solely to high latitudes. Under what circumstances might the ET climate be found in more equatorward locations?

17. Where are EF climates most extensively developed?

18. The Arizona cities of Flagstaff and Phoenix are relatively close to one another yet have contrasting climates (see Figure 18.12). Briefly explain why the differences occur.

19. Why has the atmosphere's carbon dioxide level been rising for more than 130 years?

20. How are temperatures in the lower atmosphere likely to change as carbon dioxide levels continue to increase? Why?

21. Is carbon dioxide the only gas that may contribute to a future global temperature change?

22. List three possible consequences of a global warming.

Examining the Earth System

1. The Köppen climate classification is based on the fact that there is an excellent association between natural vegetation (biosphere) and climate (atmosphere). Briefly describe the climate conditions (temperature and precipitation) and natural vegetation associated with each of the following Köppen climates: Af, BWh, Dfc, ET.

2. How might the burning of fossil fuels, such as the gasoline to run your car, influence global temperature? If such a temperature change occurs, how might sea level be affected? How might the intensity of hurricanes change? How might these changes impact people who live on a beach or barrier island along the Atlantic or Gulf coasts?

(You may find these Environmental Protection Agency (EPA) Web sites interesting and helpful: **http://www. epa.gov/oppeoee1/globalwarming/** and **http://www. gcrio.org/EPA/sealevel/seatitle.html**)

3. How does the building of cities influence air temperature? Why are city temperatures higher than those in the surrounding rural countryside? What spheres of the Earth system might be involved or influenced? Try to indicate two or three specific examples.

4. What are climatic feedback mechanisms? Give an example of a positive feedback mechanism and a negative feedback mechanism (see Box 18.4).

Web Resources

The following are informative and interesting World Wide Web sites that address topics related to those presented in the chapter:

Environmental Protection Agency (EPA), Office of Air & Radiation:

- **http://www.epa.gov/oar/oarhome.html**

National Climate Data Center (National Oceanic and Atmospheric Administration):

- **http://www.ncdc.noaa.gov/**

For direct links to these sites and other appropriate Earth science Web locations, chapter review exercises, and relevant, topical investigations that utilize Web resources, contact the *Earth Science*, 9th Edition Home Page at:

- **http://www.prenhall.com/tarbuck.**

Part Four

Astronomy

These eerie, dark pillar-like structures are actually columns of cool interstellar hydrogen gas and dust that are also incubators for new stars. They are part of the Eagle Nebula, a nearby star-forming region 7,000 light-years away. This image was made by the Hubble Space Telescope. (Courtesy of J. Hester and P. Scowen/Arizona State University/NASA)

Chapter 19

Earth's Place in the Universe

Ancient Astronomy

The Birth of Modern Astronomy

Constellations

Positions in the Sky

Motions of Earth

Motions of the Earth–Moon System

■ *Box 19.1*   Proof of Earth's Rotation
■ *Box 19.2*   Astrology

Stonehenge, an ancient observatory in England. (Photo by Robert Llewellyn)

Earth is one of nine planets and numerous smaller bodies that orbit the Sun. The Sun is part of a much larger family of perhaps 100 billion stars that make up the Milky Way, which in turn is only one of billions of galaxies in an incomprehensibly large universe. This view of Earth's position in space is considerably different from that held only a few hundred years ago, when Earth was thought to occupy a privileged position as the center of the universe. This chapter unfolds with some events that led to modern astronomy. In addition, it examines Earth's place in time and space.

Long before recorded history, which began about 5000 years ago, people were aware of the close relationship between events on Earth and the positions of heavenly bodies, the Sun in particular. People noted that changes in the seasons and floods of great rivers like the Nile in Egypt occurred when the celestial bodies, including the Sun, Moon, planets, and stars, reached a particular place in the heavens. Early agrarian cultures, which were dependent on the weather, believed that if the heavenly objects could control the seasons, they must also strongly influence all Earthly events. This belief undoubtedly was the reason that early civilizations began keeping records of the positions of the celestial objects. The Chinese, Egyptians, and Babylonians in particular are noted for this.

These cultures recorded the locations of the Sun, Moon, and the five planets visible to the unaided eye as these objects moved slowly against the background of "fixed" stars. In addition, the Chinese kept quite accurate records of comets and "guest stars" (Figure 19.1). Today we know that a "guest star" is really a normal star, usually too faint to be visible, that increases its brightness as it explosively ejects gases from its surface, a phenomenon we call a *nova*.

A study of Chinese archives shows that they recorded every appearance of the famous Halley's comet for at least ten centuries. However, because this comet appears only once in a lifetime—once every 76 years—they were unable to link these appearances to establish that what they saw was the same object each time. Thus, like most ancients, the Chinese considered comets to be mystical. Comets were seen as bad omens and were blamed for a variety of disasters, from wars to plagues (Figure 19.2).

Ancient Astronomy

The "Golden Age" of early astronomy (600 B.C.–A.D. 150) was centered in Greece. The early Greeks have been criticized, and rightly so, for using philosophical arguments to explain natural phenomena. However, they did rely on observational data as well. The basics of geometry and trigonometry, which they had developed, were used to measure the sizes and distances of the largest-appearing bodies in the heavens—the Sun and the Moon.

Early Greeks

Many astronomical discoveries have been credited to the Greeks. They held the **geocentric** ("Earth-centered") view, believing that Earth was a sphere that stayed motionless at the center of the universe. Orbiting Earth were the Moon, Sun, and the known planets—Mercury, Venus, Mars, and Jupiter. Beyond the planets was a transparent, hollow sphere (**celestial sphere**) on which the stars traveled daily around Earth (this is how it looks, but of course, the effect is actually caused by Earth's rotation about its axis). Some early Greeks realized that the motion of the stars could be explained just as easily by a rotating Earth, but they rejected that idea, because Earth exhibited no sense of motion and seemed too large to be movable. In fact, proof of Earth's rotation was not demonstrated until 1851, a topic we consider in Box 19.1.

To the Greeks, all of the heavenly bodies, except seven, appeared to remain in the same relative position to one another. These seven wanderers (*planetai* in Greek)

Figure 19.1 Comet Hale-Bopp. The two tails are about 10 million to 15 million miles long. (A Peoria Astronomical Society, Inc. photograph by Eric Clifton and Greg Neaveill)

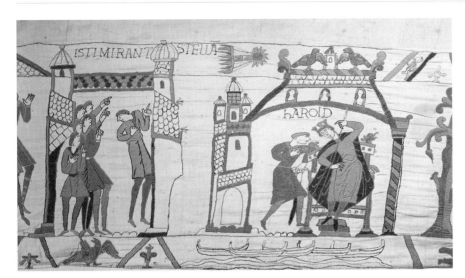

Figure 19.2 The Bayeux Tapestry that hangs in Bayeux, France, shows the apprehension caused by Halley's comet in A.D. 1066. This event preceded the defeat of King Harold by William the Conqueror. (Photo by Giraudon/Art Resource with special authorization of the City of Bayeux)

Box 19.1   Proof of Earth's Rotation

Every schoolchild learns that Earth rotates on its axis once each day to produce periods of daylight and darkness. However, day and night and the apparent motions of the stars can be accounted for equally well by a Sun and celestial sphere that revolve around a stationary Earth. Copernicus realized that a rotating Earth greatly simplified the existing model of the universe and strongly advocated this as the correct view. He was unable, however, to *prove* that Earth rotates. The first substantial proof was presented 300 years after his death by the French physicist Jean Foucault.

In 1851 Foucault used a free-swinging pendulum to demonstrate that Earth does, in fact, turn on its axis. To envision Foucaults's experiments, imagine a large pendulum swinging over the North Pole (Figure 19.A). Keep in mind that once a pendulum is put into motion, it continues swinging in the same plane unless acted upon by some outside force. Assume that a sharp stylus is attached to the bottom of this pendulum, marking the snow as it oscillates. When we observe the marks made by the stylus, we note that the pendulum is slowly but continually changing position. At the end of 24 hours it has returned to the starting position (Figure 19.A).

Because no outside force acted on the pendulum to change its position, what we observed must have been

Figure 19.A Apparent movement of a pendulum at the North Pole caused by Earth rotating beneath it.

Earth rotating beneath it. Foucault conducted a similar experiment when he suspended a long pendulum from the dome of the Pantheon in Paris. Today, Foucault pendulums can be found in a number of museums to re-create this famous scientific experiment (Figure 19.B).

Figure 19.B Foucault pendulum housed in Museum of Science and Industry, Chicago. (Courtesy Museum of Science and Industry)

included the Sun, the Moon, Mercury, Venus, Mars, Jupiter, and Saturn. Each was thought to have a circular orbit around Earth. Although this system was incorrect, the Greeks refined it to the point that it explained the apparent movements of all celestial bodies.

As early as the fifth century B.C., the Greeks understood what causes the phases of the Moon. Anaxagoras reasoned that the Moon shines by reflected sunlight, and because it is a sphere, only half is illuminated at one time. As the Moon orbits Earth, that portion of the illuminated half that is visible from Earth is always changing. Anaxagoras also realized that an eclipse of the Moon occurs when it moves into Earth's shadow.

The famous Greek philosopher Aristotle (384–322 B.C.) concluded that Earth is spherical because it always casts a curved shadow when it eclipses the moon. Although most of the teachings of Aristotle were passed along and were considered infallible by many, his belief in a spherical Earth was abandoned during the Middle Ages.

The first Greek to profess a Sun-centered, or **heliocentric,** universe was Aristarchus (312–230 B.C.). Aristarchus also used simple geometric relations to calculate the relative distances from Earth to the Sun and the Moon. He later used these data to calculate their sizes. As a result of observational errors beyond his control, he came up with measurements that were much too small. However, he did learn that the Sun was many times more distant than the Moon and many times larger than Earth. The latter fact may have

prompted him to suggest a Sun-centered universe. Nevertheless, because of the strong influence of Aristotle, the Earth-centered view dominated Western thought for nearly 2000 years.

The first successful attempt to establish the size of Earth is credited to Eratosthenes (276–194 B.C.). Eratosthenes observed the angles of the noonday Sun in two Egyptian cities that were roughly north and south of each other—Syene (presently Aswan) and Alexandria (Figure 19.3). Finding that the angles differed by 7 degrees, or 1/50 of a complete circle, he concluded that the circumference of Earth must be 50 times the distance between these two cities. The cities were 5000 *stadia* apart, giving him a measurement of 250,000 *stadia*. Many historians believe the *stadia* was 157.6 meters (517 feet), which would make Eratosthenes' calculation of Earth's circumference—39,400 kilometers (24,428 miles)—a measurement very close to the modern value of 40,075 kilometers (24,902 miles).

Probably the greatest of the early Greek astronomers was Hipparchus (second century B.C.), best known for his star catalogue. Hipparchus determined the location of almost 850 stars, which he divided into six groups according to their brightness. He measured the length of the year to within minutes of the modern value and developed a method for predicting the times of lunar eclipses to within a few hours.

Although many of the Greek discoveries were lost during the Middle Ages, the Earth-centered view that the Greeks proposed became established in

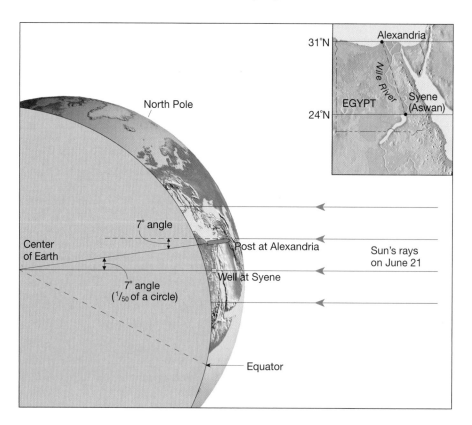

Figure 19.3 Orientation of the Sun's rays at Syene (Aswan) and Alexandria in Egypt on June 21 when Eratosthenes calculated Earth's circumference.

Europe. Presented in its finest form by Cladius Ptolemy, this geocentric outlook became known as the **Ptolemaic system.**

The Ptolemaic System

Much of our knowledge of Greek astronomy comes from a 13-volume treatise, *Almagest* ("the great work"), which was compiled by Ptolemy in A.D. 141 and survived thanks to the work of Arab scholars. In this work, Ptolemy is credited with developing a model of the universe that accounted for the observable motions of the planets (Figure 19.4). The precision with which his model was able to predict planetary motion is attested to by the fact that it went virtually unchallenged, in principle if not in detail, for nearly 13 centuries.

In the Greek tradition, the Ptolemaic model had the planets moving in circular orbits around a motionless Earth. (The circle was considered the pure and perfect shape by the Greeks.) However, the motion of the planets, as seen against the background of stars, is not so simple. Each planet, if watched night after night, moves slightly eastward among the stars. Periodically, each planet appears to stop, reverse direction for a period of time, and then resume an eastward motion. The apparent westward drift is called **retrograde motion.** This rather odd *apparent* motion results from the combination of the motion of Earth and the planet's own motion around the Sun.

Figure 19.5 illustrates the retrograde motions of Mars. Earth has a faster orbital speed than Mars so it overtakes its neighbor. While doing so, Mars *appears* to be moving backward, in retrograde motion. This is analogous to what a race-car driver sees out the side window when passing a slower car. The slower planet, like the slower car, appears to be going backward, although its actual motion is in the same direction as the faster-moving body.

A.

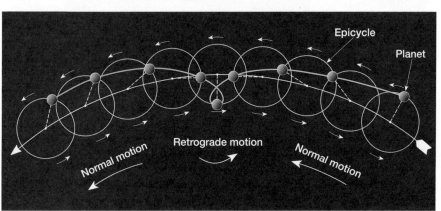

B.

Figure 19.4 The universe according to Ptolemy, second century A.D. **A.** Ptolemy believed that the star-studded celestial sphere made a daily trip around a motionless Earth. In addition he proposed that the Sun, Moon, and planets made trips of various lengths along individual orbits. **B.** Retrograde motion as explained by Ptolemy.

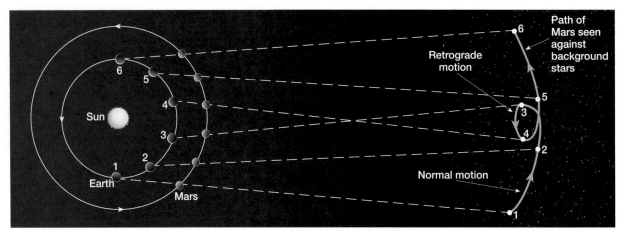

Figure 19.5 Retrograde (backward) motion of Mars as seen against the background of distant stars. When viewed from Earth, Mars moves eastward among the stars each day, then periodically appears to stop and reverse direction. This apparent westward drift is a result of the fact that Earth has a faster orbital speed than Mars and overtakes it. As this occurs, Mars appears to be moving backward; that is, it exhibits retrograde motion.

Figure 19.5 shows how retrograde motion works. It is much more difficult to accurately represent retrograde motion using the incorrect Earth-centered model, but Ptolemy was able to do so (Figure 19.4B). Rather than using a simple circle for each planet's orbit, he showed that the planets orbited on small circles (*epicycles*), revolving along large circles (*deferents*). By trial and error, he found the right combinations of circles to produce the amount of retrograde motion observed for each planet. (An interesting note is that almost any closed curve can be produced by the combination of two circular motions, a fact that can be verified by anyone who has used the Spirograph® design-drawing toy.)

It is a tribute to Ptolemy's genius that he was able to account for the planets' motions as well as he did, considering that he used an incorrect model. Some suggest that he did not mean his model to represent reality, but only to be used for calculating the positions of heavenly bodies. We probably will never know his intentions. However, the Roman Catholic Church, which dominated European thought for centuries, accepted Ptolemy's theory as the correct representation of the heavens, and this created problems for those who found fault with it.

The Birth of Modern Astronomy

Modern astronomy was not born overnight. Its development involved a break from deeply entrenched philosophical and religious views and, during the sixteenth and seventeenth centuries, by the founding of a "new and greater universe" governed by discernible laws. Let us now look at the work of five noted scientists involved in this transition: Nicolaus Copernicus, Tycho Brahe, Johannes Kepler, Galileo Galilei, and Sir Isaac Newton.

Nicolaus Copernicus

For almost 13 centuries after the time of Ptolemy, very few astronomical advances were made in Europe. The first great astronomer to emerge after the Middle Ages was Nicolas Copernicus (1473–1543) from Poland (Figure 19.6). Copernicus became convinced that

Figure 19.6 Polish astronomer Nicolaus Copernicus (1473–1543) believed that Earth was just another planet. (Yerkes Observatory Photograph/University of Chicago)

Earth is a planet, just like the other five then-known planets. The daily motions of the heavens, he reasoned, could be better explained by a rotating Earth. To counter the Ptolemaic objection that Earth would fly apart if it rotated, Copernicus suggested that the much-larger celestial sphere would be even more likely to fly apart if it rotated!

Having concluded that Earth is a planet, Copernicus reconstructed the solar system with the Sun at the center and the planets Mercury, Venus, Earth, Mars, Jupiter, and Saturn orbiting around it. This was a major break from the ancient idea that a motionless Earth lies at the center of all movement. However, Copernicus retained a link to the past and used circles, which were considered to be the perfect geometric shape, to represent the orbits of the planets. Although these circular orbits were close to reality, they didn't quite match what people saw. Unable to get satisfactory agreement between predicted locations of the planets and their observed positions, Copernicus found it necessary to add epicycles like those used by Ptolemy. The discovery that the planets have *elliptical* orbits would wait another century for the insights of Johannes Kepler.

Also, like his predecessors, Copernicus used philosophical justifications, such as the following, to support his point of view.

> …in the midst of all stands the Sun. For who could in this most beautiful temple place this lamp in another or better place than that from which it can at the same time illuminate the whole?

Copernicus's monumental work, *De Revolutionibus, Orbium Coelestium (On the Revolution of the Heavenly Spheres)*, which set forth his controversial ideas, was published as he lay on his deathbed. Hence, he never suffered the criticisms that fell on many of his followers.

The Copernican system challenged the primacy of Earth in the universe and was considered heretical by the Catholic Church. Expounding the idea cost at least one person his life. Giordano Bruno was seized by the Inquisition, a Church tribunal, in 1600 and, refusing to denounce the Copernican theory, was tied to a stake and burned alive.

Tycho Brahe

Tycho Brahe (1546–1601) was born of Danish nobility 3 years after the death of Copernicus. Reportedly, Tycho became interested in astronomy while viewing a solar eclipse that had been predicted by astronomers. He persuaded King Frederick II to establish an observatory, which he headed, near Copenhagen. There he designed and built pointers (the telescope would not be invented for a few more decades), which he used for 20 years to systematically measure the locations of the heavenly bodies (Figure 19.7). These observations, particularly of Mars, were far more precise than any made previously and are his legacy to astronomy.

Tycho did not believe in the Copernican (Sun-centered) system, because he was unable to observe an apparent shift in the position of stars that would be caused by Earth's motion. His argument went like this: If Earth does revolve along an orbit around the Sun, the position of a nearby star, when observed from extreme points in Earth's orbit 6 months apart, should shift with respect to the more distant stars. His *idea* was correct, and this apparent shift of the stars is called *stellar parallax* (see Figure 22.2, p. 603).

The principle of parallax is easy to visualize: Close one eye, and with your index finger vertical, use your eye to line up your finger with some distant object. Now, without moving your finger, view the object with your other eye and notice that the object's position appears to change. The farther away you hold your finger, the less the object's position seems to shift. Herein lay the flaw in Tycho's argument. He was right about parallax, but, because the distance to even the nearest stars is enormous compared to the width of Earth's orbit, the shift that occurs is too small to be noticed by using the first primitive telescopes, let alone the unaided eye.

With the death of his patron, the King of Denmark, Tycho was forced to leave his observatory. It was probably his arrogant and extravagant nature that caused a conflict with the next ruler, so Tycho moved to Prague in the Czech Republic. Here, in the last year of his life, he acquired an able assistant, Johannes Kepler. Kepler retained most of the observations made by Tycho and put them to exceptional use. Ironically, the data Tycho collected to refute the Copernican view would later be used by Kepler to support it.

Johannes Kepler

If Copernicus ushered out the old astronomy, Johannes Kepler (1571–1630) ushered in the new (Figure 19.8). Armed with Tycho's data, a good mathematical mind, and, of greater importance, a strong faith in the accuracy of Tycho's work, Kepler derived three basic laws of planetary motion. The first two laws resulted from his inability to fit Tycho's observations of Mars to a circular orbit. Unwilling to concede that the discrepancies were a result of observational error, he searched for another solution. This endeavor led him to discover that the orbit of Mars is not a perfect circle but is elliptical (Figure 19.9). About the same time, he realized that the orbital speed of Mars varies in a predictable way. As it approaches the Sun, it speeds up, and as it moves away from the Sun, it slows down.

In 1609, after almost a decade of work, Kepler proposed his first two laws of planetary motion:

Figure 19.7 Tycho Brahe (1546–1601) in his observatory, Uraniborg, on the Danish island of Hveen. Tycho (central figure) and the background are painted on the wall of the observatory within the arc of the sighting instrument called a quadrant. In the far right, Tycho can be seen "sighting" a celestial object through the "hole" in the wall. Tycho's accurate measurements of Mars enabled Johannes Kepler to formulate his three laws of planetary motion. (Courtesy of Thomas Clarke, McLaughlin Planetarium/Royal Ontario Museum copyright ROM)

1. The path of each planet around the Sun is an ellipse, with the Sun at one focus (Figure 19.9). The other focus is symmetrically located at the opposite end of the ellipse.

2. Each planet revolves so that an imaginary line connecting it to the Sun sweeps over equal areas in equal intervals of time (Figure 19.10). This law of equal areas expresses geometrically the variations in orbital speeds of the planets.

Figure 19.10 illustrates the second law. Note that in order for a planet to sweep equal areas in the same amount of time, it must travel more rapidly when it is nearer the Sun and more slowly when it is farther from the Sun.

Kepler was very religious and believed that the Creator made an orderly universe. The uniformity he tried to find eluded him for nearly a decade. Then in 1619, Kepler published his third law in *The Harmony of the Worlds.*

3. The orbital periods of the planets and their distances to the Sun are proportional.

Figure 19.8 German astronomer Johannes Kepler (1571–1630) helped establish the era of modern astronomy by deriving three laws of planetary motion. (Smithsonian Institution Photo #56123)

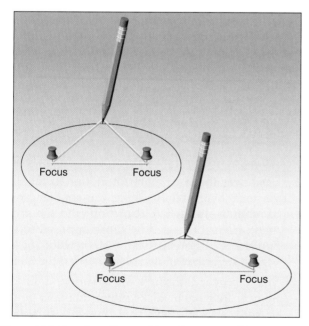

Figure 19.9 Drawing ellipses with various eccentricities. Using two straight pins for foci and a loop of string, trace out a curve while keeping the string taut, and you will have drawn an ellipse. The farther the pins (the foci) are moved apart, the more flattened (more eccentric) is the resulting ellipse.

In its simplest form, the orbital period of revolution is measured in Earth years, and the planet's distance to the Sun is expressed in terms of Earth's mean distance to the Sun. The latter "yardstick" is called the **astronomical unit (AU)** and averages about 150 million kilometers (93 million miles).

Using these units, Kepler's third law states that the planet's orbital period squared is equal to its mean solar distance cubed ($p^2 = d^3$). Consequently, the solar distances of the planets can be calculated when their periods of revolution are known. For example, Mars has a period of 1.88 years, which squared equals 3.54. The cube root of 3.54 is 1.52, and that is the distance to Mars in astronomical units (Table 19.1).

Kepler's laws assert that the planets revolve around the Sun and therefore support the Copernican theory. Kepler, however, did fall short of determining the *forces* that act to produce the planetary motion he had so ably described. That task would remain for Galileo Galilei and Sir Isaac Newton.

Galileo Galilei

Galileo Galilei (1564–1642) was the greatest Italian scientist of the Renaissance (Figure 19.11). He was a contemporary of Kepler and, like Kepler, strongly supported the Copernican theory of a Sun-centered solar system. Galileo's greatest contributions to science were his descriptions of the behavior of moving objects. These he derived from experimentation. The method of using experiments to determine natural laws had essentially been lost since the time of the early Greeks.

An example of such experimentation is the legend that Galileo climbed to the top of the Leaning Tower of Pisa, Italy, where he dropped two objects of differing masses. To the astonishment of the observers, they hit the ground at the same time. (Galileo probably did not attempt this experiment, although a colleague might

Table 19.1 Period of revolution and solar distances of planets.

Planet	Solar Distance (AU)*	Period (years)
Mercury	0.39	0.24
Venus	0.72	0.62
Earth	1.00	1.00
Mars	1.52	1.88
Jupiter	5.20	11.86
Saturn	9.54	29.46
Uranus	19.18	84.01
Neptune	30.06	164.80
Pluto	39.44	247.70

*AU = astronomical unit.

Figure 19.10 Kepler's law of equal areas. A line connecting a planet (Earth) to the Sun sweeps out an area in such a manner that equal areas are swept out in equal times. Thus, Earth revolves slower when it is farther from the Sun (aphelion) and faster when it is closest (perihelion). The eccentricity of Earth's orbit is greatly exaggerated in this diagram.

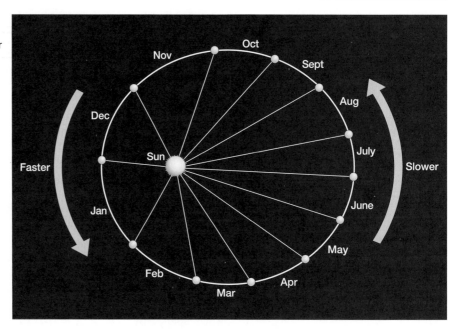

have.) Galileo correctly concluded that variations in the rate of fall of very light objects, like feathers, are due to air resistance, not differences in mass. This same experiment was performed most dramatically on the airless Moon nearly four centuries later when David Scott, an *Apollo 15* astronaut, showed that a feather and a hammer fall at the same rate.

All astronomical discoveries before Galileo's time were made without the aid of a telescope. In 1609, Galileo heard that a Dutch lens maker had devised a system of lenses that magnified objects. Apparently without ever seeing a telescope, Galileo constructed his own, which magnified distant objects to three times the size seen by the unaided eye. He immediately made others, the best having a magnification of about 30.

With the telescope, Galileo was able to view the universe in a new way. He made many important discoveries that supported the Copernican view of the universe, such as:

1. The discovery of four satellites, or moons, orbiting Jupiter. Galileo accurately determined their periods of revolution, which range from 2 to 17 days (Figure 19.12). This find dispelled the old idea that Earth was the only center of motion in the universe; for here, plainly visible, was another center of motion—Jupiter. It also countered the argument, frequently used by those opposed to the Sun-centered system, that the Moon would be left behind if Earth really revolved around the Sun.

2. The discovery, through observation, that the planets are circular disks rather than just points of light, as was previously thought. This indicated that the planets must be Earth-like.

3. The discovery that Venus has phases just like the Moon, demonstrating that Venus orbits its source of light—the Sun. Galileo saw that Venus appears smallest when it is in full phase, and

Figure 19.11 Italian scientist Galileo Galilei (1564–1642) used a new invention, the telescope, to observe the Sun, Moon, and planets in more detail than ever before. (Yerkes Observatory Photograph/University of Chicago)

4. The discovery that the Moon's surface is not a smooth glass sphere, as the ancients had suspected and the Church had decreed. Rather, Galileo saw mountains, craters, and plains. He thought the plains might be bodies of water, and this idea was strongly promoted by others, as we can tell from the names given to these features (Sea of Tranquility, Sea of Storms, and so forth).

5. The discovery that the Sun (the viewing of which may have caused the eye damage that later blinded him) had sunspots (dark regions caused by slightly lower temperatures). He tracked the movement of these spots and estimated the rotational period of the Sun as just under a month. Hence, another heavenly body was found to have both "blemishes" and rotational motion.

In 1616, the Church condemned the Copernican theory as contrary to Scripture, and Galileo was told to abandon it. Unwilling to accept this verdict, Galileo began writing his most famous work, *Dialogue of the Great World Systems*. Despite poor health, he completed the project and in 1630 went to Rome, seeking permission from Pope Urban VIII to publish. Because the book was a dialogue that expounded both the Ptolemaic and Copernican systems, publication was allowed. However, Galileo's enemies were quick to realize that he was promoting the Copernican view at the expense of the Ptolemaic system. Sales of the book were quickly halted, and Galileo was called before the Inquisition. Tried and convicted of proclaiming doctrines contrary to religious doctrine, he was sentenced to permanent house arrest, under which he remained for the last 10 years of his life.

Despite this restriction, and his age, and his grief after the death of his eldest daughter, Galileo continued to work. In 1637 he became totally blind, yet during the next few years he completed his finest scientific work, a book on the study of motion. When Galileo died in 1642, the Grand Duke of Tuscany wanted to erect a monument in his honor, but fear that it might offend the Church prevailed, and it was never built. It was not until 1992 that Galileo was finally exonerated by the Catholic Church.

Sir Isaac Newton

Sir Isaac Newton (1642–1727) (Figure 19.14) was born in the year of Galileo's death. His many accomplishments in mathematics and physics led a successor to say that "Newton was the greatest genius that ever existed."

Although Kepler and those who followed attempted to explain the forces involved in planetary motion, their explanations were less than satisfactory. Kepler believed that some force pushed the planets along in their orbits. Galileo, however, correctly reasoned that no force is required to keep an object in

Figure 19.12 Sketch by Galileo of how he saw Jupiter and its four largest satellites through his telescope. The positions of Jupiter's four largest Moons (drawn as stars) change nightly. You can observe these same changes with binoculars. (Yerkes Observatory Photograph/University of Chicago)

thus is farthest from Earth (Figure 19.13). In the Ptolemaic system, as shown in Figure 19.4, the orbit of Venus lies between Earth and the Sun, which means that only the crescent phase of Venus could be seen from Earth.

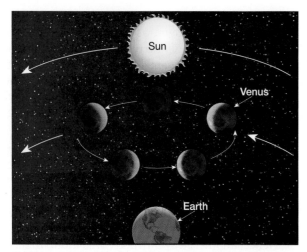

A. Phases of Venus as seen from
Earth in the Earth-centered model.

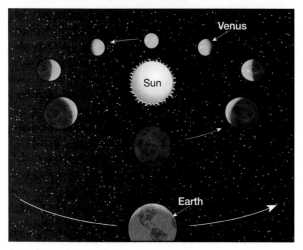

B. Phases of Venus as seen from
Earth in the sun-centered model.

Figure 19.13 Using a telescope, Galileo discovered that Venus has phases just like the Moon. **A.** In the Ptolemaic (Earth-centered) system, the orbit of Venus lies between the sun and Earth as shown in Figure 19.4A. Thus, in an Earth-centered solar system, only the crescent phase of Venus would be visible from Earth. **B.** In the Copernican (Sun-centered) system, Venus orbits the Sun and hence all of the phases of Venus should be visible from Earth. **C.** As Galileo observed, Venus goes through a series of Moonlike phases. Further, Venus appears smallest during the full phase when it is farthest from Earth and largest in the crescent phase when it is closest to Earth. This verified Galileo's belief that the Sun was the center of the solar system. (Photo courtesy of Lowell Observatory)

C.

motion. Galileo proposed that the natural tendency for a moving object (that is unaffected by an outside force) is to continue moving at a uniform speed and in a straight line. This concept, *inertia*, was later formalized by Newton as his first law of motion.

The problem, then, was not to explain the force that keeps the planets moving but rather to determine the force that *keeps them from going in a straight line out into space*. It was to this end that Newton conceptualized the force of *gravity*. At the early age of 23, he envisioned a force that extends from Earth into space and holds the Moon in orbit around Earth. Although others had theorized the existence of such a force, he was the first to formulate and test the *law of universal gravitation*. It states:

Every body in the universe attracts every other body with a force that is directly proportional to their masses and inversely proportional to the square of the distance between them.

Thus, the gravitational force decreases with distance, so that two objects 3 kilometers apart have 3^2, or 9, times less gravitational attraction than if the same objects were 1 kilometer apart.

The law of gravitation also states that the greater the mass of the object, the greater its gravitational force. For example, the large mass of the Moon has a gravitational force strong enough to cause ocean tides on Earth, whereas the tiny mass of a communications satellite has no measurable effect on Earth. The mass of an object is a measure of the total amount of matter it contains. But more often mass is measured by determining the resistance an object exhibits in response to any effort made to change its state of motion.

Often we confuse the concept of *mass* with our notion of *weight*. Specifically, weight is the force of gravity acting upon an object. Therefore, weight varies when gravitational forces change. An object weighs less on the Moon than on Earth because the Moon is much less massive than Earth. However, unlike

Figure 19.14 English scientist Sir Isaac Newton (1642–1727) explained gravity as the force that holds planets in orbit around the Sun. (Yerkes Observatory Photograph/University of Chicago)

moves forward in its orbit about 30 kilometers (18.5 miles) each second, and during the same second, the force of gravity pulls it toward the sun about 0.5 centimeter (1/8 inch). Therefore, as Newton concluded, it is the combination of Earth's forward motion and its "falling" motion that defines its orbit (Figure 19.15). If gravity were somehow eliminated, Earth would move in a straight line out into space. Conversely, if Earth's forward motion suddenly stopped, gravity would pull it directly toward the Sun.

Up to this point, we have discussed Earth as if the only forces involved in its motion were caused by its gravitational relationship with the Sun. However, all bodies in the solar system have gravitational effects on Earth and on each other. For this reason, the orbit of Earth is not the perfect ellipse determined by Kepler. Any variance in the orbit of a body from its predicted path is called **perturbation.** For example, Jupiter's gravitational pull on Saturn reduces Saturn's orbital period by nearly one week from the predicted period. As we shall see, the application of this concept led to the discovery of the planet Neptune, because of Neptune's gravitational effect on the orbit of Uranus.

Newton used the law of universal gravitation to redefine Kepler's third law, which states the relationship between the orbital periods of the planets and their solar distances. When restated, Kepler's third law takes into account the masses of the bodies involved and thereby provides a method for determining the mass of a body when the orbit of one of its satellites is known. For example, the mass of the Sun is known from Earth's orbit, and Earth's mass has been determined from the orbit of the Moon. In fact, the mass of any body with a satellite can be determined. The masses of bodies that do not have satellites can be determined only if the bodies noticeably affect the orbit of a neighboring body or of a nearby artificial satellite.

weight, the mass of an object does not change. For example, a person weighing 120 pounds on Earth weighs 1/6 as much, or 20 pounds, on the Moon, but the person's mass remains unchanged.

With his laws of motion, Newton proved that the force of gravity—combined with the tendency of a planet to remain in straight-line motion—results in the elliptical orbits discovered by Kepler. Earth, for example,

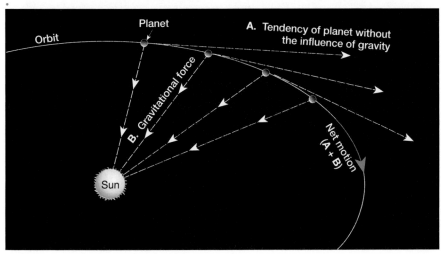

Figure 19.15 Orbital motion of Earth and other planets.

Constellations

As early as 5000 years ago, people became fascinated with the star-studded skies and began to name the patterns they saw (see Box 19.2). These configurations, called **constellations,** were named in honor of mythological characters or great heroes, such as Orion. It takes a good bit of imagination to make out the intended subjects, as most constellations were probably not thought of as likenesses in the first place. Although we inherited many of the constellations from Greek mythology, it is believed that Greek astronomers acquired most of theirs from the Babylonians, Egyptians, and Mesopotamians.

Although the stars that make up a constellation all appear to be the same distance from Earth, this is not the case. Some are many times farther away than others. Thus, the stars in a particular constellation are not associated with each other in any physical way. In addition, various cultural groups, including Native Americans, attached their own names, pictures, and stories to the constellations. For example, the constellation Orion, the Hunter, was known as the White Tiger to ancient Chinese astronomers.

Today, 88 constellations are recognized, and they are used to divide the sky into units, just as state boundaries divide the United States. Every star in the sky is in, but is not necessarily part of, one of these constellations. Constellations therefore enable astronomers to roughly identify the area of the heavens they are observing. For the student, the constellations provide a good way to become familiar with the night sky (see Appendix D).

Some of the brightest stars were given proper names, such as Sirius, Arcturus, and Betelgeuse. In addition, the brightest stars in a constellation are generally named in order of their brightness by the letters of the Greek alphabet—alpha (α), beta (β), and

---

Box 19.2 Astrology

Many people confuse astrology and astronomy to the point of believing these terms to be synonymous. Nothing can be further from the truth. *Astronomy* is a scientific probing of the universe to derive the properties of celestial objects and the laws under which the universe operates. *Astrology,* on the other hand, is based on ancient superstitions that hold that an individual's actions and personality are based on the positions of the planets and stars now, and at the person's birth. Scientists do not accept astrology, regarding it as a pseudoscience ("false science"). It is hoped that most people read horoscopes only as a pastime and do not let them influence daily living.

Apparently astrology had its origin more than 3000 years ago when the positions of the planets were plotted as they regularly migrated against the background of the "fixed" stars. Because the solar system is "flat" like a whirling Frisbee, the planets orbit the Sun along nearly the same plane. Therefore, the planets, Sun, and Moon all appear to move along a band around the sky known as the *zodiac.* Because Earth's Moon cycles through its phases about 12 times each year, the Babylonians divided the zodiac into 12 constellations (Figure 19.C). Thus, each

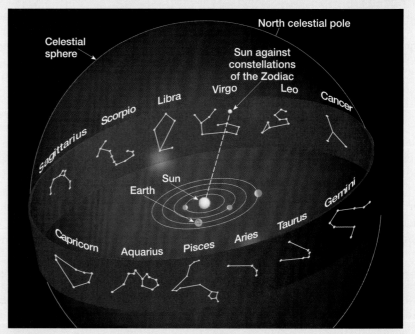

Figure 19.C The 12 constellations of the zodiac. Earth is shown in its autumn (September) position in orbit, from which the Sun is seen against the background of the constellation Virgo.

successive full Moon can be seen against the backdrop of the next constellation.

The dozen constellations of the zodiac ("Zone of Animals," so named because some constellations represent animals) are Aries, Taurus, Gemini, Cancer, Leo, Virgo, Libra, Scorpio, Sagittarius, Capricorn, Aquarius, and Pisces. These names may be familiar to you as the astrological signs of the zodiac. When first established, the vernal equinox (first day of spring) occurred when the Sun was viewed against the constellation Aries. However, during each succeeding vernal equinox, the position of the Sun shifts very slightly against the background of stars. Now, over 2000 years later, the vernal equinox occurs when the Sun is in Pisces (Figure 19.C). In several years, it will occur when the Sun appears against Aquarius. (Hence, the "Age of Aquarius" is coming.)

Although astrology is not a science and has no basis in fact, it did contribute to the science of astronomy. The positions of the Moon, Sun, and planets at the time of a person's birth (sign of the zodiac) were considered to have great influence on that person's life.* Even the great astronomer Kepler was required to make horoscopes as part of his duties. To make horoscopes for the future, astrologers attempted to predict the future positions of the celestial bodies. Consequently, some of the improve-

*It is interesting to note that 2000 years ago a person born on July 28 was considered a Leo because the Sun was in that constellation. During modern times the Sun appears in the constellation Cancer on this date, but individuals born during this time frame are still dubbed Leos.

Figure 19.D Stonehenge, an ancient observatory in England. On June 21–22 (summer solstice), the Sun can be observed rising above the heel stone. (Robin Scagell/Science Photo Library/Photo Researchers, Inc.)

ments in astronomical instruments were made because of the desire for more accurate predictions of events such as eclipses, which were considered highly significant in a person's life.

Even prehistoric people built observatories. The structure known as Stonehenge, in England, was undoubtedly an attempt at better solar predictions (Figure 19.D). At the

time of midsummer in the Northern Hemisphere (June 21–22—the summer solstice), the rising Sun emerges directly above the heel stone of Stonehenge. Besides keeping this calendar, Stonehenge may also have provided a method of determining eclipses. The remnants of other early observatories exist elsewhere in the Americas, Europe, Asia, and Africa.

so on—followed by the name of the parent constellation. For example, Sirius, the brightest star in the constellation Canis Major (Larger Dog), is called Alpha (α) Canis Majoris.

Positions in the Sky

If you gaze away from the city lights on a clear night, you will get the distinct impression that the stars produce a spherical shell surrounding Earth. This impression seems so real that it is easy to understand why many early Greeks regarded the stars as being fixed to a crystalline celestial sphere. Although we realize that no such sphere exists, it is convenient to use this concept for locating stars.

One method for doing this, called the **equatorial system,** divides the celestial sphere into a coordinate system. It is very similar to the latitude-longitude system used for locations on Earth's surface (Figure 19.16). Because the celestial sphere appears to rotate around an imaginary line extending from Earth's axis, the north and south celestial poles are in line with the

terrestrial North Pole and South Pole. The north celestial pole happens to be very near the bright star whose various names reflect its location: "pole star," Polaris, and North Star. To an observer in the Northern Hemisphere, the stars appear to circle Polaris, because it, like the North Pole, is in the center of motion (Figure 19.17). Figure 19.18 shows how to locate the North Star by using two stars in the very visible constellation called the Big Dipper.

Now, imagine a plane through Earth's equator, a plane that extends outward from Earth and intersects the celestial sphere. The intersection of this plane with the celestial sphere is the *celestial equator* (see Figure 19.16). In the equatorial system, the terms *declination,* which is analogous to latitude, and *right ascension,* which is analogous to longitude, are used (Figure 19.16). **Declination,** like latitude, is the angular distance north or south of the celestial equator. **Right ascension** is the angular distance measured eastward along the celestial equator from the position of the vernal equinox. (The *vernal equinox* is at the point in the sky where the Sun crosses the celestial equator, at

Figure 19.16 Astronomical coordinate system on the celestial sphere.

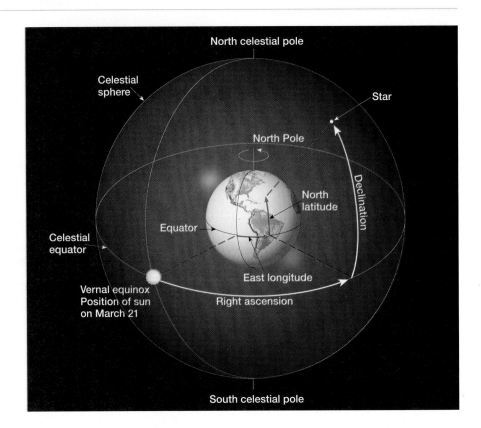

the onset of spring.) While declination is expressed in degrees, right ascension is usually expressed in hours, where each hour is equivalent to 15 degrees.

To visualize distances on the celestial sphere, it helps to remember that the Moon and Sun have an apparent width of about 0.5 degree. The declination and right ascension of some of the brightest stars (in other words, their locations) can be determined from the star charts in Appendix D.

Figure 19.17 Star trails in the region of Polaris (north celestial pole) on a time exposure. (Courtesy of National Optical Astronomy Observatories)

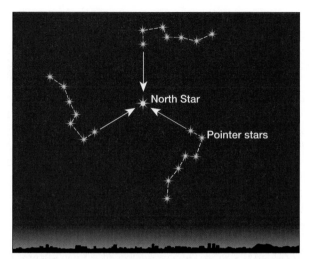

Figure 19.18 Locating the North Star (Polaris) from the pointer stars in the Big Dipper, which is part of the constellation Ursa Major. The Big Dipper is shown soon after sunset in December (lower figure), April (upper figure), and August (left). Refer to the star charts in Appendix D to find the current position of the Big Dipper.

Motions of Earth

The two primary motions of Earth are rotation and revolution. **Rotation** is the turning, or spinning, of a body on its axis. **Revolution** is the motion of a body, such as a planet or moon, along a path around some point in space. For example, Earth *revolves* around the Sun, and the Moon *revolves* around Earth. Earth also has another very slow motion known as **precession,** which is the slight movement, over a period of 26,000 years, of Earth's axis.

Rotation

The main consequences of Earth's rotation are day and night. Earth's rotation has become a standard method of measuring time because it is so dependable and easy to use. Each rotation equals about 24 hours. You may be surprised to learn that we can measure Earth's rotation in two ways, making two kinds of days. Most familiar is the **mean solar day,** the time interval from one noon to the next, which averages about 24 hours. Noon is when the Sun has reached its zenith (highest point in the sky).

The **sidereal day,** on the other hand, is the time it takes for Earth to make one complete rotation (360 degrees) with respect to a star other than our Sun. The sidereal day is measured by the time required for a star to reappear at the identical position in the sky where it was observed the day before. The sidereal day has a period of 23 hours, 56 minutes, and 4 seconds (measured in solar time), which is almost 4 minutes shorter than the mean solar day. This difference results because the direction to distant stars changes only infinitesimally owing to Earth's slow revolution along its orbit, whereas the direction to the Sun changes by almost 1 degree each day. This difference is shown in Figure 19.19.

If it is not apparent why we use the mean solar day rather than the sidereal day as a measurement of our day, consider the fact that in sidereal time, "noon" occurs 4 minutes earlier each day. Therefore, after a span of 6 months, "noon" occurs at "midnight." Astronomers use sidereal time because the stars appear in the same position in the sky every 24 sidereal hours. Usually, an observatory will begin its sidereal day when the position of the vernal equinox is directly overhead, that is, over the meridian on which the observatory is located.

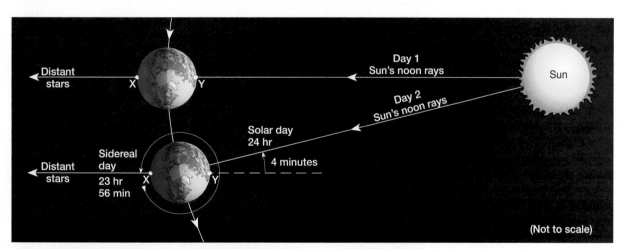

Figure 19.19 The difference between a solar day and a sidereal day. Locations X and Y are directly opposite each other. It takes Earth 23 hours and 56 minutes to make one rotation with respect to the stars (sidereal day). However, notice that after Earth has rotated once with respect to the stars, point Y is not yet returned to the "noon position" with respect to the Sun. Earth has to rotate another 4 minutes to complete the solar day.

Therefore, when the observatory's sidereal clock is the same as the star's right ascension, the star will be overhead, or at its highest point. For example, the brightest star in the heavens, Sirius, has a right ascension of 6 hours, 42 minutes, and 56 seconds and will be overhead when the clock at the observatory indicates that time.

Revolution

Earth revolves around the Sun in an elliptical orbit at an average speed of 107,000 kilometers (66,000 miles) per hour. Its average distance from the Sun is 150 million kilometers (93 million miles), but because its orbit is an ellipse, Earth's distance from the Sun varies. At **perihelion** (closest to the Sun), it is 147 million kilometers (91.5 million miles) distant, which occurs about January 3rd each year. At **aphelion** (farthest from the Sun), Earth is 152 million kilometers (94.5 million miles) distant, which occurs about July 4th.

Because of Earth's annual movement around the Sun, each day the Sun appears to be displaced among the constellations at a distance equal to about twice its width, or 1 degree. The apparent annual path of the Sun against the backdrop of the celestial sphere is called the **ecliptic** (Figure 19.20). Generally, the planets and the Moon travel in nearly the same plane as Earth.

Hence, their paths on the celestial sphere lie near the ecliptic. (The most notable exception is Pluto, which has an orbit that is tilted 17 degrees to the plane of Earth's orbit.)

The imaginary plane that connects Earth's orbit with the celestial sphere is called the **plane of the ecliptic.** From the reference plane, Earth's axis of rotation is tilted about $23\frac{1}{2}$ degrees. Because of Earth's tilt, the apparent path of the Sun (ecliptic) and the celestial equator intersect each other at an angle of $23\frac{1}{2}$ degrees (Figure 19.20). This angle is very important to Earth's inhabitants. Because of the inclination of Earth's axis to the plane of the ecliptic, Earth exhibits its yearly cycle of seasons, a topic discussed in detail in Chapter 14.

When the *apparent* position of the Sun is plotted on the celestial sphere over a period of a year's time, its path intersects the celestial equator at two points (Figure 19.21). From a Northern Hemisphere point of view, these intersections are called the vernal (spring) equinox (March 20–21) and autumnal equinox (September 22–23). On June 21–22, the date of the summer solstice, the Sun appears $23\frac{1}{2}$ degrees north of the celestial equator, and 6 months later, on December 21–22, the date of the winter solstice, the Sun appears $23\frac{1}{2}$ degrees south of the celestial equator.

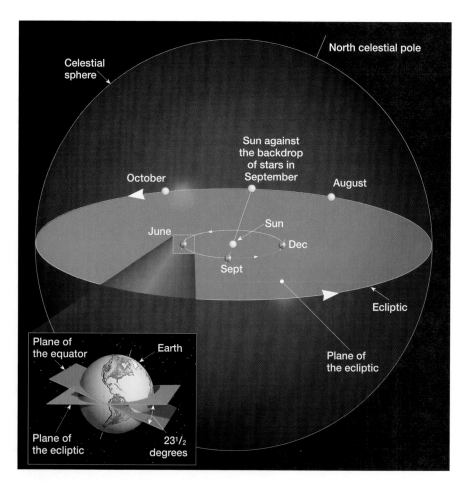

Figure 19.20 Earth's orbital motion causes the apparent position of the Sun to shift about 1 degree each day on the celestial sphere.

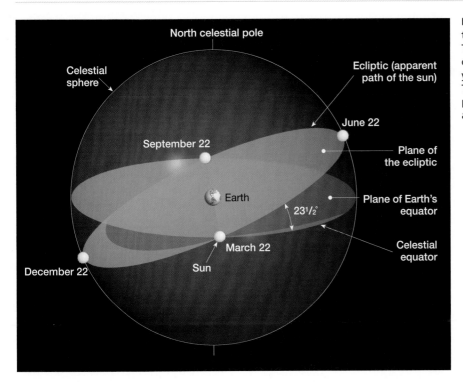

Figure 19.21 The apparent position of the Sun plotted on the celestial sphere. The path of the Sun (ecliptic) crosses the celestial equator on two occasions each year, March 20–21 and September 22–23. These are known as the equinox positions because the lengths of daylight and darkness on Earth are equal.

Precession

A third and very slow movement of Earth is called *precession*. Although Earth's axis maintains approximately the same angle of tilt, the direction in which the axis points continually changes. As a result, the axis traces a circle on the sky. This movement is very similar to the movement (wobble) of a spinning top (Figure 19.22A). At the present time, the axis points toward the bright star Polaris. In A.D. 14,000, it will point toward the bright star Vega, which will then become the North Star for a few thousand years (Figure 19.22B). The period of precession is 26,000 years. By the year 28,000, Polaris will once again be the North Star.

Precession has only a minor effect on the seasons, because the angle of tilt changes only slightly. It does, however, cause the positions of the seasons (equinox and solstice) to move slightly each year among the constellations.

In addition to its own movements, Earth shares numerous motions with the sun. It accompanies the Sun as the entire solar system speeds in the direction of the bright star Vega at 20 kilometers (12 miles) per second. Also, the Sun, like other nearby stars, revolves around the galaxy, a trip that requires 230 million years to traverse at speeds approaching 250 kilometers (150 miles) per second. In addition, the galaxies themselves

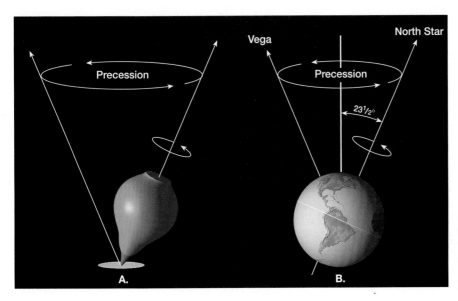

Figure 19.22 A. Precession illustrated by a spinning top. **B.** Precession of Earth causes the North Pole to point to different parts of the sky during a 26,000-year cycle. Today, the North Pole points to Polaris (North Star). In 13,000 years, Vega will be the North Star.

are in motion. We are presently approaching one of our nearest galactic neighbors, the Great Galaxy in Andromeda. In summary, the motions of Earth are many and complex, and its speed in space is very great.

Motions of the Earth–Moon System

Earth has one natural satellite, the Moon. In addition to accompanying Earth in its annual trek around the sun, our Moon orbits Earth within a period of about 1 month. When viewed from above the North Pole, the direction of this motion is counterclockwise (eastward). Because the Moon's orbit is elliptical, its distance to Earth varies by about 6 percent, averaging 384,401 kilometers (238,329 miles).

The motions of the Earth–Moon system constantly change the relative positions of the Sun, Earth, and Moon. The results are some of the most obvious of astronomical phenomena, namely the *phases of the Moon* and the occasional *eclipses of the Sun and Moon.*

Phases of the Moon

The first astronomical phenomenon to be understood was the regular cycle of the **phases of the Moon.** On a monthly basis, we observe the phases as a systematic change in the amount of the Moon that appears illuminated (Figure 19.23). We will choose the "new–Moon" position in the cycle as a starting point. About 2 days after the new Moon, a thin sliver (*crescent phase*) appears low in the western sky just after sunset. During the following week, the illuminated portion of the Moon visible from Earth increases (*waxing*) to a half circle (*first-quarter phase*) and can be seen from about noon to midnight. In another week, the complete disk (*full-Moon phase*) can be seen rising in the east as the Sun is sinking in the west. During the next 2 weeks, the percentage of the Moon that can be seen steadily declines (*waning*), until the Moon disappears altogether (*new-Moon phase*). The cycle soon begins anew with the reappearance of the crescent Moon.

The lunar phases are a consequence of the motion of the Moon and the sunlight that is reflected from its surface (Figure 19.23B). Half of the Moon is illuminated at all times (note the inner group of Moon sketches in Figure 19.23A). But to an Earthbound observer, the percentage of the bright side that is visible depends on the location of the Moon with respect to the Sun and Earth. When the Moon lies *between* the Sun and Earth, none of its bright side faces Earth, so we see the new-Moon ("no-Moon") phase. Conversely, when the Moon lies on the side of Earth opposite the Sun, all of its lighted side faces Earth, so we see the full Moon. At all positions between these extremes, an intermediate amount of the Moon's illuminated side is visible from Earth.

Lunar Motions

The cycle of the Moon through its phases requires $29\frac{1}{2}$ days, a time span called the **synodic month.** This cycle was the basis for the first Roman calendar. However, this is the *apparent period* of the Moon's revolution around Earth and not the true period, which takes only $27\frac{1}{3}$ days and is known as the **sidereal month.** The reason for the difference of nearly 2 days each cycle is shown in Figure 19.24. Note that as the Moon orbits Earth, the Earth–Moon system also moves in an orbit around the Sun. Consequently, even after the Moon has made a complete revolution around Earth, it has not yet reached its starting position, which was directly between the Sun and Earth (new-Moon phase). The additional motion to reach the starting point takes another 2 days.

An interesting fact concerning the motions of the Moon is that its period of rotation about its axis and its revolution around Earth are the same—$27\frac{1}{3}$ days. Because of this, the same lunar hemisphere always faces Earth. All of the manned *Apollo* missions were confined to the Earth-facing side. Only orbiting satellites and astronauts have seen the "back"side of the Moon.

Because the Moon rotates on its axis only once every $27\frac{1}{3}$ days, any location on its surface experiences periods of daylight and darkness lasting about 2 weeks. This, along with the absence of an atmosphere, accounts for the high surface temperature of 127°C (261°F) on the day side of the Moon and the low surface temperature of –173°C (–280°F) on its night side.

Eclipses

Along with understanding the Moon's phases, the early Greeks also realized that eclipses are simply shadow effects. When the Moon moves in a line directly between Earth and the Sun, which can occur only during the new-Moon phase, it casts a dark shadow on Earth, producing a **solar eclipse** (Figure 19.25). Conversely, the Moon is eclipsed (**lunar eclipse**) when it moves within Earth's shadow, a situation that is possible only during the full-Moon phases (Figure 19.26).

Why then does a solar eclipse not occur with every new-Moon phase and a lunar eclipse with every full-Moon phase? They would, if the orbit of the Moon lay exactly along the plane of Earth's orbit (the plane of the ecliptic). However, the Moon's orbit is inclined about 5 degrees to the plane that contains Earth and the Sun (Figure 19.27). Thus, during most new-Moon phases, the shadow of the Moon misses Earth (passes above or below); and during most full-Moon phases, the shadow of Earth misses the Moon. Only when a new- or full-Moon phase occurs where

Figure 19.23 Phases of the Moon. **A.** The outer figures show the phases as seen from Earth. **B.** Compare these photographs with the diagram. (Courtesy of UCO/Lick Observatory Image)

the Moon's orbit crosses the plane of the ecliptic can an eclipse take place.

Because these conditions are normally met only twice a year, the usual number of eclipses is four. These occur as a set of one solar and one lunar eclipse, followed 6 months later with another set (Figure 19.27). Occasionally the alignment is such that three eclipses can be squeezed into a 1-month period. These occur as a solar eclipse flanked by two lunar eclipses, or vice versa. Furthermore, it occasionally happens that the first set of eclipses occurs at the very beginning of a year, and a third set occurs before the year ends, resulting in six eclipses in that year. More rarely, if one of these sets is a three-eclipse kind, the total number of eclipses in a year can reach seven, which is the maximum.

During a total lunar eclipse, Earth's circular shadow can be seen moving slowly across the disk of the full Moon. When totally eclipsed, the Moon is completely

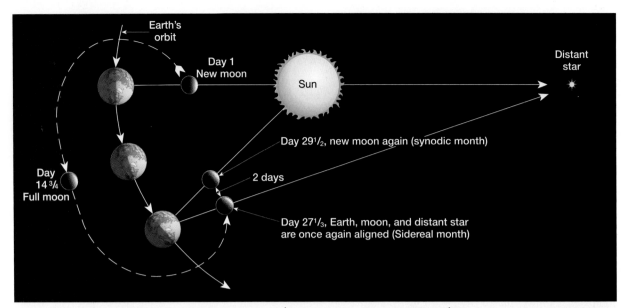

Figure 19.24 The difference between the sidereal month ($27\frac{1}{3}$ days) and the synodic month ($29\frac{1}{2}$ days). Distances and angles are not shown to scale.

Figure 19.25 Solar eclipse. Observers in the zone of the umbral shadow see a total solar eclipse. Those in the penumbra see a partial eclipse. The path of the solar eclipse moves eastward across the globe.

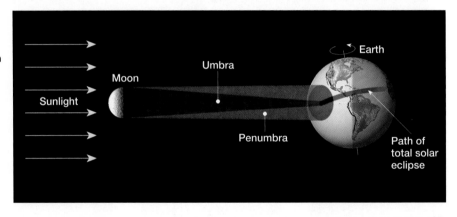

Figure 19.26 Lunar eclipse. During a total lunar eclipse the Moon's orbit carries it into the dark shadow of Earth (umbra). During a partial eclipse only a portion of the Moon enters the umbra.

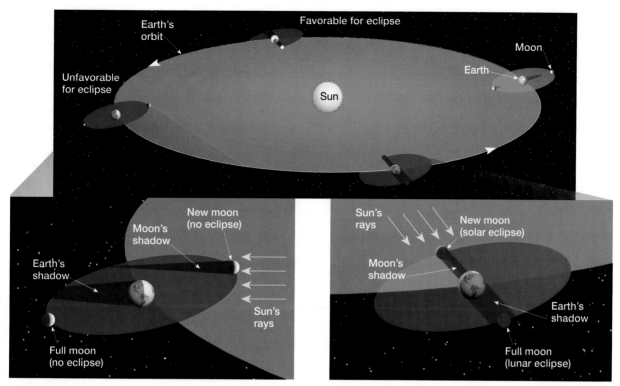

Figure 19.27 The Moon's orbit is inclined about 5 degrees to the plane that contains the Sun and Earth. Thus, during most new-Moon phases, the shadow of the Moon misses Earth (passes above or below), and during most full-Moon phases, the shadow of Earth misses the Moon. Only when a new- or full-Moon phase occurs where the Moon's orbit crosses the Earth–Sun plane can an eclipse take place. These conditions are met at roughly 6-month intervals.

within Earth's shadow but still is visible as a coppery disk, because Earth's atmosphere bends and transmits some long-wavelength light (red) into its shadow. A total eclipse of the Moon can last up to 4 hours and is visible to anyone on the side of Earth facing the Moon.

During a total solar eclipse, the Moon casts a circular shadow that is never wider than 275 kilometers (170 miles), about the size of South Carolina. Anyone observing in this region will see the Moon slowly block the Sun from view and the sky darken (Figure 19.28).

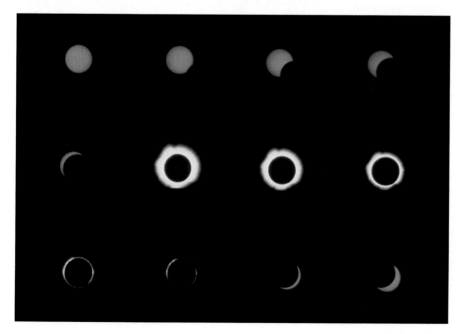

Figure 19.28 This sequence of photos starting from the upper left to the lower right shows the stages of a total solar eclipse. (From *Foundations of Astronomy, Third Edition,* by Michael Seeds, reprinted by permission of the author and Wadsworth Publishing Company)

Near totality, a sharp drop in temperature of a few degrees is experienced. The solar disk is completely blocked for at most 7 minutes, because the Moon's shadow is so small, and then one edge reappears.

At totality, the dark Moon is seen covering the complete solar disk, and only the Sun's brilliant white outer atmosphere is visible (see Figure 21.21). Total solar eclipses are visible only to people in the dark part of the Moon's shadow (*umbra*), while a partial eclipse

is seen by those in the light portion (*penumbra*) (see Figure 19.25).

Partial solar eclipses are more common in the polar regions, because it is this zone that the penumbra covers when the dark umbra of the Moon's shadow just misses Earth. A total solar eclipse is a rare event at any given location. The next one that will be visible from the contiguous United States will take place on August 21, 2017.

Chapter Summary

- Early Greeks held the *geocentric* ("Earth–centered") view of the universe, believing that Earth was a sphere that stayed motionless at the center of the universe. Orbiting Earth were the seven wanderers (*planetai* in Greek), which included the Moon, Sun, and the known planets—Mercury, Venus, Mars, Jupiter, and Saturn. To the early Greeks, the stars traveled daily around Earth on a transparent, hollow sphere called the *celestial sphere*. In A.D. 141, *Claudius Ptolemy* presented the geocentric outlook of the Greeks in its most sophisticated form in a model that became known as the *Ptolemaic system*. The Ptolemaic model had the planets moving in circular orbits around a motionless Earth. To explain the *retrograde motion* of planets (the apparent westward or opposite motion that planets exhibit for a period of time as Earth overtakes and passes them) Ptolemy proposed that the planets orbited in small circles (*epicycles*), revolving along large circles (*deferents*).

- In the fifth century B.C., the Greek *Anaxagoras* reasoned that the Moon shines by reflected sunlight, and because it is a sphere, only half is illuminated at one time. *Aristotle* (384–322 B.C.) concluded that Earth is spherical. The first Greek to profess a Sun-centered, or *heliocentric*, universe was *Aristarchus* (312–230 B.C.). The first successful attempt to establish the size of Earth is credited to *Eratosthenes* (276–194 B.C.). The greatest of the early Greek astronomers was *Hipparchus* (second century B.C.), best known for his star catalogue.

- Modern astronomy evolved through the work of many dedicated individuals during the sixteenth and seventeenth centuries. *Nicolaus Copernicus* (1473–1543) reconstructed the solar system with the Sun at the center and the planets orbiting around it, but erroneously continued to use circles to represent the orbits of planets. *Tycho Brahe's* (1546–1601) observations were far more precise than any made previously and are his legacy to astronomy. *Johannes Kepler* (1571–1630) ushered in the new astronomy with his three laws of planetary motion. After constructing his own telescope, *Galileo Galilei*

(1564–1642) made many important discoveries that supported the Copernican view of a Sun-centered solar system. *Sir Isaac Newton* (1642–1727) was the first to formulate and test the law of universal gravitation, develop the laws of motion, and prove that the force of *gravity*, combined with the tendency of an object to move in a straight line (*inertia*), results in the elliptical orbits discovered by Kepler.

- As early as 5000 years ago people began naming the configurations of stars, called *constellations*, in honor of mythological characters or great heroes. Today, 88 constellations are recognized that divide the sky into units, just as state boundaries divide the United States.

- One method for locating stars, called the *equatorial system*, divides the celestial sphere into a coordinate system similar to the latitude-longitude system used for locations on Earth's surface. *Declination*, like latitude, is the angular distance north or south of the *celestial equator*. *Right ascension* is the angular distance measured eastward from the position of the *vernal equinox* (the point in the sky where the Sun crosses the celestial equator at the onset of spring).

- The two primary motions of Earth are *rotation* (the turning, or spinning, of a body on its axis) and *revolution* (the motion of a body, such as a planet or moon, along a path around some point in space). Another very slow motion of Earth is *precession* (the slow motion of Earth's axis that traces out a cone over a period of 26,000 years). Earth's rotation can be measured in two ways, making two kinds of days. The *mean solar day* is the time interval from one noon to the next, which averages about 24 hours. In contrast, the *sidereal day* is the time it takes for Earth to make one complete rotation with respect to a star other than the Sun, a period of 23 hours, 56 minutes, and 4 seconds. Earth revolves around the Sun in an elliptical orbit at an average distance from the sun of 150 million kilometers (93 million miles). At *perihelion* (closest to the Sun), which occurs in January, Earth is 147 million kilometers from the Sun. At *aphelion* (farthest from the Sun), which occurs in July, Earth is 152 million kilometers distant. The imaginary plane that connects Earth's orbit with the celestial sphere is called the *plane of the ecliptic*.

- One of the first astronomical phenomena to be understood was the regular cycle of the phases of the Moon. The cycle of the Moon through its phases requires $29\frac{1}{2}$ days, a time span called the *synodic month*. However, the true period of the Moon's revolution around Earth takes $27\frac{1}{3}$ days and is known as the *sidereal month*. The difference of nearly 2 days is due to the fact that as the Moon orbits Earth, the Earth–Moon system also moves in an orbit around the Sun.

- In addition to understanding the Moon's phases, the early Greeks also realized that eclipses are simply shadow effects. When the Moon moves in a line directly between Earth and the Sun, which can occur only during the new-Moon phase, it casts a dark shadow on Earth, producing a *solar eclipse*. A *lunar eclipse* takes place when the Moon moves within the shadow of Earth during the full-Moon phase. Because the Moon's orbit is inclined about 5 degrees to the plane that contains the Earth and Sun (the plane of the ecliptic), during most new- and full-Moon phases no eclipse occurs. Only if a new- or full-Moon phase occurs as the Moon crosses the plane of the ecliptic can an eclipse take place. The usual number of eclipses is four per year.

Key Terms

aphelion (p. 536)
astronomical unit (AU) (p. 527)
celestial sphere (p. 520)
constellations (p. 532)
declination (p. 533)
ecliptic (p. 536)
equatorial system (p. 533)
geocentric (p. 520)
heliocentric (p. 522)

lunar eclipse (p. 538)
mean solar day (p. 535)
perihelion (p. 536)
perturbation (p. 531)
phases of the Moon (p. 538)
plane of the ecliptic (p. 536)
precession (p. 535)
Ptolemaic system (p. 523)
retrograde motion (p. 523)

revolution (p. 535)
right ascension (p. 533)
rotation (p. 535)
sidereal day (p. 535)
sidereal month (p. 538)
solar eclipse (p. 538)
synodic month (p. 538)

Review Questions

1. Why did the ancients believe that celestial objects had some control over their lives?

2. Describe what produces the retrograde motion of Mars. What geometric arrangements did Ptolemy use to explain this motion?

3. What major change did Copernicus make in the Ptolemaic system? Why was this change philosophically significant?

4. What was Tycho Brahe's contribution to science?

5. Does Earth move faster in its orbit near perihelion (January) or near aphelion (July)? Keeping your answer to the previous question in mind, is the solar day longest in January or July?

6. Use Kepler's third law ($p^2 = d^3$) to determine the period of a planet whose solar distance is

   (a) 10 AU

   (b) 1 AU

   (c) 0.2 AU

7. Use Kepler's third law to determine the distance from the Sun of a planet whose period is

   (a) 5 years.

   (b) 10 years.

   (c) 10 days.

8. Did Galileo invent the telescope?

9. Explain how Galileo's discovery of a rotating Sun supported the Copernican view of a Sun-centered universe.

10. Using a diagram, explain why the fact that Venus appears full when it is smallest supports the Copernican view and is inconsistent with the Ptolemaic system.

11. Newton learned that the orbits of the planets are the result of two actions. Explain these actions.

12. Of what value are constellations to modern-day astronomers?

13. Express the declination and right ascension of the star Arcturas (see Appendix D).

14. Explain the difference between the mean solar day and the sidereal day.

15. What is the approximate length of the cycle of the phases of the Moon?

16. What is different about the crescent phase that precedes the new-Moon phase and that which follows the new-Moon phase?

17. What phase of the Moon occurs approximately 1 week after the new Moon? Two weeks?

18. When you observe the crescent phase early in the evening, is the visible Moon waxing (growing) or waning (declining)? (See Figure 19.23.)

19. What phenomenon results from the fact that the Moon's period of rotation and revolution are the same?

20. The Moon rotates very slowly (once in $27\frac{1}{3}$ days) on its axis. How does this affect the lunar surface temperature?

21. Describe the locations of the Sun, Moon, and Earth during a solar eclipse and during a lunar eclipse.

22. How many eclipses normally occur each year?

23. Solar eclipses are slightly more common than lunar eclipses. Why, then, is it more likely that your region of the country will experience a lunar eclipse?

24. How long can a total eclipse of the Moon last? How about a total eclipse of the Sun?

Examining the Earth System

1. Currently, Earth is closest to the Sun (perihelion) in January (147 million kilometers/91.5 million miles) and farthest from the Sun in July (152 million kilometers/94.5 million miles). As the result of the precession of Earth's axis, 13,000 years from now perihelion (closest) will occur in July and aphelion (farthest) will take place in January. Assuming no other changes, how might this change *average* summer temperatures for your location? What about *average* winter temperatures? What might the impact be on the biosphere and hydrosphere? (To aid your understanding of the affect of Earth's orbital parameters on the seasons, you may want to review the section on "Variations in Earth's Orbit" in Chapter 5 and also investigate this National Oceanic and Atmospheric Administration (NOAA) Web site at **http://www.ngdc. noaa.gov/paleo/milankovitch.html**)

2. In what ways does the interaction of the Earth and Moon influence the Earth system? If Earth did not have a Moon, would the atmosphere, hydrosphere, solid Earth, and biosphere be any different? Explain.

Web Resources

The following are informative and interesting World Wide Web sites that address topics related to those presented in the chapter:

Index of Biographies:
- **http://www-groups.dcs.stand.ac.uk/history/ BiogIndex.html**

Star Chart (National Geographic Society):
- **http://www.nationalgeographic.com/stars/chart/ index.html**

Virtual Reality Moon Phase Pictures:
- **http://tycho.usno.navy.mil/vphase.html**

For direct links to these sites and other appropriate Earth science Web locations, chapter review exercises, and relevant, topical investigations that utilize Web resources, contact the *Earth Science*, 9th Edition Home Page at:

- **http://www.prenhall.com/tarbuck.**

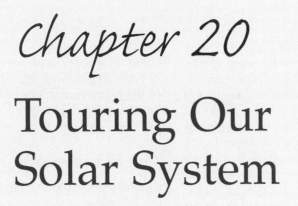

Chapter 20

Touring Our Solar System

The Planets: An Overview
Evolution of the Planets
Earth's Moon
The Planets: A Brief Tour
Minor Members of the Solar System

■ *Box 20.1* Earth as a System: Venus and the Runaway Greenhouse Effect—
A Lesson for Planet Earth?

■ *Box 20.2* Pathfinder: The First Geologist on Mars

■ *Box 20.3* Is Pluto Really a Planet?

■ *Box 20.4* Earth as a System: Is Earth on a Collision Course?

■ *Box 20.5* Comet Shoemaker-Levy Impacts Jupiter

Meteor Crater, near Winslow, Arizona. This cavity is about 1.2 kilometers (0.75 mile) across and 170 meters (560 feet) deep. The solar system is cluttered with meteoroids and other objects that can strike Earth with explosive force. (Photo by Michael Collier)

When people first recognized that the planets resembled Earth more than the stars, excitement grew. Could intelligent life exist on these other planets, or elsewhere in the universe? Space exploration has rekindled this interest. So far, no evidence of extraterrestrial life within our solar system has emerged. Nevertheless, we study the other planets to learn about Earth's formation and early history. Recent space explorations have been organized with this goal in mind. To date, Mercury, Venus, Mars, Jupiter, Saturn, Uranus, and Neptune have been explored by space probes.

The Sun is the hub of a huge rotating system of nine planets, their satellites, and numerous small asteroids, comets, and meteoroids. An estimated 99.85 percent of the mass of our solar system is contained within the Sun. The planets collectively make up most of the remaining 0.15 percent. The planets, traveling outward from the Sun, are Mercury, Venus, Earth, Mars, Jupiter, Saturn, Uranus, Neptune, and Pluto (Figure 20.1).

Under control of the Sun's gravitational force, each planet is tethered in an elliptical orbit, and all of them travel in the same direction. The nearest planet to the Sun, Mercury, has the fastest orbital motion, 48 kilometers per second, and the shortest period of revolution around the Sun, 88 Earth-days. By contrast, the most distant planet, Pluto, has an orbital speed of 5 kilometers per second and requires 248 Earth-years to complete one revolution.

Imagine a planet's orbit drawn on a flat sheet of paper. The paper represents the planet's *orbital plane.* The orbital planes of seven planets lie within 3 degrees of the plane of the Sun's equator. The other two, the innermost and outermost, Mercury and Pluto, are inclined 7 and 17 degrees, respectively.

The Planets: An Overview

Careful examination of Table 20.1 shows that the planets fall quite nicely into two groups: the **terrestrial (Earth-like) planets** (Mercury, Venus, Earth, and Mars) and the **Jovian (Jupiter-like) planets** (Jupiter, Saturn, Uranus, and Neptune). Pluto is not included in either category because its great distance from Earth and its small size make this planet's true nature a mystery.*

The most obvious difference between the terrestrial and the Jovian planets is their size (Figure 20.2). The largest terrestrial planets (Earth and Venus) have diameters only one-quarter as great as the diameter of the smallest Jovian planet (Neptune). Also, their masses are only 1/17 as great as Neptune's. Hence, the Jovian planets are often called *giants.* Because of their relative locations, the four Jovian planets are referred to as the *outer planets,* whereas the terrestrial planets are called the *inner planets.* As we shall see, there appears to be a correlation between the positions of these planets and their sizes.

---

*For more on this, see Box 20.3, p. 563.

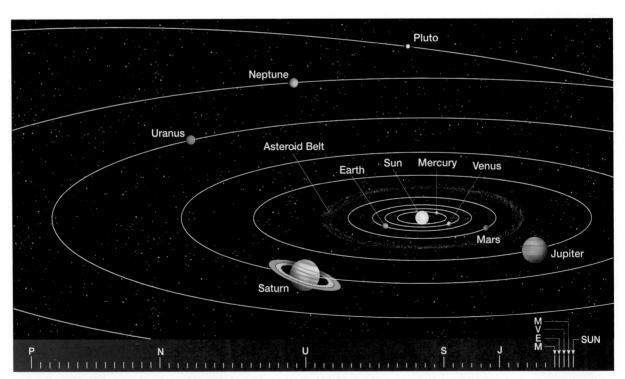

Figure 20.1 Orbits of the planets. Scale shown below.

Table 20.1 Planetary data.

Planet	Symbol	AU*	Mean Distance from Sun		Period of Revolution	Inclination of Orbit	Orbital Velocity	
			Millions of Miles	Millions of Kilometers			mi/s	km/s
Mercury	☿	0.39	36	58	88^d	7°00′	29.5	47.5
Venus	♀	0.72	67	108	225^d	3°24′	21.8	35.0
Earth	⊕	1.00	93	150	365.25^d	0°00″	18.5	29.8
Mars	♂	1.52	142	228	687^d	1°51′	14.9	24.1
Jupiter	♃	5.20	483	778	12^{yr}	1°18′	8.1	13.1
Saturn	♄	9.54	886	1427	29.5^{yr}	2°29′	6.0	9.6
Uranus	♅	19.18	1783	2870	84^{yr}	0°46′	4.2	6.8
Neptune	♆	30.06	2794	4497	165^{yr}	1°46′	3.3	5.3
Pluto	♇	39.44	3666	5900	248^{yr}	17°12′	2.9	4.7

Planet	Period of Rotation	Diameter		Relative Mass (Earth = 1)	Average Density (g/cm^3)	Polar Flattening (%)	Eccentricity	Number of Known Satellites
		Miles	Kilometers					
Mercury	59^d	3015	4878	0.06	5.4	0.0	0.206	0
Venus	244^d	7526	12,104	0.82	5.2	0.0	0.007	0
Earth	23^{h}56^{m}04^s	7920	12,756	1.00	5.5	0.3	0.017	1
Mars	24^{h}37^{m}23^s	4216	6794	0.11	3.9	0.5	0.093	2
Jupiter	9^{h}50^m	88,700	143,884	317.87	1.3	6.7	0.048	16
Saturn	10^{h}14^m	75,000	120,536	95.14	0.7	10.4	0.056	21
Uranus	17^{h}14^m	29,000	51,118	14.56	1.2	2.3	0.047	15
Neptune	16^{h}03^m	28,900	50,530	17.21	1.7	1.8	0.009	8
Pluto	6.4^d	~1425	~2300	0.002	1.8	0.0	0.250	1

*AU = astronomical unit, Earth's mean distance from the Sun.

Other dimensions along which the two groups markedly differ include density, chemical makeup, and rate of rotation. The densities of the terrestrial planets average about 5 times the density of water, whereas the Jovian planets have densities that average only 1.5 times that of water. One of the outer planets, Saturn, has a density only 0.7 times that of water, which means that Saturn would float if placed in a large enough water tank! Variations in the chemical compositions of the planets are largely responsible for these density differences.

The substances that make up both groups of planets are divided into three groups—*gases, rocks,* and *ices*—based on their melting points.

1. The gases, hydrogen and helium, are those with melting points near absolute zero (0° Kelvin or –273°C), the lowest possible temperature.

2. The rocks are principally silicate minerals and metallic iron, which have melting points exceeding 700°C.

3. The ices include ammonia (NH_3), methane (CH_4), carbon dioxide (CO_2), and water (H_2O). They have intermediate melting points (for example, H_2O has a melting point of 0°C).

The terrestrial planets are mostly rock. They are dense, rocky, and metallic, with minor amounts of gases. The Jovian planets, on the other hand, contain a large percentage of gases (hydrogen and helium), with varying amounts of ices (mostly water, ammonia, and methane). This accounts for their low densities. (The outer planets may contain as much rocky and metallic material as the terrestrial planets, but this material would be concentrated in their small central cores.)

The Jovian planets have very thick atmospheres of hydrogen, helium, methane, and ammonia. By contrast, the terrestrial planets, including Earth, have meager atmospheres at best. The reason is that a planet's ability to retain an atmosphere depends on its mass and temperature.

Simply stated, a gas molecule can "evaporate" from a planet if it reaches a speed known as the **escape velocity.** For Earth, this velocity is 11 kilometers (7 miles) per second. Any material, including a rocket, must reach this speed before it can escape Earth's gravity and go into space.

The Jovian planets, because of their greater surface gravities, have higher escape velocities (21 to 60 kilometers per second) than the terrestrial planets.

Figure 20.2 The planets drawn to scale.

of the Jovian planets even the lightest gases are unlikely to acquire the speed needed to escape.

In contrast, a comparatively warm body with a small surface gravity, like our Moon, is unable to hold even heavy gases, like carbon dioxide and radon, and thus lacks an atmosphere. The slightly larger terrestrial planets of Earth, Venus, and Mars retain some heavy gases such as carbon dioxide, but even their atmospheres make up only an infinitesimally small portion of their total mass.

The primordial cloud of dust and gas from which all the planets are thought to have condensed may have had a composition similar to Jupiter's. However, unlike Jupiter, the terrestrial planets today are nearly devoid of light gases and ices. Were the terrestrial planets once much larger? Did they contain these materials but lose them because of heating from the nearby Sun? In the following section we will consider the evolutionary histories of the terrestrial and Jovian planets in an attempt to answer these questions.

Evolution of the Planets

Recall from Chapter 11 the nebular hypothesis, which suggests that the entire solar system condensed from a gaseous nebula. The denser metallic elements (iron and nickel) in the forming planets sank toward their centers, whereas the lighter substances (silicate minerals, oxygen, hydrogen) migrated toward their surfaces.

During this period of *chemical differentiation*, gaseous materials escaped from the planet's interiors, much like what happens during a volcanic event on Earth. The hottest and second-smallest planet, Mercury, was unable to retain even the heaviest of these gases. Mars, on the other hand, being 40 percent larger and considerably cooler than Mercury, retained a thin layer of carbon dioxide and some water in the form of ice.

The largest of the terrestrial planets, Venus and Earth, have surface gravities strong enough to retain a substantial amount of the heavier gases, like nitrogen, oxygen, and carbon dioxide. However, when compared to the four Jovian planets, even the atmospheres of these two terrestrial planets must be looked upon as meager.

As the terrestrial planets were forming, the larger Jovian planets, along with their extensive satellite systems, were also developing. However, because of the frigid temperatures so far from the Sun, the fragments from which these planets formed contained a high percentage of ices—water, carbon dioxide, ammonia, and methane. Perhaps by random choice, two of the outer planets, Jupiter and Saturn, grew many times larger (by mass) than Uranus and Neptune. (For comparison, Jupiter is 318 times and Saturn is 95 times more massive than Earth. However, Uranus and Neptune have masses only 15 and 17 times greater than Earth.)

Consequently, it is more difficult for gases to "evaporate" from them. Also, because the molecular motion of a gas is temperature-dependent, at the low temperatures

When Jupiter and Saturn reached a certain size, estimated to be about 10 Earth masses, their surface gravity became sufficient to attract and hold even the lightest materials—hydrogen and helium. Thus, much of their size is attributable to the large envelope of light elements. They exist as a dense liquid below a thick, hydrogen-rich atmosphere. Jupiter and Saturn therefore consist of a central core of ices and rock, and a much larger outer envelope of mostly hydrogen and helium.

By contrast, the smaller Jovian planets, Uranus and Neptune, are proposed to have a small rocky-iron core and a large mantle of water, ammonia, and methane surrounded by a thin ocean of liquid hydrogen. These planets structurally resemble Jupiter and Saturn, but without their large hydrogen-helium envelopes.

In many respects, the development of the outer planets with their large satellite systems roughly parallels the events that formed the solar system. Like their parent planets, the satellites of the outer planets are primarily ices, with lesser amounts of rock. However, their small masses could not retain appreciable hydrogen and helium.

In the remainder of this chapter, we will consider each planet briefly, plus the minor members of the solar system. First, however, let us visit Earth's companion in space, our own Moon.

Earth's Moon

Earth now has hundreds of satellites, but only one natural satellite, the Moon, accompanies us on our annual journey around the Sun. Although other planets have moons, our planet-satellite system is unique in the solar system, because Earth's Moon is unusually large compared to its parent planet. The diameter of the Moon is 3475 kilometers (2150 miles), about one-fourth of Earth's 12,751 kilometers.

From calculation of the Moon's mass, its density is 3.3 times that of water. This density is comparable to that of *mantle* rocks on Earth but is considerably less than Earth's average density, which is 5.5 times that of water. Geologists have suggested that this difference can be accounted for if the Moon's iron core is small. The gravitational attraction at the lunar surface is one-sixth of that experienced on Earth's surface (a 100-pound person on Earth weighs only 17 pounds on the Moon). This difference allows an astronaut to carry a "heavy" life-support system with relative ease. If not

Figure 20.3 Telescopic view from Earth of the lunar surface. The major features are the dark "seas" (maria) and the light highly cratered highlands. (UCO/Lick Observatory Image)

burdened with such a load, an astronaut could jump six times higher than on Earth.

The Lunar Surface

When Galileo first pointed his telescope toward the Moon, he saw two different types of terrain—dark lowlands and bright, cratered highlands (Figure 20.3). Because the dark regions resembled seas on Earth, they were later named **maria,** Latin for "sea" (singular **mare**). This name is unfortunate, because the Moon's surface is totally devoid of water. Figure 20.4 shows typical features of the lunar surface.

Today we know that the Moon has no atmosphere or water. Therefore, the weathering and erosion that continually modify Earth's surface are virtually lacking on the Moon. In addition, tectonic forces are not active on the Moon, so earthquakes and volcanic eruptions no longer occur. However, because the Moon is unprotected by an atmosphere, a different kind of erosion occurs: tiny particles from space (micrometeorites) continually bombard its surface and ever-so-gradually smooth the landscape. Moon rocks become slightly rounded on top if they are long exposed at the lunar surface. Nevertheless, it is unlikely that the Moon has changed appreciably in the last 3 billion years, except for a few craters created by large meteorites.

Craters. The most obvious features of the lunar surface are craters. They are so profuse that craters-within-craters are the rule! The larger ones in the lower portion

of Figure 20.3 are about 250 kilometers (150 miles) in diameter, roughly the width of Indiana. Most craters were produced by the impact of rapidly moving debris (meteoroids), a phenomenon that was considerably more common in the early history of the solar system than it is today.

By contrast, Earth has only about a dozen easily recognized impact craters. This difference can be attributed to Earth's atmosphere. Friction with the air burns up small debris before it reaches the ground. In addition, evidence for most of the craters that formed in Earth's history has been obliterated by erosion or tectonic processes.

The formation of an impact crater is illustrated in Figure 20.5. Upon impact, the high-speed meteoroid compresses the material it strikes, then almost instantaneously the compressed rock rebounds, ejecting material from the crater. This process is analogous to the splash that occurs when a rock is dropped into water, and it often results in the formation of a central peak, as seen in the large crater in Figure 20.6. Most of the ejected material (*ejecta*) lands near the crater, building a rim around it. The heat generated by the impact is sufficient to melt some of the impacted rock. Astronauts have brought back samples of glass beads produced in this manner, as well as rock formed when angular fragments and dust were welded together by the impact.

A meteoroid only 3 meters (10 feet) in diameter can blast out a 150-meter (500-foot) wide crater. A few of the large craters such as Kepler and Copernicus,

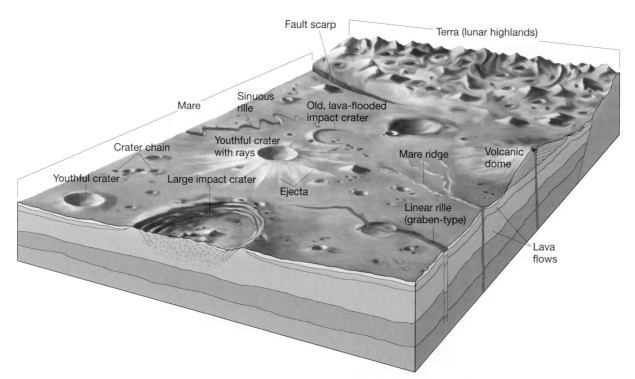

Figure 20.4 Block diagram illustrating major topographic features on the lunar surface.

Meteorite impact

Compressional wave

Fused rock and meteorite

Compressional wave

Rebound

Ejecta

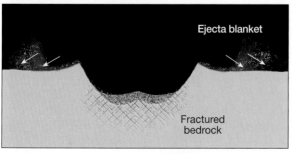

Ejecta blanket

Fractured bedrock

Figure 20.5 Formation of an impact crater. The energy of the rapidly moving meteoroid is transformed into heat energy and compressional waves. The rebound of the compressed rock causes debris to be ejected from the crater, and the heat melts some material, producing glass beads. Small secondary craters are formed by the material "splashed" from the impact crater. (After E. M. Shoemaker)

shown in Figure 20.3 formed from the impact of bodies 1 kilometer or more in diameter. These two large craters are thought to be relatively young because of the bright *rays* ("splash" marks) that radiate outward for hundreds of kilometers.

Highlands. Densely pockmarked highland areas make up most of the lunar surface. In fact, all of the "back" side of the Moon is characterized by such topography. (Only astronauts have seen the "back" side, because the Moon rotates on its axis once with each revolution around Earth, always keeping the same side facing Earth.) Within the highland regions are mountain ranges. The highest lunar peaks reach elevations approaching 8 kilometers, only 1 kilometer lower than Mount Everest.

Maria. The "seas" of basaltic lava originated when asteroids punctured the lunar surface, letting basaltic magma "bleed" out (Figure 20.7). Apparently the craters were flooded with layer upon layer of very fluid basaltic lava somewhat resembling the Columbia Plateau in the northwestern United States. The lava flows are often over 30 meters (100 feet) thick, and the total thickness of the material that fills the maria must approach thousands of meters.

Regolith. All lunar terrains are mantled with a layer of gray, unconsolidated debris derived from a few billion years of meteoric bombardment (Figure 20.8). This soil-like layer, properly called **lunar regolith,** is composed of igneous rocks, breccia, glass beads, and fine *lunar dust.* In the maria that have been explored by *Apollo* astronauts, the lunar regolith is apparently just over 3 meters (10 feet) thick.

Lunar History

Although the Moon is our nearest planetary neighbor and astronauts have sampled its surface, much is still unknown about its origin. The most widely accepted model for the origin of the Moon is that during the formative period of the solar system, a Mars-sized body impacted Earth. The impact would have liquefied Earth's surface and ejected huge quantities of crustal and mantle rock from an infant Earth. A portion of this ejected debris would have entered an orbit around Earth where it coalesced to for the Moon.

The giant-impact hypothesis is consistent with a number of facts known about the Moon. The ejected material would have been mostly iron-poor mantle and crustal rocks, which would account for the lack of a sizable iron core on the Moon. Further, the ejected material would have remained in orbit long enough to have lost the volatile materials (water) that the Moon lacks. Despite this supporting evidence, some questions remain unanswered.

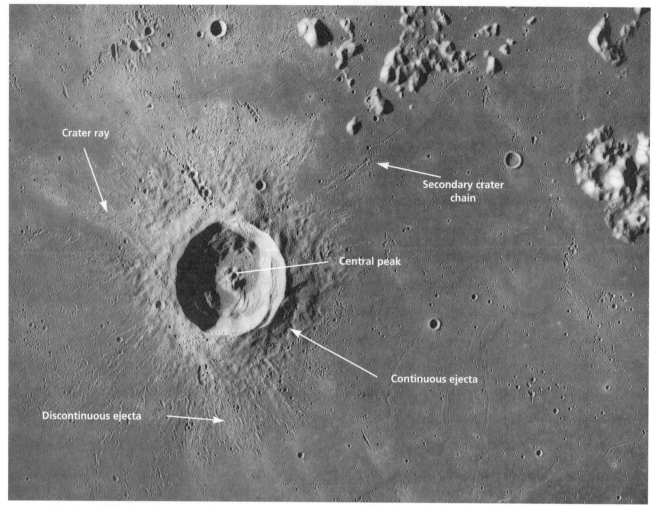

Figure 20.6 The 20-kilometer-wide lunar crater Euler in the southwestern part of Mare Imbrium. Clearly visible are the bright rays, central peak, secondary craters, and the large accumulation of ejecta near the crater rim. (Courtesy of NASA)

Figure 20.7 Formation of lunar maria. **A.** Impact of an asteroid-sized mass produced a huge crater hundreds of kilometers in diameter and disturbed the lunar crust far beyond the crater. **B.** Filling of the impact area with fluid basalts, perhaps derived from partial melting deep within the lunar mantle.

Figure 20.8 Astronaut Harrison Schmitt sampling the lunar surface. Notice the footprints (inset) in the lunar "soil." (Courtesy of NASA)

Nevertheless, planetary geologists have worked out the basic details of the Moon's later history. One of their methods is to observe variations in crater density (quantity per unit area). The greater the crater density, the longer the topographic feature must have existed. From such evidence, scientists concluded that the Moon evolved in three phases: the original crust (highlands), maria basins, and rayed craters.

During its early history the Moon was continually impacted as it swept up debris. This continuous bombardment, and perhaps radioactive decay, generated enough heat to melt the Moon's outer shell, and quite possibly the interior as well. Remnants of this original crust occupy the densely cratered highlands, which have been estimated to be as much as 4.5 billion years old—about the same age as Earth.

The second major event in the Moon's evolution was the formation of maria basins (see Figure 20.7). Radiometric dating of the maria basalts puts their age between 3.2 billion and 3.8 billion years, roughly a billion years younger than the initial crust. In places, the lava flows overlap the highlands, another testimonial to the younger age of the maria deposits.

The last prominent features to form were the rayed craters, as exemplified by the Copernicus crater (see Figure 20.3). Material ejected from these "young" depressions is clearly seen blanketing the surface of the maria and many older rayless craters. Even a relatively young crater like Copernicus must be millions of years old. Had it formed on Earth, erosional forces would have long since obliterated it.

If photos of the Moon taken several hundreds of millions of years ago were available, they would reveal that the Moon changed little in the intervening years. By all measures, the Moon is a dead body wandering through space and time.

The Planets: A Brief Tour

Mercury: The Innermost Planet

Mercury, the innermost and second smallest planet, is hardly larger than Earth's Moon and is smaller than three other moons in the solar system. Like our own Moon, it absorbs most of the sunlight that strikes it, reflecting only 6 percent into space. This is characteristic of terrestrial bodies that have no atmosphere. (Earth reflects about 30 percent of the light that strikes it, most of it from clouds.)

Mercury's close proximity to the Sun makes viewing from telescopes on Earth difficult because of the glare. The first good glimpse of this planet came in 1974 when *Mariner 10* passed within 800 kilometers (500 miles) of its surface. Its striking resemblance to the Moon was immediately evident from the high-resolution images that were sent back (Figure 20.9).

Mercury has cratered highlands, much like the Moon, and vast smooth terrains that resemble maria. However, unlike the Moon, Mercury is a very dense planet, which implies that it contains a very large iron core for its size. Also, Mercury has very long scarps that cut across the plains and craters alike. These scarps may have resulted from crustal shortening as the planet cooled and shrank.

Mercury revolves quickly but rotates slowly. One full day–night cycle on Earth takes 24 hours but on

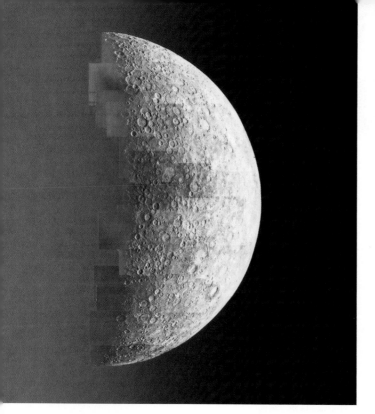

Figure 20.9 Photomosaic of Mercury. This view of Mercury is remarkably similar to the "far side" of the moon. (Courtesy of NASA)

Mercury it requires 179 Earth-days. Thus, a night on Mercury lasts for about 3 months and is followed by 3 months of daylight. Nighttime temperatures drop as low as −173°C (−280°F) and noontime temperatures exceed 427°C (800°F), hot enough to melt tin and lead. Mercury has the greatest extremes of any planet. The odds of life as we know it existing on Mercury are nil.

Venus: The Veiled Planet

Venus, second only to the Moon in brilliance in the night sky, is named for the goddess of love and beauty. It orbits the Sun in a nearly perfect circle once every 255 Earth-days. Venus is similar to Earth in size, density, mass, and location in the solar system. Thus, it has been referred to as "Earth's twin." Because of these similarities, it is hoped that a detailed study of Venus will provide geologists with a better understanding of Earth's evolutionary history.

Venus is shrouded in thick clouds impenetrable to visible light. Nevertheless, radar mapping by unmanned spacecraft and by instruments on Earth has revealed a varied topography with features somewhat between those of Earth and Mars. Simply, radar pulses in the microwave range are sent toward the Venusian surface, and the heights of plateaus and mountains are measured by timing the return of the radar echo. These data have confirmed that basaltic volcanism and tectonic deformation are the dominant processes operating on Venus. Further, based on the low density of impact craters, volcanism and tectonic deformation must have been very active during the recent geologic past (Figure 20.10).

Over 80 percent of the Venusian surface consists of subdued plains that are mantled by volcanic flows. Some lava channels extend hundreds of kilometers; one meanders 6800 kilometers across the planet. Thousands of volcanic structures have been identified, mostly small shield volcanoes, although over 1500 volcanoes greater than 20 kilometers across have been mapped. One is Sapas Mons, 400 kilometers (250

Figure 20.10 Computer-generated image of Venus. On the horizon is Maat Mons, a large volcano. Below it is a volcanic cone, Sapas Mons, from which light-colored lava flows extend hundreds of kilometers. (Courtesy of NASA)

Figure 20.11 These domelike volcanic structures on Venus average 25 kilometers in diameter and are less than 1 kilometer high. They are interpreted as very thick lava flows. (Photo courtesy of Jet Propulsion Laboratory)

miles) across and 1.5 kilometers (0.9 mile) high. Many flows from this volcano erupted from its flanks rather than the summit, in the manner of Hawaiian shield volcanoes. Other volcanic structures discovered on Venus are circular, pancake-shaped domes about 25 kilometers (15 miles) in diameter and nearly 1 kilometer high (Figure 20.11). These domes are thought to be the result of outpouring of very viscous lava, much like volcanic domes on Earth.

Only 8 percent of the Venusian surface consists of highlands that may be likened to continental areas on Earth. Tectonic activity on Venus seems to be driven by upwelling and downwelling of material in the planet's interior. Although mantle convection still operates on Venus, the processes of plate tectonics, which recycle rigid lithosphere, do not appear to have contributed to the present Venusian topography.

Before the advent of space vehicles, Venus was considered to be a potentially hospitable site for living organisms. However, evidence from *Mariner* fly-by space probes and Russian *Venera* landers indicates differently. The surface of Venus reaches temperatures of 475°C (900°F), and the Venusian atmosphere is 97 percent carbon dioxide (see Box 20.1). Only scant water vapor and nitrogen have been detected. The Venusian atmosphere contains an opaque cloud deck about 25 kilometers thick, which begins approximately 70 kilometers from the surface. Although the unmanned *Venera 8* survived less than an hour on the Venusian surface, it determined that the atmospheric pressure on that planet is 90 times that on Earth's surface. This hostile environment makes it unlikely that life as we know it exists on Venus.

Mars: The Red Planet

Mars has evoked greater interest than any other planet, for both scientists and nonscientists (see Box 20.2). When we imagine intelligent life on other worlds, "little green Martians" may come to mind. Interest in Mars stems mainly from this planet's accessibility to observation. All other planets within telescopic range have their surfaces hidden by clouds, except for Mercury, whose nearness to the Sun makes viewing difficult. Through the telescope, Mars appears as a reddish ball interrupted by some permanent dark regions that change intensity during the Martian year. The most prominent telescopic features of Mars are its brilliant white polar caps, resembling Earth's.

The Martian Atmosphere. The Martian atmosphere has only 1 percent the density of Earth's and it is primarily carbon dioxide with tiny amounts of water vapor. Data from Mars probes confirm that the polar caps of Mars are made of water ice, covered by a thin layer of frozen carbon dioxide. As winter nears in either hemisphere, we see the equatorward growth of that hemisphere's ice cap as temperatures drop to –125°C (–193°F) and additional carbon dioxide is deposited.

Although the atmosphere of Mars is very thin, extensive dust storms occur and may cause the color changes observed from Earth-based telescopes. Hurricane-force winds up to 270 kilometers (170 miles) per hour can persist for weeks. Images from *Viking 1* and *Viking 2* revealed a Martian landscape remarkably similar to a rocky desert on Earth (Figure 20.12), with abundant sand dunes and impact craters partially filled with dust.

Box 20.1 Earth as a System: Venus and the Runaway Greenhouse Effect— A Lesson for Planet Earth?

Carl Sagan, one of the foremost experts on extraterrestrial life, called Earth "the Heaven of the solar system" and Venus "the hell." Why should environments of two planets that are nearly the same size and are located in close proximity to one another be so dramatically different? The primary reason is the blistering temperature of the Venusian atmosphere, which is caused by a runaway greenhouse effect.

The greenhouse effect occurs when a planet's atmosphere acts like the glass in a greenhouse. It allows visible solar energy to penetrate the surface, but, like the glass in the greenhouse, certain gases in the planet's atmosphere are quite opaque to outgoing heat radiation. Thus, as in a greenhouse, the surface temperatures of these planets become warmer than expected, because the heat is temporarily trapped.

Both Earth and Venus experience a greenhouse effect. So why do the surface temperatures of Venus reach 475°C (900°F), whereas Earth is very hospitable? The answer is that carbon dioxide, a main contributor to the greenhouse effect, makes up 97 percent of the Venusian atmosphere, but less than 1 percent of Earth's (Figure 20.A).

The primary source of carbon dioxide is outgassing during volcanic eruptions. Why does Earth, which has over 600 active volcanoes, have an atmosphere containing only 0.035 percent carbon dioxide? The answer can be found in Earth's abundant plant life.

Plants use carbon dioxide and water in photosynthesis to generate organic matter, while oxygen is released as a by-product. Thus, over millions of years, plant life has altered our atmosphere, making it carbon dioxide-poor (0.035 percent) and oxygen-rich (21 percent).

In addition, carbon dioxide dissolved in seawater is used by a vast number of organisms for the production of carbonate shells. These shells are eventually deposited as sediment on the ocean floor. Consequently, huge quantities of carbon dioxide from Earth's atmosphere are continually being converted into organic matter and carbonate sediments. Apparently, Venus does not possess living organisms or a chemical mechanism capable of removing atmospheric carbon dioxide. Hence, our neighboring planet has a carbon dioxide concentration that has reached extreme proportions.

The discovery of a hostile Venusian atmosphere provides a good example of the practical benefits that people derive from space exploration. These data clearly reveal what can happen to a planet's environment when the greenhouse effect runs rampant. Some climatologists believe that a similar situation may be in the making on Earth. Through the burning of fossil fuels, we are converting increasingly larger amounts of oxygen and organic materials into carbon dioxide and water. In this manner, we are altering Earth's atmosphere by rapidly undoing what plant life has taken millions of years to accomplish.

We expect that further studies of the Venusian atmosphere will provide insights into how Earth's atmosphere evolved and how it will respond to changes induced by human activity. Such insights could keep our planet from becoming, as some have suggested, "hothouse Earth."

Figure 20.A Venus is shrouded in a hot, cloud-filled atmosphere composed mainly of carbon dioxide. (Courtesy of NASA)

Mars' Dramatic Surface. Mariner 9, the first artificial satellite to orbit another planet, reached Mars in 1971 amid a raging dust storm. When the dust cleared, images of Mars' northern hemisphere revealed numerous large volcanoes. The biggest, Mons Olympus, is the size of Ohio and 23 kilometers (75,000 feet) high.

This gigantic volcano and others resemble Hawaiian shield volcanoes on Earth (Figure 20.13). Their extreme size is thought to result from the absence of plate movements on Mars. Therefore, rather than a chain of smaller volcanoes forming as we find in Hawaii, large single cones developed.

Figure 20.12 This picture of the Martian landscape by the *Viking* 1 lander shows a dune field with features remarkably similar to many seen in the deserts of Earth. The dune crests indicate that recent wind storms were capable of moving sand over the dunes in the direction from lower right to upper left. The large boulder at the left is about 10 meters from the spacecraft and measures 1 by 3 meters. (Courtesy of NASA)

Box 20.2   *Pathfinder:* **The First Geologist on Mars**

On July 4, 1997, the *Mars Pathfinder* bounced onto the rock-littered surface of Mars and deployed its wheeled companion, Sojourner. For the next 3 months the lander sent back to Earth three gigabits of data, including 16,000 images and 20 chemical analyses. The landing site was a vast rolling landscape carved by ancient floods. The flood-deposit locale was selected in hope that a variety of rock types would be available for the rover Sojourner to examine.

Sojourner carried an alpha photon X-ray spectrometer (APXS) used to determine the chemical composition of rocks and Martian "soil" (regolith) at the landing site (Figure 20.B). In addition, the rover was able to take closeup images of the rocks. From these images, researchers concluded that the rocks were igneous. However, one hard, white, flat object named Scooby Doo was originally thought to be sedimentary rock, but the APXS data suggest its chemistry is like that of the soil found at the site. Thus, Scooby Doo is probably a well-cemented soil.

During its first week on Mars, Sojourner's APXS obtained data for a patch of windblown soil and a medium-sized rock, known affectionately as Barnacle Bill. Preliminary evaluation of the APXS data on Barnacle Bill shows that it contains over 60 percent silica. If these data are confirmed, it could indicate that Mars contains the

Figure 20.B Pathfinder's rover Sojourner obtaining data on the chemical composition of a Martian rock named "Yogi." (Photo courtesy of NASA)

volcanic rock andesite. Researchers had expected that most volcanic rocks on Mars would be basalt, which is lower in silica (less than 50 percent). On Earth, andesites are associated with tectonically active regions where oceanic crust is subducted into the mantle. Examples include the volcanoes of South America's Andes Mountains and the Cascades of North America.

Sojourner analyzed eight rocks and seven soils. The results thus far are only preliminary. Because these rocks are covered with a reddish dust that is high in sulfur, some controversy has

arisen as to the exact composition of these Martian rocks. Some researchers believe they are all of the same composition. The differences in measurements, they claim, are the result of varying thickness of dust.

In 2002, NASA plans to land another rover on Mars. It will be bigger than Sojourner and will explore what is thought to be an ancient sedimentary environment for up to a year. This rover will be capable of drilling a 5-centimeter bore into rocks and extracting a core for possible later return to Earth.

Figure 20.13 Image of Mons Olympus, an inactive shield volcano on Mars that covers an area about the size of the state of Ohio. (Courtesy of the U.S. Geological Survey)

Less-abundant impact craters indicate that at least some of the volcanic topography formed more recently, following the early period of heavy bombardment. Nevertheless, most Martian surface features are old by Earth standards. The highly cratered Martian southern hemisphere is probably similar in age to comparable lunar highlands (3.5 billion to 4.5 billion years old). Even the relatively fresh-appearing volcanic features of the northern hemisphere may be older than 1 billion years. This fact and the absence of "Marsquake" recordings by *Viking* seismographs point to a tectonically dead planet.

Another surprising find made by *Mariner 9* was the existence of several canyons that dwarf even Earth's Grand Canyon of the Colorado River. One of the largest, Valles Marineris, is thought to have formed by slippage of material along huge faults in the crustal layer. In this respect, it would be comparable to the rift valleys of Africa (Figure 20.14).

Water on Mars? Not all Martian valleys have a tectonic origin. Many have tributaries in a pattern similar to that of stream valleys on Earth. In addition, Viking orbiter images have revealed unmistakable ancient islands in what is now a dry stream bed. When these streamlike channels were first discovered, some observers speculated that a thick water-laden atmosphere capable of generating torrential downpours once existed on Mars. If so, what happened to this water? The present Martian atmosphere contains only traces. Moreover, the environment of Mars is far too harsh to let water exist as a liquid. Despite these difficulties, the work of

flowing water still remains the most acceptable explanation for many Martian channels.

Many planetary geologists do not accept the premise that Mars once had an active water cycle similar to Earth's. Rather, they believe that many of the large streamlike valleys were created by the collapse of surface material caused by the slow melting of subsurface ice. If this is the case, these large valleys would be more akin to features formed by mass wasting processes on Earth.

The Martian Satellites. Tiny Phobos and Deimos, the two satellites of Mars, were not discovered until 1977 because they are only 24 and 15 kilometers, respectively, in diameter. Phobos is nearer to its parent than any other natural satellite in the solar system—only 5500 kilometers—and requires just 7 hours and 39 minutes for one revolution. *Mariner 9* revealed that both satellites are irregularly shaped and have numerous impact craters, much like their parent (Figure 20.15).

Undoubtedly, these moons are asteroids captured by Mars. A most interesting coincidence in astronomy and literature is the close resemblance between Phobos and Deimos and two fictional satellites of Mars described by Jonathan Swift in *Gulliver's Travels*, written about 150 years before these satellites were actually discovered.

Jupiter: Lord of the Heavens

Jupiter, truly a giant among planets, has a mass 2 1/2 times greater than the combined mass of all the remaining planets, satellites, and asteroids. In fact, had Jupiter been about 10 times larger, it would have

Figure 20.14 This image shows the entire Valles Marineris canyon system, over 5000 kilometers long and up to 8 kilometers deep. The dark red spots on the left edge of the image are huge volcanoes, each about 25 kilometers high. (Courtesy of U.S. Geological Survey)

evolved into a small star. Despite its great size, however, it is only 1/800 as massive as the Sun. Jupiter also rotates more rapidly than any other planet, completing one rotation in slightly less than 10 Earth-hours. The effect of this fast spin is to make the equatorial region

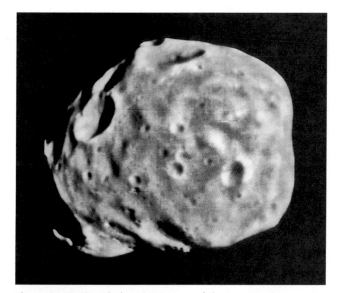

Figure 20.15 Tiny Phobos, innermost of the two Martian satellites. (Courtesy of NASA)

bulge and to make the polar dimension flatten (see "Polar Flattening" column in Table 20.1).

When viewed through a telescope or binoculars, Jupiter appears to be covered with alternating bands of multicolored clouds aligned parallel to its equator (Figure 20.16). The most striking feature is the *Great Red Spot* in the southern hemisphere (Figure 20.16). The Great Red Spot has been a prominent feature since is was first discovered more than three centuries ago. When *Voyager 2* swept by Jupiter in 1979, it was the size of two Earth-sized circles placed side by side. On occasion, it has grown even larger.

Images from *Pioneer 11* as it moved within 42,000 kilometers of Jupiter's cloud tops in 1974 indicated that the Great Red Spot is a counterclockwise-rotating (cyclonic) storm. It is caught between two jetstream-like bands of atmosphere flowing in opposite directions. This huge hurricane-like storm rotates once every 12 Earth-days. Although several smaller storms have been observed in other regions of Jupiter's atmosphere, none have survived for more than a few days.

Structure of Jupiter. Jupiter's hydrogen-helium atmosphere also has methane, ammonia, water, and sulfur compounds as minor constituents. The wind systems generate the light- and dark-colored bands that

Figure 20.16 Artist's view of Jupiter with Great Red Spot visible in its southern hemisphere. Earth for scale.

Atmospheric pressure at the top of the clouds is equal to sea-level pressure on Earth. Because of Jupiter's immense gravity, the pressure increases rapidly toward its surface. At 1000 kilometers below the clouds, the pressure is great enough to compress hydrogen gas into a liquid. Consequently, Jupiter's surface is thought to be a gigantic ocean of liquid hydrogen. Less than halfway into Jupiter's interior, extreme pressures cause the liquid hydrogen to turn into *liquid metallic* hydrogen. Jupiter is also believed to contain as much rocky and metallic material as is found in the terrestrial planets, probably in a central core.

Jupiter's Moons. Jupiter's satellite system, consisting of 16 moons discovered so far, resembles a miniature solar system. The four largest satellites, discovered by Galileo, travel in nearly circular orbits around the parent with periods of from 2 to 17 Earth-days (Figure 20.18). The two largest Galilean satellites, Callisto and Ganymede, surpass Mercury in size, while the two smaller ones, Europa and Io, are about the size of Earth's Moon. These Galilean moons can be observed with binoculars or a small telescope and are interesting in their own right. Because their orbits are along Jupiter's equatorial plane, and because they all have the same orbital direction, these moons most probably formed from "leftover" debris in much the same way as the planets did.

By contrast, Jupiter's four outermost satellites are very small (20 kilometers in diameter), revolve in a direction that is opposite the other moons, and have orbits that are steeply inclined to the Jovian equator. These satellites appear to be asteroids that passed near enough to be captured gravitationally by Jupiter.

Images from *Voyagers 1* and 2 in 1979 revealed, to the surprise of almost everyone, that each of the four Galilean satellites has a character all its own, as

encircle this giant (Figure 20.17). Unlike the winds on Earth, which are driven by solar energy, Jupiter itself gives off nearly twice as much heat as it receives from the Sun. Thus, it is the *interior* heat from Jupiter that produces huge convection currents in the atmosphere.

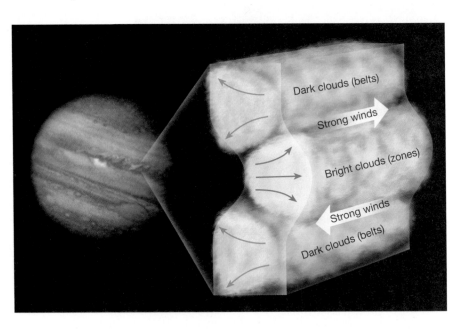

Figure 20.17 The structure of Jupiter's atmosphere. The areas of light clouds (zones) are regions where gases are ascending and cooling. Sinking dominates the flow in the darker cloud layers (belts). This convective circulation, along with the rapid rotation of the planet, generates the high-speed winds observed between the belts and zones.

A. Callisto

B. Europa

C. Ganymede

D. Io

Figure 20.18 Jupiter's four largest moons called the Galilean moons because they were discovered by Galileo. **A.** Callisto, the outermost of the Galilean satellites, is densely cratered, much like Earth's moon. **B.** Europa, smallest of the Galilean moons, has an icy surface that is criss-crossed by many linear features. **C.** Ganymede, the largest Jovian satellite, contains cratered areas, smooth regions, and areas covered by numerous parallel grooves. **D.** The innermost moon, Io, is one of only three volcanically active bodies in the solar system. (Courtesy of NASA)

Figure 20.19 A volcanic eruption on Io. This plume of volcanic gases and debris is rising over 100 kilometers (60 miles) above Io's surface. (Courtesy of NASA)

shown in Figure 20.18. The innermost of the Galilean moons, Io (Figure 20.19), is the only volcanically active body discovered in our solar system other than Earth and Neptune's moon Triton. To date, eight active sulfurous volcanic centers have been discovered. Umbrella-shaped plumes have been seen rising from the surface of Io to heights approaching 200 kilometers (Figure 20.19). The heat source for Io's volcanic activity is thought to be tidal energy generated by a relentless "tug of war" between Jupiter and the Galilean satellites. Because Io is gravitationally locked to Jupiter, the same side always faces the giant planet, like Earth's Moon. The gravitational power of Jupiter and the other nearby satellites pulls and pushes on Io's tidal bulge as its slightly eccentric orbit takes it alternately closer to and farther from Jupiter. This gravitational flexing of Io is transformed into heat energy.

One of the most interesting discoveries made by *Voyager 1* is Jupiter's thin ring system, believed to be different from that encircling Saturn. Rather than particles held in planetary-type orbits, the particles that make up Jupiter's rings appear to be temporarily entrapped by the planet's intense magnetic field. The ring material may be sulfur particles from Io's volcanoes.

Saturn: The Elegant Planet

Requiring 29.46 Earth-years to make one revolution, Saturn is almost twice as far from the Sun as Jupiter, yet its atmosphere, composition, and internal structure are thought to be remarkably similar to Jupiter's.

Saturn's Rings. The most prominent feature of Saturn is its system of rings (Figure 20.20), discovered by Galileo in 1610. Because he could not resolve them with his primitive telescope, they appeared to him as two smaller bodies adjacent to the planet. Their ring nature was revealed 50 years later by the Dutch astronomer Christian Huygens. Until the recent discovery that Jupiter, Uranus, and Neptune also have very faint ring systems, this phenomenon was thought to be unique to Saturn. When viewed from Earth, Saturn's rings appear as distinct bands.

Our view of Saturn slowly changes as both Saturn and Earth proceed along their orbits, continually shifting relative positions. This changes our angle of view of Saturn's rings. Once every 15 years, we see them edge-on, and they appear as an extremely fine line.

Saturn Close-Up. In 1980 and 1981, fly-by missions of the nuclear-powered *Voyagers* 1 and 2 space vehicles came within 100,000 kilometers of Saturn. More information was gained in a few days than had been acquired since Galileo first viewed this elegant planet telescopically (Figure 20.21):

1. Saturn's atmosphere is very dynamic with winds roaring at up to 1500 kilometers (930 miles) per hour.

2. Large cyclonic "storms" similar to Jupiter's Great Red Spot, although much smaller, occur in Saturn's atmosphere.

3. Eleven additional moons were discovered.

4. The icy rings of Saturn were discovered to be more complex than expected. Each of the seven rings is made of numerous ringlets, resembling the grooves on a phonograph record. The ringlets in one ring are intertwined in a braid-like configuration.

5. Satellite images reveal the thickness of the ring system to be no more than a few hundred meters. We easily see the thin rings from more than a billion kilometers distance because they are highly reflective.

No image obtained so far can resolve the fine structures of the rings, but they undoubtedly are composed of small particles (moonlets) that orbit the planet much like any other satellite. Radar observations indicate particles no larger than 10 meters, and the more abundant particles down to 10 centimeters (4 inches).

Figure 20.20 A view of the dramatic ring system of Saturn.

Figure 20.21 Montage of the Saturnian satellite system. The moon Dione is in foreground; Tethys and Mimas are at lower right; Enceladus and Rhea are off ring's left; and Titan is upper right. (Photo courtesy of NASA)

Beyond the outermost bright ring (A ring), some moonlets have accreted to form very small satellites having diameters on the order of 100 kilometers (60 miles). Five of these asteroid-sized moons have been discovered orbiting within the faint outer rings, and others probably exist. Planetary geologists are very interested in the gravitational interaction in Saturn's ring system. They hope this will reveal how material from the primordial cloud of dust and gases condensed to produce the planets.

Saturn's Moons. The Saturnian satellite system consists of at least 21 bodies (Figure 20.21). The largest, Titan, is bigger than Mercury and is the second-largest satellite in the solar system (after Jupiter's Ganymede). It is the only satellite in the solar system known to have a substantial atmosphere. Because of its dense gaseous cover, the atmospheric pressure at Titan's surface is about 1.5 times that at Earth's surface.

Data from *Voyager 1* reveal Titan's atmosphere to be roughly 80 percent nitrogen and perhaps 6 percent methane. This planet-sized moon appears to have polar ice caps that show seasonal variations in size. Its surface, if unfrozen, would be an ocean of liquid nitrogen.

Uranus and Neptune: The Twins

Earth and Venus have similar traits, but Uranus and Neptune are nearly twins. Only 1 percent different in diameter, both appear a pale greenish-blue, attributable to the methane in their atmospheres. Their structure and composition are believed to be similar. Neptune, however, is colder, because it is half again as distant from the Sun's warmth as is Uranus.

Uranus, the Sideways Planet. A unique feature of Uranus is that it rotates "on its side" (Figure 20.22). Its axis of rotation, instead of being generally perpendicular to the plane of its orbit, like the other planets, lies only 8 degrees from the plane of its orbit. Its rotational motion, therefore, has the appearance of rolling, rather than the toplike spinning of the other planets. Because the axis of Uranus is inclined almost 90 degrees, the Sun is nearly overhead at one of its poles once each revolution, and then half a revolution later, it is overhead at the other pole.

A surprise discovery in 1977 revealed that Uranus has rings, much like those encircling Jupiter. This find occurred as Uranus passed in front of a distant star and blocked its view, a process called *occultation* (the word occult means "hidden"). Observers saw the star

Figure 20.22 Painting of *Voyager 2* as it might have appeared when it encountered Uranus on January 24, 1986. (By Don Davis, courtesy of NASA)

"wink" briefly five times (meaning five rings) before the primary occultation and again five times afterward. Later studies indicate that Uranus has at least nine distinct belts of debris orbiting its equatorial region.

Spectacular views from *Voyager 2* of the five largest moons of Uranus show quite varied terrains. Some have long, deep canyons and linear scars, whereas others possess large, smooth areas on otherwise crater-riddled surfaces. The Jet Propulsion Laboratory described Miranda, the innermost of the five largest moons, as having a greater variety of landforms than any body yet examined in the solar system.

Neptune. Even when the most powerful telescope is focused on Neptune, it appears as a bluish fuzzy disk. Until the 1989 *Voyager 2* encounter, astronomers knew very little about this planet. However, the 12-year, nearly 3-billion-mile journey of *Voyager 2* provided investigators with so much new information about Neptune and its satellites that years will be needed to analyze it all.

Neptune has a dynamic atmosphere, much like those of Jupiter and Saturn (Figure 20.23). Winds exceeding 1000 kilometers per hour (600 miles per hour) encircle the planet, making it one of the windiest places in the solar system. It also has an Earth-sized blemish called the *Great Dark Spot* that is reminiscent of Jupiter's Great Red Spot, and is assumed to be a large rotating storm.

Perhaps most surprising are white, cirruslike clouds that occupy a layer about 50 kilometers above the main cloud deck, probably frozen methane. Six new satellites were discovered in the *Voyager* images, bringing Neptune's family to eight. All of the newly discovered moons orbit the planet in a direction opposite that of the two larger satellites. *Voyager* images also revealed a ring system around Neptune.

Triton, Neptune's largest moon, is a most interesting object (Figure 20.24). Its diameter is nearly that of our own Moon. Triton is the only large moon in the solar system that orbits its parent in the direction opposite to the direction in which all the planets travel. This indicates that Triton formed independently of Neptune and was gravitationally captured.

Triton also has the lowest surface temperature of any body in the solar system, –200°C (–391°F). Its very thin atmosphere is mostly nitrogen with a little methane. The surface of Triton apparently is largely water ice, covered with layers of solid nitrogen and methane.

Pluto: Planet X

Pluto lies on the fringe of the solar system, almost 40 times farther from the Sun than Earth (see Box 20.3). It is 10,000 times too small to be visible to the unaided eye. Because of its great distance and slow orbital speed, it takes Pluto 248 Earth-years to orbit the Sun. Ever since its discovery in 1930, it has completed about one-fourth of a revolution. Pluto's orbit is noticeably elongated (highly eccentric), causing it to occasionally travel inside the orbit of Neptune, where it currently resides. There is no chance that Pluto and Neptune will ever collide, because their orbits are inclined to each other and do not actually cross (see Figure 20.1).

In 1978, the moon Charon was discovered orbiting Pluto. Because of its close proximity to the planet, the best ground-based images of Charon show it only as an elongated bulge (Figure 20.25, left). In 1990, the Hubble Space Telescope produced an image that clearly resolves the separation between these two icy worlds (Figure 20.25, right). Charon orbits Pluto once every 6.4 Earth-days at a distance 20 times closer to Pluto than our Moon is to Earth.

Figure 20.23 This image of Neptune shows the Great Dark Spot (left center). Also visible are bright cirruslike clouds that travel at high speed around the planet. A second oval spot is at 54° south latitude on the east limb of the planet. (Courtesy of the Jet Propulsion Laboratory)

Figure 20.24 A photomosaic of Neptune's largest moon, Triton. The large south polar cap is on the right half of the image. Seasonal ice (probably nitrogen) covers the region. Because spring in Triton's southern hemisphere extends from 1960 to the year 2000, some of the polar cap has evaporated. (Courtesy of the Jet Propulsion Laboratory)

The discovery of Charon greatly altered earlier estimates of Pluto's size. Current data indicate that Pluto has a diameter of 2300 kilometers, about one-fifth the size of Earth, making it the smallest planet in the solar system. Charon is about 1300 kilometers across, exceptionally large in proportion to its parent.

The average temperature of Pluto is estimated at –210°C, cold enough to solidify most gases that might be present. Thus, Pluto might best be described as a dirty iceball of frozen gases with lesser amounts of rocky substances.

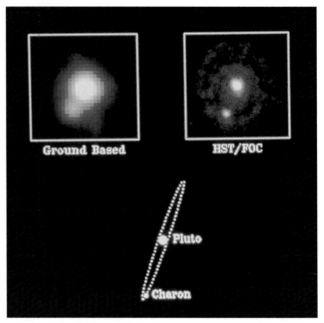

Figure 20.25 Pluto and its moon Charon. The Hubble Space Telescope produced the first image (upper right) that resolved these two icy worlds into separate objects. The image in the upper left is the best ground-based photo produced to date. (Courtesy of NASA)

Minor Members of the Solar System

Asteroids

Asteroids are smaller bodies that have been likened to "flying mountains." The largest, Ceres, is about 1000 kilometers (600 miles) in diameter, but most of the 50,000 that have been observed are only about 1 kilometer across. The smallest asteroids are assumed to be no larger than grains of sand. Most lie between the orbits of Mars and Jupiter and have periods of 3 to 6 years (Figure 20.26). Some have very eccentric orbits and travel very near the Sun, and a few larger ones regularly pass close to Earth and the Moon. Many of the most recent impact craters on the Moon and Earth were probably caused by

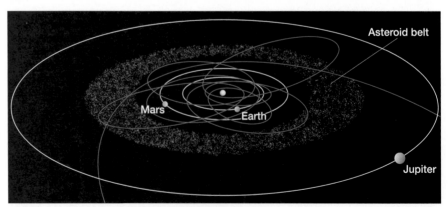

Figure 20.26 The orbits of most asteroids lie between Mars and Jupiter. Also shown are the orbits of a few known near-Earth asteroids. Perhaps a thousand or more asteroids have near-Earth orbits. Luckily, only a few dozen are thought to be larger than 1 kilometer in diameter.

Box 20.3   Is Pluto Really a Planet?

Ever since Pluto's discovery in 1930, it has been a mystery on the edge of the solar system. At first, Pluto was thought to be about as large as Earth, but as better images were obtained, Pluto's diameter was estimated to be a little less than one-half that of Earth. Then, in 1978, astronomers discovered that Pluto has a moon (Charon), whose brightness combined with its parent made Pluto appear much larger than it really was (Figure 20.C). Recent images obtained by the Hubble Space Telescope established the diameter of Pluto at only 2300 kilometers. This is about one-fifth that of Earth, and less than half that of Mercury, long considered the runt of the solar system. In fact, seven moons in the solar system are larger than Pluto.

In addition to its diminutive size, during the last few years over 60 cometlike objects have been discovered in a belt that extends beyond the orbit of Neptune. One of these objects approaches 500 kilometers (300 miles) in diameter. It appears that Pluto and its moon Charon are just two of thousands of icy worlds orbiting in the far reaches of the solar system.

A growing number of astronomers assert that Pluto's small size and its location within a swarm of similar icy objects means that it should be reclassified as a "minor planet," like asteroids and comets. Others insist that regardless of how Pluto's identity changes, demoting Pluto to a minor planet would dishonor astronomical history and confuse the public.

For now it seems that the International Astronomical Union, a group that has the power to vote on whether or not Pluto is a planet, is content with the status quo. Nevertheless, Pluto's planetary status will never be the same. It is now clear that Pluto is unique among the planets, being very different from the four, rocky innermost planets, and unlike the four gaseous giants. Perhaps Pluto is best described as the largest member of a belt of thousands of small, icy worlds that orbit in the outer reaches of our solar system.

Figure 20.C Pluto and its moon Charon. Earth is shown for scale.

collisions with asteroids. Inevitably, future Earth–asteroid collisions will occur—see Box 20.4.

Because many asteroids have irregular shapes, planetary geologists first speculated that they might be fragments of a broken planet that once orbited between Mars and Jupiter (Figure 20.27). However, the total mass of the asteroids is estimated to be only 1/1000 that of Earth, which itself is not a large planet. What happened to the remainder of the original planet? Others have hypothesized that several larger bodies once coexisted in close proximity and that their collisions produced numerous smaller ones. The existence of several "families" of asteroids has been used to support this explanation. However, no conclusive evidence has been found for either hypothesis.

Comets

Comets are among the most interesting and unpredictable bodies in the solar system (see Box 20.5). They have been compared to dirty snowballs, because they

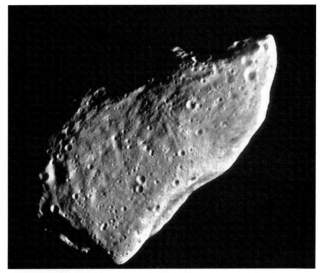

Figure 20.27 Image of asteroid 951 (Gaspra) obtained by the Jupiter-bound *Galileo* spacecraft. Like other asteroids, Gaspra is probably a collision-produced fragment of a larger body. (Courtesy of NASA)

Box 20.4   Earth as a System: Is Earth on a Collision Course?

The solar system is cluttered with meteoroids, asteroids, active comets, and extinct comets. These fragments travel at great speeds and can strike Earth with the explosive force of a powerful nuclear weapon.

In the last few decades, it has become increasingly clear that comets and asteroids have collided with Earth far more frequently than was previously known. The evidence is giant impact structures called *astroblemes*. More than 100 have been identified (Figure 20.D). Many were once misunderstood to result from some volcanic process. Although most astroblemes are so old that they no longer resemble impact craters, evidence of their intense impact remains (Figure 20.E). One notable exception is a very fresh-looking crater near Winslow, Arizona, known as Meteor Crater (see Figure 20.30).

Evidence is mounting that about 65 million years ago a large asteroid about 10 kilometers (6 miles) in diameter collided with Earth. This impact may have caused the extinction of the dinosaurs, as well as nearly 50 percent of all plant and animal species (see Box 11.2).

More recently, a spectacular explosion has been attributed to the collision of our planet with a comet or asteroid. In 1908, in a remote region of Siberia, a "fire-ball" that appeared more brilliant than the Sun exploded with a violent force. The shock waves rattled windows and triggered reverberations heard up to 1000 kilometers away. The "Tunguska event," as it is called, scorched, delimbed, and flattened trees to 30 kilometers from the epicenter. But expeditions to the area found

Figure 20.D World map of major impact structures (àstroblemes). Others are being identified every year. (Data from Griffith Observatory).

no evidence of an impact crater, nor any metallic fragments. Evidently, the explosion, which equaled at least a 10-megaton nuclear bomb, occurred a few kilometers above the surface. Most likely it was the demise of a comet or perhaps a stony asteroid. Why it exploded prior to impact is uncertain.

The dangers of living with these small but deadly objects from space again came to public attention in 1989 when an asteroid nearly 1 kilometer across shot past Earth. It was a near miss, about twice the distance to the Moon. Traveling at 70,000 kilometers (44,000 miles) per hour, it could have produced a crater 10 kilometers (6 miles) in diameter and perhaps 2 kilometers (1.2 miles) deep. As an observer noted, "Sooner or later it will be back." As it was, it crossed our orbit just 6 hours ahead of Earth. Statistics show that collisions of this tremendous magnitude should take place every few hundred million years and could have drastic consequences for life on Earth.

Figure 20.E Manicouagan, Quebec, is a 200-million-year-old eroded impact crater. The lake outlines the crater remnant, which is 70 kilometers (42 miles) across. Fractures related to this event extend outward for an additional 30 kilometers. (Courtesy of U.S. Geological Survey)

---

are made of frozen gases (water, ammonia, methane, carbon dioxide, and carbon monoxide) that hold together small pieces of rocky and metallic materials. Many comets travel very elongated orbits that carry them beyond Pluto.

When first observed, a comet appears very small; but as it approaches the Sun, solar energy begins to vaporize the frozen gases, producing a glowing head called the **coma** (Figure 20.28). The size of the coma varies greatly from one comet to another. Extremely rare ones exceed the size of the Sun, but most approximate the size of Jupiter. Within the coma, a small glowing nucleus with a diameter of only a few kilometers can sometimes be detected. As comets approach the Sun, some, but not all, develop a tail that extends for millions of kilometers. Despite the enormous size of their tails and comas, comets are relatively small members of the solar system.

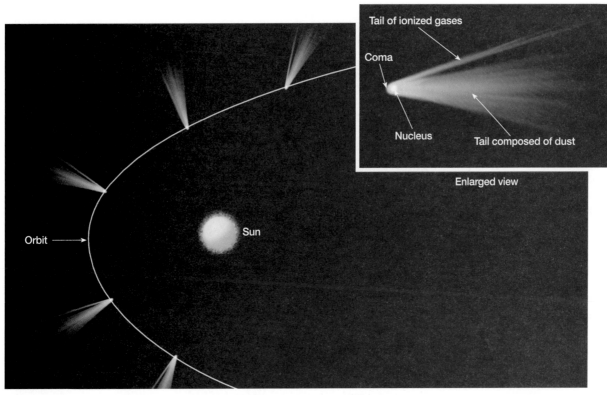

Figure 20.28 Orientation of a comet's tail as it orbits the sun.

A comet's tail is not a *trailing* tail, but a plume composed of dust and ionized gases repelled by solar forces. This explains why the tail always points away from the Sun (Figure 20.28). The solar forces are *radiation pressure*, which pushes dust particles away from the coma, and the *solar wind*, which moves the ionized gases, particularly carbon monoxide (Figure 20.28). Sometimes a single tail composed of both dust and ionized gases is produced, but often two tails are observed (Figure 20.29).

As a comet moves away from the Sun, the gases forming the coma recondense, the tail disappears, and the comet returns to "cold storage." Material that was blown from the coma to form the tail is lost from the comet forever. Consequently, it is believed that

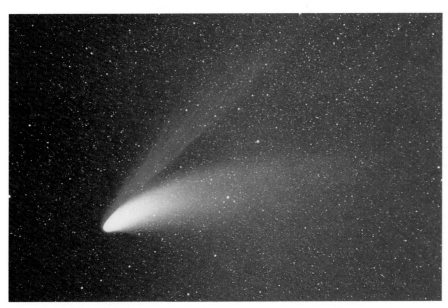

Figure 20.29 Comet Hale-Bopp. The two tails seen in the photograph are between 10 million and 15 million miles long. (A Peoria Astronomical Society Photograph by Eric Clifton and Craig Neaveill)

Box 20.5   Comet Shoemaker-Levy Impacts Jupiter

In July 1994, Comet Shoemaker-Levy impacted Jupiter with the force equal to 6 million megatons of energy. (A megaton is the equivalent of a million tons of the explosive, TNT). Clearly, this was the most dramatic event in the solar system ever observed by humans. Concern that a similar event might occur on Earth led NASA to establish the Near-Earth Object Search Committee to detect asteroids or comets with trajectories that might cross Earth's orbit.

It is possible that a few small fragments of a comet may have actually impacted Earth in 1908. That year, a strong explosion flattened more than 1000 square kilometers of a remote Siberian forest.

Comet Shoemaker-Levy was discovered at California's Mount Palomar Observatory barely a year before it impacted Jupiter. Careful observation showed that it consisted of two dozen fragments. Researchers concluded that the comet had broken up during an earlier pass by Jupiter. As the larger fragments penetrated Jupiter's outer atmosphere, they produced brilliant impact flashes and debris plumes

Figure 20.F Dark blemishes on Jupiter produced by the impact of fragments of Comet Shoemaker-Levy in July 1994. (Courtesy of NASA)

soaring thousands of kilometers above the planet. The result of these fiery impacts were dark zones in Jupiter's atmosphere that exceeded Earth in size. The largest of these dark blemishes lasted for months (Figure 20.F). Investigators hope to learn more about the dynamics of Jupiter's atmosphere by observing how these blemishes disperse.

---

most comets cannot survive more than a few hundred close orbits of the Sun. Once all the gases are expelled, the remaining material—a swarm of unconnected metallic and stony particles—continues the orbit without a coma or a tail.

Little is known about the origin of comets. Millions are believed to orbit the Sun beyond Neptune, and it is proposed that the gravitational effect of nearby stars sends some of them into the highly eccentric orbits that carry them toward the center of our solar system. Here, the gravitational pull of the larger planets, particularly Jupiter, alters a comet's orbit and accelerates its period of revolution.

The most famous short-period comet is Halley's comet. Its orbital period averages 76 years, and every one of its 29 appearances since 240 B.C. has been recorded by Chinese astronomers. This record is a testimonial to their dedication as astronomical observers and to the endurance of their culture. When seen in 1910, Halley's comet had developed a tail nearly 1.6

million kilometers (1 million miles) long and was visible during the daylight hours.

In 1986, the unspectacular showing of Halley's comet was a disappointment to many people in the Northern Hemisphere. Yet it was during this most recent visit to the inner solar system that a great deal of new information was learned about this most famous of comets. The new data were gathered by space probes sent to rendezvous with the comet. Most notably, the European probe *Giotto* approached to within 600 kilometers of the comet's nucleus and obtained the first images of this elusive structure.

We now know that the nucleus is potato-shaped, 16 kilometers by 8 kilometers. The surface is irregular and full of craterlike pits. Gases and dust that vaporize from the nucleus to form the coma and tail appear to gush from its surface as bright jets or streams. Only about 10 percent of the comet's total surface was emitting these jets at the time of the rendezvous. The remaining surface area of the comet appeared to be covered with a dark layer that may consist of organic material.

In 1997, the comet Hale-Bopp made for spectacular viewing around the globe. As comets go, the nucleus of Hale-Bopp was unusually large, about 40 kilometers (25 miles) in diameter. As shown in Figure 20.29, two tails nearly 15 million miles long extended from this comet. The bluish gas-tail is composed of positively charged ions, and it points almost directly away from the Sun. The yellowish tail is composed of dust and other rocky debris. Because the rocky material is more massive than the ionized gases, it is less affected by the solar wind and follows a different trajectory away from the comet.

Meteoroids

Nearly everyone has seen a **meteor,** popularly (but inaccurately) called a "shooting star." This streak of light lasts from an eyeblink to a few seconds and occurs when a small solid particle, a **meteoroid,** enters Earth's atmosphere from interplanetary space. The friction between the meteoroid and the air heats both and produces the light we see. Although a rare meteoroid is as large as an asteroid, most are the size of sand grains and weigh less than 1/100 gram. Consequently, they vaporize before reaching Earth's surface. Some, called *micrometeorites,* are so tiny that their rate of fall becomes too slow to cause them to burn up, so they drift down as "space dust." Each day, the number of meteoroids that enter Earth's atmosphere must reach into the thousands. After sunset, a half dozen or more are bright enough to be seen with the naked eye each hour from anywhere on Earth.

Occasionally, meteor sightings increase dramatically to 60 or more per hour. These displays, called **meteor showers,** result when Earth encounters a swarm of meteoroids traveling in the same direction and at nearly the same speed as Earth. The close association of these swarms to the orbits of some short-term comets strongly suggests that they represent material lost by these comets (Table 20.2). Some swarms not associated with orbits of known comets are probably the remains of the nucleus of a long-defunct comet. The notable Perseid meteor shower that occurs each year around August 12 is believed to be the remains of the Comet 1862 III, which has a period of 110 years.

Meteoroids associated with comets are small and not known to reach the ground. Most meteoroids large enough to survive the heated fall are thought to originate among the asteroids, where chance collisions modify their orbits and send them toward Earth. Earth's gravitational force does the rest.

The remains of meteoroids, when found on Earth, are referred to as **meteorites.** A few very large

Table 20.2 Major meteor showers.

Shower	Approximate Dates	Associated Comet
Quadrantids	January 4–6	—
Lyrids	April 20–23	Comet 1861 I
Eta Aquarids	May 3–5	Halley's comet
Delta Aquarids	July 30	—
Perseids	August 12	Comet 1862 III
Draconids	October 7–10	Comet Giacobini-Zinner
Orionids	October 20	Halley's comet
Taurids	November 3–13	Comet Encke
Andromedids	November 14	Comet Biela
Leonids	November 18	Comet 1866 I
Geminids	December 4–16	—

meteoroids have blasted out craters on Earth's surface that strongly resemble those on the lunar surface. The most famous is Meteor Crater in Arizona (Figure 20.30). This huge cavity is about 1.2 kilometers (0.75 miles) across, 170 meters (560 feet) deep, and has an upturned rim that rises 50 meters (165 feet) above the surrounding countryside. Over 30 tons of iron fragments have been found in the immediate area, but attempts to locate the main body have been unsuccessful. Based on erosion, the impact likely occurred within the last 20,000 years.

Prior to Moon rocks brought back by lunar explorers, meteorites were the only extraterrestrial materials that could be directly examined (Figure 20.31). Meteorites are classified by their composition: (1) **irons**—mostly iron with 5 percent to 20 percent nickel, (2) **stony**—silicate minerals with inclusions of other minerals, and (3) **stony-irons**—mixtures. Although stony meteorites are probably more common, people find mostly irons. This is understandable, for irons withstand the impact better, weather more slowly, and are much easier for a layperson to distinguish from terrestrial rocks. Iron meteorites are probably fragments of once-molten cores of large asteroids or small planets.

One rare kind of meteorite, called a *carbonaceous chondrite,* was found to contain simple amino acids and other organic compounds, which are the basic building blocks of life. This discovery confirms similar findings in observational astronomy, which indicate that numerous organic compounds exist in the frigid realm of outer space.

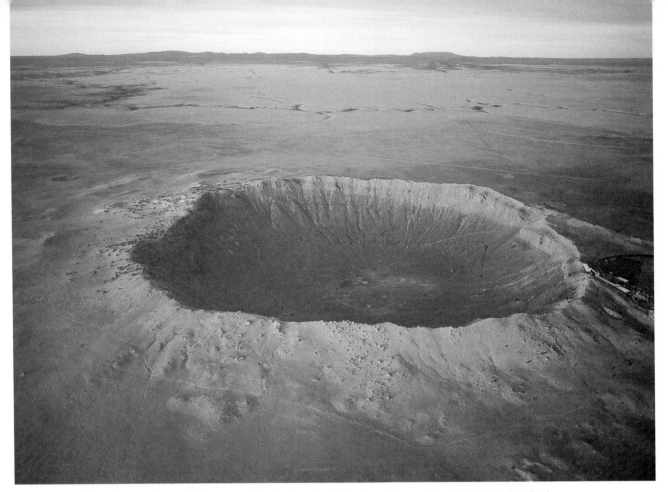

Figure 20.30 Meteor Crater, about 32 kilometers (20 miles) west of Winslow, Arizona. (Photo by Michael Collier)

Figure 20.31 Iron meteorite found near Meteor Crater, Arizona. (Courtesy of Meteor Crater, Northern Arizona, USA)

If meteorites represent the makeup of Earth-like planets, as some planetary geologists suggest, then Earth must contain a much larger percentage of iron than is indicated by surface rocks. This is one reason why geologists suggest that Earth's core may be mostly iron and nickel. In addition, meteorite dating indicates that our solar system's age certainly exceeds 4.5 billion years. This "old age" has been confirmed by data from lunar samples.

Chapter Summary

- The planets can be arranged into two groups: the *terrestrial* (Earth-like) *planets* (Mercury, Venus, Earth, and Mars) and the *Jovian* (Jupiter-like) *planets* (Jupiter, Saturn, Uranus, and Neptune). Pluto is not included in either group. *When compared to the Jovian planets, the terrestrial planets are smaller, more dense, contain proportionally more rocky material, have slower rates of rotation, and lower escape velocities.*

- The *nebular hypothesis* (discussed in Chapter 11) describes the formation of the solar system. Because of *chemical differentiation*, as the terrestrial planets formed, the denser metallic elements (iron and nickel) sank toward their centers, whereas the lighter substances (silicate minerals, oxygen, hydrogen) migrated toward their surfaces. Owing to their surface gravities, Earth and Venus were able to retain the heavier gases, such as nitrogen, oxygen, and carbon dioxide. Because of the very cold temperatures existing far from the Sun, the fragments from which the Jovian planets formed contained a high percentage of ices—water, carbon dioxide, ammonia, and methane.

- The lunar surface exhibits several types of features. Most *craters* were produced by the impact of rapidly moving interplanetary debris (*meteoroids*). Bright, densely cratered *highlands* make up most of the lunar surface. The dark, fairly smooth lowlands are called *maria* (singular, *mare*). Maria basins are enormous impact craters that have been flooded with layer-upon-layer of very fluid basaltic lava. All lunar terrains are mantled with a soil-like layer of gray, unconsolidated debris, called *lunar regolith,* which has been derived from a few billion years of meteoric bombardment. Much is still unknown about the Moon's origin. One hypothesis suggests that a giant asteroid collided with Earth to produce the Moon. Scientists conclude that the *lunar surface evolved in three phases: (1) the original crust (highlands); (2) maria basins;* and (3) *youthful rayed craters.*

- *Mercury* is a small, dense planet that has no atmosphere and exhibits the greatest temperature extremes of any planet. *Venus,* the brightest planet in the sky, has a thick, heavy atmosphere composed of 97 percent carbon dioxide, a surface of relatively subdued plains and inactive volcanic features, a surface atmospheric pressure 90 times that of Earth's, and surface temperatures of 475°C (900°F). *Mars,* the "Red Planet," has a carbon dioxide atmosphere only 1 percent as dense as Earth's, extensive dust storms, numerous inactive volcanoes, many large canyons, and several valleys of debatable origin exhibiting drainage patterns similar to stream valleys on Earth. *Jupiter,* the largest planet, rotates rapidly, has a banded appearance caused by huge convection currents driven by the planet's interior heat, a *Great Red Spot* that varies in size, a thin ring system, and at least 16 moons (one of the moons, *Io,* is a volcanically active body). *Saturn* is best known for its system of rings. It also has a dynamic atmosphere with winds up to 930 miles per hour and "storms" similar to Jupiter's Great Red Spot. *Uranus* and *Neptune* are often called "the twins" because of similar structure and composition. A unique feature of Uranus is the fact that it rotates "on its side." Neptune has white, cirruslike clouds above its main cloud deck and an Earth-sized *Great Dark Spot,* assumed to be a large rotating storm similar to Jupiter's Great Red Spot. *Pluto,* is a small frozen world with one relatively large moon (Charon). Pluto's noticeably elongated orbit causes it to occasionally travel inside the orbit of Neptune, but with no chance of collision.

- The minor members of the solar system include the *asteroids, comets,* and *meteoroids.* Most asteroids lie between the orbits of Mars and Jupiter. No conclusive evidence has been found to explain their origin. Comets are made of frozen gases (water, ammonia, methane, carbon dioxide, and carbon monoxide) with small pieces of rocky and metallic material. Many travel in very elongated orbits that carry them beyond Pluto and little is known about their origin. Meteoroids, small solid particles that travel through interplanetary space, become *meteors* when they enter Earth's atmosphere and vaporize with a flash of light. *Meteor showers* occur when Earth encounters a swarm of meteoroids, probably material lost by a comet. *Meteorites* are the remains of meteoroids found on Earth. The *three types of meteorites* (classified by their composition) are (1) *irons,* (2) *stony,* and (3) *stony-irons.* One rare kind of meteorite, called a *carbonaceous chondrite,* was found to contain amino acids and other organic compounds.

Key Terms

asteroid (p. 568)
coma (p. 570)
comet (p. 569)
escape velocity (p. 549)
iron meteorite (p. 573)
Jovian planet (p. 548)

lunar regolith (p. 553)
maria (p. 551)
meteor (p. 573)
meteorite (p. 573)
meteoroid (p. 573)
meteor shower (p. 573)

stony-iron meteorite (p. 573)
stony meteorite (p. 573)
terrestrial planet (p. 548)

Review Questions

1. By what criteria are the planets placed into either the Jovian or terrestrial group?

2. What are the three types of materials thought to make up the planets? How are they different? How does their distribution account for the density differences between the terrestrial and Jovian planetary groups?

3. Explain why different planets have different atmospheres.

4. How is crater density used in the relative dating of features on the Moon?

5. Briefly outline the history of the Moon.

6. How are the maria of the Moon thought to be similar to the Columbia Plateau?

7. Why has Mars been the planet most studied telescopically?

8. What surface features does Mars have that are also common on Earth?

9. Although Mars has valleys that appear to be products of stream erosion, what fact makes it unlikely that Mars has had a water cycle like that on Earth?

10. The two "moons" of Mars were once suggested to be artificial. What characteristics do they have that would cause such speculation?

11. What is thought to be the nature of Jupiter's Great Red Spot?

12. Why are the Galilean satellites of Jupiter so named?

13. What is distinctive about Jupiter's satellite Io?

14. Why are the four outer satellites of Jupiter thought to have been captured?

15. What evidence indicates that Saturn's rings are composed of individual moonlets rather than consisting of solid disks?

16. What is unique about Saturn's satellite Titan?

17. What three bodies in the solar system exhibit volcanic-like activity?

18. What do you think would happen if Earth passed through the tail of a comet?

19. Describe the origin of comets according to the most widely accepted hypothesis.

20. Compare a meteoroid, meteor, and meteorite.

21. Why are meteorite craters more common on the Moon than on Earth, even though the Moon is a much smaller target?

22. It has been estimated that Halley's comet has a mass of 100 billion tons. Further, this comet is thought to lose 100 million tons of material during the few months that its orbit brings it close to the Sun. With an orbital period of 76 years, what is the maximum remaining life span of Halley's comet?

Examining the Earth System

1. On Earth the four major spheres (atmosphere, hydrosphere, solid Earth, and bioshpere) interact as a system with occasional influence from our near-space neighbors. Which of these spheres are absent, or nearly absent, on the Moon? Because the Moon lacks these spheres, list at least five processes that operate on Earth but are absent on the Moon.

2. Among the planets in our solar system, Earth is unique because water exists in all three states (solid, liquid, and gas) on its surface. In what state(s) of matter is water found on Mercury, Venus, and Mars? How would Earth's hydrologic cycle be different if (a) its orbit was inside the orbit of Venus? (b) its orbit was outside the orbit of Mars?

3. If a large meteorite were to strike Earth in the near future, what affect might this event have on the atmosphere (in particular, average temperatures and climate)? Assuming the conditions persisted, how would the changes impact the biosphere?

Web Resources

The following are informative and interesting World Wide Web sites that address topics related to those presented in the chapter:

National Aeronautics and Space Administration (NASA):

• **http://www.nasa.gov/**

Observatorium (NASA):

• **http://observe.ivv.nasa.gov/nasa/core.shtml**

Planetary Tour Guide (National Aeronautics and Space Administration):

• **http://ranier.oact.hq.nasa.gov/Sensors_page/Planets.html**

Solar System Tour:

• **http://seds.lpl.arizona.edu/billa/tnp/nineplanets.html**

For direct links to these sites and other appropriate Earth science Web locations, chapter review exercises, and relevant, topical investigations that utilize Web resources, contact the *Earth Science,* 9th Edition Home Page at:

• **http://www.prenhall.com/tarbuck.**

Chapter 21

Light, Astronomical Observations, and the Sun

Dawn view of some of the astronomical observatories on Mauna Kea in Hawaii, with a full moon in the sky. (Photo by David Nunuk/ Science Photo Library/Photo Researchers, Inc.)

Astronomers are in the business of gathering and studying light. Almost everything that is known about the universe beyond Earth comes by analyzing light from distant sources. Consequently, an understanding of the nature of light is basic to modern astronomy. This chapter deals with the study of light and the tools used by astronomers to gather light in order to probe the universe. In addition, we will examine the nearest source of light, our Sun. By understanding the processes that operate on the Sun, astronomers can better grasp the nature of more distant celestial objects.

The Study of Light

The vast majority of our information about the universe is obtained from the study of the light emitted from celestial bodies (Figure 21.1). Although visible light is most familiar to us, it constitutes only a small part of an array of energy generally referred to as **electromagnetic radiation.** Included in this array are gamma rays, X-rays, ultraviolet light, visible light, infrared radiation, and radio waves (see Figure 14.14, p. 384). All radiant energy travels through the vacuum of space in a straight line at the rate of 300,000 kilometers (186,000 miles) per second.* Over a 24-hour day, this equals a staggering 26 billion kilometers.

Nature of Light

Experiments have demonstrated that light can be described in two ways. In some instances, light behaves like waves, and in others, like discrete particles. In the wave sense, light is analogous to swells in the ocean. This motion is characterized by the wavelength, which is the distance from one wave crest to the next. Wavelengths vary, from several kilometers for radio waves to less than a billionth of a centimeter for gamma rays. Most of these waves are either too long or too short for our eyes to detect.

---

*Light rays are "bent" slightly when they pass nearby a very massive object such as the Sun.

Figure 21.1 The Trifid Nebula, in the constellation Sagittarius. This colorful nebula is a cloud of dust, plus hydrogen and helium gases. These gases are excited by the radiation of the hot, young stars within and produce a pink glow. (Courtesy of National Optical Astronomy Observatories)

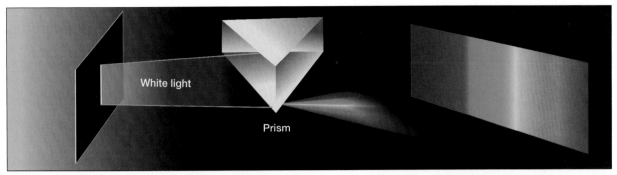

Figure 21.2 A spectrum is produced when sunlight (white light) is passed through a prism, which bends each wavelength at different angles.

The narrow band of electromagnetic radiation we can see is sometimes referred to as *white light.* However, white light consists of an array of waves having various wavelengths, a fact easily demonstrated with a prism (Figure 21.2). As white light passes through a prism, the color with the shortest wavelength, violet, is bent more than blue, which is bent more than green, and so forth (Table 21.1). Thus, white light can be separated into its component colors in the order of their wavelengths, producing the familiar rainbow of colors (Figure 21.2).

Wave theory, however, cannot explain some effects of light. In some cases, light acts like a stream of particles, analogous to infinitesimally small bullets fired from a machine gun. These particles, called **photons,** can exert a pressure ("push") on matter, which is called **radiation pressure.** Photons from the Sun are responsible for "pushing" material away from a comet to produce its tail. Each photon has a specific amount of energy, which is related to its wavelength in a simple way: *Shorter wavelengths* correspond to *more energetic* photons. Thus, blue light has more energetic photons than does red light.

Which theory of light—the wave theory or the particle theory—is correct? Both, because each will predict the behavior of light for certain phenomena. As George Abell, a prominent astronomer, stated about all scientific laws, "The mistake is only to apply them to situations that are outside their range of validity."

Table 21.1 Colors and corresponding wavelengths.

Color	Wavelength (nanometers*)
Violet	380–440
Blue	440–500
Green	500–560
Yellow	560–590
Orange	590–640
Red	640–750

*One nanometer is 10^{-9} meter.

Spectroscopy

When Sir Isaac Newton used a prism to disperse white light into its component colors, he unknowingly initiated the field of **spectroscopy,** which is the study of properties of light that depend on wavelength. The rainbow of colors Newton produced is called a "continuous spectrum," because all wavelengths of light are included. It was later learned that two other types of spectra exist, and that all three are generated under somewhat different conditions (Figure 21.3).

1. A **continuous spectrum** is produced by an incandescent solid, liquid, or gas under high pressure. It consists of an uninterrupted band of color (Figure 21.3A). One example would be light generated by a common light bulb. (*Incandescent* means "to emit light when hot.")

2. A **dark-line spectrum** (**absorption spectrum**) is produced when white light is passed through a comparatively cool gas under low pressure. The gas absorbs selected wavelengths of light, so the spectrum that is produced appears as a continuous spectrum, but with a series of dark lines running through it (Figure 21.3B).

3. A **bright-line spectrum** (**emission spectrum**) is produced by a hot (incandescent) gas under low pressure. It is a series of bright lines of particular wavelengths, depending on the gas that produces them (Figure 21.3C). These bright lines appear in the exact location as the dark lines that are produced by this gas in a dark-line (absorption) spectrum.

The spectra of most stars are of the dark-line type. The importance of these spectra is that each element or compound that is in gaseous form (in a star, material is usually in the gaseous form) produces a unique set of spectral lines. When the spectrum of a star is studied, the spectral lines act as "fingerprints," which identify the elements present.

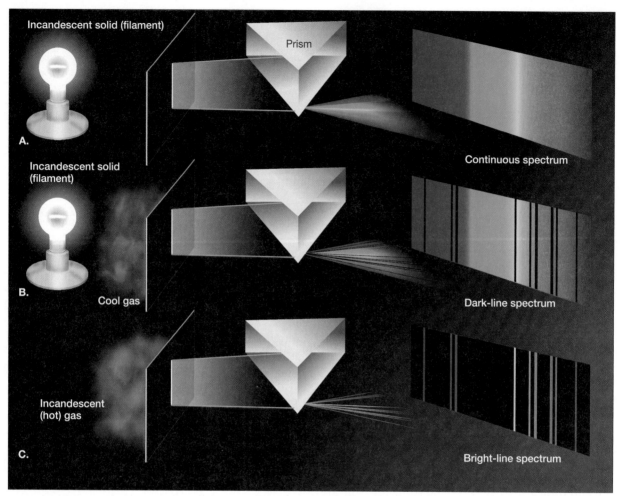

Figure 21.3 Formation of the three types of spectra. **A.** Continuous spectrum. **B.** Dark-line spectrum. **C.** Bright-line spectrum.

In an admittedly oversimplified manner, we can imagine how the Sun and other stars create a dark-line spectrum. A continuous spectrum is produced in the interior of the Sun, where the gases are under very high pressure. When this light passes through the less dense gases of the solar atmosphere, they absorb selected wavelengths, which appear as dark lines in the spectrum.

When Newton studied solar light, he obtained a continuous spectrum. However, when a prism is used in conjunction with lenses, the solar spectrum can be dispersed even further. An instrument that does this is called a **spectroscope.** The spectrum of the Sun contains thousands of dark lines. Over 60 elements have been identified by matching these lines with those of elements known on Earth.

Two other facts concerning a radiating body are important. First, if the temperature of a radiating surface is increased, the total amount of energy emitted increases. The rate of increase is stated by the Stefan-Boltzmann law: The energy radiated by a body is directly proportional to the fourth power of its absolute temperature. For example, if the temperature of a star is doubled, the total radiation emitted increases by 2^4 ($2 \times 2 \times 2 \times 2$), or 16 times. Second, as the temperature of an object increases, a larger proportion of its energy is radiated at shorter wavelengths. To illustrate this, imagine a metal rod that is heated slowly. The rod first appears dull red (longer wavelengths), and later bluish white (shorter wavelengths). From this, it follows that blue stars are hotter than yellow stars, which are hotter than red stars.

The Doppler Effect

You may have heard the change in pitch of a car horn or ambulance siren as it passes by. When it is approaching, the sound seems to have a higher-than-normal pitch, and when it is moving away, the pitch sounds lower than normal. This effect, which occurs for both sound and light waves, was first explained by Christian Doppler in 1842 and is called the **Doppler effect.** The reason for the difference in pitch is that it takes time for the wave to be emitted. If the source is moving away, the beginning of the wave is emitted nearer to you than the end, which

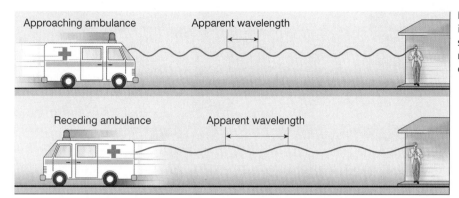

Figure 21.4 The Doppler effect, illustrating the apparent lengthening and shortening of wavelengths caused by the relative motion between a source and an observer.

"stretches" the wave, that is, gives it a longer wavelength (Figure 21.4). The opposite is true for an approaching source.

In the case of light, when a source is moving away, its light appears redder than it actually is, because its waves appear lengthened. Objects approaching have their light waves shifted toward the blue (shorter wavelength). Thus, if a source of red light approached you at a very high speed (near the speed of light), it would actually appear blue. The same effect would be produced if you moved and the light were stationary.

Therefore, the Doppler effect reveals whether Earth is approaching or receding from a star or another celestial body. In addition, the amount of shift allows us to recalculate the rate at which the relative movement is occurring. Larger Doppler shifts indicate higher velocities; smaller Doppler shifts indicate slower velocities. Doppler shifts are generally measured from the dark lines in the spectra of stars by comparing them with a standard spectrum produced in the laboratory (Figure 21.5).

A. Standard sodium lines

B. Red-shifted sodium lines

C. Large red-shifted sodium lines

D. Blue-shifted sodium lines

Figure 21.5 A. An illustration of standard sodium lines produced in a laboratory compared to sodium lines as they would appear when the source is receding (red shift, **B.** and **C.**) or approaching (blue shift, **D.**).

Astronomical Tools

Having examined the nature of light, we will now turn our attention to the tools astronomers use to intercept and study the energy emitted by distant objects in the universe (Figure 21.6). Because the basic principles of detecting radiation were originally developed through visual observations, we will consider optical telescopes first.

Refracting Telescopes

Galileo is considered to be the first person to use telescopes for astronomical observations. Having learned about the newly invented instrument, Galileo built one of his own design that was capable of magnifying objects 30 times. Because this early instrument, as well as its modern counterparts, used a lens to bend or refract light, it is known as a **refracting telescope.**

The most important lens in a refracting telescope, the **objective lens,** produces an image by bending light from a distant object in such a way that the light converges at an area called the **focus** (Figure 21.7). For an object such as a star, the image appears as a point of light, but for a nearby objects, it appears as an inverted replica of the original.

You can easily demonstrate the latter case by holding a lens in one hand and with the other hand placing a white card behind the lens. Now, vary the distance between them until an image appears on the card. The distance between the focus (where the image appears) and the lens is called the **focal length** of the lens.

Astronomers usually study an image from a telescope by first photographing the image. However, if a telescope is used to examine an image directly, a second lens, called an **eyepiece,** is required (Figure 21.7). The eyepiece magnifies the image produced by the objective lens. In this respect, it is similar to a magnifying glass. Thus, the objective lens produces a very small, bright image of an object, and the eyepiece enlarges the image so that details can be seen.

Figure 21.6 The Mayall 4-meter telescope at Kitt Peak National Observatory near Tucson, Arizona. (Photo by Tom Bean/DRK Photo)

Although used extensively in the nineteenth century, refracting telescopes suffer a major optical defect. As light passes through any lens, the shorter wavelengths of light are bent more than the longer wavelengths. (Recall the effect of a prism in separating the colors of the spectrum.) Consequently, when a refracting telescope is in focus for red light, blue and violet light are out of focus. The troublesome effect, known as **chromatic** ("color") **aberration,** weakens the image and produces a halo of color around it. When blue light is in focus, a reddish halo appears, and vice versa. Although this effect cannot be eliminated completely, it is reduced by using a second lens made of a different type of glass.

Reflecting Telescopes

Newton was so bothered by chromatic aberration that he built and used telescopes that reflected light from a shiny surface (mirror). Because reflected light is not dispersed into its component colors, the problem is avoided. **Reflecting telescopes** use a concave mirror that focuses the light in front of the objective (the mirror), rather than behind it, like a lens (Figure 21.8). The mirror is generally made of glass that is finely ground. In the case of the 5-meter (200-inch) Hale Telescope, the grinding is accurate to about 1 millionth of a centimeter. The surface is then coated with a highly reflective material, usually an aluminum compound.

To focus parallel incoming rays onto one spot, the mirror is ground to a special curved surface called a *paraboloid.* This is the same shape as that used for the reflector in the headlights of automobiles. In the case of the auto bulb, however, the light source is at the focus, and the light goes out in parallel rays, rather than coming in.

Because the focus of a reflecting telescope is in front of the mirror, provisions have to be made to view the image without blocking too much of the incoming light. Figure 21.9 illustrates the most common arrangements. Most large telescopes employ more than one type. When using a very large reflecting telescope, the observer enters a viewing cage positioned at the focus to make the observations. The viewing cage blocks only about 10 percent of the total incoming light, and this is more than compensated for by the large objective (mirror) that is used.

In the "good old days," an astronomer would spend numerous long nights outdoors in the cold mountain air, perched in a viewing cage. But advances in photographic materials and computer-enhancing technologies now allow indoor work and reduce the time required to obtain an image.

Nearly all large optical telescopes built today are reflectors. Among the reasons is the monumental task of producing a large piece of high-quality, bubble-free

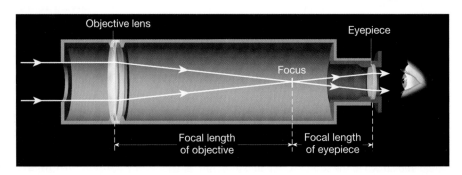

Figure 21.7 Simple refracting telescope.

A.

B.

Figure 21.8 A. Diagram illustrating how concave mirrors, like those used in reflector telescopes, gather light. **B.** Preparation of the 2.4-meter mirror for the Hubble Space Telescope. (Courtesy of Space Telescope Science Institute)

Large, reflecting telescopes of about 4 meters (150 inches) are located at Kitt Peak, Arizona (Figure 21.10); Mauna Kea, Hawaii; Cerro Tololo, Chile; and Siding Spring, Australia (see Box 21.1). By comparison, the largest refractor in the world is the 1-meter (40 inch) telescope at Yerkes Observatory in Williams Bay, Wisconsin. This instrument was built before the turn of the twentieth century.

Figure 21.9 Viewing methods used with reflecting telescopes. **A.** Prime focus method (only used in very large telescopes). **B.** Cassegrain method (most common). **C.** Newtonian method.

glass for refracting telescopes. Because light does not pass through a mirror, the glass for a reflecting telescope does not have to be of optical quality, nor does the instrument suffer from chromatic aberration. In addition, a lens can be supported only around the edge, so it sags. Mirrors, on the other hand, can be supported fully from behind.

Figure 21.10 The 3.9-meter Anglo-Australian telescope. This modern reflecting telescope is one of the world's most advanced. (Royal Observatory, Edinburgh/Photo Researchers, Inc.)

Properties of Optical Telescopes

Telescopes have three properties that aid astronomers in their work. They provide the observer with light-gathering power, resolving power, and magnifying power.

Light-Gathering Power. Because most celestial objects are very faint sources of light, astronomers are *most* interested in improving the *light-gathering* power of their instruments. As seen in Figure 21.11, a telescope with a large lens (or mirror) intercepts more light from distant objects and thereby produces brighter images. As very distant stars appear very dim as well, a great deal of light must be collected before the image is bright enough to be seen. Consequently, telescopes with large objectives "see" farther into space than do those with small objectives.

Resolving Power. Another advantage of telescopes with large-diameter objectives is their greater *resolving power* (Figure 21.12), which allows for sharper images and finer detail. For example, with the unaided eye, the Milky Way appears as a vague band of light in the night sky, but even a small telescope is capable of resolving (separating it into) individual stars. Even so, the condition of Earth's atmosphere (which is known as *seeing*) greatly limits the resolving power of telescopes on Earth. On a night when the stars twinkle, the seeing is poor because the air is moving rapidly. This causes the image to move about, blurring it. Conversely, when the stars shine steadily, the seeing is described as good. Even under ideal conditions, however, some blurring occurs, eliminating the fine details. Thus, even the largest telescopes cannot photograph lunar features less than 0.5 kilometer (0.3 mile) in size.

To eliminate the problems of Earthbound viewing, the United States built the Hubble Space Telescope, which was put into Earth orbit in April 1990 (Figure 21.13). This 2.4-meter (94-inch) space telescope has 10 billion times more light-gathering power than the human eye. As the technical problems that hampered the early operation of this instrument have been corrected, many spectacular images have been received. For example, as shown in Figure 20.25 (p. 562), the Hubble Space Telescope has provided images that

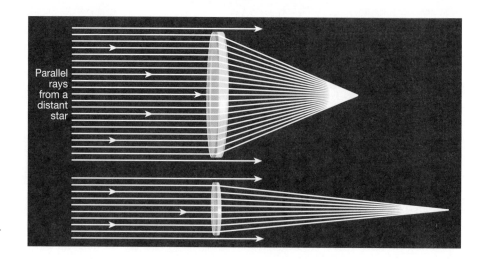

Figure 21.11 Comparison of the light-gathering ability of two lenses.

Parallel rays from a distant star

Figure 21.12 Appearance of the galaxy in the constellation Andromeda using telescopes of different resolution. (Courtesy of Leiden Observatory)

clearly resolve the separation between Pluto and its moon, Charon. Pluto is at the center of the photo while Charon is the faint object in the lower left. For comparison, the best ground-based image taken to date is shown in the upper left of the figure.

Magnifying Power. When you think of the power of a telescope, you probably think of its *magnifying power,* the ability to make an object larger. Magnification is calculated by dividing the focal length of the objective by the focal length of the eyepiece. Thus, the magnification of a telescope can be changed by simply

Figure 21.13 The deployment of the Hubble Space Telescope in Earth orbit, April 24, 1990, from Space Shuttle *Discovery*. (Courtesy of NASA)

changing the eyepiece. However, increased magnification does not necessarily improve the clarity of the image. What can be viewed telescopically is limited by atmospheric conditions and the resolving power of the telescope. Any part of an image that is not clear at low magnification will appear only as a larger blur at higher magnification. Furthermore, increasing magnification spreads out the light and decreases the brightness of the object. Thus, astronomers describe telescopes not in terms of their magnification but by the diameter of the objective mirror or lens, because it is this factor that determines both the light-gathering power and the resolving power of a telescope.

Most modern telescopes have supplemental devices that enhance the image. A simple, but important, example is a photographic plate that can be exposed for long periods of time, thereby collecting enough light from a star to make an image that otherwise would be undetectable. One of the latest advances makes use of high-speed computers that adjust the optics to partially remove the distortion caused by the atmosphere. This process greatly enhances the image sharpness.

Detecting Invisible Radiation

As we said, sunlight is made up of more than just the radiation that is visible to our eyes. Gamma rays, X-rays, ultraviolet radiation, infrared radiation, and radio waves are also produced by stars. Photographic film that is sensitive to ultraviolet and infrared radiation has been developed, thereby extending the limits of our vision. However, most of this radiation cannot penetrate our atmosphere, so balloons, rockets, and satellites must transport cameras "above" the atmosphere to record it.

Of great importance is a narrow band of radio radiation that does penetrate the atmosphere. One particular wavelength is the 21-centimeter (8-inch) line

Box 21.1 The Largest Telescopes

The world's largest conventional telescope is the 6-meter (236-inch) reflector in the Caucasus Mountains in Russia. However, because of problems with location and optics, it has been less useful than many smaller telescopes. Just behind it in size is the 5-meter (200-inch) Hale Telescope on Mount Palomar in California. The Hale Telescope is a 500-ton steel and glass marvel that floats into position on oil bearings.

Casting and grinding a single large mirror is time-consuming and expensive. For example, the production of the 5-meter mirror for the Hale Telescope began in 1934 and was finished 14 years later. Thus, astronomers driven by the desire to obtain even greater light-gathering and resolving power have looked for innovative techniques for telescope building.

A new technology for casting an 8-meter mirror is currently being employed under the football stadium at the University of Arizona. This new oven turns like a merry-go-round, so that molten glass is forced outward in the mold, thereby producing a concaved outer surface. Once cooled, this glass casting has nearly the correct shape, which reduces the grinding time considerably. Twin 8-meter telescopes are planned. One will be located in the Northern Hemisphere, on Hawaii's Mauna Kea at an altitude of nearly 4200 meters (13,800 feet).

The other is to be located in Chile, on the western slopes of the Andes Mountains.

Another effort is the Multiple Mirror Telescope (MMT), which uses six 1.8-meter mirrors linked together to focus at a single point (Figure 21.A). The result is a telescope with a light-gathering capability of a mirror 4.5 meters in diameter.

A different innovative design is the recently completed 10-meter Keck Telescope located on Mauna Kea. This instrument uses a mosaic of 36 six-sided mirrors carefully positioned by computer to give the optical effect of a 10-meter (400-inch) mirror.

Even more ambitious proposals are being considered. Plans are under way to build four 8.4-meter telescopes located on the same mountaintop. The four telescopes will be able to observe independently, or by combining their light they will have the light-gathering power of a single 16-meter instrument. When completed, this instrument will have 10 times the light gathering capacity of the 5-meter Hale Telescope.

Figure 21.A The Multiple Mirror Telescope Observatory. This unique telescope consists of six 1.8-meter mirrors that are brought to a common focus. (Courtesy of MMT Observatory/Smithsonian Inst./Univ. of Arizona)

produced by neutral hydrogen (hydrogen atoms that lack an electrical charge). Measurement of this radiation has permitted us to map the galactic distribution of hydrogen—the material from which stars are made.

The detection of radio waves is accomplished by "big dishes" called **radio telescopes** (Figure 21.14A). In principle, the dish of one of these telescopes operates in the same manner as the mirror of an optical telescope. It is parabolic in shape and focuses the incoming radio waves on an antenna, which absorbs and transmits these waves to an amplifier, just like any radio antenna.

Because radio waves are about 100,000 times longer than visible radiation, the surface of the dish need not be as smooth as a mirror. In fact, except for

the shortest radio waves, a wire mesh is a good reflector. On the other hand, because radio signals from celestial sources are very weak, large dishes are necessary to intercept an adequate signal. The largest radio telescope is a bowl-shaped antenna hung in a natural depression in Puerto Rico (Figure 21.15). It is 300 meters (1000 feet) in diameter and has some directional flexibility in its movable antenna. The largest steerable types have about 100-meter (330-foot) dishes like that currently being constructed at the National Radio Astronomy Observatory in Green Bank, West Virginia.

Radio telescopes also have rather poor resolution, making it difficult to pinpoint the radio source. Pairs or groups of telescopes are used to reduce this prob-

A.

B.

Figure 21.14 A. The 43-meter (140-foot) steerable radio telescope at Green Bank, West Virginia. The dish acts like the mirror of a reflector-type optical telescope to focus radio waves onto the antenna. The antenna is the small round object supported on four "legs" above the dish. **B.** Twenty-seven identical antennas operate together to form the Very Large Array near Socorro, New Mexico. (Courtesy of National Radio Astronomy Observatory)

lem. When several radio telescopes are wired together, the resulting network is called a **radio interferometer** (Figure 21.14B).

Radio telescopes have some advantages over optical telescopes. They are much less affected by turbulence in the atmosphere, clouds, and the weather in general. No protective dome is required, which reduces the cost of construction, and "viewing" is possible 24 hours a day. More importantly, radio telescopes can "see" through interstellar dust clouds that obscure our view at visible wavelengths. Radio signals from distant points in the universe pass unhindered through the dust, giving us an unobstructed view. Furthermore, radio telescopes can

detect clouds of gases too cool to emit visible light. These cold gas clouds are important because they are the sites of star formation.

Radio telescopes are, however, hindered by human-made radio interference. Thus, while optical telescopes are placed on remote mountaintops to reduce interference from city lights, radio telescopes are often hidden in valleys to block human-made radio interference.

Radio telescopes have revealed such spectacular events as the collision of two galaxies. Of even greater interest was the discovery of *quasars* (quasi-stellar radio sources). These perplexing objects are the most distant things known in the universe and will be examined further in Chapter 22.

Figure 21.15 The 300-meter (1000-foot) radio telescope at Arecibo, Puerto Rico. (Courtesy of National Astronomy and Ionosphere Center's Arecibo Observatory, operated by Cornell University under contract with the National Science Foundation)

The Sun

The Sun is one of the 200 billion stars that make up the Milky Way Galaxy. Although the Sun is of no significance to the universe as a whole, to us on Earth it is the primary source of energy (see Box 21.2). Everything from the fossil fuels we burn in our automobiles and power plants to the food we eat is ultimately derived from solar energy (Figure 21.16). The Sun is also important to astronomers, since it is the only star close enough to permit study of the surface. Even with the largest telescopes, other stars appear only as points of light.

Because of the Sun's brightness and its damaging radiation, it is not safe to observe it directly. However, a small telescope will project its image on a piece of cardboard held behind the telescope's eyepiece, and the Sun can be studied safely in this manner. This basic method is used in several telescopes around the world, which keep a constant vigil of the Sun. One of the finest is at the Kitt Peak National Observatory in

southern Arizona (Figure 21.17). It consists of a 150-meter sloped enclosure that directs sunlight to a mirror situated below ground. From the mirror, an 85-centimeter (33-inch) image of the Sun is projected to an observing room, where it is studied.

Compared to other stars of the universe, many of which are larger, smaller, hotter, cooler, more red, or more blue, the Sun is an "average star." However, on the scale of our solar system, it is truly gigantic, having a diameter equal to 109 Earth diameters (1.35 million kilometers) and a volume of 1.25 million times as great as that of Earth. Yet, because of its gaseous nature, the density is only one-quarter that of solid Earth, nearly the density of water.

Structure of the Sun

For convenience of discussion, we divide the Sun into four parts: the solar interior; the visible surface, or photosphere; and the two layers of its atmosphere, the chromosphere and the corona (Figure 21.18). Because the Sun is gaseous throughout, no sharp boundaries exist between these layers. The Sun's interior makes up all but a tiny fraction of the solar mass, and unlike the outer three layers, it is not accessible to direct observation. We shall discuss the visible layers first.

Photosphere. The **photosphere** ("sphere of light") is aptly named, for it radiates most of the sunlight we see and therefore appears as the bright disk of the Sun. Although it is considered to be the Sun's "surface," it

Figure 21.16 The Sun is the source of more than 99 percent of all energy on Earth. (Photo by Thomas Dimock/the Stock Market)

Figure 21.17 The unique Robert J. McMath Solar Telescope at Kitt Peak, near Tucson, Arizona. Movable mirrors at the top follow the Sun, reflecting its light down the sloping tunnel. (Photo by Kent Wood/Photo Researchers, Inc.).

is unlike most surfaces we are accustomed to. The photosphere consists of a layer of incandescent gas less than 500 kilometers (300 miles) thick, having a pressure less than 1/100 of our atmosphere. Furthermore, it is neither smooth nor uniformly bright as the ancients had imagined. It has numerous blemishes.

When viewed through a telescope under ideal conditions, the photosphere's grainy texture is apparent. This is the result of numerous, relatively small, bright markings called **granules,** which are surrounded by narrow, dark regions (Figure 21.19). Granules are typically the size of Texas, and they owe their brightness to hotter gases that are rising from below. As this gas spreads laterally, cooling causes it to darken and sink back into the interior. Each granule survives only 10 to 20 minutes, while the combined motion of new

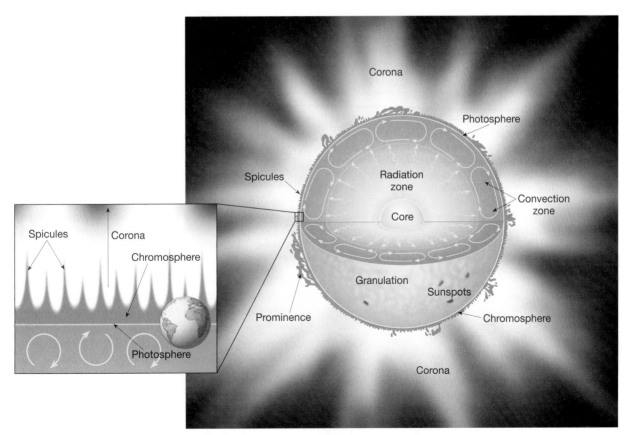

Figure 21.18 Diagram of solar structure in cutaway view. Earth is shown for scale.

Box 21.2 Earth as a System: Variable Sun and Climatic Change

Among the most persistent hypotheses of climatic change have been those based on the idea that the Sun is a variable star and that its output of energy varies through time. The effect of such changes would seem direct and easily understood: Increases in solar output would cause the atmosphere to warm, and reductions would result in cooling. This notion is appealing because it can be used to explain climatic changes of any length or intensity. Still, there is at least one major drawback: No major long-term variations in the total intensity of solar radiation have yet been measured outside the atmosphere. Such measurements were not even possible until modern satellite technology became available. Now that it is possible, we will need many years of records before we begin to sense how variable (or invariable) energy from the Sun really is.

Several proposals for climatic change based on a variable Sun relate to *sunspot cycles*. The most conspicuous and best-known features on the surface of the Sun are the dark blemishes called *sunspots*. Although their origin is uncertain, it has been established that sunspots are huge magnetic storms that extend from the Sun's surface deep into the interior. Moreover, these spots are associated with the Sun's ejection of huge masses of particles that, on reaching Earth's atmosphere, interact with gases there to produce auroral displays.

Along with other solar activity, the number of sunspots increases and decreases on a regular basis, creating a cycle of about 11 years. A curve of the annual number of sunspots, beginning in the early eighteenth century, appears to be very regular (Figure 21.B).

Studies indicate that there have been prolonged periods when sunspots were absent, or nearly so. Moreover, it was found that these events corresponded closely with cold periods in Europe and North America. Conversely, periods characterized by plentiful sunspots were found to correlate well with warmer times in these regions. Referring to these excellent matches, the solar astronomer John Eddy stated: "These early results in comparing solar history with climate make it appear that changes on the Sun are the dominant agent of climatic changes lasting between 50 and several hundred years."

But other scientists seriously questioned this conclusion. Their hesitation stems in part from subsequent investigations, using different climate records from around the world, that failed to find a significant correlation between variations in solar activity and climate. Even more troubling seems to be the fact that no testable physical mechanism exists to explain the purported effect.

Figure 21.B Mean annual sunspot numbers.

Figure 21.19 Granules of the solar photosphere. The granules appear as yellowish-orange patches, while the dark areas shown are small sunspots. Each granule is about the size of Texas and lasts for only 10 to 20 minutes before being replaced by a new granule. (Courtesy of National Optical Astronomy Observatories)

granules replacing old ones gives the photosphere the appearance of boiling. This up-and-down movement of gas is called *convection*. Besides producing the grainy appearance of the photosphere, convection is believed to be responsible for the transfer of energy in the uppermost part of the Sun's interior (Figure 21.18).

The composition of the photosphere is revealed by the dark lines of its absorption spectrum (see Figure 21.3). When these "fingerprints" are compared to the spectra of known elements, they indicate that most of the elements found on Earth also occur on the Sun. When the strengths of the absorption lines are analyzed, the relative abundance of the elements can be determined. These studies reveal that 90 percent of the Sun's surface atoms are hydrogen, almost 10 percent are helium, and only minor amounts of the other detectable elements are present. Other stars also indicate similar disproportionate percentages of these two lightest elements, a fact we shall consider later.

Chromosphere. Just above the photosphere lies the **chromosphere** ("color sphere"), a relatively thin layer of hot, incandescent gases a few thousand kilometers thick. The chromosphere is observable for a few moments during a total solar eclipse or by using a special instrument that blocks out the light from the photosphere. Under such conditions, it appears as a thin red rim around the Sun. Because the chromosphere consists of hot, incandescent gases under low pressure, it produces a bright-line spectrum that is nearly the reverse of the dark-line spectrum of the photosphere. One of the bright lines of hydrogen contributes a good portion of its total light and accounts for this sphere's red color.

A study of the chromospheric spectrum conducted in 1868 revealed the existence of an unknown element. It was named helium, from *helios*, the Greek word for "sun." Originally, helium was thought to be an element unique to the stars, but 27 years later, it was discovered in natural-gas wells on Earth.

The top of the chromosphere contains numerous **spicules,** flamelike structures that extend upward as much as 10,000 kilometers into the lower corona, almost like trees that reach into our atmosphere (Figure 21.20). Spicules are produced by the turbulent motion of the granules below. As we shall see later in the chapter, chromospheric gases trapped in the Sun's strong magnetic field may shoot upward 500,000 kilometers in a few hours. These structures, called *prominences*, are 50 times bigger than spicules.

Figure 21.21 Solar corona photographed during a total eclipse. (Photo by Lawrence Burr/Gamma Liaison)

Figure 21.20 Spicules of the chromosphere on the edge of the solar disk. (Courtesy of Sacramento Peak Observatory, Air Force Cambridge Research Laboratories)

Corona. The outermost portion of the solar atmosphere, the **corona,** is very tenuous and, as with the chromosphere, is visible only when the brilliant photosphere is covered (Figure 21.21). This envelope of ionized gases normally extends a million kilometers from the Sun and produces a glow about half as bright as the full Moon.

At the outer fringe of the corona, the ionized gases have speeds great enough to escape the gravitational pull of the Sun. The streams of protons and electrons that "boil" from the corona constitute the **solar wind.** They travel outward through the solar system at very high speeds (250 to 800 kilometers a second) and eventually are lost to interstellar space. During its journey, the solar wind interacts with the bodies of the solar system, continually bombarding lunar rocks and altering their appearance. Although Earth's magnetic field prevents the solar winds from reaching our surface, these winds do affect our atmosphere, as we shall discuss later.

Studies of the energy emitted from the photosphere indicate that its temperature averages about 6000 K (10,000°F). Upward from the photosphere, the temperature unexpectedly increases, exceeding 1 million K at the top of the corona. It should be noted that, although the coronal temperature exceeds that of the photosphere by many times, it radiates much less energy overall because of its very low density.

Surprisingly, the high temperature of the corona is probably caused by sound waves generated by the convective motion of the photosphere. Just as boiling water makes noise, the energetic sound waves generated in the photosphere are believed to be absorbed by the gases of the corona and thereby raise their temperatures.

The Active Sun

Sunspots. The most conspicuous features on the surface of the Sun are the dark blemishes called **sunspots** (Figure 21.22A). Although large sunspots were occasionally observed before the advent of the telescope, they were generally regarded as opaque objects located somewhere between the Sun and Earth. In 1610 Galileo concluded that they were residents of the solar surface, and from their motion, he deduced that the Sun rotates on its axis about once a month.

Observations made later indicated that not all parts of the Sun rotate at the same speed. The Sun's equator rotates once in 25 days, while a place located 70 degrees from the solar equator, whether north or south, requires 33 days for one rotation. If Earth rotated in a similar disjointed manner, imagine the consequences! The Sun's nonuniform rotation is a testimonial to its gaseous nature.

Sunspots begin as small dark pores about 1600 kilometers (1000 miles) in diameter. Although most pores last for only a few hours, some grow into blemishes many times larger than Earth and last for a month or more. The largest spots often occur in pairs surrounded by several smaller spots. An individual spot contains a black center, the *umbra,* which is rimmed by a lighter region, the *penumbra* (Figure 21.22B). Sunspots appear dark only by contrast with the brilliant photosphere, a fact accounted for by their temperature, which is about 1500 K less than that of solar surface. If these dark spots could be observed away from the Sun, they would appear many times brighter than the full moon.

During the early nineteenth century, it was believed that a tiny planet named Vulcan orbited between Mercury and the Sun. In the search for Vulcan an accurate record of sunspot occurrences was kept. Although the planet was never found, the sunspot data collected did reveal that the number of sunspots observable on the solar disk varies in an 11-year cycle.

First, the number of sunspots increases to a maximum, with perhaps a hundred or more visible at a given time. Then, over a period of 5 to 7 years, their numbers decline to a minimum, when only a few or even none are visible. At the beginning of each cycle, the first sunspots form about 30 degrees from the solar equator, but as the cycle progresses and their numbers increase, they form nearer the equator. During the period when sunspots are most abundant, the majority form about 15 degrees from the equator. They rarely occur more than 40 degrees away from the Sun's equator, or within 5 degrees of it.

Another interesting characteristic of sunspots was discovered by astronomer George Hale, for whom the Hale Telescope is named. Hale deduced that the large

A.

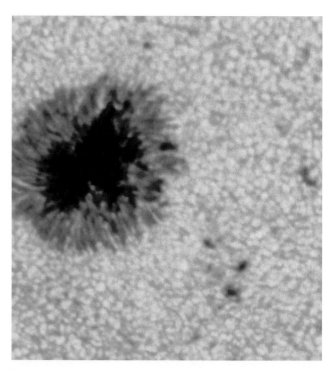

B.

Figure 21.22 A. Large sunspot group on the solar disk. (Celestron 8 photo courtesy of Celestron International) **B.** Sunspots having visible umbra (dark central area) and penumbra (lighter area surrounding umbra). (Courtesy of National Optical Astronomy Observatories)

spots are strongly magnetized, and when they occur in pairs, they have opposite magnetic poles. For instance, if one member of the pair is a north magnetic pole, then the other member is a south magnetic pole, as with the north and south poles of Earth's magnetic field. Also, every pair located in the same hemisphere is magnetized in the same manner. However, all pairs in the *other* hemisphere are magnetized in the opposite manner. At the beginning of each sunspot cycle, the situation reverses, and the polarity of these sunspot pairs is opposite those of the previous cycle. The cause of this change in polarity—in fact, the cause of sunspots themselves—is not fully explained. However, other solar activity varies in the same cyclic manner as sunspots do, indicating a common origin.

Prominences. Among the more spectacular features of the active sun are **prominences.** These huge cloudlike structures, consisting of concentrations of chromospheric gases, are best observed when they are on the edge, or limb, of the Sun, where they often appear as great arches that extend well into the corona (Figure 21.23). Many prominences have the appearance of a fine tapestry and seem to hang motionless for days at a time, but motion pictures reveal that the material within them is continually falling like luminescent rain.

On the other hand, eruptive prominences rise almost explosively away from the Sun. These active prominences reach velocities up to 1000 kilometers (620 miles) per second and may leave the Sun entirely. Whether eruptive or quiescent, prominences are ionized chromospheric gases trapped by magnetic fields that extend from regions of intense solar activity.

Solar Flares. These are the most explosive events associated with sunspots. **Solar flares** are brief outbursts that normally last an hour or so and appear as a sudden brightening of the region above a sunspot cluster. During their existence, enormous quantities of energy are released, much of it in the form of ultraviolet, radio, and X-ray radiation. Simultaneously, fast-moving atomic particles are ejected, causing the solar wind to intensify noticeably. Although a major flare could conceivably endanger a manned space flight, they are relatively rare. About a day after a large outburst, the ejected particles reach Earth and disturb the ionosphere,[†] affecting long-distance radio communications.

The most spectacular effects of solar flares, however, are the **auroras,** also called the Northern and Southern Lights (Figure 21.24). Following a strong solar flare, Earth's upper atmosphere near its magnetic poles is set aglow for several nights. The auroras appear in a wide variety of forms. Sometimes the display consists of vertical streamers with considerable movement. At other times, the auroras appear as a series of luminous expanding arcs or as a quiet, almost foglike, glow. Auroral displays, like other solar activities, vary in intensity with the 11-year sunspot cycle.

The Solar Interior

The interior of the Sun cannot be observed directly. For that reason, all we know about it is based on information acquired from the energy it radiates and from

[†]The ionosphere is a complex atmospheric zone of ionized gases extending between about 80 and 400 kilometers (50 and 250 miles) above Earth's surface.

Figure 21.23 A huge solar prominence. (Sacramento Peak Observatory, Air Force Cambridge Research Laboratory)

Figure 21.24 Aurora borealis (Northern Lights) as seen from Alaska. The same phenomenon occurs toward the South Pole, where it is called the Aurora australis (Southern Lights). (Photo by Michio Hoshino/Minden Pictures)

theoretical studies. The source of the Sun's energy, **nuclear fusion,** was not discovered until the late 1930s.

Deep in its interior, a nuclear reaction called the **proton-proton chain** converts four hydrogen nuclei (protons) into the nucleus of a helium atom. The energy released from the proton-proton reaction results because some of the matter involved is actually converted to energy. This can be illustrated by noting that four hydrogen atoms have a combined atomic mass of 4.032 (4 × 1.008), whereas the atomic mass of helium is 4.003, or 0.029 less than the combined mass of the hydrogen. The tiny missing mass is emitted as energy according to Einstein's formula $E = mc^2$, where E equals energy, m equals mass, and c equals the speed of light. Because the speed of light is very great, the amount of energy released from even a small amount of mass is enormous.

The conversion of just one pinhead's worth of hydrogen to helium generates more energy than burning thousands of tons of coal. Most of this energy is in the form of high-energy photons that work their way toward the solar surface, being absorbed and reemitted many times until they reach an opaque layer just below the photosphere. Here, convection currents serve to transport this energy to the solar surface, where it radiates through the transparent chromosphere and corona (see Figure 21.18).

Only a small percentage (0.7 percent) of the hydrogen in the proton-proton reaction is actually converted to energy. Nevertheless, the Sun is consuming an estimated 600 million tons of hydrogen each second, with about 4 million tons of it being converted to energy. As hydrogen is consumed, the product of this reaction, helium, forms the solar core, which continually grows in size.

Just how long can the Sun produce energy at its present rate before all of its fuel (hydrogen) is consumed? Even at the enormous rate of consumption, the Sun has enough fuel to last easily another 100 billion years. However, evidence from other stars indicates that the Sun will grow dramatically and engulf Earth long before all of its hydrogen is gone. It is thought that a star the size of the Sun can exist in its present stable state for 10 billion years. As the sun is already 5 billion years old, it is "middle-aged."

To initiate the proton-proton reaction, the Sun's internal temperature must have reached several million degrees. What was the source of this heat? As previously noted, the solar system is believed to have formed from an enormous cloud of dust and gases (mostly hydrogen) that condensed gravitationally. The consequence of squeezing (compressing) a gas is to increase its temperature. Although all of the bodies in the solar system were compressed, the Sun was the only one, because of its size, that became hot enough to trigger the proton-proton reaction. Astronomers currently estimate its internal temperature at 15 million K.

The planet Jupiter is basically a hydrogen-rich gas ball; if it were about 10 times more massive, it too might have become a star. The idea of one star orbiting another may seem odd, but recent evidence indicates that about 50 percent of the stars in the universe probably occur in pairs or multiples!

Chapter Summary

- Visible light constitutes only a small part of an array of energy generally referred to as *electromagnetic radiation*. Light, a type of electromagnetic radiation, can be described in two ways: (1) as waves and (2) as a stream of particles, called *photons*. The wavelengths of electromagnetic radiation vary from several kilometers for *radio waves* to less than a billionth of a centimeter for *gamma rays*. The shorter wavelengths correspond to more energetic photons.

- *Spectroscopy* is the study of the properties of light that depend on wavelength. When a prism is used to disperse visible light into its component parts (wavelengths), one of three possible types of *spectra* is produced (a *spectrum*, the singular form of *spectra*, is the light pattern produced by passing light through a prism). The three types of spectra are (1) *continuous spectrum*, (2) *dark-line (absorption) spectrum*, and (3) *bright-line (emission) spectrum*. The spectra of most stars are of the dark-line type. Spectroscopy can be used to determine (1) the state of matter of an object (solid, liquid, high- or low-pressure gas; (2) the composition of gaseous objects; (3) the temperature of a radiating body; and (4) the motion of an object. Motion (direction toward or away and velocity) is determined using the *Doppler effect*—the apparent change in the wavelength of radiation emitted by an object caused by the relative motions of the source and the observer.

- There are two types of optical telescopes: (1) the *refracting telescope*, which uses a *lens* as its *objective* to bend or refract light so that it converges at an area called the *focus*; and (2) the *reflecting telescope*, which uses a *concave mirror* to focus (gather) the light. When examining an image directly, both types of telescopes require a second lens, called an *eyepiece*, which magnifies the image produced by the objective.

- Telescopes have three properties that aid astronomers: (1) *light-gathering power*, which is a function of the size of the objective—large objectives gather more light and therefore "see" farther into space; (2) *resolving power*, which allows for sharper images and finer details, is the ability of a telescope to separate objects that are close together, e.g., Pluto and its moon, Charon; and (3) *magnifying power*, the ability to make an object larger. Most modern telescopes have supplemental devices that enhance the image.

- Invisible radio wave radiation is detected by "big dishes" called *radio telescopes*. A parabolic-shaped dish, often consisting of a wire mesh, operates in the same manner as the mirror of a reflecting telescope. Radio telescopes have poor resolution, making it difficult to pinpoint a radio source. To reduce this problem, several can be wired together into a network called a *radio interferometer*. The advantages of radio telescopes over optical telescopes are that radio telescopes are less affected by the weather, they are less expensive to construct, "viewing" is possible 24 hours a day, they can detect material in the universe too cool to emit visible radiation, and they can "see" through interstellar dust clouds.

- The *Sun* is one of the 200 billion stars that make up the Milky Way Galaxy. The Sun can be divided into four parts: (1) the *solar interior*, (2) the *photosphere* (visible surface), and the two layers of its atmosphere, (3) the *chromosphere* and (4) *corona*. The photosphere radiates most of the light we see. Unlike most surfaces, it consists of a layer of incandescent gas less than 500 kilometers (300 miles) thick with a grainy texture consisting of numerous, relatively small, bright markings called *granules*. Just above the photosphere lies the chromosphere, a relatively thin layer of hot, incandescent gases a few thousand kilometers thick. At the edge of the uppermost portion of the solar atmosphere, called the *corona*, ionized gases escape the gravitational pull of the Sun and stream toward Earth at high speeds, producing the *solar wind*.

- Numerous features have been identified on the active Sun. *Sunspots* are dark blemishes with a black center, the *umbra*, which is rimmed by a lighter region, the *penumbra*. The number of sunspots observable on the solar disk varies in an 11-year cycle. *Prominences*, huge cloudlike structures best observed when they are on the edge, or limb, of the Sun, are produced by ionized chromospheric gases trapped by magnetic fields that extend from regions of intense solar activity. The most explosive events associated with sunspots are *solar flares*. Flares are brief outbursts that release enormous quantities of energy that appear as a sudden brightening of the region above sunspot clusters. During the event, radiation and fast-moving atomic particles are ejected, causing the solar wind to intensify. When the ejected particles reach Earth and disturb the ionosphere, radio communication is disrupted and the *auroras*, also called the Northern and Southern Lights, occur.

- The source of the Sun's energy is *nuclear fusion*. Deep in the solar interior, at a temperature of 15 million K, a nuclear reaction called the *proton-proton chain* converts four hydrogen nuclei (protons) into the nucleus of a helium atom. During the reaction some of the matter is converted to the energy of the Sun. A star the size of the Sun can exist in its present stable state for 10 billion years. As the Sun is already 5 billion years old, it is a "middle-aged" star.

Key Terms

aurora (p. 595)
bright-line (emission) spectrum
  (p. 581)
chromatic aberration (p. 584)
chromosphere (p. 593)
continuous spectrum (p. 581)
corona (p. 593)
dark-line (absorption) spectrum
  (p. 581)
Doppler effect (p. 582)
electromagnetic radiation (p. 580)

eyepiece (p. 583)
focal length (p. 583)
focus (p. 583)
granules (p. 591)
nuclear fusion (p. 596)
objective lens (p. 583)
photon (p. 581)
photosphere (p. 590)
prominence (p. 595)
proton-proton chain (p. 596)
radiation pressure (p. 581)

radio interferometer (p. 589)
radio telescope (p. 588)
reflecting telescope (p. 584)
refracting telescope (p. 583)
solar flare (p. 595)
solar wind (p. 593)
spectroscope (p. 582)
spectroscopy (p. 581)
spicule (p. 593)
sunspot (p. 594)

Review Questions

1. What term is used to describe the collection that includes gamma rays, X-rays, ultraviolet light, visible light, infrared radiation, and radio waves?

2. Which color has the longest wavelength? The shortest?

3. What is spectroscopy?

4. Describe a continuous spectrum. Give an example of a natural phenomenon that exhibits a continuous spectrum.

5. What produces emission lines (bright lines) in a spectrum?

6. What can be learned about a star (or other celestial objects) from a dark-line (absorption) spectrum?

7. How can astronomers determine whether a star is moving toward or away from Earth?

8. The primary objective of a refracting telescope is a mirror: True or false?

9. What three properties do telescopes have that aid astronomers?

10. Of the three properties listed in Question 9, which is the least important to astronomers?

11. What is the advantage of a space telescope over a similar instrument on Earth?

12. Why do astronomers seek to design telescopes with larger and larger objectives?

13. Why do all large optical telescopes use mirrors to collect light rather than lenses?

14. With reflecting telescopes, why are special viewing systems needed?

15. Explain the following statement: "Photography has extended the limits of our vision."

16. Why would the Moon make a good site for an observatory?

17. Why are radio telescopes much larger than optical telescopes?

18. What are some of the advantages of radio telescopes over optical telescopes?

19. Compare the diameter of the Sun to that of Earth.

20. Describe the photosphere, chromosphere, and corona.

21. List the features associated with the active Sun and describe each.

22. Explain how a sunspot can be very hot and yet appear dark.

23. What is the solar wind?

24. What "fuel" does the Sun consume?

25. What happens to the matter that is consumed in the proton-proton chain reaction?

Examining the Earth System

1. Of the two sources of energy that power the Earth system, the Sun is the main driver of Earth's external processes. If the Sun increased its energy output by 10%, what would happen to global temperatures? What affect would this temperature change have on the percentage of water that exists as ice? What would be the impact on the position of the ocean shoreline (see Box 5.2)? Would the change in temperature produce an increase or decrease in the amount of surface vegetation? In turn, what impact would the change in vegetation have on the level of atmospheric carbon dioxide? How would the change in the amount of carbon dioxide in the atmosphere affect global temperature?

Web Resources

The following are informative and interesting World Wide Web sites that address topics related to those presented in the chapter:

Hubble Space Telescope (Space Telescope Science Institute):

- **http://oposite.stsci.edu/**

Mt. Wilson Observatory:

- **http://www.mtwilson.edu/**

National Solar Observatory:

- **http://argo.tuc.noao.edu/**

For direct links to these sites and other appropriate Earth science Web locations, chapter review exercises, and relevant, topical investigations that utilize Web resources, contact the *Earth Science*, 9th Edition Home Page at:

- **http://www.prenhall.com/tarbuck.**

Chapter 22

Beyond Our Solar System

This image from the Hubble Space Telescope shows a column of molecular hydrogen gas and dust in the Eagle Nebula. (Courtesy of NASA)

The star Proxima Centauri is about 4.3 light-years away, roughly 100 million times farther than the Moon. Yet, other than our own Sun, it is the closest star to Earth. This fact suggests that the universe is incomprehensibly large. What is the nature of this vast cosmos beyond our solar system? Are the stars distributed randomly, or are they organized into distinct clusters? Do stars move, or are they permanently fixed features, like lights strung out against the black cloak of outer space? Does the universe extend infinitely in all directions, or is it bounded? To consider these questions, this chapter will examine the universe by taking a census of the stars—the most numerous objects in the night sky.

Properties of Stars

The Sun is the only star whose surface we can observe. Yet a great deal is known about the universe beyond our solar system. In fact, more is known about the stars than about our outermost planet, Pluto. This knowledge hinges on the fact that stars, and even gases in the "empty" space between stars, radiate energy in all directions into space (Figure 22.1). The key to understanding the universe is to collect this radiation and unravel the secrets it holds. Astronomers have devised many ingenious methods to do just that. We will begin by examining stellar distances and some intrinsic properties of stars, including *color, brightness, mass, temperature,* and *size.*

Measuring Distances to the Stars

Measuring the distance to a star is very difficult. Obviously, we cannot journey to the star. Nevertheless, astronomers have developed some *indirect* methods of measuring stellar distances. The most basic of these measurements is called stellar parallax.

Recall from Chapter 19 that **stellar parallax** is the extremely slight back-and-forth shifting in the apparent position of a nearby star due to the orbital motion of Earth. To review, the principle of parallax is easy to visualize. Close one eye, and with your index finger in a vertical position, use your eye to line up your finger with some distant object. Now, without moving your finger, view the object with your other eye and notice that its position appears to have changed. The farther away you hold your finger, the less its position seems to shift.

In practice, parallax is determined by photographing a nearby star against the background of distant stars. Then, 6 months later, when Earth has moved halfway around its orbit, a second photograph is taken. When these photographs are compared, the position of the nearby star appears to have shifted with respect to the background stars. Figure 22.2 illustrates this shift and the parallax angle determined from it. The nearest stars have the largest parallax angles, while those of distant stars are too slight to measure. Recall that the sixteenth-century astronomer Tycho Brahe was unable to detect stellar

Figure 22.1 Lagoon Nebula. It is in glowing clouds like these that gases and dust particles become concentrated into stars. (Courtesy of National Optical Astronomy Observatories)

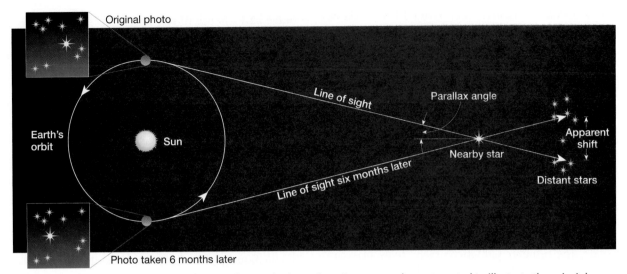

Original photo

Photo taken 6 months later

Earth's orbit

Sun

Line of sight

Line of sight six months later

Parallax angle

Nearby star

Apparent shift

Distant stars

Figure 22.2 Geometry of stellar parallax. The parallax angle shown here is enormously exaggerated to illustrate the principle. Because the distance to even the nearest stars is thousands of times greater than the Earth–Sun distance, the triangles that astronomers work with are quite long and narrow, making the angles that are measured very small.

parallax, leading him to reject the idea that Earth orbits the Sun.

It should be emphasized that parallax angles are *very* small. The parallax angle to the nearest star, Proxima Centauri, is less than 1 second of arc, which equals 1/3600 of a degree. To put this in perspective, fully extend your arm and raise your little finger. Your finger is roughly 1 degree wide. Try doing this on a moonlit night, covering the Moon with your finger. The Moon is only about 1/2 degree wide. Now, imagine detecting a movement that is only 1/3600 as wide as your finger. It should be apparent why Tycho Brahe, without the aid of a telescope, was unable to observe stellar parallax.

The distances to stars are so large that the conventional units, such as kilometers or astronomical units, are often too cumbersome to use. A better unit to express stellar distance is the **light-year**, which is the distance light travels in one Earth year—about 9.5 trillion kilometers (5.8 trillion miles).

In principle, the method used to measure stellar distances is elementary and was known to the ancient Greeks. But in practice, measurements are greatly complicated because of the tiny angles involved and because the Sun, as well as the star being measured, also have actual motion through space. The first accurate stellar parallax was not determined until 1838. Even today, parallax angles for only a few thousand of the nearest stars are known with certainty.

Most stars have such small parallax shifts that accurate measurement is not possible. Fortunately, a few other methods have been derived for estimating distances to these stars. Also, the Hubble Space Telescope, which is not hindered by Earth's turbulent, light-distorting atmosphere, is expected to obtain accurate parallax distances for many additional remote stars.

Stellar Brightness

Three factors control the apparent brightness of a star as seen from Earth: *how big* it is, *how hot* it is, and *how far away* it is. The stars in the night sky are a grand assortment of sizes, temperatures, and distances, so their brightnesses vary widely.

Apparent Magnitude. Stars have been classified according to their brightness since the second century B.C., when Hipparchus placed about 1000 of them into six categories. The measure of a star's brightness is called its **magnitude.** Some stars may appear dimmer than others only because they are farther away. Therefore, a star's brightness, *as it appears when viewed from Earth,* has been termed its **apparent magnitude.**

When numbers are employed to designate relative brightness, the larger the magnitude number, the dimmer the star. *Stars that appear the brightest are of the first magnitude, whereas the faintest stars visible to the unaided eye are of the sixth magnitude.* With the invention of the telescope, many stars fainter than the sixth magnitude were discovered.

In the mid-1800s, a method was developed for comparing the brilliance of stars using magnitude. Just as we can compare the brightness of a 50-watt light bulb to that of a 100-watt bulb, we can compare the brightness of stars having different magnitudes. It was determined that a first-magnitude star was about 100 times brighter than a sixth-magnitude star. Therefore, on the scale that was devised, two stars that differ by 5 magnitudes have a ratio in brightness of 100 to 1.

Hence, a seventh-magnitude star is 100 times brighter than a twelfth-magnitude star.

It follows, then, that the brightness ratio of two stars differing by only one magnitude is about 2.5.* Thus, a star of the first magnitude is about 2.5 times brighter than a star of the second magnitude. Table 22.1 shows some magnitude differences and the corresponding brightness ratios.

Because some stars are brighter than first-magnitude stars, zero and negative magnitudes were introduced. The brightest star in the night sky, Sirius, has an apparent magnitude of –1.4, about 10 times brighter than a first-magnitude star. On this scale, the Sun has an apparent magnitude of –26.7. At its brightest, Venus has a magnitude of –4.3. At the other end of the scale, the 5-meter (200-inch) Hale Telescope can view stars with an apparent magnitude of 23, approximately 100 million times dimmer than stars that are visible to the unaided eye.

Absolute Magnitude. Astronomers are also interested in the "true" brightness of stars, called their **absolute magnitude.** Stars of the same luminosity, or brightness, usually do not have the same *apparent* magnitude, because their distances from us are not equal. To compare their true, or intrinsic, brightness, astronomers determine what magnitude the stars would have if they were at a standard distance of about 32.6 light-years. For example, the Sun, which has an apparent magnitude of –26.7, would, if located at a distance of 32.6 light-years, have an absolute magnitude of about 5. Thus, stars with absolute magnitudes greater than 5 (smaller numerical value) are intrinsically brighter than the Sun, but because of their distance, they appear much dimmer.

Table 22.2 lists the absolute and apparent magnitudes of some stars as well as their distances from Earth. Most stars have an absolute magnitude between –5 and 15, which puts the Sun near the middle of this range.

Stellar Color and Temperature

The next time you are outdoors on a clear night, take a good look at the stars and note their color (Figure 22.3). Some that are quite colorful can be found in the constellation Orion (see Appendix E, winter chart). Of the two brightest stars in Orion, Betelgeuse (α Orionis) is definitely red, whereas Rigel (β Orionis) appears blue.

Very hot stars with surface temperatures above 30,000 K emit most of their energy in the form of short-wavelength light and therefore appear blue. Red stars, on the other hand, are much cooler, generally less than 3000 K, and most of their energy is

Table 22.1 Ratios of star brightness

Difference in Magnitude	Brightness Ratio
0.5	1.6:1
1	2.5:1
2	6.3:1
3	16:1
4	40:1
5	100:1
10	10,000:1
20	100,000,000:1

emitted as longer-wavelength red light. Stars with temperatures between 5000 and 6000 K appear yellow, like the Sun. Because color is primarily a manifestation of a star's temperature, this characteristic provides the astronomer with useful information about a star.

Binary Stars and Stellar Mass

One of the night sky's best known constellations, the Big Dipper, appears at first glance to consist of seven stars.[†] But anyone with good eyesight can resolve the second star in the handle of the Big Dipper (Mizar) as two stars. During the eighteenth century, astronomers used their new tool, the telescope, to discover numerous such star pairs. One of the stars in the pair is usually fainter than the other, and for this reason, it was considered to be farther away. In other words, the stars were not considered true pairs but were thought only to lie in the same line of sight.

In the early nineteenth century, careful examination of numerous star pairs by William Herschel revealed that many stars actually orbit one another. The two stars are in fact united by their mutual gravitation. These pairs of stars, in which the members are far enough apart to be

---

[†]For the sky location, see Appendix D, first star chart, top.

Table 22.2 Distance, apparent magnitude, and absolute magnitude of some stars.

Name	Distance (Light years)	Apparent Magnitude*	Absolute Magnitude*
Sun	NA	–26.7	5.0
Alpha Centauri	4.27	0.0	4.4
Sirius	8.70	–1.4	1.5
Arcturus	36	–0.1	–0.3
Betelgeuse	520	0.8	–5.5
Deneb	1600	1.3	–6.9

Calculations: 2.512 x 2.512 x 2.512 x 2.512 x 2.512, or 2.512 raised to the fifth power, equals 100.

*The more negative, the brighter; the more positive, the dimmer.

Binary stars are used to determine the star property most difficult to calculate—its mass. The mass of a body can be established if it is gravitationally attached to a partner, which is the case for any binary star system. Binary stars orbit each other around a common point called the *center of mass* (Figure 22.4). For stars of equal mass, the center of mass lies exactly halfway between them. If one star is more massive than its partner, their common center will be located closer to the more massive one. Thus, if the sizes of their orbits can be observed, a determination of their individual masses can be made. You can experience this relationship on a seesaw by trying to balance a person who has a much greater mass.

For illustration, when one star has an orbit half the size (radius) of its companion, it is twice as massive. If their combined masses are equal to three times the mass of the Sun, then the larger will be twice as massive as the Sun, and the smaller will have a mass equal to that of the Sun. Most stars have masses that range between 1/10 and 50 times the mass of the Sun.

Hertzsprung-Russell Diagram

Early in the twentieth century, Einar Hertzsprung and Henry Russell independently studied the relation between the true brightness (absolute magnitude) and temperature of stars. From this research each developed a graph, now called a **Hertzsprung-Russell diagram, (H-R diagram),** that exhibits these intrinsic stellar properties. By studying H-R diagrams, we learn a great deal about the sizes, colors, and temperatures of stars.

To produce an H-R diagram, astronomers survey a portion of the sky and plot each star according to its luminosity (brightness) and temperature (Figure 22.5). Notice that the stars in Figure 22.5 are not uniformly distributed. Rather, about 90 percent of all stars fall along a band that runs from the upper-left corner to the lower-right corner of an H-R diagram. These "ordinary" stars are called **main-sequence stars.** As shown in Figure 22.5, *the hottest main-sequence stars are intrinsically the brightest, and vice versa.*

The luminosity of the main-sequence stars is also related to their mass. The hottest (blue) stars are about 50 times more massive than the Sun, whereas the coolest (red) stars are only 1/10 as massive. Therefore, on the H-R diagram, the main-sequence stars appear in a decreasing order, from *hotter, more massive* blue stars to *cooler, less massive* red stars.

Note the location of the Sun in Figure 22.5. The Sun is a yellow, main-sequence star with an absolute magnitude of about 5. Because the magnitude of a vast majority of main-sequence stars lie between –5 and 15, and because the Sun falls midway in this range, the Sun is often considered an average star. However, more main-sequence stars are cooler and less massive than our Sun.

Figure 22.3 Time-lapse photograph of stars in the constellation Orion. These star trails show some of the various star colors. It is important to note that the eye sees color somewhat differently than does photographic film. (Courtesy of National Optical Astronomy Observatories)

resolved telescopically, are called *visual binaries.* The idea of one star orbiting another may seem unusual, but evidence indicates that more than 50 percent of the stars in the universe occur in pairs or multiples.

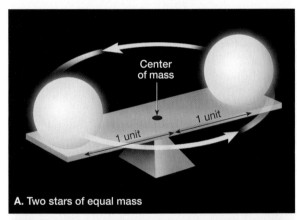

A. Two stars of equal mass

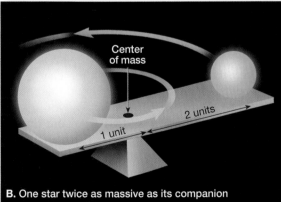

B. One star twice as massive as its companion

Figure 22.4 Binary stars orbit each other around their common center of mass. **A.** For stars of equal mass, the center of mass lies exactly halfway between them. **B.** If one star is twice as massive as its companion, it is twice as close to their common center of mass. Therefore, more massive stars have proportionately smaller orbits than do their less massive companions.

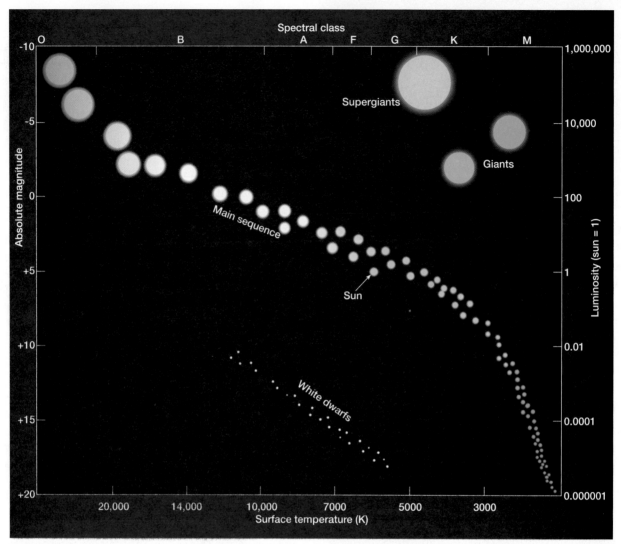

Figure 22.5 Idealized Hertzsprung-Russell diagram on which stars are plotted according to temperature and absolute magnitude.

Just as all humans do not fall into the normal size range, some stars do not fit in with the main-sequence stars. Above and to the right of the main sequence in the H-R diagram in Figure 22.5 lies a group of very luminous stars called *giants,* or, on the basis of their color, **red giants.** The size of these giants can be estimated by comparing them with stars of known size that have the same surface temperature. We know that objects having equal surface temperatures radiate the same amount of energy per unit area. Therefore, any difference in the brightness of two stars having the same surface temperature is attributable to their relative sizes.

As an example, let us compare the Sun, which has been assigned a luminosity of 1, with another yellow star that has a luminosity of 100. Because both stars have the same surface temperature, they both radiate the same amount of energy per unit area. Therefore, for the more luminous star to be 100 times brighter than

the Sun, it must have 100 times more surface area. It should be clear why stars whose plots fall in the upper-right position of an H-R diagram are called *giants.*

Some stars are so large that they are called **supergiants.** Betelgeuse, a bright-red supergiant in the constellation Orion (Appendix E, winter chart), has a radius about 800 times that of the Sun. If this star were at the center of our solar system, it would extend beyond the orbit of Mars, and Earth would find itself inside the star! Other red giants that are easy to locate in our sky are Arcturus in the constellation Bootes and Antares in Scorpius (Appendix E, summer chart).

In the lower-central portion of the H-R diagram, the opposite situation occurs. These stars are much fainter than main-sequence stars of the same temperature, and by using the same reasoning as before, they must be much smaller. Some probably approximate Earth in size. This group has come to be called **white dwarfs,** although not all are white.

March 10, 1935

May 6, 1935

Figure 22.6 Photographs of Nova Herculis (a nova in the constellation Hercules), taken about 2 months apart, showing the decrease in brightness. (Courtesy of Lick Observatory)

Soon after the first H-R diagrams were developed, astronomers realized their importance in interpreting stellar evolution. Just as with living things, a star is born, ages, and dies. Owing to the fact that almost 90 percent of the stars lie on the main sequence, we can be relatively certain that stars spend most of their active years as main-sequence stars. Only a few percent are giants, and perhaps 10 percent are white dwarfs.

After considering some variable stars and the nature of interstellar matter, we will return to the topic of stellar evolution.

Variable Stars

Stars that fluctuate in brightness are known as variables. Some, called **pulsating variables,** fluctuate regularly in brightness by expanding and contracting in size. The importance of one member of this group (cepheid variables) in determining stellar distances is discussed in Box 22.1. Other pulsating variables have irregular periods and are of no value for this purpose.

The most spectacular variables belong to a group known as **eruptive variables.** When one of these explosive events associated with these stars occurs, it appears as a sudden brightening of a star, called a **nova** (Figure 22.6). The term *nova,* meaning "new," was used by the ancients because these stars were unknown to them before their abrupt increase in luminosity.

During the outburst, the outer layer of the star is ejected at high speed. The "cloud" of ejected material occasionally can be captured photographically

(Figure 22.7). A nova generally reaches maximum brightness in a couple of days, remains bright for only a few weeks, then slowly returns in a year or so to its original brightness. Because the star returns to its prenova brightness, we can assume that only a small amount of its mass is lost during the flareup. Some stars have experienced more than one such event. In fact, the process probably occurs repeatedly.

Figure 22.7 Expanding "cloud" around Nova Persei. (Courtesy of Hale Observatories. Copyright by the California Institute of Technology and Carnegie Institution of Washington)

Box 22.1 Determining Distance From Magnitude

For a star too distant for parallax measurements, knowing its absolute and apparent brightness provides astronomers with a tool for approximating its distance. The apparent magnitude is measured with a photometer (light meter) attached to a telescope. If we also know a star's true brightness we can determine just how far away that star would have to be for it to have the brightness we observe.

You use this same principle when you drive at night and estimate the distance to an oncoming car from the brightness of its headlights. You can do this because you know the true brightness of an automobile headlight. But how do astronomers determine the intrinsic brightness of a star? Fortunately, some stars have characteristics that provide the necessary data.

One important star group is called "cepheid variables." These are pulsating stars that get brighter and fainter in a rhythmic fashion. The interval between two successive occurrences of maximum brightness of a pulsating variable is termed its *light period.* Most cepheid variables pulsate with periods of between 2 and 50 days. For example, the North Pole Star (Polaris) varies about 10 percent in brightness over a period of 4 days. In general, the longer the light period of a cepheid, the greater is its absolute magnitude (Figure 22.A). Thus, by determining the light period of a cepheid, its absolute magnitude can be calculated. When this absolute magnitude is compared to the apparent magnitude, a good approximation of its distance can be made.

Figure 22.A Relationship between the light period (two successive occurrences of maximum brightness) and absolute magnitude of pulsating stars (cepheid variables).

The modern explanation of novae proposes that they occur in binary systems consisting of an expanding red giant and a hot white dwarf that are in close proximity. Hydrogen-rich gas from the oversized giant encroaches near enough to the white dwarf to be gravitationally transferred. Eventually, enough of the hydrogen-rich gas is added to the dwarf to cause it to ignite explosively. Such a thermonuclear reaction rapidly heats and expands the outer gaseous envelope of the hot dwarf to produce a nova event. In a relatively short time, the white dwarf returns to its prenova state, where it remains inactive until the next buildup occurs.

Interstellar Matter

Between existing stars is "the vacuum of space." However, it is far from being a pure vacuum, for it is populated with accumulations of dust and gases. The name applied to these concentrations of interstellar matter is **nebula** (Latin for "cloud"). If this interstellar matter is close to a very hot (blue) star it will glow and is called a **bright nebula.** The two main types of bright nebulae are known as *emission nebulae* and *reflection nebulae.*

Emission nebulae are gaseous masses that consist largely of hydrogen. They absorb *ultraviolet radiation* emitted by an embedded or nearby hot star. Because these gases are under very low pressure, they reradiate, or emit, this energy as *visible light.* This conversion of ultraviolet light to visible light is known as *fluorescence,* an effect you observe daily in fluorescent lights. A well-known emission nebula easily seen with binoculars is the sword of the hunter in the constellation Orion (Figure 22.8).

Reflection nebulae, as the name implies, merely reflect the light of nearby stars (Figure 22.9). Reflection nebulae are thought to be composed of rather dense clouds of large particles called **interstellar dust.** This view is supported by the fact that atomic gases with low densities could not reflect light sufficiently to produce the glow observed.

When a dense cloud of interstellar material is not close enough to a bright star to be illuminated, it is referred to as a **dark nebula.** Exemplified by the Horsehead nebula in Orion, dark nebulae appear as opaque objects silhouetted against a bright background (Figure 22.10). Dark nebulae can also easily be seen as starless regions—"holes in the heavens"—when viewing the Milky Way.

Although nebulae appear very dense, they actually consist of very thinly scattered matter. Because of their enormous size, however, the total mass of

Figure 22.8 The Orion Nebula is a well-known emission nebula. Bright enough to be seen by the naked eye, the Orion Nebula is located in the sword of the hunter in the constellation of the same name. (Courtesy of National Optical Astronomy Observatories)

rarefied particles and molecules may be many times that of the Sun. Interstellar matter is of great interest to astronomers because it is from this material that stars and planets are formed.

Figure 22.9 A faint blue reflection nebula, in the Pleiades star cluster, is caused by the reflection of starlight from dust in the nebula. The Pleiades star cluster, just visible to the naked eye in the constellation Taurus, is spectacular when viewed through binoculars or a small telescope. (Palomar Observatories/California Institute of Technology [Caltech])

Stellar Evolution

The idea of describing how a star is born, ages, and then dies may seem a bit presumptuous, for many of these objects have life spans that surely exceed billions of years. However, by studying stars of different ages, astronomers have been able to piece together a plausible model for stellar evolution. The method that was used to create this model is analogous to what an alien being, upon reaching Earth, might do to determine the developmental stages of human life. By examining a large number of humans, this stranger would be able to observe the birth of human life, the activities of children and adults, and the death of the elderly. From this information, the alien would attempt to put the stages of human development into their proper sequence. Based on the relative abundance of humans in each stage of development, it would even be possible to conclude that humans spend more of their lives as adults than as toddlers. In a similar fashion, astronomers have pieced together the story of the stars.

Simply, stars exist because of gravity. The mutual gravitational attraction of particles in a thin, gaseous cloud causes the cloud to collapse. As the cloud is squeezed to unimaginable pressures, its temperature rises, igniting its nuclear furnace, and a star is born. A star is a ball of very hot gases, caught between the opposing forces of gravity trying to contract it and thermal nuclear energy trying to expand it. Eventually, all of a star's nuclear fuel will be exhausted and gravity takes over, collapsing the stellar remnant into a small, dense body.

Star Birth

The birthplaces of stars are dark, cool interstellar clouds, which are comparatively rich in dust and gases

Figure 22.10 The Horsehead Nebula, a dark nebula in a region of glowing nebulosity in Orion. (© Anglo-Australian Observatory/Royal Observatory Edinburgh, photography by David Malin)

(Figure 22.11). In the neighborhood of the Milky Way, these gaseous clouds consist of 92 percent hydrogen, 7 percent helium, and less than 1 percent of the remaining heavier elements. By some mechanism not yet fully understood, these thin gaseous clouds become concentrated enough to begin to contract gravitationally. One

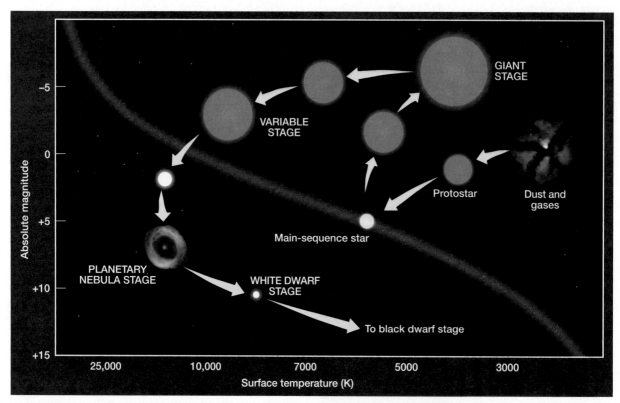

Figure 22.11 Diagram of stellar evolution of H-R diagram for a star about as massive as the Sun.

proposal to explain the triggering of stellar formation is a shock wave traveling from a catastrophic explosion (supernova) of a nearby star. But, regardless of the force that initiates the concentration of interstellar matter, once it is accomplished, mutual gravitational attraction of the particles squeeze the cloud, pulling every particle toward the center. As the cloud shrinks, gravitational energy (potential energy) is converted into energy of motion, or heat energy.

The initial contraction spans a million years or so. With the passage of time, the temperature of this gaseous body slowly rises, eventually reaching a temperature sufficiently high to cause it to radiate energy from its surface in the form of long-wavelength red light. Because this large red object is not hot enough to engage in nuclear fusion, it is not yet a star. The name **protostar** is applied to these bodies.

Protostar Stage

During the protostar phase, gravitational contraction continues, slowly at first, then much more rapidly (Figure 22.11). This collapse causes the core of the developing star to heat much more intensely than the outer envelope. When the core has reached at least 10 million K, the pressure within is so great that groups of four hydrogen nuclei are fused together into single helium nuclei. Astronomers refer to this nuclear reaction as **hydrogen burning** because an enormous amount of energy is released. However, keep in mind that thermonuclear "burning" is not burning in the usual chemical sense.

Heat from hydrogen fusion causes the stellar gases to increase their motion. This in turn results in an increase in the outward gas pressure. At some point, this outward pressure exactly balances the inward force of gravity. When this balance is reached, the star becomes a stable main-sequence star (Figure 22.11). Stated another way, a stable main-sequence star is balanced between two forces: *gravity,* which is trying to squeeze it into a smaller sphere, and *gas pressure,* which is trying to expand it.

Main-Sequence Stage

From this point in the evolution of a main-sequence star until its death, the internal gas pressure struggles to offset the unrelenting force of gravity. Typically, hydrogen burning continues for a few billion years and provides the outward pressure required to support the star from gravitational collapse.

Different stars age at different rates. Hot, massive blue stars radiate energy at such an enormous rate that they substantially deplete their hydrogen fuel in only a few million years. By contrast, the very smallest (red) main-sequence stars may remain stable for hundreds of billions of years. A yellow star, such as the Sun, remains a main-sequence star for about 10 billion

years. As the solar system is about 5 billion years old, it is comforting to know that the Sun will remain stable for another 5 billion years.

An average star spends 90 percent of its life as a hydrogen-burning main-sequence star. Once the hydrogen fuel in the star's core is depleted, it evolves rapidly and dies. However, with the exception of the least-massive (red) stars, a star can delay its death by burning other fuels and becoming a giant.

Red Giant Stage

The evolution to the red giant stage results because the zone of hydrogen burning continually migrates outward, leaving behind an inert helium core. Eventually, all of the hydrogen in the star's core is consumed. While hydrogen fusion is still progressing in the star's outer shell, no fusion is taking place in the core. Without a source of energy, the core no longer has enough pressure to support itself against the inward force of gravity. As a result, the core begins to contract.

Although the core cannot generate nuclear energy, it does grow hotter by converting gravitational energy into heat energy. Some of this energy is radiated outward, initiating a more vigorous level of hydrogen fusion in the star's outer shell. This energy in turn heats and enormously expands the star's outer envelope, producing a giant body hundreds to thousands of times its main-sequence size (Figure 22.11).

As the star expands, its surface cools, which explains the star's reddish appearance. Eventually the star's gravitational force will stop this outward expansion. Once again, the two opposing forces, gravity and gas pressure, will be in balance, and this gaseous mass will be a stable but much larger star. Some red giants overshoot the equilibrium point and rebound like an overextended spring. Such stars continue to oscillate in size, becoming variable stars.

While the envelope of a red giant expands, the core continues to collapse and heat until it reaches 100 million K. At this incredible temperature, it is hot enough to initiate a nuclear reaction in which helium is converted to carbon. Thus, a red giant consumes both hydrogen and helium to produce energy. In stars more massive than the Sun, still other thermonuclear reactions occur that generate all the elements on the periodic table up to number 26, iron. Nuclear "burning" of elements heavier than iron requires an additional source of energy to keep the reaction progressing. Hence, these elements are not produced in ordinary stars.

Eventually, all the usable nuclear fuel in these giants will be consumed. The Sun, for example, will spend less than a billion years as a giant, and the more massive stars will pass through this stage even more rapidly. The force of gravity will again control the star's destiny as it squeezes the star into the smallest, most dense piece of matter possible.

Burnout and Death

Most of the events of stellar evolution discussed thus far are well documented. What happens to a star after the red giant phase is more speculative. We do know that a star, regardless of its size, must eventually exhaust all of its usable nuclear fuel and collapse in response to its immense gravitational force. With this in mind, we will now consider the final stage of stars in three different mass categories.

Death of Low-Mass Stars. Stars less than one-half the mass of the Sun (0.5 solar mass) consume their fuel at a comparatively slow rate (Figure 22.12A). Consequently, these small, *cool red stars* may remain on the main sequence for up to 100 billion years. Because the interior of a low-mass star never attains sufficiently high temperatures and pressures to fuse helium, its only energy source is hydrogen fusion. Thus, low-mass stars never evolve to become bloated red giants. Rather, they remain as stable main-sequence stars until they consume their hydrogen fuel and collapse into a hot, dense *white dwarf*. As we shall see, white dwarfs are small, compact objects unable to support nuclear burning.

Death of Medium-Mass (Sunlike) Stars. Main-sequence stars with masses ranging between half that of the Sun and three times that of the Sun evolve in essentially the same way (Figure 22.12B). During their giant phase, sunlike stars fuse hydrogen and helium fuel at an accelerated rate. Once this fuel is exhausted, these stars (like low-mass stars) collapse into an Earth-sized body of great density—a white dwarf. The density of a white dwarf is as great as physics will allow short of destroying protons and electrons.

Figure 22.12 The evolutionary stages of stars having various masses.

Figure 22.13 The Helix Nebula, the nearest planetary nebula to our solar system. A planetary nebula is the ejected outer envelope of a sunlike star that formed during the star's collapse from a red giant to a white dwarf. (© Anglo-Australian Observatory, photography by David Malin)

The gravitational energy supplied to a collapsing white dwarf is reflected in its high surface temperature. However, without a source of nuclear energy, a white dwarf becomes cooler and dimmer as it continually radiates its remaining thermal energy into space.

During their collapse from red giants to white dwarfs, medium-mass stars are thought to cast off their bloated outer atmosphere, creating an expanding spherical cloud of gas. The remaining hot, central white dwarf heats the gas cloud, causing it to glow. These often beautiful, gleaming spherical clouds are called **planetary nebulae.** A good example of a planetary nebula is the Helix Nebula in the constellation Aquarius (Figure 22.13). This nebula appears as a ring because our line of sight through the center traverses less gaseous material than at the nebula's edge. It is, nevertheless, spherical in shape.

Death of Massive Stars. In contrast to sunlike stars, which expire gracefully, stars exceeding 3 solar masses have relatively short life spans and terminate in a brilliant explosion called a **supernova** (Figure 22.12C). During a supernova event, a star becomes millions of times brighter than its prenova stage (see Box 22.2). If one of the nearest stars to Earth produced such an outburst, its brilliance would surpass that of the Sun. Supernovae are rare; none have been observed in our galaxy since the advent of the telescope, although Tycho Brahe and Galileo each recorded one about 30 years apart. An even larger supernova was recorded in A.D.

1054 by the Chinese. Today, the remnant of this great outburst is the Crab nebula, shown in Figure 22.14.

A supernova event is thought to be triggered when a massive star consumes most of its nuclear fuel. Without a heat engine to generate the gas pressure required to balance its immense gravitational field, it collapses. This implosion is of cataclysmic proportion, resulting in a shock wave that moves out from the star's interior. This energetic shock wave destroys the star and blasts the outer shell into space, generating the supernova event.

Theoretical work predicts that during a supernova, the star's interior condenses into a very hot object, possibly no larger than 20 kilometers in diameter (see Box 22.3). These incomprehensibly dense bodies have been named *neutron stars.* Some supernovae events are thought to produce even smaller and most intriguing objects called *black holes.* We will consider the nature of neutron stars and black holes in the "Stellar Remnants" section.

H-R Diagrams and Stellar Evolution

The Hertzsprung-Russell diagrams have been very helpful in formulating and testing models of stellar evolution. They are also useful for illustrating the changes that take place in an individual star during its life span. Figure 22.11 shows on an H-R diagram the evolution of a star about the size of the Sun. Keep in mind that the

Box 22.2 Supernova 1987A

The first naked-eye supernova in 383 years was discovered in the southern sky in February 1987 (Figure 22.B). This stellar explosion was officially named SN 1987A (SN stands for "supernova" and 1987A indicates that it was the first supernova observed in 1987). Naked-eye supernovae are rare. Only a few have been recorded in historic times. Arab observers saw one in 1006, and the Chinese recorded one in 1054 at the present location of the Crab Nebula. In addition, the astronomer Tycho Brahe observed a supernova in 1572 and Kepler saw one shortly thereafter in 1604.

Prior to this event, researchers could only test their hypotheses on dim supernovae seen in distant galaxies. Thus, when SN 1987A occurred, astronomers quickly focused every available telescope in the Southern Hemisphere on this spectacular event. As one astronomer remarked, "This supernova is better studied by far than any supernova in history." More importantly, this event has allowed astronomers to use observational data to test their theoretical models of stellar evolution.

Supernova 1987A occurred about 170,000 light-years away in the Large Magellanic Cloud, a satellite galaxy to our own Milky Way. As expected, the supernova rapidly increased in brightness to a peak magnitude of 2.4, outshining all the other stars in the Large Magellanic Cloud. Also as predicted, within a few weeks it began to fade. However, SN 1987A did provide some surprises.

From old photographs taken of the area, researchers identified the exploded star as Sanduleak −69°202. Astronomers were surprised to find that the parent star was a hot blue star about 15 times the mass of our Sun. Recall that only cool red giants are thought to die as a supernova event. Further, the Hubble Space Telescope made another unexpected discovery. Its camera revealed a very large shell of gas that predates the supernova explosion by about 40,000 years.

Astronomers now think that Sanduleak −69°202 was once a red supergiant that had blown away its outer shell, exposing a hot blue core. It is this ejected outer shell that appears in the image produced by the Hubble Space Telescope. Then, some 40,000 years later, the remaining hot core of the red supergiant collapsed, producing the supernova of 1987.

Despite these twists, the theory of stellar evolution has held up very well. Theory predicts that the expanding remnants to Supernova 1987A will be large enough to be observed by the turn of the twenty-first century. Thus, astronomers continue to monitor SN 1987A to unravel its secrets and to confirm or refute their ideas about the final stages of stellar evolution.

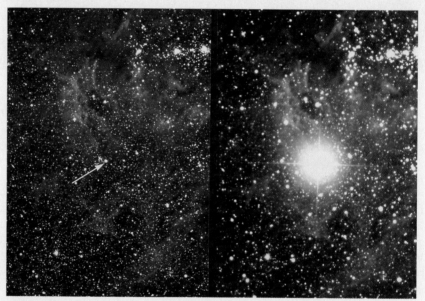

Figure 22.B The great Supernova 1987A. The photo on the left was made prior to the supernova and the one on the right was made following the event. (© Anglo-Australian Observatory, photography by David Malin)

star does not physically move along this path, but rather that its position on the H-R diagram represents the color (temperature) and absolute magnitude (brightness) of the star at various stages in its evolution.

For example, on the H-R diagram, a protostar would be located to the right and above the main sequence (Figure 22.11). It is formed to the right because of its relatively cool surface temperature (red color), and above because it would be more luminous than a main-sequence star of the same color, a fact attributable to its large size. Careful examination of Figure 22.11 should help you visualize the evolutionary

changes experienced by a star the size of our Sun. In addition, Table 22.3 provides a summary of the evolutionary history of stars having various masses.

Stellar Remnants

Eventually, all stars consume their nuclear fuel and collapse into one of three final states—white dwarf, neutron star, or black hole. Although different in some ways, these small, compact objects are all composed of incomprehensibly dense material and all have extreme surface gravity.

Figure 22.14 Crab Nebula in the constellation Taurus: the remains of the supernova of A.D. 1054. (UCO Lick Observatory Image)

White Dwarfs

White dwarfs are extremely small stars with densities greater than any known terrestrial material. It is believed that white dwarfs once were low-mass or medium-mass stars whose internal heat was able to keep these gaseous bodies from collapsing under their own gravitational force.

Although some white dwarfs are no larger than Earth, the mass of such a dwarf can equal 1.4 times that of the Sun. Thus, their densities may be a million times greater than water. A spoonful of such matter would weigh several tons. Densities this great are possible only when electrons are displaced inward from their regular orbits, around an atom's nucleus, allowing the atoms to take up less than the "normal" amount of space. Material in this state is called **degenerate matter.**

In degenerate matter, the atoms have been squeezed together so tightly that the electrons are displaced much nearer to the nucleus. Degenerate matter uses electrical repulsion rather than molecular motion to support itself from total collapse. Although atomic particles in degenerate matter are much closer together than in normal Earth matter, they still are not packed as tightly as possible. Stars made of matter that has an even greater density are thought to exist.

As a star contracts into a white dwarf, its surface becomes very hot, sometimes exceeding 25,000 K. Even so, without a source of energy, it can only become cooler and dimmer. Although none have been observed, the terminal stage of a white dwarf must be a small, cold, nonluminous body called a *black dwarf.*

Neutron Stars

A study of white dwarfs produced what might at first appear to be a surprising conclusion. The smallest white dwarfs are the most massive, and the largest are the least massive. The explanation for this is that a more massive star, because of its greater gravitational force, is able to squeeze itself into a smaller, more densely packed object than can a less massive star. Thus, the smallest white dwarfs were produced from the collapse of larger, more massive stars than were the larger white dwarfs.

This conclusion led to the prediction that stars smaller and more massive than white dwarfs must exist. Named **neutron stars,** these objects are thought to be the remnants of supernova events. In a white

Table 22.3   Summary of evolution for stars of various masses.

Initial Mass of Interstellar Cloud (Sun = 1)	Main-Sequence Stage	Giant Phase	Evolution After Giant Phase	Terminal State (Final Mass*)
0.001	None (Planet)	No	None	Planet (0.001)
0.1	Red	No	None	White dwarf (0.1)
1–3	Yellow	Yes	Planetary nebula	White dwarf (< 1.4)
6	White	Yes	Supernova	Neutron star (1.4–3)
20	Blue	Yes (Supergiant)	Supernova	Black hole (> 3.0)

*These mass numbers are estimates.

Box 22.3   Earth as a System: From Stardust to You

During a supernova implosion, the internal temperature of a star may reach 1 billion K, a condition thought to produce very heavy elements such as gold and uranium. These heavy elements, plus the debris of novae and the planetary nebulae, are continually returned to interstellar space where they are available for the formation of other stars (Figure 22.C).

Astronomers believe that the earliest stars were made of nearly pure hydrogen. Fusion during the life and death of stars in turn produced heavier elements, some of which were returned to space. Because the Sun contains some heavy elements but has not yet reached the stage in its evolution where it could have produced them, it must be at least a second-generation star. Thus, our Sun, as well as the rest of the solar system, is believed to have formed from debris scattered from preexisting stars. If this is the case, the atoms in your body were produced billions of years ago inside a star, and the gold in your jewelry was formed during a supernova event that occurred trillions of kilometers away. Without such events, the development of life on Earth would not have been possible.

Figure 22.C Eagle Nebula in the constellation Serpens. This gaseous nebula is the site of a recent star formation. (Courtesy of National Optical Astronomy Observatories)

---

dwarf, the electrons are pushed close to the nucleus, whereas in a neutron star, the electrons are forced to combine with protons to produce neutrons (hence the name). If Earth were to collapse to the density of a neutron star, it would have a diameter equivalent to the length of a football field. A pea-sized sample of this matter would weigh 100 million tons. This is approximately the density of an atomic nucleus; thus, neutron stars can be thought of as large atomic nuclei.

During a supernova implosion, the envelope of the star is ejected (Figure 22.15), while the core collapses into a very hot neutron star about 20 kilometers (12.4 miles) in diameter. Although neutron stars have high surface temperatures, their small size would greatly limit their luminosity. Consequently, locating one visually would be extremely difficult.

However, theory predicts that a neutron star would have a very strong magnetic field. Further, as a star collapses, it will rotate faster, for the same reason ice skaters rotate faster as they pull in their arms. If the Sun were to collapse to the size of a neutron star, it would increase its rate of rotation from once every 25 days to nearly 1000 times per second. Radio waves generated by these rotating stars would be concentrated into two narrow zones that would align with the star's magnetic poles. Consequently, these stars

Figure 22.15 Veil Nebula in the constellation Cygnus is the remnant of an ancient supernova implosion. (Palomar Observatories/California Institute of Technology [Caltech])

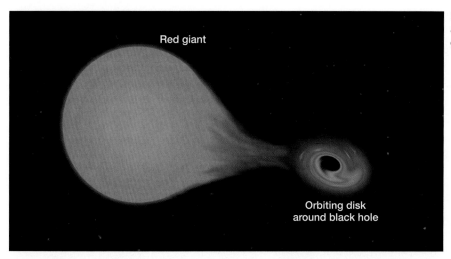

Figure 22.16 This illustration shows how astronomers believe a binary pair (red giant/black hole) might function.

would resemble a rapidly rotating beacon emitting strong radio waves. If Earth happened to be in the path of these beacons, the star would appear to blink on and off, or pulsate, as the waves swept past.

In the early 1970s, a source that radiates short pulses of radio energy, called a **pulsar** (pulsating radio source), was discovered in the Crab Nebula. Visual inspection of this radio source revealed it to be a small star centered in the nebula. The pulsar found in the Crab Nebula is undoubtedly the remains of the supernova of A.D. 1054 (see Figure 22.14). Thus, the first neutron star had been discovered.

Black Holes

Are neutron stars made of the most dense materials possible? No. During a supernova event, remnants of stars greater than 3 solar masses apparently collapse into objects even smaller and denser than neutron stars. Even though these objects would be very hot, their surface gravity would be so immense that even light could not escape the surface. Consequently, they would literally disappear from sight. These incredible bodies have appropriately been named **black holes.** Anything that moved too near a black hole would be swept in by its irresistible gravity and be devoured forever.

How can astronomers find an object whose gravitational field prevents the escape of all matter and energy? One strategy is to seek evidence of matter being rapidly swept into a region of apparent nothingness. Theory predicts that as matter is pulled into a black hole, it should become very hot and emit a flood of X-rays before being engulfed. Because isolated black holes would not have a source of matter to engulf, astronomers first looked at binary star systems.

A likely candidate for a black hole is Cygnus X-1, a strong X-ray source in the constellation Cygnus. In this case, the X-ray source can be observed orbiting a supergiant companion with a period of 5.6 days. It appears that gases are pulled from this companion and spiral into the disc-shaped structure around the black hole (Figure 22.16). The result is a stream of X-rays. Because X-rays cannot penetrate our atmosphere efficiently, the existence of black holes was not confirmed until recently. The first X-ray sources were discovered in 1971 by detectors on satellites. Cygnus X-1 is such a source.

The Milky Way Galaxy

On a clear and moonless night away from city lights, you can see a truly marvelous sight—our own Milky Way Galaxy (Figure 22.17). With his telescope, Galileo discovered that this band of light was produced by countless individual stars that the unaided eye is unable to resolve. Today, we realize that the Sun is actually a part of this vast system of stars, which number about 100 billion (Figure 22.18A). The "milky" appearance of our galaxy results because the solar system is located within the flat *galactic disk.* Thus, when it is viewed from the "inside," a higher concentration of stars appears in the direction of the galactic plane than in any other direction. You can see this in the edge-on view in Figure 22.18B.

When astronomers began telescopically to survey the stars located along the plane of the Milky Way, it appeared that equal numbers lay in every direction. Could Earth actually be at the center of the galaxy? A better explanation was put forth. Imagine that the trees in an enormous forest represent the stars in the galaxy. After hiking into this forest a short distance, you look around. What you see is an equal number of trees in every direction. Are you really in the center of the forest? Not necessarily; anywhere in the forest, except at the very edge, you will seem to be in the middle.

Structure of the Milky Way Galaxy

Attempts to visually inspect the Milky Way Galaxy are hindered by the large quantities of interstellar matter that lie in our line of sight. Nevertheless, with

Figure 22.17 Panorama of our galaxy, the Milky Way. Notice the dark bands caused by the presence of interstellar dark nebulae. (Lund Observatory photograph)

the aid of radio telescopes, the gross structure of our galaxy has been determined. The Milky Way is a rather large spiral galaxy whose disk is about 100,000 light-years wide and about 10,000 light-years thick at the nucleus (Figure 22.18). As viewed from Earth, the center of the galaxy lies beyond the constellation Sagittarius.

Radio telescopes reveal the existence of at least three distinct *spiral arms,* with some showing splintering (Figure 22.19). The Sun is positioned in one of

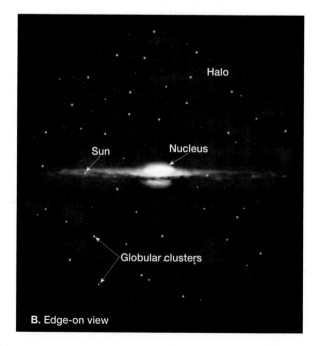

Figure 22.18 Structure of the visible portion of the Milky Way Galaxy.

Figure 22.19 If the Milky Way were photographed from a distance, it might appear like the spiral galaxy NGC 2997. (© Anglo-Australian Observatory, photography by David Malin)

these arms about two-thirds of the way from the center, at a distance from the hub of about 30,000 light-years. The stars in the arms of the Milky Way rotate around the *galactic nucleus*, with the most outward ones moving the slowest, such that the ends of the arms appear to trail. The Sun and the arm it is in require about 200 million years for each orbit around the nucleus.

Surrounding the galactic disk is a nearly spherical *halo* made of very tenuous gas and numerous globular clusters. These star clusters do not participate in the rotating motion of the arms, but rather have their own orbits that carry them through the disk. Although some clusters are very dense, they pass among the stars of the arms with plenty of room to spare.

Galaxies

In the mid-1700s, German philosopher Immanuel Kant proposed that the telescopically visible fuzzy patches of light scattered among the stars were actually distant galaxies like the Milky Way. Kant described them as "island universes." Each galaxy, he believed, contained billions of stars and, as such, was a universe in itself. The weight of opinion, however, favored the hypothesis that they were dust and gas clouds (nebulae) within our galaxy.

This matter was not resolved until the 1920s, when American astronomer Edwin Hubble was able to locate, within one of these fuzzy patches, some unique stars that are known to be intrinsically very bright. Because these very bright stars appeared only very faintly in the telescope, Hubble believed they must lie outside the Milky Way.

This fuzzy patch, which lies over a million light-years away, was named the Great Galaxy in Andromeda (Figure 22.20). Hubble had extended the universe far beyond the limits of our imagination, to include hundreds of billions of galaxies, each containing hundreds of billions of stars. It has been said that a million galaxies are found in that portion of the sky bounded by the cup of the Big Dipper. There are more stars in the heavens than grains of sand in all the beaches on Earth.

Figure 22.20 Great Galaxy, a spiral galaxy, in the constellation Andromeda. The two bright spots to the left and right are dwarf elliptical galaxies. (Palomar Observatories/California Institute of Technology [Caltech])

Figure 22.21 Two views illustrating the idealized structure of spiral galaxies. (Courtesy of Hansen Planetarium/U.S. Naval Observatory)

Four Types of Galaxies

From the hundreds of billions of galaxies, several basic types have been identified: spiral, elliptical, and irregular.

The Milky Way and the Great Galaxy in Andromeda are examples of fairly large **spiral galaxies** (Figure 22.21). Andromeda can be seen with the unaided eye as a fuzzy fifth-magnitude object. Typically, spiral galaxies are disk-shaped with a somewhat greater concentration of stars near their centers, but there are numerous variations. Viewed broadside, arms are often seen extending from the central nucleus and sweeping gracefully away. The outermost stars of these arms rotate most slowly, giving the galaxy the appearance of a fireworks pinwheel.

One type of spiral galaxy, however, has the stars arranged in the shape of a bar, which rotates as a rigid system. This requires that the outer stars move faster than the inner ones, a fact not easy for astronomers to reconcile with the laws of motion. Attached to each end of these bars are curved spiral arms. These have become known as **barred spiral galaxies** (Figure 22.22). Spiral galaxies are generally quite large, ranging from 20,000 to about 125,000 light-years in diameter. About 10 percent of all galaxies are thought to be barred spirals and another 20 percent are regular spiral galaxies like the Milky Way.

The most abundant group, making up 60 percent of the total, is the **elliptical galaxies.** These are generally smaller than spiral galaxies. Some are so much smaller,

Figure 22.22 Barred spiral galaxy. (Palomar Observatories/California Institute of Technology [Caltech])

in fact, that the term *dwarf* has been applied. Because these dwarf galaxies are not visible at great distances, a survey of the sky reveals more of the conspicuous large spiral galaxies. Although most elliptical galaxies are small, the very largest known galaxies (200,000 light-years in diameter) are also elliptical. As their name implies, elliptical galaxies have an ellipsoidal shape that ranges to nearly spherical, and they lack spiral arms. The two dwarf companions of Andromeda shown in Figure 22.20 are elliptical galaxies.

Only 10 percent of the known galaxies lack symmetry and are classified as **irregular galaxies.** The best-known irregular galaxies, the Large and Small Magellanic Clouds in the Southern Hemisphere, are easily visible with the unaided eye. Named after the explorer Ferdinand Magellan, who observed them when he circumnavigated Earth in 1520, they are our nearest galactic neighbors—only 150,000 light-years away.

One of the major differences among the galactic types is the age of the stars that make them up. The irregular galaxies are composed mostly of young stars, whereas the elliptical galaxies contain old stars. The Milky Way and other spiral galaxies consist of both young and old stars, with the youngest located in the arms.

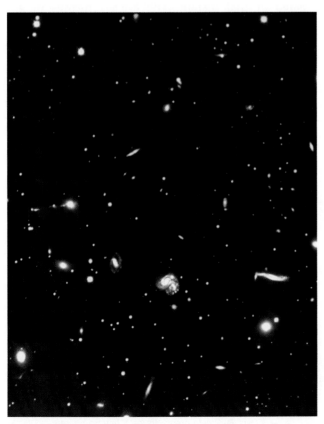

Figure 22.23 Numerous galaxies grouped in the constellation Hercules. (Courtesy of National Optical Astronomy Observatories)

Galactic Clusters

Once astronomers discovered that stars were associated in groups, they set out to determine whether galaxies also were grouped or just randomly distributed throughout the universe. They found that, like stars, galaxies are grouped in **galactic clusters** (Figure 22.23). Some abundant clusters contain thousands of galaxies. Our own, called the **Local Group,** contains at least 28 galaxies. Of these, 3 are spirals, 11 are irregulars, and 14 are ellipticals. Galactic clusters also reside in huge swarms called superclusters. From visual observations, it appears that superclusters may be the largest entities in the universe.

Red Shifts

You probably have noticed the change in pitch of a car horn or ambulance siren as it passes by. When it is approaching, the sound seems to have a higher-than-normal pitch, and when it is moving away, the pitch sounds lower than normal. This effect, which occurs for all wave motion, including sound and light waves, was first explained by Christian Doppler in 1842 and is called the *Doppler effect.* The reason for the difference in pitch is that it takes time for the wave to be emitted. If the source of the wave is moving away, the beginning of the wave is emitted nearer to you than the end of the wave, effectively "stretching" the wave. This gives it a longer wavelength (see Figure 21.4, p. 583). The opposite is true for an approaching source.

In the case of light, when a source is moving away, its light appears redder than it actually is, because its waves appear lengthened. Objects approaching have their light waves shifted toward the blue (shorter wavelength). Therefore, the Doppler effect reveals whether Earth and another celestial body are approaching or leaving one another. In addition, the amount of shift allows us to calculate the *rate* at which the relative movement is occurring. Large Doppler shifts indicate higher velocities; smaller Doppler shifts indicate lower velocities.

Expanding Universe

One of the most important discoveries of modern astronomy was made in 1929 by Edwin Hubble. Observations completed several years earlier revealed that most galaxies have Doppler shifts toward the red end of the spectrum. Recall that red shift occurs because the light waves are "stretched," indicating that Earth and the source are moving away from each other. Hubble set out to explain the predominance of red shift.

Hubble realized that dimmer galaxies were probably farther away than were brighter galaxies. Thus, he tried to determine whether a relation existed between the distances to galaxies and their red shifts. Using estimated distances based on relative brightness and the observed Doppler red shifts, Hubble discovered that galaxies that exhibit the greatest red shifts are the most distant.

A consequence of the universal red shift is that it predicts that most galaxies (except for a few nearby) are receding from us. Recall that the amount of Doppler red shift is dependent on the velocity at which the object is moving away. Greater red shifts indicate faster recessional velocities. Because more distant galaxies have greater red shifts, Hubble concluded that they must be retreating from us at greater velocities. This idea is currently termed **Hubble's law** and states that galaxies are receding from us at a speed that is proportional to their distance.

Hubble was surprised at this discovery because it implied that the most distant galaxies are moving away from us many times faster than those nearby. What type of cosmological theory can explain this fact? It was soon realized that an *expanding universe* can adequately account for the observed red shifts.

To help visualize the nature of this expanding universe, we will employ a popularly used analogy. Imagine a loaf of raisin bread dough that has been set out to rise for a few hours (Figure 22.24). As the dough doubles in size, so does the distance between all of the raisins. However, the raisins that were originally farther apart traveled a greater distance in the same time span than those located closer together. We therefore conclude that, in an expanding universe, as in our analogy, those objects located farther apart move away from each other more rapidly.

Another feature of the expanding universe can be demonstrated using the raisin bread analogy. No matter which raisin you select, it will move away from all the other raisins. Likewise, no matter where one is located in the universe, every other galaxy (except those in the same cluster) will be receding. Edwin Hubble had indeed advanced our understanding of the universe. The Hubble Space Telescope is named in his honor.

The Big Bang

The universe—did it have a beginning? Will it have an end? Cosmologists are trying to answer these questions, and that makes them a rare breed.

First and foremost, any viable theory regarding the origin of the universe must account for the fact that all galaxies (except for the very nearest) are moving away from us. Because all galaxies appear to be moving away from Earth, is our planet in the center of the universe? Probably not, because if we are not even in the center of our own solar system, and our solar system is not even in the center of the galaxy, it seems unlikely that we could be in the center of the universe.

A more probable explanation exists: Imagine a balloon with paper-punch "dots" glued to its surface. When the balloon is inflated, each dot spreads apart from every other dot. Similarly, if the universe is expanding, every galaxy would be moving away from every other galaxy.

This concept of an expanding universe led to the widely accepted **big bang** theory. According to this theory, the entire universe was at one time confined to a dense, hot, supermassive ball. Then, about 20 billion years ago, a cataclysmic explosion occurred, hurling this material in all directions. The big bang marks the inception of the universe; all matter and space were created at that instant. The ejected masses of gas cooled and condensed, forming the stellar systems we now observe fleeing from their birthplace.

Will the stars eventually dim from view and invisible galaxies travel on forever? It has been suggested that after a certain point, perhaps 20 billion years in the future, the galaxies will slow and eventually stop their outward flight. Gravitational contraction would follow. The galaxies would then collide and coalesce, and a new fireball would be born. For this event to occur, the universe must have an average density of one atom for every cubic meter of space. But present estimates indicate that the concentration of matter in the universe is far less than this amount, so this hypothesis remains unsubstantiated.

Figure 22.24 Illustration of the raisin bread analogy of an expanding universe. As the dough rises, raisins originally farther apart travel a greater distance in the same time span as those located closer together. Thus, raisins (like galaxies in a uniform expanding universe) that are located farther apart move away from each other more rapidly than those located nearer to each other.

A. Raisin bread dough before it rises.

B. Raisin bread dough a few hours later.

However, this subject does not end here. It has been proposed that heretofore undetected matter exists in large quantities in the universe. For example, numerous black holes may occupy many of the voids in the universe. If this is true, the galaxies could, in fact, collapse upon themselves.

"Absence of evidence is not evidence of absence."

—Anonymous

Chapter Summary

- One method for determining the distance to a star is to use a measurement called *stellar parallax,* the extremely slight back-and-forth shifting in a nearby star's position due to the orbital motion of Earth. *The farther away a star is, the less its parallax.* A unit used to express stellar distance is the *light-year,* which is the distance light travels in a year—about 9.5 trillion kilometers (5.8 trillion miles).

- The intrinsic properties of stars include *brightness, color, temperature, mass,* and *size.* Three factors control the brightness of a star as seen from Earth: how big it is, how hot it is, and how far away it is. *Magnitude* is the measure of a star's brightness. *Apparent magnitude* is how bright a star appears when viewed from Earth. *Absolute magnitude* is the "true" brightness were a star at a standard distance of about 32.6 light-years. The difference between the two magnitudes is directly related to a star's distance. Color is a manifestation of a star's temperature. Very hot stars (surface temperatures above 30,000 K) appear blue; red stars are much cooler (surface temperatures generally less than 3000 K). Stars with surface temperatures between 5000 and 6000 K appear yellow, like our Sun. The center of mass of orbiting *binary stars* (two stars revolving around a common center of mass under their mutual gravitational attraction) is used to determine the mass of the individual stars in a binary system.

- A *Hertzsprung-Russell diagram* is constructed by plotting the absolute magnitudes and temperatures of stars on a graph. A great deal about the sizes of stars can be learned from H-R diagrams. Stars located in the upper-right position of an H-R diagram are called *giants,* luminous stars of large radius. *Supergiants* are very large. Very small *white dwarf* stars are located in the lower-central portion of an H-R diagram. Ninety percent of all stars, called *main-sequence stars,* are in a band that runs from the upper-left corner to the lower-right corner of an H-R diagram.

- *Variable stars* fluctuate in brightness. Some, called *pulsating variables,* fluctuate regularly in brightness by expanding and contracting in size. When a star explosively brightens, it is called a *nova.* During the outburst, the outer layer of the star is ejected at high speed. After reaching maximum brightness in a few days, the nova slowly returns in a year or so to its original brightness.

- New stars are born out of enormous accumulations of dust and gases, called *nebula,* that are scattered between existing stars. A *bright nebula* glows because the matter is close to a very hot (blue) star. The two main types of bright nebulae are *emission nebulae* (which derive their visible light from the fluorescence of the ultraviolet light from a star in or near the nebula) and *reflection nebulae* (relatively dense dust clouds in interstellar space that are illuminated by reflecting the light of nearby stars). When a nebula is not close enough to a bright star to be illuminated, it is referred to as a *dark nebula.*

- Stars are born when their nuclear furnaces are ignited by the unimaginable pressures and temperatures in collapsing nebulae. New stars not yet hot enough for nuclear fusion are called *protostars.* When collapse causes the core of a protostar to reach a temperature of at least 10 million K, the fusion of hydrogen nuclei into helium nuclei begins in a process called *hydrogen burning.* The opposing forces acting on a star are *gravity* trying to contract it and *gas pressure* (*thermal nuclear energy*) trying to expand it. When the two forces are balanced, the star becomes a stable *main-sequence star.* When the hydrogen in a star's core is consumed, its outer envelope expands enormously and a *red giant* star, hundreds to thousands of times larger than its main-sequence size, forms. When all the usable nuclear fuel in these giants is exhausted and gravity takes over, the stellar remnant collapses into a small dense body.

- The *final fate of a star is determined by its mass.* Stars with less than one-half the mass of the Sun collapse into hot, dense *white dwarf* stars. Medium-mass stars (between 0.5 and 3.0 times the mass of the Sun) become red giants, collapse, and end up as white dwarf stars, often surrounded by expanding spherical clouds of glowing gas called *planetary nebulae.* Stars more than three times the mass of the Sun terminate in a brilliant explosion called a *supernova.* Supernovae events can produce small, extremely dense *neutron stars,* composed entirely of subatomic particles called neutrons; or even smaller and more dense *black holes,* objects that have such immense gravity that light cannot escape their surface.

- The *Milky Way Galaxy* is a large, disk-shaped *spiral galaxy* about 100,000 light-years wide and about 10,000 light-years thick at the center. There are three distinct *spiral arms* of stars, with some showing splintering. The Sun is positioned in one of these arms about two-thirds of the way from the galactic center, at a distance of about 30,000 light-years. Surrounding the galactic disk is a nearly spherical halo made of very tenuous gas and numerous *globular clusters* (nearly spherically shaped groups of densely packed stars).

- The various types of galaxies include (1) *irregular galaxies,* which lack symmetry and account for only 10 percent of the known galaxies; (2) *spiral galaxies,* which are typically disk-shaped with a somewhat greater concentration of stars near their centers, often containing arms of stars extending from their central nucleus; and (3) *elliptical galaxies,* the most abundant type, which have an ellipsoidal shape that ranges to nearly spherical, and that lack spiral arms.

- Galaxies are not randomly distributed throughout the universe. They are grouped in *galactic clusters,* some containing thousands of galaxies. Our own, called the *Local Group,* contains at least 28 galaxies.

- By applying the *Doppler effect* (the apparent change in wavelength of radiation caused by the motions of the source and the observer) to the light of galaxies, galactic motion can be determined. Most galaxies have Doppler shifts toward the red end of the spectrum, indicating increasing distance. The amount of Doppler shift is dependent on the velocity at which the object is moving. Because the most distant galaxies have the greatest red shifts, Edwin Hubble concluded in the early 1900s that they were retreating from us with greater recessional velocities than were more nearby galaxies. It was soon realized that an *expanding universe* can adequately account for the observed red shifts.

- The belief in the expanding universe led to the widely accepted *Big Bang theory.* According to this theory, the entire universe was at one time confined in a dense, hot, supermassive concentration. About 20 billion years ago, a cataclysmic explosion hurled this material in all directions, creating all matter and space. Eventually the ejected masses of gas cooled and condensed, forming the stellar systems we now observe fleeing from their place of origin.

Key Terms

absolute magnitude (p. 604)
apparent magnitude (p. 603)
barred spiral galaxy (p. 620)
big bang (p. 622)
black hole (p. 617)
bright nebula (p. 608)
dark nebula (p. 608)
degenerate matter (p. 615)
elliptical galaxy (p. 620)
emission nebula (p. 608)
eruptive variables (p. 607)
galactic cluster (p. 621)

Hertzsprung-Russell (H-R) diagram
   (p. 605)
Hubble's law (p. 622)
hydrogen burning (p. 611)
interstellar dust (p. 608)
irregular galaxy (p. 621)
light-year (p. 603)
Local Group (p. 621)
magnitude (p. 603)
main-sequence stars (p. 605)
nebula (p. 608)
neutron star (p. 615)

nova (p. 607)
planetary nebula (p. 613)
protostar (p. 611)
pulsar (p. 617)
pulsating variables (p. 607)
red giant (p. 606)
reflection nebula (p. 608)
spiral galaxy (p. 620)
stellar parallax (p. 602)
supergiant (p. 606)
supernova (p. 613)
white dwarf (p. 606)

Review Questions

1. How far away in light-years is our nearest stellar neighbor, Proxima Centauri? Convert your answer to kilometers.

2. What is the most basic method of determining stellar distances?

3. Explain the difference between a star's apparent and absolute magnitudes. Which one is an intrinsic property of a star?

4. What is the ratio of brightness between a twelfth-magnitude and fifteenth-magnitude star?

5. What information about a star can be determined from its color?

6. What color are the hottest stars? Medium-temperature stars? Coolest stars?

7. Which property of a star can be determined from binary star systems?

8. Make a generalization relating the mass and luminosity of main-sequence stars.

9. The disk of a star cannot be resolved telescopically. Explain the method that astronomers have used to estimate the size of stars.

10. Where on an H-R diagram does a star spend most of its lifetime?

11. How does the Sun compare in size and brightness to other main-sequence stars?

12. Why is interstellar matter important to stellar evolution?

13. Compare a bright nebula and a dark nebula.

14. What element is the fuel for main-sequence stars? Red giants?

15. What causes a star to become a giant?

16. Why are less massive stars thought to age more slowly than more massive stars, even though they have much less "fuel"?

17. Enumerate the steps thought to be involved in the evolution of sunlike stars.

18. What is the final state of a low-mass (red) main-sequence star?

19. What is the final state of a medium-mass (sunlike) star?

20. How do the "lives" of the most massive stars end? What are the two possible products of this event?

21. Describe the general structure of the Milky Way Galaxy.

22. Compare the three general types of galaxies.

23. Explain why astronomers consider elliptical galaxies more abundant than spiral galaxies, even though more spiral galaxies have been sighted.

24. How did Edwin Hubble determine that the Great Galaxy in Andromeda is located beyond our galaxy?

25. What evidence supports the big bang theory?

Examining the Earth System

1. Briefly describe how the atmosphere, hydrosphere, solid Earth, and biosphere are each related to the death of a star(s) that occurred billions of years ago.

2. If a supernova explosion were to occur within the immediate vicinity of our solar system, what might be some possible consequences of the intense X-ray and gamma radiation that would reach Earth?

3. Scientists are continuously searching the Milky Way Galaxy for other stars that may have planets. What types of stars would most likely have a planet or planets suitable for life as we know it? If you would like to investigate extra-solar planets on line, you might find these two Web sites helpful: *NOVA Online* at **http://www.pbs.org/wgbh/nova/worlds/** and from the *Electronic Universe Project* at **http://zebu.uoregon.edu/newplanet.html**

4. Based upon your knowledge of the Earth system, the planets in our solar system, and the cosmos in general, speculate about the likelihood that extra-solar planets exist with atmospheres, hydrospheres, lithospheres, and biospheres, similar to Earth's? Explain your speculation.

Web Resources

The following are informative and interesting World Wide Web sites that address topics related to those presented in the chapter:

Imagine the Science (NASA):
- **http://imagine.gsfc.nasa.gov/docs/science/science.html**

Our Hierarchical Universe (University of Illinois):
- **http://www.ncsa.uiuc.edu/Cyberia/Cosmos/HierarchUni.html**

Windows to the Universe (University of Michigan):
- **http://windows.engin.umich.edu/**

For direct links to these sites and other appropriate Earth science Web locations, chapter review exercises, and relevant, topical investigations that utilize Web resources, contact the *Earth Science,* 9th Edition Home Page at:
- **http://www.prenhall.com/tarbuck.**

Appendix A

Metric and English Units Compared

Units

1 kilometer (km) = 1000 meters (m)
1 meter (m) = 100 centimeters
1 centimeter (cm) = 0.39 inch (in.)
1 mile (mi) = 5280 feet (ft)
1 foot (ft) = 12 inches (in.)
1 inch (in.) = 2.54 centimeters (cm)
1 square mile (mi^2) = 640 acres (a)
1 kilogram (kg) = 1000 grams (g)
1 pound (lb) = 16 ounces (oz)
1 fathom = 6 feet (ft)

Conversions

When you want
to convert: multiply by: to find:

Length

inches	2.54	centimeters
centimeters	0.39	inches
feet	0.30	meters
meters	3.28	feet
yards	0.91	meters
meters	1.09	yards
miles	1.61	kilometers
kilometers	0.62	miles

Area

square inches	6.45	square centimeters
square centimeters	0.15	square inches
square feet	0.09	square meters
square meters	10.76	square feet
square miles	2.59	square kilometers
square kilometers	0.39	square miles

Volume

cubic inches	16.38	cubic centimeters
cubic centimeters	0.06	cubic inches
cubic feet	0.028	cubic meters
cubic meters	35.3	cubic feet
cubic miles	4.17	cubic kilometers
cubic kilometers	0.24	cubic miles
liters	1.06	quarts
liters	0.26	gallons
gallons	3.78	liters

to convert: multiply by: to find:

Masses and Weights

ounces	20.33	grams
grams	0.035	ounces
pounds	0.45	kilograms
kilograms	2.205	pounds

Temperature

When you want to convert degrees Fahrenheit (°F) to degrees Celsius (°C), subtract 32 degrees and divide by 1.8.

When you want to convert degrees Celsius (°C) to degrees Fahrenheit (°F), multiply by 1.8 and add 32 degrees.

When you want to convert degrees Celsius (°C) to kelvins (K), delete the degree symbol and add 273. When you want to convert kelvins (K) to degrees Celsius (°C), add the degree symbol and subtract 273.

Figure A.1 Temperature scales.

Appendix B
Earth's Grid System

A glance at any globe reveals a series of north-south and east-west lines that together make up Earth's grid system, a universally used scheme for locating points on Earth's surface. The north-south lines of the grid are called **meridians** and extend from pole to pole (Figure B.1). All are halves of great circles. A **great circle** is the largest possible circle that may be drawn on a globe; if a globe were sliced along one of these circles, it would be divided into two equal parts called **hemispheres**. By viewing a globe or Figure B.1, it can be seen that meridians are spaced farthest apart at the equator and converge toward the poles. The east-west lines (circles) of the grid are known as **parallels**. As their name implies, these circles are parallel to one another (Figure B.1). While all meridians are parts of great circles, all parallels are not. In fact, only one parallel, the equator, is a great circle.

Latitude and Longitude

Latitude may be defined as distance, measured in degrees, *north* and *south* of the equator. Parallels are used to show latitude. Because all points that lie along the same parallel are an identical distance from the equator, they all have the same latitude designation. The latitude of the equator is 0 degrees, while the north and south poles lie 90 degrees N and 90 degrees S, respectively.

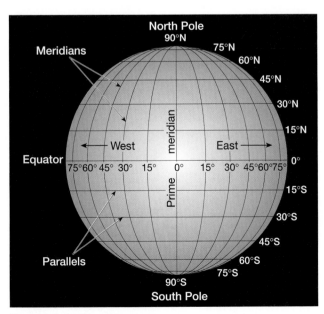

Figure B.1 Earth's grid system.

Longitude is defined as distance, measured in degrees, *east* and *west* of the zero or *prime meridian*. Because all meridians are identical, the choice of a zero line is obviously arbitrary. However, the meridian that passes through the Royal Observatory at Greenwich, England, is universally accepted as the reference meridian. Thus, the longitude for any place on the globe is measured east or west from this line. Longitude can vary from 0 degrees along the prime meridian to 180 degrees, halfway around the globe.

It is important to remember that when a location is specified, directions must be given, that is, north or south latitude and east or west longitude (Figure B.2). If this is not done, more than one point on the globe is being designated. The only exceptions, of course, are places that lie along the equator, the prime meridian, or the 180-degree meridian. It should also be noted that while it is not incorrect to use fractions, a degree of latitude or longitude is usually divided into minutes and seconds. A minute (′) is $1/60$th of a degree, and a second (″) is $1/60$th of a minute. When locating a place on a map, the degree of exactness will depend on the scale of the map. When using a small-scale world map or globe, it may be difficult to estimate latitude and longitude to the nearest whole degree or two. On the other hand, when a large-scale map of an area is used, it is often possible to estimate latitude and longitude to the nearest minute or second.

Distance Measurement

The length of a degree of longitude depends on where the measurement is taken. At the equator, which is a great circle, a degree of east-west distance is equal to approximately 111 kilometers (69 miles). This figure is found by dividing Earth's circumference—40,075 kilometers (24,900 miles)—by 360. However, with an increase in latitude, the parallels become smaller, and the length of a degree of longitude diminishes (see Table B.1). Thus, at about latitude 60 degrees N and S, a degree of longitude has a value equal to about half of what it was at the equator.

As all meridians are halves of great circles, a degree of latitude is equal to about 111 kilometers (69 miles), just as a degree of longitude along the equator is. However, Earth is not a perfect sphere but is slightly flattened at the poles and bulges slightly at the equator. Because of this, there are small differences in the length of a degree of latitude.

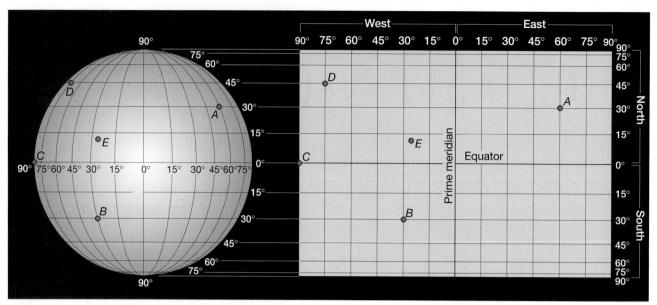

Figure B.2 Locating places using the grid system. For both diagrams: Point *A* is latitude 30 degrees N, longitude 60 degrees E; Point *B* is latitude 30 degrees S, longitude 30 degrees W; Point *C* is latitude 0 degrees, longitude 90 degrees W; Point *D* is latitude 45 degrees N, longitude 75 degrees W; Point *E* is approximately latitude 10 degrees N, longitude 25 degrees W.

Determining the shortest distance between two points on a globe can be done easily and fairly accurately using the "globe and string" method. It should be noted here that the arc of a great circle is the shortest distance between two points on a sphere. To determine the great circle distance (as well as observe the great circle route) between two places, stretch the string between the locations in question. Then, measure the length of the string along the equator (since it is a great circle with degrees marked on it) to determine the number of degrees between the two points. To calculate the distance in kilometers or miles, simply multiply the number of degrees by 111 or 69, respectively.

Table B.1   Longitude as distance.

°Lat.	Length of 1° Long.		°Lat.	Length of 1° Long.		°Lat.	Length of 1° Long.	
	km	miles		km	miles		km	miles
0	111.367	69.172	30	96.528	59.955	60	55.825	34.674
1	111.349	69.161	31	95.545	59.345	61	54.131	33.622
2	111.298	69.129	32	94.533	58.716	62	52.422	62.560
3	111.214	69.077	33	93.493	58.070	63	50.696	31.488
4	111.096	69.004	34	92.425	57.407	64	48.954	30.406
5	110.945	68.910	35	91.327	56.725	65	47.196	29.314
6	110.760	68.795	36	90.203	56.027	66	45.426	28.215
7	110.543	68.660	37	89.051	55.311	67	43.639	27.105
8	110.290	68.503	38	87.871	54.578	68	41.841	25.988
9	110.003	68.325	39	86.665	53.829	69	40.028	24.862
10	109.686	68.128	40	85.431	53.063	70	38.204	23.729
11	109.333	67.909	41	84.171	52.280	71	36.368	22.589
12	108.949	67.670	42	82.886	51.482	72	34.520	21.441
13	108.530	67.410	43	81.575	50.668	73	32.662	20.287
14	108.079	67.130	44	80.241	49.839	74	30.793	19.126
15	107.596	66.830	45	78.880	48.994	75	28.914	17.959
16	107.079	66.509	46	77.497	48.135	76	27.029	16.788
17	106.530	66.168	47	76.089	47.260	77	25.134	15.611
18	105.949	65.807	48	74.659	46.372	78	23.229	14.428
19	105.337	65.427	49	73.203	45.468	79	21.320	13.242
20	104.692	65.026	50	71.727	44.551	80	19.402	12.051
21	104.014	64.605	51	70.228	43.620	81	17.480	10.857
22	103.306	64.165	52	68.708	42.676	82	15.551	9.659
23	102.565	63.705	53	67.168	41.719	83	13.617	8.458
24	101.795	63.227	54	65.604	40.748	84	11.681	7.255
25	100.994	62.729	55	64.022	39.765	85	9.739	6.049
26	100.160	62.211	56	62.420	38.770	86	7.796	4.842
27	99.297	61.675	57	60.798	37.763	87	5.849	3.633
28	98.405	61.121	58	59.159	36.745	88	3.899	2.422
29	97.481	60.547	59	57.501	35.715	89	1.950	1.211
30	96.528	59.955	60	55.825	34.674	90	0.000	0.000

Appendix C

Topographics Maps

A map is a representation on a flat surface of all or a part of Earth's surface drawn to a specific scale. Maps are often the most effective means for showing the locations of both natural and human structures, their sizes, and their relationships to one another. Like photographs, maps readily display information that would be impractical to express in words.

Although most maps show only the two horizontal dimensions, geologists, as well as other map users, often require that the third dimension, elevation, be shown on maps. Maps that show the shape of the land are called **topographic maps**. Although various techniques may be used to depict elevations, the most accurate method involves the use of contour lines.

Contour Lines

A **contour line** is a line on a map representing a corresponding imaginary line on the ground that has the same elevation above sea level along its entire length. While many map symbols are pictographs, resembling the objects they represent, a contour line is an abstraction that has no counterpart in nature. It is, however, an accurate and effective device for representing the third dimension on paper.

Some useful facts and rules concerning contour lines are listed as follows. This information should be studied in conjunction with Figure C.1.

1. Contour lines bend upstream or upvalley. The contours form Vs that point upstream, and in the upstream direction the successive contours represent higher elevations. For example, if you were standing on a stream bank and wished to get to the point at the same elevation directly opposite you on the other bank, without stepping up or down, you would need to walk upstream along the contour at that elevation to where it crosses the stream bed, cross the stream, and then walk back downstream along the same contour.

2. Contours near the upper parts of hills form closures. The top of a hill is higher than the highest closed contour.

3. Hollows (depressions) without outlets are shown by closed, hatched contours. Hatched contours are contours with short lines on the inside pointing downslope.

4. Contours are widely spaced on gentle slopes.

5. Contours are closely spaced on steep slopes.

6. Evenly spaced contours indicate a uniform slope.

7. Contours usually do not cross or intersect each other, except in the rare case of an overhanging cliff.

8. All contours eventually close, either on a map or beyond its margins.

9. A single high contour never occurs between two lower ones, and vice versa. In other words, a change in slope direction is always determined by the repetition of the same elevation either as two different contours of the same value or as the same contour crossed twice.

10. Spot elevations between contours are given at many places, such as road intersections, hill summits, and lake surfaces. Spot elevations differ from control elevation stations, such as bench marks, in not being permanently established by permanent markers.

Relief

Relief refers to the difference in elevation between any two points. *Maximum relief* refers to the difference in elevation between the highest and lowest points in the area being considered. Relief determines the **contour interval**, which is the difference in elevation between succeeding contour lines that is used on topographic maps. Where relief is low, a small contour interval, such as 10 or 20 feet, may be used. In flat areas, such as wide river valleys or broad, flat uplands, a contour interval of 5 feet is often used. In rugged mountainous terrain, where relief is many hundreds of feet, contour intervals as large as 50 or 100 feet are used.

Scale

Map **scale** expresses the relationship between distance or area on the map to the true distance or area on Earth's surface. This is generally expressed as a ratio or fraction, such as 1:24,000 or $\frac{1}{24,000}$. The numerator, usually 1, represents map distance, and the denominator, a large number, represents ground distance. Thus, 1:24,000 means that a distance of 1 unit on the map represents a distance of 24,000 such units on the surface of Earth. It does not matter what the units are.

Figure C.1 Perspective view of an area and a contour map of the same area. These illustrations show how features are depicted on a topographic map. The upper illustration is a perspective view of a river valley and the adjoining hills. The river flows into a bay, which is partly enclosed by a hooked sandbar. On either side of the valley are terraces through which streams have cut gullies. The hill on the right has a smoothly eroded form and gradual slopes, whereas the one on the left rises abruptly in a sharp precipice, from which it slopes gently, and forms an inclined plateau traversed by a few shallow gullies. A road provides access to a church and the two houses situated across the river from a highway that follows the seacoast and curves up the river valley. The lower illustration shows the same features represented by symbols on a topographic map. The contour interval (vertical distance between adjacent contours) is 20 feet. (After U.S. Geological Survey)

Often, the graphic or bar scale is more useful than the fractional scale, because it is easier to use for measuring distances between points. The graphic scale (Figure C.2) consists of a bar divided into equal segments, which represent equal distances on the map. One segment on the left side of the bar is usually divided into smaller units to permit more accurate estimates of fractional units.

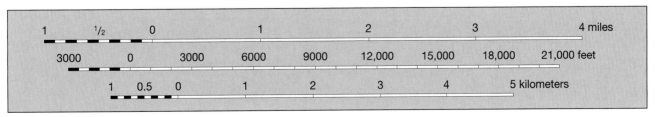

Figure C.2 Graphic scale.

Topographic maps, which are also referred to as *quadrangles*, are generally classified according to publication scale. Each series is intended to fulfill a specific type of map need. To select a map with the proper scale for a particular use, remember that large-scale maps show more detail and small-scale maps show less detail. The sizes and scales of topographic maps published by the U.S. Geological Survey are shown in Table C.1.

Color and Symbol

Each color and symbol used on U.S. Geological Survey topographic maps has significance. Common topographic map symbols are shown in Figure C.3. The meaning of each color is as follows:

Blue — water features
Black — works of humans, such as homes, schools, churches, roads, and so forth
Brown — contour lines
Green — woodlands, orchards, and so forth
Red — urban areas, important roads, public land subdivision lines

Table C.1 National Topographic Maps

Series	Scale	1 inch Represents	Standard Quadrangle Size (latitude-longitude)	Quadrangle Area (square miles)	Paper Size E-W N-S Width Length (inches)
7½ minute	1:24,000	2000 feet	7½′ × 7½′	49–70	22 × 27
Puerto Rico 7½ minute	1:20,000	about 1667 feet	7½′ × 7½′	71	29½ × 32½
15-minute	1:62,500	nearly 1 mile	15′ × 15′	197–282	17 × 21
Alaska 1:63,360	1:63,360	1 mile	15′ × 20′–36′	207–281	18 × 21
U.S. 1:250,000	1:250,000	nearly 4 miles	1° × 2°	4580–8669	34 × 22
U.S. 1:1,000,000	1:1,000,000	nearly 16 miles	4° × 6°	73,734–102,759	27 × 27

SOURCE: U.S. Geological Survey

TOPOGRAPHIC MAP SYMBOLS

VARIATIONS WILL BE FOUND ON OLDER MAPS

Primary highway, hard surface

Secondary highway, hard surface

Light-duty road, hard or improved surface

Unimproved road

Road under construction, alinement known

Proposed road

Dual highway, dividing strip 25 feet or less

Dual highway, dividing strip exceeding 25 feet

Trail

Railroad: single track and multiple track

Railroads in juxtaposition

Narrow gage: single track and multiple track

Railroad in street and carline

Bridge: road and railroad

Drawbridge: road and railroad

Footbridge

Tunnel: road and railroad

Overpass and underpass

Small masonry or concrete dam

Dam with lock

Dam with road

Canal with lock

Buildings (dwelling, place of employment, etc.)

School, church, and cemetery

Buildings (barn, warehouse, etc.)

Power transmission line with located metal tower

Telephone line, pipeline, etc. (labeled as to type)

Wells other than water (labeled as to type)   oOil   oGas

Tanks: oil, water, etc. (labeled only if water)   • • •   ⊘Water

Located or landmark object; windmill   o

Open pit, mine, or quarry; prospect   x

Shaft and tunnel entrance   Y

Horizontal and vertical control station:

    Tablet, spirit level elevation   BM △ 5653

    Other recoverable mark, spirit level elevation   △ 5455

Horizontal control station: tablet, vertical angle elevation   VABM △ 95/9

    Any recoverable mark, vertical angle or checked elevation   △ 3775

Vertical control station: tablet, spirit level elevation   BM × 957

    Other recoverable mark, spirit level elevation   × 954

Spot elevation   × 7369   × 7369

Water elevation   670   670

Boundaries: National

    State

    County, parish, municipio

    Civil township, precinct, town, barrio

    Incorporated city, village, town, hamlet

    Reservation, National or State

    Small park, cemetery, airport, etc.

    Land grant

Township or range line, United States land survey

Township or range line, approximate location

Section line, United States land survey

Section line, approximate location

Township line, not United States land survey

Section line, not United States land survey

Found corner: section and closing

Boundary monument: land grant and other

Fence or field line

Index contour   Intermediate contour

Supplementary contour   Depression contours

Fill   Cut

Levee   Levee with road

Mine dump   Wash

Tailings   Tailings pond

Shifting sand or dunes   Intricate surface

Sand area   Gravel beach

Perennial streams   Intermittent streams

Elevated aqueduct   Aqueduct tunnel

Water well and spring o   o⌒   Glacier

Small rapids   Small falls

Large rapids   Large falls

Intermittent lake   Dry lake bed

Foreshore flat   Rock or coral reef

Sounding, depth curve   10   Piling or dolphin   o

Exposed wreck: Sunken wreck

Rock, bare or awash; dangerous to navigation   *   ⊛

Marsh (swamp)   Submerged marsh

Wooded marsh   Mangrove

Woods or brushwood   Orchard

Vineyard   Scrub

Land subject to controlled inundation   Urban area

Figure C.3 U.S. Geological Survey topographic map symbols. (Variations will be found on older maps.)

Appendix D

Star Charts

The star charts on the next four pages can be used to locate the brighter stars and prominent constellations on the dates and times indicated on each chart. To use these charts, face southward and hold the chart overhead with the top toward the north.

[*]From Robert Dixon, *Dynamic Astronomy*, 6th ed., Englewood Cliffs, NJ: Prentice Hall, 1992.

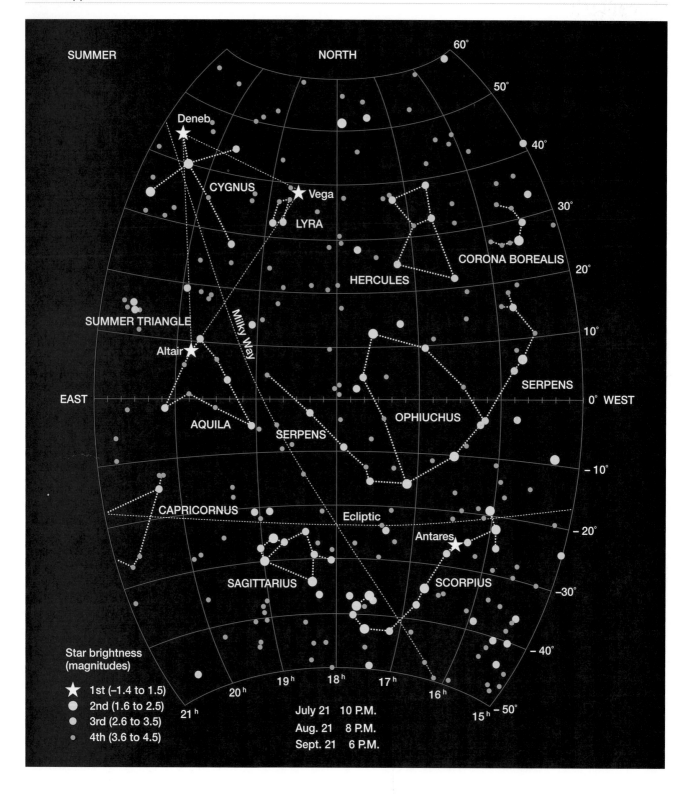

SUMMER

NORTH

60°

50°

Deneb

40°

CYGNUS

Vega

30°

LYRA

CORONA BOREALIS

20°

HERCULES

SUMMER TRIANGLE

10°

Altair

Milky Way

SERPENS

EAST

OPHIUCHUS

0° WEST

AQUILA

SERPENS

– 10°

CAPRICORNUS

Ecliptic

– 20°

Antares

SAGITTARIUS

SCORPIUS

–30°

– 40°

Star brightness
(magnitudes)

1st (–1.4 to 1.5)

20ʰ

19ʰ

18ʰ

17ʰ

16ʰ

15ʰ – 50°

2nd (1.6 to 2.5)

21ʰ

July 21   10 P.M.

3rd (2.6 to 3.5)

Aug. 21   8 P.M.

4th (3.6 to 4.5)

Sept. 21   6 P.M.

Appendix E
World Soils

Figure E.1, on pages 640–1, shows the generalized pattern of global soil orders according to the *Comprehensive Soil Classification System* (CSCS). It should be examined in conjunction with Table E.1, which briefly describes each of the soil orders depicted on the map. To avoid subjective decisions as to classification (a problem that plagued earlier systems), the CSCS defined its classes strictly in terms of soil characterisitics. That is, it is based on features that can be observed or inferred.

The CSCS uses a hierarchy of six categories, or levels. The system recognizes 10 major global *orders* that can be further subdivided into *suborders, great groups, subgroups, families,* and *series.* Note, however, that on the scale of a world map such as Figure E.1, only the largest units (soil orders) can be shown and then only in an extremely generalized way. Although the distribution pattern of major soil orders is more complex than can be shown in Figure E.1, the major distinguishing regional properties of world soils are depicted.

Figure E.1 Soil distribution. The pattern of global soil orders is remarkably similar to the pattern of major climates. Terminology employed is from the *Comprehensive Soil Classification System*. (Adapted from E. Willard Miller, *Physical Geography*, Columbus, Ohio: Merril, 1985, Plate 4.)

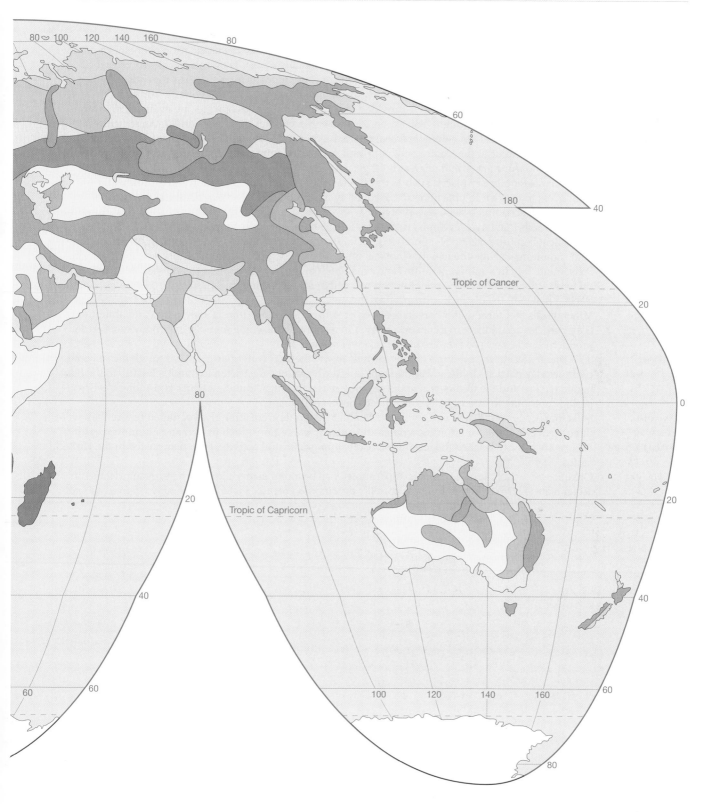

Table E.1   World soil orders.

Entisols	Youngest soils on the Earth. Just beginning to develop in response to the weathering phenomena in the environment. Do not display natural horizons. Found in all climates. They weather slowly over thousands of years; consequently, volcanic ash deposits or sand deposits form the basis for entisols.
Vertisols	Soils containing large amounts of clay, which shrink upon drying and swell with the addition of water. Found in subhumid to arid climates, provided that adequate supplies of water are available to saturate the soil after periods of drought. Soil expansion and contraction exert stresses on human structures.
Inceptisols	Young soils that reveal developmental characteristics (horizons) in response to climate and vegetation. Exist from the Arctic to the tropics on young land surfaces. Common in alpine areas, on river floodplains, in stables and dune areas, and in areas once glaciated.
Aridsols	Soils that develop in dry places, such as the desert, where water—precipitation and groundwater—is insufficient to remove soluble minerals. Frequently irrigated for intensive agricultural production, although salt accumulation poses a problem.
Mollisols	Dark, soft soils that have developed under grass vegetation, generally found in prairie areas. Soil fertility is excellent because potential evaporation generally exceeds precipitation. Also found in hardwood forests with significant earthworm activity. Climatic range is boreal or alpine to tropical. Dry seasons are normal.
Spodosols	Soils found only in humid regions on sandy material. Range from the boreal coniferous forests into tropical forests. Beneath the dark upper horizon of weathered organic material lies a light-colored horizon of leached material, the distinctive property of this soil.
Alfisols	Mineral soils that form under boreal forests or broadleaf deciduous forests, rich in iron and aluminum. Clay particles accumulate in a subsurface layer in response to leaching in moist environments. Fertile, productive soils, because they are neither too wet nor too dry.
Ultisols	Soils that represent the products of long periods of weathering. Water percolating through the soil concentrates clay particles in the lower horizons (argillic horizons). Restricted to humid climates in the temperate regions and the tropics where the growing season is long. Abundant water and a long frost-free period contribute to extensive leaching, hence poorer soil quality.
Oxisols	Soils that occur on old land surfaces unless parent materials were strongly weathered before they were deposited. Generally found in the tropics and subtropical regions. Rich in iron and aluminum oxides, oxisols are heavily leached; hence are poor soils for agricultural activity. Few, if any, exist in the United States.
Histosols	Organic soils with little or no climatic implications. Can be found in any climate where organic debris can accumulate to form a "bog soil." Dark, partially decomposed organic material commonly referred to as *peat*.

Source: Robert E. Norris et al., *Geography: An Introductory Perspective*, Columbus, Ohio: Merrill, 1982.

Appendix F
Landforms of the Conterminous United States

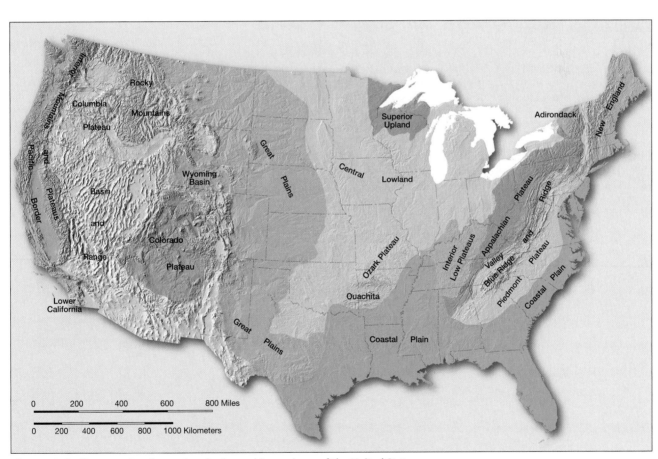

Figure F.1 Outline map showing major physiographic provinces of the United States.

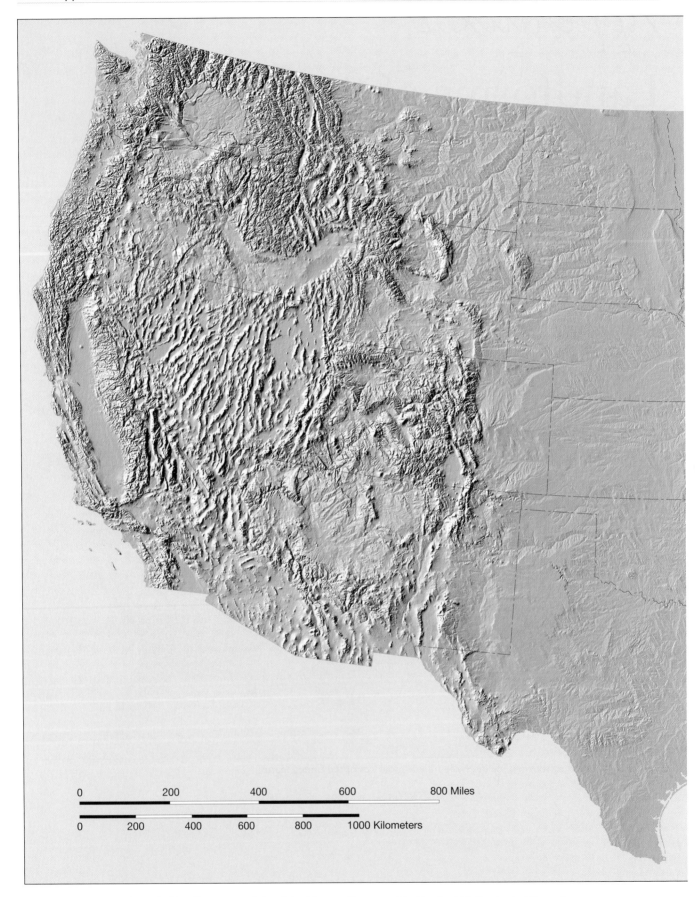

Figure F.2 Digital shaded relief landform map of the United States. (Data provided by the U.S. Geological Survey)

Landforms of the Conterminous United States

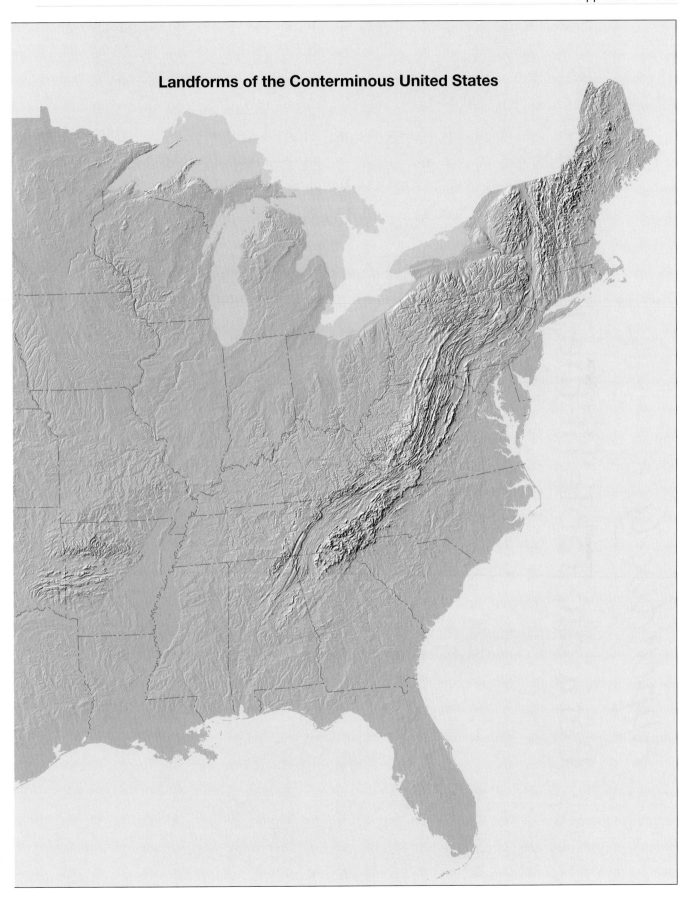

Appendix G
Periodic Table of the Elements

Legend (example cell):
- 2 — Atomic number
- He — Symbol of element
- 4.003 — Atomic weight
- Helium — Name of element

Key:
- Metals
- Transition metals
- Nonmetals
- Noble gases
- Lanthanide series
- Actinide series

IA	IIA	IIIB	IVB	VB	VIB	VIIB	VIIIB			IB	IIB	IIIA	IVA	VA	VIA	VIIA	VIIIA
1 **H** 1.0080 Hydrogen																	2 **He** 4.003 Helium
3 **Li** 6.939 Lithium	4 **Be** 9.012 Beryllium											5 **B** 10.81 Boron	6 **C** 12.011 Carbon	7 **N** 14.007 Nitrogen	8 **O** 15.9994 Oxygen	9 **F** 18.998 Fluorine	10 **Ne** 20.183 Neon
11 **Na** 22.990 Sodium	12 **Mg** 24.31 Magnesium											13 **Al** 26.98 Aluminum	14 **Si** 28.09 Silicon	15 **P** 30.974 Phosphorus	16 **S** 32.064 Sulfur	17 **Cl** 35.453 Chlorine	18 **Ar** 39.948 Argon
19 **K** 39.102 Potassium	20 **Ca** 40.08 Calcium	21 **Sc** 44.96 Scandium	22 **Ti** 47.90 Titanium	23 **V** 50.94 Vanadium	24 **Cr** 52.00 Chromium	25 **Mn** 53.94 Manganese	26 **Fe** 55.85 Iron	27 **Co** 58.93 Cobalt	28 **Ni** 58.71 Nickel	29 **Cu** 63.54 Copper	30 **Zn** 65.37 Zinc	31 **Ga** 69.72 Gallium	32 **Ge** 72.59 Germanium	33 **As** 74.92 Arsenic	34 **Se** 78.96 Selenium	35 **Br** 79.909 Bromine	36 **Kr** 83.80 Krypton
37 **Rb** 85.47 Rubidium	38 **Sr** 87.62 Strontium	39 **Y** 88.91 Yttrium	40 **Zr** 91.22 Zirconium	41 **Nb** 92.91 Niobium	42 **Mo** 95.94 Molybdenum	43 **Tc** (99) Technetium	44 **Ru** 101.1 Ruthenium	45 **Rh** 102.90 Rhodium	46 **Pd** 106.4 Palladium	47 **Ag** 107.87 Silver	48 **Cd** 112.40 Cadmium	49 **In** 114.82 Indium	50 **Sn** 118.69 Tin	51 **Sb** 121.75 Antimony	52 **Te** 127.60 Tellurium	53 **I** 126.90 Iodine	54 **Xe** 131.30 Xenon
55 **Cs** 132.91 Cesium	56 **Ba** 137.34 Barium	57 TO 71	72 **Hf** 178.49 Hafnium	73 **Ta** 180.95 Tantalum	74 **W** 183.85 Tungsten	75 **Re** 186.2 Rhenium	76 **Os** 190.2 Osmium	77 **Ir** 192.2 Iridium	78 **Pt** 195.09 Platinum	79 **Au** 197.0 Gold	80 **Hg** 200.59 Mercury	81 **Tl** 204.37 Thallium	82 **Pb** 207.19 Lead	83 **Bi** 208.98 Bismuth	84 **Po** (210) Polonium	85 **At** (210) Astantine	86 **Rn** (222) Radon
87 **Fr** (223) Francium	88 **Ra** 226.05 Radium	89 TO 103															

Lanthanide series:

57 **LA** 138.91 Lanthanum	58 **Ce** 140.12 Cerium	59 **Pr** 140.91 Praseodymium	60 **Nd** 144.24 Neodymium	61 **Pm** (147) Promethium	62 **Sm** 150.35 Samarium	63 **Eu** 151.96 Europium	64 **Gd** 157.25 Gadolinium	65 **Tb** 158.92 Terbium	66 **Dy** 162.50 Dysprosium	67 **Ho** 164.93 Holmium	68 **Er** 167.26 Erbium	69 **Tm** 168.93 Thulium	70 **Yb** 173.04 Ytterbium	71 **Lu** 174.97 Lutetium

Actinide series:

89 **Ac** (227) Actinium	90 **Th** 232.04 Thorium	91 **Pa** (231) Protactinium	92 **U** 238.03 Uranium	93 **Np** (237) Neptunium	94 **Pu** (242) Plutonium	95 **Am** (243) Americium	96 **Cm** (247) Curium	97 **Bk** (249) Berkelium	98 **Cf** (251) Californium	99 **Es** (254) Einsteinium	100 **Fm** (253) Fermium	101 **Md** (256) Mendelevium	102 **No** (254) Nobelium	103 **Lw** (257) Lawrencium

Glossary

Aa A type of lava flow that has a jagged blocky surface.

Abrasion The grinding and scraping of a rock surface by the friction and impact of rock particles carried by water, wind, or ice.

Absolute humidity The weight of water vapor in a given volume of air (usually expressed in grams/m^3).

Absolute instability Air that has a lapse rate greater than the dry adiabatic rate.

Absolute magnitude The apparent brightness of a star if it were viewed from a distance of 10 parsecs (32.6 light-years). Used to compare the true brightness of stars.

Absolute stability Air with a lapse rate less than the wet adiabatic rate.

Absorption spectrum A continuous spectrum with dark lines superimposed.

Abyssal plain Very level area of the deep-ocean floor, usually lying at the foot of the continental rise.

Abyssal zone A subdivision of the benthic zone characterized by extremely high pressures, low temperatures, low oxygen, few nutrients, and no sunlight.

Accretionary wedge A large wedge-shaped mass of sediment that accumulates in subduction zones. Here sediment is scraped from the subducting oceanic plate and accreted to the overriding crustal block.

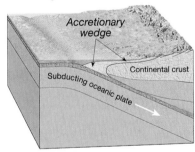

Acid precipitation Rain or snow with a pH value that is less than the pH of unpolluted precipitation.

Active continental margin Usually narrow and consisting of highly deformed sediments. They occur where oceanic lithosphere is being subducted beneath the margin of a continent.

Adiabatic temperature change Cooling or warming of air caused when air is allowed to expand or is compressed, not because heat is added or subtracted.

Advection Horizontal convective motion, such as wind.

Advection fog A fog formed when warm, moist air is blown over a cool surface.

Aerosols Tiny solid and liquid particles suspended in the atmosphere.

Aftershocks Smaller earthquakes that follow the main earthquake.

Air A mixture of many discrete gases, of which nitrogen and oxygen are most abundant, in which varying quantities of tiny solid and liquid particles are suspended.

Air mass A large body of air that is characterized by a sameness of temperature and humidity.

Air-mass weather The conditions experienced in an area as an air mass passes over it. Because air masses are large and fairly homogenous, air-mass weather will be fairly constant and may last for several days.

Air pollutants Airborne particles and gases that occur in concentrations that endanger the health and well-being of organisms or disrupt the orderly functioning of the environment.

Air pressure The force exerted by the weight of a column of air above a given point.

Albedo The reflectivity of a substance, usually expressed as a percentage of the incident radiation reflected.

Alluvial fan A fan-shaped deposit of sediment formed when a stream's slope is abruptly reduced.

Alluvium Unconsolidated sediment deposited by a stream.

Alpine glacier A glacier confined to a mountain valley, which in most instances had previously been a stream valley.

Altitude (of the Sun) The angle of the Sun above the horizon.

Anemometer An instrument used to determine wind speed.

Aneroid barometer An instrument for measuring air pressure that consists of evacuated metal chambers very sensitive to variations in air pressure.

Angle of repose The steepest angle at which loose material remains stationary without sliding downslope.

Angular unconformity An unconformity in which the strata below dip at an angle different from that of the beds above.

Annual mean temperature An average of the 12 monthly temperature means.

Annual temperature range The difference between the highest and lowest monthly temperature means.

Anthracite A hard, metamorphic form of coal that burns clean and hot.

Anticline A fold in sedimentary strata resembling an arch.

Anticyclone A high-pressure center characterized by a clockwise flow of air in the Northern Hemisphere.

Aphelion The place in the orbit of a planet where the planet is farthest from the Sun.

Aphotic zone That portion of the ocean where there is no sunlight.

Apparent magnitude The brightness of a star when viewed from Earth.

Aquifer Rock or soil through which groundwater moves easily.

Aquitard Impermeable beds that hinder or prevent groundwater movement.

Archean eon The second eon of Precambrian time, following the Hadean and preceding the Proterozoic. It extends between 3.8 billion and 2.5 billion years before the present.

Arête A narrow knifelike ridge separating two adjacent glaciated valleys.

Arid *See* Desert.

Arkose A feldspar-rich sandstone.

Artesian well A well in which the water rises above the level where it was initially encountered.

Asteroids Thousands of small planet-like bodies, ranging in size from a few hundred kilometers to less than a kilometer, whose orbits lie mainly between those of Mars and Jupiter.

Asthenosphere A subdivision of the mantle situated below the lithosphere. This zone of weak material exists below a depth of about 100 kilometers and in some regions extends as deep as 700 kilometers. The rock within this zone is easily deformed.

Astronomical theory A theory of climatic change first developed by the Yugoslavian astronomer Milankovitch. It is based upon changes in the shape of Earth's orbit, variations in the obliquity of Earth's axis, and the wobbling of Earth's axis.

Astronomical Unit (AU) Average distance from Earth to the Sun; 1.5×10^8 km or 93×10^6 miles.

Astronomy The scientific study of the universe; it includes the observation and interpretation of celestial bodies and phenomena.

Atmosphere The gaseous portion of a planet; the planet's envelope of air. One of the traditional subdivisions of Earth's physical environment.

Atoll A continuous or broken ring of coral reef surrounding a central lagoon.

Atoll
Lagoon

Atom The smallest particle that exists as an element.

Atomic number The number of protons in the nucleus of an atom.

Atomic weight The average of the atomic masses of isotopes for a given element.

Aurora A bright display of ever-changing light caused by solar radiation interacting with the upper atmosphere in the region of the poles.

Autumnal equinox The equinox that occurs on September 21–23 in the Northern Hemisphere and on March 21–22 in the Southern Hemisphere.

Backswamp A poorly drained area on a floodplain that results when natural levees are present.

Barchan dune Solitary sand dune shaped like a crescent with its tips pointing downward.

Wind

Barchan dune

Barchanoid dune Dunes forming scalloped rows of sand oriented at right angles to the wind. This form is intermediate between isolated barchans and extensive waves of transverse dunes.

Barograph A recording barometer.

Barometer An instrument that measures atmospheric pressure.

Barometric tendency *See* Pressure tendency.

Barred spiral A galaxy having straight arms extending from its nucleus.

Barrier island A low, elongate ridge of sand that parallels the coast.

Barrier island

Basalt A fine-grained igneous rock of mafic composition.

Base level The level below which a stream cannot erode.

Basin A circular downfolded structure.

Batholith A large mass of igneous rock that formed when magma was emplaced at depth, crystallized, and subsequently exposed by erosion.

Baymouth bar A sandbar that completely crosses a bay, sealing it off from the open ocean.

Baymouth bar

Beach drift The transport of sediment in a zigzag pattern along a beach caused by the uprush of water from obliquely breaking waves.

Beach nourishment Large quantities of sand are added to the beach system to offset losses caused by wave erosion.

Bed load Sediment that is carried by a stream along the bottom of its channel.

Benioff zone Zone of inclined seismic activity that extends from a trench downward into the asthenosphere.

Benthic zone The marine life zone that includes *any* seabottom surface regardless of its distance from shore.

Bergeron process A theory that relates the formation of precipitation to supercooled clouds, freezing nuclei, and the different saturation levels of ice and liquid water.

Big bang theory The theory that proposes that the universe originated as a single mass, which subsequently exploded.

Binary stars Two stars revolving around a common center of mass under their mutual gravitational attraction.

Biogenous sediment Seafloor sediments consisting of material of marine-organic origin.

Biosphere The totality of life on Earth; the parts of the solid Earth, hydrosphere, and atmosphere in which living organisms can be found.

Bituminous The most common form of coal, often called soft, black coal.

Black dwarf A final state of evolution for a star, in which all of its energy sources are exhausted and it no longer emits radiation.

Black hole A massive star that has collapsed to such a small volume that its gravity prevents the escape of all radiation.

Blowout (deflation hollow) A depression excavated by the wind in easily eroded deposits.

Bode's law A sequence of numbers that approximates the mean distances of the planets from the Sun.

Body waves Seismic waves that travel through Earth's interior.

Braided stream A stream consisting of numerous intertwining channels.

Breakwater A structure protecting a near-shore area from breaking waves.

Breccia A sedimentary rock composed of angular fragments that were lithified.

Bright-line spectrum The bright lines produced by an incandescent gas under low pressure.

Bright nebula A cloud of glowing gas excited by ultraviolet radiation from hot stars.

Cactolith A quasi-horizontal chonolith composed of anastomosing ductoliths, whose distal ends curl like a harpolith, thin like a sphenolith, or bulge discordantly like an akmolith or ethmolith.

Caldera A large depression typically caused by collapse or ejection of the summit area of a volcano.

Calorie The amount of heat required to raise the temperature of one gram of water 1°C.

Calving Wastage of a glacier that occurs when large pieces of ice break off into water.

Capacity The total amount of sediment a stream is able to transport.

Cassini division A wide gap in the ring system of Saturn between the A ring and the B ring.

Catastrophism The concept that Earth was shaped by catastrophic events of a short-term nature.

Cavern A naturally formed underground chamber or series of chambers most commonly produced by solution activity in limestone.

Celestial sphere An imaginary hollow sphere upon which the ancients believed the stars were hung and carried around Earth.

Cenozoic era A time span on the geologic calendar beginning about 66 million years ago following the Mesozoic era.

Cepheid variable A star whose brightness varies periodically because it expands and contracts. A type of pulsating star.

Chemical sedimentary rock Sedimentary rock consisting of material that was precipitated from water by either inorganic or organic means.

Chemical weathering The processes by which the internal structure of a mineral is altered by the removal and/or addition of elements.

Chinook A wind blowing down the leeward side of a mountain and warming by compression.

Chromatic aberration The property of a lens whereby light of different colors is focused at different places.

Chromosphere The first layer of the solar atmosphere found directly above the photosphere.

Cinder cone A rather small volcano built primarily of pyroclastics ejected from a single vent.

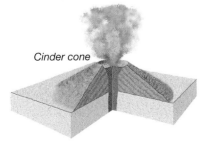

Cinder cone

Circle of illumination The great circle that separates daylight from darkness.

Cirque An amphitheater-shaped basin at the head of a glaciated valley produced by frost wedging and plucking.

Cirrus One of three basic cloud forms; also one of the three high cloud types. They are thin, delicate ice crystal clouds often appearing as veil-like patches or thin, wispy fibers.

Clastic rock A sedimentary rock made of broken fragments of preexisting rock.

Cleavage The tendency of a mineral to break along planes of weak bonding.

Climate A description of aggregate weather conditions; the sum of all statistical weather information that helps describe a place or region.

Climate System The exchanges of energy and moisture that occur among the atmosphere, hydrosphere, solid earth, biosphere, and cryosphere.

Climatic feedback mechanism One of the several different outcomes that may result if one of the many elements in the atmosphere's extremely complex interactive system is altered.

Climatology The scientific study of climate.

Cloud A form of condensation best described as a dense concentration of suspended water droplets or tiny ice crystals.

Clouds of vertical development A cloud that has its base in the low height range but extends upward into the middle or high altitudes.

Cluster (star) A large group of stars.

Coarse-grained texture An igneous rock texture in which the crystals are roughly equal in size and large enough so that individual minerals can be identified with the unaided eye.

Col A pass between mountain valleys where the headwalls of two cirques intersect.

Cold front A front along which a cold air mass thrusts beneath a warmer air mass.

Collision-coalescence process A theory of raindrop formation in warm clouds

(above 0°C) in which large cloud droplets ("giants") collide and join together with smaller droplets to form a raindrop. Opposite electrical charges may bind the cloud droplets together.

Column A feature found in caves that is formed when a stalactite and stalagmite join.

Columnar joints A pattern of cracks that form during cooling of molten rock to generate columns that are generally six-sided.

Coma The fuzzy, gaseous component of a comet's head.

Comet A small body that generally revolves about the Sun in an elongated orbit.

Competence A measure of the largest particle a stream can transport; a factor dependent on velocity.

Composite cone A volcano composed of both lava flows and pyroclastic material.

Compound A substance formed by the chemical combination of two or more elements in definite proportions and usually having properties different from those of its constituent elements.

Condensation The change of state from a gas to a liquid.

Condensation nuclei Tiny bits of particulate matter that serve as surfaces on which water vapor condenses.

Conditional instability Moist air with a lapse rate between the dry and wet adiabatic rates.

Conduction The transfer of heat through matter by molecular activity. Energy is transferred through collisions from one molecule to another.

Cone of depression A cone-shaped depression in the water table immediately surrounding a well.

Conformable Layers of rock that were deposited without interruption.

Conglomerate A sedimentary rock composed of rounded gravel-sized particles.

Constellation An apparent group of stars originally named for mythical characters. The sky is presently divided into 88 constellations.

Contact metamorphism Changes in rock caused by the heat from a nearby magma body.

Continental (c) air mass An air mass that forms over land; it is normally relatively dry.

Continental drift theory A theory that originally proposed that the continents are rafted about. It has essentially been replaced by the plate tectonics theory.

Continental margin That portion of the sea floor adjacent to the continents. It may include the continental shelf, continental slope, and continental rise.

Continental rise The gently sloping surface at the base of the continental slope.

Continental shelf The gently sloping submerged portion of the continental margin extending from the shoreline to the continental slope.

Continental slope The steep gradient that leads to the deep-ocean floor and marks the seaward edge of the continental shelf.

Continental volcanic arc Mountains formed in part by igneous activity associated with the subduction of oceanic lithosphere beneath a continent.

Continuous spectrum An uninterrupted band of light emitted by an incandescent solid, liquid, or gas under pressure.

Convection The transfer of heat by the movement of a mass or substance. It can take place only in fluids.

Convergence The condition that exists when the distribution of winds within a given area results in a net horizontal inflow of air into the area. Because convergence at lower levels is associated with an upward movement of air, areas of convergent winds are regions favorable to cloud formation and precipitation.

Convergent boundary

Convergent boundary A boundary in which two plates move together, causing one of the slabs of lithosphere to be consumed into the mantle as it descends beneath on overriding plate.

Coral reef Structure formed in a warm, shallow, sunlit ocean environment that consists primarily of the calcite-rich remains of corals as well as the limy secretions of algae and the hard parts of many other small organisms.

Core Located beneath the mantle, it is the innermost layer of Earth. The core is divided into an outer core and an inner core.

Coriolis force (effect) The deflective force of Earth's rotation on all free-moving objects, including the atmosphere and oceans. Deflection is to the right in the Northern Hemisphere and to the left in the Southern Hemisphere.

Corona The outer, tenuous layer of the solar atmosphere.

Correlation Establishing the equivalence of rocks of similar age in different areas.

Crater The depression at the summit of a volcano, or that which is produced by a meteorite impact.

Creep The slow downhill movement of soil and regolith.

Crevasse A deep crack in the brittle surface of a glacier.

Cross-bedding Structure in which relatively thin layers are inclined at an angle to the main bedding. Formed by currents of wind or water.

Cross-cutting A principle of relative dating. A rock or fault is younger than any rock (or fault) through which it cuts.

Crust The very thin outermost layer of Earth.

Crystal An orderly arrangement of atoms.

Crystal form The external appearance of a mineral as determined by its internal arrangement of atoms.

Crystallization The formation and growth of a crystalline solid from a liquid or gas.

Cumulus One of three basic cloud forms; also the name given one of the clouds of vertical development. Cumulus are billowy individual cloud masses that often have flat bases.

Cup anemometer *See* Anemometer

Curie point The temperature above which a material loses its magnetization.

Cutoff A short channel segment created when a river erodes through the narrow neck of land between meanders.

Cyclone A low-pressure center characterized by a counterclockwise flow of air in the Northern Hemisphere.

Daily mean The mean temperature for a day that is determined by averaging the 24 hourly readings or, more commonly, by averaging the maximum and minimum temperatures for a day.

Daily temperature range The difference between the maximum and minimum temperatures for a day.

Dark-line spectrum *See* Absorption spectrum.

Dark nebula A cloud of interstellar dust that obscures the light of more distant stars and appears as an opaque curtain.

Daughter product An isotope resulting from radioactive decay.

Debris flow A relatively rapid type of mass wasting that involves a flow of soil and regolith containing a large amount of water. Also called *mudflows*.

Declination (stellar) The angular distance north or south of the celestial equator denoting the position of a celestial body.

Deep-ocean trench A narrow, elongated depression on the floor of the ocean.

Deep-sea fan A cone-shaped deposit at the base of the continental slope. The sediment is transported to the fan by turbidity currents that follow submarine canyons.

Deflation The lifting and removal of loose material by wind.

Delta An accumulation of sediment formed where a stream enters a lake or ocean.

Dendritic pattern A stream system that resembles the pattern of a branching tree.

Dendritic pattern

Density The weight per unit volume of a particular material.

Deposition The process by which water vapor is changed directly to a solid without passing through the liquid state.

Desalination The removal of salts and other chemicals from seawater.

Desert One of the two types of dry climate; the driest of the dry climates.

Desert pavement A layer of coarse pebbles and gravel created when wind removed the finer material.

Detrital sedimentary rock Rock formed from the accumulation of material that originated and was transported in the form of solid particles derived from both mechanical and chemical weathering.

Dew point The temperature to which air has to be cooled in order to reach saturation.

Differential weathering The variation in the rate and degree of weathering caused by such factors as mineral makeup, degree of jointing, and climate.

Diffused light Solar energy scattered and reflected in the atmosphere that reaches Earth's surface in the form of diffuse blue light from the sky.

Dike A tabular-shaped intrusive igneous feature that cuts through the surrounding rock.

Dip-slip fault A fault in which the movement is parallel to the dip of the fault.

Discharge The quantity of water in a stream that passes a given point in a period of time.

Disconformity A type of unconformity in which the beds above and below are parallel.

Disseminated deposit Any economic mineral deposit in which the desired mineral occurs as scattered particles in the rock but in sufficient quantity to make the deposit an ore.

Dissolved load That portion of a stream's load carried in solution.

Distributary A section of a stream that leaves the main flow.

Diurnal tide Tides characterized by a single high and low water height each tidal day.

Divergence The condition that exists when the distribution of winds within a given area results in a net horizontal outflow of air from the region. In divergence at lower levels the resulting deficit is compensated for by a downward movement of air from aloft; hence, areas of divergent winds are unfavorable to cloud formation and precipitation.

Divergent boundary A region where the rigid plates are moving apart, typified by the mid-oceanic ridges.

Divergent boundary

Divide An imaginary line that separates the drainage of two streams; often found along a ridge.

Dome A roughly circular upfolded structure similar to an anticline.

Doppler effect The apparent change in wavelength of radiation caused by the relative motions of the source and the observer.

Doppler radar In addition to the tasks performed by conventional radar, this new generation of weather radar can detect motion directly and hence greatly improve tornado and severe storm warnings.

Drainage basin The land area that contributes water to a stream.

Drawdown The difference in height between the bottom of a cone of depression and the original height of the water table.

Drift The general term for any glacial deposit.

Drumlim A streamlined asymmetrical hill composed of glacial till. The steep side of the hill faces the direction from which the ice advanced.

Dry adiabatic rate The rate of adiabatic cooling or warming in unsaturated air. The rate of temperature change is 1°C per 100 meters.

Dry climate A climate in which yearly precipitation is not as great as the potential loss of water by evaporation.

Dry-summer subtropical climate A climate located on the west sides of continents between latitudes 30 and 45°. It is the only humid climate with a strong winter precipitation maximum.

Dune A hill or ridge of wind-deposited sand.

Earthflow The downslope movement of water-saturated, clay-rich sediment. Most characteristic of humid regions.

Earthquake The vibration of Earth produced by the rapid release of energy.

Ebb current The movement of a tidal current away from the shore.

Eccentricity The variation of an ellipse from a circle.

Echo sounder An instrument used to determine the depth of water by measuring the time interval between emission of a sound signal and the return of its echo from the bottom.

Eclipse The cutting off of the light of one celestial body by another passing in front of it.

Ecliptic The yearly path of the Sun plotted against the background of stars.

Elastic rebound The sudden release of stored strain in rocks that results in movement along a fault.

Electromagnetic radiation *See* Radiation.

Electromagnetic spectrum The distribution of electromagnetic radiation by wavelength.

Electron A negatively charged subatomic particle that has a negligible mass and is found outside an atom's nucleus.

Element A substance that cannot be decomposed into simpler substances by ordinary chemical or physical means.

Elements of weather and climate Those quantities or properties of the atmosphere that are measured regularly and that are used to express the nature of weather and climate.

Elliptical galaxy A galaxy that is round or elliptical in outline. It contains little gas and dust, no disk or spiral arms, and few hot, bright stars.

El Niño The name given to the periodic warming of the ocean that occurs in the central and eastern Pacific. A major El Niño episode can cause extreme weather in many parts of the world.

Eluviation The washing out of fine soil components from the A horizon by downward-percolating water.

Emergent coast A coast where land that was formerly below sea level has been exposed either because of crustal uplift or a drop in sea level or both.

Emission nebula A gaseous nebula that derives its visible light from the fluorescence of ultraviolet light from a star in or near the nebula.

End moraine A ridge of till marking a former position of the front of a glacier.

Environmental lapse rate The rate of temperature decrease with increasing height in the troposphere.

Eon The largest time unit on the geologic time scale, next in order of magnitude above era.

Ephermeral stream A stream that is usually dry because it carries water only in response to specific episodes of rainfall. Most desert streams are of this type.

Epicenter The location on Earth's surface that lies directly above the focus of an earthquake.

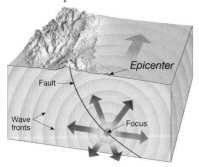

Epoch A unit of the geologic calendar that is a subdivision of a period.

Equatorial low A belt of low pressure lying near the equator and between the subtropical highs.

Equatorial system A method of locating stellar objects much like the coordinate system used on Earth's surface.

Equinox The time when the vertical rays of the Sun are striking the equator. The length of daylight and darkness is equal at all latitudes at equinox.

Era A major division on the geologic calendar; eras are divided into shorter units called periods.

Erosion The incorporation and transportation of material by a mobile agent, such as water, wind, or ice.

Eruptive variable A star that varies in brightness.

Escape velocity The initial velocity an object needs to escape from the surface of a celestial body.

Esker Sinuous ridge composed largely of sand and gravel deposited by a stream flowing in a tunnel beneath a glacier near its terminus.

Estuary A partially enclosed coastal water body that is connected to the ocean. Salinity here is measurably reduced by the freshwater flow of rivers.

Euphotic zone The portion of the photic zone near the surface where light is bright enough for photosynthesis to occur.

Evaporation The process of converting a liquid to a gas.

Evaporite A sedimentary rock formed of material deposited from solution by evaporation of the water.

Evolution, (Theory of) A fundamental theory in biology and paleontology that sets forth the process by which members of a population of organisms come to differ from their ancestors. Organisms evolve by means of mutations, natural selection, and genetic factors. Modern species are descended from related but different species that lived in earlier times.

Exfoliation dome Large, dome-shaped structure, usually composed of granite, formed by sheeting.

Exotic stream A permanent stream that traverses a desert and has its source in well-watered areas outside the desert.

External process Process such as weathering, mass wasting or erosion that is powered by the Sun and transforms solid rock into sediment.

Extrusive Igneous activity that occurs outside the crust.

Eye A zone of scattered clouds and calm averaging about 20 kilometers in diameter at the center of a hurricane.

Eyepiece A short-focal-length lens used to enlarge the image in a telescope. The lens nearest the eye.

Eye wall The doughnut-shaped area of intense cumulonimbus development and very strong winds that surrounds the eye of a hurricane.

Fall A type of movement common to mass wasting processes that refers to the free falling of detached individual pieces of any size.

Fault A break in a rock mass along which movement has occurred.

Fault-block mountain A mountain formed by the displacement of rock along a fault.

Fault creep Displacement along a fault that is so slow and gradual that little seismic activity occurs.

Felsic The group of igneous rocks composed primarily of feldspar and quartz.

Fetch The distance that the wind has traveled across the open water.

Filaments Dark, thin streaks that appear across the bright solar disk.

Fine-grained texture A texture of igneous rocks in which the crystals are too small for individual minerals to be distinguished with the unaided eye.

Fiord A steep-sided inlet of the sea formed when a glacial trough was partially submerged.

Fissure eruption An eruption in which lava is extruded from narrow fractures or cracks in the crust.

Flare A sudden brightening of an area on the Sun.

Flood basalts Flows of basaltic lava that issue from numerous cracks or fissures and commonly cover extensive areas to thicknesses of hundreds of meters.

Flood current The tidal current associated with the increase in the height of the tide.

Floodplain The flat, low-lying portion of a stream valley subject to periodic inundation.

Flow A type of movement common to mass wasting processes in which water-saturated material moves downslope as a viscous fluid.

Fluorescence The absorption of ultraviolet light, which is reemitted as visible light.

Focal length The focal length of a lens is the distance from the lens to the point where it focuses parallel rays of light.

Focus (earthquake) The zone within Earth where rock displacement produces an earthquake.

Focus (light) The point where a lens or mirror causes light rays to converge.

Fog A cloud with its base at or very near Earth's surface.

Fold A bent rock layer or series of layers that were originally horizontal and subsequently deformed.

Foliated A texture of metamorphic rocks that gives the rock a layered appearance.

Foreshocks Small earthquakes that often precede a major earthquake.

Fossil fuel General term for any hydrocarbon that may be used as a fuel, including coal, oil, and natural gas.

Fossils The remains or traces of organisms preserved from the geologic past.

Fossil succession Fossil organisms succeed one another in a definite and determinable order, and any time period can be recognized by its fossil content.

Fracture Any break or rupture in rock along which no appreciable movement has taken place.

Freezing The change of state from a liquid to a solid.

Freezing nuclei Solid particles that serve as cores for the formation of ice crystals.

Front The boundary between two adjoining air masses having contrasting characteristics.

Frontal fog Fog formed when rain evaporates as it falls through a layer of cool air.

Frontal wedging Lifting of air resulting when cool air acts as a barrier over which warmer, lighter air will rise.

Frost wedging The mechanical breakup of rock caused by the expansion of freezing water in cracks and crevices.

Galactic cluster A system of galaxies containing from several to thousands of member galaxies.

Geocentric The concept of an Earth-centered universe.

Geologic time scale The division of Earth history into blocks of time—eons, eras, periods, and epochs. The time scale was created using relative dating principles.

Geology The science that examines Earth, its form and composition, and the changes it has undergone and is undergoing.

Geostrophic wind A wind, usually above a height of 600 meters (2000 feet), that blows parallel to the isobars.

Geothermal energy Natural steam used for power generation.

Geyser A fountain of hot water ejected periodically.

Giant (star) A luminous star of large radius.

Glacial erratic An ice-transported boulder that was not derived from bedrock near its present site.

Glacial striations Scratches and grooves on bedrock caused by glacial abrasion.

Glacial trough A mountain valley that has been widened, deepened, and straightened by a glacier.

Glacier A thick mass of ice originating on land from the compaction and recrystallization of snow that shows evidence of past or present flow.

Glassy A term used to describe the texture of certain igneous rocks, such as obsidian, that contain no crystals.

Glaze A coating of ice on objects formed when super-cooled rain freezes on contact.

Globular cluster A nearly spherically shaped group of densely packed stars.

Globule A dense, dark nebula thought to be the birthplace of stars.

Gondwanaland The southern portion of Pangaea consisting of South America, Africa, Australia, India, and Antarctica.

Graben A valley formed by the downward displacement of a fault-bounded block.

Graben

Graded bed A sediment layer that is characterized by a decrease in sediment size from bottom to top.

Gradient The slope of a stream; generally measured in feet per mile.

Granules The fine structure visible on the solar surface caused by convective cells below.

Greenhouse effect The transmission of short-wave solar radiation by the atmosphere coupled with the selective absorption of longer-wavelength terrestrial radiation, especially by water vapor and carbon dioxide.

Groin A short wall built at a right angle to the shore to trap moving sand.

Ground moraine An undulating layer of till deposited as the ice front retreats.

Groundwater Water in the zone of saturation.

Guyot A submerged flat-topped seamount.

Gyre The large circular surface current pattern found in each ocean.

Hadean eon The first eon on the geologic time scale; this eon ended 3.8 billion years ago and preceded the Archean eon.

Hail Nearly spherical ice pellets having concentric layers and formed by the successive freezing of layers of water.

Half-life The time required for one-half of the atoms of a radioactive substance to decay.

Halocline A layer of water in which there is a high rate of change in salinity in the vertical dimension.

Hanging valley A tributary valley that enters a glacial trough at a considerable height above its floor.

Hanging valley

Hardness The resistance a mineral offers to scratching.

Heliocentric The view that the Sun is at the center of the solar system.

Heat The kinetic energy of random molecular motion.

Hertzsprung-Russell diagram *See* H-R diagram.

High A center of high pressure characterized by anticyclonic winds.

High cloud A cloud that normally has its base above 6000 meters; the base may be lower in winter and at high-latitude locations.

Highland climate Complex pattern of climate conditions associated with mountains. Highland climates are characterized by large differences that occur over short distances.

Hogback A narrow, sharp-crested ridge formed by the upturned edge of a steeply dipping bed of resistant rock.

Horizon A layer in a soil profile.

Horn A pyramidlike peak formed by glacial action in three or more cirques surrounding a mountain summit.

Horst An elongate, uplifted block of crust bounded by faults.

Horst

Hot spot A concentration of heat in the mantle capable of producing magma which, in turn, extrudes onto Earth's surface. The intraplate volcanism that produced the Hawaiian Islands is one example.

Hot spring A spring in which the water is 6°–9°C (10°–15°F) warmer than the mean annual air temperature of its locality.

H-R diagram A plot of stars according to their absolute magnitudes and spectral types.

Hubble law Relates the distance to a galaxy and its velocity.

Humid continental climate A relatively severe climate characteristic of broad continents in the middle latitudes between approximately 40 and 50 degrees north latitude. This climate is not found in the Southern Hemisphere, where the middle latitudes are dominated by the oceans.

Humidity A general term referring to water vapor in the air but not to liquid droplets of fog, cloud, or rain.

Humid subtropical climate A climate generally located on the eastern side of a continent and characterized by hot, sultry summers and cool winters.

Humus Organic matter in soil produced by the decomposition of plants and animals.

Hurricane A tropical cyclonic storm having winds in excess of 119 kilometers (74 miles) per hour.

Hydrogen burning The conversion of hydrogen through fusion to form helium.

Hydrogenous sediment Seafloor sediments consisting of minerals that crystallize from seawater. An important example is manganese nodules.

Hydrosphere The water portion of our planet; one of the traditional subdivisions of Earth's physical environment.

Hydrothermal solution The hot, watery solution that escapes from a mass of magma during the later stages of crystallization. Such solutions may alter the surrounding country rock and are frequently the source of significant ore deposits.

Hygrometer An instrument designed to measure relative humidity.

Hygroscopic nuclei Condensation nuclei having a high affinity for water, such as salt particles.

Hypothesis A tentative explanation that is tested to determine if it is valid.

Ice cap A mass of glacial ice covering a high upland or plateau and spreading out radially.

Ice cap climate A climate that has no monthly means above freezing and supports no vegetative cover except in a few scattered high mountain areas. This climate, with its perpetual ice and

snow, is confined largely to the ice sheets of Greenland and Antarctica.

Ice sheet A very large, thick mass of glacial ice flowing outward in all directions from one or more accumulation centers.

Igneous rock A rock formed by the crystallization of molten magma.

Immature soil A soil lacking horizons.

Incised meander Meandering channel that flows in a steep, narrow valley. They form either when an area is uplifted or when base level drops.

Inclination of the axis The tilt of Earth's axis from the perpendicular to the plane of Earth's orbit.

Inclusion A piece of one rock unit contained within another. Inclusions are used in relative dating. The rock mass adjacent to the one containing the inclusion must have been there first in order to provide the fragment.

Index fossil A fossil that is associated with a particular span of geologic time.

Inertia A property of matter that resists a change in its motion.

Infiltration The movement of surface water into rock or soil through cracks and pore spaces.

Infrared Radiation with a wavelength from 0.7 to 200 micrometers.

Inner core The solid innermost layer of Earth, about 1300 kilometers (800 miles) in radius.

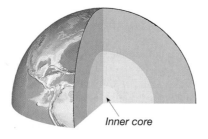
Inner core

Inselberg An isolated mountain remnant characteristic of the late stage of erosion in an arid region.

Intensity (earthquake) An indication of the destructive effects of an earthquake at a particular place. Intensity is affected by such factors as distance to the epicenter and the nature of the surface materials.

Interior drainage A discontinuous pattern of intermittent streams that do not flow to the ocean.

Intermediate composition The composition of igneous rocks lying between felsic and mafic.

Interstellar matter Dust and gases found between stars.

Intertidal zone The area where land and sea meet and overlap; the zone between high and low tides.

Intrusive rock Igneous rock that formed below Earth's surface.

Ion An atom or molecule that possesses an electrical charge.

Ionosphere A complex zone of ionized gases that coincides with the lower portion of the thermosphere.

Iron meteorite One of the three main categories of meteorites. This group is composed largely of iron with varying amounts of nickel (5–20 percent). Most meteorite finds are irons.

Irregular galaxy A galaxy that lacks symmetry.

Isobar A line drawn on a map connecting points of equal atmospheric pressure, usually corrected to sea level.

Isostacy The concept that Earth's crust is "floating" in gravitational balance upon the material of the mantle.

Isotherms Lines connecting points of equal temperature.

Isotopes Varieties of the same element that have different mass numbers; their nuclei contain the same number of protons but different numbers of neutrons.

Jet stream Swift (120–240-kilometer per hour), high-altitude winds.

Jetties A pair of structures extending into the ocean at the entrance to a harbor or river that are built for the purpose of protecting against storm waves and sediment deposition.

Jetties

Joint A fracture in rock along which there has been no movement.

Jovian planet The Jupiter-like planets Jupiter, Saturn, Uranus, and Neptune. These planets have relatively low densities.

Kame A steep-sided hill composed of sand and gravel originating when sediment collected in openings in stagnant glacial ice.

Karst A topography consisting of numerous depressions called sinkholes.

Kettle holes Depressions created when blocks of ice became lodged in glacial deposits and subsequently melted.

Köppen classification A system for classifying climates devised by Wladimir Koeppen that is based on mean monthly and annual values of temperature and precipitation.

Laccolith A massive igneous body intruded between pre-existing strata.

Laccolith

Lahar Mudflows on the slopes of volcanoes that result when unstable layers of ash and debris become saturated and flow downslope, usually following stream channels.

Lake-effect snow Snow showers associated with a cP air mass to which moisture and heat are added from below as the air mass traverses a large and relatively warm lake (such as one of the Great Lakes), rendering the air mass humid and unstable.

Land breeze A local wind blowing from land toward the water during the night in coastal areas.

Lapse rate (normal) The average drop in temperature (6.5°C per kilometer; 3.5°F per 1000 feet) with increased altitude in the troposphere.

Latent heat The energy absorbed or released during a change in state.

Lateral moraine A ridge of till along the sides of an alpine glacier composed primarily of debris that fell to the glacier from the valley walls.

Laterite A red, highly leached soil type found in the tropics that is rich in oxides of iron and aluminum.

Laurasia The northern portion of Pangaea consisting of North America and Eurasia.

Lava Magma that reaches Earth's surface.

Law of conservation of angular momentum The product of the velocity of an object around a center of rotation (axis) and the distance squared of the object from the axis is constant.

Leaching The depletion of soluble materials from the upper soil by downward-percolating water.

Lightning A sudden flash of light generated by the flow of electrons between oppositely charged parts of a cumulonimbus cloud or between the cloud and the ground.

Light-year The distance light travels in a year; about 6 trillion miles.

Liquefaction A phenomenon, sometimes associated with earthquakes, in which soils and other unconsolidated materials containing abundant water are turned into a fluidlike mass that is not capable of supporting buildings.

Lithification The process, generally cementation and/or compaction, of converting sediments to solid rock.

Lithosphere The rigid outer layer of Earth, including the crust and upper mantle.

Local group The cluster of 20 or so galaxies to which our galaxy belongs.

Loess Deposits of wind-blown silt, lacking visible layers, generally buff colored, and capable of maintaining a nearly vertical cliff.

Longitudinal (seif dunes) Long ridges of sand oriented parallel to the prevailing wind; these dunes form where sand supplies are limited.

Longitudinal dunes

Longshore current A near-shore current that flows parallel to the shore.

Low A center of low pressure characterized by cyclonic winds.

Low cloud A cloud that forms below a height of 2000 meters.

Low-velocity zone *See* Asthenosphere.

Luminosity The brightness of a star. The amount of energy radiated by a star.

Lunar breccia A lunar rock formed when angular fragments and dust are welded together by the heat generated by the impact of a meteoroid.

Lunar eclipse An eclipse of the Moon.

Lunar regolith A thin, gray layer on the surface of the Moon, consisting of loosely compacted, fragmented material believed to have been formed by repeated meteoritic impacts.

Luster The appearance or quality of light reflected from the surface of a mineral.

Mafic Igneous rocks with a low silica content and a high iron-magnesium content.

Magma A body of molten rock found at depth, including any dissolved gases and crystals.

Magmatic differentiation The process of generating more than one rock type from a single magma.

Magnitude (earthquake) The total amount of energy released during an earthquake.

Magnitude (stellar) A number given to a celestial object to express its relative brightness.

Main sequence A sequence of stars on the Hertzsprung-Russell diagram, containing the majority of stars, that runs diagonally from the upper left to the lower right.

Manganese nodules Rounded lumps of hydrogenous sediment scattered on the ocean floor consisting mainly of manganese and iron, and usually containing small amounts of copper, nickel, and cobalt.

Mantle The 2900-kilometer (1800-mile)-thick layer of Earth located below the crust.

Maria The Latin name for the smooth areas of the Moon formerly thought to be seas.

Marine west coast climate A climate found on windward coasts from latitudes 40 to 65 degrees and dominated by maritime air masses. Winters are mild and summers are cool.

Maritime (m) air mass An air mass that originates over the ocean. These air masses are relatively humid.

Mass number The number of neutrons and protons in the nucleus of an atom.

Mass wasting The downslope movement of rock, regolith, and soil under the direct influence of gravity.

Meander A looplike bend in the course of a stream.

Meander

Mean solar day The average time between two passages of the Sun across the local celestial meridian.

Mechanical weathering The physical disintegration of rock, resulting in smaller fragments.

Medial moraine A ridge of till formed when lateral moraines from two coalescing alpine glaciers join.

Melt The liquid portion of magma excluding the solid crystals.

Melting The change of state from a solid to a liquid.

Mercalli intensity scale A 12-point scale originally developed to evaluate earthquake intensity based on the amount of damage to various structures.

Mercury barometer A mercury-filled glass tube in which the height of the mercury column is a measure of air pressure.

Mesocyclone An intense, rotating wind system in the lower part of a thunderstorm that precedes tornado development.

Mesopause The boundary between the mesophere and the thermosphere.

Mesosphere The layer of the atmosphere immediately above the stratosphere and characterized by decreasing temperatures with height.

Mesozoic era A time span on the geologic calendar between the Paleozoic and Cenozoic eras—from about 245 million to 66.4 million years ago.

Metamorphic rock Rocks formed by the alteration of preexisting rock deep within Earth (but still in the solid state) by heat, pressure, and/or chemically active fluids.

Metamorphism The changes in mineral composition and texture of a rock subjected to high temperature and pressure within Earth.

Meteor The luminous phenomenon observed when a meteoroid enters Earth's atmosphere and burns up; popularly called a "shooting star."

Meteorite Any portion of a meteoroid that survives its traverse through Earth's atmosphere and strikes Earth's surface.

Meteoroid Small solid particles that have orbits in the solar system.

Meteorology The scientific study of the atmosphere and atmospheric phenomena; the study of weather and climate.

Meteor shower Many meteors appearing in the sky caused when Earth intercepts a swarm of meteoritic particles.

Middle cloud A cloud occupying the height range from 2000 to 6000 meters.

Middle-latitude cyclone Large center of low pressure with an associated cold front and often a warm front. Frequently accompanied by abundant precipitation.

Mid-ocean ridge A continuous mountainous ridge on the floor of all the major ocean basins and varying in width from 500 to 5000 kilometers (300 to 3000 miles). The rifts at the crests of these ridges represent divergent plate boundaries.

Mineral A naturally occurring, inorganic crystalline material with a unique chemical composition.

Mineral resource All discovered and undiscovered deposits of a useful mineral that can be extracted now or at some time in the future.

Mixed tide A tide characterized by a large inequality in successive high water heights, low water heights, or both.

Mixing depth The height to which convectional movements extend above Earth's surface. The greater the mixing depth, the better the air quality.

Mohorovicic discontinuity (Moho) The boundary separating the crust from the mantle, discernible by an increase in seismic velocity.

Mohs scale A series of 10 minerals used as a standard in determining hardness.

Monsoon Seasonal reversal of wind direction associated with large continents, especially Asia. In winter, the wind blows from land to sea; in summer, from sea to land.

Monthly mean temperature The mean temperature for a month that is calculated by averaging the daily means.

Mountain breeze The nightly downslope winds commonly encountered in mountain valleys.

Natural leeves The elevated landforms that parallel some streams and act to confine their waters, except during floodstage.

Neap tide Lowest tidal range, occurring near the times of the first- and third-quarter phases of the Moon.

Nebula A cloud of interstellar gas and/or dust.

Nebular hypothesis The basic idea that the sun and planets formed from the same cloud of gas and dust in interstellar space.

Negative-feedback mechanism As used in climatic change, any effect that is opposite of the initial change and tends to offset it.

Neritic zone The marine life zone that extends from the low tideline out to the shelf break.

Neutron A subatomic particle found in the nucleus of an atom. The neutron is electrically neutral and has a mass approximately that of a proton.

Neutron star A star of extremely high density composed entirely of neutrons.

Nonconformity An unconformity in which older metamorphic or intrusive igneous rocks are overlain by younger sedimentary strata.

Nonfoliated Metamorphic rocks that do not exhibit foliation.

Nonmetallic mineral resource Mineral resource that is not a fuel or processed for the metals it contains.

Nonrenewable resource Resource that forms or accumulates over such long time spans that it must be considered as fixed in total quantity.

Normal fault A fault in which the rock above the fault plane has moved down relative to the rock below.

Normal fault

Normal polarity A magnetic field that is the same as that which exists at present.

Nova A star that explosively increases in brightness.

Nucleus The small heavy core of an atom that contains all of its positive charge and most of its mass.

Nuée ardente Incandescent volcanic debris buoyed up by hot gases that moves downslope in an avalanche fashion.

Numerical date Date that specifies the actual number of years that have passed since an event occurred.

Objective lens In a refracting telescope, the long-focal-length lens that forms an image of the object viewed. The lens closet to the object.

Oblique fault A fault having both vertical and horizontal movement.

Obliquity The angle between the planes of Earth's equator and orbit.

Obsidian A volcanic glass of felsic composition.

Occluded front A front formed when a cold front overtakes a warm front. It marks the beginning of the end of a middle-latitude cyclone.

Occultation An eclipse of a star or planet by the Moon or a planet.

Oceanic zone The marine life zone beyond the continental shelf.

Oceanography The scientific study of the oceans and oceanic phenomena.

Open cluster A loosely formed group of stars of similar origin.

Orbit The path of a body in revolution around a center of mass.

Ore Usually a useful metallic mineral that can be mined at a profit. The term is also applied to certain nonmetallic minerals such as fluorite and sulfur.

Original horizontality Layers of sediments are generally deposited in a horizontal or nearly horizontal position.

Orogenesis The processes that collectively result in the formation of mountains.

Orographic lifting Mountains acting as barriers to the flow of air force the air to ascend. The air cools adiabatically, and clouds and precipitation may result.

Outer core A layer beneath the mantle about 2200 kilometers (1364 miles) thick that has the properties of a liquid.

Outgassing The escape of gases that had been dissolved in magma.

Outwash plain A relatively flat, gently sloping plain consisting of materials deposited by meltwater streams in front of the margin of an ice sheet.

Oxbow lake A curved lake produced when a stream cuts off a meander.

Oxbow lake

Ozone A molecule of oxygen containing three oxygen atoms.

Pahoehoe A lava flow with a smooth-to-ropy surface.

Paleomagnetism The natural remnant magnetism in rock bodies. The permanent magnetization acquired by rock that can be used to determine the location of the magnetic poles and the latitude of the rock at the time it became magnetized.

Paleontology The systematic study of fossils and the history of life on Earth.

Paleozoic era A time span on the geologic calendar between the Precambrian and Mesozoic eras—from about 570 million to 245 million years ago.

Pangaea The proposed supercontinent that 200 million years ago began to break apart and form the present landmasses.

Parabolic dunes The shape of these dunes resembles barchans except their tips point into the wind; they often form along coasts that have strong on-shore winds, abundant sand, and vegetation that partly covers the sand.

Parabolic dunes

Parallax The apparent shift of an object when viewed from two different locations.

Parent material The material upon which a soil develops.

Parsec The distance at which an object would have a parallax angle of 1 second of arc (3.26 light-years).

Partial melting The process by which most igneous rocks melt. Since individual minerals have different melting points, most igneous rocks melt over a temperature range of a few hundred degrees. If the liquid is squeezed out after some melting has occurred, a melt with a higher silica content results.

Passive continental margin Margins that consist of a continental shelf, continental slope, and continental rise. They are *not* associated with plate boundaries and therefore experience little volcanism and few earthquakes.

Pedalfer Soil of humid regions characterized by the accumulation of iron oxides and aluminum-rich clays in the B horizon.

Pedocal Soil associated with drier regions and characterized by an accumulation of calcium carbonate in the upper horizons.

Pelagic zone Open ocean of *any* depth. Animals in this zone swim or float freely.

Penumbra The portion of a shadow from which only part of the light source is blocked by an opaque body.

Penumbra

Perched water table A localized zone of saturation above the main water table created by an impermeable layer (aquiclude).

Peridotite An igneous rock of ultra-mafic composition thought to be abundant in the upper mantle.

Perihelion The point in the orbit of a planet where it is closest to the sun.

Period A basic unit of the geologic calendar that is a subdivision of an era. Periods may be divided into smaller units called epochs.

Permeability A measure of a material's ability to transmit water.

Perturbation The gravitational disturbance of the orbit of one celestial body by another.

Phanerozoic eon That part of geologic time represented by rocks containing abundant fossil evidence. The eon extending from the end of the Proterozoic eon (570 million years ago) to the present.

Phases of the Moon The progression of changes in the Moon's appearance during the month.

Pheoncryst Conspicuously large crystals imbedded in a matrix of finer-grained crystals.

Photic zone The upper part of the ocean into which any sunlight penetrates.

Photochemical reaction A chemical reaction in the atmosphere that is triggered by sunlight, often yielding a secondary pollutant.

Photon A discrete amount (quantum) of electromagnetic energy.

Photosphere The region of the sun that radiates energy to space. The visible surface of the Sun.

pH scale A common measure of the degree of acidity or alkalinity of a solution, it is a logarithmic scale ranging from 0 to 14. A value of 7 denotes a neutral solution, values below 7 indicate greater acidity, and numbers above 7 indicate greater alkalinity.

Piedmont glacier A glacier that forms when one or more valley glaciers emerge from the confining walls of mountain valleys and spread out to create a broad sheet in the lowlands at the base of the mountains.

Pipe A vertical conduit through which magmatic materials have passed.

Placer Deposit formed when heavy minerals are mechanically concentrated by currents, most commonly streams and waves. Placers are sources of gold, tin, platinum, diamonds, and other valuable minerals.

Plane of the ecliptic The imaginary plane that connects Earth's orbit with the celestial sphere.

Planetary nebula A shell of incandescent gas expanding from a star.

Plate One of numerous rigid sections of the lithosphere that moves as a unit over the material of the asthenosphere.

Plate tectonics The theory that proposes that Earth's outer shell consists of individual plates which interact in various ways and thereby produce earthquakes, volcanoes, mountains, and the crust itself.

Playa A flat area on the floor of an undrained desert basin. Following heavy rain, the playa becomes a lake.

Playa lake A temporary lake in a playa.

Pleistocene epoch An epoch of the Quaternary period beginning about 1.6 million years ago and ending about 10,000 years ago. Best known as a time of extensive continental glaciation.

Plucking (quarrying) The process by which pieces of bedrock are lifted out of place by a glacier.

Pluvial lake A lake formed during a period of increased rainfall. During the Pleistocene epoch this occurred in some nonglaciated regions during periods of ice advance elsewhere.

Polar (P) air mass A cold air mass that forms in a high-latitude source region.

Polar easterlies In the global pattern of prevailing winds, winds that blow from the polar high toward the subpolar low. These winds, however, should not be thought of as persistent winds, such as the trade winds.

Polar front The stormy frontal zone separating air masses of polar origin from air masses of tropical origin.

Polar front

Polar high Anticyclones that are assumed to occupy the inner polar regions and are believed to be thermally induced, at least in part.

Polar wandering As the result of paleomagnetic studies in the 1950s, researchers proposed that either the magnetic poles migrated greatly through time or the continents had gradually shifted their positions.

Population I Stars rich in atoms heavier than helium. Nearly always relatively young stars found in the disk of the galaxy.

Population II Stars poor in atoms heavier than helium. Nearly always relatively old stars found in the halo, globular clusters, or nuclear bulge.

Porosity The volume of open spaces in rock or soil.

Porphyritic An igneous texture consisting of large crystals embedded in a matrix of much smaller crystals.

Positive-feedback mechanism As used in climatic change, any effect that acts to reinforce the initial change.

Precambrian All geologic time prior to the Paleozoic era.

Precession A slow motion of Earth's axis which traces out a cone over a period of 26,000 years.

Precipitation fog Fog formed when rain evaporates as it falls through a layer of cool air.

Pressure gradient The amount of pressure change occurring over a given distance.

Pressure tendency The nature of the change in atmospheric pressure over the past several hours. It can be a useful aid in short-range weather prediction.

Prevailing wind A wind that consistently blows from one direction more than from another.

Primary pollutants Those pollutants emitted directly from identifiable sources.

Primary (P) wave A type of seismic wave that involves alternating compression and expansion of the material through which it passes.

Prominence A concentration of material above the solar surface that appears as a bright archlike structure.

Proterozoic eon The eon following the Archean and preceding the Phanerozoic. It extends between 2500 million and 570 million years ago.

Proton A positively charged subatomic particle found in the nucleus of an atom.

Proton-proton chain A chain of thermonuclear reactions by which nuclei of hydrogen are built up into nuclei of helium.

Protostar A collapsing cloud of gas and dust destined to become a star.

Psychrometer A device consisting of two thermometers (wet bulb and dry bulb) that is rapidly whirled and, with the use of tables, yields the relative humidity and dew point.

Ptolemaic system An Earth-centered system of the universe.

Pulsar A variable radio source of small size that emits radio pulses in very regular periods.

Pulsating variable A variable star that pulsates in size and luminosity.

Pyroclastic An igneous rock texture resulting from the consolidation of individual rock fragments that are ejected during a violent eruption.

Pyroclastic flow A highly heated mixture, largely of ash and pumice fragments, traveling down the flanks of a volcano or along the surface of the ground.

Pyroclastic material The volcanic rock ejected during an eruption, including ash, bombs, and blocks.

Radial pattern A system of streams running in all directions away from a central elevated structure, such as a volcano.

Radial pattern

Radiation The transfer of energy (heat) through space by electromagnetic waves.

Radiation fog Fog resulting from radiation heat loss by Earth.

Radiation pressure The force exerted by electromagnetic radiation from an object such as the Sun.

Radioactivity The spontaneous decay of certain unstable atomic nuclei.

Radiocarbon (carbon-14) The radioactive isotope of carbon, which is produced continuously in the atmosphere and is used in dating events as far back as 40,000 years.

Radio interferometer Two or more radio telescopes that combine their signals to achieve the resolving power of a larger telescope.

Radiometric dating The procedure of calculating the absolute ages of rocks and minerals that contain radioactive isotopes.

Radio telescope A telescope designed to make observations in radio wavelengths.

Rainshadow desert A dry area on the lee side of a mountain range. Many middle-latitude deserts are of this type.

Rainshadow desert

Rapids A part of a stream channel in which the water suddenly begins flowing more swiftly and turbulently because of an abrupt steepening of the gradient.

Ray (lunar) Any of a system of bright elongated streaks, sometimes associated with a crater on the Moon.

Recessional moraine An end moraine formed as the ice front stagnated during glacial retreat.

Rectangular pattern A drainage pattern characterized by numerous right angle bends that develops on jointed or fractured bedrock.

Red giant A large, cool star of high luminosity; a star occupying the upper right portion of the Hertzsprung-Russell diagram.

Reflecting telescope A telescope that concentrates light from distant objects by using a concave mirror.

Reflection nebula A relatively dense dust cloud in interstellar space that is illuminated by starlight.

Refracting telescope A telescope that employs a lens to bend and concentrate the light from distant objects.

Refraction The process by which the portion of a wave in shallow water slows, causing the wave to bend and tend to align itself with the underwater contours.

Regional metamorphism Metamorphism associated with the large-scale mountain-building processes.

Regolith The layer of rock and mineral fragments that nearly everywhere covers Earth's surface.

Relative dating Rocks are placed in their proper sequence or order. Only the chronological order of events is determined.

Relative humidity The ratio of the air's water vapor content to its water vapor capacity.

Renewable resource A resource that is virtually inexhaustible or that can be replenished over relatively short time spans.

Reserve Already identified deposits from which minerals can be extracted profitably.

Residual soil Soil developed directly from the weathering of the bedrock below.

Resolving power The ability of a telescope to separate objects that would otherwise appear as one.

Retrograde motion The apparent westward motion of the planets with respect to the stars.

Reverse fault A fault in which the material above the fault plane moves up in relation to the material below.

Reverse fault

Reverse polarity A magnetic field opposite to that which exists at present.

Revolution The motion of one body about another, as Earth about the Sun.

Richter scale A scale of earthquake magnitude based on the motion of a seismograph.

Rift zone A region of Earth's crust along which divergence is taking place.

Right ascension An angular distance measured eastward along the celestial equator from the vernal equinox. Used with declination in a coordinate system to describe the position of celestial bodies.

Rime A thin coating of ice on objects produced when supercooled fog droplets freeze on contact.

Rock A consolidated mixture of minerals.

Rock cycle A model that illustrates the origin of the three basic rock types and the interrelatedness of Earth materials and processes.

Rock flour Ground-up rock produced by the grinding effect of a glacier.

Rockslide The rapid slide of a mass of rock downslope along planes of weakness.

Rockslide

Rotation The spinning of a body, such as Earth, about its axis.

Runoff Water that flows over the land rather than infiltrating into the ground.

Salinity The proportion of dissolved salts to pure water, usually expressed in parts per thousand (%).

Saltation Transportation of sediment through a series of leaps or bounces.

Santa Ana The local name given a chinook wind in southern California.

Saturation The maximum possible quantity of water vapor that the air can hold at any given temperature and pressure.

Scoria Hardened lava that has retained the vesicles produced by escaping gases.

Sea arch An arch formed by wave erosion when caves on opposite sides of a headland unite.

Sea breeze A local wind blowing from the sea during the afternoon in coastal areas.

Seafloor spreading The process of producing new seafloor between two diverging plates.

Seamount An isolated volcanic peak that rises at least 1000 meters (3000 feet) above the deep-ocean floor.

Sea stack An isolated mass of rock standing just offshore, produced by wave erosion of a headland.

Seawall A barrier constructed to prevent waves from reaching the area behind the wall. Its purpose is to defend property from the force of breaking waves.

Secondary enrichment The concentration of minor amounts of metals that are scattered through unweathered rock into economically valuable concentrations by weathering processes.

Secondary pollutants Pollutants that are produced in the atmosphere by chemical reactions that occur among primary pollutants.

Secondary (S) wave A seismic wave that involves oscillation perpendicular to the direction of propagation.

Sediment Unconsolidated particles created by the weathering and erosion of rock, by chemical precipitation from solution in water, or from the secretions of organisms and transported by water, wind, or glaciers.

Sedimentary rock Rock formed from the weathered products of preexisting rocks that have been transported, deposited, and lithified.

Seismic sea wave A rapidly moving ocean wave generated by earthquake activity capable of inflicting heavy damage in coastal regions.

Seismogram The record made by a seismograph.

Seismograph An instrument that records earthquake waves.

Seismology The study of earthquakes and seismic waves.

Semiarid *See* Steppe.

Semidiurnal tide The predominant type of tide throughout the world, with two high waters and two low waters each tidal day.

Shadow zone The zone between 104 and 143 degrees distance from an earthquake epicenter in which direct waves do not arrive because of refraction by Earth's core.

Sheeting A mechanical weathering process characterized by the splitting off of slablike sheets of rock.

Shelf break The point where a rapid steepening of the gradient occurs, marking the outer edge of the continental shelf and the beginning of the continental slope.

Shield A large, relatively flat expanse of ancient metamorphic rock within the stable continental interior.

Shield volcano A broad, gently sloping volcano built from fluid basaltic lavas.

Shield volcano

Sidereal day The period of rotation of Earth with respect to the stars.

Sidereal month A time period based on the revolution of the Moon around Earth with respect to the stars.

Silicate Any one of numerous minerals that have the oxygen and silicon tetrahedron as their basic structure.

Silicon-oxygen tetrahedron A structure composed of four oxygen atoms surrounding a silicon atom that constitutes the basic building block of silicate minerals.

Sill A tabular igneous body that was intruded parallel to the layering of preexisting rock.

Sink hole A depression produced in a region where soluble rock has been removed by groundwater.

Sinkhole

Sleet Frozen or semifrozen rain formed when raindrops freeze as they pass through a layer of cold air.

Slide A movement common to mass wasting processes in which the material moving downslope remains fairly coherent and moves along a well-defined surface.

Slip face The steep, leeward slope of a sand dune; it maintains an angle of about 34 degrees.

Slump The downward slipping of a mass of rock or unconsolidated material moving as a unit along a curved surface.

Slump

Snow A solid form of precipitation produced by sublimination of water vapor.

Snowfield An area where snow persists year-round.

Snowline Lower limit of perennial snow.

Soil A combination of mineral and organic matter, water, and air; that portion of the regolith that supports plant growth.

Soil horizon A layer of soil that has identifiable characteristics produced by chemical weathering and other soil-forming processes.

Soil profile A vertical section through a soil showing its succession of horizons and the underlying parent material.

Soil texture The relative proportions of clay, silt, and sand in a soil. Texture strongly influences the soil's ability to retain and transmit water and air.

Solar constant The rate at which solar radiation is received outside Earth's atmosphere on a surface perpendicular to the Sun's rays when Earth is at an average distance from the Sun.

Solar eclipse An eclipse of the Sun.

Solar flare A sudden and tremendous eruption in the solar chromosphere.

Solar winds Subatomic particles ejected at high speed from the solar corona.

Solifluction Slow, downslope flow of water-saturated materials common to permafrost areas.

Solstice The time when the vertical rays of the Sun are striking either the Tropic of Cancer or the Tropic of Capricorn. Solstice represents the longest or shortest day (length of daylight) of the year.

Solum The O, A, and B horizons in a soil profile. Living roots and other plant and animal life are largely confined to this zone.

Sorting The process by which solid particles of various sizes are separated by moving water or wind. Also the degree of similarity in particle size in sediment or sedimentary rock.

Source region The area where an air mass acquires its characteristic properties of temperature and moisture.

Specific gravity The ratio of a substance's weight to the weight of an equal volume of water.

Specific humidity The weight of water vapor compared with the total weight of the air, including the water vapor.

Spectral class A classification of a star according to the characteristics of its spectrum.

Spectroscope An instrument for directly viewing the spectrum of a light source.

Spectroscopy The study of spectra.

Spheroidal weathering Any weathering process that tends to produce a spherical shape from an initially blocky shape.

Spicule A narrow jet of rising material in the solar chromosphere.

Spiral galaxy A flattened, rotating galaxy with pinwheel-like arms of interstellar material and young stars winding out from its nucleus.

Spiral galaxy

Spit An elongate ridge of sand that projects from the land into the mouth of an adjacent bay.

Spring A flow of groundwater that emerges naturally at the ground surface.

Spring equinox The equinox that occurs on March 21–22 in the Northern Hemisphere and on September 21–23 in the Southern Hemisphere.

Spring tide Highest tidal range that occurs near the times of the new and full moons.

Stalactite The iciclelike structure that hangs from the ceiling of a cavern.

Stalagmite The columnlike form that grows upward from the floor of a cavern.

Star dune Isolated hill of sand that exhibits a complex form and develops where wind directions are variable.

Star dune

Stationary front A situation in which the surface position of a front does not move; the flow on either side of such a boundary is nearly parallel to the position of the front.

Steam fog Fog having the appearance of steam; produced by evaporation from a warm water surface into the cool air above.

Stellar parallax A measure of stellar distance.

Steppe One of the two types of dry climate. A marginal and more humid variant of the desert that separates it from bordering humid climates.

Stony-iron meteorite One of the three main categories of meteorites. This group, as the name implies, is a mixture of iron and silicate minerals.

Stony meteorite One of the three main categories of meteorites. Such meteorites are composed largely of silicate minerals with inclusions of other minerals.

Storm surge The abnormal rise of the sea along a shore as a result of strong winds.

Strata Parallel layers of sedimentary rock.

Stratified drift Sediments deposited by glacial meltwater.

Stratopause The boundary between the stratosphere and the mesosphere.

Stratosphere The layer of the atmosphere immediately above the troposphere characterized by increasing temperatures with height owing to the concentration of ozone.

Stratovolcano *See* Composite cone

Stratus One of three basic cloud forms; also the name given one of the flow clouds. They are sheets or layers that cover much or all of the sky.

Streak The color of a mineral in powdered form.

Striations (glacial) Scratches or grooves in a bedrock surface caused by the grinding action of a glacier and its load of sediment.

Strike-slip fault A fault along which the movement is horizontal.

Stromatolite Structures that are deposited by algae and consist of layered mounds of calcium carbonate.

Subarctic climate A climate found north of the humid continental climate and south of the polar climate and characterized by bitterly cold winters and short cool summers. Places within this climatic realm experience the highest annual temperature ranges on Earth.

Subduction The process of thrusting oceanic lithosphere into the mantle along a convergent boundary.

Subduction

Subduction zone A long, narrow zone where one lithospheric plate descends beneath another.

Sublimation The conversion of a solid directly to a gas without passing through the liquid state.

Submarine canyon A seaward extension of a valley that was cut on the continental shelf during a time when sea level was lower, or a canyon carved into the outer continental shelf, slope, and rise by turbidity currents.

Submergent coast A coast with a form that is largely the result of the partial drowning of a former land surface either because of a rise of sea level or subsidence of the crust or both.

Subpolar low Low pressure located at about the latitudes of the Arctic and

Antarctic circles. In the Northern Hemisphere the low takes the form of individual oceanic cells; in the Southern Hemisphere there is a deep and continuous trough of low pressure.

Subsoil A term applied to the B horizon of a soil profile.

Subtropical high Not a continuous belt of high pressure but rather several semipermanent, anticyclonic centers characterized by subsidence and divergence located roughly between latitudes 25 and 35 degrees.

Summer solstice The solstice that occurs on June 21–22 in the Northern Hemisphere and on December 21–22 in the Southern Hemisphere.

Sunspot A dark spot on the Sun, which is cool by contrast to the surrounding photosphere.

Supercooled The condition of water droplets that remain in the liquid state at temperatures well below 0°C.

Supergiant A very large star of high luminosity.

Supernova An exploding star that increases in brightness many thousands of times.

Superposition In any undeformed sequence of sedimentary rocks, each bed is older than the layers above and younger than the layers below.

Surf A collective term for breakers; also the wave activity in the area between the shoreline and the outer limit of breakers.

Surface soil The uppermost layer in a soil profile: the A horizon.

Surface waves Seismic waves that travel along the outer layer of Earth.

Suspended load The fine sediment carried within the body of flowing water.

Swells Wind-generated waves that have moved into an area of weaker winds or calm.

Syncline A linear downfold in sedimentary strata; the opposite of anticline.

Synodic month The period of revolution of the Moon with respect to the Sun, or its cycle of phases.

Talus An accumulation of rock debris at the base of a cliff.

Tarn A small lake in a cirque.

Tectonics The study of the large-scale processes that collectively deform Earth's crust.

Temperature A measure of the degree of hotness or coldness of a substance; a measure of the *average* kinetic energy of individual atoms or molecules in a substance.

Temperature inversion A layer in the atmosphere of limited depth where the temperature increases rather then decreases with height.

Temporary (local) base level The level of a lake, resistant rock layer, or any other base level that stands above sea level.

Terminal moraine The end moraine marking the farthest advance of a glacier.

Terrace A flat, benchlike structure produced by a stream, which was left elevated as the stream cut downward.

Terrane A crustal block bounded by faults, whose geologic history is distinct from the histories of adjoining crustal blocks.

Terrestrial planet Any of the Earth-like planets including Mercury, Venus, Mars, and Earth.

Terrigenous sediment Seafloor sediments derived from terrestrial weathering and erosion.

Texture The size, shape, and distribution of the particles that collectively constitute a rock.

Theory A well-tested and widely accepted view that explains certain observable facts.

Thermal gradient The increase in temperature with depth. It averages 1°C per 30 meters (1–2°F per 100 feet) in the crust.

Thermocline A layer of water in which there is a rapid change in temperature in the vertical dimension.

Thermohaline circulation Movements of ocean water caused by density differences brought about by variations in temperature and salinity.

Thermosphere The region of the atmosphere immediately above the meso-sphere and characterized by increasing temperatures due to absorption of very short-wave solar energy by oxygen.

Thrust fault A low-angle reverse fault.

Thrust fault

Thunder The sound emitted by rapidly expanding gases along the channel of lightning discharge.

Thunderstorm A storm produced by a cumulonimbus cloud and always accompanied by lightning and thunder. It is of relatively short duration and usually accompanied by strong wind gusts, heavy rain, and sometimes hail.

Tidal current The alternating horizontal movement of water associated with the rise and fall of the tide.

Tidal delta A delta-like feature created when a rapidly moving tidal current emerges from a narrow inlet and slows, depositing its load of sediment.

Tidal flat A marshy or muddy area that is covered and uncovered by the rise and fall of the tide.

Tide Periodic change in the elevation of the ocean surface.

Till Unsorted sediment deposited directly by a glacier.

Tombolo A ridge of sand that connects an island to the mainland or to another island.

Tombolo

Tornado A small, very intense cyclonic storm with exceedingly high winds, most often produced along cold fronts in conjunction with severe thunderstorms.

Tornado warning A warning issued when a tornado has actually been sighted in an area or is indicated by radar.

Tornado watch A warning issued for areas of about 65,000 square kilometers (25,000 square miles) indicating that conditions are such that tornadoes may develop; it is intended to alert people to the possibility of tornadoes.

Trade winds Two belts of winds that blow almost constantly from easterly directions and are located on the equatorward sides of the subtropical highs.

Transform boundary A boundary in which two plates slide past one another without creating or destroying lithosphere.

Transpiration The release of water vapor to the atmosphere by plants.

Transported soil Soils that form on unconsolidated deposits.

Transverse dunes A series of long ridges oriented at right angles to the prevailing wind; these dunes form where vegetation is sparse and sand is very plentiful.

Travertine A form of limestone ($CaCO_3$) that is deposited by hot springs or as a cave deposit.

Trellis pattern A system of streams in which nearly parallel tributaries occupy valleys cut in folded strata.

Trench An elongated depression in the seafloor produced by bending of oceanic crust during subduction.

Tropical depression By international agreement, a tropical cyclone with maximum winds that do not exceed 61 kilometers (38 miles) per hour.

Tropical rain forest A luxuriant broadleaf evergreen forest; also the name given the climate associated with this vegetation.

Tropical storm By international agreement, a tropical cyclone with maximum winds between 61 and 119 kilometers (38 and 74 miles) per hour.

Tropical wet and dry A climate that is transitional between the wet tropics and the subtropical steppes.

Tropic of Cancer The parallel of latitude, 23-1/2 degrees north latitude, marking the northern limit of the Sun's vertical rays.

Tropic of Capricorn The parallel of latitude, 23-1/2 degrees south latitude, marking the southern limit of the Sun's vertical rays.

Tropopause The boundary between the troposphere and the stratosphere.

Troposphere The lowermost layer of the atmosphere. It is generally characterized by a decrease in temperature with height.

Tsunami The Japanese word for a seismic sea wave.

Tundra climate Found almost exclusively in the Northern Hemisphere or at high altitudes in many mountainous regions. A treeless climatic realm of sedges, grasses, mosses, and lichens that is dominated by a long, bitterly cold winter.

Turbidite Turbidity current deposit characterized by graded bedding.

Turbidity current A downslope movement of dense, sediment-laden water created when sand and mud on the continental shelf and slope are dislodged and thrown into suspension.

Turbidity current

Ultimate base level Sea level; the lowest level to which stream erosion could lower the land.

Ultramafic Igneous rocks composed mainly of iron and magnesium-rich minerals.

Ultraviolet Radiation with a wavelength from 0.2 to 0.4 micrometer.

Umbra The central, completely dark part of a shadow produced during an eclipse.

Unconformity A surface that represents a break in the rock record, caused by erosion or nondeposition.

Uniformitarianism The concept that the processes that have shaped Earth in the geologic past are essentially the same as those operating today.

Upslope fog Fog created when air moves up a slope and cools adiabatically.

Upwelling The rising of cold water from deeper layers to replace warmer surface water that has been moved away.

Urban heat island Refers to the fact that temperatures within a city are generally higher than in surrounding rural areas.

Valley breeze The daily upslope winds commonly encountered in a mountain valley.

Valley glacier *See* Alpine glacier.

Valley train A relatively narrow body of stratified drift deposited on a valley floor by meltwater streams that issue from a valley glacier.

Vapor pressure That part of the total atmospheric pressure attributable to water vapor content.

Vein deposit A mineral filling a fracture or fault in a host rock. Such deposits have a sheet-like, or tabular form.

Ventifact A cobble or pebble polished and shaped by the sandblasting effect of wind.

Vesicular A term applied to igneous rocks that contain small cavities called vesicles, which are formed when gases escape from lava.

Viscosity A measure of a fluid's resistance to flow.

Visible light Radiation with a wavelength from 0.4 to 0.7 micrometer.

Volcanic bomb A streamlined pyroclastic fragment ejected from a volcano while molten.

Volcanic island arc A chain of volcanic islands generally located a few hundred kilometers from a trench where active subduction of one oceanic slab beneath another is occurring.

Volcanic neck An isolated, steep-sided, erosional remnant consisting of lava that once occupied the vent of a volcano.

Volcanic neck

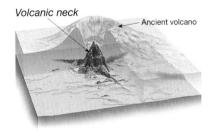

Ancient volcano

Volcano A mountain formed of lava and/or pyroclastics.

Warm front A front along which a warm air mass overrides a retreating mass of cooler air.

Wash A common term for a desert stream course that is typically dry except for brief periods immediately following a rain.

Water table The upper level of the saturated zone of groundwater.

Water table

Wave-cut cliff A seaward-facing cliff along a steep shoreline formed by wave erosion at its base and mass wasting.

Wave-cut platform A bench or shelf in the bedrock at sea level, cut by wave erosion.

Wave height The vertical distance between the trough and crest of a wave.

Wavelength The horizontal distance separating successive crests or troughs.

Wave of oscillation A water wave in which the wave form advances as the water particles move in circular orbits.

Wave of translation The turbulent advance of water created by breaking waves.

Wave period The time interval between the passage of successive crests at a stationary point.

Wave refraction *See* Refraction.

Weather The state of the atmosphere at any given time.

Weathering The disintegration and decomposition of rock at or near Earth's surface.

Welded tuff A pyroclastic rock composed of particles that have been fused together by the combination of heat still contained in the deposit after it has come to rest and by the weight of overlying material.

Well An opening bored into the zone of saturation.

Westerlies The dominant west-to-east motion of the atmosphere that characterizes the regions on the poleward side of the subtropical highs.

Wet adiabatic rate The rate of adiabatic temperature change in saturated air. The rate of temperature change is variable, but it is always less than the dry adiabatic rate.

White dwarf A star that has exhausted most or all of its nuclear fuel and has collapsed to a very small size; believed to be near its final stage of evolution.

White frost Ice crystals instead of dew that form on surfaces when the dew point is below freezing.

Wind Air flowing horizontally with respect to Earth's surface.

Wind vane An instrument used to determine wind direction.

Winter solstice The solstice that occurs on December 21–22 in the Northern Hemisphere and on June 21–22 in the Southern Hemisphere.

Yazoo tributary A tributary that flows parallel to the main stream because a natural levee is present.

Zodiac A band along the ecliptic containing the 12 constellations of the zodiac.

Zone of accumulation The part of a glacier characterized by snow accumulation and ice formation. Its outer limit is the snowline.

Zone of aeration Area above the water table where openings in soil, sediment, and rock are not saturated but filled mainly with air.

Zone of fracture The upper portion of a glacier consisting of brittle ice.

Zone of saturation Zone where all open spaces in sediment and rock are completely filled with water.

Zone of wastage The part of a glacier beyond the zone of accumulation where all of the snow from the previous winter melts, as does some of the glacial ice.

Index

Installation Instructions

Macintosh Information

Hardware and software requirements:

- Macintosh with 256 color graphics
- 13" or larger display (640×480 or higher)
- CD-ROM drive (double-speed)
- 6 MB of free RAM memory
- System 7.0 or newer

Refer to the Read Me file on the CD-ROM if you need information regarding how to run the program.

Windows Information

Hardware and software requirements:

- IBM compatible 486 or better
- SVGA (256 colors in 640×480)
- CD-ROM drive (double-speed)
- 16 MB RAM suggested
- Windows 3.1 (enhanced mode) or Windows 95/98/NT

Refer to the README.TXT file on the CD-ROM if you need information regarding how to run the program.

For technical support, please e-mail Tasa Graphic Arts, Inc. at:

support@tasagraphicarts.com

GEODe II

Geologic Explorations on Disk

YOU SHOULD CAREFULLY READ THE TERMS AND CONDITIONS BEFORE USING THE DISKETTE PACKAGE. USING THIS DISKETTE PACKAGE INDICATES YOUR ACCEPTANCE OF THESE TERMS AND CONDITIONS.

Prentice-Hall, Inc. provides this program and licenses its use. You assume responsibility for the selection of the program to achieve your intended results, and for the installation, use, and results obtained from the program. This license extends only to use of the program in the United States or countries in which the program is marketed by authorized distributors.

License Grant

You hereby accept a nonexclusive, nontransferable, permanent license to install and use the program ON A SINGLE COMPUTER at any given time. You may copy the program solely for backup or archival purposes in support of your use of the program on the single computer. You may not modify, translate, disassemble, decompile, or reverse engineer the program, in whole or in part.

Term

The License is effective until terminated. Prentice-Hall, Inc. reserves the right to terminate this License automatically if any provision of the License is violated. You may terminate the License at any time. To terminate this License, you must return the program, including documentation, along with a written warranty stating that all copies in your possession have been returned or destroyed.

Limited Warranty

THE PROGRAM IS PROVIDED "AS IS" WITHOUT WARRANTY OF ANY KIND, EITHER EXPRESSED OR IMPLIED, INCLUDING, BUT NOT LIMITED TO, THE IMPLIED WARRANTIES OR MERCHANTABILITY AND FITNESS FOR A PARTICULAR PURPOSE. THE ENTIRE RISK AS TO THE QUALITY AND PERFORMANCE OF THE PROGRAM IS WITH YOU. SHOULD THE PROGRAM PROVE DEFECTIVE, YOU (AND NOT PRENTICE-HALL, INC. OR ANY AUTHORIZED DEALER) ASSUME THE ENTIRE COST OF ALL NECESSARY SERVICING, REPAIR, OR CORRECTION. NO ORAL OR WRITTEN INFORMATION OR ADVICE GIVEN BY PRENTICE-HALL, INC., ITS DEALERS, DISTRIBUTORS, OR AGENTS SHALL CREATE A WARRANTY OR INCREASE THE SCOPE OF THIS WARRANTY.

SOME STATES DO NOT ALLOW THE EXCLUSION OF IMPLIED WARRANTIES, SO THE ABOVE EXCLUSION MAY NOT APPLY TO YOU. THIS WARRANTY GIVES YOU SPECIFIC LEGAL RIGHTS AND YOU MAY ALSO HAVE OTHER LEGAL RIGHTS THAT VARY FROM STATE TO STATE.

Prentice-Hall, Inc. does not warrant that the functions contained in the program will meet your requirements or that the operation of the program will be uninterrupted or error-free.

However, Prentice-Hall, Inc. warrants the diskette(s) on which the program is furnished to be free from defects in material and workmanship under normal use for a period of ninety (90) days from the date of delivery to you as evidenced by a copy of your receipt.

The program should not be relied on as the sole basis to solve a problem whose incorrect solution could result in injury to person or property. If the program is employed in such a manner, it is at the user's own risk and Prentice-Hall, Inc. explicitly disclaims all liability for such misuse.

Limitations of Remedies

Prentice-Hall, Inc.'s entire liability and your exclusive remedy shall be:

1. the replacement of any diskette not meeting Prentice-Hall, Inc.'s "LIMITED WARRANTY" and that is returned to Prentice-Hall, or

2. if Prentice-Hall is unable to deliver a replacement diskette that is free of defects in materials and workmanship, you may terminate this agreement by returning the program.

IN NO EVENT WILL PRENTICE-HALL, INC. BE LIABLE TO YOU FOR ANY DAMAGES, INCLUDING ANY LOST PROFITS, LOST SAVINGS, OR OTHER INCIDENTAL OR CONSEQUENTIAL DAMAGES ARISING OUT OF THE USE OR INABILITY TO USE SUCH PROGRAM EVEN IF PRENTICE-HALL, INC. OR AN AUTHORIZED DISTRIBUTOR HAS BEEN ADVISED OF THE POSSIBILITY OF SUCH DAMAGES, OR FOR ANY CLAIM BY ANY OTHER PARTY.

SOME STATES DO NOT ALLOW FOR THE LIMITATION OR EXCLUSION OF LIABILITY FOR INCIDENTAL OR CONSEQUENTIAL DAMAGES, SO THE ABOVE LIMITATION OR EXCLUSION MAY NOT APPLY TO YOU.

General

You may not sublicense, assign, or transfer the License of the program. Any attempt to sublicense, assign or transfer any of the rights, duties, or obligations hereunder is void.

This Agreement will be governed by the laws of the State of New York.

Should you have any questions concerning this Agreement, you may contact Prentice-Hall, Inc. by writing to:

Director of Multimedia Development
Higher Education Division
Prentice-Hall, Inc.
1 Lake Street
Upper Saddle River, NJ 07458

Should you have any questions concerning technical support, you may write to:

Tasa Graphic Arts, Inc.
9301 Indian School Road NE
Suite 208
Albuquerque, NM 87112-2861

YOU ACKNOWLEDGE THAT YOU HAVE READ THIS AGREEMENT, UNDERSTAND IT, AND AGREE TO BE BOUND BY ITS TERMS AND CONDITIONS. YOU FURTHER AGREE THAT IT IS THE COMPLETE AND EXCLUSIVE STATEMENT OF THE AGREEMENT BETWEEN US THAT SUPERSEDES ANY PROPOSAL OR PRIOR AGREEMENT, ORAL OR WRITTEN, AND ANY OTHER COMMUNICATIONS BETWEEN US RELATING TO THE SUBJECT MATTER OF THIS AGREEMENT.